The Biology and Chemistry of the Compositae

A joint symposium sponsored by the Linnean Society of London
and
The Phytochemical Society

The Biology and Chemistry of the Compositae

Volume I

edited by

V. H. HEYWOOD and J. B. HARBORNE

Department of Botany, University of Reading, Reading, England

and

B. L. TURNER

Department of Botany, University of Texas, Austin, Texas, U.S.A.

1977

ACADEMIC PRESS

London   New York   San Francisco

A Subsidiary of Harcourt Brace Jovanovich, Publishers

ACADEMIC PRESS INC. (LONDON) LTD.
24/28 Oval Road
London NW1

United States edition published by
ACADEMIC PRESS INC.
111 Fifth Avenue
New York, New York 10003

QK
495
.C74B53
Vol.1

Copyright © 1977 by
ACADEMIC PRESS INC. (LONDON) LTD.

All Rights Reserved

No part of this book may be reproduced in any form by photostat, microfilm, or any other means, without written permission from the publishers

Library of Congress Catalog Card Number: 76–24428
ISBN: 0–12–346801–9

*Printed in Great Britain by William Clowes & Sons Limited,
London, Beccles and Colchester*

Contributors

J. BAAGØE, Botanical Museum of the University, Copenhagen, Denmark
B. A. BOHM, Botany Department, University of British Columbia, Vancouver B.C., Canada
F. BOHLMANN, Institute of Organic Chemistry, Technical University of Berlin, Germany
B. L. BURTT, Royal Botanic Garden, Edinburgh, Scotland
A. L. CABRERA, Museo de la Plata and Consejo Nacional de Investigaciónes, La Plata, Argentina
M. DITTRICH, Conservatoire et Jardin botaniques, Geneva, Switzerland
X. A. DOMINGUEZ, Department of Chemistry, Institute of Technology, Monterrey, Mexico
A. G. GONZALEZ, Department of Organic and Biochemistry, University of La Laguna, Tenerife, Canary Islands
J. GRAU, Botanische Staatssamlung, München, Germany
H. GREGER, Botanical Institute, University of Vienna, Austria
L. R. G. VALADON, Department of Botany, Royal Holloway College, University of London, England
J. B. HARBORNE, Department of Botany, University of Reading, Whiteknights, Reading, England
R. HEGNAUER, Laboratorium voor Experimentele Plantensystematiek, Leiden, Netherlands
W. HERZ, Department of Chemistry, The Florida State University, Tallahassee, Florida, U.S.A.
V. H. HEYWOOD, Department of Botany, University of Reading, Whiteknights, Reading, England
C. J. HUMPHRIES, Department of Botany, British Museum (Natural History), London, England
C. JEFFREY, The Herbarium, Royal Botanic Gardens, Kew, Surrey
S. B. JONES, Botany Department, University of Georgia, Athens, Georgia, U.S.A.
R. M. KING, National Museum of Natural History, Smithsonian Institute, Washington D.C., U.S.A.
G. LEDYARD STEBBINS, Department of Genetics, University of California, Davis, U.S.A.
P. LEINS, Botanical Institute, University of Bonn, Germany
E. E. LEPPICK, Plant Genetics and Germplasm Institute, U.S. Department of Agriculture, Beltsville, Maryland, U.S.A.

CONTRIBUTORS

T. J. MABRY, Department of Botany, University of Texas, Austin, Texas, U.S.A.
H. MERXMULLER, Botanische Staatssamlung, München, Germany
B. NORDENSTAM, Botany Section, Swedish Museum of Natural History, Stockholm, Sweden
T. NORLINDH, Botany Section, Swedish Museum of Natural History, Stockholm, Sweden
V. C. PATEL, Department of Botany, University of Texas, Austin, Texas, U.S.A.
A. M. POWELL, Department of Biology, Sul Ross State University, Alpine, Texas, U.S.A.
D. J. ROBINS, Department of Chemistry, The University, Glasgow, Scotland
H. ROBINSON, National Museum of Natural History, Smithsonian Institute, Washington D.C., U.S.A.
E. RODRIGUEZ, Department of Botany and the Cell Research Institute, University of Texas, Austin, Texas, U.S.A.
H. ROESSLER, Botanische Staatssamlung, München, Germany
J. J. SKVARLA, Department of Botany and Microbiology, University of Oklahoma, Norman, Oklahoma, U.S.A.
O. T. SOLBRIG, Department of Biology, Harvard University, Cambridge, Massachusetts, U.S.A.
N. A. SØRENSEN, Norges Tekniske Hogskole, Trondheim, Norway
J. L. STROTHER, University Herbarium, Department of Botany, University of California, Berkeley, U.S.A.
T. F. STUESSY, Department of Botany, The Ohio State University, Columbus, Ohio, U.S.A.
T. SWAIN, Boston University, Biological Science Center, Boston, Massachusetts, U.S.A.
A. S. TOMB, Department of Biology, Kansas State University, Manhattan, Kansas, U.S.A.
C. A. WILLIAMS, Botany Department, University of Reading, England
B. L. TURNER, Department of Botany, University of Texas, Austin, Texas, U.S.A.
H. WAGNER, Institut für Pharmazeutische Arzneimittellehre, University of Munich, Germany

Preface

These volumes are based on the papers presented at an international symposium on the Biology and Chemistry of the Compositae held at the University of Reading on July 14–18, 1975, under the auspices of the Linnean Society of London and the Phytochemical Society.

The Compositae or Asteraceae is one of the largest and most familiar families of Flowering Plants, and has been the subject of research for centuries. Although not of major economic importance it has attracted a great deal of attention from botanists largely because of its characteristic aggregations of reduced flowers into specialized inflorescences called capitula. These capitula are highly specialized structures and have been subject to many evolutionary processes. Their structure, biology and evolution have long fascinated botanists and at the same time the overall uniformity imposed on the family by the common possession of capitula has had the effect of making the classification of tribes and genera exceptionally difficult, a phenomenon shared with other "natural" families possessing highly characteristic inflorescences, such as the Umbelliferae and the Gramineae.

The sheer size of the family, about 1310 genera and 13 000 species, has meant inevitably that no overall revision of the family has been attempted in recent times. On the other hand, scores if not hundreds, of specialists have revised genera or generic groups within individual tribes but no overall synthesis of this revisionary activity has been available.

This symposium was in many ways an experiment. No attempt has been made in previous symposia on the biology and chemistry of large families to revise the component tribes systematically down to generic level. In this case, as is explained in the first chapter, for each tribe one or two specialists were invited to prepare a systematic report, including a description of the tribe, a list of valid or accepted genera and approximate number of species they contain, and their general distribution, as well as a discussion of significant data on floral biology, karyology, cytogenetics, palynology, embryology, anatomy, morphology and dispersal mechanisms where appropriate. We were fortunate that most of the invited reviewers felt able to comply with this request and provided a preliminary version of their report for inclusion in the several-hundred page volume of Abstracts and Reports circulated privately to the symposium participants.

These reports were extensively debated during the meeting and the authors were able to make appropriate revisions of their texts before they were submitted for publication. The result is the first account this century of the family which includes a comprehensive listing down to generic level, and an assessment of all the genera published since the times of Bentham and Hoffmann.

In addition the Compositae has attracted many chemists and biochemists and a very substantial body of research has built up over the past four decades on the chemical constituents of individual species and genera. Accordingly specialists were also invited to review each tribe from a chemical point of view, covering such topics as the chemical pattern of the tribe, low molecular weight constituents such as terpenoids, flavonoids, alkaloids and other compounds, macromolecular constituents, the relevance of chemistry to the systematics of the tribe, and economic and pharmaceutical uses. These reports were included in the preliminary volume of Abstracts and Reports and like the systematics reports, revised during and after the meeting.

To complete the survey of the family, twelve review lectures on various topics of biological or chemical importance were commissioned. All these papers, with the exception of that by Dr. A. Cronquist which, regretfully, it was not possible to include,* are included in these two volumes. Together with the tribal reports they represent a vast and unrivalled source of information and ideas on this important family and are likely to serve as a major reference work for several decades.

The editors are most grateful to all the contributors for agreeing so willingly to co-operate in this enterprise.

In presenting the papers for publication we have placed the general review papers at the beginning, followed by the systematic and chemical reviews, tribe by tribe, and the work concludes with two summary chapters, the first by Mabry and Bohlmann on the chemistry and the second by Turner on the systematics and biology. For ease of reference we have included an alphabetical list of all genera (plus key synonyms) recognized by the tribal reviewers together with an indication of the tribe to which each belongs. This has been compiled by Mr. Stephen Jury from the proofs of the various systematic reviews. An earlier list of genera published since 1896, compiled by Dr. B. Nordenstam and Dr. T. Stuessy, has been of great assistance as a cross check.

The symposium was attended by over 130 scientists from many parts of the world. We are grateful to The Royal Society, Tate and Lyle, Shell International Petroleum Company Limited, Imperial Chemical Industries and the Phytochemical Society for financial contributions towards the costs of some of the invited speakers. The Philip Lyle Memorial Laboratory very generously invited all the participants to a buffet reception on the

* Now published in *Brittonia* **29**, 137–153 (1977).

last evening and our thanks are due to Professor A. J. Vlitos, the Chief Executive. The University of Reading kindly offered a reception on Thursday 17 July. Many members of the staff of the Department of Botany at Reading assisted in making arrangements for the meeting and in particular we would like to acknowledge the generous and cheerful assistance of Mrs Abigail Gillett who acted as secretary to the symposium and did so much to make it a successful occasion.

The production department of Academic Press has greatly eased the task of the editors in preparing the book for the press.

September 1977

V. H. Heywood
J. B. Harborne
B. L. Turner

Contents

VOLUME I

Contributors v
Preface vii

1. An overture to the Compositae 1
   V. H. Heywood, Jeffrey B. Harborne and B. L. Turner
2. Fossil history and geography 21
   B. L. Turner
3. Aspects of diversification in the capitulum 41
   B. L. Burtt
4. The evolution of capitulum types of the Compositae in the light of insect–flower interaction 61
   Elmar E. Leppick
5. Developmental and comparative anatomy of the Compositae . 91
   G. Ledyard Stebbins
6. Corolla forms in Compositae—some evolutionary and taxonomic speculations 111
   C. Jeffrey
7. Microcharacters in the ligules of the Compositae . . . 119
   Jette Baagøe
8. Pollen morphology in the Compositae and in morphologically related families 141
   John J. Skvarla, B. L. Turner, Varsha C. Patel and A. Spencer Tomb

   Appendix: Principal works on the pollen morphology of the Compositae 249
   C. Thanikaimoni
9. Chromosomal cytology and evolution in the family Compositae 267
   Otto T. Solbrig
10. The chemistry of the Compositae 283
    R. Hegnauer
11. Sesquiterpene lactones in the Compositae 337
    Werner Herz

12. Flavonoid profiles in the Compositae 359
    Jeffrey B. Harborne
13. Polyacetylenes and conservatism of chemical characters in the Compositae 385
    N. A. Sørensen
14. Pharmaceutical and economic uses of the Compositae . . 411
    H. Wagner
15. Eupatorieae—systematic review 437
    H. Robinson and *R. M. King*
16. Eupatorieae—chemical review 487
    Xorge A. Domínguez
17. Vernonieae—systematic review 503
    Samuel B. Jones
18. Vernonieae—chemical review 523
    Jeffrey B. Harborne and *Christine A. Williams*
19. Astereae—systematic review 539
    J. Grau
20. Astereae—chemical review 567
    Werner Herz
21. Inuleae—systematic review 577
    H. Merxmüller, P. Leins and *H. Roessler*
22. Inuleae—chemical review 603
    Jeffrey B. Harborne

CONTENTS OF VOLUME II

Contributors v
Preface vii

23. Heliantheae—systematic review 621
    Tod F. Stuessy
24. Heliantheae—chemical review 673
    Tony Swain and *Christine A. Williams*
25. Helenieae—systematic review 699
    B. L. Turner and *A. M. Powell*
26. Heliantheae—chemical review 739
    Bruce A. Bohm
27. Tageteae—systematic review 769
    John L. Strother
28. Tageteae—chemical review 785
    Eloy Rodríguez and *Tom J. Mabry*
29. Senecioneae and Liabeae—systematic review . . . 799
    Bertil Nordenstam
30. Senecioneae—chemical review 831
    David J. Robins
31. Anthemideae—systematic review 851
    V. H. Heywood and *C. J. Humphries*
32. Anthemideae—chemical review 899
    H. Greger
33. Arctoteae—systematic review 943
    Tycho Norlindh
34. Calenduleae—systematic review 961
    Tycho Norlindh
35. Arctoteae and Calenduleae—chemical review . . . 989
    L. R. Guy Valadon
36. Cynareae—systematic review 999
    M. Dittrich
37. Cynareae—chemical review 1017
    H. Wagner

38. Mutisieae—systematic review 1039
    Angel L. Cabrera
39. Lactuceae—systematic review 1067
    A. Spencer Tomb
40. Lactuceae—chemical review 1081
    Antonio G. González
41. Summary of the chemistry of the Compositae . . . 1097
    Tom J. Mabry and *Ferdinand Bohlmann*
42. Summary of the biology of the Compositae 1105
    B. L. Turner

  Appendix: List of genera 1119
  Subject index 1141
  Organism index 1155
  Chemical compound index 1179

Chapter 1
An overture to the Compositae

V. H. HEYWOOD, JEFFREY B. HARBORNE

Department of Botany, University of Reading, England

and

B. L. TURNER

Department of Botany, University of Texas at Austin, U.S.A.

Abstract. The scene is set for the Symposium proceedings which follow this introductory chapter. A brief history of the taxonomy is followed by an outline of previously accepted subdivisions within the family. Morphological variation is summarized and discussed in terms of adaptation to habitat and to pollinators. The chemistry is considered in relation to the economic importance of the family and to the ability of these plants to protect themselves from grazing by herbivores.

CONTENTS

Systematic introduction	1
Subdivision of the family	4
Structure of the Symposium	5
Morphology and anatomy of the family	5
Ecological diversification	5
Habitual diversification	6
Capitular diversification	7
Floral diversification	8
Chemistry of the family	9
General pattern	9
Sesquiterpene lactones	13
Polyacetylenes	15
Insecticidal principles	15
Rubber production	16
Inulin accumulation	17
Conclusion	18
References	19

SYSTEMATIC INTRODUCTION

The family Compositae or Asteraceae has attracted, fascinated and even repelled botanists for over two centuries. There have been few heroes—

Cassini and Bentham, certainly—while hundreds of other botanists have worked away at parts of the family with little visible or lasting effect, although the cumulative result of their efforts has modified much of the detail.

The basic classification of the family as recognized today is little different from that established by Bentham in his treatment for *Genera Plantarum* (1873a) and further explained in his celebrated essay *Notes on the classification, history and geographical distribution of the Compositae* (1873b). Of course Bentham's classification, especially into tribes (see Table I), was largely based on the painstaking work of earlier authors, notably Cassini, De Candolle and Lessing, but it represented the first comprehensive account assessing and summating all this earlier research. Bentham's account was a personal one and he studied for himself representatives of most of the genera as represented in the Herbarium at the Royal Gardens, Kew.

TABLE I. The tribes of the Compositae as recognized by Bentham (1873). The original spellings for the tribes are given in parentheses

Tribe 1	Vernonieae (Vernoniaceae)
Tribe 2	Eupatorieae (Eupatoriaceae)
Tribe 3	Astereae (Asteroideae)
Tribe 4	Inuleae (Inuloideae)
Tribe 5	Heliantheae (Helianthoideae)
Tribe 6	Helenieae (Helianthoideae)
Tribe 7	Anthemideae
Tribe 8	Senecioneae (Senecionideae)
Tribe 9	Calenduleae (Calendulaceae)
Tribe 10	Arctotideae
Tribe 11	Cynareae (Cynaroideae)
Tribe 12	Mutisieae (Mutisaceae)
Tribe 13	Cichorieae (Cichoriaceae)

Cassini's extensive researches on the Compositae, largely published in various parts of Cuvier's 60-volume *Dictionnaire des Sciences Naturelles*, have tended to be neglected until recent years or even castigated because of the large number of new genera (324) which he put forward. Some of Cassini's genera have found favour with taxonomists who have adopted a narrower generic concept than is normally favoured but it has been difficult to assess his work adequately because of the scattered way in which it was published. Botanists have, therefore, every reason to be grateful to R. M. King and Helen W. Dawson for assembling Cassini's contributions on the family in Cuvier's *Dictionnaire* and arranging to have them published together, with an introduction and index, in a three-volume facsimile. As Cronquist (1977) shrewdly observes, "In spite of Cassini's perspicacity in recognizing relationships and detecting useful technical characters, his work was not highly influential. His idiosyncratic approach and numerous new names alienated those who might have profited from his more fundamental ideas". As we have already remarked, however, Bentham took full cognisance of Cassini's work in arriving at his own treatment of the family.

Bentham's tribal classification has stood the test of time, and with some modifications such as those introduced by Hoffmann in his account of the family in Engler and Prantl's *Die natürlichen Pflanzenfamilien* (1889–94), Dalla Torre and Harms in their *Genera Siphonogamarum* (1907) and Melchior in the sixth edition of Engler's *Syllabus* forms the basis of most current work.

It is generally accepted that the Compositae are a "natural" family with well established limits and a basic uniformity of floral structure imposed on all members by the common possession of characters such as the aggregation of the flowers into capitula and the special features of the stamens and corolla. Most of the various names applied to the group such as the Synanthereae, Androtomeae, Nevramphipetalae, Aggregatae, etc. refer to such features. Bentham said of the family, "Ordo omnium vastissimus et quam maxime naturalis, genere nullo hucusque cognito ad limites ambiguo".

In common with other natural families such as the Cruciferae, Umbelliferae and Gramineae, "the uniformity in inflorescence, floral and fruit structure imposed by the family characters tends to make recognition of tribes and genera difficult" (Heywood, 1971). For the recognition and circumscription of tribes and genera, use has often to be made of small-scale or "trivial" characters whose validity and significance is often placed in doubt. There has been much research into the genetical basis and adaptive significance of many such features, especially those that appear to be related to floral mechanisms and fruit dispersal, but a vast field remains open for further research. Even features such as cypsela structure and anatomy, the potential importance of which has been known since the times of Gaertner and Schultz Bipontinus, have been studied in detail in only a few groups such as the Anthemideae and Cardueae where they have been shown to be taxonomically valuable. It is difficult to believe that carpological features will prove to be of lesser value in all the remaining tribes.

It seems to be a characteristic which the Compositae shares with other large natural families that the tribes or subtribes contain a small number (one to four) large genera which make up the bulk of the group in terms of species, together with a larger number of small genera, often peripheral in terms of relationships, many of which contain only one or two species.

Even after attempts are made to dismember the large core genera, this basic pattern remains and it has also been found, for example, in the Cruciferae and Umbelliferae. Whether this reflects the underlying evolutionary history of the families concerned or is the consequence of recent taxonomic practice, or more likely a combination of both, is not yet clear.

SUBDIVISION OF THE FAMILY

Although the 13 tribes recognized by Bentham and by Hoffmann have been largely accepted to the present day, some more recent proposals for modification to this system have been made. There is, for example, a growing recognition that the Helenieae is an unnatural assemblage as first indicated by Cronquist (1955). The position of the genus *Liabum* in the Vernonieae (Cassini) or Senecioneae (Bentham) is anomalous and Rydberg's proposal to create a separate tribe for it, the Liabeae, is now gaining acceptance. The genus *Echinops* which Cassini recognized as forming a separate tribe, placed between the Cardueae and Arctotideae, but which Bentham placed in the Cardueae is also puzzling.

The positioning of a considerable number of genera within the thirteen or so generally accepted tribes is, however, uncertain as will be seen from the following chapters. This is not surprising in view of the vast amount of additional information of all kinds that has become available since the times of Bentham and Hoffmann.

Recently Wagenitz (1976) has surveyed the distribution of certain characters in the tribes of the Compositae and has revealed "the existence of two rather distinct groups of tribes inside the Asteroideae, each characterized by several common characters and common tendencies". The two groups are:

    I Vernonieae, Liabeae, Mutisieae, Cardueae, Echinopeae, Arctotideae.
   II Eupatorieae, Heliantheae, Helenieae, Senecioneae, Calenduleae, Astereae, Inuleae, ? Anthemideae.

He notes that the affinities of the Cichorioideae are closer to group I of the Asteroideae than to group II. This would agree with Carlquist's recent (1966) division of the whole family into two phyletic groups, the first of which contains the Cichorieae along with the tribes of Wagenitz's group I apart from the Eupatorieae:

   Subfamily Cichorioideae: Mutisieae, Vernonieae, Cardueae, Arctotideae, Cichorieae, Eupatorieae.
   Subfamily Asteroideae: Heliantheae, Astereae, Inuleae, Calenduleae, Senecioneae, Anthemideae.

Already one can detect a trend towards acceptance of such classifications in which one of the two subfamilies recognized includes the Cichorioid group as just one of the tribes instead of making it the basis of its own subfamily. Or, following the implications of Wagenitz's proposal, three subfamilies can be recognized of which one is the Cichorioideae. Jeffrey (1978) goes further and splits the family into two subfamilies in a way similar to Carlquist and then suggests that the tribes Eupatorieae and Senecioneae might warrant segregation into a third subfamily, as they are in some ways intermediate between the other two subfamilies.

STRUCTURE OF THE SYMPOSIUM

This symposium is the first attempt to tackle the Compositae down to generic level since Bentham. In the absence of such a review, the Compositae has grown by accretion of species and genera and, in common with other such families which are too large to monograph as a whole, the basic structure has been accepted and bits added on without full regard to the consequences. The number of papers published in the systematics of the Compositae since Bentham runs into thousands and the *Kew Record of Taxonomic Literature* for the years 1971–74 gives over 1300 references for the family.

In planning the review of the family for this symposium, ideally we should have attempted to start with a *tabula rasa* and build up a new classification from below by grouping together related species and genera to form into subtribes and tribes. This was impossible in practical terms and we have been forced to accept the given tribes, in the full knowledge that several of them are unsatisfactory, and have used them as the units within which different specialists would operate. Such a method has its dangers in that no uniformity of generic treatment is possible between tribes (is it ever?) and there was the likelihood of dispute, as will be seen in some of the following chapters, between different tribal reviewers as to their inclusion or not of particular genera in their tribe. We have, in fact, a few cases where genera have been excluded by one tribal reviewer but not accepted by any of the others!

Nonetheless the systematic reviews of each tribe should provide for the first time a virtually complete generic listing, with tribal affiliation, for the whole family. Likewise in view of the large amount of chemical investigations that have been made into the family, a review of the relevant literature has also been made for each tribe by an appropriate specialist. In addition several major review papers on general aspects of the biology, systematics and chemistry of the Compositae have been included.

MORPHOLOGY AND ANATOMY OF THE FAMILY
Ecological diversification

As noted by several contributors to this symposium, the Compositae, in spite of its relatively uniform capitular and floral features, occupies a wide range of habitat types and is found in abundance on every continent except Antarctica. The family is most abundant, as to diversity, in montane subtropical or tropical latitudes, this diversity being most evident in mountainous areas that border semi-arid or desert regions.

Ecological diversification must have begun very early in the development of the family and, when successful, persisted for relatively long periods. Thus there is a tendency towards tribal and subtribal specialization such that, for example, the phyletically remote tribes Vernonieae and Eupatorieae have largely adapted to mainly mesic, tropical or subtropical habitats; the Astereae, Inuleae and Lactuceae, among themselves also phyletically remote, to largely temperate, montane or high latitudinal habitats etc. Within the larger tribes considerable ecological diversification may be found, but habitat predelictions usually hold: wholly aquatic genera (e.g. *Gymnocoronis, Trichocoronis* and *Shinneria*) and epiphytic taxa are more often found in the mesically disposed Eupatorieae, while extreme xerophytic adaptation is more often found in the more xerically disposed Astereae, Inuleae and Helantheae.

Habital diversification

The Compositae are so plastic as regards habitat accommodation that extreme xerophytes or mesophytes often occur in the same genus (e.g. *Erigeron* and *Coreopsis*, both with aquatic and highly adapted desert species; *Senecio* with a wide range of ecological types, etc.). This ecological plasticity has been made possible largely through habitat diversifications, which includes everything from highly reduced, minute, montane annuals less than 1 cm high (e.g. *Cuchamatanea* — Heliantheae) to relatively large (albeit soft-wooded) tropical trees up to 20 meters tall (e.g. *Brachylaena* — Inuleae and *Montanoa* — Heliantheae). Trees occur but sporadically in phyletically remote tribes, these habits presumably having evolved independently in such groups. Slow-growing, cold-enduring, hard-wooded shrubs, such as *Artemisia*, also occur, along with rapid-growing, cold-sensitive, herbaceous genera, which are common in warm deserts, as well as perennial, often succulent, saline taxa such as *Borrichia* and *Varilla*.

Which of these habital types is primitive in the family is debatable. A woody forbearer is favoured by Cronquist (1977), this being a reversal from his earlier view (1955) on the subject, while a perennial herbaceous or suffrutescent habit is championed by yet others. This question is not easily resolved, but what does seem apparent is that perennial herbs and suffruticose species predominate in regions where the greatest phyletic diversity occurs: the more xeric montane communities of subtropical and tropical North and South America, regions from which most acknowledged workers would derive the Compositae (cf. Chapter 2). Trees or tree-like taxa nearly always occur in regions where phyletic diversification is low, such as mid to low tropical or subtropical rain forests or else as insular endemics where woodiness is clearly a secondary phenomenon (Carlquist, 1966).

Plant communities *dominated* by tree-like or "woody" Compositae are also relatively rare but are perhaps best developed in the planalto regions of east-central Brazil where fire-resistant, short trees such as *Piptocarpha* (Vernonieae) tend to dominate seasonally burned landscapes. The wood in such species is usually quite soft, easily ruptured, and probably not commercially durable. Only the occasional tropical forest tree is reported to have sufficient hard wood to be commercially productive, although it may be conceded that the woody base of a persistent herbaceous annual, such as *Helianthus annuus* might be readily sawn and shaved into shingles (Cronquist, 1977) but hardly as quarted wood for construction purposes.

Capitular diversification

As may be noted in the several papers dealing with the subject in this symposium, one of the most neglected areas of experimental study within the Compositae is that of the adaptive value of characters, or character suites, associated with the inflorescence (capitulum) and floral features. While, as noted above, habital modifications seem fairly explicable in terms of ecological strategies (e.g. annuals, spines and succulents in xeric areas; aquatic habit in mesic areas; rapid-growing, r-strategy, weedy "trees" in tropical regions, etc.), floral modifications are less open to interpretation, except where obvious correlations exist such as in the dioecious, wind-pollinated genera such as *Ambrosia*.

Leppick (Chapter 4) discusses several such strategies or morphological adaptations for the capitulum *per se* and Burtt (Chapter 3) discusses some of the over-all problems relating to floral modifications *within* the capitulum, but relatively few detailed, experimental, studies have been made. With the recent emergence of quantitative methods in the field of population ecology, more meaningful studies can be expected, especially as regards adaptive strategies (i.e. morphological accommodations) for such things as pollination, dispersal and propagule survival.

A wide range of pollinators has been reported for the Compositae, from beetles to humming birds. Indeed, as noted by Leppick (Chapter 4), the capitulum itself seems to have evolved such that it stimulates an individual flower with respect to pollinator attraction. Somehow floral aggregation with a single involucre served an adaptive function and this must have occurred at an early time within the family, so universal is the structure. Once formed, the involucre itself took on various adaptive modifications, some of these having to do with attraction (e.g. coloured, persistent involucral bracts in *Helipterium* and *Helichysum*, where these have essentially replaced ray florets in this function) and dispersal (e.g. cockle burs in *Xanthium* spp.).

Whatever the adaptive value of floral aggregation might have been, it has occasionally extended to the secondary aggregation of heads into heads. This phenomenon, termed syncephaly, has formed independently in a number of tribes. It would seem that, while there is safety in numbers, in some Compositae an important adaptive strategy is organization and cohesion in the ranks, from the standpoint of both pollination and dissemination.

Floral diversification

Relatively little is known about adaptive strategies in floral and fruit modifications, although the significance of pappus structure seems apparent in many groups such as, for example, the tribe Lactuceae where a "parachute"-type pappus enhances dispersal, and in *Bidens*, etc., where the pappus acts as an adhering agent (Pijl, 1972).

Adaptive strategies within this large family are bound to become better known as more and better attempts at statistical correlations are made and as more attention is given to the evolutionary forces which affect adaptational strategy.

Venable (in preparation), using a cosmopolitan range of floristic treatments as well as herbarium sheets, obtained data from 5451 species distributed among all tribes of the Compositae in an attempt to correlate habitat and habit with dispersal adaptations. He ascertained the different dispersal mechanisms found among these species and by correlating this with habitat types noted, among other things, that (1) species of plains and prairies tend to have capillary pappus less frequently than species of deciduous forests; (2) species of mountainous regions tend to have capillary pappus more frequently, and "no obvious" dispersal adaptation less frequently, than do those of lower elevations within the same region; (3) species of desert regions tend to have more adhesive structures and less wind-dispersed pappus types than those of non-desert regions; and (4) very mesic, riverine or sea-coast habitats, tend to have a high percentage of species with no obvious modifications of pappus for dispersal.

One of the more interesting results of Venable's study is that found in the correlation of pappus structure and habit. In the progression from annual to perennial herbs to shrubs he found a concomitant increase in the proportion of capillary or plumose pappus (wind dispersal) and a decreasing proportion of both adhesive or "no obvious" dispersal adaptations.

A more speculative, but essentially correlative, study by Levin and Turner (1977) entitled "Clutch size in the Compositae" may be mentioned here. They sampled herbarium sheets of 1007 species distributed among

110 genera of the tribe Heliantheae in an attempt to ascertain if there was any correlation between the number of propagules within an involucre (the "clutch" unit) and (1) the weight of individual propagules, (2) general habitat of the species or (3) life-habit of the species itself. Statistical analysis of their data revealed that annuals tended to have smaller clutch sizes and smaller propagules, the mean annual seed weighing 1·68 mg *vs* 4·45 mg in perennials. Other things being equal, this undoubtedly makes for greater dispersibility of annuals, most of which are weedy, *r*-strategy species (Whittaker, 1975). In short, they claim to have demonstrated that clutch size differences are related to ecological sites and growth form, the differences being shaped by biotic selective agents as well as physiological and morphological adaptations.

A point not discussed by these authors is the phyletic import or insight that might be derived from such studies. Most systematists tend to focus upon easily observed characters and use these in their classificatory schemes. This was especially true of early workers who did not use evolutionary theory as a working motif. Thus, some of the better scholars working in the Compositae were inclined to let certain selected characters "make" their genera (and generic clusterings!) rather than the other way about (i.e. good systematics revolves about the recognition of genera using a multiplicity of characters out of which the critical or "key" characters are selected; Cronquist, 1968).

As already noted, ecological and coevolutionary adaptive strategies among plant groups are only now receiving the kind of experimental studies and statistical analysis needed to make them meaningful within the framework of systematics generally. Character evolution is the backbone of phyletic thinking, however many characters we might employ in the construction of our classification. Interpretative thinking as to the direction that selection took in arriving at these characters must also play a part in our phyletic thinking.

CHEMISTRY OF THE FAMILY

General pattern

A study of the early plant herbals reveals that a surprisingly large number of plants of the Compositae were used for their curative properties. Thus, leaf extracts of the dandelion, *Taraxacum officinale*, have long been known to be diuretic, hence the names 'piss-a-bed' and 'pissenlit' (in France) for this plant. Sneezewort, *Achillea ptarmica*, apart from its sneeze-inducing smell, was chiefly employed against the toothache, according to Gerard. Even the unlikely daisy, *Bellis perennis*, was reported by Culpepper to be a wound-herb.

Undoubtedly the wide medicinal use of many composites inspired the early organic chemists at the turn of the century to explore their chemistry in order to identify the active constituents. Since those days, the chemistry of the family has continued to be widely investigated for a variety of reasons and, as a result, we now have an enormous amount of information on the organic constituents present. Several classes of plant compound are characteristic of this family, notably the terpenoid-based sesquiterpene lactones, the fatty acid derived polyacetylenes and the polysaccharide fructans. Composites, in fact, are exceptionally rich, both in the range of secondary compounds present and also in the numbers of complex structures known of any one class. Futhermore, as Hegnauer (Chapter 10) points out, the family is very distinctive in its chemical attributes. Although no single class of constituent is unique to the family, the Compositae are unlike any other family in the array of characteristic constituents (Table II).

The distribution of the different classes of chemical constituent at the tribal level within the family is indicated in Table I, but information on

TABLE II. Chemical pattern of the Compositae

Class of compound	Location and biological activity
PRESENT IN ALL TRIBES	
1. Inulin-type fructans	in storage organs
2. Characteristic fatty acids	in seed oils
3. Sesquiterpene lactones	mainly in leaves; bitter-tasting
4. Pentacyclic triterpene alcohols	as esters in fruit pericarps, and in lipids generally
5. Caffeic acid esters	in leaves; cynarin is diuretic
6. Methylated flavonoids	in leaves and flowers (as yellow pigments)
PRESENT IN MOST TRIBES	
7. Acetylenic compounds	in roots and leaves; antimicrobial activity
8. Essential oils, including phenolic monoterpenes	in leaves and fruits
9. Cyclitols	in leaves
10. Coumarins	in leaves and flowers
PRESENT IN FEW TRIBES	
11. Rubber (polyisoprene)	in roots and stems
12. Pyrrolizidine alkaloids	in leaves; highly toxic
13. Triterpene acids	free in flowers; combined with sugar (as saponin) in leaves
14. Diterpenes	in all tissues
15. Cyanogenic glycosides	in leaves and fruits; toxic
16. Anthochlor pigments	in yellow flowers
17. Chromenes	in leaves and roots; insecticidal
18. Fatty acid amides	in roots; insecticidal

these points is still limited and the outline may have to be revised as new results come in. As will be apparent from the chemical reviews of the different tribes which appear later in this volume, the degree of chemical ascertainment is quite low, as compared with morphological or anatomical ascertainment; it is probably less than 10% of species in most tribes. In the case of the Mutisieae, chemical information is so meagre that there is no report on this group of plants.

Clearly, more surveys are needed before a complete picture becomes available of the chemistry of the family. One valuable outcome of the Compositae Symposium, the proceedings of which are reported here, has been a significant increase in chemical studies and many new results have already appeared in the primary literature, the details of which have come in too late for consideration in the present volume. One might ask at this stage: is the present picture of the chemistry likely to be altered drastically as new data pour in? Fortunately, we do have an answer to this question in the review of Hegnauer (Chapter 10). This author has summarized the chemistry and its systematic implications both in 1964 (in *Chemotaxonomie der Pflanzen*, Volume 4) and in 1976 (this volume). Although the phytochemical information doubled in the interim, he found that the broad generalizations made in 1964 stood the test of time and were still true 13 years later.

Many of the substances elaborated by the family are toxic or show other significant physiological activity, and this may be one reason why plants of the Compositae are rarely used in human diets or for animal fodder. Some of the more important economic plants are listed in Table III. It will be seen from this list that there is only one plant which is a regular item in human diets—the lettuce. Even this plant is not eaten primarily for its nutritional properties and indeed, it is low in protein or carbohydrate. It is used because of its crispness and crunchiness as a basic salad ingredient. Alternatives to lettuce as a salad material are chicory and endive. There are also several minor vegetables in the family, especially the Jerusalem artichoke *Helianthus* and the globe artichoke *Cynara*, of which the inflorescence is the part consumed.

The most generally useful economic plant of the Compositae, however, must be the sunflower, *Helianthus annuus*. It is widely cultivated as an oil seed crop, but is also a good source of seed protein and the leaves provide animal fodder. The oil is a valuable food, with a good balance of dietary fatty acids, and has even been employed beneficially in the treatment of human cancer. In 1970, the sunflower ranked second only to the soyabean as a world oil crop, largely due to its extensive cultivation in the U.S.S.R. (Heiser, 1976). Another composite with potential as an oil seed crop is *Carthamus tinctorius*, better known as a source of a yellow dye in the flowers.

The rich accumulation of essential oils and other terpenoids in certain composites is responsible for the use of various members such as tansy, *Tanacetum vulgare*, and wormwood, *Artemisia absinthium*, for flavouring foods or liqueurs. Terpenoids and certain phenolic constituents are also responsible for the value of many composites in pharmacy and medicine (see Wagner, Chapter 14).

When considering the economic value of plants of the Compositae (Table III), it must be pointed out that the useful plants are to a considerable extent counter-balanced by the large number of weeds in the family. Indeed, there are few families with such an abundance of weedy

TABLE III. Economically important members of the Compositae[a]

FOOD PLANTS
Cichorium endivia Willd.—endive
Cichorium intybus L.—chicory
Cirsium oleraceum (L.) Scop.—meadow cabbage
Cynara cardunculus L.—cardoon
Cynara scolymus L.—globe artichoke
Helianthus tuberosus L.—Jerusalem artichoke
Lactuca sativa L.—garden lettuce
Scorzonera hispanica L.—black salsify
Tragopogon porrifolius L.—salsify

MEDICINAL PLANTS
Achillea ptarmica L.—sneezewort
Antennaria dioica (L.) Gaertner—cat's foot
Arnica montana L.—arnica
Chamaemelum nobile (L.) All.—chamomile
Cnicus benedictus L.—blessed thistle
Petasites hybridus (L.) Gaertner—butterbur
Silybum marianum (L.) Gaertner—milk-thistle
Tussilago farfara L.—coltsfoot

FLAVOURING PLANTS
Artemisia absinthium L.—wormwood
Artemisia dracunculus L.—tarragon
Inula helenium L.—elecampane
Tanacetum vulgare L.—tansy

RUBBER PLANTS
Parthenium argentatum Gray—guayule
Taraxacum bicorne L.—kok-saghuz

OIL PLANTS
Helianthus annuus L.—sunflower

INSECTICIDAL PLANTS
Anacyclus pyrethrum DC.
Echinacea angustifolia DC.
Pulicaria dysenterica (L.) Bernh.—fleabane
Tanacetum cinerariifolium (Trev.) Schultz Bip.—pyrethrum

DYE PLANTS
Carthamus tinctorius L.—safflower
Centaurea cyanus L.—cornflower
Solidago virgaurea L.—golden rod

ORNAMENTAL PLANTS
Calendula officinalis L.—pot marigold
Chrysanthemum carinatum Schousboe—annual chrysanthemum
Dahlia variabilis L.—dahlia
Dendranthema indica (L.) Desmoulins—autumn chrysanthemum
Tagetes indica L.—African marigold
Tagetes patula L.—French marigold

[a] The above list is only a selection. Many plants serve several purposes, e.g. sunflower, tansy, safflower, etc.

1. AN OVERTURE TO THE COMPOSITAE

TABLE IV. Some weedy members of the Compositae

Achillea millefolium L.—yarrow
Ambrosia artemisifolia L.—roman ragweed
Anthemis cotula L.—stinking mayweed
Bellis perennis L.—daisy
Centaurea nigra L.—lesser knapweed
Centaurea scabiosa L.—greater knapweed
Chrysanthemum segetum L.—corn marigold
Cirsium spp.—thistles
Cotula coronopifolia L.—brass buttons
Crepis spp.—hawk's-beards
Hieracium spp.—hawkweeds
Leontodon spp.—hawkbits
Matricaria matricarioides (Less.) Porter—pineapple weed
Parthenium hysterophorus L.—wild feverfew
Senecio jacobaea L.—ragwort
Senecio vulgaris L.—groundsel
Sonchus spp.—sow thistles
Taraxacum officinale L.—dandelion
Xanthium strumarium L.—cockle bur

members (Table IV), many of which are extremely successful and have spread throughout the temperate areas of the world. The success of these weeds stems mainly from the development of biological features which ensure both survival under adverse environmental conditions and also a high reproductive rate. For example, the common ragwort has the ability to develop a massive root system so that it can grow on poor soils and it also has an enormous yield of seeds. An average plant produces between 50 000 and 60 000 fruits, with a germination frequency of over 80%. Chemical factors are, nevertheless, important in Compositae weeds in providing protection from over-grazing. The ragwort, for example, is well protected from the majority of potential herbivores by the presence of toxic levels of pyrrolizidine alkaloids in the leaf tissue. Indeed, pyrrolizidine alkaloids are so effective as mammalian toxins that about 50% of all cattle deaths due to plant poisoning are the result of ingestion of these particular alkaloids.

In the following paragraphs, some of the major classes of chemical constituents are discussed in relation to economic importance and to their possible value to the plants themselves.

Sesquiterpene lactones

Sesquiterpene lactones are colourless, often bitter-tasting, lipophilic constituents, which are the most characteristic single group of chemicals known in the Compositae. They are present mainly in leaf tissues and can constitute up to 5% of the dry weight. They have been detected in all the tribes except the Tageteae. A profusion of different lactone structures have been characterized, over 600 in number according to Herz (Chapter 11). Although regular constituents in composites, these lactones do occur sporadically in 10 other angiosperm families and also in a few liverworts. The only related family where they are found—but only in some 12 taxa—is the Umbelliferae.

Until recently, only the chemistry of the sesquiterpene lactones was studied, their structural identification, biogenesis and stereochemistry presenting many interesting problems. Attention has turned in the last few years to their biological importance and they have been found to be significantly active in a variety of tests (Rodriguez et al., 1976). A number are toxic to livestock and their major role in the ecology of the family seems to be as a deterrent to mammalian herbivores. They also have insecticidal activity (see section below), but this may be less important than their deterrent effects on rabbits, deer and other browsing animals. Furthermore, the lactones are not only feeding toxins in the case of mammals but they also cause allergic contact dermatitis.

Over 80 sesquiterpene lactones have been examined in patch tests in man and the results of these experiments show that the γ-lactone grouping with the exocyclic α-methylene function is a structural requirement for dermatitic activity. This grouping is present in practically all the common lactones, so that it is not surprising that the list of composites capable of causing contact dermatitis is a long and impressive one (see Rodriguez et al., 1976). The list includes plants used as food flavouring (lettuce, sunflower, elecampane, globe artichoke), several ornamentals (dahlia, *Rudbeckia*) and the common dandelion, *Taraxacum*. The effectiveness of the lactones in causing dermatitis in humans is illustrated by the recent outbreak in India of many cases, due to the accidental introduction of *Parthenium hysterophorus* into this subcontinent. This weed contains the active compound, parthenin. The severity of the dermatitis produced can occasionally lead to the death of the victim.

The presence of sesquiterpene lactones in composites is often associated with a bitter taste, and it is likely that this repellent taste response acts as a signal to protect the plants from being heavily grazed. It has been necessary to remove this bitter taste from those composites used as salad materials. While the wild relatives of *Lactuca sativa* contain two bitter compounds, lactucin and lactupicrin, these two substances are largely absent from cultivated varieties.

The toxicity of these lactones can be an important cause of livestock poisoning. *Hymenoxys odorata* growing in Texas, U.S.A. is said to be responsible in the State for an annual loss of sheep and goats worth several millions of dollars. Toxicity of hymenovin, the active principle, may be due to the alterations it produces in the microbial composition of the animal's rumen and these alterations affect vital metabolic functions. Although the lactones have useful anti-tumour and cytotoxic activity, their toxicity as indicated above will probably limit their application in treatment of human cancer.

Polyacetylenes

These reactive substances have been found in roots and/or leaves of the great majority of the composites that have been surveyed. Acetylenic compounds are much more labile than most other plant substances and they can only be isolated successfully from fresh plant material. In spite of this limitation, their distribution in the family is well documented (Sørensen, Chapter 13). They occur widely in 11 of the 13 tribes, but are rare in the Cichorieae and Senecioneae.

Hardly any economic use has been made of the polyacetylenes, although one such compound, ichthyothereol, is the active principle of *Ichthyothera terminalis*, extracts of which have been used as a fish poison. Perhaps, the most important function of polyacetylenes in the Compositae is as antifungal agents, although this has yet to be proved. At least acetylenic compounds have been identified as phytoalexins with certainty in the safflower *Carthamus tinctorius* and there is some evidence that they have a similar role in *Dahlia*. It may also be significant that acetylenes also have been identified as phytoalexins in a different family, in the genus *Vicia* of the Leguminosae.

Insecticidal principles

It has long been recognized that members of the Compositae contain insecticidal components. Such properties are reflected in the common names of such plants as fleabane which was recommended in early herbals for its ability to repel midges and fleas. Again, the leaves of tansy have a long history as insect repellents. The generic name for tansy, *Tanacetum*, the "deathless" plant, is derived from the early gruesome practice of rubbing the leaves on to human corpses with the express purpose of repelling the attack of various corpse worms.

The most familiar insecticidal plant in the family is undoubtedly pyrethrum, which has been used commercially for many years, with the flower heads being employed as the basis of the pyrethrin compounds (see Chapter 14). Six constituents provide the insecticidal "knockdown" of pyrethrum extracts. The relationship between activity and structure has been explored and used as a basis for synthesizing even more active analogues. The pyrethrins are not the only insecticidal agents in the Compositae, since long-chain fatty acid amides are responsible for the insecticidal action of root extracts of many Anthemideae and Heliantheae (see Chapter 10).

Other classes of secondary compound may also be important repellents to insect feeding in the family. There is some evidence that sesquiterpene lactones are effective in preventing polyphagous Lepidoptera from feeding on *Vernonia*. The lactone glaucolide A, which occurs on *Vernonia* leaves, is rejected in laboratory tests by insect larvae given a choice of leaf containing it or lacking it. Further, larvae deliberately fed on a glaucolide A-containing diet suffered as a result and were not as fit as similar insects reared on a glaucolide A-free diet (see Jones, Chapter 17). The toxicity of these lactones to insects is underlined by the recent report that eremanthine, also a substance from the Vernonieae (Harborne and Williams, Chapter 18), is very effective against the cercarieae of the trematode *Schistosoma mansonii*, a serious insect parasite in man.

One other class of composite constituent which has excited interest as insecticidal material is that based on 2,2-dimethylchromene. Two such compounds, called precocenes 1 and 2, have been extracted from the insect-resistant *Ageratum houstonianum* (Eupatorieae). These compounds are toxic to insects because they interfere with juvenile hormone activity in such a way that precocious metamorphosis occurs. The nymphs of the milkweed bug, for example, miss out one or more larval stages to become imperfect adults, the females being sterile (Maugh, 1976). These chromenes are thus very subtle insecticides and act by reducing the insects' reproductive capacity.

One final group of insecticidal constituents of taxonomic significance in the family are the sulphur-containing thiophene derivatives. One such compound, α-terthienyl, occurs in *Eclipta prostrata* and is known to be active against the nematode *Pratylenchus penetrans*. Similar sulphur-based nematocides have been detected in 70 of 175 composite species tested, and their distribution, interestingly enough, follows taxonomic divisions within the family (Gommers and Voor in Tholt, 1976).

Rubber production

The ability to synthesize latex, usually in special laticifer-producing tubes and cells, is widely present in angiosperms, occurring as it does in at least 20 families. In the Compositae, milky sap is present characteristically in the Cichorieae and is virtually universal throughout the tribe. Latex does appear in individual members of the tribes Cynareae, Vernonieae, Arctotideae and Mutisieae, so that it is by no means absolutely restricted to the one tribe. A related family where plants may have laticiferous cells is the Campanulaceae.

The principal chemical of the latex is polyisoprene or rubber and this is the main constituent that is of commercial value. There are often many other minor components in the latex, including essential oils and diterpenoid resins and the presence of such contaminants may limit the exploitation of a given plant as a source of rubber. While the rubber tree *Hevea brasiliensis* (Euphorbiaceae) is an unrivalled source of natural rubber, other plants have been examined as possible alternatives: several composites (see Table III) show considerable promise and indeed have been used for commercial rubber production. The woody shrub guayule (*Parthenium argentatum*) is of special interest since rubber is secreted in parenchymatous cells which occur in most parts of the plant. In favourable conditions, the yield of rubber from mature guayule plants reaches 20%. By contrast, rubber occurs in *Taraxacum bicorne* only in the roots and the yield here never reaches more than 10% dry weight. This is considerably better than most Cichorieae, where the yield of rubber rarely exceeds 1%.

The presence of milky sap, or the ability to synthesize polyisoprene, is a key chemical character which has been used, together with the ligulate corolla, to raise the Cichorieae to subfamilial or even familial status. As mentioned above, latex does occur in other tribes, so that the character does not provide real grounds for putting these plants into their own family. Indeed, there is much other chemical evidence linking the Cichorieae with the rest of the Compositae. Many characteristic composite chemicals occur in the Cichorieae, including sesquiterpene lactones, polyacetylenes, phytosterols, carotenoids, fatty acids and coumarins (see González, Chapter 40).

Recently, we have had the opportunity of surveying the Cichorieae for yellow flower pigments and have discovered as a result further links with other composite tribes. Chalcones, for example, which are found especially in Heliantheae flowers, occur in several taxa among Cichorieae. Coreopsin, first isolated from *Coreopsis* flowers, occurs in the genus *Pyrrhopappus* (Harborne, 1977) and marein, a major yellow pigment of *Coreopsis maritima*, occurs in yellow ligules of *Malacothrix* and *Calycoseris*. Furthermore, the yellow flavonol gossypetin, which has been found in yellow rays of *Chrysanthemum segetum* (Anthemideae—see Harborne, Chapter 12) also occurs in the yellow ligules of the desert annual *Glyptopleura marginata* (J. B. Harborne and D. M. Smith, unpublished results). The ability to synthesize more than one type of yellow pigment as revealed here may be important in the attraction of the special composite inflorescence to animal pollinators (see Leppik, Chapter 4).

Inulin accumulation

One of the most distinctive features in the biochemistry of the Compositae is the production of storage polysaccharides based on fructose instead of glucose. These unusual polysaccharides are known as fructans or inulins, the latter name being derived from *Inula helenium*, the roots of which are a particularly rich source. Fructans also occur in the Campanulaceae but are otherwise rare in the angiosperms. Structurally different branched-chain fructans, called levans, do, however, occur in certain grasses.

Starches, which are the common storage carbohydrates of higher plants, consist of the related components: amylose with $\alpha 1 \rightarrow 4$ linked glucose units and the branched polymer amylopectin containing glucose linked both $\alpha 1 \rightarrow 4$ and $\alpha 1 \rightarrow 6$. By contrast, inulins appear to be simple straight chain polymers containing fructofuranose units linked $\alpha 1 \rightarrow 2$ with a single glucose unit attached at the non-reducing terminal end. Inulins differ from starches in their relatively low molecular weight ($< 10\,000$), water solubility and inability to stain with iodine.

Inulins are principally root constituents, so that they are found in quantity in biennial or perennial species. The nutritive value of composite vegetables is based on the presence of large amounts of inulin. As a carbohydrate source for humans, fructans are just as useful as glucans. The best known inulin-containing vegetable is the Jerusalem artichoke, but several other species with storage roots have had similar use (see Table III). The sweet taste of Jerusalem artichoke after cooking may be due in part to the presence of fructose, released from inulin during boiling, which is sweeter (on a molar basis) than glucose, the breakdown product of the more usual starch. The flavour of artichoke roots is probably due to other (terpenoid?) constituents but these have not as yet been fully identified. As mentioned by Wagner (Chapter 14), inulins have a special medicinal value in diabetic diets, since diabetic sufferers can tolerate fructose much better than glucose.

Conclusion

In the present volume, the broad chemistry of the Compositae is discussed in Chapter 10 and this is followed by three chapters dealing at greater length with the sesquiterpene lactones, the flavonoids and the polyacetylenes. This section concludes with a review of pharmaceutical and economic uses (Chapter 14). All the more detailed information on individual compounds and their natural distributions are contained in the chemical tribal reports which appear in turn following each systematic report on a particular tribe. A final summary of the chemistry can be found in Chapter 41.

The absence of a general account on nitrogen compounds in the family may appear to need some explanation. However, there is at present rather little known about such compounds. Non-protein amino acids, which are so well represented in other angiosperm groups, such as the Leguminosae, are not known in the Compositae. They may well be present. Several composites are selenium accumulators and it is likely that they may contain similar selenium-based amino acid analogues as have been obtained from the legume genus *Astragalus* (Shrift, 1972). Alkaloids may similarly be thought to have been overlooked. However, the only alkaloids to have been thoroughly investigated in the family are those from *Senecio* and these are covered by Robins under the Senecioneae in Chapter 30.

Finally, protein patterns in the Compositae seem to be another neglected area. The situation here is being rectified by Boulter, who earlier published two cytochrome c amino acid sequences for the family (see Ramshaw and Boulter, 1975). Very recently, N-terminal amino acid sequences for the plastocyanins of 10 composites have been determined (Haslett et al., 1977). The plants investigated include four members of the Cichorieae (*Hieracium*, *Lactuca*, *Taraxacum*, *Tragopogon*), two Cynareae (*Centaurea*, *Cirsium*), two Heliantheae (*Helianthus*, *Guizotia*), one Anthemideae (*Tanacetum*) and one Senecioneae (*Senecio*). These results provide new information with important systematic implications and there is little doubt that nitrogenous compounds will not be neglected in future chemical investigations of this vast and fascinating plant family.

REFERENCES

CARLQUIST, S. (1976). Tribal interrelationships and phylogeny of the Asteraceae. *Aliso* **8**, 465–492.

CARLQUIST, S. (1966). Wood anatomy of Compositae. A summary, with comments on factors controlling wood evolution. *Aliso* **6**, 25–44.

CRONQUIST, A. (1955). Phylogeny and taxonomy of the Compositae. *Am. Midl. Nat.* **53**, 478–511.

CRONQUIST, A. (1968). "The Evolution and Classification of Flowering Plants". Houghton Mifflin, Boston.

CRONQUIST, A. (1977). The Compositae revisited. *Brittonia* **29**, 137–240.

GOMMERS, F. J. and VOOR IN THOLT, D. J. M. (1976). Chemotaxonomy of Compositae related to their host suitability for *Pratylenchus penetrans*. *Neth. J. Pl. Pathol.* **82**, 1–8.

HARBORNE, J. B. (1977). Variations in pigment patterns in *Pyrrhopappus* and related taxa of the Cichorieae. *Phytochemistry* **16**, 927–928.

HASLETT, B. G., GLEAVES, T. and BOULTER, D. (1977). N-Terminal sequences of plastocyanins from various members of the Compositae. *Phytochemistry* **16**, 363–365.

HEISER, C. B. (1976). "The Sunflower". University of Oklahoma Press, Norman, U.S.A.

HEYWOOD, V. H. (1971). Systematic survey of Old World Umbelliferae. *In* "The Biology and Chemistry of the Umbelliferae" (V. H. Heywood, ed.), 31–41. Academic Press, London and New York.

JEFFREY, C. (1978). Compositae. *In* "Flowering Plants of the World by Families" (V. H. Heywood, ed.), Elsevier International Projects, Oxford (in press).

LEVIN, D. A. and TURNER, B. L. (1977). Clutch size in the Compositae. *In* "Evolutionary Ecology" (B. Stonehouse and E. Perrin, eds.), pp. 215–222. MacMillan, London.

MAUGH, T. S. (1976). Plant biochemistry: two new ways to fight pests. *Science* **192**, 874–877.

PIJL, L. VAN DER (1972). "Principles of Dispersal in Higher Plants", 2nd edition. Springer-Verlag, New York.

RAMSHAW, J. A. M. and BOULTER, D. (1975). The amino acid sequence of cytochrome c from *Guizotia abyssinica*. *Phytochemistry* **14**, 1945–1950.

RODRIGUEZ, E., TOWERS, G. H. N. and MITCHELL, J. C. (1976). Biological activities of sesquiterpene lactones. *Phytochemistry* **15**, 1573–1580.

SHRIFT, A. (1972). Selenium Toxicity. *In* "Phytochemical Ecology" (J. B. Harborne, ed.), pp. 145–162. Academic Press, London.

SMITH, A. C. and KOCH, M. F. (1935). The genus *Espeletia*: a study in phylogenetic taxonomy. *Brittonia* **1**, 479–532.

VENABLE, L. (in preparation). Habitat and habit correlations to dispersal adaptations in Compositae.

WHITTAKER, R. H. (1975). "Communities and Ecosystems", 2nd edition. MacMillan, New York.

WAGENITZ, G. (1976). Systematics and phylogeny of the Compositae. *Plant Syst. Evol.* **125**, 29–46.

Chapter 2

Fossil history and geography

B. L. TURNER

*The University of Texas,
Austin, Texas, U.S.A.*

Abstract. The fossil and distributional data bearing upon the origin of the Compositae is reviewed and discussed in the light of plate tectonics. Contrary to what many recent workers believe, the family is accepted as a very old phyletic line whose origin stems back to at least the Cretaceous. Its world-wide distribution is largely accounted for by Continental Drift, its relative abundance in the fossil record since Miocene time being due to adaptive mechanisms in the exploitation of short-term, mostly disturbed habitats which have presumably become more characteristic of the continental land masses since that period. It is conjectured that the family arose in western Gondwanaland from perennial herbaceous populations which gave rise to both the Compositae and Calyceraceae.

> No fossil pollen of the vast family Asteraceae [Compositae] is known prior to the uppermost Oligocene, despite extensive search . . . . and pre-Miocene records are exceedingly few . . . . No Asteraceae have what appears to be a distribution achieved by direct migration between Australia and southern South America, suggesting with other evidence no more than a mid-Oligocene age for the family.
>
> (Raven and Axelrod, 1974)

CONTENTS

Introduction	22
Fossil record to 1919	22
Fossil record after 1919	24
Fossil record interpreted	25
Tribal distributions	27
Introduction	27
Vernonieae	28
Eupatorieae	28
Astereae	29
Inuleae	29
Heliantheae	29
Coreopsideae	30
Tageteae	30

Senecioneae 30
Anthemideae 31
Calenduleae 31
Arctoteae 31
Cynareae 31
Mutiseae 31
Lactuceae 32
Origin and dispersal of Compositae 32
  Place and time of origin 32
  Continental drift 34
  Protoasteroids and phyletic considerations 37
  Man and weeds 37
Acknowledgements 38
References 38

INTRODUCTION

As will be noted below, my views, as well as those of others concerned with the problem, differ substantially from those expressed by Raven and Axelrod in their excellent account of *Angiosperm Biogeography and Past Continental Movements* (quoted above) who briefly summarized fossil and biogeographical data bearing upon the distribution and origin of the Compositae. Bentham (1873, p. 390) would presumably also disagree, to judge from his sagacious consideration of the fossil record to that date which, incidentally, except for microfossils, has not been added to appreciably by subsequent workers. In the light of his remarks (quoted below) it would have been interesting to obtain Bentham's opinion on the origin and subsequent dispersal of the family had he knowledge of Continental Drift as we know this today. I would like to think that the views expressed below would be much like his were he commissioned with the paper to be presented here.

FOSSIL RECORD TO 1919

Small (1919) presented "an extensive but not quite exhaustive" account of the paleobotanical literature up to that date. Most of this related to macrofossils, mainly leaf or fruit impressions. It *is* an extensive, excellent presentation but has, unfortunately, been largely ignored by subsequent workers (e.g. by Raven and Axelrod, 1974; in fact these authors, presumably independently, came to more or less the same conclusion as Small regarding the place and origin of the angiosperms generally and the Compositae specifically: Andean regions of northern South America or what was once western Gondwanaland). This neglect has undoubtedly been due, in part, to Small's phyletic bias which arose out of his preoccupation with the Age and Area hypothesis of Willis (1919): he simply insisted that the large genus *Senecio* must be the most primitive member of the Compositae, a view shared by few, if any, subsequent workers. Nevertheless, Small did point out several important facts and interpretations to be made from the fossil record and these may be summarized as follows:

(1) Altogether, macrofossils representing approximately 100 species among 10 tribes are reported from 40 different geological sites, mostly in England.

(2) Macrofossils of asteroid taxa similar to forms existing today are found in the Lower Oligocene of Europe.

(3) Macrofossils, however scant, "of most, if not all", of the major tribes, including the Lactuceae, are found in increasing numbers from the Oligocene onward.

(4) Late Cretaceous or early Eocene are indicated as the time of probable origin of the Compositae, and the paleobotanical evidence shows that the origin of the family took place at approximately the same date as the first upheaval of the Andes.

(5) "Although the American tertiary has as yet yielded practically no Compositae, and the European forms, even from the Lower Oligocene, cannot be accepted as the earliest types, [paleobotanical] evidence, as far as it goes, supports the view that the setose type of pappus is the primitive form."

Inferences drawn from these macrofossils are only as good as the reliability of the identifications. Many, if not most, of the leaf impressions probably do not belong to the Compositae. However, fully 50 or more of the fossils are fruit impressions, nearly all from the Lower Oligocene and Miocene and, pictorially speaking, they look as much like fruits from the family they are supposed to represent as do most macrofossils, identity of which is not especially quibbled over by paleobotanists. At least, I believe that many of the species referred to the form genus *Cypselites* by Heer and listed by Small (1919) probably are asteroid cypselas.

Bentham's (1873, p. 390) evaluation of the macrofossils, most of which were the same as those cited by Small, seem pertinent.

> The geological record is remarkably scanty; but in the case of the members of this order, the absence of their remains is no proof of their non-existence at various geological periods. They are very rarely aquatic; and a comparatively small number only are to be met with on the borders of such waters as are wont to accumulate stores of organic remains; nor yet do they shed a profusion of leaves likely to be carried to any such hoarding-places. The great mass of them live, die, and are thoroughly consumed, without leaving a single fragment to serve as evidence or indication to future generations. It is only here and there that the winds appear to have carried an achene, by means of its pappus, to some place of deposit; and thus it is that Oswald Heer found in the upper miocene tertiary deposits of central Europe various impressions which he refers, on plausible grounds, to Compositae. He is also probably justified in his conjecture that the great majority of them belong to Cichoriaceae, two or three to Cynaroideae, and that one is probably the

achene of an aquatic *Bidens*. All this, if well founded, would show that at that tertiary epoch Compositae existed in Europe of the same general character as those which are now to be met with. It would seem to prove that they had then already attained that highly differentiated character they now possess, and consequently must have been already of very old date, although they had left no previous record of their existence which has as yet been exposed to our observation ... large herbaceous Compositae are not in the habit of casting their leaves unwithered, so as to have become encased in mud unaltered in shape.... And even Heer's above-mentioned Miocene Compositae achenes are doubted by some palaeontologists, who contend that they are seeds of Apocyneae. Some, indeed, of Heer's figures show the pappus not to be strictly terminal, but to proceed from an oblique or somewhat lateral notch, which is unusual though not unknown in Compositae (e.g. *Tourneuxia*); but many of the figures might be identified with more than one recent achene and pappus.

FOSSIL RECORD AFTER 1919

Relatively few significant *additions* have been made to the fossil record of the Compositae since 1919. Small himself deplored the fact that, at the time when his paper was published, little or nothing was known about the Compositae of the New World, remarking (p. 134), "This field for research is still very open and investigations on the early Tertiary Compositae of America would prove very interesting."

Probably the most significant fossil find, at least as to credibility is *Viguiera cronquistii* from the lowermost Miocene of Montana (Becker, 1969). The fossil consists of an intact head which the author states "closely resembles *Viguiera laciniata* H.B.K.... from California. It is suggested that composites of such a high level of development during the early Miocene must have had a considerably earlier evolution". Cronquist (pers. comm.) agrees that the fossil is a composite but probably not a *Viguiera*. Not having seen the fossil personally, I cannot be as certain, but from the published figures, it does indeed appear to be an involucrate receptacle with attached florets, presumably belonging to the tribe Heliantheae and perhaps closest to *Viguiera*. Considering the generic names assigned to the fossils reported upon (e.g. *Berberis, Cassia, Cercocarpus, Parkinsonia, Paulownia, Platanus, Potentilla, Quercus*, etc.) the genus would not be unexpected. *Viguiera*, even today, occurs somewhat south of the Montana fossil locales; in fact, if the fossil *is* a *Viguiera* I would relate the taxon to the extant *V. multiflora*, an annual species of the southern Rocky Mountains (Arizona and New Mexico) with a base chromosome number of $x=8$ which is sometimes segregated, along with a few other species, as the genus *Heliomeris*.

Most other macrofossils described since 1919 have been given specific names under form-genera and without the fossils in hand it is difficult, if not impossible, to relate them to known genera. An interesting exception is a fossil achene from the Pliocene of western Kansas which Segal (1965) has described as *Achaenites kansanum*. This is probably a species of *Lagenifera* (tribe Astereae) belonging to a group of genera largely confined at the present time to South America and Australia.

A very interesting discovery has been that of *Achaenites cichorioides* described from Eocene beds of Colorado by Knowlton (1925). I have personally examined the fossil, which is housed in the paleobotanical collection at the Natural History Museum, Washington, D.C. (specimen No. 36594) and take this to be an achene from the tribe Heliantheae (subtribe Galinsoginae) or Coreopsideae (subtribe Jaumeinae). The body of the impression is about 8 mm long, 2·3 mm wide, obpyramidal and presumably four-sided in the manner of *Helianthus*, being crowned at the apex with about 12 linear-lanceolate pappus scales about 7 mm long, in the manner of *Jaumea* or *Hypericophyllum*. It can of course be argued that this is wrongly identified but it *is* certainly a fruiting body of what most of us would call a composite achene were we to have encountered this in a herbarium packet, and is about as good a fossil as might be expected from the organic debris of this family.

In general, the most important recent addition to the fossil record of the Compositae has been from palynological studies. What it certainly shows is that pollen from the family became increasingly abundant in the fossil record from the Miocene onward and, as summarized by Raven and Axelrod (1974), is not known in the record before the Miocene.* Since such pollen appears to be derived from taxa referable to both the subfamilies Lactucoideae and Helianthoideae it must be assumed that the family as a whole was well differentiated at that time. Also noteworthy in the fossil record is the appearance of pollen during the Miocene in both the Caribbean Islands and the northwestern United States referable to the relatively advanced genus *Ambrosia* (Graham, 1963; Germeraad *et al.*, 1968), a taxon still found in these areas today, but certainly not centered in either of these regions.

FOSSIL RECORD INTERPRETED

So what does the fossil record of the Compositae, as currently known, tell us about the time and place of origin of this large, phyletically compact, family? Not much, really. What is perhaps most impressive is that the family was already well developed by Oligocene time, and that from Miocene on the family became increasingly frequent in the fossil record.

*Kemp and Harris (1975: *Nature* **258**, 303–37) have recently reported Compositae-like pollen from Paleocene sediments of the Indian Ocean and unpublished data confirm such pollen for the Cretaceous of North America (Tomb, pers. comm.).

The microfossil record *does not* tell us that the Compositae differentiated or developed as a family during the Oligocene or shortly thereafter; it only tells us that its pollen became increasingly frequent and more widely dispersed in the post-Oligocene sediments. Germeraad et al. (1968), in their palynological study of tertiary sediments from tropical areas, perhaps put this fact in proper perspective: "The conspicuously sudden appearance and subsequent rapid increase of Asteraceae (Compositae) pollen types in the Lower Miocene of northern South America, Nigeria and the Far East could reflect the adaptation to easy dispersal of Asteraceae seeds." A similar cautionary statement was also echoed by Muller (1970) in his major review of the subject, "It appears likely therefore that the fossil record [of the Compositae] reflects a real increase in their numbers and diversification during the Mio-Pliocene. The possibility still remains, however, that a fairly long period of macromorphological evolution in small populations, without any simultaneous morphological pollen differentiation, preceded this phase."

Lacking, then, any strong conclusions to be drawn from the fossil record, we must turn to other sources for evidence bearing on the origin of the Compositae. This fact has been appropriately commented upon by Bentham and I can do no better than quote him directly (1873, p. 391):

> In the absence of all direct evidence we are left to judge of the antiquity and origin of Compositae from their comparative structure and from their geographical distribution, as to both of which we have still much to learn, and in both which respects several of the boldest of modern hypothesists have neglected or been ignorant of much that is known.
>
> A general notion is prevalent, especially among French botanists, that Compositae are at the summit of the scale of progression in the vegetable kingdom—that DeCandolle's idea that the greatest perfection was to be sought for where, as in some Thalamiflorae, the essential parts of the flower, the petals, stamens, and carpels are the most distinct from each other, is altogether erroneous—that these Thalamiflorae are, in fact, the nearest to the Monochlamydeae, which commence from the base of the Dicotyledonous scale—and that the high degree of consolidation in the floral organs of Compositae is a strong proof of perfection and thence of a comparatively recent origin. It seems very probable that these views may be correct; yet, on the other hand, we must bear in mind that the numerous monotypic or oligotypic highly distinct genera confined respectively to the widely distant centres of preservation of the Mediterranean region, tropical and Southern Africa, Southern and Western Australia, Chili, the Mexican region, etc., *point to a very wide dispersion of the original stock of the order at a very early period, when the physical configuration of the surface of the globe must have been very different from what it is now,—that this dispersion appears, indeed, to have been so early as to give time for the absolute fixation of secondary characters, which in most orders are very inconstant—and that, moreover, previously to this dispersion the stock must have existed long enough to give absolute permanence and*

an otherwise unexampled constancy to those essential characters of primary importance which I shall recur to in detail...

But although Compositae must thus have existed, in some shape or other, but yet with all these essential characters, at an early geological period, the differentiation of the larger groups probably took place after the isolation of the actual centres of preservation. Of the thirteen tribes adopted, two only, Asteroideae and Senecionideae, may be said to be cosmopolitan or nearly so. Cichoriaceae, Cynaroideae, and Anthemideae belong to the northern hemisphere with chief centres in the Mediterranean and Central Asiatic regions; a few, but those (except some Cichoriaceae) either forming part of or closely allied to Europaeo-Asiatic genera, have spread over North America and even down the Andes to extra-tropical South America. Calendulaceae and Arctotideae are African, extending sparingly into Europe. Vernoniaceae, Eupatoriaceae, Helianthoideae, Helenioideae, and Mutisiaceae are essentially American, but with a few types which may have arisen in the tropical and subtropical regions of Africa and Asia. The great tribe of Inuloideae is for the most part Old-World, although the subtribes Plucheineae and Gnaphalieae have been long enough in America to have there formed a very few generic types.

With the above as an introduction, Bentham proceeded to discuss the distribution of the 13 tribes known to him. This is what I propose to do here, but with much greater brevity since this is discussed in considerably more detail in other chapters. What I want to do then is set the stage for a creative synthesis of the fossil and distributional data, bringing up to date any new significant facts or generalizations about the tribes themselves and proceed from there to a likely formulation as to their time and center of origin. But before I leave this section I cannot help but call the reader's attention to the passage which is italicized in the above quote from Bentham, especially "when the physical configuration of the surface of the globe must have been very different from what it is now..."! And hence my earlier allusion to what he might have written about the origin of the family had he a knowledge of plate tectonics and Continental Drift.

TRIBAL DISTRIBUTIONS

Introduction

At the time Bentham (1873) wrote his masterful account of the Compositae, approximately 10 000 species distributed among 760 genera were recognized, these being accommodated among 13 tribes. In the century since, the number of species recognized has more than tripled (c. 25 000) and the number of genera approximately doubled (c. 1400). In spite of this proliferation of specific and generic names (the latter being mainly elevations of subgenera and sections) little in the way of new information has become available so as to alter significantly the tribal groupings.

The most far-reaching taxonomic change affecting tribal circumscriptions and generic realignments is that proposed by Turner and Powell (Chapter 25) in which the tribe Helenieae is discarded and the approximately 67 genera realigned with groups distributed among eight tribes, mostly Heliantheae and Senecioneae.

The tribal accounts given below take into consideration these redispositions as well as other significant changes within the groups which have come to the fore since Bentham's treatment. Basically, however, these are thumbnail sketches largely taken from Bentham, the purpose being to acquaint the reader briefly with both his and my own thinking preparatory to the conclusions which follow.

Vernonieae (Jones, Chapter 17)

As noted by Bentham (1873: 393), this tribe "consists primarily of one large genus [*Vernonia*] with a number of smaller ones closely connected with it...". Altogether the tribe is comprised of about 55 genera, including taxa of the subtribe Liabinae which was assigned to the Senecioneae by Bentham. The Liabinae belongs to or near the Vernonieae on both palynological and megamorphic grounds (Turner and Powell, Chapter 25). It is conjectured that the Vernonieae as early as the Eocene was centered in northern South America with a well developed secondary center in central Africa. Except for the rare exploitation of warm temperate habitats of North America in relatively recent times, the tribe is essentially confined to the tropics or subtropics.

Eupatorieae (Robinson and King, Chapter 15)

Bentham did not have available to him the rich collections of material from the American tropics and subtropics which have accrued in herbaria over the past century. Nevertheless, his comment, that the "Eupatoriaceae may be regarded as one large and natural essentially American group or genus in an extended sense of the term; for multifarious and distinct as it is, some of the last mentioned small groups or monotypic genera of Vernoniaceae rank as high in the latter respect", seems appropriate today as then.

In the classical sense the tribe comprises perhaps 60 genera, although if one accepts the propositions of King and Robinson (1970, *ad abundicum*), and any ultimate treatment along such lines, this figure should be tripled. Like the Vernonieae, the tribe is largely tropical or subtropical, being centered in Central America and the immediate regions on either side. Bentham (1873, p. 400), noting its predominantly American occurrence, commented that the tribe must "either not be so ancient as some other groups of Compositae, or some other reason must have interfered with their early dispersion". I accept both Bentham's points; it presumably evolved out of an ancestral complex which gave rise to the Vernonieae, not having become widely dispersed prior to the break-up of Gondwanaland. The relatively few, weedy-type, species which occur in the Old World tropics today presumably reflect introductions since the Pliocene.

Astereae (Grau, Chapter 19)

This tribe comprises about 3000 species in approximately 110 genera. It is distributed throughout the world but is largely centered, as to relictual, morphologically remote types, in western North America, South America and South Africa.

Inuleae (Merxmüller *et al.*, Chapter 21)

This is one of the larger tribes of the Compositae with perhaps 2800 species distributed among approximately 160 genera. Except for the transfer of elements of the small subtribe Tarchonanthinae to the Anthemideae (Skvarla and Turner, in mss.) and the segregation of Filaginae as a distinct tribe of doubtful affinity by Poljakov (1967), the Inuleae has been but little tampered with since Bentham's time. It is largely an Old World taxon with centers of diversity in Africa and Australia, the widespread genera *Gnaphalium, Antennaria*, etc. having spread to the New World from the north in Neocene time.

Heliantheae (Stuessy, Chapter 23)

According to Bentham (1873) this tribe is second in size only to the Inuleae but must rank first in diversity. As currently circumscribed (excluding the Coreopsideae, but including the Gaillardinae and other helenoid elements; Turner and Powell, Chapter 25) it is comprised of perhaps 2500 species distributed among about 150 genera. It is largely a New World element with centers of diversity in the lower montane tropical and subtropical regions.

Bentham (1873, p. 482) reckoned the tribe to be among the more primitive of the Compositae, as did Cronquist (1955). For much the same reasons given by these authors I too would reckon the tribe to house some of the more primitive elements of the family. Because of this the distribution of the taxon, especially its relict genera, is important to any considerations of the origin of the family. As Bentham noted, these primal elements are largely centered in the northern Andes and the montane regions of Central Mexico. Apparently the tribe was well developed (but not common as to numbers, or at least confined to habitats not conducive to fossil entrapment) in the western portion of Gondwanaland *and* southwestern-most Lauratia in the early Tertiary (pre-Oligocene) and presumably the Cretaceous, as will be discussed below.

Coreopsideae (Stuessy, Chapter 23)

This taxon has been treated as a subtribe of the Heliantheae by nearly all workers, but for reasons given elsewhere (Turner and Powell, Chapter 25) it seems deserving of tribal status. It is a well marked tribe with perhaps 1000 species and 30 or more genera if treated so as to include the subtribe Jaumeinae. The tribe is centered in tropical America and Africa with relictual elements on both continents. Further, the taxa included seem related to, but about equally distant from, the tribes Senecioneae and Heliantheae, suggesting a widespread well developed plexus within the proasteroid elements of Gondwanaland as early as the Cretaceous.

Tageteae (Strother, Chapter 27)

The number of taxa and distribution of this previously helenoid element is neatly discussed by Strother and it need only be noted here that it too is largely centred in the tropical and subtropical regions of North and South America.

Senecioneae (Nordenstam, Chapter 29)

In spite of the loss of the Liabinae and Neurolaeninae, with the transfer to this tribe of elements previously included in the Helenieae (Stuessy, Chapter 23; Turner and Powell, Chapter 25) the Senecioneae becomes larger and more diverse than previously thought. It comprises perhaps 6000 species distributed among some 100 or more genera. The tribe is cosmopolitan in distribution suggesting a very early development over a large region, but the more ancestral, relictual, types seem largely confined to Africa and South America.

Some of the North American genera such as *Cacalia* and *Arnica* are undoubtedly derived from more temperate or circumboreal prototypes, but many of the Mexican and Californian elements such as Peritylinae and Eriophyllinae have presumably developed *in situ* from ancestral senecioid prototypes no longer extant. It should be noted, however, that the subtribe Madinae of the Heliantheae is possibly related to the Senecioneae, thus strengthening the case for a strong developmental center of the latter tribe in North America. Nevertheless, Africa must be considered the center of diversity of the tribe since the small, well marked, tribe Calenduleae is restricted to that continent, being quite closely related to the Senecioneae "and might have been enumerated amongst the subtribes of Senecionideae (with which it has much more affinity than with Cynaroideae, under which it is usually classed), but that there is a tendency to produce appendages or tails to the anther auricles, and there is never any pappus" (Bentham, 1873, p. 463).

Anthemideae (Heywood and Humphries, Chapter 31)

This appears to be a very natural tribe of perhaps 1800 species distributed among some 60, often poorly marked, genera. I would transfer to this group the Tarchonanthinae, a small taxon of two or three genera heretofore assigned to the Inuleae where they appear to be out of place. Megamorphically and in habit features, the subtribe approaches the Vernonieae, but palynological evidence suggests a closer relationship with the Anthemideae (Skvarla and Turner, in prep.). The latter is essentially an Old World group, being about equally well developed in Eurasia and South Africa, especially as regards diversity. It is poorly developed in North America, most of the taxa being temperate or boreal elements extending into the region from Eurasian terminals.

Calenduleae (Norlindh, Chapter 34)

A small tribe of perhaps 150 species and 10 or so genera largely confined to Africa. As noted above, the taxon seems quite close to the Senecioneae and perhaps should be treated as a subtribe of the latter.

Arctoteae (Norlindh, Chapter 34)

Also a small tribe with perhaps 300 species and 15 or so genera largely confined to Africa. According to Bentham (1873, p. 464) they pass rather gradually into the Cynareae and since the tribe is centered in South-Africa "they may perhaps be considered the southern representatives" of the Cynareae and might profitably be combined with this tribe in spite of some authors' desire to divide the group into yet further miniscule tribal elements (e.g. Ursineae; Robinson and Brettel, 1973).

Cynareae (Dittrich, Chapter 36)

This tribe comprises perhaps 2500 species distributed among approximately 80 genera. It is largely centered in Eurasia, the New World taxa being few in number, presumably extending (concomitantly evolving) into the region in Neocene times from wide-ranging genera (e.g. *Cirsium*).

Mutiseae (Cabrera, Chapter 38)

This is a seemingly very natural but highly variable tribe with perhaps 1200 species distributed among approximately 60 genera. It is largely centered in South America and is of interest because it approaches both the Cynareae and the Inuleae in some of its characters (Bentham, 1873, p. 471). Because of its bilabiate corollas some workers have suggested a close relationship with the Lactuceae but this appears more technical than natural. A few peculiar genera are found in the tribe, the most notable being *Fitchia*, which Poljakov (1967) segregates as the sole member of the tribe Fitcheae, although it might more appropriately be positioned as an anomalous or relictual member of the tribe Heliantheae much as treated by Stuessy (Chapter 23).

Lactuceae (Tomb, Chapter 39)

This tribe comprises perhaps 2200 species distributed among approximately 100 genera being largely centered in the more temperate regions of Eurasia with a well developed secondary center in temperate North America, thus marking the group as perhaps having developed in the early Tertiary largely on the Lauratian land mass. The tribe has subsequently diversified and entered the southern realms presumably as montane or long-distance immigrants.

ORIGIN AND DISPERSAL OF COMPOSITAE

Relatively few significant papers have been written on the distribution of the Compositae since Bentham's opus (1873). Probably the most comprehensive has been that of Small (1919) but he did little more than reiterate what Bentham had to say. More recent, largely regional, accounts have been published for Asia by Hu (1958) and for Mexico by Rzedowski (1972). Probably the best presentation of purely distributional information for the various tribes is to be found among the contributions to the present book.

Place and time of origin

As indicated by Raven and Axelrod (1974), any consideration of the origin of a group believed to be relatively old, geologically speaking, must take into account the phenomenon of plate tectonics or Continental Drift. Except for these two authors, most previous phytogeographers have accepted the heretofore well established dogma of stable, but isostatically variable, continents. Even so, the fact of Continental Drift has not led to notably differing views as to the place of origin of the Compositae, for Bentham (1873), Small (1919), Raven and Axelrod (1974) and the present author all reckon the family to have had its origin in the southern hemisphere, more specifically in the montane northwestern portions of South America. Rzedowski (1972) and Hu (1958) have pointed out the montane predilection of the family, the latter in particular noting that it presumably reached its current world-wide distribution by dispersal along the major mountain chains of the various land masses (assuming stable continents).

This was also the surmise of Small (1919) and must be inferred from the statement of Raven and Axelrod (1974) that the migrations of Compositae must "be seen in the light of present geography". Since the continental masses were essentially in the positions they occupy today by Miocene time, it follows that speciation and dispersal of this family so as to produce so large a number of species (*and genera, and subtribes, and tribes*) over such an expansive region by the end of Miocene must have been truly explosive! Indeed, a mere reading of the geopalynological record might suggest this, but contrary to what Raven and Axelrod state, *not* "with other evidence".

Except for the assumption of recent age because of their morphologically advanced flowers (as Bentham noted), the "other evidence" consists of macrofossils which suggest that the family existed in about the same morphological form (tribally speaking) in about the same regions during Oligocene as it does today. Bentham astutely noted this fact, commenting that this must mean the family had a much earlier beginning. He also wisely noted that were the family to have had a relatively recent origin, presumably explosive, one would not expect the extraordinary uniformity of floral structure found throughout this large family. I agree; for where such recent, explosive, radiation and dispersal exists, especially if this arises out of a diverse populational matrix as suggested for the Compositae (the main tribes, indeed even *advanced* members, e.g. Ambrosinae, having been present during Miocene), a broad range of intergrading floral types is to be expected. In the Compositae this would mean the presence of inflorescence types ranging from relatively open spikes with solitary flowers (elongate "receptacles") and poorly developed basal bracts ("involucres"), to say nothing of variable floral parts, bilocular ovaries, etc. Indeed, the only significant floral variation found in the family is that of syncephalization, loss of function (e.g. sterile florets), development of asymmetry and relatively minor quantitative elaborations upon these variables; in short, the capitulum structure and floral plan are remarkably the same throughout the family.

"Other evidence" also includes distributional data, especially of relictual groups. This was pointed out by Bentham (1873), Small (1919), Stebbins (1941) and Cronquist (1955): there are disjunctional elements (both continental and extra-continental) within Compositae which attest to their relatively old occupancy on the various continents. Such examples may be found among numerous taxa of the well marked tribes Vernonieae, Astereae, Inuleae, Coreopsideae, Senecioneae, Anthemideae, Mutiseae and Lactuceae and have been commented on by several workers noted above and by numerous monographers.

Apart from fossil and purely morphological and distributional data bearing upon the place and time of origin, there is chemical evidence bearing upon the problem. Apart from the micromolecular data (Herout, 1971), this consists of amino acid sequence data from cytochrome *c* of two genera (*Guizotia* and *Helianthus*) of the Compositae, strongly suggesting that the family is a very old one among angiosperms generally (*Boulter et al.*, 1972) and, more important, that the divergence between what are thought to be at least tribally related genera must date back to at least 60 million years (Turner, 1975). This places primary divergence of the family back to at least the early Tertiary, which presupposes at least a Cretaceous origin for the family. In fact, relative to the 20 other angiosperm families sequenced to date, including both monocots and dicots, the Compositae branches off the main phyletic line earlier than most (Boulter *et al.*, 1972). Krassilov (1973), in discussing his possible proto-asteroid, Jurassic fossil disseminule, *Problematospermum*, also comments upon the significance of the cytochrome c data. I refer the reader to a recent paper by Kimura and Ohta (1974) for a lucid presentation of the significance of such data for phyletic purposes. Sequence data, while sparse in the extreme, are sufficient to suggest that the Compositae is a very old group and, along with the fact that the family is not *readily* related to any extant family makes both the morphological and chemical data corroborative.

Finally, as a frill among the "other evidence" discussed above, it should be noted that van der Pijl (1961) calls attention to the fact that the Compositae are fundamentally adapted to pollination by Coleoptera which he believes to be among the most primitive forms of insect pollination, suggesting to Krassilov (1973) that the family antedated the rise of more specialized pollinators such as Hymenoptera and Lepidoptera.

Continental Drift

If one can accept a medial Cretaceous age for the development of the Compositae (i.e. about 100 million years before the present; Raven and Axelrod, 1974), which, I submit, seems reasonable in view of the evidence discussed above, then the world-wide dispersal of the family seems more logically bound to the phenomenon of Continental Drift than it is to explosive evolution and adaptive radiation up and down the world's mountain chains since Miocene time. No doubt intercontinental *dispersal*, concomitantly with speciation, occurred in many asteroid genera (e.g. *Eupatorium*, *Vernonia*, *Antennaria*) in the post-Miocene up to the recent but it is my contention that the great tribal divisions of the family attained their continental positions of relative concentration as a result of plate tectonics.

Raven and Axelrod (1974) based their premise largely on palynological findings (presumably in the belief that the fossil disseminules were not sufficient documentation for an early Tertiary beginning) but did note, "No Asteraceae have what appears to be a distribution achieved by direct migration between Australia and southern America . . .", a statement with which I take issue. In fact, from the world-wide, basically radial (Fig. 1),

2. FOSSIL HISTORY AND GEOGRAPHY

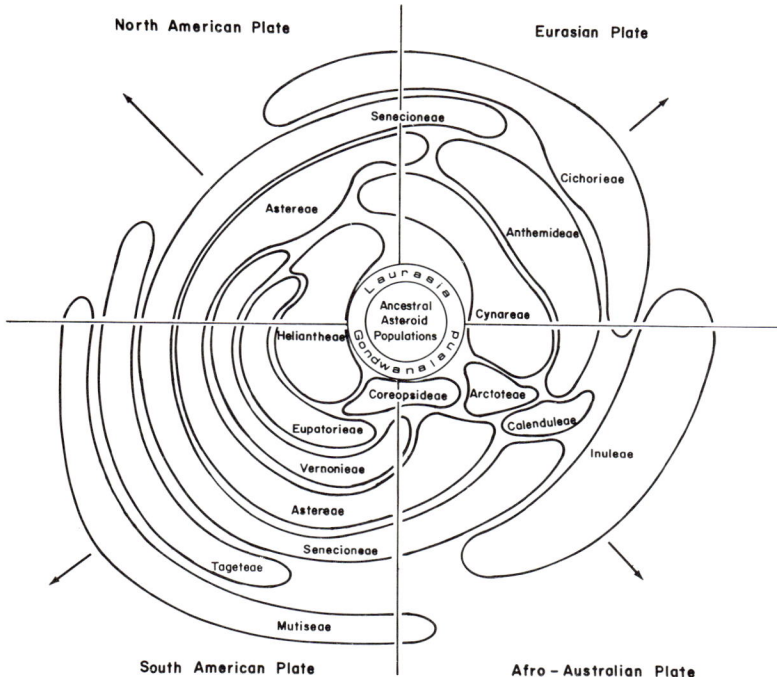

Fig. 1. Idealized, hypothetical diagram showing generalized present-day distribution of Bentham's tribal groupings (the Helenieae excepted) as these might have been affected by Continental Drift. Additional explanation in text.

distribution of the family I submit that the Eurasian concentration of the tribes Lactuceae, Anthemideae and Cynareae was largely due to their peripheral development (with respect to a western Gondwanaland center) in northeastern Lauratia, and subsequent rafting. This might also be said to be true of the Inuleae: it appears to have had peripheral development in southeastern Gondwanaland and the present strong concentration of this tribe in both Australia and South Africa no doubt reflects their development in the lower hinterlands of that ancestral continent. Further, if the Compositae arose in what was once western-most Gondwanaland, as most workers seem to agree, then the more ancestral tribal members such as the Vernonieae, Heliantheae and Mutiseae ought to be centered there: they need not have been widely distributed or frequent, as indeed they were not, to judge from the palynological record. It is as if the family developed in the form of a half-target, the center of which lay in what is now north-western South America, the more highly developed tribes (Inuleae, Anthemideae and Lactuceae) being more remote and peripheral to this source, and becoming more isolated and remote with post-Paleocene rafting. This would also account for the fact that Africa is a strong secondary center of the family: that continent houses many types connecting this or that taxon to its more southwestern or northwestern (with respect to Pangaea) elements, such as the Arctoteae, which connects the Cynareae to the Anthemideae, and the Calenduleae, which connects the Senecioneae with the Cynareae (Bentham, 1873). This leaves South America with the core of the family but does not account for the strong development of the Compositae in the southern parts of North America, unless the latter continent and South America were in much closer proximity during the Cretaceous than is apparent from the data assembled so far (cf. Raven and Axelrod, 1974: Fig. 2). Considering the distribution of extant members of the Compositae, as well as other angiosperm families, it will be surprising if new geological data do not, in fact, suggest a much closer early Cretaceous "fit" for these two continents than has heretofore been assumed.

In considering the above I cannot help but once again call to the reader's attention the remarkable insight brought to bear on the comparative antiquity of the Compositae by Bentham (1873, p. 481). After discussing the possible origin of the composite flower he states, "we might further be led to imagine that several of these changes had taken place at a very early period, previous to the disruption or stoppage of communication between what are now the tropical regions of the globe, that, besides the parent form above supposed, Compositae existed showing several important modifications . . ." and he goes on more or less to communicate what I have reiterated above about tribes and continents. Clearly, had Bentham been convinced of Continental Drift he would have taken much the same view as I have taken here, for he concludes (1873, p. 576):

> Africa, west America, and possibly Australia possessed the order at the earliest recognizable stage, Africa showing the greatest variety of individual isolated remnants of extinct races, Andine America and some of the scattered islands showing a few [e.g. *Fitchia* and *Dendroseris*] of what may be deemed the nearest approach to what we have conjectured to have been the primitive form of the order; that at this early period there must have been some means of reciprocal interchange of races between these regions; that, since the disruption of this intercourse between the two great divisions of the globe, there must have been for a time a certain continuity of Compositae races across the tropics from south to north, a continuity which was probably further prolonged in America than in the Old World. . . .

How he might have stated all this had he known what we now know can only be presumed; but, as stated above, I would like to think that his analysis of the fossil history and distribution of the family would be much like mine, given the same facts, which, after all, have not changed appreciably from his day, except for the fact of Continental Drift, which of course makes all the difference in *how* things came to be.

Protoasteriods and Phyletic Considerations

From what phyletic source came the Asteraceae? There are many contenders, but it seems to me that the evidence points to the South American herbaceous family Calyceraceae, as noted by Bentham, among others. The families are very similar palynologically (Skvarla *et al.*, Chapter 8), occupy the same continent and have a seemingly homologous capitular structure and similar floral features. In accepting the Compositae as primitively herbaceous but perennial (at most suffruticose) I ascribe to the views of Stebbins (1965) and Stuessy (Chapter 23): there is simply no compelling evidence to suggest that they were woody or possessed of considerable xylar development. Indeed, extant genera with well developed xylem are nearly always soft-wooded or else their floral structures are relatively highly derived types such as occur in *Tarchonanthus*, *Eupatorium*, *Artemisia* and *Polymnia*.

In summary, the populations ancestral to the Asteraceae must have been montane perennial suffruticose herbs with phyletic relationships relatively close to that line leading to the Calyceraceae. Such populations undoubtedly existed in the lower-most Cretaceous and, with the morphological radiation characteristic of the angiosperms of that era generally, gave rise to the first asteroid taxa consisting of relatively rare helianthoid-like, beetle-pollinated, perennial herbs with alternate leaves, eradiate heads and yellow flowers. These were at first (early Cretaceous) confined to relatively high or possibly more restricted xeric habitats of western Pangaea, especially westernmost Gondwanaland; from this focal point the family became widespread and diversified by the end of the Cretaceous. By the end of the Oligocene most, if not all, of the major tribes had evolved, having become increasingly successful occupants (or survivors) along the frontal mountain chains which formed along the leading edges of rafting continents. In Miocene the adaptive strategies of these montane populations had become sufficiently successful to permit a rapid seemingly explosive, increase in their numbers, both as to individuals and species. This increase is reflected in the fossil record and appears to have been due to world-wide, mainly fluctuating, short-term changes in habitats, reflecting increasing global aridity, interspersed with variable fluvial periods and, more recently, climatic fluctuations due to global glacial advances. These physiologically diverse populations, which presumably occupied diverse Oligocene montane habitats (which must have varied from warm xeric to cold humid) were well programmed for the global situation that began to develop in the Miocene and holds today.

MAN AND WEEDS

No account of the distribution of the Compositae would be complete without mention of man as a vector in the dispersal of certain weedy taxa of this large family. This has been appropriately emphasized by Hu (1958) who suggested that many specific, seemingly "natural" (i.e. established without man's aid) disjunctions in the group have resulted from dispersal by early man. What is certain is that the family contains many species which are prone to become weedy, given a disturbed habitat. Some of the more notable, widely distributed taxa, include *Conyza canadensis* (L.) Cronq., *Bidens pilosa* L., *Galinsoga parviflora* Cav., *Parthenium hysterophorus* L., *Taraxacum officinale* Weber, *Tridax procumbens* L., *Verbesina encelioides* (Cav.) Gray, *Vernonia cinerea* L. and *Xanthium strumarium* L., to name but a few. No doubt these and similar weeds will become increasingly significant components of the world's ravished ecosystems.

ACKNOWLEDGEMENTS

I am grateful to Dr Leo J. Hickey, for helping me locate literature and fossil material deposited at the United States National Museum, and to Drs Beryl Simpson and James Walker for stimulating discussion relating to the problem of Continental Drift and dispersal of the Asteraceae. Supported in part by N.S.F. Grant BMS 71-01088.

REFERENCES

BECKER, H. F. (1969). Fossil plants of the Tertiary Beaverhead Basins in southwestern Montana. *Paleontographica* **127**, 1-142.

BENTHAM, G. (1873). Notes on the classification, history and geographical distribution of Compositae. *J. Linn. Soc. (Bot).* **13**, 335-577.

BOULTER, D., RAMSHAW, J. A. M., THOMPSON, E. W., RICHARDSON, M and BROWN, R. H. (1972). A phylogeny of higher plants based on the amino acid sequence of cytochrome *c* and its biological implications. *Proc. R. Soc.* B **181**, 441-455.

CRONQUIST, A. 1955. Phylogeny and taxonomy of the Compositae. *Am. Midl. Nat.* **53**, 478-511.

GERMERAAD, J. H., HOPPING, C. A. and MULLER, J. (1968). Palynology of Tertiary sediments from tropical areas. *J. Palaeobot. Palynol.* **6**, 189-348.

GRAHAM, A. (1963). Systematic revision of the Sucker Creek and Trout Creek Miocene floras of southeastern Oregon. *Am. J. Bot.* **50**, 921-936.

HEROUT, V. (1971). Chemotaxonomy of the family Compositae (Asteraceae). *In* "Pharmacognosy and Phytochemistry" (H. Wagner, and L. Hörhammer, eds), pp. 93-110.

HU, S. (1958). Statistics of Compositae in relation to the flora of China. *J. Arnold Arbor.* **39**, 347-419.

KIMURA, M. and OHTA, T. (1974). On some principles governing molecular evolution. *Proc. natn. Acad. Sci. U.S.A.* **71**, 2848-2852.

KING, R. M. and ROBINSON, H. (1970). A monograph of the genus *Trichocoronis*. *Phytologia* **19**, 497-500.

KNOWLTON, F. H. (1925). Flora of the Green River. *U.S. Geol. Surv.* **131**, 173.

Krassilov, V. (1973). Mesozoic plants and the problem of angiosperm ancestry. *Lethaia* **6**, 163–178.

Muller, J. (1970). Palynological evidence on early differentiation of angiosperms. *Biol. Rev.* **45**, 417–450.

van der Pijl, L. (1969). "Principles of Dispersal in Higher Plants". Springer-Verlag, New York.

Poljakov, P. P. (1967). ["Systematics and Origin of the Compositae".] (in Russian). Acad. Sci. Kazakh S.S.R. Alma-Ata.

Raven, P. H. and Axelrod, D. I. (1974). Angiosperm biogeography and past continental movements. *Ann. Mo. bot. Gdn* **61**, 539–673.

Robinson, H. and Brettell, R. D. (1973). Tribal revisions in the Asteraceae. VIII. A new tribe, Ursinieae. *Phytologia* **26**, 76–85.

Rzedowski, J. (1972). Contribuciones a la fitogeografía floristica e histórica de México. III. Algunas tendencias en la distribución geográfica y ecológica de las Compositae mexicanas. *Ciencia Mex.* **27**, 123–132.

Segal, R. (1965). New fossil fruit (Compositae) from the Pliocene of western Kansas. *Am. Midl. Nat.* **73**, 430–432.

Small, J. (1919). The origin and development of the Compositae. *New Phytol.* **18**, 129–176.

Stebbins, G. L. (1941). Additional evidence for a holarctic dispersal of flowering plants in the Mesozoic era. *Proc. 6th Pacific Sci. Congr.* **6**, 649–660.

Stebbins, G. L. (1965). The probable growth habit of the earliest flowering plants. *Ann. Mo. bot. Gdn* **52**, 457–468.

Turner, B. L. (1975). *In* "The Basis of Angiosperm Phylogeny" (J. R. Walker, ed.). *Ann. Mo. bot. Gdn* **62**, 765–800.

Willis, J. C. (1918). The Age and Area hypothesis. *Science, N.Y.* **47**, 626.

Chapter 3
Aspects of diversification in the capitulum

B. L. Burtt

Royal Botanic Garden,
Edinburgh, Scotland

Abstract. The constantly racemose structure of the capitulum of Compositae is well suited to centripetal diversification of different types of flower and fruit. The situation is contrasted with that in Dipsacaceae, where the capitulum has a cymose basis. Probably the first form of diversification in Compositae was the restriction of the peripheral flowers to a female role. This confers a functional protogyny on the capitulum. In general, the two main types of female flower, radiate or filiform, are respectively correlated with large and small achenes throughout the head. The more complex elaborations of capitular structure tend to be found in the radiate, larger-fruited group. The varying functions of the involucre, from bud protection to fruit dispersal, form a major link between flowering and fruiting phases. Some special aspects of dispersal are considered.

Although fruit diversification within the capitulum is much less widespread than diversification of the flowers, it is nevertheless very important. Heteromorphic achenes and their role are discussed in relation to some general biological ideas. Certain capitula with the central flowers male, including some that have been interpreted as compound, are thought to have reached this stage of evolution by passing through one with heteromorphic achenes. The tendency towards loss of pappus is often first expressed in the peripheral flowers. The curious situation in which the central male flowers are the only ones with a pappus is considered. In Compositae there has been a transfer of variability from flowers to capitulum. This has been an important factor in the success of the family.

CONTENTS

Introduction	42
Peripheral unisexuality	43
Functional protogyny of the capitulum	43
Comparison with Dipsacaceae	43
Self-incompatibility	44
Hybrids between radiate and discoid plants	44
Capitula with filiform female flowers	45
Peripheral male flowers	46
Size of achenes	46

The involucre.	47
Functions linking flowering and fruiting phases	47
Development to set free achenes	47
Development to enclose achenes	48
Heteromorphic capitula	48
Heteromorphic achenes	49
Functional differences and their correlations	49
Heteromorphic achenes in relation to some modern biological theories	51
Achaenial heteromorphy in Calenduleae	52
Central unisexuality	53
Female sterility in simple capitula	53
Some allegedly compound capitula	54
The pappus of male flowers	55
Conclusion	56
Acknowledgements	57
References	57

INTRODUCTION

The purpose of this contribution is to discuss the organization of the capitulum, both in flowering and in fruiting phases, and in particular to examine some ways in which the requirements of these two phases interact. My treatment of so vast a topic is of necessity highly selective.

In the dawn of Compositae the evolutionary passage from a more open inflorescence into an integrated capitulum must have been a result of selection acting on the flowering phase. Mere aggregation of the flowers into a capitulum gives immediate opportunity for the development of certain cooperative features of floral biology. After flowering, however, a capitulum contains a store of developing fruits, which offers a well stocked larder for the larval stages of various insects. Thus selection pressures in the fruiting phase may have moulded such features as the type of involucre or size of head, features we normally record when the plant is in flower. Pest pressure on the fruits may also be the underlying reason for the massive evolution of bitter chemicals in the family.

The capitulum of Compositae is a racemose inflorescence showing consistently centripetal development. It is, in fact, simply a condensed shoot system developing from its apical meristem; thus centripetal diversification within the capitulum (for instance, the peripheral flowers radiate, the central ones tubular) requires nothing different in the system of developmental control from that which produces the normal sequence of leaf types and axillary buds on the vegetative shoot. It is this sequential differentiation of the flowers and fruits from the periphery to the centre of the head that gives the capitulum its great versatility. At some point or other in the family we find sex of flower, form of corolla, of pappus and of achene all varying within the head in this centripetal manner.

It is within the range of this versatility that the topics to be discussed are best illustrated. The tribes that show less diversification, notably

Vernonieae, Eupatorieae, Cynareae and Mutisieae, receive small attention. They must not be forgotten for they comprise a large and important part of the family. They show slight diversity within the head (e.g. *Stokesia* and *Elephantopus* in Vernonieae or *Centaurea* in Cynareae) and where the whole head is concerned they show the same features as the rest of the family: variations in size of head, secondary aggregations, elaboration of the involucre. In a fuller treatment they would take their rightful place.

PERIPHERAL UNISEXUALITY

Functional protogyny of the capitulum

The first important step in diversification within the capitulum was, most probably, the restriction of the peripheral flowers to a female role, producing a capitulum that is functionally protogynous. The temporal separation of the male and female stages of the individual flower is, of course, very frequent in angiosperms. Earlier maturation of the stamens (protandry) is the commonest condition. Protogyny, though widely distributed, is rarer, but is important in congested inflorescences, where protandry would mean that there was always pollen about before the first stigmas were receptive. Thus protogyny is found in such a genus as *Plantago*; without it, the stigmas of the first flowers at the bottom of the spike would be showered with pollen from the anthers of the next flower above just as they were becoming receptive. In Compositae this protogyny of the capitulum gives a very important opportunity for cross-pollination to take place.

Comparison with Dipsacaceae

The contrast with Dipsacaceae, another family with flowers in a capitulum, emphasizes the value of centripetal development and peripheral protogyny. The capitulum here has a cymose basis and does not develop in a sequential centripetal manner, as shown by the order of opening of the flowers. Some peripheral differentiations may occur in the form of enlarged rays or, much more rarely, in the development of the fruiting calyx. It does not affect fundamental structure and appears to be restricted to development late in the ontogeny of the individual flower. Sexual differentiation may occur between plants, as gynodioecism (cf. Jaeger, 1938; Assoud, 1971), but is not found regularly within the capitulum.

The non-racemose capitulum of Dipsacaceae does not show peripheral protogyny. The outcrossing mechanism is different. The individual flowers are protandrous, but in most genera (*Dipsacus* is an exception) none of the stigmas becomes receptive until the male stage of all the flowers in the head is over. Then the whole head becomes receptively female at the same time (Jaeger, 1938). Thus Dipsacaceae does not have the characteristic feature of Compositae, that the female stage of the flowers in a capitulum is spread over a long period.

Self-incompatibility

In throwing emphasis on the importance of peripheral female flowers, I am ignoring one difficult problem. This is the possibility that Compositae are basically self-incompatible. If selfing cannot take place, why do we find peripheral female flowers? I cannot answer this question. In many genera the peripheral flowers are female, and some of these plants certainly are self-incompatible; but if this were the condition in the basic stock of Compositae, then one would expect to find peripheral flowers developed as attractants only. This is what happens, for instance, in *Centaurea montana* L. and other species whose radiant flowers are neuter.

Unfortunately knowledge of self-incompatibility in the family is very inadequate. Too little is known about its distribution, too little on the degree of effectiveness, which is not always absolute. The mechanism of self-incompatibility has been the subject of very elegant studies; but as a factor in evolution its importance is still poorly known. The laboratory has out-stripped field studies. Some species with female rays are self-compatible and it seems likely that most of those with filiform female flowers are so. My remarks on the advantages of this effective protogyny apply at least to these.

Hybrids between radiate and discoid plants

We can also get some hints on the importance of female ray flowers from a study of hybrids. Recently I was looking at hybrid swarms between two species of *Senecio* in Natal; one parent was discoid and grew on the drier ground round the edge of a marsh; the other parent grew in the marsh itself and was radiate. The hybrids showed a wide range of intermediate floral types. It was very noticeable that they all grew in the marsh, closely mixed with the radiate parent. There may, of course, have been an ecological factor responsible, but it is at least possible that this pattern was produced because it was the female peripheral flowers that were being cross-pollinated. The discoid species, without a female phase, would have been acting only as pollen parent. Both the species concerned proved to be undescribed: the radiate one has been named *S. parentalis*, the discoid *S. submontanus* (Hilliard and Burtt, 1976). Hybrids have also recently been found between radiate and discoid red-flowered senecios of the *S. erubescens* Ait.–*S. polyodon* DC. group. Here too plants with intermediate, bilabiate, marginal flowers were a feature of the populations. Trow (1912) recorded similar flower forms in his artificial crosses between radiate and discoid

Senecio vulgaris L. It has recently been suggested that the spread of a radiate form of *S. vulgaris* in Scotland is due to the fact that the plant is constantly formed by introgression from *S. squalidus* L. (Hull, 1974a, b, 1975; Richards, 1975). In this case, however, first-generation crosses between the species are apparently so rare as to escape detection; there seem to be no records of the intermediate corolla forms that might be expected.

On the Essex coast I have seen apparent crosses between discoid and radiate plants of *Aster tripolium* L. Here the intermediate capitula do not show bilabiate corollas but have only one or two rays. These presumed hybrids were clustered at the margin of the habitat of the radiate plants. The discoids were at a distinctly lower level in the salt marsh and did not have any hybrids near them, again suggesting that the female ray flowers may have been important in hybridization. A complete catalogue of known hybrids between radiate and discoid Compositae might be instructive.

Of course not all Compositae with 1-rayed heads are of hybrid origin. The South African *Senecioadnatus* DC. and its ally *S. hygrophilus* Dyer & Smith normally have one ray, rarely two.

Capitula with filiform female flowers

That conspicuous rays are associated with cross-pollination is the central theme of Leppik's paper (Chapter 4). It must not be thought, however, that the head without ray flowers is necessarily associated predominantly with self-pollination, or fails to attract insect visitors. Such heads can be very conspicuous by mere size (*Centaurea*), by having coloured and exserted styles (*Cynara*), by being massed together (*Helichrysum*), by having an involucre of coloured bracts (*Helichrysum, Helipterum*) or by being strongly scented (*Raoulia*).

Disciform capitula often have several peripheral rows of inconspicuous filiform female flowers and thus have a marked female stage before any pollen is shed by the hermaphrodite flowers of the same capitulum. This stage may vary considerably within a genus. For instance in *Blumea, B. belangeriana* DC. has a ratio of two female flowers to one hermaphrodite (*c.*50:25), whereas in *B. hieraciifolia* (Don) DC. the ratio is 26:1 (*c.*180:7) and in *B. obliqua* (L.) DC. it is 20:1 (*c.* 740:37). Clearly the female phase, and therefore the chance of cross-pollination, is much better developed in the last two species than in the first.

Helichrysum is a much larger and more variable genus than *Blumea*. Data for the 122 species found in Natal, South Africa, show that heads range from the homogamous 2–5-flowered head of *H. infaustum* Wood & Evans to the heterogamous one of *H. ruderale* Hilliard & Burtt with 811–1053 flowers (121–226 female, 681–754 hermaphrodite). But distribution of sexes has little to do with size of head. *H. stenopterum* DC. with 4–6-flowers is heterogamous, *H. bellidiastrum* Moeser with 217–277 is homogamous. The condition in which the female flowers considerably outnumber the male does not occur in *Helichrysum*, such species being placed at present in the neighbouring genus *Gnaphalium*. In the same area *G. oligandrum* (DC.) Hilliard & Burtt may have 282 female and 16 hermaphrodite flowers in a head; further north in Zambia (Barotseland) the figures reach 421 female to 17 hermaphrodite, giving these capitula an extended female phase. Thus, some plants in this affinity behave in the same way as species of *Blumea*, having marked but varying female phases in each capitulum; but the functional significance, if any, of 1–few female flowers in the species with small heads is difficult to understand.

Peripheral male flowers

The peripheral flowers are commonly hermaphrodite, female, or neuter (acting as attractants only). The development of male flowers at the periphery of the head is not a normal pattern, but it is worth recording here that it does occur exceptionally. The two instances known to me are in *Centaurodendron dracaenoides* Johow of Juan Fernandez (Skottsberg, 1938), and in the western Asiatic *Gundelia*. *Gundelia* has compound capitula of which the individual component heads consist of a single central hermaphrodite flower surrounded by 4–5 which are functionally male. The resultant fruiting unit is reminiscent of the fruiting umbellules of Umbelliferae—Echinophoreae rather than the capitulum of other Compositae!

SIZE OF ACHENES

It seems that in groups where filiform female flowers are important, the achenes are usually small. This is shown in its most extreme form in Inuleae-Gnaphaliinae. One feature of such heads is that they ripen fast. Small achenes that are quickly dispersed offer scant opportunity for exploitation as a food supply.

The somewhat larger seeds associated with more conspicuous flower-heads offer attractive nutriment which is available for a longer time because of the slower development of the achenes. This food supply is exploited by various insects which lay their eggs in the flower heads (cf. Burtt, 1961). Here I want to do no more than draw attention to the fact that the insects fall into two main groups. The larvae of Coleoptera and Diptera are relatively inactive: the grubs hatch amongst the young achenes and eat some or all of them according to the size of the head and the number of grubs in it. Decrease in the size of capitula and increase in their number is likely to protect many heads from attack: increase in size of head may ensure a surfeit of food so that there are always some achenes left for reproduction. A higher order of insects, the Lepidoptera, have more mobile larvae (caterpillars). These may move from one head to another, sometimes spinning a web and drawing several heads together, and against their attacks a small capitulum offers no insurance. In the first volume of a monograph on British Tortricoid Moths (Bradley *et al.*, 1973), Compositae figure conspicuously in the list of food plants, and in many cases it is the capitula that are attacked. The whole interaction of the family with its seed-predators needs much more observation, bearing in mind that the situations that we normally see in nature are those that are in balance. The species so restricts the range of its parasites by morphological or chemical defences, or evades them by its mobility, that enough seed for the perpetuation of the species always survives their depredations. This is not to deny that pest pressure may occasionally be a major factor in the extermination of individual populations.

THE INVOLUCRE

Functions linking flowering and fruiting phases

The involucre is manifestly the organ that maintains continuity between the various phases of the capitulum, for it has functions in all of them. These functions may be roughly listed: (i) protection of young flowers; (ii) protection of mature flowers by reclosure during unfavourable weather or at night; (iii) attraction of pollinators; (iv) protection of developing fruits; (v) release of ripe achenes, or their enclosure and dispersal. Not every involucre performs all these functions, but all perform some of them. In consequence there is often change in the anatomical structure of the involucre during its life; and at some stage it dies. Both these features are very poorly documented in descriptions.

Development to set free achenes

The involucre of *Senecio* and allied genera demonstrates not only a very efficient protective mechanism, but the importance of the death of the involucre. The involucre is uniseriate and the bracts are held together in the bud and in the flowering phase by a very neat interlocking mechanism, the margin of one bract locking into a groove under the dorsal ridge of its neighbour. This position is firmly held because the living cells are turgid and exert pressure. When fruiting sets in the involucral bracts wither, the cells shrink and the interlocking no longer holds. During the growth and maturing of the head, special tissue has developed at the base of the bracts: at maturity this shrinks and the withered bracts, freed from one another, become reflexed and the achenes are set free (for this latter phase see Guttenberg, 1971: 100, quoting Goebel, 1924). The uniseriate interlocking bracts, their precisely timed and rapid death, and the basal reflexion tissue (cohesion tissue: Fahn, 1967, p. 447), all mark the senecioid involucre as advanced in comparison with the multiseriate, scarcely changing structure of many Vernonieae, Heliantheae etc.

Most often the involucre does simply spread and set free the ripe achenes, though usually the mechanism is less specialized than in *Senecio*. Some capitula open in dry weather; others, usually in arid regions, when it is damp; these aspects are discussed briefly by Zohary (1950).

Development to enclose achenes

More rarely the involucre retains the ripe achenes within it, becomes more or less indurated and effects their dispersal. Examples of this are found in the South African *Cuspidia* and *Gorteria*; in *Didelta* the involucre and receptacle break up into four several-seeded segments (see Burtt, 1961). This is the condition that Murbeck (1920, 1943) called synaptospermy; it is developed by a number of families in semi-arid regions. I need not dwell further on these, but recent availability of fruiting heads of the Australian *Erodiophyllum* does tempt me to refer to it. The genus belongs to Astereae and is commonly (but I suspect wrongly) placed in subtribe Bellidineae because of its reduced pappus. The flowers are trimorphic: the outermost female and radiate, the next series with a very reduced filiform tubular corolla and also female, while those in the centre of the head are functionally male flowers. The receptacle is paleaceous. In fruit the female part of the receptacle elongates so that the paleae, with the fruits between them, are now held horizontally. The whole base of the capitulum, the paleae and the receptacle become woody. There is here no possibility of the radicles of germinating seeds escaping through the scar of the peduncle, as in *Gorteria* or *Cuspidia*. Achenes germinate freely when removed from the head, but a capitulum sown intact has not, in 6 months, produced any seedlings. One must presume that in nature the achenes are eventually shaken free, or that the capitulum itself rots away. However, it is clear that the whole capitulum acts as a protection and as the primary disseminule. The plant grows on land subject to flooding and the fruiting capitula are certainly buoyant and probably water-dispersed.

HETEROMORPHIC CAPITULA

Heteromorphic capitula are characteristic of a few members of the Compositae, and like those just mentioned that show synaptospermy, they are found in semi-arid regions. Two examples may be mentioned. Perhaps the best known is *Catananche lutea* L. (Battandier 1883; Trotter 1910); the normal aerial capitula are borne on long peduncles and stand some 20-30 cm above ground level; down at the base of the plant, the lower part often in the soil, are some reduced axillary capitula containing 2-3 flowers each. These have a reduced pappus and larger achenes. The other example is

Gymnarrhena micrantha Desf. (Koller and Roth, 1964); the whole plant is dwarf, so that the aerial capitula are only 2–3 cm above soil level; the basal ones are partly underground, only the tops of the flowers being exserted. The achenes of the basal heads are more or less epappose and germinate quickly, those of the aerial heads are pappose and show some delay in germination. It is more usual, when distinct dispersive and non-dispersive fruits are produced, to find that the dispersive type shows the quicker germination (see below). The reversal here may be connected with the fact that in *Gymnarrhena* the non-dispersive achene is already sown in the soil. As a parallel we may look at *Aellenia autrani* (Post) Zohary (Chenopodiaceae: see Werker and Many, 1974) in which upper exposed flowers produce dispersive fruits with non-dormant embryo and basal (but not buried) flowers produce persistive fruits with dormant embryo. This agrees with the behaviour of heteromorphic achenes from a single head (see below). Other examples are given by Zohary (1937). Amongst amphicarpous Cyperaceae, *Bulbostylis humilis* (Kunth) C.B.Cl. has hermaphrodite basal flowers which produce nutlets similar to the aerial ones; some other species of *Bulbostylis* C.B.Cl. and of *Scirpus* L. sect. *Actaeogeton* Reichenb. have female basal flowers, and the basal nutlets are larger than the aerial ones (Haines, 1971).

HETEROMORPHIC ACHENES

Functional differences and their correlations
The diversification of fruits within the capitulum is much less widespread than that of flowers, but it is a recurrent specialization throughout the family, although the number of species affected must be only a small proportion of the total. In its simplest form it is behavioural rather than morphological; the peripheral achenes are retained, pressed against the inner surface of the involucre, when the other achenes are blown away (for instance in *Senecio jacobaea* L. or *Picris echioides* L.). More often the pappus is lacking on the peripheral flowers (as in *Galinsoga* Ruiz & Pavon, *Felicia* (*Charieis*) *heterophylla* (Cass.) Grau). Diversification in these cases is clearly a matter of dispersal, the peripheral flowers being retained close to the parent plant. The effect, therefore, is the same as that achieved in other families by the development of some basal, non-dispersive, fruits (e.g. *Scrophularia* spp.) or by some fruits being produced underground (e.g. *Commelina benghalensis* L. or by some seeds not being shed from the capsule (members of the *Mesembryanthemum* group such as *Apatesia* N.E. Br.).

In those heads where the peripheral flowers lack a pappus, the distinction may be sharp between pappus absent and pappus present, or there may be a gradual increase of pappus development from the periphery to the centre of the head (e.g. *Hedypnois cretica* (L.) Willd.: see Zohary, 1950). It is a feature of heteromorphisms, that their taxonomic value is not necessarily very great. For example the African *Gnaphalium declinatum* L.f. has copious pappus on all flowers throughout the greater part of its range; but, especially in the south-east, there are also forms which differ only in having the female flowers epappose (these are the plants described as *Amphidoxa gnaphalodes* DC.), and various intermediate conditions can be found. This situation is paralleled in some species of *Helichrysum* (see Hilliard and Burtt, 1973, 336, 341).

The small achenes of *Gnaphalium* and *Helichrysum* are not known to show any further diversification within the capitulum other than that associated with the presence or absence of pappus. However, it would be unwise to say that it does not occur, for dormancy is known in the equally small seeds of the South African Rhenoster bush (*Elytropappus rhinocerotis* (L.f.) Less.: see Levyns, 1972) and there is no reason why differential dormancy should not be found in association with the pappose and epappose achenes.

Where heteromorphism is associated with larger achenes there are morphological differences in the achenes and therefore differences in germination behaviour are much more easily observed and have been known for a long time. Becker (1912) carried out studies on 46 species and this is still the most wide-ranging investigation that has been made; it was supplemented by anatomical studies of some of the species by Grimbach (1913). There are three aspects to be considered: dispersability, germination characteristics and the types of flower from which the heteromorphic fruits originated. A few examples have been put together in Table I. Where the number of types of fruit is shown in the dispersal column, this indicates lack of information on whether there is, in fact, truly differential dispersal.

In *Crepis* L. achaenial dimorphism occurs both in self-compatible species (e.g. *C. foetida* L. subsp. *foetida*) and in self-incompatible species (e.g. *C. pulchra* L.). The range of dimorphism expressed varies from very slight (e.g. *C. vesicaria* subsp. *stellata* (Ball) Babc.) to marked (e.g. *C. dioscoridis* L.), and there is a distinct tendency towards larger and less dispersive peripheral achenes and for the involucre to play an increasing part in their retention (Babcock, 1947).

As yet there is, to my knowledge, little information available on any chemical differences in heteromorphic achenes, but I can say that preliminary work at Delhi (K. M. M. Dakshini, in litt.) has shown chromatographic differences between the two types of fruits of *Bidens bipinnata*, whose distinctive germination characteristics have already been published (Dakshini and Aggarwal, 1974). There is also one marked difference in the chromatograms for the two types of achene in *Dimorphotheca*, although both germinate without any dormancy.

The table suggests a link between the outer fruits of the head, near-dispersal and delayed germination; it can be added that these outer fruits

3. DIVERSIFICATION IN THE CAPITULUM

TABLE I. Characteristics of heteromorphic achenes
(Dispersal: N, near; F, far. Germination: Q, quick; D, delayed)

♀ ray/⚥ disc	Dispersal		Germination	
	Outer	Inner	Outer	Inner
Dimorphotheca pluvialis	N	F	Q	Q
Felicia heterophylla	N	F	Q	Q
Bidens bipinnata	(similar)		D	Q
Synedrella nodiflora	N	F	D	Q
Chrysanthemum segetum	N	F	D	Q
Galinsoga parviflora	N	F	Q	slight D
♀ ray				
Calendula officinalis	3 types		Q	Q
Osteospermum spp.	2 types		?	
⚥ ligule				
Crepis dioscoridis	2 types		D	Q
Crepis aspera	2 types		D	Q
Crepis zacintha	2 types		D	Q
Tragopogon glaber	2 types		D	Q
Leontodon taraxacoides	N	F	D	Q

are usually somewhat larger than the central ones. The outer fruits are developed from the first-opening flowers of the head and, where these are female, they will therefore have a good chance of being cross-fertilized. Because most of these plants are annuals, and annuals tend to be self-compatible, the inner flowers will tend to be self-fertilized. Even in the Lactuceae, where all the flowers are hermaphrodite, if any flowers of a self-compatible species are going to be out-crossed they are likely to be the first to open, and it has recently been shown that there is a very good chance of this in *Pyrrhopappus carolinianus* (Walt.) DC. (Estes and Thorp, 1975), a species with homomorphic achenes.

Heteromorphic achenes in relation to some modern biological theories

The linkages just mentioned, though far from absolute, seem sufficiently probable to justify enquiring how they fit with general ideas about breeding and competition. It has been suggested, from computer simulation studies, that outbreeding is increasingly important as intraspecific competition becomes more intense. If the non-dispersive achenes of the peripheral flowers are largely the result of cross-pollination, then that would be in harmony with the theoretical study, since non-dispersal necessarily implies competition between plants of the same species.

Weeds that lack the power of vegetative spread are characteristically inbreeders with effective dispersal mechanisms. Tried genetic combinations are reproduced and dispersed, as it were, in search of new but similar habitats. This would fit with the probability that in plants with heteromorphic achenes the central, dispersive ones are self-pollinated.

Valen (1971) has pointed out that within-population selection is inevitably against dispersive mechanisms, but that these are just as inevitably selected when the duration of a population is short. Many plants with heteromorphic achenes are either annuals or perennials of open ground; the dual type of diaspore should permit them to exploit a suitable habitat for the maximum time without losing the mobility necessary to survive frequent population extinctions. Such extinctions will be most often due to normal succession in the habitat, but a semi-permanent population is also much more vulnerable to insect pests, and especially in those Compositae where intraspecific competition may well lead to an increase in the size of seed. Outbreeding will increase the chances of development of pest-resistant genotypes (Levin, 1975), but an alternative dispersive diaspore may be important in the survival of the species.

In recent years recognition has been given to the differences between organisms that devote a high proportion of resources to reproduction and those that devote a high proportion to vegetative growth. Gadgil and Solbrig (1972) have been able to differentiate apomicts of *Taraxacum* on this basis: those devoting a higher proportion to reproduction (and bearing more capitula) are the more mobile types of open situations. In a very general sense all the Composites with heteromorphic fruits belong to the group concentrating on reproduction, simply because they are mostly annuals. However we may, perhaps, think of the production of small dispersive achenes as satisfying the preponderantly reproductive aspect and that of larger, more static achenes, giving stronger vegetative growth in the seedlings, as satisfying the competitive aspect. These two types of selection have been designated r-selection, favouring increased reproduction, and K-selection favouring ability to survive increasing density of population (MacArthur and Wilson, 1967).

These speculations are not to be read as an attempt to provide glib explanations of achaenial heteromorphism. My intention is simply to bring the structure of capitula into close proximity with work on breeding systems and selection patterns.

Achaenial heteromorphy in the Calenduleae

The tribe Calenduleae needs separate consideration. Here, alone in the family, achaenial heteromorphy is basically a tribal character, for the groups with homomorphic achenes are marked by female-sterilization of either rays or disc. *Castalis* has the ray flowers neuter, and is clearly derived from *Dimorphotheca* in which the female rays have one sort of achene, the hermaphrodite disc another. This is not to say that achaenial heteromorphy in this tribe is always associated with different flower-types. In *Dimorphotheca* it is, and follows the familiar pattern that the fruits of the ray lack adaptation for dispersal, while those of the disc, being flat and more or less winged may certainly be ranked as wind-dispersed. In *Calendula* itself the heteromorphic fruits are all developed from ligulate female flowers, which are in more than one series. In *Osteospermum* most species have homomorphic achenes, for the opposite reason to that found in *Castalis*, for in *Osteospermum* it is only the uniseriate ray flowers that produce fruit; the more numerous disc flowers are functionally male. However, a few species of *Osteospermum* may have some fruits winged and others unwinged even in this single series of female flowers. Thus it seems that *Castalis* and most species of *Osteospermum* have secondarily homomorphic achenes, owing to the loss of fruit-forming capacity in one type of flower; on the other hand when heteromorphy is found within the uniseriate rays of *Osteospermum* it is secondary heteromorphy, developed after the primary heteromorphy was lost by the female-sterilization of the disc flowers. It may also be noted that the achenes of this tribe are always fairly large and show a tendency towards an increase in size as the number of fruit-bearing flowers in the head is reduced. All these data are available in Norlindh's excellent monograph (Norlindh, 1943). In functional terms the change from *Dimorphotheca* to *Osteospermum* subgen. *Osteospermum* represents loss of the dispersive achene but an increase in size of the non-dispersive type. Then some achenes have regained a certain degree of dispersive power in a few species of this subgenus and throughout the subgenus *Tripteris*. The plants with heteromorphic achenes may be either annual or perennial and some at least are self-incompatible.

CENTRAL UNISEXUALITY

Female sterility in simple capitula

The flowers at the centre of the capitulum are most often hermaphrodite; however, there are many genera in which the ovules are aborted and the flowers are functionally male. This condition can usually be recognized at flowering time not only by the lack of an ovule in the ovary (more rarely by the complete abortion of the ovary) but by the notched rather than divided style. When a female-sterile disc is associated with filiform female flowers in the periphery of the head (as in *Anaphalis* L.) it seems that fruit production is reduced, but that structure is not changed in comparison with related species with fully fertile heads (e.g. in *Helichrysum* Mill.). However, in other groups where the female flowers at the periphery are radiate, there does seem to be a tendency for the reduction in numbers of fruit to be accompanied by an increase in their size (e.g. in the tribe Calenduleae).

In the previous section the advantages in having two types of diaspore, one dispersive and the other persistent, have been suggested. The conditions accompanying the loss of duality are therefore of interest, and the explanation may lie in the increase in size of the peripheral achenes. This may be associated with a change in conditions of growth increasing the permanence of populations, and rendering the more highly dispersive achenes at least temporarily unnecessary; or by the normally larger and non-dispersive peripheral achenes developing a dispersal mechanism, such as the fruit wings of *Osteospermum* subgen. *Tripteris* or the succulent exocarp of the genus *Chrysanthemoides* Medik.

Some allegedly compound capitula

Compound capitula are well known in Compositae and there is no need to discuss them in general terms. However, there are several genera where the nature of the capitulum, whether simple or compound, is in dispute; and these are relevant here because the central flowers in all of them are functionally male. I myself (see Hilliard and Burtt, 1971) once suggested that the capitulum of *Ifloga* (Inuleae) might be compound, and others have recently suggested the same interpretation of the heads of *Acanthospermum* (Tiagi and Manilal, 1964), *Melampodium* (Manilal, 1973, 1975) both in Heliantheae, and *Moscharia* (Crisci, 1974) in Mutiseae. In the capitula of all these plants the achenes are protected by bracts, open on one side or completely enclosed. One genus related to *Moscharia* does have compound heads; this is *Polyachyrus*. *Moscharia*, however, has a much greater resemblance to some species of *Leucheria*, in which the inner involucral bracts may be decidedly concave on the inner surface and embrace an outer achene. *Moscharia* might well have developed from such a species of *Leucheria*.

The arrangement of capitula in a compound inflorescence is relevant here. It is usually found that the head terminating the central axis opens first, followed by lateral ones. When the heads of such an inflorescence are aggregated to form a compound capitulum it does not, therefore, open in a regular centripetal manner. In *Echinops*, for instance, the first individual capitula to open are those at the top of the secondary head. In these doubtful genera, however, the flowers open from the periphery to the centre. In other words, the indications are that the underlying structure is still racemose. Racemes of capitula are rare in Compositae, but they do occur; they are developed from a cymose panicle by reduction of the lateral branches to a single head, inversion of flowering order and eventual loss of the terminal capitulum of the main axis. Maresquelle (1964) calls this process racemization. Good examples are found in *Ligularia przewalskii* Maxim. and its allies (cf. also Kunze, 1969). Condensation of these inflorescences would, of course, give a racemose compound capitulum but there is no indication that this occurs; in fact the extended length of these inflorescences is their main feature. To my knowledge racemes of capitula do not occur in the taxonomic affinity of *Ifloga* or *Moscharia*.

The situation with regard to *Acanthospermum* and *Melampodium* is somewhat different, for the genus *Ambrosia* belongs to the same tribe, Heliantheae. In *Ambrosia* the male capitula form a raceme which has short basal branches bearing the female heads. Payne (1964) has shown that the derivation of this pattern can be traced through the intermediate state of *A. divaricata* (Brandegee) Payne, in which the basal part of the male inflorescence is also branched. However, inflorescence structure in *Ambrosia* seems to be closely connected with wind pollination and it remains to be shown that the capitulum of *Melampodium* or *Acanthospermum* is derived from a compound inflorescence of this pattern. I am still inclined to regard all these genera as having simple capitula. It has just been suggested that the relegation of the disc flowers in Calenduleae to a purely male function may have been sometimes associated with the development of dispersive mechanisms by the usually non-dispersive ray. A feature of *Acanthospermum* and *Melampodium* is that the fertile peripheral achenes are large, and particularly well protected; some species are endowed with hooks as aids to dispersal. It is possible that these genera are derived from plants having heteromorphic achenes by a similar series of steps.

There is a comparable situation in *Zinnia* and *Tragoceros*, now *Zinnia* sect. *Tragoceros* (Torres, 1963a,b; Olorode and Torres, 1970). *Zinnia*, in the narrow sense, has all the flowers producing achenes and the ray-corollas are persistent on the fruits. In sect. *Mendezia* the peripheral fruits are also hooked. In *Tragoceros* (reduced to a section of *Zinnia* on the evidence of hybrids with sect. *Mendezia*) there are strongly developed hooked horns on the peripheral achenes (whence the name) and the disc flowers are female-sterile. Again the reduction of the disc to a male function is linked to the development of a more precise dispersal mechanism for the peripheral achenes.

The pappus of male flowers

It is, at first, rather baffling to dissect a capitulum of *Ifloga* and find the fertile achenes without a pappus, but the male flowers amply provided with one. This pappus has setae subplumose at the tip, is opaquely white and is conspicuous in the flowering head. Northern botanists will remember that *Anaphalis dioica* L. has a thin setose pappus on the female heads, but on the males the setae are expanded with a flat blade. In these plants the pappus of the male flowers may be taking on a new role in making the head conspicuous.

In heads where the pappus is not uniform it is absent or weakly developed on the peripheral flowers. It seems that this is brought about by delayed expression of the pappus in the ontogeny of the head. The rather complex situation in *Ifloga* may be a further development from this state: that is to say it may have developed, just as *Acanthospermum* and *Melampodium* may have developed, from a preceding condition with heteromorphic fruits. In *Psychrogeton* Boiss. the peripheral female flowers have more pappus setae than the central, functionally male, ones (Grierson, 1967): a notable exception to the more normal condition.

CONCLUSION

At several points in the foregoing discussion it has been necessary to distinguish between plants with filiform female flowers and those with radiate heads. Despite the fact that radiate and filiform female flowers can occur in the same capitulum (*Erigeron* sect. *Trimorphaea*, *Nanopappus*, *Erodiophyllum*), these two divisions do represent a major divergence in the structure of the capitulum. This is carried through to the fruiting phase where filiform flowers are generally associated with small, quick-ripening achenes and a pappus, if any, composed of slender setae. All the more considerable structural elaborations within the capitulum have been associated with radiate female flowers and larger achenes. Nevertheless it seems that both pathways lead ultimately to a capitulum in which a few peripheral female flowers surround a cluster of functionally male flowers. The two end-points differ mainly in the size of the fertile achenes and their dispersal mechanisms.

Diversification of fruits often follows the pattern laid down for the flowers, ray and disc showing achaenial differences; but it is not dependent on this floral differentiation. The heteromorphic achenes of the homogamous Lactuceae and those found amongst the ray flowers of Calenduleae show that fruit diversification can occur independently.

Annuals appear at the end point of many different evolutionary lines in Compositae. They are the potential weeds, and in that role many special features are needed. Here they have attracted our attention in fruit protection, fruit heteromorphism and in small, rapidly maturing fruits. It is well to remember, therefore, that they comprise only a small proportion of this vast family, and that the features that have caught our interest are not wholly restricted to annual plants.

In a sense we have been looking at the potential of the capitulum; so that the selection of plants mentioned is by no means an average sample of the family. In this selection it is clear that development of the potential has frequently taken place under an arid climate.

The structure of the capitulum must meet the demands of both the flowering and fruiting phases. There are four major factors: in the flowering phase, the efficiency of pollination and the balance of the breeding system (ratio of inbreeding to outbreeding); in the fruiting phase, protection of the maturing achenes and then their adequate dispersal. In all plants the demands of flowering and fruiting phases interact (cf. Burtt, 1975) and study of the co-evolution of these two parts of the life history has been rather neglected. Compositae offer a particularly promising field for its investigation because so many trends are paralleled in different parts of the family, of which the basic classification is well founded. I have tried to indicate a few of the more interesting points, and to suggest where they link on to general topics of biology.

Compositae is not the only family in which a centripetal capitulum occurs, but it is unique in having attained this, together with an efficient unspecialized floral mechanism, before ecological specialization set in. To take a contrasting case: the capitulum of *Protea* is clearly a late development in a family where both vegetative and floral ecology are already highly specialized. The capitulum of Compositae, with an underlying pattern of development that has an immense potential, was virtually a new structure. It has become the main seat of variation and therefore of response to changing pressures. In Wardlaw's phrase, it is the main reaction system of Compositae and has, as it were, acted as a shield for the flowers. That is one reason why this group is outstandingly successful: variation being largely in the capitulum, the characters of the individual flowers, by which the taxonomist defines the family, remain rather constant.

ACKNOWLEDGEMENTS

It is a pleasure to thank Dr Prithipal Singh (University of Delhi) for the numerical data on *Blumea*, Dr O. M. Hilliard (University of Natal) for those on *Helichrysum* and *Gnaphalium*, Dr K. M. M. Dakshini (University of Delhi) for chromatograms of dimorphic fruits of *Bidens* and *Dimorphotheca*, and Mr R. W. Phelps (South Australia) for the fruiting heads of *Erodiophyllum*.

REFERENCES

Assouad, M. W. (1971). Observations et experiences préliminaires sur la variabilité de la forme sexuelle chez Cephalaria leucantha Schr. *Natur. Monsp.* **22**, 5–11.

Babcock, E. B. (1947). "The Genus *Crepis*". California University Press, Berkeley and Los Angeles.

Battandier, M. A. (1883). Sur quelques cas d'hétéromorphisme. *Bull. Soc. bot. Fr.* **30**, 238–244, tab. 3.

Becker, W. (1912). Über die Keimung verschiedenartiger Früchte und Samen derselben Spezies. *Beih. bot. Zbl.* **29**, 21–143.

BRADLEY, J. D., TREMEWAN, W. G. and SMITH, A. (1973). "British Tortricoid Moths". The Ray Society, London.
BURTT, B. L. (1961). Compositae and the study of functional evolution. *Trans. bot. Soc. Edinb.* **39**, 216–232.
BURTT, B. L. (1975). Patterns of structural change in the flowering plants. *Trans. bot. Soc. Edinb.* **42**, 133–142.
CRISCI, J. V. (1974). Revision of the genus *Moscharia* (Compositae: Mutisieae) and a reinterpretation of its inflorescence. *Contr. Gray Herb.* **205**, 163–173.
DAKSHINI, K. M. M. and AGGARWAL, S. K. (1974). Intracapitular cypsele dimorphism and dormancy in *Bidens bipinnata*. *Biologia Pl. (Praha)*, **16**, 469–471.
ESTES, J. R. and THORP, R. W. (1975). Pollination ecology of *Pyrrhopappus carolinianus* (Compositae). *Am. J. Bot.* **62**, 148–149.
FAHN, A. (1967). "Plant Anatomy". Pergamon, Oxford.
GADGIL, M. and SOLBRIG, O. T. (1972). The concept of r- and K-selection: evidence from wild flowers and some theoretical considerations. *Am. Nat.* **106**, 14–31.
GOEBEL, K. (1924). "Die Entfaltungsbewegungen der Pflanzen und deren teleologische Deutung". Fischer, Jena.
GRIERSON, A. J. C. (1967). The genus *Psychrogeton* (Compositae). *Notes R. bot. Gdn Edinb.* **27**, 101–147.
GRIMBACH, P. (1913). Vergleichende Anatomie verschiedenartiger Früchte und Samen bei derselben Spezies. *Engl. Bot. Jahrb.* **51**, Beibl. no. 113, 1–52.
GUTTENBERG, H. VON (1971). "Bewegungsgewebe und Perzeptionsorgane". (Handbuch der Pflanzenanatomie 5, teil 5). Borntraeger, Berlin & Stuttgart.
HAINES, R. W. (1971). Amphicarpy in East African Cyperaceae. *Mitt. Bot. Staatss. München*, **10**: 534–538.
HILLIARD, O. M. and BURTT, B. L. (1971). Notes on some plants of southern Africa, chiefly from Natal: II. *Notes R. bot. Gdn Edinb.* **31**, 1–33.
HILLIARD, O. M. and BURTT, B. L. (1973). Notes on some plants of southern Africa, chiefly from Natal: III. *Notes R. bot. Gdn Edinb.* **32**, 303–387.
HILLIARD, O. M. and BURTT, B. L. (1976). Notes on some plants of southern Africa, chiefly from Natal: V. *Notes R. bot. Gdn Edinb.* **34**, 253–286.
HULL, P. (1974a). Self-fertilization and the distribution of the radiate form of *Senecio vulgaris* L. in Scotland. *Watsonia* **10**, 69–75.
HULL, P. (1974b). Differences in esterase distribution detected by electrophoresis as evidence of continuing interspecific hybridization in the genus *Senecio*. *Ann. Bot.*, n.s. **38**, 697–700.
HULL, P. (1975). Selection and hybridisation as possible causes of changes in the frequency of alleles controlling capitulum-type in *Senecio vulgaris* L. *Watsonia* **10**, 395–402.
JAEGER, P. (1938). "Morphologie et biologie florales chez les Dipsacacées". Imprimerie "Alsatia", Colmar.
KOLLER, D. and ROTH, N. (1964). Studies in the ecological and physiological significance of amphicarpy in *Gymnarrhena micrantha* (Compositae). *Am. J. Bot.*, **51**: 26–35.
KUNZE, H. (1969). Vergleichend-morphologische Untersuchungen an Komplexen Compositen Blütenstand. *Beitr. Biol. Pfl.* **46**, 97–154.

LEVIN, D. A. (1975). Pest pressure and recombination systems in plants. *Am. Nat.* **109**, 437-451.
LEVYNS, M. R. (1972). The Rhenosterbush. *Veld & Flora* **2**, 7-9.
MACARTHUR, R. H. and WILSON, E. O. (1967). "The Theory of Island Biogeography". Princeton University Press, New Jersey.
MANILAL, K. S. (1973). Morphology of the capitulum of *Melampodium divaricatum* (Rich.) DC. *Curr. Sci.* **42**, 578-580.
MANILAL, K. S. (1975). A compound capitulum in *Melampodium* L. (Compositae). *J. Linn. Soc. (Bot.)* **70**, 70-74.
MARESQUELLE, H. J. (1964). Sur la filiation des inflorescences. IV, La notion de racemisation en morphologie vegetale. *Mém. Soc. bot. Fr.* 90-100.
MURBECK, S. (1920). Beiträge zur Biologie der Wüstenpflanze: II. Synaptospermie. *Lunds Univ. Årrskr.* N.F. Avd. 2, **17**, 3-53.
MURBECK, S. (1943). Weitere Beobachtungen über Synaptospermie. *Lunds Univ. Årrskr.* N.F. Avd. 2, 39, nr. **10**, 3-24.
NORLINDH, T. (1943). "Studies in the Calenduleae". Gleerup, Lund.
OLORODE, O., and TORRES, A. M. (1970). Artificial hybridization of the genera *Zinnia* (sect. *Mendezia*) and *Tragoceros* (Compositae-Zinninae). *Brittonia* **22**, 359-369.
PAYNE, W. W. (1964). The significance of the organization of the inflorescence of *Ambrosia divaricata* (Compositae). *Pap. Mich. Acad. Sci. Arts Letters* **49**, 41-46.
RICHARDS, A. J. (1975). The inheritance and behaviour of the rayed gene complex in *Senecio vulgaris*. *Heredity* **34**, 95-104.
SKOTTSBERG, C. (1938). On Mr C. Bock's collection of plants from Masatierra (Juan Fernandez), with remarks on the flowers of *Centaurodendron*. *Acta Horti. gothoburg.* **12**, 361-373.
TIAGI, Y. D. and MANILAL, K. S. (1964). The morphology and anatomy of the capitulum of *Acanthospermum hispidum* DC. *Proc. natn. Acad. Sci. India* **34**, 291-305.
TORRES, A. M. (1963a). Taxonomy of *Zinnia*. *Brittonia* **15**, 1-25.
TORRES, A. M. (1963b). Revision of *Tragoceros* (Compositae). *Brittonia* **15**, 290-302.
TROTTER, A. (1910). Intorno alla anficarpia di *Catananche lutea* L. *Bull. Soc. bot. Ital.* **1910**, 150-154.
TROW, A. H. (1912). On the inheritance of certain characters in the common groundsel—*Senecio vulgaris* Linn.—and its segregates. *J. Genet.* **2**, 239-276.
VALEN, L. VAN (1971). Group selection and the evolution of dispersal. *Evolution* **25**, 591-598.
WERKER, E. and MANY, T. (1974). Heterocarpy and its ontogeny in *Aellenia autrani* (Post) Zoh.; light and electron microscope study. *Israel J. Bot.* **23**, 132-144.
ZOHARY, M. (1937). Die verbreitungsökologischen Verhältnisse der Pflanzen Palästinas. *Beih. bot. Zbl.* **56A**, 1-155.
ZOHARY, M. (1950). Evolutionary trends in the fruiting head of Compositae. *Evolution* **4**, 103-109.

Chapter 4

The evolution of capitulum types of the Compositae in the light of insect–flower interaction*

ELMAR E. LEPPIK
*Plant Genetics and Germplasm Institute,
U.S. Department of Agriculture, Beltsville, Maryland, U.S.A.*

Abstract. The form and function of diverse capitulum types of the Compositae were studied and related to the sensory reaction and frequency of attendance of plant pollinators. No capitulum types specific to the Compositae were found. All the types studied imitated some floral pattern extant in solitary flowers. Differentiation of heads appears to be a recapitulation of the evolutionary sequence of solitary flowers. Four capitulum classes may be recognized among the Compositae: haplomorphic, actinomorphic, pleomorphic, stereomorphic. No example of the zygomorphic class has been found. The development of pseudo-flowers from multi-flowered inflorescences of the Compositae may be plausibly explained by the directing action and selective pressure of pollinators. This recapitulatory process and its underlying mechanism were found to be correlated with the inherited sensory reaction of food-searching insects and birds, and the selective pressure which they exercise upon plants.

CONTENTS

Introduction	62
Brief review of earlier studies	62
Concept of pseudanthic recapitulation	65
Capitulum evolution and its bearing on the origin of the Compositae	66
Phylogenetic origin	66
Correlation between the phylogenetic trend and the sequence of floral evolution	66
Observable evolutionary trends	67
Evolution of corolla types in the Compositae	67
Evolution of capitulum types in the Compositae	68
Differentiation of numeral patterns in the capitula	70
Phyllotaxis and semataxis	72
Current development of flower heads of the Compositae	73

* Contribution from the Agricultural Research Service, United States Department of Agriculture, Beltsville, Maryland. Plant Introduction and Genetics Resources Investigation Paper No. 38. All drawings are credited to the Agricultural Research Service, USDA.

Asymmetric trend of floral evolution 76
Inflorescence types 77
   Simple capitulum 78
   Compound capitula 78
   Conflorescences 79
Pollination and fertilization of the Compositae compared with the
   Umbelliferae 80
   Pollinators of the Compositae 81
   Statistical data on flower visitors observed on the Compositae . . 82
Coenogenetic approach 83
Practical application of the sequence of floral evolution . . . 84
Discussion and conclusion 85
References 87

INTRODUCTION

Some years ago, a new theory of evolutionary recapitulation in the differentiation of flower heads of the Compositae was proposed (Leppik, 1960a, b). This attempted to clarify the remarkable process and mechanism of the pseudanthous trend in the evolution of the capitula. New evidence has accumulated confirming the earlier findings that the pseudanthous inflorescence or head of the Compositae not only imitates the form and shape of solitary flowers, but also follows the main evolutionary trend of flower-types. Thus the general evolutionary sequence of solitary flowers— amorphic → haplomorphic → actinomorphic → pleomorphic → stereomorphic—is repeated in the development of flower heads of the Compositae (Figs 1, 2).

Common examples of this remarkable phenomenon include flower heads of *Aster, Centaurea, Leucanthemum, Taraxacum, Solidago, Tagetes* and *Helianthus*. These false flowers imitate true flowers of some phylogenetically remote plant groups, mainly the Ranunculaceae, which have no close relationship with the Compositae. The evolutionary relationship between entomophilous plants and anthophilous insects has been fully discussed in earlier papers (Leppik, 1956–1972). Phylogenetic and taxonomic consideration related to floral evolution in this paper are based on the studies by Cassini (1826), Darwin (1859), Bentham (1873), Hoffmann (1894), Hutchinson (1916), Melchior (1964), Cronquist (1955, 1968), Takhtajan (1966), Polyakov (1967), Thorne (1968), and Mayr (1972).

Brief review of earlier studies

The main challenge in understanding the Compositae capitulum lies in the apparent paradox of their pseudanthic inflorescences. In the capitular inflorescence, bracts may function as sepals and ray-florets imitate petals, whereas the closely aggregated disk-florets resemble the central part of an individual flower. The similarity of the capitulum to a solitary flower is so striking that it has long been a source of confusion and controversy among botanists (Small, 1915, 1919; Troll, 1928; Good, 1931, 1956; Cronquist, 1951, 1955) and has caused lively debates among recent workers in this field (Leppik, 1960a; Burtt, 1961).

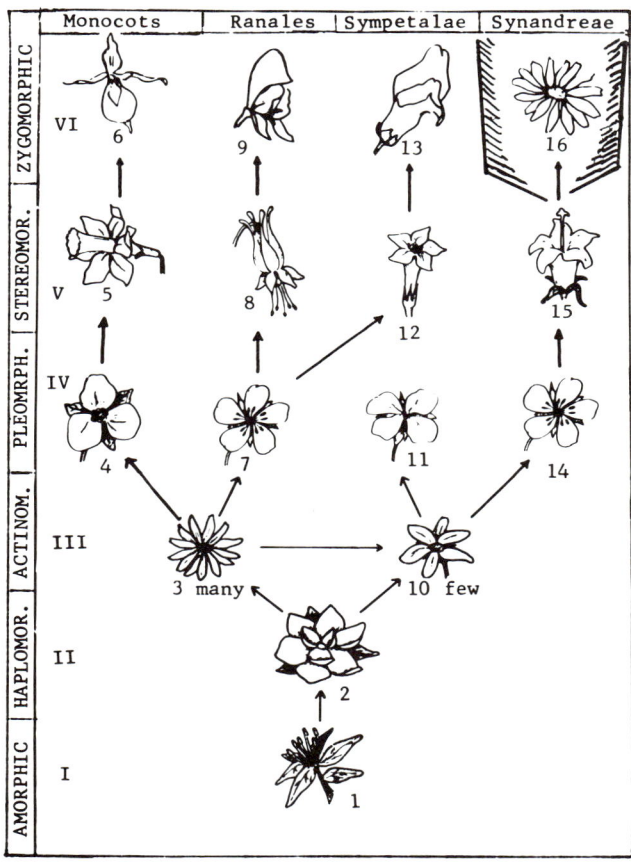

FIG. 1. Assumed historical development of flower types during the evolution of angiosperms in the last 100 million years: Level I.: amorphic flower types; Level II.: haplomorphic; Level III.: actinomorphic; Level IV.: pleomorphic; Level V.: stereomorphic; Level VI.: zygomorphic flowers. Arrows indicate the sequence of floral evolution in the main phylogenetic groups of angiosperms (Monocots, Ranales, Sympetalae, Synandrae). At the upper right corner (16), recapitulation of the main sequence of floral evolution in the differentiation of the capitula. Rearranged from Leppik (1957).

Although it is well known that the capitulum functions as an alluring showpiece for pollinators, its morphogenic development and evolutionary significance have never been satisfactorily explained. Without knowledge of the decisive role of pollinators in the process of floral evolution, it was difficult to suggest any satisfactory explanation of this pseudanthous trend.

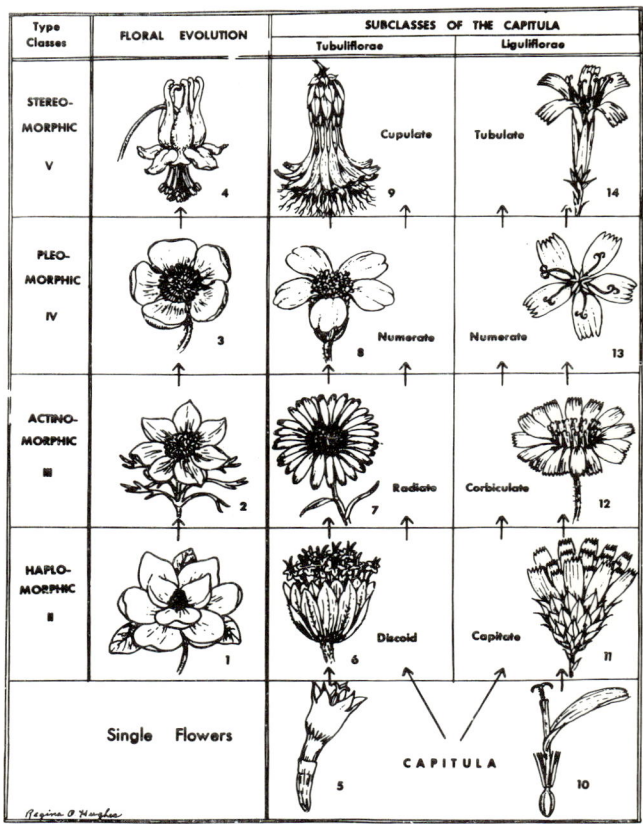

FIG. 2. Morphogenic differentiation of flower heads of the Compositae compared with the evolutionary sequence of flower-types, selected examples. Ranunculaceae: 1. *Magnolia*, 2. *Adonis*, 3. *Ranunculus*, 4. *Aquilegia*. Compositae, subfam. Tubuliflorae. 5. *Tanacetum alpinum* Sch. Bip., disk (tubulate) floret in the female stage, 6. *Adenostemma viscosum* Forst. (= *A. lavenia* (L.) Kuntze), discoid haplomorphic capitulum, 7. *Aster alpinus* L., radial (actinomorphic) capitulum, 8. *Achillea millefolium* L., numerate (pleomorphic) capitulum, 9. *Mutisia grandiflora* Humb. & Bonpl., cupulate (stereomorphic) capitulum. Compositae subfamily Linguliflorae, 10. *Lactuca virosa* L. zymorphic ligulate floret, 11. *Catananche caerulea* L., capitate (haplomorphic) head, 12. *Cichorium intybus* L., corbiculate (actinomorphic) head, 13. *Lactuca muralis* (L.) Gaertn., numerate (pleomorphic) capitulum. 14. *Lygodesmia juncea* (Pursh) D. Don. tubulate (stereomorphic) head (after Leppik, 1970b).

Thus, the whole problem remained unsolved until discoveries were made concerning the sensory physiology of insects.

A new theory is emerging that correlates floral evolution with the sensory behavior and selective pressure of insect pollinators. According to this theory, certain stages in the sensory development of anthophilous insects, particularly their ability to distinguish definite flower types, must concur with corresponding evolutionary levels in floral differentiation. Floral evolution of entomophilous plants may consequently be considered to be a gradual adaptation of evolving flower types to the increasing sensory ability and selective activity of certain plant pollinators. The majority of composites are unquestionably entomophilous, and therefore this family cannot be an exception to this fundamental process of floral evolution.

The discovery of the sign language of the honeybees, the sensitivity of the insect eye to polarized light, their metabolic time sense, their food-sharing habits, and other findings in the field of sensory physiology enable us to comprehend better the evolution of flower types. The ability of insects, for instance, to distinguish form and symmetry as well as other floral characteristics provides opportunities for experiments in floral ecology (Leppik, 1956, 1957, 1964, 1966).

Concept of pseudanthic recapitulation

Flower heads evolve continuously in response to the sensory abilities and selective activities of their exploiter-pollinators, which are mainly insects and sometimes birds or other animals. Earlier workers tried to explain the evolution of pseudoflowers from the morphological viewpoint only, thus overlooking or ignoring the important selective role of pollinating insects.

When the image of a floral pattern is deeply engraved in the inherited sensory behaviour of certain pollinators, this pattern may reappear, even among flowers of various phyletic origins, as long as they are visited by this particular group of pollinators. Frequently, similar flower patterns are formed from entirely different morphological organs, such as petals, sepals, bracts, upper leaves, or inflorescences. These patterns appear as distinctive semaphylls, or food markers, for nectar-searching insects, birds, and bats.

For instance, haplomorphic and actinomorphic floral patterns existed among Mesozoic Bennettitatae (Leppik, 1960b, 1963b) about 200 million years ago. Although these plants possessed neither petals nor true sepals, they had semaphylls formed from bracts and sporophylls. These semaphylls were arranged into showy floral patterns, which attracted contemporary pollinators. The resemblance of such semaphylls to true flowers was so striking that early palaeontologists mistook them for angiosperms (Leppik, 1963b). Fossil evidence indicates that the primary pollinators of these ancient plants were beetles and other primitive insects whose sensory abilities were less developed than those of the more advanced pollinators of the present day.

Thus, the general trend of floral evolution is recapitulated in various phylogenetically unrelated plant groups, including the Compositae (Figs 1, 2). Depending on the selective activity of certain specialized pollinators, the general evolutionary trend of the capitula may progress, regress, or remain constant. Sometimes deviations in the general evolutionary trend of capitula are caused by other pollination agencies, such as the wind (*Ambrosia, Artemisia*) and birds (*Mutisia, Barnadesia, Cnicothamnus, Chuquiragua,* and others), as described by Vogel (1954).

CAPITULUM EVOLUTION AND ITS BEARING ON THE ORIGIN OF THE COMPOSITAE

Phylogenetic origin

Cronquist (1955), believes that the Rubiales–Dipsacales complex, rather than the Campanulales, has many of the characters necessary for it to be a near ancestor of the Compositae. Yet, both the Rubiaceae and Dipsacaceae are too far advanced to be considered ancestors of the Compositae. These families could have undergone similar changes as they developed from a common ancestral stock. Further evidence is presented by Koch (1930), Leonhardt (1949), Zohary (1950), Augier and Mérac (1951), Takhtajan (1966), and Polyakov (1967), but see also Turner (chapter 2).

Correlation between phylogenetic trend in the Compositae and sequence of floral evolution

The sequence of floral evolution can be used as criterion for the estimation of the age and ancestry of the Compositae. Since the family is characterized and defined basically by the capitulum, which is a secondary aggregation of the primary inflorescence, the age of the capitulum itself cannot be very old geologically. Thus the structure of individual florets of the capitulum must serve as the best guide to the origin and ancestry of the Compositae.

Cronquist (1968) lists putative primitive characters of the family. It is clear that the five-lobed gamopetalous corolla, the epipetalous stamens, and the inferior, bicarpellate, uniloculate ovary with a single basal ovule must have been well established in the ancestral stock. These characteristics have remained constant during the evolution of all tribes. The dry fruit, an achene (cypsela), bearing a single seed without endosperm, is also characteristic. Such well advanced flower-types appeared in the stereomorphic evolutionary level in the middle of the Tertiary period. Only from the aggregation of such solitary flowers could the capitulum of the Compositae develop. The aggregation of solitary flowers into a capitulum has drastically changed their further evolution. The capitulum continued to evolve as an ecological unit for attraction of pollinators, whereas the differentiation of ligulate, bilabiate and tubulate corollas took place entirely in the framework of the capitulum as a whole.

4. EVOLUTION OF CAPITULUM TYPES

OBSERVABLE EVOLUTIONARY TRENDS

The classical subfamilies, Tubuliflorae and Liguliflorae, exhibit similar evolutionary trends in the differentiation of their capitula. They are actually a repetition of the general evolutionary course of the solitary flowers of the Magnoliaceae, Ranunculaceae, Campanulaceae, and other groups of phylogenetically older angiosperms (Leppik, 1963a, 1964). Observable evolutionary stages of the capitula, arranged in phylogenetic sequence of the Compositae, are depicted in Fig. 10.

Evolution of corolla types in the Compositae

Evolution of the primary flowers of the Compositae as morphological units has been described and discussed by Leppik (1960a). A pentamerous gamopetalous corolla with five syngenesious stamens, and an inferior bicarpellate uniloculate ovary, usually with a pappus, are typical of all florets of present-day composites. These are the basic characteristics of a campanuloid or rubioid type at the stereomorphic level of the precomposite evolution of solitary flowers. In the adaptation to pseudanthic capitula, changes occurred only in the corolla lobes of the ray-florets.

There are four basic corolla types among the present-day Compositae: tubulate, radiate, labiate, and ligulate (Fig. 3). Radiate and labiate corollas could have been derived from tubulate structures, whereas the ligulate florets may have evolved independently.

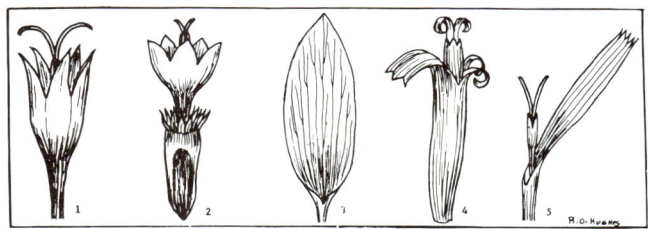

FIG. 3. Corolla-types of the Compositae: 1. tubulate corolla of *Petasites*; 2. tubulate disk-floret of *Grangea*; 3. ray-floret of a radiate type, *Helianthus*; 4. bilabiate corolla of *Trixis*; 5. ligulate corolla of *Tragopogon*. After Leppik, 1970b.

Tubulate, five-lobed corollas (Fig. 3: 1, 2) occur predominantly in the subfamily Tubuliflorae. In ray-florets (Fig. 3: 3), the corolla-lobes develop unequally, usually forming a large three-lobed ray extending outside, frequently, but not always, supplemented by two rudimentary inside lobes (Fig. 3: 4).

Bilabiate corollas (Fig. 3: 4) are characteristic of the tribe Mutisieae. They are found in the peripheral as well as the central part of the capitulum. Typical ligulate (Fig. 3: 5) corollas occur almost exclusively in the subfamily Liguliflorae. (For more detailed description of corolla types, see Leppik, 1967, 1970b, 1971 and 1972.)

Evolution of capitulum types in the Compositae

After aggregation into a capitulum, single florets tend to lose their individual shapes and to evolve further as definite parts of a pseudanthium. At the lower amorphic and haplomorphic evolutionary levels, all the florets of a capitulum, in spite of their tight clustering, still look alike. But from the actinomorphic level onwards, gradual differentiations of rays and central florets takes place. The ray-florets become bilateral and zygomorphic, whereas the central florets tend to keep their original structure.

The capitulum-types described in the following can be observed in the present-day Compositae (Figs 2, 4). In the Tubuliflorae, they are com-

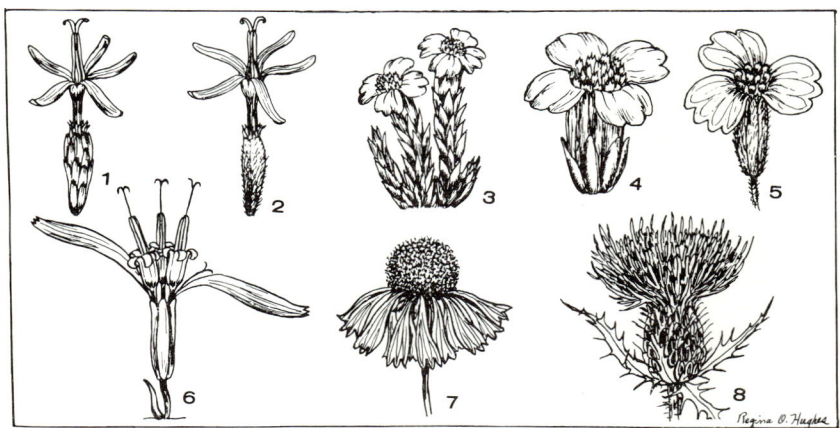

FIG. 4. Capitulum types: 1. one-flowered capitulum of *Echinops sphaerocephalus* L.; 2. the same: single floret; 3. tetramerous capitulum of *Bryomorphe zeyheri* Harvey; 4. the same, single floret; 5. trimerous capitulum of *Riddellia tagetina* Nutt.; 6. bimerous head of *Ligularia przewalskii* (Maxim.) Diels; 7. actinomorphic head of *Helenium autumnale* L.; 8. a secondary haplomorphic head of *Cirsium discolor* (Mühl.) Spreng.

posed of radiate and tubulate florets, and in the Liguliflorae, of ligulate florets only.

Discoid is a frequently used term for haplomorphic capitula of the Tubuliflorae without showy rays (Fig. 2: 6). *Capitate* is an equivalent type of the Liguliflorae (Fig. 2: 11), exclusively composed of ligulate florets. *Radiate* (Fig. 2: 7) and *corbiculate* (Fig. 2: 12) capitula show a similar actinomorphic shape and are composed of radiate (Tubuliflorae) and ligulate (Liguliflorae) florets.

4. EVOLUTION OF CAPITULUM TYPES

Numerate capitula are characterized by iconic numerals—8, 5, 3, 2—in both subfamilies (Fig. 2: 8, 13). Differentiation of numerate capitula is demonstrated in the ray-florets of *Bidens bipinnata* L. (Fig. 5). Five-rayed heads are dominant, whereas a few five-, and even three-rayed heads are found on different individuals of the same plant population, frequently in the same place. Well established numerical patterns (iconic numerals), such as 2, 3, 4, 5, and 8, are shown in Fig. 6.

Cupulate or cup-shaped capitula with recessed nectar are found among some ornithophilous composites (Fig. 2: 9).

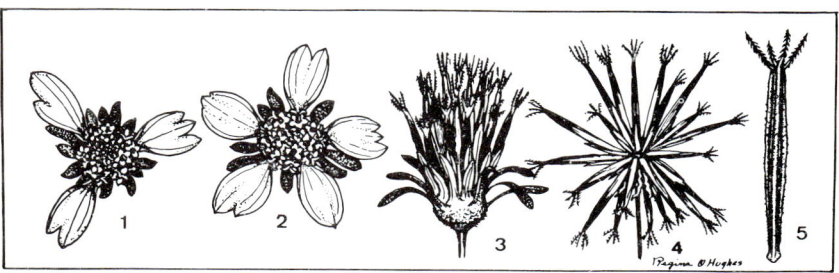

FIG. 5. Differentiation of the inflorescence and infructescence of *Bidens bipinnata* L.: 1. flower head with three ray florets; 2. head with five rays; 3. young infructescence; 4. dry infructescence; 5. achene.

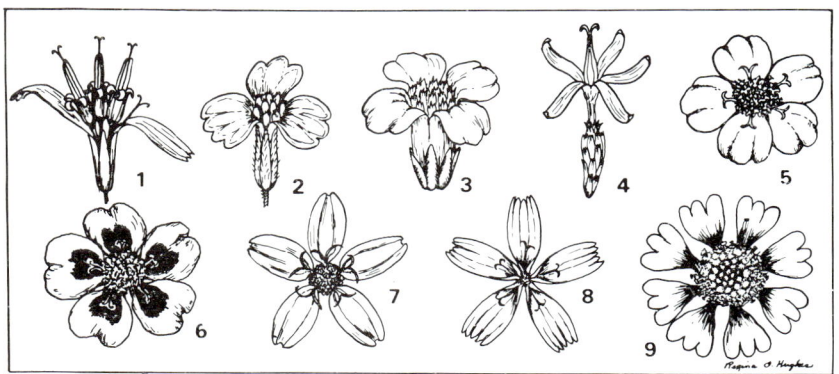

FIG. 6. Differentiation of numeral patterns (iconic numerals) in the flower heads of the Compositae: 1. bimerous head of *Ligularia przewalskii* (Maxim.) Diels; 2. trimerous head of *Riddellia tagetina* Nutt.; 3. tetramerous head of *Bryomorphe zeyheri* Harvey; 4. pentamerous head of *Echinops sphaerocephalus* L. with a single floret; 5. pentamerous head of *Achillea millefolium* L. with five rays, but numerous tubulate florets in the centre; 6. pentamerous head of *Tagetes tenuifolia* Cav. with ornate rays; 7. simple pentamerous flower of *Baltimora recta* L.; 8. simple pentamerous flower head of *Lactuca muralis* (L.) Gaertn., with serrate rays; 9. octomerous flower head of *Actinella odorata* A. Gray with eight serrate rays.

Tubulate capitula are represented in some genera of the Liguliflorae (Fig. 2: 14).

Reduction and secondary aggregation of simple capitula into compound capitula and conflorescences is a further interesting trend in the evolution of the Compositae. The phenomenon is thoroughly discussed by Burtt (1961 and Chapter 3) and a few examples are depicted in Figs 14, 15. The one-flowered capitula of *Echinops* (Fig. 4: 1, 2) are secondarily aggregated into large compound heads (Fig. 11: 1), sometimes consisting of as many as 200 primary capitula.

Reduction of ray-florets to a particular iconic (pictorial) numeral pattern is demonstrated by *Bryomorphe* (Fig. 4: 3, 4), *Riddellia* (Fig. 4: 5), *Ligularia* (Fig. 4: 6), and *Achillea* (Fig. 15: 1). The secondary aggregation of the pentamerous capitula of *Achillea millefolium* L. into a corymboid conflorescence is shown in Fig. 15: 2.

Differentiation of numeral patterns in the capitula

The constant number of ray-florets in many composites has long attracted attention. The capitula ordinarily have 8, 6, 5, 4, 3, or 2 rays, seldom a single one. The number varies in different genera, but usually remains constant in the capitula of the same species (Fig. 7).

Expressed in definite size, shape and colours, these numerical patterns

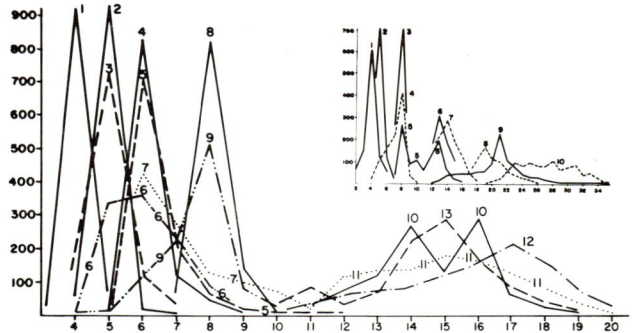

FIG. 7. Total numbers of counted heads (ordinate) arranged according to the number of rays (abscissa)—Note that the species with constant number of rays have high peaks (1, 2, 3, 4): 1. *Solidago flexicaulis* L., Minnesota; 2. *Baltimora recta* L., Central America; 3. *Rudbeckia laciniata* L., Minnesota; 4. *Ratibida columnifera* (Nutt) W. & S., Minnesota; 5. *Rudbeckia hirta* L., Ames, Iowa; 6. *Helianthus laetiflorus* Pers., Ames, Iowa; 7. *Helianthus grosseserratus* Mart., Ames, Iowa; 8. *Lactuca sativa* L., cult., Ames, Iowa; 9. *Chrysanthemum leucanthemum* L., Europe (average from 3000 counts according to Ludwig 1883, 1887, 1897); 10. *C. leucanthemum*, cult., Ames, Iowa. At the upper right corner a reduced scheme of similar grouping of the solitary flowers of the Ranunculaceae, according to Leppik, 1964, p. 41.

4. EVOLUTION OF CAPITULUM TYPES

imitate the corresponding number and arrangement of petals in solitary flowers. The perfect radial symmetry, fine configuration of rays, variegated colours, and pleasant odours of these pseudoflowers may even surpass the aesthetic appeal of solitary flowers. To the pollinators, these numerical patterns serve as semaphylls, or food markers, helping insects locate their food plants and restricting their flights to certain species.

Birds, bats, and beetles are less adapted to numeral patterns in flowers and are not able to distinguish *iconic numerals*. In ornithophilous and cantharophilous flowers, therefore, iconic numerals are reduced or may disappear completely.

Several earlier workers, not understanding the purpose of iconic numerals, tried to derive the number of ray-florets from the progressive phyllotactic order of phyllaries and cauline leaves. They succeeded only with sunflowers, daisies, and other plants with multi-rayed capitula. They failed when the number of rays was below 10. Ludwig (1884, 1890, 1897a, b), for instance, showed that the ray-florets in a sunflower head normally occur only at the ends of long spirals (Figs 8, 9). The theoretically expected 13, 21, or 34 rays in a capitulum are not always found.

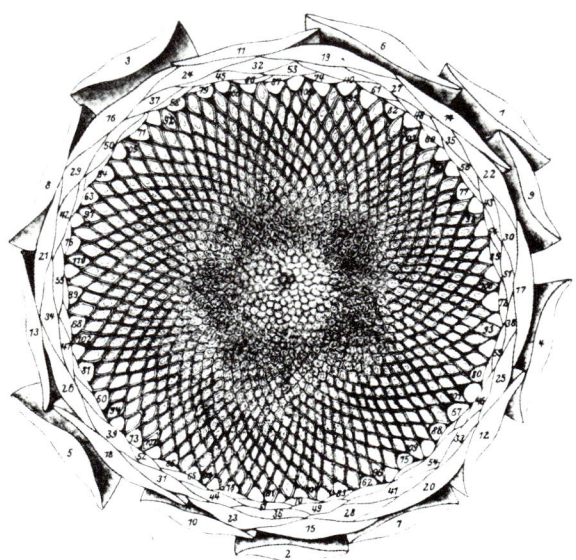

FIG. 8. *Helianthus annuus* L.: capitulum with fruits in a phyllotactic arrangement (after Weisse, 1897).

Extensive observations of Ludwig (1890, 1897a, b), Weisse (1897), Church (1904, 1920), Troll (1928), and Leppik (1960a, 1961) have shown that the most frequent number among few-rayed capitula is 5. The next most frequent is 8, whereas 6, 4, 3, and 2 occur in only special cases. Single-rayed capitula are known in *Barnadesia, Myripnois, Ainsliaea, Flaveria,* and *Schkuhria*. In all of these, the rays are reduced and have lost their importance as semaphylls. The differentiation of numeral patterns, in the capitula, consequently, follows the evolutionary sequence of solitary flowers with 5, 6, 4 and 8 petals dominating at the higher evolutionary levels.

Phyllotaxis and semataxis

Beautiful phyllotactic configurations are produced by the symmetrical arrangements of florets in the flower heads of the Compositae. In the capitula, definite curves and logarithmic spirals appear. These can be depicted as geometric figures and expressed by mathematical formulae (Figs 8, 9). The exact radial symmetry of the heads of composites has long been the object of admiration and scientific curiosity. Earlier literature has been reviewed by Small (1915, 1919), Church (1904, 1920), and more recently by Leppik (1961).

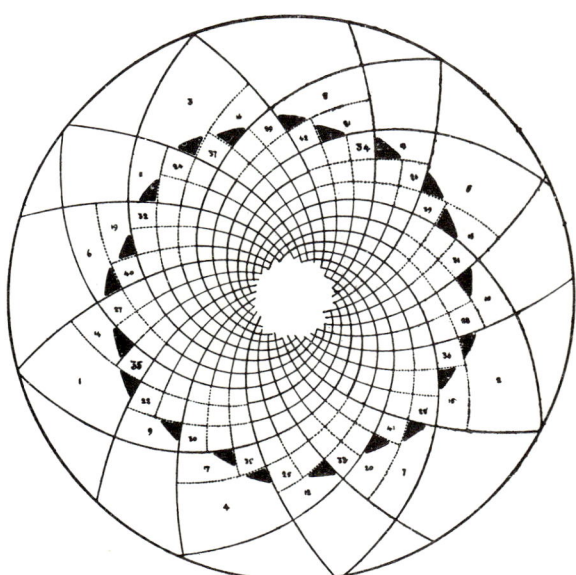

FIG. 9. Sketch showing complicated phyllotaxis order of sunflower with many intersecting spirals (after Weisse, 1897).

Extensive observations have shown that the most frequent patterns in the arrangement of the florets in the capitula are those that fall into the so-called Schimper-Braun series of divergences as follows: 2/5, 3/8, 5/13, 8/21,

13/34, or 5/8, 8/13, 13/21, 21/34, and so on. For example, a sunflower head, according to Weisse (1897), shows a phyllotaxis ratio that increases from 8/13 to 21/34. There are 21 long spirals intersecting 34 short ones (Figs 8, 9). Since ray-florets (black triangles) occur at the ends of long spirals only, there are normally 21 rays in a capitulum. These ray-florets may furthermore be seen in Fig. 9 to be grouped as 2-1-2-1-2-2-1-2-1-2-2-1-2. Thus, the number and position of the ray-florets follow the geometrical construction of the capitulum.

Using mathematical formulae, Church (1904, 1920) worked out a complete phyllotactic order for plants, including the Compositae. However, insects do not know phyllotactic orders and have no interest in logarithmic spirals. If they prefer some species as a source of food, insects automatically select flowers with familiar syndromes for their consecutive visits.

Under the continuous selective pressure of insect pollinators, phyllotactic orders in the capitula are gradually disintegrated such that more familiar numerical patterns are restored. Church was greatly confused about this phenomenon. He could not understand how and why such an advanced phyllotactic order, which served its purpose so well, could revert to elementary numerical patterns. He did not know, of course, much about the role of pollinating insects in such phenomena.

Earlier authors tried to generalize these phenomena and to explain the arrangement of florets in the capitula by the conventional phyllotaxis theory. Later studies revealed that, depending on their function, four different systems—phyllotaxis, anthotaxis, semataxis, and carpotaxis—are involved (Leppik, 1961). During its maturation, a plant passes from the vegetative to the flowering and fruiting stages, and the leaf arrangement systems change correspondingly from phyllotaxis to anthotaxis, semataxis, and carpotaxis (Leppik, 1970b, p. 335).

Current development of flower heads of the Compositae

The capitulum may be depicted as having developed, as have flowers in many phylogenetic groups, in response to sensory mechanisms pollinators (Fig. 10).

The *amorphic* level, in general, represents the first transient stage in the evolution of pseudanthous capitula. In this stage, many solitary flowers of the previously loose inflorescence are clustered into a compact head, which functions as a unit. Even without definite form and symmetry, such elementary capitula are more conspicuous than solitary flowers and may be visited more frequently by insects. On a head, insects can walk around easily and pollinate more florets in less time than they can when flying from flower to flower. Therefore, the chance for florets in a head to become fertilized is considerably higher than for solitary blossoms.

In their current development, amorphic heads are gradually changing to the *haplomorphic* stage, which is more conspicuous and attractive to insect pollinators. This is obviously the reason why amorphic types are rare or absent among the primary flower heads of the present-day Compositae. More frequently, amorphic capitula appear secondarily in compound heads, such as the inflorescences of *Leontopodium, Ageratum, Lagascea, Solidago,* and others.

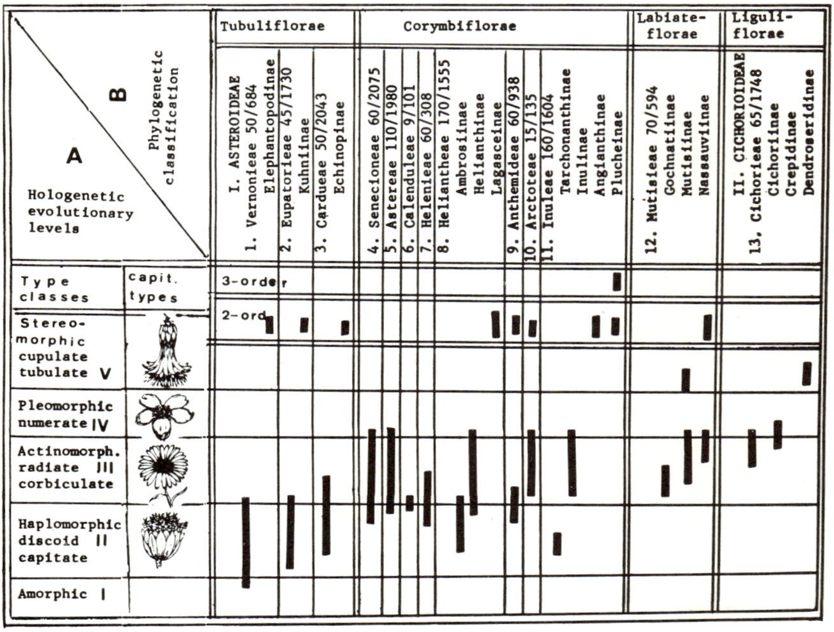

FIG. 10. Observable evolutionary stages of the capitula, arranged in a phylogenetic sequence. Numbers indicate genera and species in a tribe.

At the *haplomorphic* level, both the *discoid* and *capitate* subclasses are frequently represented by the present-day Compositae (Fig. 2:6, 2:11). They are characterized by haplomorphic shape and symmetry and by simple colours such as yellow, blue, or white—rarely pink or red. On the same haplomorphic level *sphaeromorphic* types can be placed, such as the capitulum of *Echinops* (Fig. 11:1). Some members of these subclasses, such as *Vernonia* and *Arnoseris*, still have loose involucres and easily discernible florets in the least regular heads.

The *actinomorphic* type with *radiate* and *corbiculate* subtypes is the commonest pattern in existence today (Fig. 2:7, 2:12). In this type class, the conspicuousness of the capitula is considerably increased by special ray-florets in the subfamily Tubuliflorae and by elongated ligulate florets in the Liguliflorae. Their pseudanthous capitula imitate single flowers and are recognized as such by their pollinators. Most present-day Compositae bear actinomorphic capitula.

In the *pleomorphic* type class the number of ray-florets is reduced to certain numerical patterns (iconic numerals), most frequently 8 or 5, infrequently 4, 3, or 2 (Fig. 2:8, 13). This marked tendency towards specialization is frequently accompanied by elongation of the corolla-tubes. Pleomorphic capitula are preferred by Hymenoptera and other similar pollinators.

FIG. 11. Compound capitula: 1. *Echinops ritro* L.; 2. *Syncephalantha decipiens* Bartl., capitulum of second order.

At the *stereomorphic* level, the *cupulate* and *tubulate* subclasses exhibit their relatively specialized flower heads to insect visitors (Fig. 2: 9, 14). This is the highest evolutionary level so far achieved by the present-day Compositae. Highest specialization of the cupulate capitula is exhibited by the ornithophilous Mutisiinae. The robust pendent heads of *Mutisia* and *Barnadesia* are formed from a deep flask-shaped receptacle with a single large nectary at the bottom. A narrow opening to the nectar eliminates most insect visitors. Only hummingbirds can reach the nectar (Fig. 2: 9). *Lygodesmia juncea* (Pursh) D. Don (Fig. 2: 14), has a similarly shaped capitulum. With its elongated pink flower heads, this plant imitates the true flowers of *Dianthus* and is visited mainly by Hymenoptera.

The *zygomorphic* type class is the highest stage in the evolution of true flowers achieved by the most advanced groups of angiosperms, such as Orchidaceae, Scrophulariaceae, Labiatae, and Leguminosae. However, none of the known capitula types among the Compositae has yet reached the zygomorphic level, except some bilateral capitula that show the tendency toward zygomorphism.

Although both the bilabiate ray-florets of the Tubuliflorae and the ligulate florets of the Liguliflorae are typically zygomorphic in their individual structure, in the flower heads of the Compositae these ray-florets serve only to heighten the radiate symmetry of the capitula.

Asymmetric trend of floral evolution

It would be illogical to think that the zygomorphic type class is the end-product of floral evolution and that the future will only recapitulate earlier evolutionary sequences. On the contrary, a new class is already actively evolving. Surprisingly, this new trend is so different from anything we have yet seen, so entirely removed from the classical symmetrical patterns, that we could call this new development the asymmetric trend.

The asymmetric floral arrangement is not primitive but represents the highest level of floral evolution and is comparable in complexity to the zygomorphic trend of orchids. All asymmetric flower types may be derived morphologically from some symmetrical type class.

The adaptive significance of asymmetric flowers may be related to sophisticated pollination mechanisms, which have coevolved with the sensory systems of insects and birds. Field observations show that birds are less dependent upon symmetrical patterns and iconic numerals in flowers than are insects. In the ornithophilous flowers of the Zingiberales, for instance, regular floral symmetry is irreversibly lost, whereas in the entomophilous flowers it is gradually replaced by extra floral semaphylls, formed from ornamented bracts and leaves.

This new asymmetric trend is evolving most rapidly among ornithophilous plants. *Strelitzia reginae* Banks, (the bird of paradise, Musaceae, South Africa) has completely departed from the symmetrical patterns and developed a distinctive asymmetric flower, with rare orange and blue coloration that is strange and unusual to our eyes and to insects but familiar to birds (Fig 12). Morphologically, these flowers still have three sepals,

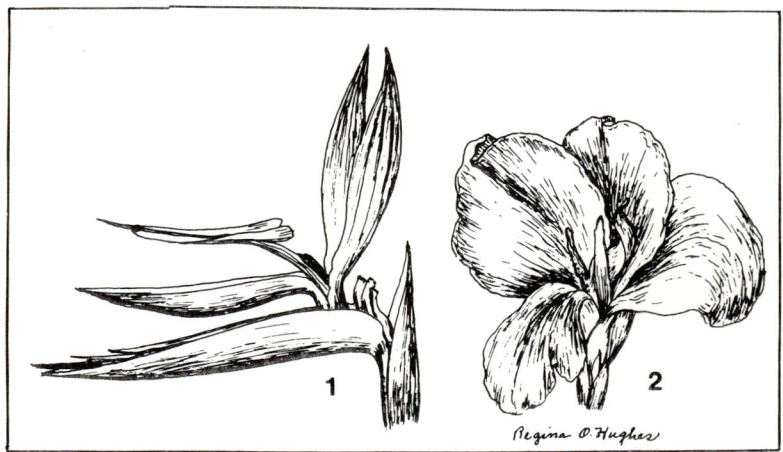

FIG. 12. Asymmetric ornithophilous flowers. 1. *Strelitzia reginae* Banks. 2. *Canna indica* L. (both ¼ natural size).

three petals, six stamens, and three ovaries, but these numbers are arranged into an asymmetric pattern.

Another example of asymmetric bird flowers is the genus *Heliconia* (Heliconiaceae), where the flowers are asymmetric and characterized by vivid colors but no odors, abundant nectar but no nectar guides, heteranthery, hard outside floral parts, and protected ovaries. Flower colors include fiery red, scarlet, deep orange, or yellow. Further ornithophilous syndromes in these flowers are the lack of iconic numerals and the irregular arrangement of perianth parts, petals, sepals, and bracts.

An example of entomophilous asymmetric flowers is shown in Fig. 13—*Marantha leuconeura* Morr., with irregular, bisexual flowers.

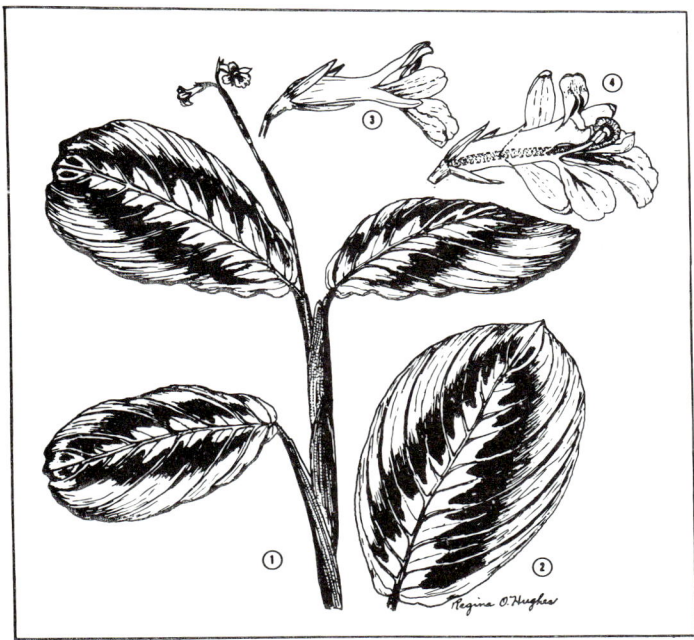

FIG. 13. Asymmetric entomophilous flowers and ornamented leaves of *Maranta leuconeura* Morr. The diminishing colour effect of tiny flowers is restored by showy vegetative leaves. 1. A branchlet with flowers ($\frac{1}{2}$ natural size). 2. leaf blade ($\times \frac{1}{2}$). 3-4. Magnified flowers ($\times 2$).

INFLORESCENCE TYPES

The evolution of the flower heads of the Compositae displays two parallel progressive trends. In the first trend, the single capitula are composed of tubular and bilabiate florets, in the second, of ligulate florets. These typological characteristics of the pseudanthous capitula are sufficiently persistent and distinctive to have been used by earlier taxonomists in the systematic classification of the Compositae, hence, the division of the family into two subfamilies: Tubuliflorae (Asteroideae) and Liguliflorae (Cichorioideae).

Simple capitulum

In current botanical literature, the terms *discoid* and *radiate* are used to designate different types of capitula of the subfamily Tubuliflorae. The discoid capitula are homogeneous, being composed of identically colored tubular florets (Fig. 2: 6). The radiate capitula have tubular florets in the central disk, surrounded by bilabiate rays on their margins (Fig. 2: 7). The rays are not always the same color as the disk, thus increasing the variability and attractiveness of the capitulum types.

Because liguliflorous capitula are composed only of ligulate florets, there is no sharp difference between the central part of the capitulum and its marginal area. Thus, the liguliflorous capitula different from the tubuliflorous heads. For instance dandelion capitula look different from those of asters, daisies, or marigolds. They also look different to their insect visitors.

In the subfamily Liguliflorae, the capitular subclasses can be called *capitate* and *corbiculate* with dandelion and chicory heads as examples (Fig. 2: 11, 12). In capitate heads, numerous florets are distributed evenly over the whole surface of the capitular area and provide a broad, brushlike surface for the visitors (*Taraxacum, Hieracium, Sonchus,* (Fig 2: 11). In corbiculate heads, the florets are less numerous and their lobes are directed outward, becoming better developed toward the margin of the capitulum. Being almost empty in the central part, such a head resembles a low and broad basket (corbicula), from which the name corbiculate is derived (*Cichorium, Lactuca,* Fig 2: 12). For a complete list of capitulum-types, see Leppik, 1970b: 339-342.

Compound capitula

The conspicuousness and attractiveness of small heads are considerably heightened by their aggregation into compound capitula or synflorescences. In such clusters, the heads not only are closely aggregated in space, but show a definite pseudanthic pattern. The general sequence of floral evolution is once more recapitulated at the syncapitulum level.

The next stage of this cycle is formed by haplomorphic compound heads, such as the capitula of *Echinops* (Fig. 11: 1), *Lagascea, Elephantopus, Vanillosmopsis,* and *Syncephalantha* (Fig. 11: 2). In these perfect synflorescences, numerous single-flowered heads, each with its own involucre, are aggregated into separately involucrated secondary heads. In *Triplocephalum* (Inuleae) compound capitula are aggregated to form a syncapitulum of the third order. Compound capitula are found exclusively in the subfamily Tubuliflorae, being absent from the Liguliflorae.

Conflorescences

In addition to compound capitula or synflorescences, heads are sometimes assembled into loose clusters imitating inflorescences, such as spikes, racemes, panicles, corymbs, and cymes. These loose assemblages of capitula without a general cover can be called conflorescences. Thus, inflorescences, synflorescences (capitula with an involucral cover), and conflorescences (without such a cover) represent an evolutionary sequence of reproductive clusters with observable progressive and regressive trends.

Examples of this interesting phenomenon are numerous. The capitular clusters of *Solidago, Aster, Anthemis, Ligularia, Liatris,* and other genera exhibit various inflorescent forms, such as thyrsoid, paniculoid, corymboid, racemoid, and subcapitate (Fig. 14). In *Achillea millefolium* L. capitula clusters with numerous (often over 100) five-rayed capitula are assembled in a compact corymb (Fig. 15) that provides a continuous flat surface and walking ground for insects. Using extended ray-florets as bridges, insects can pass from one head to another without having to fly. In this way, a single visitor can quickly pollinate numerous florets.

Another remarkable conflorescence is found in *Ligularia przewalskii* (Maxim.) Diels from China and Mongolia, described by Good (1956, p. 321). Each capitulum of this plant contains only five florets, three disk- and two ray-florets (Fig 4: 6). A large number (300–400) of such well advanced bilabiate capitula are assembled into a multiradiate raceme on an almost leafless axis (Fig 14: 4). The florets are visited by bumblebees and

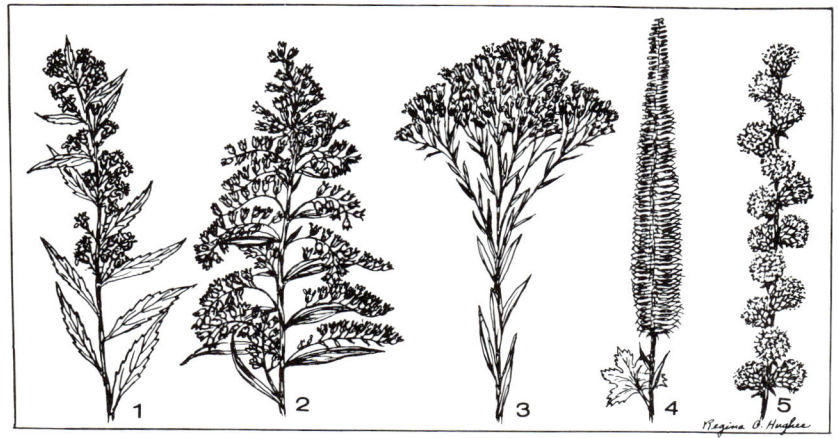

FIG. 14. Conflorescences: 1. *Solidago caesia* L.; 2. *Solidago shortii* T. & G.; 3. *Solidago chinensis* Ridd.; 4. *Ligularia przewalskii* (Maxim.) Diels; 5. *Liatris regiomontis* (Small) K. Schum.

other hymenopterous pollinators. *L. veitchiana* (Hemsl.) Greenm. from western China has a similar conflorescence.

POLLINATION AND FERTILIZATION OF THE COMPOSITAE COMPARED WITH THE UMBELLIFERAE

The discussion of pollinating insects in this Chapter is of necessity very brief. In the literature, there are several papers that deal with the sensory behavior of anthophilous insects (Sprengel, 1793; Müller, 1883; Knuth, 1905, 1908; Kugler, 1955, 1963, 1966; Knoll, 1956; Werth, 1956; Leppik, 1956, 1957; Pijl, 1960–1961; Meeuse, 1961; Kullenberg, 1961; Faegri and Pijl, 1966; Grant and Grant, 1967; Percival, 1965; Proctor and Yeo, 1973). Although the advantageous pollination systems and variety of insect visitors that most Compositae share with the Umbelliferae are recognized, the following points should be considered here:

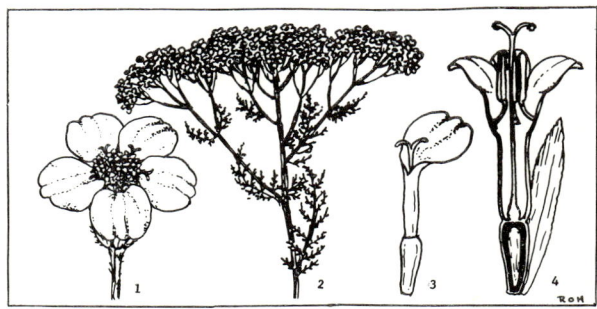

FIG. 15. Corymboid conflorescence of *Achillea millefolium* L.; 1. pentamerous (pleomorphic) head; 2. conflorescence; 3. single ray-floret; 4. single disk-floret with chaff.

1. The close association of many florets renders the head more conspicuous to insect visitors. In most cases conspicuousness is further increased by outwardly directed rays, whose colour frequently differs from that of the disk florets. The disk florets usually form a flat surface consisting of many protruding reproductive organs, over which insect visitors crawl, thus quickly fertilizing many florets. All these characters are clearly displayed in most apioid Umbelliferae, but in a less perfect form than in the Compositae.

2. The free accessibility of nectar to all visitors, short-lipped or long-lipped, ensures successful pollination at all times of the season. In the Compositae, nectar is secreted by a ring of glands surrounding the style at the base of the narrow tubular corolla. As it accumulates, it rises up into the wider part of the corolla, where it is protected from rain. In the Umbelliferae, nectar lies upon the fleshy disk where it is fully exposed to the rain.

3. The anthers of the Compositae cohere to form a hollow cylinder, and dehisce introrsely, filling the cylinder with pollen before the floret opens. The growing style brushes the pollen out of the anther-cylinder by means of the hairs on its outer surface. Thus in the first stage of pollination, the pollen is exposed to the visitors. In the second stage, the stigmas separate and expose their papillar surface, to catch the pollen that is brought in. Such a pollen presentation mechanism corresponds essentially to that of the Lobeliaceae and has some similarities to that of papilionaceous flowers In all these families, the mechanisms serve the same purpose. In the Umbelliferae, the stamens are free, arising from an epigynous disc, the anthers dehiscing longitudinally.

4. The capitular inflorescences and pseudanthous floral patterns of the Compositae greatly favour both pollination and seed production in this family. The well developed pappus of some Compositae and the wings found in some Umbelliferae favour the dissemination of seeds by wind and air currents.

5. The inferior ovary of both Compositae and Umbelliferae is less likely to be damaged by phytophagous insects that frequently act as pollen carriers.

6. Each single-seeded fruit in the Compositae and each of the two single-seeded mericarps in the Umbelliferae, is the product of separate selection and fertilization by individual pollinators. This ensures the genetic variability of the numerous offspring produced in the same capitulum or umbel and preserves the morphological stability of genetically-fixed characteristics.

Pollinators of the Compositae

Numerous insect pollinators have been observed visiting Compositae in various geographic areas. They are listed by Delpino (1868-1875), Müller (1883), Knuth (1905, 1908), Robertson (1928) and others. The pollen presentation mechanisms are described by Sprengel (1793), Müller (1883), Small (1915, 1919), Troll (1928), Vogel (1954), Kugler (1955, 1966), Werth (1958), Knoll (1956), Jaeger (1961), Kullenberg (1961), Meeuse (1961), Percival (1965), Faegri and Pijl (1966), Free (1970), Leppik (1970, p. 346), and Proctor and Yeo (1973).

The more elementary capitula of the haplomorphic and actinomorphic levels are freely accessible to all floral visitors. They are mostly exploited by beetles, flies, bugs, and other less specialized pollinators.

The numerate, cupulate, and elongate capitula of the pleomorphic levels attract bees, bumblebees, wasps, moths, and even hummingbirds in the tropics. No bats are among the known regular flower visitors of the Compositae. Birds, relative newcomers among the pollinators of the Compositae, are observed mainly on Mutisieae and Cardueae and mostly in the tropics. Knuth (1905, p. 233) and Vogel (1954, pp. 308-309) listed the following

82   ELMAR E. LEPPIK

ornithophilous genera of Mutisieae from South America and South Africa: *Barnadesia, Chuquiragua, Cnicothamnus, Mutisia,* and *Echinops* (Cardueae). These genera represent the most advanced stereomorphic pseudanthia among the Compositae. Their capitula form a deep and robust campanuloid gullet with a single large nectary deep in the centre. Nectar production is so rich that occasionally it drips out of the gullet.

Grant and Grant (1967, p. 109) observed four species of hummingbirds, visiting the thistle, *Cirsium neomexicanum* Gray, in Arizona. Pollen adhered to the birds' bills on some floral visits, but not on others. The authors concluded that hummingbirds are active, but relatively inefficient, pollinators of this thistle.

Wind pollination occurs in several genera, such as *Artemisia, Ambrosia, Franseria, Gnaphalium, Iva,* and *Xanthium*. All these genera belong to the Tubuliflorae.

Statistical data on flower visitors observed on the Compositae

Extensive statistical data on the visitors of the Compositae have been published by many classic and modern anthecologists, such as Delpino (1868–1875), Müller (1883), Knuth (1908), Robertson (1928), Leppik (1960a), Grant and Grant (1967), and Proctor and Yeo (1973). Data from the literature, including the personal observations of the writer, are summarized in Fig. 16. The solid columns in Fig. 16 indicate the approximate number of observed visitors in each type class (II to V), and the historical sequence of the evolution of the capitula is given at the left.

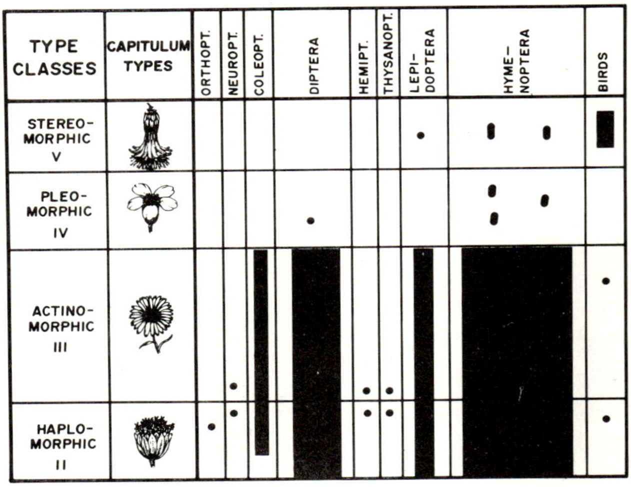

TYPE CLASSES	CAPITULUM TYPES	ORTHOPT.	NEUROPT.	COLEOPT.	DIPTERA	HEMIPT.	THYSANOPT.	LEPI-DOPTERA	HYMENOPTERA	BIRDS
STEREO-MORPHIC V					•			●	●	■
PLEO-MORPHIC IV					•				● ●	
ACTINO-MORPHIC III				■	■	•	•	■	■	•
HAPLO-MORPHIC II		•	•			•	•	■	■	•

Fig. 16. Frequency of insect visitors registered on Compositae of different type-classes. Black columns indicate the total number of visitors observed on the flower head of the Compositae. Hymenoptera dominate at all levels.

The greater part of the Compositae, particularly those whose capitula are at lower evolutionary levels, appear to be visited by a large number of unspecialized pollinators, such as beetles, bugs, wasps, and flies. In this respect, the Compositae closely resemble the Umbelliferae. Yet, the umbel of the latter can be considered morphologically and ecologically only a pre-stage of the capitulum of the Compositae.

At the upper evolutionary levels, the capitula are more advanced and are visited by specialized pollinators, such as bees, bumblebees, butterflies, and birds. The selective activity of these pollinators has produced a variety of capitula types, which resemble the solitary flowers of the pleomorphic and stereomorphic levels (Fig. 4). But the number of species bearing pleomorphic or stereomorphic capitula is relatively small in comparison with the great mass of haplomorphic and actinomorphic flower heads of the Compositae.

COENOGENETIC APPROACH

Two different trends, *phylogenetic* and *coenogenetic* (Leppik, 1974), caused by two independent groups of factors, are readily distinguishable in the evolution of the flower heads of the Compositae. The *morphological phylogenetic* trend, is controlled by internal, hereditary factors, and the *typological coenogenetic* trend, is controlled by external selective agencies (Fig. 17). Both approaches can be used in the study of the capitulum.

FIG. 17. Presumed interaction of internal and external factors on the course of evolution of the flower heads of the Compositae, showing the diagonal course of evolution between phylogenetic and coenogenetic trends. Figs 2, 4–17 after Leppik, 1970a; courtesy of the *Societas Biologica Fennica Vanamo*

Considering the capitulum as a morphological unit tends to explain the anatomical structure and phylogenetic origin of single organs, such as sepals, petals, bracts, and phyllaries, while viewing it as a typological entity deals with the functional adaptation of these organs to form a capitulum of a definite ecological type. Knowledge of both these trends and the application of both methods of study are essential to understanding the evolution of the capitulum as a whole.

The flower head of the Compositae provides particularly good illustrations of both phylogenetic and coenogenetic evolutionary trends. Interpreted morphologically, the capitulum is a dense inflorescence assembled from a great number of single florets and supported by a common involucre. Yet, from the standpoint of insect attraction, the same capitulum is a pseudanthium, biologically and functionally equivalent to a single flower. After the completion of its morphological development, further changes in the capitulum appear to have a coenogenetic direction. (Leppik, 1963c, 1970a, 1974).

PRACTICAL APPLICATION OF THE SEQUENCE OF FLORAL EVOLUTION

Having the above-described natural sequence of floral evolution at hand, we can use it for many practical purposes.

1. On the basis of this evolutionary sequence, we can now construct a reliable classification system for the enormous number of flower-types. However, such a system is actually typological which is not identical with, but is complementary to, a taxonomic-phylogenetic classification (Figs 1, 2, 17).

2. This arrangement of flower types supplies a natural test of the ability-grades of pollinators and classifies them according to their evolutionary stages (Fig. 18). This finding proved a surprising and very valuable aid

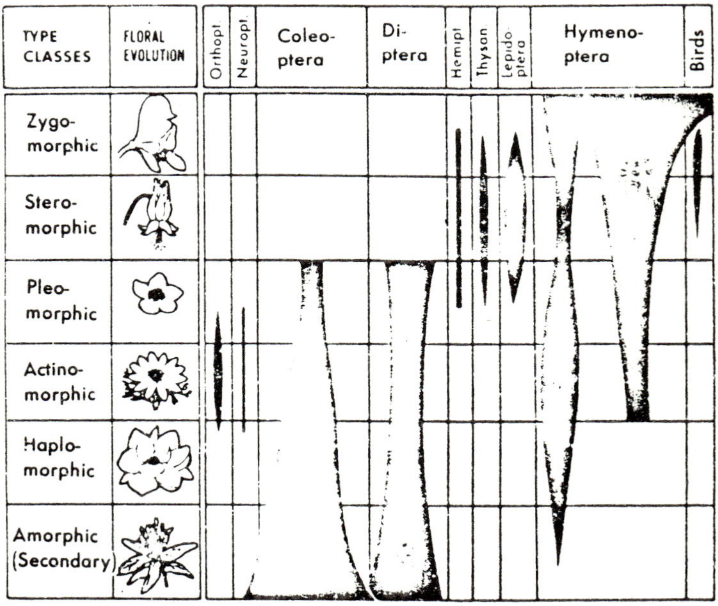

FIG. 18. Relative ability of pollinating insects and birds to exploit various type classes of flowers. In left columns are type classes and representative types, as pictured in Fig. 1. The black areas at right roughly indicate the number of visitors registered on the flowers of a particular type class. (After Leppik, 1963b.)

to entomologists who can, with this simple method, now trace and test the sensory ability of present pollinating insect groups and also of paleontological insects from as far back as 100 million years.

3. With fossil imprints of flower types, we can connect their whole evolutionary sequence with the geological timetable and estimate their geological age.

4. With the floral evolution, we can correlate the metabolic, biochemical, and nutritional evolution of flowering plants, although such a procedure is still very general. Under the persistent selective pressure of pollinators, the amount of sugar (nectar), protein, seed oils, vitamins, and other nutritive substances has been constantly increasing in the zoophilous plants.

5. With this standard evolutionary scheme, we can establish the position of families, orders, and other plant groups in their general evolutionary sequence (Fig. 19).

6. Finally, in this sequence of floral evolution, we have a natural scale for aesthetic judgment. We know which flower-types are elementary, which are primitive, or progressive, and which may be regarded as at the highest level of floral beauty.

DISCUSSION AND CONCLUSION

Controlled by inborn instincts and inherited sensory abilities, pollinators tend to select capitula with a high content of sugar (nectar), and other food substances. Pollinators cannot see these food qualities from a distance, but have to associate them with colour, size, shape, iconic numerals, odour, and other floral characteristics. The attractiveness of capitula to the insect eye is demonstrated by the numerous visitors to the Compositae. In their successive flights, pollinators are attracted to better-looking capitula with a higher content of nectar or pollen. In this way, the evolution of the capitula is directed toward better-looking types with higher food qualities. Evolution has progressed in many directions, but mainly towards the morphological perfection and biochemical and metabolic improvement of flowers and fruits. This improvement is attested by the high content and quality of the seed oils of many Compositae, such as *Carthamus, Helianthus,* and *Dimorphotheca.*

Recapitulation of the historical sequence of floral evolution by the differentiation of the capitula of the Compositae is a striking example of insect–flower coevolution and coexistence. It shows that the insects and flowers are mutually interrelated in evolution as reciprocal selective factors, adapted to each other and to their particular environments. Distinctive entomophilous syndromes can be recognized in fossil angiosperm flowers from the Cretaceous period, about 100 million years ago, and in the pre-angiospermous flowers and flower-like structures, about 400 million years ago.

During this long time, most attractive floral patterns and specialized pollination mechanisms have evolved in correlation with the sensory development of their pollinators. In this synagonistic interplay, insects with their advanced senses and inherited instincts became undisputable leaders. They directed the floral evolution towards nutritional and visual improvement and perfected their own sensory faculty and physical effi-

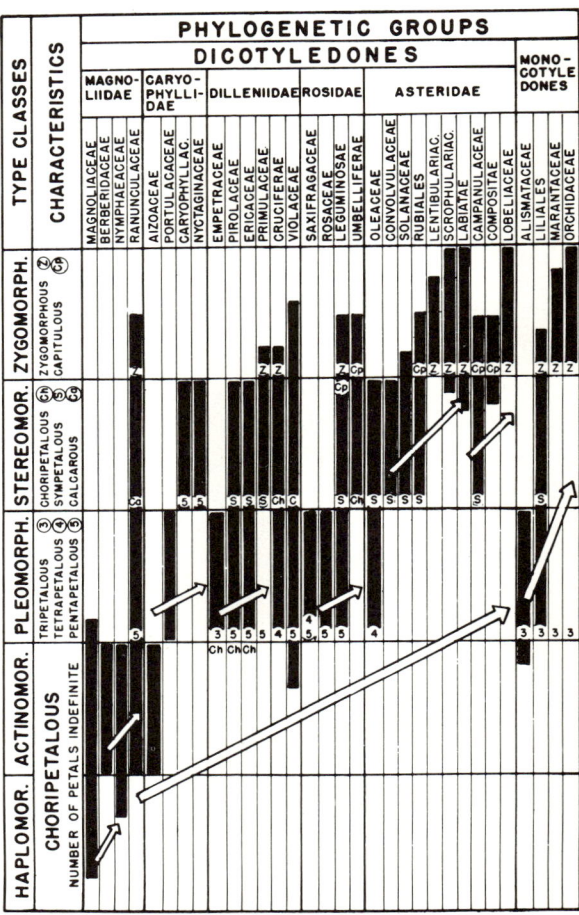

FIG. 19. Main orders of angiosperms, arranged according to hologenetic (typological) and phylogenetic (taxonomic) systems. White arrows point to the evolutionary trends in the main phylogenetic groups. (After Leppik, 1968.)

ciency. In genetically responsive plant groups, the change was fast; in less responsive plants, slow. Thus, we still have most evolutionary levels, beginning from amorphic and haplomorphic and ending with the zygomorphic, represented in present day floras (Fig. 1).

REFERENCES

Augier, J., and du Mérac, M. L. (1951). La phylogénie des Composées. *Rev. scient.* **89**, (3), 167–182.
Bentham, G. (1873). Notes on the classification, history, and geographical distribution of Compositae. *J. Linn. Soc. (Bot.)*, **13**, 335–577.
Burtt, B. L. (1961). Compositae and the study of functional evolution. *Trans. bot. Soc. Edinb.* **39**, (2), 216–232.
Cassini, H. (1826–1834). "Opuscules Phytologiques" Vols I–III.
Church, A. H. (1904). Principles of phyllotaxis. *Ann. Bot.* **18**, 227–243.
Church, A. H. (1920). On the interpretation of phenomena of phyllotaxis. *Oxford bot. Mem.* **6**, 1–58. (Reprint by Hafner, New York, 1968).
Cronquist, A. 1951. Orthogenesis in evolution. *Res. Stud. State Coll. Washington* **19**, 3–18.
Cronquist, A. (1955). Phylogeny and taxonomy of the Compositae. *Am. Mid. Nat.* **53**, 478–511.
Cronquist, A. (1968). "The evolution and Classification of Flowering plants" (Riverside Studies in Biology). Houghton Miffin, Boston.
Darwin, C. (1859). "On the Origin of Species by Natural Selection". John Murray, London.
Delpino, F. (1868–1875). Ulteriori osservazioni e considerazioni sulla dicogamia nel regno vegetale. Milano, Pt. I, 1868, 1869. Pt. II (fasc. 1) 1870; *Atti Soc. Ital. Sci. nat., Milano* **11**, 1868: 265–332; **12**, 1869: 21–141, 179–233; **16**, 1873: 151–349.
Faegri, K. and van der Pijl L. (1966). "The Principles of Pollination Ecology". Pergamon Press, Oxford (2nd edition 1972).
Free, T. B. (1970). "Insect Pollination of Crops". Academic Press, London and New York.
Good, R. D. O. (1931). Some evolutionary problems presented by certain members of Compositae. *J. Bot. Lond.* **99**, 299–395.
Grant, C. and Grant, K. A. (1967). Records of hummingbird pollination in the Western American Flora. III. Arizona records. *Aliso* **6**(3), 107–110.
Hoffmann, O. (1894). Compositae. *In* "Die natürlichen Pflanzenfamilien" (Engler and Prantl, eds) **4** (5), 87–387. Leipzig.
Hutchinson, J. (1916). Aquatic Compositae. *Gdnrs' Chron.* **59**, 305.
Jaeger, P. (1961). "The Wonderful Life of Flowers". Harrap, New York.
Knoll, F. (1956). "Die Biologie der Blüte". Springer Verlag, Berlin.
Knuth, P. (1905). "Handbuch der Blütenbiologie". III Band. 2 Teil. Leipzig.
Knuth, P. (1908). "Handbook of Flower Pollination", Volume 2. Clarendon Press. Oxford.
Koch, M. F. (1930). Studies in the anatomy and morphology of the Composite flower. I. The corolla; II. The corollas of the Heliantheae and Mutisieae. *Am. J. Bot.* **17**, 938–952; 995–1010.
Kugler, H. (1955). "Einführung in die Blütenökologie". Fischer, Stuttgart.
Kugler, H. (1963). UV-Musterungen auf Blüten und ihr Zustandekommen. *Planta* **59**, 296–329.
Kugler, H. (1966). UV-Male auf Blüten. *Ber. dt. bot. Ges.* **79**(2), 57–70.
Kullenberg, B. (1961). Studies in *Ophrys* pollination. *Zool. Bidr. Upps.* **34**, 1–340.

Leonhardt, T. (1949). Phylogenetisch-systematische Betrachtungen. I. Betrachtungen zur Systematik der Compositen. *Öst. bot. Z.* **96**, 293–324.

Leppik, E. E. (1956). The form and function of numeral patterns in flowers. *Am. J. Bot.* **43**, 445–455.

Leppik, E. E. (1957). Evolutionary relationship between entomophilous plants and anthophilous insects. *Evolution* **11**, 466–481.

Leppik, E. E. (1960a). Evolutionary differentiation of the flower head of the Compositae. *Arch. Soc. zool. bot. fenn.* **14**, 466–481.

Leppik, E. E. (1960b). Early evolution of flower types. *Lloydia* **23**, 72–92.

Leppik, E. E. (1961). Phyllotaxis, anthotaxis, semataxis. *Acta biotheor.* **14**, 1–28.

Leppik, E. E. (1963a). Reconstruction of a Cretaceous *Magnolia* flower. *Adv. Front. Plant Sci.* **4**, 79–94.

Leppik, E. E. (1963b). Fossil evidence for floral evolution. *Lloydia* **26**, 91–115.

Leppik, E. E. (1963c). Evolutionary correlation between plants, insects, and soils, *Ann. Soc. Litt. Est. Am.* **3**, 28–50.

Leppik, E. E. (1964). Floral evolution in the Ranunculaceae. *Iowa State J. Sci.* **39**, 1–101.

Leppik, E. E. (1966). Floral evolution and pollination of the Leguminosae. *Ann. bot. fenn.* **3**, 299–308.

Leppik, E. E. (1967). Morphogenetic classification of flower types. *Phytomorphology* **18**(4), 451–466. New Delhi, India.

Leppik, E. E. (1968). Directional trend of floral evolution. *Acta biotheor.* **18**, 87–102.

Leppik, E. E. (1970a). Evolutionary correlations between plants, animals, and their environments. *Adv. Front. Pl. Sci.* **25**, 1–32.

Leppik, E. E. (1970b). Evolutionary differentiation of the flower head of the Compositae II. *Ann. bot. fenn.* **7**, 325–352.

Leppik, E. E. (1971). Paleotological evidence on the morphogenic development of flower types. *Phytomorphology* **21**, 164–174.

Leppik, E. E. (1972). Origin and evolution of bilateral symmetry in flowers. *Evol. Biol.* **5**, 49–85.

Leppik, E. E. (1974). Phylogeny, hologeny and coenogeny, basic concepts of environmental biology. *Acta biotheor.* **23**, 170–193.

Ludwig, F. (1884). Die Anzahl der Strahlen-blüthen bei *Leucanthemum vulgare* u. anderen Kompositen. *Dt. bot. Monatsschr.* **V** 52.

Ludwig, F. (1890). Die constant. Strahlenkurv. der Compositen u. ihre Maxima. *Schrift. Naturf. Ges. Danzig.* **7**, 177.

Ludwig, F. (1897a). Das Gesetz der Variabilität der Zahl der Zungenblüten v. *Chrysanthemum leucanthemum*. *Mitt. thüring. bot. Ver.* **10**, 20.

Ludwig, F. (1897b). Beitrage zur Phytarithmetik. *Bot. Zentbl.* **71**, 257–265.

Mayr, E. (1972). The nature of the Darwinian revolution. *Science, N.Y.* **176**, 981–989.

Meeuse, I. D. (1961). "The Story of Pollination". New York.

Melchior, H. (1964). Compositae. *In Engler* and *Prantl.* 'Syllabus der Pflanzenfamilien" (Engler and Prantl, eds) (2nd edition). II Band: Angiospermen. Gebr. Borntraeger, Berlin.

Müller, H. (1883). "The Fertilization of Flowers". Macmillan, London.

Percival, M. S. (1965). "Floral Biology". Pergamon Press, Oxford.

Pijl, L. van der (1960–1961). Ecological aspects of flower evolution. I–II. *Evolution* **14**, 403–416; **15**, 44–59.

Polyakov, P. P. (1967). "Taxonomy and Origin of the Compositae" [in Russian]. Alma-Ata, USSR.

Proctor, M. and Yeo, P. (1973). "The Pollination of Flowers". Collins, London.

Robertson, C. (1928). "Flowers and Insects, List of Visitors of 453 Flowers". Carlinville, Illinois.

Small, F. (1915). The pollen-presentation mechanism in the Compositae. *Ann. Bot.* **29**, 457–470.

Small, F. (1919). The origin and development of the Compositae. *New Phytol.* reprint **11**, 334 pp.

Solecki, R. S. (1971). "Shanidar. The First Flower People". A. A. Knopf, New York.

Sprengel, C. K. (1793). "Das entdeckte Geheimhis der Natur in Bau und in der Befruchtung der Blumen". Berlin.

Takhtajan, A. (1966). "System and Phylogeny of Flowering Plants" [in Russian]. Nauka, Moscow and Leningrad.

Thorne, R. F. (1968). Synopsis of a putatively phylogenetic classification of the flowering plants. *Aliso* **6**(4), 57–66.

Troll, W. (1928). "Organization und Gestalt in Bereich der Blüte". Springer, Berlin.

Vogel, S. (1954). "Blütenbiologische Typen als Elemente der Sippengliederung, dargestellt anhand der Flora Südafrikas". Fischer, Jena.

Weisse, A. (1897). Die Zahl der Randblüten an Compositen-Köpfchen. *J. wiss. Bot.* **30**, 453–483.

Werth, E. (1956). "Bau und Leben der Blumen". Enke, Stuttgart.

Zohary, M. (1950). Evolutionary trends in the fruiting head of Compositae. *Evolution* **4**, 103–109.

Chapter 5

Developmental and Comparative anatomy of the Compositae

G. LEDYARD STEBBINS

Department of Genetics, University of California, Davis, U.S.A.

Abstract. The vegetative and floral anatomy of the Compositae are reviewed from the comparative and developmental standpoint. The original growth form is believed to be shrubby, based upon the condition in the tribes Mutisieae, Heliantheae and Inuleae. Reduction to the herbaceous condition has been frequent, and increases to the woody habit occasional. In some of the largely herbaceous tribes (Anthemideae, Cichorieae) secondarily woody forms occur. The woody growth form is most common in the tropics and the south temperate regions. The Compositae were adapted originally to semi-arid regions, but numerous evolutionary lines have become more mesic as well as more xeric. When related groups occur sympatrically, one of them much more specialized than the other, the greater specialization has often been evolved via an intermediate form living in more severe habitat. This is called the "adaptive bottleneck hypothesis", and is exemplified by the genera *Microseris* and *Agoseris*. Arboreal Compositae living on oceanic islands are regarded as derived specializations, although some of them may also be relictual. Woody anatomy shows trends of paedomorphosis plus specialization for xeric habitats, and elongation of xylem elements in more mesic habitats. Floral anatomy exhibits many parallel trends of reduction, but in some genera (e.g. *Helianthus*) increase in ovular traces has accompanied selection for increased size of achenes. Large achenes having few traces (*Tragopogon*) are characterized by occurrence of a high proportion of their growth during late stages of development. When Compositae are compared with other families that have been placed near them in phylogenetic systems, the resemblances appear to be more superficial than fundamental, and suggest parallel and convergent evolution. The Compositae cannot be regarded as descended from or closely related to any other modern family.

CONTENTS

Introduction	92
Growth habit and vegetative structures	93
The shrub as the basic growth habit in Compositae	93
Reversals of evolutionary trends	96
Arboreal species of oceanic islands	98
Woody anatomy and the phylogeny of Compositae	99

The significance of differences in floral anatomy	99
Trends in floral anatomy	100
Relation between ovary anatomy and development	102
Anatomy of the corolla and relationships of the Compositae	103
References	107

INTRODUCTION

As a discipline for revealing phylogeny, the comparative anatomy of the vascular system is rapidly declining in favor. This new skepticism about the significance of "vestigial bundle" and similar phenomena (Carlquist, 1969) is fully justified. An earlier generation of botanists had made unjustifiable analogies between the complex vestigial organs of animals, which can be a guide to phylogeny, and individual vascular traces of plants, which can appear as a result of the differentiation of a few cells in response to their tissue environment (Wetmore and Rier, 1963). Nevertheless, the statement of Carlquist (1969): "Anatomy of flowers can be studied meaningfully only in relation to adaptation for particular modes of pollination, dispersal and allied functions," carries this skepticism too far. As I have emphasized elsewhere (Stebbins, 1974), and as is recognized by Carlquist with respect to the woody stem, adult vascular anatomy is a highly significant reflection of gene-controlled patterns of development. Evolution of form is the result of establishment of mutations and other genetic change which alter the regulator of these patterns. Adaptation to new environments usually follows the principle of adaptive modification along the lines of least resistance (Stebbins, 1974). Hence the particular form which a new adaptation will show depends to a considerable degree upon the nature of the ancestral genotype. For these reasons, comparative anatomy can make significant contributions to phylogeny only when studied in relation to functional adaptation, gene-controlled developmental patterns, or both.

Several recent contributions have set the stage for the development of this new functional and developmental comparative anatomy. The investigations of the woody anatomy of Compositae by Carlquist are reviewed in another section of this paper. An equally significant contribution is that of Kaplan (1973) to the homology of foliar appendages in monocotyledons. In the field of floral anatomy, the contributions of Leins (1971), Hiepko (1965a, b), Mayr (1969), Sattler (1972; Singh and Sattler 1972), Kam and Maze (1974) and others have considered various aspects of the new approach. Its usefulness will increase as more data are acquired and ideas become better crystallized.

A very good way of testing the value of this new approach is to apply it to a modern family that is still evolving rapidly, that displays a wide spectrum of diversity in form, and occupies an equally wide range of habitats. The Compositae fulfil these requirements to perhaps a greater degree than any other single family. The exploration of comparative functional and developmental anatomy as a guide to the evolution of this family is, therefore, the objective of the present paper.

GROWTH HABIT AND VEGETATIVE STRUCTURES

The Compositae, like some other large families of dicotyledons (Leguminosae, Rosaceae, Euphorbiaceae, Malvaceae, Umbelliferae, Scrophulariaceae, Rubiaceae) contain a wide spectrum of growth habits, from tall trees through shrubs and vine-like plants to perennial herbs and diminutive annuals. In a few genera, such as *Eupatorium* (*sensu lato*), *Vernonia*, *Senecio* and *Sonchus*, a wide variety of growth habits is displayed by a single genus. The family is, therefore, a good object for seeking answers to questions such as: "Is any particular growth habit more likely than others to be primitive?" and "Does evolutionary specialization for growth habits proceed in a straight line, towards either continuous reduction or elaboration; or does specialization consist of a series of adaptive radiations from intermediate forms in several directions?"

The Shrub as the Basic Growth Habit in Compositae

Carlquist (1966), on the basis of wood anatomy, has maintained that the shrubby habit is primitive for Compositae. His reasoning is as follows. Primitive wood ray conformation is a combination of multiseriate and uniseriate rays (Barghoorn, 1940), a condition which in Compositae is more common in shrubs than in either trees or herbs. Since relatively long ray-cells reflect persistence of a juvenile condition, or paedomorphosis, primitive wood should have only moderately long cells. Rays should also be relatively short, and should consist of both procumbent and erect cells. Since exceptionally long tracheary elements reflect paedomorphosis, and exceptionally short ones are associated with xeric specializations of other characters, an intermediate length of these cells is the most primitive. All of these characteristics are more frequent in shrubs than in either herbs or trees.

Do other lines of evidence support the hypothesis that shrubs are primitive among Compositae? Two kinds of evidence can be considered: (1) the taxonomic position of shrubs relative to herbs, trees and vines; and (2) the degree of correlation between the shrubby habit and relatively primitive characteristics of capitula, flowers and achenes.

Taxonomic position can best be considered by focusing attention on a genus that displays a great variety of growth habits. The best example is *Senecio*. It contains a few small trees, a large number of shrubs and woody-based subshrubs, and an equal or greater number of herbaceous species including many annuals. Which of these forms provide the best links between *Senecio* and other genera, particularly those marginal for the tribe Senecioneae, or connecting it with other tribes?

Arboreal species of *Senecio* are poor candidates for such connections. They occur on various continents: *S. yegua* in Chile, *S. reinoldii* in New Zealand, and the famous cabbage-like tree *Senecios* of the African mountains. These species have few or no characters in common except for those generally found in the genus. There appear to be independent origins of the tree habit on separate continents. The monotypic or ditypic genera *Brachyglottis* of New Zealand and *Bedfordia* of Australia differ from *Senecio* only in characters that indicate greater specialization (glandular achenes, loss of ray flowers, flattened achenes), and therefore appear to be specialized offshoots of the larger genus.

The strictly herbaceous species of *Senecio*, along with many of the genera of Senecioneae which contain only herbs (*Ligularia, Doronicum, Cacalia, Cacaliopsis, Tussilago, Petasites*) are also too specialized to be representative of an ancestral complex. Nearly all of them are either annuals or hemicryptophytes, having annual stems and underground rhizomes which adapt them to highly seasonal, winter-cold climates which have become widespread only in relatively recent geological epochs. Many of them have specialized inflorescences, capitula with reduced numbers of florets, loss of ray florets, unisexual disk florets. On the other hand, some of the shrubby and subshrubby genera of Senecioneae do provide links with other tribes. Most noteworthy in this connection are *Liabum, Eremothamnus* and *Newtonia* of the Liabinae, and *Neurolaena* of the Senecioninae.

Another taxonomic approach is to compare the frequency of shrubs versus other growth habits in the different tribes, to see whether those tribes that are generally regarded as the most primitive contain higher proportions of shrubs than those which are considered to be more advanced. Table I lists the percentages of genera in each tribe which contain shrubby, subshrubby or arboreal species. This was compiled from Hoffmann's (1894) treatment which, though very old, is still the only one from which this kind of information can be obtained quickly. The scoring was by genera, regardless of size: genera consisting entirely of woody species were scored as 1; those containing both woody and herbaceous species were scored as ½.

According to this table, the tribes can be roughly divided into three groups. The Mutisieae with 56% stand out as the only tribe having more than 50% of woody genera. Most of the tribes form a graded series, starting with Vernonieae (45%) and ending with Heliantheae (29%). Below this series are the Helenieae (18%), Cynareae (12%) and Cichorieae (4%). This series does not support any hypothesis which would maintain a correlation between woody habit and primitiveness with respect to floral characteristics. Although the three least woody tribes are regarded by most

TABLE I. Percent of woody or subshrubby genera in the tribes of Compositae. Compiled from Hoffmann (1894)

Tribe	% woody
Mutisieae	56
Vernonieae	45
Eupatorieae	42
Inuleae	40
Anthemideae	39
Astereae	36
Calenduleae + Arctotideae	35
Senecioneae	34
Tageteae	30
Heliantheae	29
Helenieae	18
Cynareae	12
Cichorieae	4

workers as among the most specialized (Cronquist, 1955), the Heliantheae, which are generally looked upon as having the largest number of unspecialized genera, are fourth from the bottom in percentage of woody genera. On the other hand the Mutisieae have not been selected in any phylogenetic treatment as the most primitive tribe.

If, however, one examines the data a little more carefully, the case for primitive woodiness becomes better. The Heliantheae is a highly diverse tribe, and some of the subtribes (Lagascinae, Ambrosinae, Madiinae) are highly specialized in one or more characteristics. The least specialized of the subtribes is the Verbesininae; except for the anomalous arboreal Petrobinae, the percentage of woody genera in the subtribe Verbesininae (37%) is far higher than in any of the other subtribes, which range from 11% to 24%, and this fact suggests that the original members of the Heliantheae may have been woody. Furthermore, Carlquist's (1957a, b) study of the anatomy of the Mutisieae revealed a number of possibly primitive features in the structure of the pollen grains, vascularization of the corolla and particularly of the gynoecium. The geographic distribution of the Mutisieae (Stebbins, 1940b) includes a larger proportion of relictual genera and disjunctions of range than are found in any other tribe. Consequently, Mutisieae are probably one of the older tribes of Compositae, having been derived from the original stock in a line of evolution independent of that which gave rise to the Heliantheae or any other tribe, as Bentham (1873) has suggested.

A striking correlation which emerges from Table I is between the woody habit and distribution in the tropics or the southern hemisphere. The Mutisieae are poorly represented in the northern hemisphere, and several of the northern genera, such as *Hecastocleis* in North America, *Nouelia* and *Catamixis* in Asia, and *Berardia* in Europe, are monotypic and highly localized. The Vernonieae, Eupatorieae and Inuleae are strongly represented in the tropics, the southern hemisphere or both. Since the largest and most widespread of these tribes is the Inuleae, a significant comparison can be made between floristic areas with respect to the proportion of woody genera which they contain. The result of this comparison is dramatic. Tropical and South Africa, including Madagascar, contain 72% of shrubby genera, most of them being South African. Of the Inuleae confined to tropical America, 33% are woody, while the remaining percentages are: paleotropical and pantropical, 25%; north temperate 19% and Australasian 15%. Since Africa is the center of diversity for the tribe, one could speculate that it originated there and the original Inuleae were shrubby. The lower percentages in other warm regions could reflect the fact that herbs, particularly annuals, can migrate more easily than woody species, and that members of this rather specialized tribe arrived in other continents so late that ecological niches favorable for shrubs were nearly all filled. This latter argument is particularly pertinent to Australia. One of the most remarkable features of the Australian flora is that although shrubs dominate the scene, shrubby Compositae are much more poorly represented than they are in other semiarid regions; while other families, particularly the Proteaceae, Myrtaceae and Thymeleaceae have evolved genera which mimic Compositae, often to a striking degree.

The higher proportion of shrubby Compositae in the tropics and southern hemisphere can be explained on the basis of climate. Since even in southern regions having a temperate climate, the difference between summer and winter temperatures is far less marked than in northern regions, the adaptive value of rhizomes and other subterranean organs is correspondingly lower. Winter annuals and biennials, which are frequent in northern regions, are almost unknown among plants native to the southern hemisphere. This reduces the phylogenetic significance of the correlations which have just been reported. Nevertheless, since northern climates during the early part of the Tertiary period, when the Compositae most probably originated, probably had a much lower range of variation in temperature than at present (Axelrod and Bailey, 1969), one could speculate that the early diversification and spread of the family was largely on the basis of shrubby species.

Reversals of evolutionary trends

An important question with respect to growth habit is: "If the original Compositae were woody, does this mean that woody members of this family are always more primitive than their herbaceous relatives, or have there been secondary reversions from herbaceous to woody forms?" Two examples lead me to postulate secondary reversions. The first is the genus *Artemisia*, which I have discussed in relation to the evolution of shrubs in general (Stebbins, 1972a). It is one of the most specialized genera of a relatively specialized tribe, the Anthemideae. This tribe has an intermediate percentage (39%) of woody or partly woody genera, but the three genera (*Hymenopappus, Hymenothrix, Leucampyx*) which link the Anthemidae to other tribes, particularly the largely herbaceous Helenieae, are all herbaceous. Moreover the anomalous stem anatomy of shrubby species of *Artemisia* (Diettert, 1938; Moss, 1940) would be expected in a secondary revertant from the herbaceous condition. If the phylogeny of the genus proposed by Hall and Clements (1923) is accepted, then woodiness was evolved from the herbaceous condition in three separate lines. Possibly, therefore, not only *Artemisia* but also the remaining shrubby genera of Anthemideae are secondarily woody, derived from herbaceous ancestors.

The second example, the shrubby and arboreal Cichorieae, is dealt with in the next section. Since none of them can be regarded as primitive by even the remotest stretch of the imagination, they are most probably reversions to woodiness in a predominantly herbaceous tribe. Such reversions are probably widespread in other tribes. They would be expected on the assumption that evolution in the Compositae or any other family of angiosperms has been a succession of adaptive radiations rather than a series of trends in any particular direction (Stebbins, 1974).

I have recently suggested (Stebbins, 1974) that xeric or semi-xeric adaptations of angosperms are often reversed, so that trends from drier to wetter habitats are common. The Compositae exhibits these trends more conspicuously than any other family. Examples that have already been mentioned are *Antennaria* and other northern Inuleae. Others are north temperate genera such as *Helianthus, Rudbeckia, Solidago, Aster, Agoseris, Taraxacum, Microseris, Phalacroseris, Tragopogon, Ambrosia* and the perennial species of *Madia*. Many others could be added. As a generalization, I propose the hypothesis that all or nearly all of the herbaceous groups of Compositae living in the North Temperate zone and having stems which die back each year plus winter dormant rhizomes or other underground parts, as well as many of the annuals adapted to mesic climates, are derived from more xeric ancestors. Adaptive radiation has been principally from semi-xeric habitats, which contain a mosaic of radically different plant communities existing side by side, in one direction to truly xerophytic desert plants, and in the other to specialized derivatives adapted to mesic or even hydric conditions.

A related hypothesis can be called the *adaptive bottleneck hypothesis*. It states that if two related groups exist sympatrically in a favorable habitat, and one of them is much more specialized than the other, this greater specialization has been acquired in a different, more severe habitat.

The evolutionary line leading to the more specialized group has undergone a reversal, leading from the less specialized, more mesic ancestor first to a group with intermediate specialization and adaptation to severe conditions, followed by readaptation to favorable conditions at an even higher level of specialization. The original example is the subtribe Microseridinae of the Cichorieae (Stebbins, 1972b, 1974), but the genus *Taraxacum*, as well as *Ambrosia* and *Xanthium* as compared to less specialized Heliantheae, would serve equally well. I believe that this kind of reversal is responsible for the striking examples of secondary aggregation of few-flowered capitula into capitula of a second order, as in *Elephantopus*, *Echinops* and *Moscharia*.

Arboreal species of oceanic islands

Any discussion of the evolution of growth habits in Compositae must deal with the status of arboreal species of this and other families that are endemic to oceanic islands. Are these species primitive relics or specialized adaptations to a very particular environment? Since this subject has been discussed admirably by Carlquist (1974), I shall do no more than emphasize my agreement with his strong arguments in favor of Darwin's hypothesis of derived specialization. The tribe Cichorieae definitely favor this viewpoint. Shrubby or arboreal members of this tribe occur on the Juan Fernandez Islands (*Dendroseris*), the Islas Desventuradas (*Thamnoseris*), the Channel Islands of California (*Munzothamnus* or *Stephanomeria blairii*), the Bonin Islands southeast of Japan (*Ixeris* subgen. *Crepidiastrum*) and the Canary Islands (*Sonchus*). If these endemics were the last vestiges of one or more arboreal complexes that once had a continental distribution, one would expect them to resemble each other to some degree, and to possess primitive floral characteristics. The opposite is true. Most of these groups are clearly related to and congeneric with species found on neighboring continents, and have as many as if not more specializations than, their continental relatives. The genus *Dendroseris* is anomalous, since it has no recognizable relatives in South America. The two species of which chromosome counts are available are tetraploids with $2n = 36$ (Stebbins *et al.*, 1953); in form and size, the chromosomes resemble most those of the Mediterranean genus *Tolpis*. Perhaps *Dendroseris* is derived from extinct species of *Tolpis* or a similar genus which exists in South America during the Tertiary period. Since genera having Mediterranean affinities (*Hypochaeris, Centaurea, Armeria, Valeriana, Sisymbrium, Lathyrus, Linum*, etc.) are a prominent element in the flora of Pacific South America, the former existence there of still another genus does not seem to be far fetched. At any rate, the woody anatomy of *Dendroseris* strongly supports the hypothesis that it is derived from herbaceous ancestors (Carlquist, 1967).

The genera *Fitchia* in the Pacific islands and *Petrobium* on St. Helena present greater problems, since neither of them has recognizable relatives on a neighboring continent. Both of them, however, combine primitive with specialized characteristics. The complex floral anatomy of *Fitchia* is in some respects primitive (Carlquist, 1957a), but its flattened achenes, 2-awned pappus and ligulate corollas are specializations. The specialized characteristics of *Petrobium* are dioecism, a 4-merous corolla and 2-awned pappus. These examples can be explained as specialized offshoots of ancient continental ancestors which are now extinct.

Woody anatomy and the phylogeny of Compositae

The woody anatomy of the family has been thoroughly described by Carlquist (1957c, 1958a, b, 1959, 1960a, b, 1961, 1962a, 1964, 1965a, b, 1966a, b,). He has concluded that a flexible system of wood anatomy has aided the adaptation of Compositae to a wide diversity of growth forms. From the ancestral shrubby habit, radiation has proceeded toward trees in a few lines, and repeatedly toward herbs. Among the shrubs, those having the leaves more or less evenly distributed over the branches tend to have less specialized wood, while the wood of rosette shrubs and rosette trees is highly specialized. Some groups of herbs may be unable to regain woodiness, but others have done so.

A conspicuous feature of rosette trees and rosette shrubs is paedomorphosis. This was defined by Carlquist (1962b) as the retention in secondary xylem of certain characters that are most conspicuous in the primary xylem of young stems: long vessel elements, features of pitting multiperforate perforation plates and erect ray cells. Some of these features appear also in annual and biennial species. Carlquist's evidence runs counter to the commonly accepted belief that elongate vessel elements or tracheids are always primitive, and that reduction in length is an irreversible trend. Specialized species belonging to various tribes, and secondarily arboreal forms as well as biennials or even annuals may have longer xylem elements than their shrubby or perennial relatives which possess less specialized features with respect to other wood characteristics, such as the nature of the rays. This conclusion is in harmony with much other evidence indicating that irreversible trends with respect to individual characteristics are much less prevalent than many botanists have supposed (Stebbins, 1974).

THE SIGNIFICANCE OF DIFFERENCES IN FLORAL ANATOMY

The florets of the Compositae are built upon a simple, stereotyped pattern which varies little from one genus or tribe to another. Variation in the vascular system consists chiefly of a larger or smaller number of traces passing through the ovary wall, and of more complex versus simpler patterns in the corolla. The commonest situation is as follows. A single bundle enters the base of the ovary, where it splits into six branches, five of which traverse the ovary wall to its summit. The sixth supplies the single ovule. At the summit of the ovary two of the ovary wall traces send off branches which enter the style and continue into the two stigma branches. Immediately above their branching points each of the five traces splits, sending a single branch into each of the five stamen filaments. The other branch continues into the corolla, which therefore contains five parallel traces extending in tubular corollas to the five sinuses between the lobes. There they fork and their branches continue along the margins of the corolla-lobes, so that branches from adjacent traces join at the apex of each lobe.

Deviations from this mode include both simpler and more complex patterns of vascularization. In species having flattened achenes, the ovary traces may be only four, as in *Bidens* (Koch, 1930b) or two, as in *Lactuca* (Stebbins, 1937). Reduction in the venation of the corolla is frequent in the narrow ligulate corollas found chiefly in members of the tribe Astereae (Koch, 1930a).

Greater complexity is more common, and can be evident in several ways. In *Fitchia speciosa* (Carlquist, 1957a) up to four bundles enter the floret from the receptacle. In this and several other genera (Koch, 1930a, b, Carlquist, 1957b, Stebbins, 1937, 1940a, 1973) the ovary wall is traversed by two, five or up to 30 traces in addition to the five principal ones. These branch from a smaller number of traces at the base of the ovary, and most of them either terminate below its summit or fuse with other traces at this point. Frequently, one or both of the principal stylar traces are continuations of supernumeraries, rather than branches from the principal traces. In *Dubyaea, Fitchia, Wyethia* and *Helianthus* (Carlquist, 1957a) the style is traversed by three or four traces rather than two.

In *Dubyaea, Fitchia* and other genera the ovular trace arises from a complex platform of vascular tissue, consisting of many irregular short, broad tracheary elements (Stebbins, 1940a). In *Fitchia* and *Helianthus* the trace itself may branch in a dichotomous fashion (Carlquist, 1957a). In the Heliantheae and Mutisieae (Koch, 1930a, b; Carlquist, 1957a, b), and less frequently in other tribes, a more complex venation of the corolla is found, consisting chiefly of traces which traverse the middle of the corolla lobes and terminate in their apices. They may or may not unite with the branch traces ascending from the corolla sinuses. Finally, in species of Heliantheae having a lobed or awn-like pappus, these structures may contain vascular traces (Carlquist, 1957a).

Trends in floral anatomy

According to the conventional dogmas of comparative anatomy, the extra traces just described should be interpreted as vestiges of a more complex anatomy which existed in the ancestors of the family. Carlquist (1969) has implied that they have only a functional significance in connection with the greater size of the structures involved. Which of these two is the better interpretation of the present examples?

This question is best answered by seeking other evidence which would support the primitive or derived status of the genera involved. One line of such evidence would be a greater anatomical resemblance of the forms having extra bundles to representatives of other families. Such resemblances exist. The entrance of a single vascular bundle into the base of a flower or floret is a very unusual condition among angiosperms, so that the condition described for *Fitchia speciosa* is an approach to a more usual pattern. In most angiosperms having inferior ovaries, the vascular supply to the styles and stigmas is independent of that which supplies the calyx and corolla, and this condition is realized in some Compositae that have supernumerary ovary traces. The usual vascularization of the composite corolla differs markedly from that of other sympetalous angiosperms, in which the median trace of each corolla lobe is the strongest, and traces leading to the sinuses are weak or absent. Finally, in many angiosperms having inferior ovaries, these are crowned with vascularized calyx-lobes, so that the traces found in the pappus of some Heliantheae approach this condition.

Two other lines of evidence are taxonomic and geographical. Do the genera having more complex floral anatomy possess morphological characters other than floral anatomy which suggest primitiveness, and is their geographical distribution such as to suggest a relictual condition rather than derivation from genera having the modal pattern of floral anatomy? With respect to taxonomy, the fact is noteworthy that most of the genera in which a more complex anatomy has been found—*Helianthus, Wyethia, Silphium, Petrobium, Oparanthus, Lagascea, Coulterella, Fitchia*—have been placed by at least some authors in the tribe Heliantheae, which most students of the family regard as primitive (Cronquist, 1955). This opinion is based upon such characters as the well developed receptacular paleae, which in some genera resemble the innermost involucral phyllaries; the occasional presence of a pappus which resembles calyx-lobes more than do the pappus members in any other tribe; and the relatively unspecialized style branches.

In more specialized tribes such as the Cichorieae, the few genera which have a more complex floral anatomy (*Dubyaea, Soroseris, Prenanthes, Microseris*; Stebbins, 1937, 1940a, 1973) have generalized growth habits, as well as unspecialized involucres and achenes.

The geographic distribution of many of these genera suggests that they are relictual. The genus *Fitchia* is localized in a few widely separated islands of the Pacific; it has no mainland relatives and constitutes a monogeneric subtribe of Heliantheae (Carlquist, 1957a). Another insular species without recognizable relatives on the African mainland is *Petrobium* on St

Helena. The distribution of *Microseris* is remarkably disjunct. The bulk of the genus occurs in Pacific North America, but there is a single species in Chile and another (*M. scapigera* (Sol. ex Cunn.) Sch. Bip.) in Australia and New Zealand. That of *Dubyaea* is relictual in the mountains of Asia (Stebbins, 1940a). In the case of these Compositae, therefore, the hypothesis that complex floral anatomy is primitive and that reduction has occurred in several separate lines is supported by other lines of evidence.

Relation between ovary anatomy and development

A comparative study of ovary and achene development in nine species of Compositae belonging to the tribes Heliantheae, Senecioneae and Cichorieae (Stebbins, 1973) has shown that the relationships between developmental pattern and anatomy of the mature achene are complex, but that certain regularities exist. When a cultivated variety of *Helianthus annuus* which has been selected for large seeds is compared with its wild relatives that have smaller seeds, the number of supernumerary vascular traces in the large seeded variety (26–28) is greater than in the wild relatives (19–21 and 18–24). This suggests that artificial selection for increased seed size has increased the number of supernumerary traces in the ovary, and that natural selection may have had similar results in various lines of composite evolution, as suggested by Carlquist for the large number of supernumerary traces in *Fitchia speciosa* as compared to its relatives having smaller achenes. On the other hand *Wyethia glabra*, a species generally regarded as primitive but having achenes larger than the wild species of *Helianthus*, has only 12–17 traces in the ovary at anthesis. Does this mean that during the evolution of *Helianthus* from its common ancestor with *Wyethia* an increase in number of traces accompanied a reduction in achene size? An alternative interpretation would be that the common ancestor and original member of the subtribe Verbesininae has smaller achenes and fewer traces than either *Wyethia* or *Helianthus*, and that increase in size leading to *Wyethia* added fewer supernumerary traces than that leading to *Helianthus*. More information on both adult anatomy and developmental pattern is needed before such questions can be answered.

Developmental studies have indicated a probable explanation for the existence of some achenes which in spite of very large size have a small number of ovarian traces. The most striking of these is *Tragopogon*, which has among the largest achenes known in the Compositae, but has only five vascular traces traversing the ovary and achene wall. This is associated with the fact that in *Tragopogon* nearly all of the growth in achene size occurs after anthesis, when the vascular system is fully developed. In proportion to the initial size, the growth from anthesis to maturity is 13.91 in *Tragopogon porrifolius*, as compared to 1·66, 1·97 and 2·69 in the other three species of Cichorieae investigated. The length of the young ovary at the stage of first differentiation of procambial strands is 0·29 mm in *Microseris nutans*, which has 10 strands in its mature achenes, and only 0·242 mm in *T. porrifolius*. The length of the achene at maturity is 5 mm in *M. nutans*, and 28–30 mm in *T. porrifolius*. This suggests that large achenes having a small number of vascular traces are produced by a developmental system based upon many genes for large size that act very late in development.

On the other hand, the increase in supernumerary bundles found in *Helianthus* is not due to size genes that act at the earliest developmental stages, before the first differentiation of procambium. In all four of the Heliantheae studied, ovary size at the initiation of procambium is similar (0·198–0·253 mm) and does not differ significantly from that of the Cichorieae or Senecioneae. The very large number of supernumerary bundles in *H. annuus* is associated with a greatly increased amount of growth between procambial differentiation and anthesis. The data suggest that selection for a greater amount of growth between differentiation of the first procambial initials and the earliest differentiation of xylem has been chiefly responsible for the increase in number of supernumerary bundles.

Anatomy of the corolla and relationships of the Compositae

Koch (1930a) has postulated that in the ancestors of the Compositae the petals or corolla-lobes each had three principal vascular traces, as in modern species of Gentianaceae (Eames, 1961). Reduction in vascularization involved union of lateral traces supplying adjacent lobes to produce the situation described above for some members of the Heliantheae and Mutisieae. The usual pattern for the family was the result of suppression of the median traces of the corolla-lobes.

If this hypothesis is accepted, it raises considerable difficulties with respect to the affinities of the family. The condition of 15 traces, three for each corolla lobe, is not known in any group that has an inferior ovary containing a single ovule, characters which are generally assumed to have existed in the immediate ancestors of the Compositae. The 10-stranded condition, with fused laterals, associated with a radially symmetrical 5-lobed corolla, exists only in the small family Calyceraceae, provisionally assigned to the Dipsacales. Can the Compositae have been derived from an immediate common ancestor with the Calyceraceae?

The strongest argument against this hypothesis is the position of the ovule, which in the Calyceraceae is pendant from the summit of the ovary locule, and in the Compositae is erect from its base. This would suggest that the reduction in ovule number occurred independently in the lines leading to the two families. Other differences are that in the Calyceraceae the stamen filaments are united while the anthers are free, while in the

TABLE II. Comparison of Compositae and composite-like families with respect to 10 morphological characters

	Compositae	Calyceraceae	Dipsacales	Rubiaceae Psychotriinae	Brunoniaceae
Stipules present (+) or absent (−)	−	−	−	+	−
Flowers in capitula (+) or open inflorescences (−)	+	+	+ or −	− or +	+
Number of corolla-lobes	5 or 4	5 or 4	5 or 4	4 or 5	5
Principal veins in reduced corollas	laterals	median	median	median	median
Nature of stigma	bifid	capitate	1–3 fid	bifid	asymmetric
Number of ovary-locules	1	1	5–1	2(−1)	1
Position of ovule when solitary	basal	apical	apical	basal	basal
Orientation of ovule	anatropous	anatropous	anatropous	anatropous	orthotropous
Predominant gametic (x) chromosome number	9, 8, 10	7, 8, 9, 10	9, 8, 10	11	9

Compositae the filaments are free but the anthers are united; also the stigma of the Calyceraceae is capitate and undivided, while in the Compositae it is always bifid. These two families may be distantly related, but they have evidently acquired many of their distinctive specializations independently of each other. Even if a relationship can be established between them, it has little relevance to angiosperm phylogeny in general, since the position of the Calyceraceae is as much in doubt as that of the Compositae.

Anatomical comparisons between Compositae and other families placed by Cronquist in the superorder Asteridae are no more helpful. In all of the families of Dipsacales (Caprifoliaceae, Adoxaceae, Valerianaceae, Dipsacaceae) the reduction in ovule number, when it occurs, affects chiefly the basal ovules, so that solitary ovules are always pendant in their locules. Reduction in vascularization of the corolla affects the lateral strands so that in the more reduced corollas the midvein is prominent and the laterals are absent from the corolla tube (Henslow, 1890, verified by personal observation). Finally, the inflorescence in these families is a determinate cyme or cyme-like head, while the capitulum of Compositae is indeterminate, and probably derived by reduction of a raceme.

Cronquist (1955) has maintained that the Compositae could have been derived from the Rubiaceae. In both families the style and stigma are 2-branched, and the placentation of ovules in more reduced ovaries of Rubiaceae is basal. Many genera having this condition have 4-merous corollas, while the original corolla of Compositae was probably 5-merous, but a few genera of Rubiaceae, particularly *Psychotria*, agree with Compositae in having both 5-merous corollas and ovary locules containing a single basal ovule. Personal observations of the corolla anatomy in *Psychotria* show that it is completely different from the Compositae. It is traversed by five principal traces, which end in the corolla lobes. The sinuses are supplied by laterals which branch from the principal medians, and laterals of adjacent lobes are joined only just below the sinus itself. The development of this corolla should be studied, but at present, its transformation into a vascular pattern like that of Compositae is very difficult to imagine. Furthermore, the Rubiaceae having reduced ovaries, placed by Schumann (1897) in the subfamily Coffeoideae, are all connected by a series of intermediates to genera having ovaries with numerous ovules, placed in the Cinchonoideae. In all of them, the predominant basic chromosome number is $x=11$ (Fagerlind, 1937), a number almost unknown among Compositae. Hence the origin of Compositae from Rubiaceae that resemble them in having capitate inflorescences, corollas of reduced size, and ovaries containing only one or two ovaries, is highly improbable, so that the nearest affinity having a reasonable degree of probability is descent from an unknown common ancestor. Since, however, the cinchonoid Rubiaceae appear to be connected by intermediates to the Loganiaceae, which are among the least specialized of Asteridae; the common ancestor of Compositae, Rubiaceae and Loganiaceae would have to lack all of the specialized characters found in Compositae except for the sympetalous corolla. The origin of Compositae from such a complex is made even more improbable by characters of inflorescences and leaves. The inflorescence of Rubiaceae and Loganiaceae, like that of Dipsacales, is determinate. The suggestion of Cronquist (1955), that in the derivation of Compositae from Rubiaceae a shift from a determinate cyme to an indeterminate capitulum was the first step in a general shift in phyllotaxy, which was followed later by the shift from opposite leaves, to alternate leaves, receives no support from developmental principles. Many investigations of shoot differentiation (Corson and Gifford, 1969) have shown that during the shift from the vegetative to the reproductive condition the physiology and cell metabolism of the shoot is completely altered, so that genes altering the developmental pattern of the vegetative phase would have to be completely different from those affecting the reproductive phase. Finally, leaf bases in Rubiaceae always possess stipules, while these are consistently lacking in Compositae.

If only one or two of these differences between Rubiaceae and Compositae existed, one might postulate a very unusual shift in evolutionary direction and derive the latter family from the former. When all of these points are considered, the most logical conclusion is that Compositae are no more closely related to Rubiaceae than to any other family of Asteridae.

Finally, the relationships between Compositae and the order Campanulales must be considered. These turn out to be just as weak as those with families that have already been considered. The Stylidiaceae are extremely different, and should probably be removed from the Campanulales. The monogeneric family Brunoniaceae is the only one having an almost regular corolla, a uniovulate ovary, and flowers in heads. If differs widely from Compositae in its superior ovary, asymmetrical undivided stigma, orthotropous ovule, and seeds lacking endosperm. The Goodeniaceae include genera having inferior ovaries, but their stigma is like that of Brunoniaceae, and their highly zygomorphic corolla is unlike anything in the Compositae. The Sphenocleaceae do not resemble Compositae in any important respect. This leaves the Campanulaceae (*sensu lato*, including Lobeliaceae). Most genera of this family resemble Compositae only in having sympetalous corollas, inferior ovaries, and indeterminate inflorescences. Nevertheless, each of the specializations that are combined together in Compositae— flowers in capitula, reduced corollas, united anthers, stigma-branches reduced to two, and few ovules per gynoecium—occur as more or less isolated trends in various genera or subfamilies of Campanulaceae. This suggests that the basic organization of the flower favors the same kinds of changes in Campanulaceae as have been predominant in Compositae, and might therefore be taken as evidence for an immediate or not very distant common ancestor of the two families. Even if such an ancestor is postulated, the relationship between them is not very close. The least specialized genus of Campanulaceae is *Cyananthus*, which has a superior, 5-loculate ovary containing many ovules on an axile placenta, and flowers solitary on leafy branches. Since the hypothetical common ancestor would have to be at least as unspecialized, this hypothesis implies that all of the specializations of the Compositae except for the sympetalous corolla have been evolved separately from similar specializations in other families. Since two different lines of reasoning have lead me to this same conclusion, I believe that it is the most logical one, based upon present knowledge of morphology and anatomy.

Perhaps evidence from pollen grain structure or from biochemistry will shed more light on the affinities of the family. In my opinion, the morphological and anatomical evidence is so equivocal that any solid evidence from a different line of research should take precedence over it. The Compositae illustrates to an alarming degree the dilemma of the botanist who attempts to construct angiosperm phylogeny on the basis of comparisons of morphological and anatomical characters of modern forms. Parallel trends and convergences toward superficial similarity have been so widely prevalent as to obscure almost completely many of the critical relationships between families and orders.

REFERENCES

AXELROD, D. I. and BAILEY, H. P. (1969). Paleotemperature analysis of Tertiary floras. *Palaeogeogr. Palaeoclimatol. Palaeoecol.* **6**, 163–195.

BARGHOORN, E. S. Jr. (1940). The ontogenetic development and phylogenetic specialization of rays in the xylem of Dicotyledons. I. The primitive ray structure. *Am. J. Bot.* **27**, 918–928.

BENTHAM, G. (1873). Notes on the classification; history and geographical distribution of Compositae. *J. Linn. Soc. (Bot.)* **13**, 335–577.

CARLQUIST, S. (1957a). The genus *Fitchia* (Compositae). *Univ. Calif. Publs. Bot.* **29**(1), 1–144.

CARLQUIST, S. (1957b). Anatomy of Guayana Mutisiae. *Mem. N.Y. bot. Gdn.* **9**, 441–476.

CARLQUIST, S. (1957c). Wood anatomy of Mutisieae (Compositae). *Trop. Woods* **106**, 29–45.

CARLQUIST, S. (1958a). Wood anatomy of Heliantheae (Compositae). *Trop. Woods* **108**, 1–30.

CARLQUIST, S. (1958b). The woods and flora of the Florida Keys. Compositae. *Trop. Woods* **109**, 1–37.

CARLQUIST, S. (1959). Wood anatomy of Helenieae (Compositae). *Trop. Woods* **111**, 19–39.

CARLQUIST, S. (1960a). Wood anatomy of Cichorieae (Compositae) *Trop. Woods* **112**, 65–91.

CARLQUIST, S. (1960b). Wood Anatomy of Astereae (Compositae). *Trop. Woods* **113**, 54–84.

CARLQUIST, S. (1961). Wood anatomy of Inuleae (Compositae). *Aliso* **5**, 21–37.
CARLQUIST, S. (1962a). Wood anatomy of Senecioneae (Compositae). *Aliso* **5**, 123–146.
CARLQUIST, S. (1962b). A theory of paedomorphosis in dicotyledonous woods. *Phytomorphology* **12**, 30–45.
CARLQUIST, S. (1964). Wood anatomy of Vernonieae. *Aliso* **5**, 451–467.
CARLQUIST, S. (1965a). Wood anatomy of Cynareae. *Aliso* **6**, 13–24.
CARLQUIST, S. (1965b). Wood anatomy of Eupatorieae (Compositae). *Aliso* **6**, 89–103.
CARLQUIST, S. (1966a). Wood anatomy of Anthemideae, Ambrosieae, Calenduleae and Arctotidae (Compositae). *Aliso* **6**, 1–23.
CARLQUIST, S. (1966b). Wood anatomy of Compositae: a summary with comments on factors controlling wood evolution. *Aliso* **6**, 25–44.
CARLQUIST, S. (1967). Anatomy and systematics of *Dendroseris* (sensu lato). *Brittonia* **19**, 99–121.
CARLQUIST, S. (1969). Toward acceptable evolutionary interpretations of floral anatomy. *Phytomorphology* **19**, 332–362.
CARLQUIST, S. (1974). "Island Biology". Columbia University Press, New York and London.
CORSON, G. E. Jr. and GIFFORD, E. M. Jr. (1969). Histochemical studies of the shoot apex of *Datura stramonium* during transition to flowering. *Phytomorphology* **19**, 189–196.
CRONQUIST, A. (1955). Phylogeny and taxonomy of the Compositae. *Am. Midland Nat.* **53**, 478–511.
DIETTERT, R. A. (1938). The morphology of *Artemisia tridentata* Nutt. *Lloydia* **1**, 3–74.
EAMES, A. J. (1961). "Morphology of the Angiosperms". McGraw-Hill, New York.
FAGERLIND, F. (1937). Embryologische, zytologische, und bestäubungs-experimentelle Studien in der Familie Rubiaceae nebst Bemerkungen über einige Polyploiditätsprobleme. *Acta horti bergiani* **2**, 195–470.
HAGEMANN, W. (1970). Studien zur Entwicklungsgeschichte der Angiospermenblätter. *Bot. Jahrb.* **90**, 297–413.
HALL, H. M. and CLEMENTS, F. E. (1923). The phylogenetic method in taxonomy: The North American species of *Artemisia, Chrysothamnus* and *Atriplex. Publs. Carnegie Instn* **326**, 355 pp.
HENSLOW G. (1890). On the vascular systems of floral organs and their importance in the interpretation of the morphology of flowers. *J. Linn. Soc. (Bot.)* **28**, 151–197.
HIEPKO, P. (1965a). Das zentrifugale Androeceum von *Paeonia. Ber. dt. bot. Ges.* **77**, 427–435.
HIEPKO, P. (1965b). Vergleichend-morphologische und entwicklungsgeschichtliche Untersuchungen über das Perianth bei den Polycarpicae. *Bot. Jahrb.* **84**, 359–508.
HOFFMANN, O. (1894). Compositae. *In* "Die natürliche Pflanzenfamilien" (Engler and Prantl, eds) **4** (5), 87–391.
KAM, Y. K. and MAZE, J. (1974). Studies on the relationships and evolution of supra-specific taxa utilizing developmental data. II Relationships and evolution

of *Oryzopsis hymenoides* O. *virescens*, O. *kingii*, O. *micrantha* and O. *asperifolia*. *Bot. Gaz.* **135**, 227–247.

Kaplan, D. R. (1973). The problem of leaf morphology and evolution in monocotyledons. *Q. Rev. Biol.* **48**, 437–457.

Koch, M. F. (1930a). Studies in the anatomy and morphology of the composite flower I. The corolla. *Am. J. Bot.* **17**, 938–952.

Koch, M. F. (1930b). Studies in the anatomy and morphology of the composite flower II. The corollas of the Heliantheae and Mutisieae. *Am. J. Bot.* **17**, 995–1010.

Leins, P. (1971). Das Androecium der Dicotylen. *Ber. dt. bot. Ges.* **84**, 191–193.

Leins, P. and Stadler, P. (1973). Entwicklungsgeschichtliche Untersuchungen am Androecium der Alismatales. *Öst. bot. Z.* **121**, 51–63.

Mayr, B. (1969). Ontogenetische Studien an Myrtales-Blüten. *Bot. Jahrb.* **89**, 210–271.

Moss, E. H. (1940). Interxylary cork in *Artemisia* with a reference to its taxonomic significance. *Am. J. Bot.* **27**, 762–768.

Sattler, R. (1972). Centrifugal primordial inception in floral development. *Adv. Pl. Morph.* **1972**, pp. 170–178.

Schumann, K. (1897). Rubiaceae. *In* "Die natürlichen Pflanzenfamilien" (Engler and Prantl, eds) **4** (4–5), 1–156.

Singh, V. and Sattler, R. (1972). Floral development of *Alisma triviale*. *Can. J. Bot.* **50**, 619–627.

Stebbins, G. L. (1937). Critical notes on *Lactuca* and related genera. *Am. J. Bot.* **75**, 12–18.

Stebbins, G. L. (1940a). Studies in the Cichorieae: *Dubyaea* and *Soroseris*, endemics of the Sino-Himalayan region. *Mem. Torrey bot. Club* **19**, 1–76.

Stebbins, G. L. (1940b). Additional evidence for a holarctic dispersal of flowering plants in the Mesozoic era. *Proc. 6th Pacific Sci. Congr.*, 649–660.

Stebbins, G. L. (1972a). Evolution and diversity of arid-land shrubs. *In* "Wildland Shrubs; their Biology and Utilization" (C. McKell, J. B. Blaisdell and J. R. Goodin, eds). *U.S. Dept. Agric. Forest Service Gen. Techn. Rep. INT-L*, 111–120.

Stebbins, G. L. (1972b). Ecological distribution of major centers of adaptive radiation in angiosperms. *In* "Taxonomy, Phytogeography and Evolution" (D. Valentine, ed.), pp. 7–34. Academic Press, London and New York.

Stebbins, G. L. (1973). Morphogenesis, vascularization and phylogeny in angiosperms. *Breviora, Mus. Comp. Zool. Harvard Univ.* no. **418**, 1–19.

Stebbins, G. L. (1974). "Flowering Plants: Evolution Above the Species Level". Belknap Press, Harvard, Cambridge, Mass.

Stebbins, G. L. Jr., Jenkins, J. A. and Walters, M. S. (1953). Chromosomes and phylogeny in the Compositae, tribe Cichorieae. *Univ. Calif. Publs Bot.* **26**, 401–430.

Wetmore, R. and Rier, J. P. (1963). Experimental induction of vascular tissues in callus of angiosperms. *Am. J. Bot.* **50**, 418–429.

Chapter 6
Corolla forms in Compositae—some evolutionary and taxonomic speculations

C. JEFFREY

*The Herbarium, Royal Botanic Gardens,
Kew, Surrey, England*

Abstract. Consideration of the floral biology of racemose inflorescences leads to the hypothesis that the ancestral corolla type in the Compositae was zygomorphic bilabiate and that other corolla types in the family are derivative. The implications of this hypothesis for the evolutionary relationships and classification of the Compositae are discussed, and a scheme of tribal relationships postulated.

The constantly centripetal, indeterminate order of maturation of the florets in the capitulum of Compositae is indicative of its origin from a less contracted form of indeterminate inflorescence, such as a spike or raceme. The arrangement of the capitula themselves makes it likely that each indeterminate unit was part of an overall determinate flowering region, but it is with the arrangement of the flowers in the individual racemose units that the present paper is concerned. Amongst the Compositae the daisy-like arrangement of rays and disc, as exemplified by *Leucanthemum* and *Bellis*, is very widespread and may be considered as a modal type. This is an excellent arrangement, in a capitulum which extends in a horizontal direction, for providing both attraction to, and a landing platform for, visiting insects. Such an arrangement, however, would be quite unsuited to a spike or raceme, having its main extent in a vertical direction, and that this is so is borne out by the fact that, among families in which racemes are commonly found, e.g. Leguminosae, Scrophulariaceae, Acanthaceae, Liliaceae, Orchidaceae, no spikes or racemes with such a dimorphic arrangement of flowers are known.

    Among the Compositae there is a wide range of corolla types. In the disc floret, the corolla is actinomorphic, pentamerous or sometimes tetramerous, the floret bisexual, unisexual, or neuter. In the ray floret, usually female, it is strap-shaped with three or fewer apical teeth. In the ligulate floret, it is strap-shaped with five apical teeth. Two types of bilabiate floret are known, one with a 2-lobed inner and 3-lobed outer lip, the other with a 1-lobed inner and 4-lobed outer lip. Finally, there are florets in which the corolla is much reduced or absent. Consideration of the families which have been postulated on morphological grounds as possibly similar to the ancestral Compositae, namely, Caprifoliaceae, Rubiaceae, Dipsacaceae, Calyceraceae, Brunoniaceae, Campanulaceae, Goodeniaceae, Stylidiaceae, shows they exhibit two main types of corolla—one more or less actinomorphic, usually pentamerous, the other zygomorphic, bilabiate, pentamerous, with a 3-lobed lower lip and a 2-lobed upper lip, sometimes more or less cleft adaxially. The other types of the Compositae do not occur in these families. This suggests that the inflorescence of the ancestral Compositae was probably composed of flowers of one of these two types, or, possibly, of a mixture of both.

Consideration of racemose inflorescences in modern sympetalous families shows that a biologically effective type of corolla in such an inflorescence is a zygomorphic, bilabiate type, more or less horizontally orientated, which provides both an alighting platform for a visiting insect and an attractive display in a vertically extensive inflorescence in both space and time (Stebbins, 1951). Such a corolla type, on this consideration, might well have characterized the racemose inflorescence of the protocomposites. Reinforcement of this argument is provided by consideration of the widespread occurrence of bilabiate corollas in Compositae. *A priori* they appear to be biologically not very well suited to a capitate inflorescence. The upper or inner lip of such a corolla, unless tightly rolled, would tend to interfere, both spatially and functionally, with the outer (lower) lip of the immediately inner floret on the same orthostichy. If tightly rolled, it would then be more or less functionless once the bud stage was passed. It is difficult, then, to present a logical argument in favour of selection, in a capitate inflorescence with centripetal development, for the evolution of a bilabiate corolla from another ancestral type. The bilabiate corolla form in the Compositae might better be considered as a relictual feature from a previously more elongated inflorescence.

The distribution of bilabiate forms in the Compositae appears to favour this possibility. They occur sporadically in some tribes, e.g. Vernonieae, Cynareae, Eupatorieae, rarely and then predominantly only marginally in the capitulum in others, e.g. Senecioneae, Heliantheae, Astereae, or may be completely absent, as in Lactuceae, Anthemideae, Inuleae, Arctotideae, Calenduleae; they are common only in the Mutisieae. In the last tribe, their development may in part be associated with the ornithophily exhibited by a number of genera. In most of the genera, however, they appear unassociated with any specialized pollinator, and might well be considered as relictual. In the other tribes, they appear to have been largely or wholly eliminated in favour of types more suited to the biology of a capitate organization, and their appearance is difficult to explain except as a relictual ancestral feature. It is possible to postulate as an exception only that, in the radiate tribes, the presence of strictly marginal bilabiate corollas may be wholly or partly secondary, as a result of a tendency towards the evolution of a radiate capitulum secondarily from a discoid or disciform one.

Thus it may be concluded that the ancestral corolla form in Compositae was probably a zygomorphic bilabiate pentamerous type, with the flowers arranged in a raceme or spike. The disc floret then appears as a neotonous derivative form, probably evolved in parallel with the evolution of a capitate inflorescence, perhaps in association with selection for economy in time and materials under the influence of increasing seasonal aridity and shortening of the growing season. This may well have been associated with a shift from adaptation to pollination by a larger, perhaps specialized, agent to pollination by smaller, more generalized ones. Such a shift to generalized pollination could well have been advantageous if, for the same climatic reasons, the growing seasons were becoming shorter and possible vectors less numerous. This would accord well with a change from a zygomorphic floral class organization, based on individual flowers, to an amorphic one, based on the inflorescence as a pseudanthium, biologically and functionally equivalent to a single flower (Leppik, Chapter 4). The other types of Compositae floret appear likewise derivative, by way of shifts in developmental patterns, brought about by interplay of selective forces balancing, on the one hand, the need for economy, on the other, the need for provision for the attraction and landing of pollinators.

A logically coherent reductive scheme of corolla type evolution is possible on the above hypothesis (Fig. 1). A bilabiate type in a raceme (a) leads to a similar but reduced type in a capitulum (b). Reduction of the ovary to the unilocular uniovulate condition and of the calyx to the capillary pappus most likely accompanied this markedly reductive stage. From this bilabiate 2+3 form can be derived the bilabiate 1+4 (c), the ligulate 0+5 (d), the sterile disc 5 (e), the bisexual ray 0+3 (f) and the bisexual disc 5 (g). From the former follow the female ray (h), the sterile ray (i), the reduced female ray (j), the filiform female (k) and the rudimentary female R (l). The bisexual disc floret leads to the filiform female by a different route (m), the male disc (n), the sterile disc again (e) and to similar types in the tetramerous condition.

A coherent picture of tribal inter-relationships can also be established when these progressions are considered in relation to our other knowledge of the morphology and chemistry of the family (Fig. 2). Closest to the ancestral complex stands the Mutisieae (XII), with a wide range of corolla types— 0+5, 1+4, 2+3, 5, 0+3. The Lactuceae (XIII) 0+5 stand near to it as a distinct line. Also close are the tribes Arctotideae (X) 0+3, 5, Eremothamneae /(a) 0+3, 5, Cynareae in the broad sense (XI) 0+5, 2+3, 5, Vernonieae (I) 0+5, 2+3, 5 and Liabeae (b) 0+3, 5 (Robinson and Brettell,

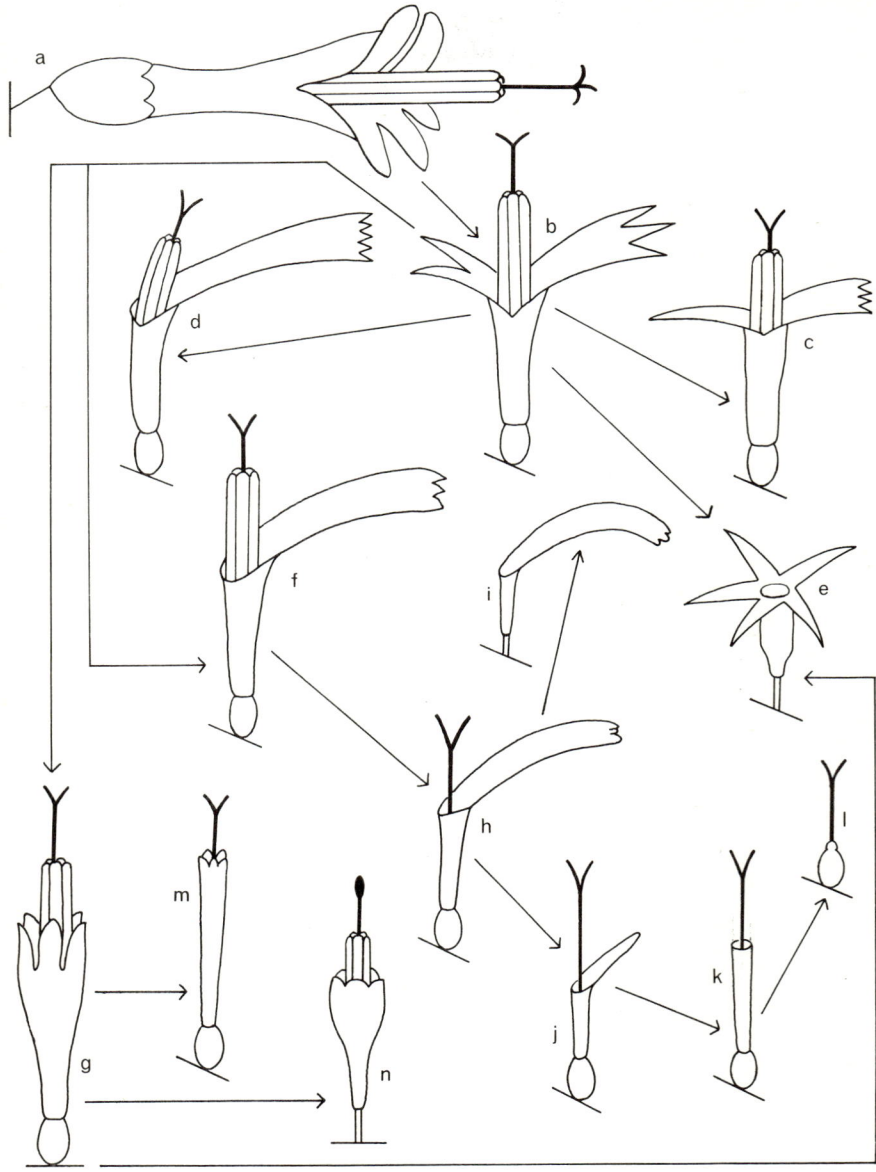

FIG. 1. Postulated scheme of evolutionary relationships of corolla types in Compositae: a, hypothetical ancestral bilabiate, horizontally orientated, in an elongated racemose inflorescence; b, bilabiate 2+3; c, bilabiate 1+4; d, ligulate 0+5; e, sterile disc 5; f, bisexual ray 0+3; g, bisexual disc, 5; h, female ray; i, sterile ray; j, reduced female ray; k, filiform female; l, vestigial or absent R; m, filiform female; n, male disc.

1973). The Eupatorieae (II) 2+3, 5 stands associated with this group of tribes though more remote from the ancestral stock. More remote still are the Heliantheae (V+VI, inclusive of the vast majority of the former Helenieae) and its associated tribes, which group contains the bulk of the typically radiate Compositae. The Heliantheae 2+3, o+5, o+3, 5, 4, R is the most diverse tribe of this alliance, with a number of evolutionary lines within it. Severally allied to it are the distinct and specialized Senecioneae (VII) 2+3, o+3, 5, 4, and Tageteae (c) o+3, 5, the Inuleae (III) o+3, 5 and the Anthemideae (VIII) in the strict sense, o+3, 5, 4. Also of this alliance are the remainder of the more specialized tribes, the Astereae (III) o+5, 2+3, o+3, 5, 4, R, the Calenduleae (IX) o+3, 5, the Cotuleae (e) o+3, 4, ?3, R and the Ursinieae (d) o+3, 5, 4. The Cotuleae are here considered provisionally to include at least the majority of genera 531–541 of Bentham (1873), although it is admitted that the relationships of some are in dispute, and to be closely allied to the grangeoid genera of the Astereae, i.e. genera 102–113, 115–116 and 124 of Bentham (1873). To the Ursinieae may be referred eventually a number of African genera hitherto considered anthemid. The relationships and classification of both Cotuleae and Ursinieae are in need of clarification.

This hypothesis is also pertinent to consideration of the possible ancestors of the Compositae and its relationship to other families. It suggests the affinities of a protocomposite are more likely to be found in racemose, often bilabiate families, such as Campanulaceae, Goodeniaceae and Stylidiaceae, rather than in cymose, predominantly actinomorphic ones, such as Rubiaceae, Caprifoliaceae and Dipsacaceae. The Brunoniaceae and Calyceraceae resemble many Compositae in having more or less actinomorphic flowers in centripetal capitate inflorescences. This suggests they may have evolved in parallel with the Compositae for a long period from the common ancestral stock; when better known developmentally and chemically, they may well prove to be the closest living relatives of the daisy family. Ovule morphology (Philipson, 1974) indicates that, among the polypetalous families, an alliance of families placed by Takhtajan (1966) in the Grossulariales (Pittosporaceae, Bruniaceae, Hydrangeaceae, Escalloniaceae) and the Aralianae (Cornaceae, Davidiaceae, Nyssaceae, Alangiaceae, Garryaceae, Araliaceae, Umbelliferae) approaches nearest to the ancestral stock of polypetalous dicotyledons from which the majority of the sympetalous families (including Compositae) have probably been derived. The striking chemical similarity between the Umbelliferae and the Compositae (Hegnauer, 1971; Harborne, Chapter 12) may indicate no more than this eventual common ancestry, if one bears in mind, first that both families have been comparatively intensively studied chemically and secondly, that both are highly derived, highly specialized families including many highly productive perennial and annual herbaceous species, on which selective pressures favouring the evolution of protective chemical sub-

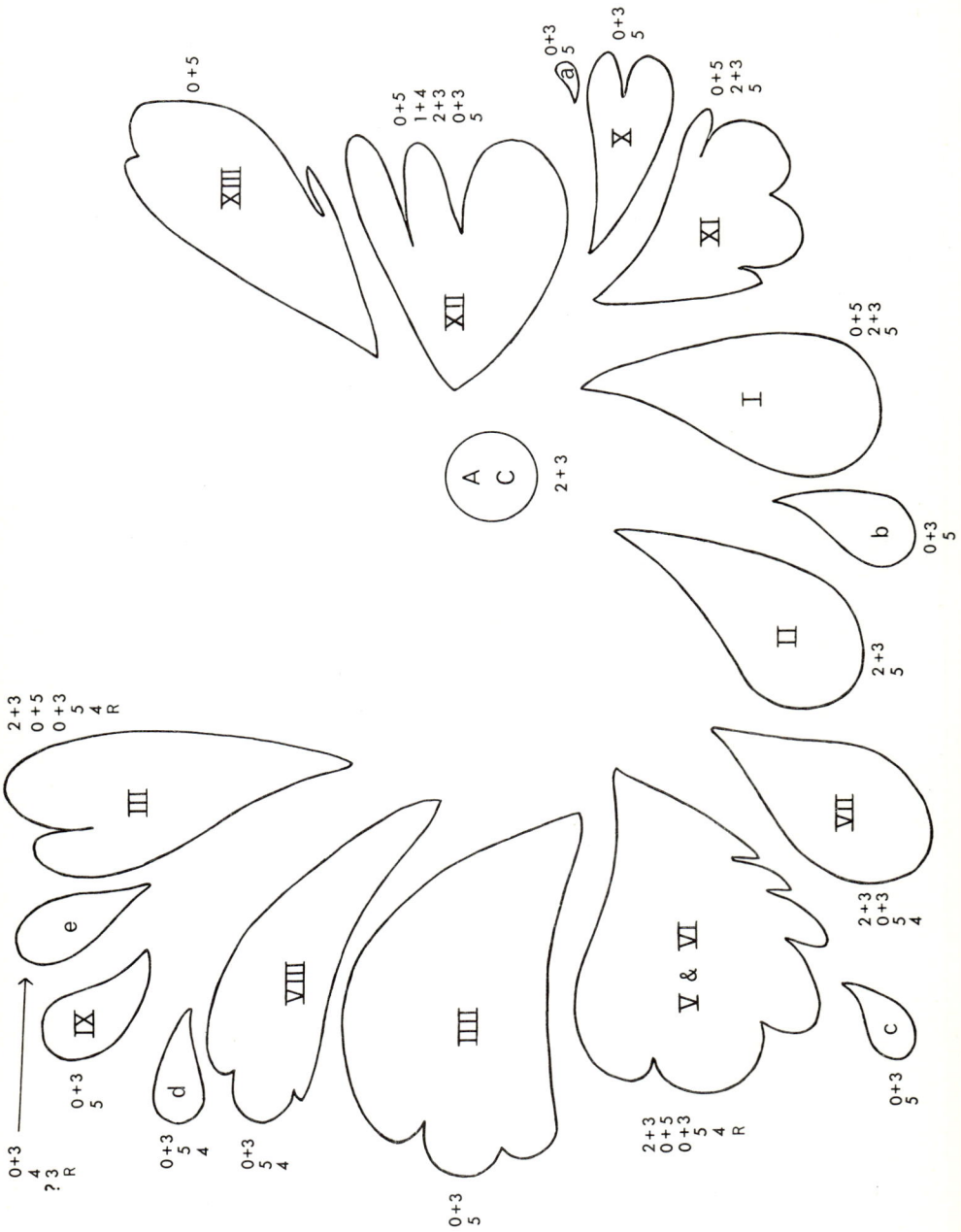

stances must have been both strong and similar. From such an ancestral complex it is possible that one line of evolution led eventually to the Gentianales-Dipsacales, a second to the Campanulales, Calycerales and Asterales. The question of the timing of this diversification is in dispute. Although no light is shed by the fossil record on the *origin* of either group, there is a marked parallel between the record of *diversification* of Silurian–Devonian–Carboniferous land plant microspores (Chaloner, 1970) and that of angiosperm pollen grains in the uppermost Jurassic, Cretaceous and Tertiary (Muller, 1970). Both records indicate single major evolutionary radiations, and provide no evidence, respectively, for a long pre-Devonian evolutionary history for the vascular land plants, or for a long pre-Cretaceous evolutionary history for the angiosperms. Pollen similar to that of modern Compositae apparently first appears comparatively late, in the post-Cretaceous, although this is questioned by some authors (cf. Turner, Chapter 2). On the other hand, the evidence from geographical distribution and cytochrome c sequences indicates a much older origin for the Compositae, and for angiosperms in general (Turner, loc. cit.). These contradictions are resolvable if it be postulated that, while the angiospermous line in itself has been a distinct one since pre-Jurassic times, only comparatively late in its evolution did its microspores acquire features characteristic of modern angiosperms. This may well be true also of some of the other organs of the latter. The evolutionary history of the Compositae indicates that it and the majority of modern angiosperm families may well have originated as distinct evolutionary lines in a pre-angiospermous stage. This conclusion should not be taken to indicate that the angiosperms are polyphyletic, rather that evolutionary advance was on a broad front, with similar specializations becoming acquired in parallel along a number of related but distinct lines. If this were so, then the similarity between Compositae and Umbelliferae in cytochrome c sequences, among the few angiosperm families so far studied, would not be unexpected. Compositae are especially characteristic of tropical wooded grassland, grassland and bushland formations and their rapid diversification in Tertiary times was no doubt associated with the spread of such formations and the evolution of herbivorous grazing mammals.

Finally, the groupings provided and relationships postulated by the present hypothesis form a basis on which the need for further evidence

FIG. 2. Postulated relationships of the main groups of Compositae. The main tribes are numbered in the order in which they are described by Bentham (1873), additional tribes are lettered. AC, Ancestral complex; I, Vernonieae; II, Eupatorieae; III, Astereae; IIII, Inuleae; V, Heliantheae; VI, Helenieae (in greater part); VII, Senecioneae (in the strict sense); VIII, Anthemideae (in the strict sense); IX, Calenduleae; X, Arctotideae; XI, Cynareae (in the broad sense); XII, Mutisieae; XIII, Lactuceae (Cichorieae); a, Eremothamneae; b, Liabeae; c, Tageteae; d, Ursinieae; e, Cotuleae; Corolla types indicated as in Fig. 1.

from micromorphology, palynology, chemotaxonomy and other disciplines can be determined, the taxonomic implications of such evidence evaluated and the necessity for taxonomic changes at subfamilial, tribal and subtribal levels assessed. I venture to predict that while groupings within the main tribes will become better defined and characterized, the latter will largely remain unchanged in both circumscription and content and that while a number of genera, at present uncertain or misplaced, will be found an appropriate tribal home, the number of tribes itself will not have to be increased to any great extent. Major changes will be needed, however, in the delimitation of many genera.

REFERENCES

BENTHAM, G. (1873). Compositae. *In* "Genera Plantarum" (G. Bentham, and J. D. Hooker, eds), **2**(1), 163–533; 536–537. Reeve & Co., London.

CHALONER, W. G. (1970). The rise of the first land plants. *In* Symposium on major evolutionary events and the geological record of plants (E. N. Willmer, ed.). *Biol. Rev.* **45**, 353–378.

HEGNAUER, R. (1971). Chemical patterns and the relationships of the *Umbelliferae*. "The Biology and Chemistry of the Umbelliferae" (V. H. Heywood, ed.), pp. 267–277. Academic Press, London and New York. (Supplement, *J. Linn. Soc. (Bot.)* **64**).

MULLER, J. (1970). Palynological evidence on early differentiation of angiosperms. *In* Symposium on major evolutionary events and the geological record of plants (E. N. Willmer, ed.). *Biol. Rev.* **45**, 417–450.

PHILIPSON, W. R. (1974). Ovular morphology and the major classification of the dicotyledons. *J. Linn. Soc. (Bot.)* **68**, 89–108.

ROBINSON, H. and BRETTELL, R. D. (1973). Tribal revisions in the Asteraceae. III. A new tribe Liabeae. *Phytologia* **25**, 404–407.

STEBBINS, G. L. (1951). Natural selection and the differentiation of angiosperm families. *Evolution* **5**, 299–324.

TAKHTAJAN, A. L. (1966). "Sistema i filogeniya tsvetkovỹkh rastenii". Nauka, Moscow and Leningrad.

Chapter 7

Microcharacters in the ligules of the Compositae

JETTE BAAGØE
*Botanical Museum of the University,
Copenhagen, Denmark*

Abstract. Ligules of the Compositae present an array of microcharacters in their epidermises, which may be taxonomically useful at different levels within the family. At least three adaxial epidermis-types may be recognized: (1) papillose, (2) longitudinally striped, and (3) tabular celled (presumably unspecialized). The anatomy of the adaxial epidermis is apparently often correlated with the distribution of visible and ultraviolet-absorbing pigments in the ligules. The significance of the epidermis-types in terms of pollination biology is known from other families of flowering plants and provides a clue to the role of the ligule in the pollination biology of the capitulum. A possible derivation of the papillose epidermis-type is suggested by the cell morphology and suggests, in combination with the functional role of the ligule, some evolutionary trends within the taxonomic groups concerned.

CONTENTS

Introduction	120
Materials and methods	120
Main epidermis types	120
Helianthoid type	121
Senecionoid type	121
Senecionoid papillose type	121
Mutisioid type	124
Functional considerations	124
Some modifications, their possible explanations and implications	129
Anatomical considerations	130
Taxonomic considerations	133
The Lactuceae	136
Acknowledgements	136
References	136
Appendix—List of genera studied	138

INTRODUCTION

The ligules of flowers in the Compositae are seldom considered in much detail in conventional systematic studies. In this paper an attempt is made to investigate the microcharacters possessed by the ligules in a wide sample of the family using light and scanning electron microscopy.

MATERIALS AND METHODS

Florets from 275 species representing 111 genera have been studied (see Appendix). Fresh material was obtained from the Botanical Garden, University of Copenhagen; voucher specimens are kept at the Botanical Museum of the University of Copenhagen (C). Herbarium specimens were borrowed from C and from K (the Herbarium, Royal Botanic Gardens, Kew).

Surface scanning was carried out on some 85 of the species. Intact ligules were stuck to the stubs with double-sided adhesive tape and, in the case of fresh material, freeze-dried in a Leybold-Heraues GT 2 freeze drier ($-18°C$, 10^{-1} mmHg). Gold coating was added by means of an Edwards Pirani Penning Model 4 evaporator at $< 10^{-4}$ mmHg. The thickness of the coating varied with the material, and especially with the height differences in it. The micrographs were exposed in a Cambridge Stereoscan Mark IIa at an accelerating voltage of 15 or 20 kV, with magnifications varying between $\times 300$ and $\times 6200$. In most cases both upper and lower surfaces were scanned.

All material was investigated by light microscopy. Before preparation, herbarium material was boiled in water. Hand-cut cross-sections and peelings of epidermises were embedded in Hoyer's solution.

The unstained preparations were photographed in a Zeiss photomicroscope.

The occurrence of ultraviolet colour patterns was studied by means of a black-ray lamp and ultraviolet photography (Kevan, 1972). Qualitative studies of the flower colours have not, however, been carried out and the colours noted refer to human colour vision.

MAIN EPIDERMIS TYPES

Most of the microcharacters studied refer to the features of the morphology and anatomy of the epidermises of the adaxial and abaxial ligules. Since studies of the outer epidermal cell wall are still incomplete, a proper distinction between the different layers is not yet possible and the term "cuticle" is used here to denote the entire cuticular complex while the rest of the wall is termed the "cell wall".

The ligule-types found fall into three main groups according to functional criteria and each of the groups is named after the tribe in which it has been found to occur. Members of the tribe Lactuceae have not been studied in detail and are excluded from the general descriptions and discussions, although they are briefly mentioned on p. 132.

Helianthoid type

Epidermis of adaxial ligules consisting of papillose, nearly isodiametric cells. Radial and tangential walls usually undulating sometimes straight with septa (Plate 2e, f), or without. Outer walls thin (c. 1·3–3 μm), of equal thickness all over the cell. Cuticle with a pattern of stripes running from a rugose area on top of the cell towards its base (Plate 1a, b). No direct connection between the cuticle patterns of adjacent cells. Length and width c. 20–75 μm (extremes excluded), height c. 15–100 μm, the height sometimes varying considerably within the same ligule (cf. p. 123).

The helianthoid type epidermis is typical of the Anthemideae studied (20 spp.) except for *Ursinia* and *Lasiospermum*, the Heliantheae *sensu* Stuessy (Chapter 23) (except *Gaillardia* and *Schkuhria*), the Tageteae *sensu* Strother (p. 765), and the subtribes Eriophyllinae and Arnicinae (cf. p. 130). The Peritylinae, Chaenactidinae, and Flaveriinae have not yet been studied. Possible explanations for the occurrence of anomalous epidermis types in the Heliantheae and Tageteae are given on p. 125.

Senecionoid type

Epidermis of adaxial ligule consisting of oblong, tabular cells with more or less convex outer wall (Plate 1c, d). Radial and tangential walls usually straight, sometimes slightly undulating. Outer wall thin (c. 1·5–3 μm), of equal thickness all over the cell, or, in the Astereae with a characteristic thickening along the median plane of the cell (Plate 2c). Cuticle with a pattern of transverse or longitudinal, more or less rugose stripes. No systematic connection between the patterns of adjacent cells. Some Liabeae show a structureless cuticle. Length c. 45–185 μm, width c. 10–35 μm, height c. 5–30 μm.

The senecionoid epidermis is typical of all Astereae studied (31 spp.), the Inuleae (8 spp. of Inulinae *s. ampl.*) except *Asteriscus* and *Buphthalmum*, the Liabeae *sensu* Nordenstam (p. 131), the Blennospermatinae, and most, but not all of the Senecioninae examined (cf. p. 120). It is present in some Arctoteae, and in *Chrysanthemoides*, *Gibbaria*, and *Osteospermum* of the Calenduleae.

Senecionoid papillose type

This can be derived from the senecionoid type described above. Cells ovoid with truncate ends as seen from above. Radial and tangential walls

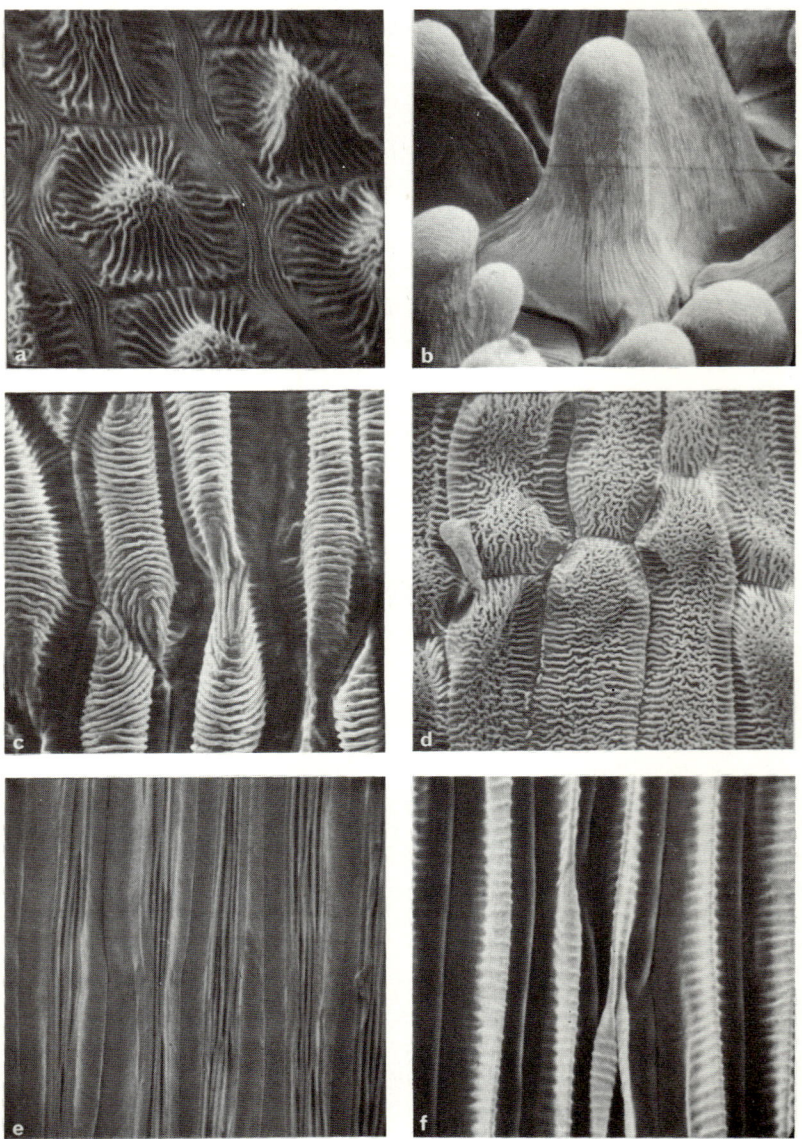

PLATE 1. Scanning electron micrographs of Compositae ligule surfaces. (a) *Lasthenia* sp., ×c. 760, (b) *Rudbeckia* sp., ×c. 530, (c) *Grindelia* sp., ×c. 620, (d) *Doronicum* sp., ×c. 780, (e) *Calendula* sp., ×c. 700, (f) *Perezia* sp., ×c. 780.

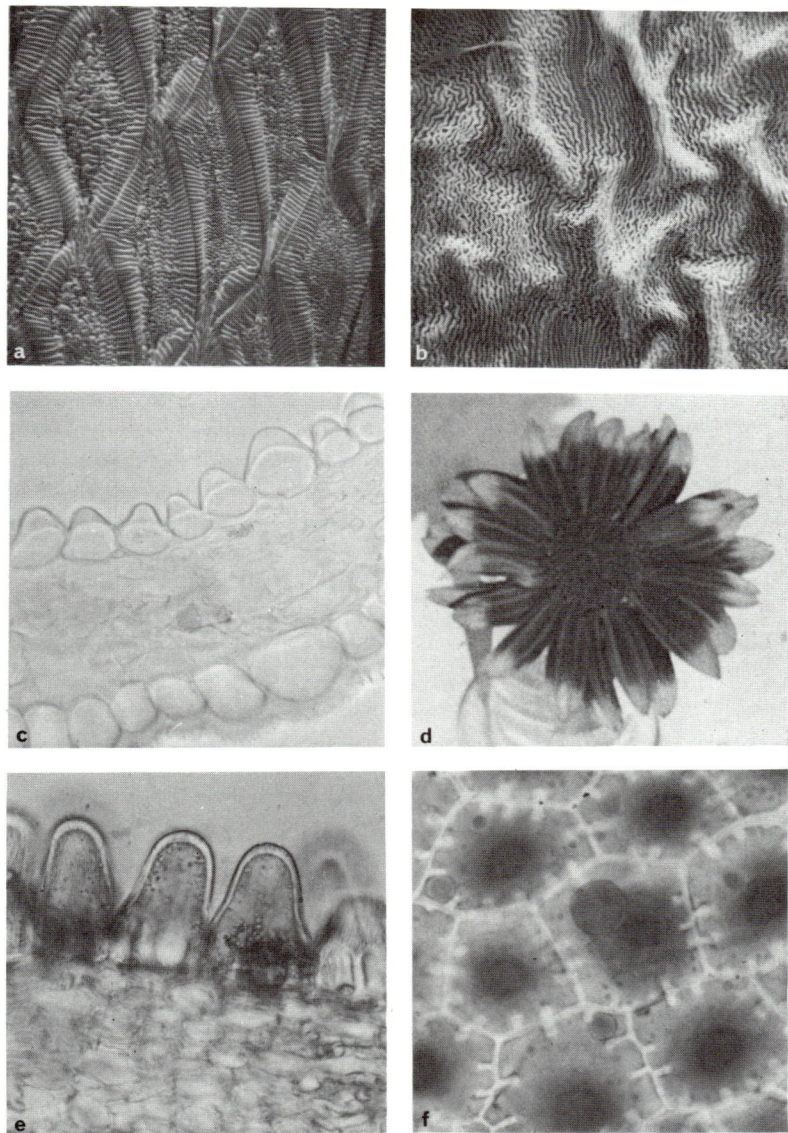

PLATE 2. (a) Scanning electron micrograph of purple *Senecio* ligule, ×c. 425, (b) Cuticle pattern in *Gaillardia pulchella* (SEM), ×c. 610, (c) *Aster* ligule. Cross section showing median wall thickening, ×c. 400, (d) Ultraviolet colour pattern in *Heliopsis scabra*, (e) *Rudbeckia* ligule. Cross section showing septa, ×c. 400, (f) *Rudbeckia* ligule. Epidermis cells with septa seen from above, ×c. 400.

straight or slightly undulating, without septa. The length: width ratio ranges from c. 1·5 to c. 4, i.e. the papillae are not typically isodiametric as are those of the helianthoid type. Cuticle pattern generally like the typical senecionoid type (Plate 2a). Length c. 55–95 μm, width c. 20–35 μm, height c. 15–35 μm.

The occurrence of this kind of cell seems to be correlated with white and purple ligule colour. It is found in purple- and white-rayed *Senecio* species, and in *Castalis*, *Dimorphotheca*, and *Osteospermum* of the Calenduleae.

Mutisioid type

Epidermis of adaxial ligules consisting of oblong, tabular cells with straight radial and tangential walls. Septa present in one case (*Onoseris*). Outer walls usually thickened in a crest 10 (–15) μm high, running longitudinally without intermissions over several cells or having slight depressions above the tangential cell walls; sometimes thin and orientated forming a pointed "roof" above the cell as seen in transverse section. Cuticle pattern variable as shown in Plate 1e, f. Length c. 55–265 μm, width c. 10–25 μm, height c. 10–25 μm.

All the Mutisieae investigated have this epidermis-type; in the Arctoteae it is found in some *Arctotis* species (*Arctotis sensu lato*), *Berkheya*, and *Gazania*; and in the Calenduleae in *Calendula*, and *Garuleum*. Apart from this I have found it only in *Asteriscus* and *Buphthalmum* of the Inuleae.

FUNCTIONAL CONSIDERATIONS

The epidermis-types of the corolla-ligules here described also occur in other angiosperm families, and their function has to some extent been explained by Kugler (1955). The effect of the papillose epidermis is illustrated by Fig. 1, which shows that only a very small fraction of the incident light is reflected directly from the epidermal surface. Most of the rays are split up on entering the cells. The light that ultimately returns to the observer's eye is consequently of low intensity, very poor in white, and the colour is perceived as very saturated. This effect, which Kugler calls the "Samt-effekt", gives the surface a velvety sheen.

Another well studied epidermal effect is that which Kugler (1955) called "Seidenglanz", where the surface has a silky sheen. This sheen is caused by the longitudinal arrangement of the cells and the crests, and/or longitudinally striped cuticle of the mutisioid cells. When a beam of light strikes an epidermal surface of this kind at an angle of 90°, it acquires its silky sheen.

The senecionoid epidermis-type represents an intermediary stage between these two extremes. Its tabular cells reflect more light than do papillose ones, but the outer cell walls and the arrangement of the cells do not give the reflected light any distinct direction; it is consequently rather diffuse and of relatively high intensity. The surface looks dull, and the colour that the observer perceives is of a relatively low saturation.

Usually the function of the cuticular complex is explained in terms of water economy and resistance to various kinds of disease, but I would suggest that at least in flowers another important function of the cuticle is to

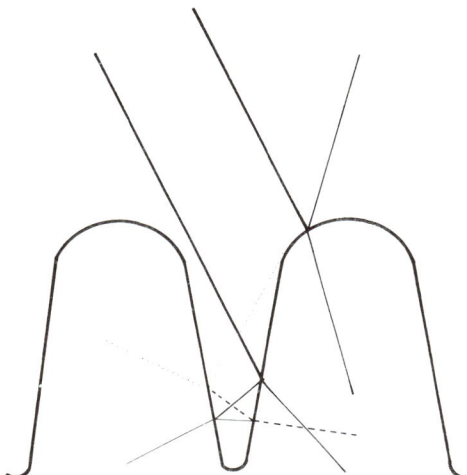

FIG. 1. Reflection of light from a papillose epidermal surface. (After Kugler, 1955.)

add to the surface sheen and colour. Recently, Brehm and Krell (1975) have suggested that a cuticle pattern like that of the helianthoid type might support the outer wall in the living papilla. This seems rather improbable since the outer wall is supported by the cell's turgor pressure.

The surface structure derived from the epidermis is of importance for the attraction of pollinators, when they come close to a flower head. Flowers or flower heads with a velvety sheen are preferred by bumblebees (*Bombus* spp.) to those without. Kugler (1955) explained this as a result of the preference of these bees for colours of a high saturation. *Bombus* spp. have also been shown to prefer flowers with a silky sheen to those without (Kugler, 1955). It seems, therefore, that, at least in relation to bumblebees, plants with a helianthoid or mutisioid epidermis will have an advantage over those with the senecionoid type.

Since colour and texture interact closely, some investigations of the ultraviolet patterns in the tribes concerned have been carried out. The chemistry of ultraviolet patterning is discussed elsewhere in this volume (see Harborne, Chapter 12). Concentric patterns like that illustrated in

Plate 2d are very common in the Heliantheae and are present in the Anthemideae, Arctoteae, Calenduleae, Senecioneae, and Tageteae. King and Krantz (1975) give a photographic report on some ultraviolet patterns in the Compositae. I have tested the persistence of such patterns in herbarium material and found that in most, but not all, cases, where real concentric patterns are concerned, they are constant within the species. However, bad drying conditions make the absorbing flavonoids deteriorate with the result that the pattern originally present may be obscured or even extinguished. As taxonomic characters these patterns should therefore be used with some caution.

The presence of both visible and ultraviolet patterns in the ligules is often correlated with differentiation of the adaxial epidermal cells as illustrated by the following examples. Figure 2 plots the height of the distal epidermal cells in some Heliantheae ligules against that of the proximal, and shows that the proximal cells are about twice the height of the distal. Furthermore, the colour and amount of chromoplastic material in these two modifications of the papilla are different. The distal cells have more light yellow chromoplasts, of which one is larger than the others, while the proximal cells are densely packed with small, more orange-yellow chromoplasts. In all the dead material I have examined the chromoplasts of the proximal cells are distributed in two layers, one at the base of the cells, the other in the apex of the papilla (cf. Brehm and Krell, 1975). However, investigation of fresh material in water indicated that the natural condition is as described above and that the stratification is an artifact caused by the death of the cell.

In *Ursinia* there is a different, although effectively similar, specialization of the cells. In *U. anthemoides* subsp. *versicolor* (DC.) Prassl. that part of the ligule which we perceive as uniformly orange absorbs ultraviolet light proximally and must appear bicoloured to an ultraviolet-perceiving insect. The centre of the capitulum in this species is, in other words, marked by three distinct circles of different colours. In *U. speciosa* DC. the situation is similar, the ligules having a distal yellow + ultraviolet zone and a proximal deep yellow zone. In both species, however, the cells in the ultraviolet-absorbing area are different from those in the ultraviolet-reflecting area. They are senecionoid papillose (most typically evolved in *U. anthemoides* subsp. *versicolor*), and their chromoplastic material is more orange (in *U. anthemoides* subsp. *versicolor*), and more densely packed. The distal cells contain, in addition to small roundish chromoplasts, a large more or less amoeboid one.

The effect of the high and/or more domed proximal cells on the incident light is evident: by refracting the light rays more efficiently than the low distal cells they cause a difference in colour saturation over the ligule. This enhances the differences in chromoplast distribution with the result that in many helianthoid species a yellow zone of higher saturation around

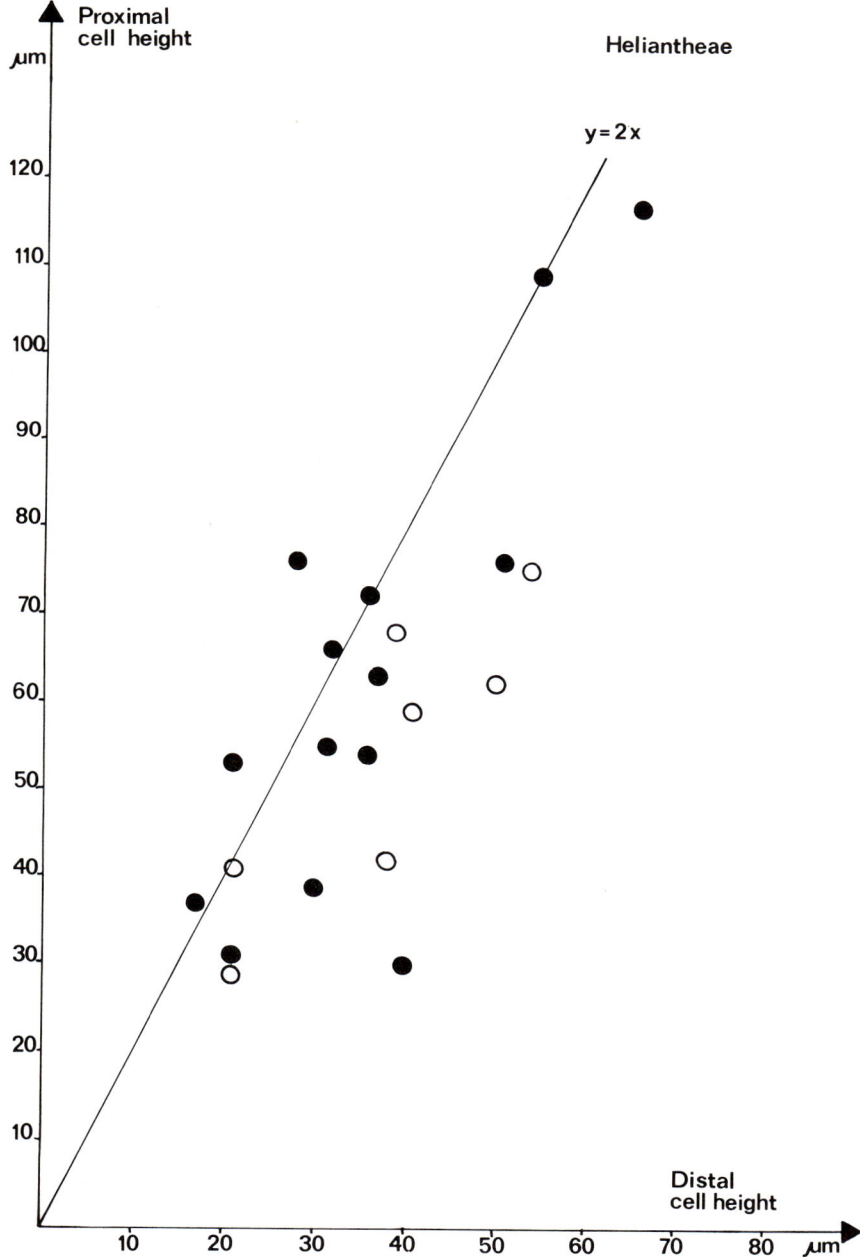

Fig. 2. Adaxial epidermal cells of ligule. Comparison of cell height in the distal and proximal part of the ligule. Solid circles: species absorbing ultraviolet light proximally.

the disc may even be visible to the human eye. Presumably a large proportion of the incident ultraviolet light is likewise caught by the basal cells, to be absorbed by the ultraviolet-absorbing compounds that are apparently dissolved in the sap of these cells.

The following simple experiment is also relevant. *Lasthenia chrysostoma* (Fisch. & Mey.) Greene absorbs ultraviolet proximally in the ligules (sometimes all over), and is known to contain polyhydroxy chalcones and aurones (cf. Bohm, Chapter 26; Ornduff, 1966). These anthochlors turn red in the presence of alkali, and upon treatment with KOH the absorbing ligule area in a partially absorbing individual all turns red, indicating, as might be expected, that the alkali-sensitive anthochlors are responsible for the ultraviolet absorption in *L. chrysostoma*. Transverse sections taken from the proximal part of the ligules, placed in water and examined in the light microscope while KOH was added, revealed that the red reaction appeared exclusively in the upper epidermal cells.

Differences in ultraviolet absorbance in the ligules are, however, not always correlated with differences in cell-shape. *Calendula* species for example, absorb some ultraviolet light all over the ligules, but some of them absorb more strongly in the proximal than in the distal part, and in these species there is no difference in the epidermal cells of the ligules. On the other hand some helianthoid species without detectable colour differences still show a differentiation into taller and shorter cells. This difference in cell-height may probably cause a slight difference in saturation of colour across the ligule detectable to an insect's eye.

The importance in terms of pollination biology of the colour patterns concerned is partly explained by Kugler (1955), who notes that bumblebees (*Bombus* spp.) prefer a yellow disc with an orange centre to a uniformly yellow one. There is, in other words, an advantage *per se* in having heterochromatic heads, and with this come other advantages. It has repeatedly been shown that insects, after alighting on a flower, turn in their search for food to places of a more saturated colour (Kugler, 1955). In a flower head of the type described, insects will therefore orientate themselves towards the centre. Kevan (1972) referred to the discriminative value of ultraviolet patterns in the high arctic flora, and in the Compositae ligules of *Rudbeckia* spp., for example, have evolved at least six different patterns ranging from completely ultraviolet-reflecting to one third reflecting, one third absorbing, and one third purple.

Burtt (1961) pointed out that "larger capitula because they are more conspicuous and produce more seed, will carry a natural advantage unless, or until, they are exposed to selection pressures which can be evaded by a decrease in size". The decreased ability of some insects to find the centre of a disc above a certain size represents such a selection pressure. Knoll has demonstrated in the moth *Macroglossa* (Sphingidae) that the critical size of a single attraction unit (although a square) was a side length of *c*. 30 mm

(Faegri and van der Pijl, 1966). Many Compositae heads are just about 30 mm in diameter, but many are much broader. In the larger heads the effect of the above-mentioned selection pressure may be diminished or even annihilated by an accentuation of the centre by differently coloured rings.

SOME MODIFICATIONS, THEIR POSSIBLE EXPLANATIONS AND IMPLICATIONS

The genus *Pectis* provides some valuable information on what may happen functionally and anatomically when the Compositae become subject to physical selection pressures against large capitula. *Pectis* comprises an array of different habit-types ranging from perennial, relatively broadleaved herbs with few, large, yellow-rayed heads at one extreme, to slender annuals with filiform leaves and many small, minutely rayed, white or purple heads in capitulescences at the other. The latter may be considered advanced in relation to the former, and may have evolved their habit in correlation with increasing aridity (J. L. Strother, pers. comm. and Chapter 27). Some of the perennial, large-headed species have helianthoid type of ligule epidermis, and the papillose cells have septa along the radial and tangential walls (cf. Plate 2e, f). Some have colour patterns (ultraviolet or visual), and present a differentiation of cells in the ligules, parallel to that mentioned for *Ursinia* and *Lasthenia*; these species have helianthoid, septate cells proximally, but senecionoid, septate cells distally. In some species with less conspicuous yellow rays the epidermis is helianthoid, in some senecionoid, and in all the small-headed species included here the epidermis is of senecionoid type. Similar differences in epidermis-type are not correlated with the differences in ligule-size in for example *Tagetes*, and are seemingly a result of the special selection pressure to which the *Pectis* species have been subject. The papillose, helianthoid epidermis has a much larger surface area than the senecionoid type, and is consequently comparatively disadvantageous in terms of water economy.

Through their adaptation to increasing aridity the derived *Pectis* species have, in other words, lost that complex of characters which, both at a distance and at close quarters, works as a stimulus to many primarily visual insects, and their success has probably partly depended upon their ability to compensate for this loss. Stebbins (1974) mentions the advantage under arid conditions of an elaborate (in this case secondary) inflorescence, whose different members mature over a long period of time. However, such a secondary inflorescence may apparently also have advantages in terms of pollination biology when other selection pressures induce a reduction in size and number of the rays. Experiments have shown that *Bombus* species are more attracted by the shape of a star than by that of a single circle of the same area, and that the attractive value of the star decreases progressively with a decrease in its number of arms. Moreover, it has been shown that bumblebees prefer the complex form pattern of several small dots to that of one big circle of the same area (Kugler, 1955). As far as visual stimuli are concerned it seems, in other words, that with *Pectis* one form-pattern of a high attractive value has been substituted for another. However, David Keil has noted a loss of self-incompatibility with reduction of ligule size in *Pectis* (*fide* J. L. Strother, pers. comm.), and although this seems to stress the importance of the ligules in the pollination biology of the heads, it does also indicate that more complicated factors than those mentioned here may be related to their reduction in size.

In *Melanthera biflora* (L.) Wild the upper epidermal cells are not truly helianthoid, but rather oblong and low, approaching the senecionoid type. This may be caused by the selection pressures to which *M. biflora* is subject in its exposed habitat close to the sea, where the desiccation by insolation combines with that of the salinity in the air.

ANATOMICAL CONSIDERATIONS

Although the shape of an epidermal cell is often dependent upon such ontogenetical factors as growth rate (Esau, 1969), these factors cannot explain the constant occurrence of highly specialized ligule epidermis-types within different taxa. The specialized epidermis-types must be interpreted as genetically fixed functional adaptations rather than as ontogenetically developed structural eventualities.

The most important specializations distinguishing the cells of a helianthoid epidermis from those of a senecionoid are increased height and decreased length : width ratio. Those distinguishing the mutisioid are longitudinal differentiation of the cuticle pattern in the individual cells, longitudinal connection of the cuticle patterns of adjacent cells, thickening of the outer wall, and increased length: width ratio. These specializations are supplemented by those in pigment content.

Figure 3 shows that, within the epidermis-types described here, increased cell height is correlated with increased undulation of anticlinal walls, and decreased length : width ratio, and that the highest cells have septa along these walls. As mentioned (p. 125), the differences in ligule size within *Pectis* are correlated with differences in epidermal cell shape, while the differences in cell shape seem to be correlated with different stages of septum-formation. The distal tabular cells in *P. canescens* H.B.K. have septate walls, as do those of the senecionoid epidermis in *P. leptocephala* (Cass.) Urb., but in the very small, thick-walled cells of *P. linifolia* L. the septae are absent.

It is not the aim of this study to decide what factors might influence the evolution of shape in the epidermal cell, but the correlations indicated in Fig. 3 do merit some interpretation. A cell's need of nutrients is

7. MICROCHARACTERS IN THE LIGULES

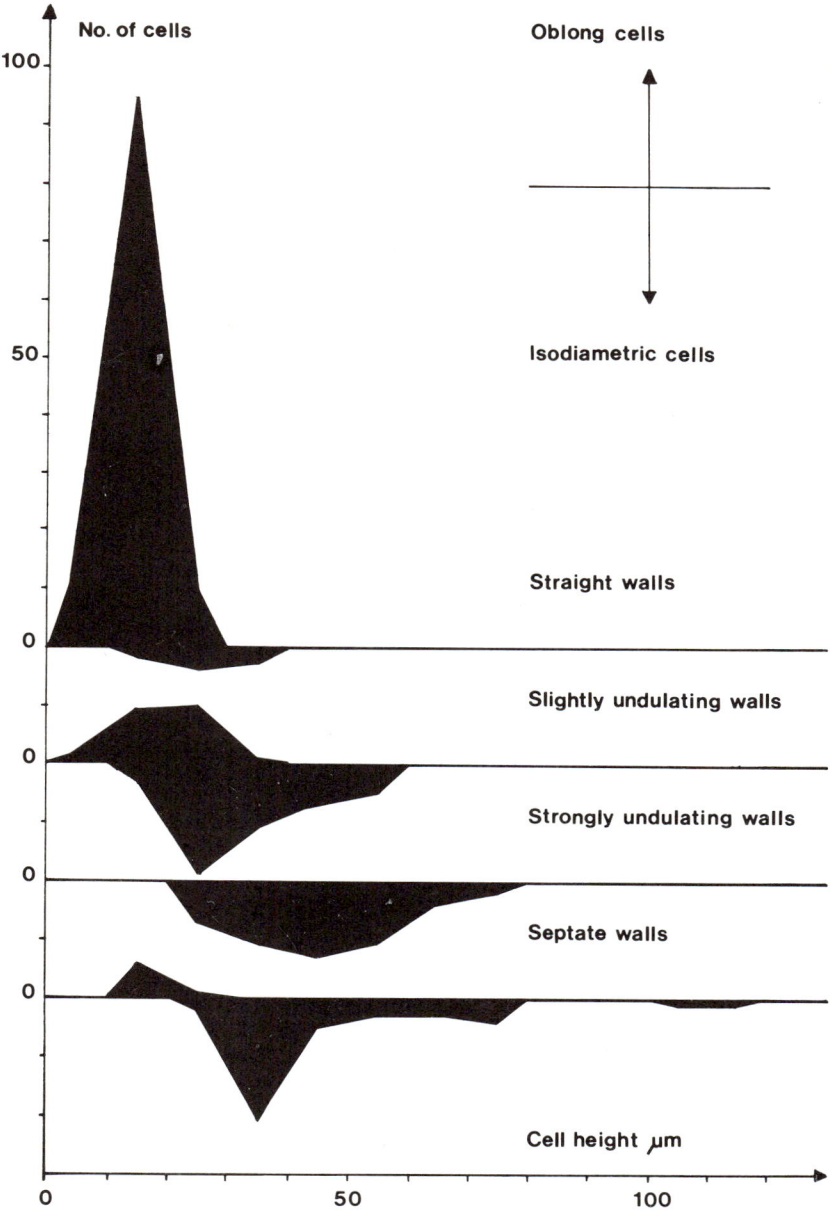

FIG. 3. Correlation between increasing cell height, decreasing length:width ratio and increasing undulation of radial and tangential walls in cells from the adaxial epidermis of Compositae ligules.

proportional to its volume, while its ability to take up nutrients is proportional to its surface area. Consequently, what limits a cell's size is primarily the ratio between its surface area and its volume (Clowes and Juniper, 1968). Greater cell size may be allowed for by polar growth, for with an approximately constant cross-sectional area, both surface area and volume will increase in proportion to length. In the papillose epidermis cells studied here the isodiametry in surface view will ensure a profitable ratio between surface area and volume. However, with increasing height the papilla exposes an increasingly larger surface area to the outside world, and a relatively smaller area to the neighbouring cells. Even if the metabolic activities of corolla epidermis cells are low, compared with those of other cells, the increase in height will, therefore, ultimately be limited by a nourishment deficiency in the cell. However, undulation of the anticlinal walls will increase their surface area and postpone the occurrence of such a deficiency, thereby allowing for a greater height of the papilla. Esau (1969) stated that the septa of epidermal cells consist of two wall-layers held together by intercellular matrix, and it seems reasonable to assume that in this material the increasing undulation of the walls may ultimately lead to the formation of septa by collapse of the sinuses. The septa provide the cell both with a large inner (membrane) surface, and with a large wall-area, and this is important since the nutrients are transported in the cell walls. It seems, therefore, that the correlation depicted in Fig. 3 may be taken as an expression in the shape of the epidermal cell of the interaction between at least two selection pressures working on the cells, namely pollination biology and physiology.

Linsbauer (1930) mentioned that, in leaves, undulating anticlinal walls are mostly found in epidermis cells facing, on their inner side, onto a mesophyll with many intercellular spaces. In such epidermises the inner cell walls are in direct contact with mesophyll cells over a very limited area only, and the cells may largely depend upon their anticlinal walls for exchange of nutrients. The occurrence of undulating anticlinal walls in these cells might, therefore, likewise be explained as a means of counteracting nutrient deficiency.

Finally it cannot completely be excluded that both undulations and septa have an additional mechanical function at least in the papillae where the cell walls are thin. This possibility is supported by the fact that the septa, as indicated by Plate 2e, extend slightly under the outer wall in the papillae.

Altogether it does not seem unreasonable to assume that in the material studied here, a papillose cell with strongly undulating or septate walls may be considered derived as compared to one with slightly undulating walls.

Apparent examples of incipient septum formation may be found. Several *Lasthenia* species have undulating walls in which the sinuses collapse towards the cell lumen. When viewed from above such walls are undulating with low sinuses, each of which ends in a two-layered point. In this connection it is interesting to note that the ultraviolet colour pattern in *L. chrysostoma* is inconstant. In *Helianthella*, on the other hand, the isodiametric cells show small points here and there along their straight walls, presumably representing reduced septa.

Esau (1969) mentions the frequent occurrence of oblong, tabular cells proximally in papillose petals. Such proximal cells are also found in the ligules of some composites such as *Helenium* and *Silphium* species. Probably these cells represent the original type from which the papillae have developed, but direct proof of this assumption is not available.

The samples described in *Pectis* indicate that the evolution from tabular to papillose cells may be reversible, and that the epidermises which are here for functional reasons collectively described a senecionoid may either be rather unspecialized or highly derived. This fact should always be taken into account if phylogenetic or taxonomic conclusions are to be drawn from the occurrence of epidermis-types with tabular cells.

Usually, however, the unspecialized type may apparently be recognized by the irregular size and shape of the cells, which very often have tapering ends and thin, slightly domed outer walls. Absence of cells with septa or specializations in the outer wall also indicate an unspecialized epidermis. There seems to be little doubt that the Senecioneae have unspecialized epidermises that probably illustrate the situation in the ancestors of the tribes with specialized epidermises. Apart from the senecionoid papilla that seems correlated with white and purple ligule-colour, some *Senecio* and *Doronicum* species show, among their highly domed oblong cells with undulating walls, some shorter ones, which may represent an early step on the path towards formation of a papilla of the helianthoid type. It is not clear whether this should be interpreted as an expression of close relationship between these taxa and the helianthoid stock or as evidence that ligules are a recent acquisition in the Senecioneae so that more specialized epidermis-types are still under formation by parallel evolution.

That the mutisioid epidermis-types may be derived by several kinds of rather simple changes in the outer walls of tabular cells seems probable. The scattered occurrence of this epidermis-type point in the same direction, but it seems that once an advantageous epidermis-type has been acquired within a group, it tends to be preserved, and some taxonomic conclusions may therefore also be drawn from the presence of the mutisioid type, though usually at a low taxonomic level.

TAXONOMIC CONSIDERATIONS

Ligule microcharacters alone cannot be used to draw major taxonomic conclusions. However, in some cases the conclusions that may be drawn from them coincide very closely with those drawn from other, better known and tested characters. Probably therefore, the microcharacters may, if used with suitable caution, help to elucidate the situation in the more problematical groups.

A simplified survey of the distribution of the epidermis-types among the tribes is given in Fig. 4. It may be seen that the Eriophyllinae and Arnicinae have been placed with the Heliantheae. Turner and Powell propose that

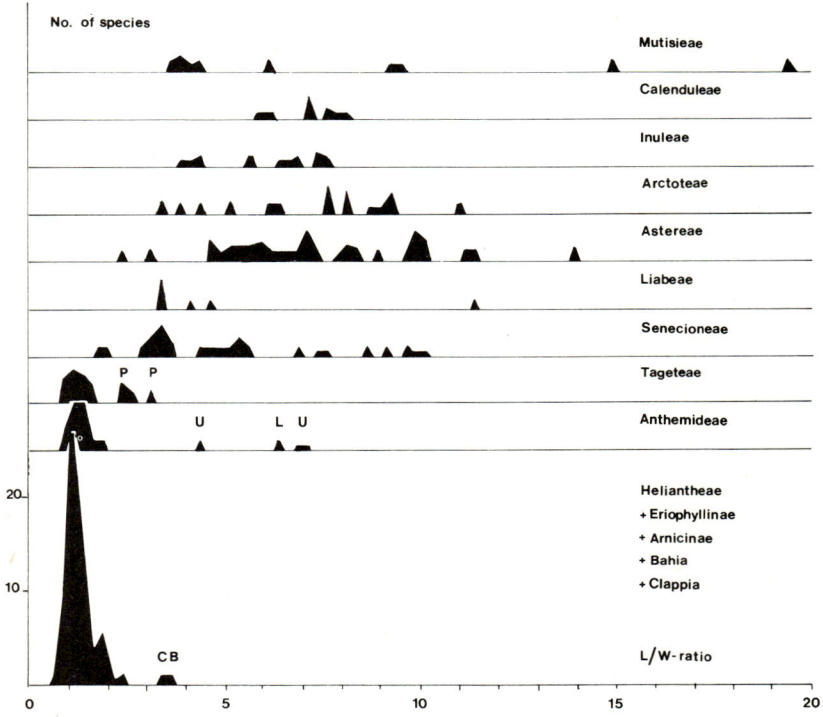

FIG. 4. Comparison of length:width ratios of cells from the distal part of Compositae ligules. B = *Bahia* sp., C = *Clappia* sp., L = *Lasiospermum* sp., P = *Pectis* sp., U = *Ursinia* sp.

they should be included in the Senecioneae (cf. Chapter 25), while Nordenstam considers them better placed with the Heliantheae (cf. Chapter 29). The evidence presented by the ligule microcharacters clearly refers these groups to the Heliantheae. *Arnica* has papillose cells, which can be referred to the helianthoid type, although they have in surface view, a squarer outline than normal. Cells with a square outline may, however, also be met with in some *Helenium* and *Psilostrophe* species.

Turner and Powell (Chapter 25) further place the Neurolaeninae and Bahiinae with the Heliantheae, but the *Clappia* and *Bahia* species studied here do not have a helianthoid epidermis. The material studied is too scanty to allow any firm conclusions, but indicates that these subtribes are intermediate between Heliantheae and Senecioneae as to epidermal characters.

Within the Heliantheae *sensu* Stuessy (Chapter 23), *Gaillardia* and *Schkuhria* deviate by having tabular cells covered by a common cuticle pattern (Plate 2b). The length : width ratios of these cells fall within the range of the helianthoid type, and they have undulating radial and tangential walls indicating that they may have a common origin with the typical helianthoid cells.

It can also be seen from Fig. 4 that the evidence provided by the microcharacters places *Ursinia* and *Lasiospermum* some distance from the Anthemideae, where they are only accepted with reluctance by Heywood and Humphries (cf. Chapter 31), and this fits in with the marked differences in their chemical profiles (see Greger, Chapter 32). *Ursinia* seems to be more naturally placed close to the Arctoteae-Calenduleae, while the relationship of *Lasiospermum* is more problematical. Judging from the material studied, the ligule-epidermis of this latter taxon strongly resembles that of the Astereae with which it has, according to F. Bohlmann (pers. comm.), nothing in common chemically. These matters will be investigated further.

In the Senecioneae the length : width ratios of around 2 are derived from the species with senecionoid papillae, and the study of more material may reveal that the microcharacters are taxonomically useful within the *Senecio* group. Material of *Ischnea* and *Abrotanella* has not been included, but *Blennosperma* and *Crocidium* are alike, and apparently epidermal characters cannot assist in separating the Blennospermatinae from the Senecioninae (cf. Turner and Powell, Chapter 25; Nordenstam, Chapter 29).

The Liabeae are unique among the tribes studied in showing structureless epidermal surfaces, and *Cacosmia* also agrees with the tribe in this aspect. Structureless epidermises are found in the corolla-lobes of the disc florets within some tribes, e.g. Arctoteae, Calenduleae, Senecioneae, and Vernonieae.

The Astereae are distinguished from other groups with a senecionoid epidermis by among other features the median thickening of the outer cell wall, and they stand out as a quite distinct and uniform group, although their cells have a rather variable length : width ratio.

Cronquist (1955) related the Inuleae to the Heliantheae through the Buphthalminae, and although the material surveyed is scanty, it is worth noticing that it has senecionoid or mutisioid, not helianthoid epidermises. A possible explanation might be that the inulean ligules are progressively losing their original function in pollination biology. The mesophyll-less, inconspicuous ligules in some *Inula* species, and the complete lack of ligules in some Inulinae support this explanation. However, the conspicuous ligules of *Buphthalmum* have mutisioid cells. *Telekia* and *Buphthalmum* may apparently be distinguished by their epidermis anatomy (cf. Merxmüller *et al.* Chapter 21).

The situation within the Calenduleae and Arctoteae is much more chaotic than Fig. 4 reveals (cf. pp. 117, 120). Among possible explanations are: (1) the mutisioid epidermis-types may have arisen several times by parallel evolution in these tribes; (2) the tribes may not be truly natural, an idea that has been advocated for example by Turner and Powell (Chapter 25). At all events both tribes comprise species or genera of either senecionoid or mutisioid type, and future investigations may reveal trends in the evolution of the mutisioid epidermis-type, which may be helpful in elucidating these problems.

The epidermis character may apparently be used to a certain extent at the generic level within the Mutisieae. *Schlechtendahlia* seems to deviate from the other genera studied by having thin outer walls.

THE LACTUCEAE

The ligule-epidermises of Lactuceae have not been adequately studied, but scanning electron microscopic spot checks in 12 genera indicate (Plate 3) that the ligule microcharacters may be taxonomically useful, though less easily applicable within the tribe. The thin, often mesophyll-less lactucaceous ligules are difficult to handle with simple light-microscopical methods and the cuticle patterns are often very fine indeed, and difficult to detect unless scanning electron microscopy is applied. None of the included genera had identical epidermises though some were alike, such as *Chondrilla* and *Cicerbita*, and it seems that several *Hieracium* species have epidermises of the type shown in Plate 3b. It may, in other words, be worthwhile for taxonomists working in this tribe to check the epidermal characters.

ACKNOWLEDGEMENTS

I am grateful to the Institute of Historic Geology and Paleontology, University of Copenhagen for placing a scanning electron microscope at my disposal; and to Professor R. Dahlgren for critically reading the manuscript.

REFERENCES

BREHM, B. G. and KRELL, D. (1975). Flavonoid localization in epidermal papillae of flower petals: A specialized adaptation for ultraviolet absorption. *Science, N.Y.* **190**, 1221–1223.

BURTT, B. L. (1961). Compositae and the study of functional evolution. *Trans. bot. Soc. Edinb.* **39**(2), 216–232.

CLOWES, F. A. L. and JUNIPER, B. E. (1968). "Plant Cells". Botanical Monographs **8**. Blackwell Scientific Publications, Oxford.

7. MICROCHARACTERS IN THE LIGULES 137

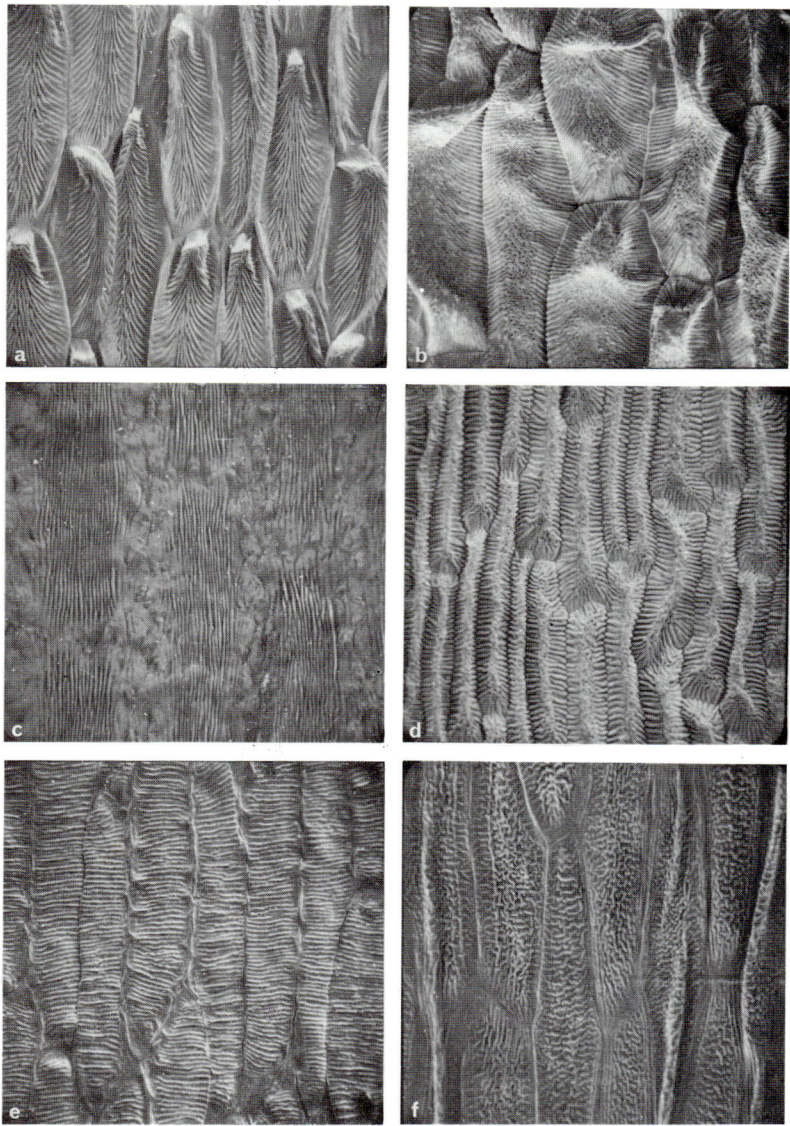

PLATE 3. Scanning electron micrographs of ligule surfaces in the Lactuceae. (a) *Leontodon* sp. ×c. 525, (b) *Hieracium* sp. ×c. 520, (c) *Cichorium* sp. ×c. 480, (d) *Scorzonera* sp. ×c. 530, (e) *Chondrilla* sp. ×c. 525, (f) *Lapsana* sp. ×c. 510.

CRONQUIST, A. (1955). Phylogeny and taxonomy of the Compositae. *Am. Midl. Nat.* **53**, 478–511.
ESAU, K. (1969). "Pflanzenanatomie". Gustav Fischer Verlag, Stuttgart.
FAEGRI, K. and PIJL, L. VAN DER. (1966) "The Principles of Pollination Ecology". Pergamon Press, Oxford.
KEVAN, P. G. (1972). Floral colors in the high Arctic with reference to insect-flower relations and pollination. *Can. J. Bot.* **50**, 2289–2316.
KING, R. M. and KRANTZ, V. E. (1975). Ultraviolet reflectance patterns in the Asteraceae. I. Local and cultivated species. *Phytologia* **31**, 66–114.
KUGLER, H. (1955). "Einführung in die Blütenökologie". Gustav Fischer Verlag, Jena.
LINSBAUER, K. (1930). "Die Epidermis". *In* "Handbuch der Pflanzenanatomie", 1. Abt. 2. Teil: Histologie. Bd. IV. Gebr. Borntraeger, Berlin,
ORNDUFF, R. (1966). A biosystematic survey of the goldfield genus *Lasthenia*. (Compositae: Helenieae). *Univ. Calif. Publs. Bot.* **40**, 1–92.
PROCTOR, M. and YEO, P. (1973). "The Pollination of Flowers". Collins, London (New Naturalist Series).

APPENDIX—LIST OF GENERA STUDIED

Achillea L.
Amellus L.
Arctotis L. sensu lato
Arnica L.
Aspilia Thouars
Aster L.
Asteriscus Mill.
Bahia Lag.
Baileya Harv. & A. Gray
Bellis L.
Berkheya Ehrh.
Bidens L.
Blennosperma Less.
Brachycome Cass.
Buphthalmum L.
Cacosmia H.B.K.
Calendula L.
Callistephus Cass.
Castalis Cass.
Catananche L.
Chaetopappa DC.
Chondrilla L.
Chrysactinia A. Gray
Chrysanthemoides Tourn. ex Medik.
Chrysanthemum L.
Cicerbita Wallr.
Cichorium L.
Cladanthus Cass.
Clappia A. Gray
Coreopsis L.
Cosmos Cav.
Crocidium Hook.
Dahlia Cav.
Dendranthema (DC.) Desmoul.
Dimorphotheca Vaill. ex Mnch
Doronicum L.
Dyssodia Cav.
Erigeron L.
Eriophyllum Lag.
Euryops Cass.
Felicia Cass.
Gaillardia Foug.
Galinsoga Ruiz & Pav.
Garuleum Cass.
Gazania Grtn.
Geigeria Griessel.
Gerbera L. ex Cass.
Gibbaria Cass.
Grindelia Cass.
Guizotia Cass.
Haplopappus Cass.
Helenium L.
Helianthella Torr. & Gray
Helianthus L.
Heliopsis Pers.
Hertia Less.
Heteranthemis Schott
Hieracium L.
Hysterionica Willd.
Inula L.
Lactuca L.
Lapsana L.
Lasiospermum Lag.
Lasthenia Cass. emend. Ornd.
Laiya Hook. & Arn.
Leucanthemum Mill.
Leontodon L.
Liabum Adans.
Ligularia Cass.
Lindheimera A. Gray & Engelm.
Madia Mol.
Matricaria L.
Melampodium L.
Melanthera Rohr
Montanoa La Llave & Lexarza
Moscharia Ruiz & Pav.
Mutisia L.f.
Nassauvia Comm. ex Juss.
Olearia Mnch
Onoseris Willd.
Osteospermum L.
Othonna L.

7. MICROCHARACTERS IN THE LIGULES

Pallenis Cass.
Pectis L.
Perezia Lag.
Polymnia L.
Prenanthes L.
Psilostrophe DC.
Pulicaria Grtn.
Rudbeckia L.
Sanvitalia Gualt. in Lam.
Schkuhria Roth
Schlechtendahlia Less.
Scolymus L.
Scorzonera L.
Senecio L.
Silphium L.
Solidago L.
Spilanthes Jacq.
Steirodiscus Less.
Tagetes L.
Taraxacum Weber
Telekia Baumg.
Tithonia Desf. ex Juss.
Tridax L.
Trixis P. Browne
Tussilago L.
Ursinia Grtn.
Xanthisma DC.
Zaluzania Pers.
Zinnia L.

Chapter 8

Pollen morphology in the Compositae and in morphologically related families

JOHN J. SKVARLA

Department of Botany and Microbiology, University of Oklahoma, Norman, U.S.A.

B. L. TURNER

Department of Botany, University of Texas, Austin, Texas, U.S.A.

VARSHA C. PATEL

Department of Botany and Microbiology, University of Oklahoma, Norman, U.S.A.

and

A. SPENCER TOMB

Department of Biology, Kansas State University, Manhattan, Kansas, U.S.A.

Abstract. Compositae pollen representing every tribe was examined by transmission electron microscopy (TEM) in order to identify the major structural patterns which occur in the family. Four were recognized: Helianthoid, Senecioid, Arctotoid and Anthemoid. These generally were restricted to a given tribe but frequently more than one pattern was noted in a tribe. The Helianthoid pattern was present exclusively in the Heliantheae, Eupatorieae, Astereae, Helenieae, and Calenduleae. It was also present in some taxa of the Inuleae, Senecioneae, Cynareae, and Anthemideae. In some of these tribes the Helianthoid structure correlated with other evidence to suggest taxonomic realignments. The Senecioid pattern characterized taxa of the Inuleae and Senecioneae. The Arctotoid structure was found in some Arctotideae and the Anthemoid occurred in the Anthemideae, Cynareae, some Inuleae and Mutisieae. Variations of these structural patterns were common: the Vernonieae and the genus *Liabum* were identical and both were considered to have a modified Anthemoid pattern; the Cichorieae was interpreted as incorporating Senecioid and Anthemoid patterns. Some genera, such as *Barnadesia*, *Dasyphyllum* and *Schlechtendalia* (Mutisieae), *Cullumia* and *Gazania* (Arctotideae) and *Catanache* (Cichorieae) were not assigned structural patterns. A limited number of scanning electron micrographs for a few genera of each tribe indicated a variety of sculptured surfaces. Internal structural patterns could not be predicted from examination of these surfaces. However, it appears that scanning electron microscopy (SEM) will be useful in distinguishing some taxa which have notable ridged-lacunar (i.e. echinolophate) morphology. This would include Vernonieae, Cichorieae, as well as the previously mentioned *Barnadesia, Cullumia* and *Gazania*. A brief survey was conducted of families which have been suggested as showing relationships to the Compositae. The Calyceraceae, with an Anthemoid pollen pattern, were indicated to be closest to the Compositae. The Valerianaceae, some Dipsacaceae, Umbelliferae, Goodeniaceae, Brunoniaceae, and possibly a few Caprifoliaceae also appeared to express the Anthemoid structural pattern to some degree. The Rubiaceae and Campanulaceae (including *Lobelia*) were found to have pollen characters unlike those of the Compositae.

CONTENTS

Introduction	143
Materials and methods	144
General pollen wall morphology of Compositae	144
Pollen wall units at ultrastructural level	147
Columellae	147
Tectum	149
Spines	151
Foot layer	153
Endexine	153
Exine patterns at ultrastructural level	154
Summary	158
Tribal considerations	158
Vernonieae	159
Eupatorieae	160
Astereae	161
Inuleae	161
Heliantheae	168
Helenieae	170
Anthemideae	171
Senecioneae	172
Calenduleae	174
Arctotideae	175
Cynareae	177
Mutisieae	178
Cichorieae	181
Related families	183
Background review and rationale	183
Cronquist system	183
Takhtajan system	184
Hutchinson system	184
Thorne system	185
Review of pollen morphology	185
Review of phytochemistry	186
Electron microscopy of related families	186
Calyceraceae	187

8. POLLEN MORPHOLOGY

Valerianaceae	188
Dipsacaceae	188
Campanulaceae	188
Brunoniaceae	189
Goodeniaceae	189
Caprifoliaceae	189
Rubiaceae	190
Umbelliferae	190
Summary of related families	191
Concluding remarks	192
Acknowledgements	193
References	193
Appendix by G. Thanikaimoni	249

INTRODUCTION

The Compositae, with more than 20 000 species, would appear to be a prime family for palynological investigation. Although numerous studies exist at the light microscope level, starting with the pollen wall interpretations of Fischer (1890), continuing with the classic investigations of Wodehouse in the 1930s and 1940s and culminating with a superb family conspectus by Stix (1960), modern day palynologists have little utilized the tools of electron microscopy (transmission and scanning) to examine and interpret the pollen of this family. Electron microscope studies are particularly crucial because of the remarkable similarity of so many taxa when viewed by light microscopy. A general statement describing the pollen of the family is that "all tribes have a similar morphology with the exception of the Vernonieae and Cichorieae". Of course, it is obvious that such a statement is far too broad because those familiar with Compositae pollen would be quick to point out numerous exceptions. Indeed, the late Professor Erdtman would have disagreed since he considered the family to be eurypalynous (i.e. showing wide diversity of morphology). Still, considerable similarity exists between such tribes as the Eupatorieae, Astereae, Inuleae, Heliantheae, Helenieae, Anthemideae, Senecioneae, Arctotideae, Calenduleae, Cynareae and Mutisieae. In the sense of Erdtman, the Vernonieae and Cichorieae are the only truly eurypalynous tribes. Wodehouse (1928b, p. 449) astutely pointed out that the "... usefulness [of the pollen] is limited only by the number of characters observable . . ." and it is with the more highly refined tools of the scanning and transmission electron microscopes that the number of characters is increased.

In this report we will attempt to provide an outline of the basic morphological patterns of Compositae pollen from a structural viewpoint based essentially on the use of transmission electron microscopy (TEM). We intend to do this by documenting by way of electron micrographs, taxa representing every tribe. The classification system we have followed is that of Bentham (1873) and, needless to say, considerable taxonomic revision has occurred in the 102 years following this work. Indeed, some of our data will reflect such revision. In order to provide a foundation for the interpretation of the structural morphology we have included some scanning electron micrographs. We will therefore present, to a limited extent, a spectrum of sculpturing or exomorphic patterns for the family.

A second part of this study will attempt to examine pollen from families of angiosperms which have been considered as showing affinities with the Compositae. This study is highly superficial at the present time and is intended only to serve as a starting point for our interest in ascertaining ancestral pollen relationships of the Compositae.

MATERIALS AND METHODS

Pollen for this study has been obtained from numerous herbaria. Table I (see end of chapter) lists the taxa examined and all relevant collecting data. After the removal of mature buds from herbarium sheets the pollen was acetolyzed by boiling in a mixture of one part concentrated H_2SO_4 and nine parts of acetic anhydride (Erdtman, 1960). Processing for TEM followed previously described methods (Skvarla, 1966, 1973). Examination and photography was done with a Philips model 200 transmission electron microscope. One sample, *Haplocarpa scaposa*, was rehydrated according to the method of Rowley and Nilsson (1972).

Acetolyzed pollen was examined with a Jeol JSM-2 scanning electron microscope. Preparatory treatment consisted of critical point drying in a Pelco model critical point dryer and rotational coating with gold-palladium.

GENERAL POLLEN WALL MORPHOLOGY OF COMPOSITAE

Morphology of Compositae pollen became best known through the investigations of Wodehouse (1926; 1928a, b, c; 1929a, b, c; 1930; 1931; 1935). These studies were directed primarily at describing the surface patterns (exomorphology) occurring throughout the family. Accordingly, three major patterns were recognized: (1) psilate — having an almost smooth, relatively unadorned surface, (2) echinate — with conspicuous spines, and (3) lophate — with the surface consisting of ridges and depressions (lacunae) that anastomose or are free and with a nearly always radio-symmetrical-hexagonal arrangement (Wodehouse, 1928c, 1935). The lophate pollen grains were grouped further into two major categories: echinolophate — having spines extending from the ridges, and psilolophate — without spines on the ridges. Most of these terms were further prefixed with "sub" in order to indicate transitional pollen grains. These patterns of surface morphology are illustrated by scanning electron micrographs in the present investigation, as for example: psilate (Plates 22A; 23A), echinate (Plates 2B, C; 3B; 4A; 5A, B: 7A, F; 8B, C; 9A, B; 12A; 13A; 15A, B; 17A, B; 19A; 20A; 21A), echinolophate (Plates 1A–E; 24A, I; 25A, B), psilophate (Plates 19C; 22E) and the various "sub" categories (Plates 3C; 5G; 19F). In addition to these sculpturing patterns Wodehouse showed that the Compositae pollen surface characteristically possessed three germinal apertures. Each was surrounded by a groove or fold, termed a germinal furrow. The furrows were considered to aid the pollen grain in response to volume changes due to conditions of moisture or aridity. In this "functional" sense the germinal furrow was considered as an "organ" or "mechanism" and was termed a harmomegathus (Wodehouse, 1935). The subject of harmomegathy, neglected since the time of Wodehouse, has received new impetus through the work of Payne (1972) whereby harmomegathal action was indicated as serving important roles for both the evolution of the pollen surface and the internal structure. Wodehouse applied the term colpate to the entire aperture mechanism (i.e. furrow and included germinal pore), using prefixes to signify the number present on a pollen grain. In more recent times the term colpate has been amended and the correct apertural term for Compositae is more accurately defined as colporate, that is, to signify a furrow (i.e. colpus) containing a definite germinal opening (i.e. pore). The publications of Erdtman (1963, 1969, 1971) should be consulted for detailed explanations of aperture terms.

The above description refers to the surface or sculpturing features of the Compositae pollen wall and for purposes of morphological description, must be distinguished from internal, structural features. The significance of recognizing these differences has been discussed by Walker (1974). The internal structure of the Compositae pollen wall, like the overwhelming number of angiosperm pollens grains, consists of two layers which are different in morphology and chemistry. The first layer, the intine, is composed of polysaccharides (pectins, cellulose) and enzymatic proteins which are mainly hydrolytic (Knox and Heslop-Harrison, 1969, 1970). The intine is the layer which immediately encircles the pollen cytoplasm, it is either uni- or bilayered, presumably reflecting different polysaccharide constituents. The second pollen wall layer, the exine, broadly defined as being composed of oxidative polymers of carotenoids and/or carotenoid esters (Shaw, 1971), consists of four basic units. Terminology applied to these units has been the subject of unending debate, as reviews by Wittman and Walker (1965), Van Campo *et al.* (1967), Manten (1970), Walker (1974) and Erdtman (1966, 1970) readily attest. We see no purpose in reviewing this subject again and only wish to direct attention to it as it applies to Compositae. This will be done by describing the basic features of the internal structure of the Compositae pollen grain wall as viewed with TEM. It should be stressed that the acetolysis treatment of the pollen (see Materials and Methods) removes the intine, so that in a correct sense it is only the exine (in deference to *pollen wall* which includes the intine) which is analyzed.

TEM shows the acetolyzed exine to consist of two major stratification layers (including four units) which can be identified by electron density. The layer which is less dense to electrons is the endexine and it is the most basal layer. (In unacetolyzed pollen, such as illustrated by *Haplocarpha scaposa*, Plate 20E, the endexine surrounds the intine.) The second exine layer, which is more dense to electrons, is located above the endexine and is termed the ektexine. It is composed of three distinct units: (1) a layer blanketing the endexine, the foot layer, (2) a series of beam- or pillar-like structures that are either partially or entirely attached to the foot layer, the

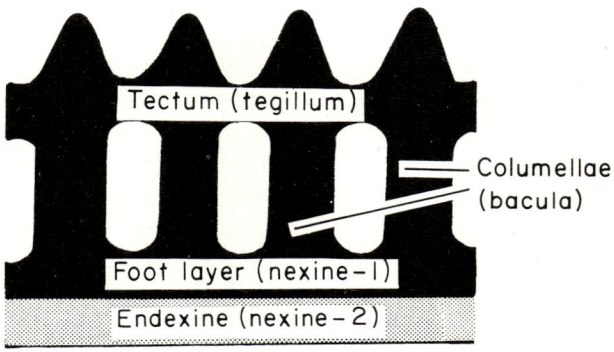

FIG. 1. Comparison of exine terminological systems of Faegri and Erdtman. The terms without parentheses are those of Faegri; terms within parentheses are those of Erdtman. Note that there is a discrepancy in these systems: (1) Faegri considers the ektexine to be composed of tectum, columellae and foot layer; Erdtman considers the sexine (= ektexine) to be composed of tegillum (= tectum) and bacula (= columellae); (2) the endexine of Faegri is unilayered whereas Erdtman interprets the nexine (= endexine) to be composed of nexine-1 (= foot layer) and nexine-2 (= endexine).

columellae (bacula), and (3) a roof-like covering of the columellae, the tectum. This terminology (i.e. ektexine and endexine) is based on chemical or electron-staining properties and is adopted from Faegri (1956). It contrasts with the morphological system of Erdtman (1952) which was employed by Stix (1960) in her survey of the family. Since Erdtman's system is used by many workers in the Compositae a brief comparison of the systems of Faegri and Erdtman is herein provided (Fig. 1). It should be noted that in place of ektexine and endexine Erdtman uses sexine and nexine. However, these two sets of terms are not equivalent. The sexine includes the tectum (and spines) and columellae; the nexine includes the foot layer and endexine.

8. POLLEN MORPHOLOGY

POLLEN WALL UNITS AT ULTRASTRUCTURAL LEVEL

In analyzing the internal morphology of Compositae pollen the ektexine is the most pleomorphic stratification layer. Of the three units which form the ektexine (i.e. tectum, columellae and foot layer) the columellae exhibit the greatest variability. This is particularly evident with respect to (1) continuity or discontinuity with the foot layer, (2) form, and (3) internal organization. These categories are interdependent; while they will be given individual discussion below, the interdependency is emphasized.

Columellae

Continuity or discontinuity with foot layer: caveate vs non-caveate exines. Many exines have columellae joined to the foot layer only at the periphery or margins of the apertures. In regions between the apertures (i.e. interapertural, mesocolpal or mesoporal regions) columellae are physically separated from the foot layer and the resultant opening or space is known as a cavus (Skvarla and Larson, 1965a). Since three apertures are typical of Compositae pollen three cavus areas are present and they impart the familiar bladdered appearance commonly observed by light microscopy. These pollen grains are designated as "caveate exines". Figure 2 depicts this interpretation. In contrast, other Compositae exines have columellae which are not separated from the foot layer. In this case the pollen wall is considered as "non-caveate".

Form. Remarkable variability of columellae form is evident in both caveate and non-caveate exines. In caveate exines the columellae display considerable variation at their basal extremities. This is expressed by the relationships of the lateral branches of the adjacent columellae. These branches, which are interpreted as a disjunct part of the foot layer, can be (1) uniform

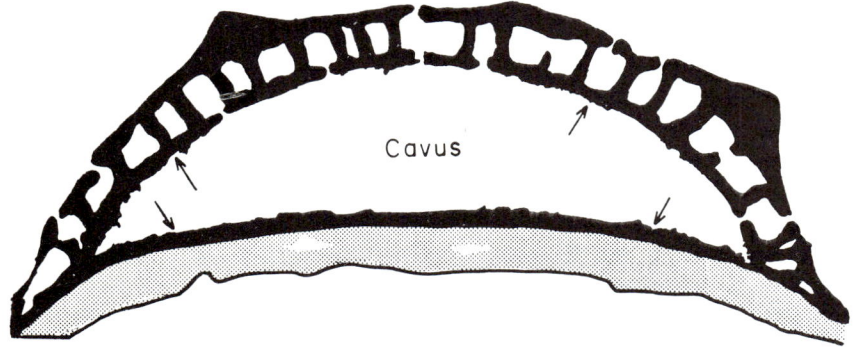

FIG. 2. Morphology of caveate exines. The arrows indicate a separation of foot layer by the cavus in area between apertures.

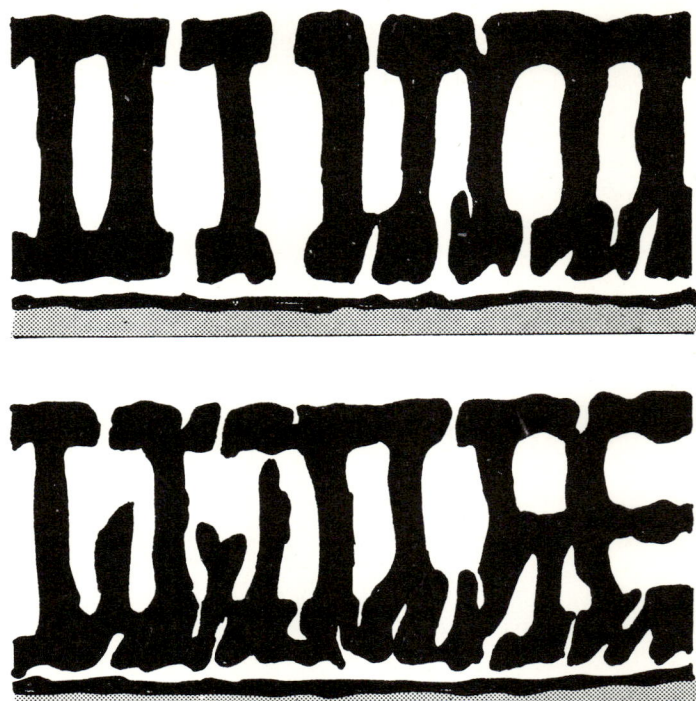

Fig. 3. Columellae variation in caveate exines. The upper diagram shows columellae bases varying from smooth, uniform connections (at left) to disrupted and more irregular ones (center and to the right). The lower diagram depicts highly ramified columellae bases; note "sub-columellae" (left of center) and internal tectum (extreme right). Both diagrams incorporate morphology of a spectrum of exines.

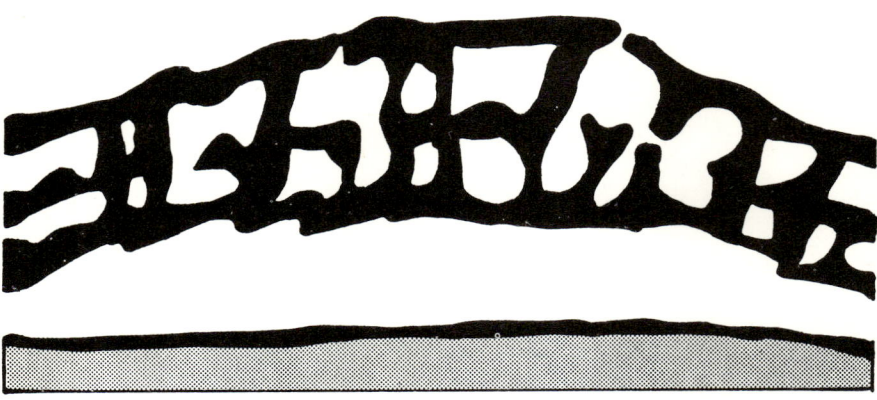

Fig. 4. Caveate exines with an internal tectum.

and continuous between adjacent columellae, (2) irregular and discontinuous, (3) ramified so that they form "sub-columellae" extending to nearly half the length of normal columellae, (4) ramified to form internal lateral branches, or (5) various combinations of the aforementioned forms. The two diagrams in Fig. 3 illustrate these features of columellae bases.

Although the columellae of caveate exines most commonly have lateral branches only at the distal and proximal ends (forming respectively the tectum and columellae bases) there are exines with columellae possessing delicate lateral branches at or near their centers. These branches unite with those of adjacent columellae and form an "internal tectum" or, according to the terminology of Stix (1960) a "supporting membrane". This is depicted in Fig. 4.

In non-caveate exines columellae are even more variable and complex than in caveate exines and a variety of internal tecta levels is evident (Fig. 5).

Exceptions to a multiplicity of internal tecta levels are (1) columellae which have a highly anastomosing net-like tectum as a result of complex distal branching, and (2) columellae which appear branched throughout so that there are no recognized levels of alternating internal tecta and columellae. These are sketched also in Fig. 5.

Internal organization. The columellae of caveate exines have been observed to be either solid or with perforations termed internal foramina (Skvarla and Larson, 1965a). These internal foramina vary in shape from circular to ellipsoidal, and in size. Internal foramina frequently occur also in the tectum and less frequently in the foot layer. In contrast, internal foramina are absent from non-caveate exines. Figure 6 illustrates these variations.

Tectum

The tectum, formed by distally branching columellae, is a more stable morphologic unit than the columellae. Tectum morphology (in combination with apertural morphology) is doubtless the major pollen wall feature utilized by the majority of investigators employing light microscopy (LM) or SEM. An integral part of the tectum are the spines which vary from long, sharp excrescences to minute granular knobs. The spines may be smooth transgressions of the surface or they may arise from bases which are highly distended. Perforations are common at the spine bases and on tectal surfaces between spines. These perforations appear to show quantitative variation among genera (Heslop-Harrison, 1969; Jones, 1970; Wells, 1971; Tomb et al., 1974). In those exines which are lophate (most Vernonieae and Cichorieae, and a few Arctotideae and Mutisieae) the tectum is composed actually of two types: (1) high ridges forming geometrical patterns, and (2) the lacunar or depressed areas enclosed by the ridges.

Fig. 5. Columellae variation in non-caveate exines. The sketches depict the great variation of columellae in non-caveate exines. Notable in the top four diagrams are the massive columellae which arise from the foot layer. The bottom diagram, which incorporates transitional ektexine forms (from left to right) resembles the top layer of the exine diagram immediately above.

FIG. 6. Internal organization of columellae. The diagrams illustrate the variation of internal foramina observed in caveate exines. Internal foramina range from large, irregular openings (top left), to elongate slits (top right), to small, roughly circular openings (bottom left). The diagram at bottom right illustrates that all exines, to a certain degree, may possess "internal foramina".

Perforations, similar to those observed in echinate grains, are less common in lophate pollen; instead, the surfaces appear "web-like".

Spines

Apart from the obvious exomorphic variability of the spines their internal organization has been a topic of considerable interest. Much taxonomic significance has been attached to two characters: (1) the presence or absence of holes or channels within the spine apices (i.e. "subapical channels" of Skvarla and Turner, 1966b), and (2) the construction of the spine bases in relation to the columellae.

Stix (1960) used the criterion of presence or absence of holes in the spine apices to help establish pollen types; Liens (1968) and Besold (1971) employed it to separate a number of closely related Inuleae taxa; and Felippe and Labouriau (1964) refined this character by including hole-number. Although subapical holes may be a reliable morphologic indicator it is our feeling that serial thin sections are required for their proper elucidation. Perusal of Plates 6E, F, 8A, 13E, 15D, 16D, 17C, 18D is illustrative in that spines on individual exines are both with and without holes. Sectioning planes obviously account for this inconsistency but it is apparent that spine-tips lacking holes are not necessarily devoid of them. This is supported by the work of Felippe and Labouriau (1964) on *Bidens gardneri*. Pollen of this taxon were shown to have spines with one hole, with two holes, and without subapical holes. We are not disqualifying this character as a taxonomic marker—certainly their presence as well as their various shapes (i.e. holes to elongate channels as indicated by Liens, 1968, 1971; Besold, 1971; and this chapter) may be of great utility—we only emphasize the need for their proper delineation so that their merits can be evaluated.

Construction of the spine-bases in relation to the columellae has been interpreted in two ways: (1) The area beneath the spine-base is devoid of columellae and is represented by an open space or large channel, (2) the spine-base is supported by columellae which can vary from the normal complement making up tectum areas to multiple and complex layers thereof. The character of large channels or cavities beneath the spine-base is very difficult to analyze. This can be put into perspective by Besold's (1971) criticisms of Stix's drawings of *Helianthus debilis* (Stix, 1960). Besold contends that the latter's diagram for the *Helianthus*-type (Fig. 16) does not correspond with her pollen diagram for the genus (Fig. A of plate I), presumably because the former does not indicate a channel beneath the spine-base, while the latter does. The problem is amplified by electron microscopy. Electron micrographs of *Galinsoga ciliata*, *Parthenium incanum*, *Viquiera dentata* (all of Heliantheae and shown in Figs 12–14 by Skvarla and Larson, 1965a), several species of *Ambrosia* (*A. artemisiifolia*, Fig. 29: *A. bidentata* Fig. 33 and *A. castanensis*, Fig. 37, by Payne and Skvarla, 1970) and different taxa in this present chapter, (Plates 3E; 10F, G; 16A–C) show, even within the same grain, presence or absence of channels beneath spine-bases.

Not to be confused with channels under the spine-bases are exines representative of the Arctotideae and Cichorieae. In these tribes large openings are evident beneath the spines (Plates 19D, E, G, I; 20D, E; 24C–H; 25D). These openings are interpreted to result, at least in part, from a compression and folding of the ektexine surface during pollen wall development.

The criterion of columellae association with the spine-bases is a more stable character than channels beneath spine apices and has been shown to be useful in taxonomy (Stix, 1960; Liens, 1971; and Besold, 1971). A number of pollen types illustrated by Stix (1960) indicate that columellae variations are of considerable importance in circumscribing members of the Senecioneae, Calenduleae, Cynareae and Arctotideae. Liens and Thyret (1971) and Besold (1971) also indicate this importance for the Arctotideae and various electron micrographs in the present study reinforce this significance.

Foot layer

The foot layer is of universal occurrence in all Compositae exines. However, as will be discussed in greater detail later, the Mutisieae appear to have some exines in which the foot layer is highly reduced. In non-caveate exines the foot layer is uniform and shows only minimal variations in thickness. Furthermore, it is of equal or greater thickness in relation to the underlying and apposed endexine. In caveate pollen we envisage the foot layer as being partially interrupted by the cavus. Accordingly, the foot layer consists of the columellae bases above the cavus and the narrow electron-dense layer immediately beneath the cavus. This conclusion is based upon similarities in electron staining and density as well as morphology. In the latter respect consideration is given to the area of the apertural mechanism (i.e. colpus): both strata are united and appear homogeneous (Fig. 2). Morphogenetic studies emphasizing the time and sequence of pollen wall deposition should help clarify such interpretations. This should be particularly instructive because if our "bifurcate foot layer" idea is correct we cannot at this time explain why, in exines proliferating in internal foramina, the foot layer commonly is without them.

Endexine

This layer, while also universal to Compositae pollen grains, frequently is not recognized (Stix, 1960; Liens, 1968, 1971; Besold, 1971; Parra and Marticorena, 1972). Similar difficulties have been recorded with TEM (Skvarla and Larson, 1965a) but are attributed to poor electron staining. In non-caveate pollen the endexine rarely exceeds the foot layer in thickness; usually it is thinner. In caveate pollen the endexine is always thicker than that part of the foot layer which is directly above it. Other work (Skvarla and Turner, 1966b) utilized this comparative thickness disparity to formulate ratios for various taxa. Although our position has changed since 1966 in that bases of the columellae (above the cavus) are now considered to be part of the foot layer, these ratio-classes still appear to be of potential use in systematics. In both caveate and non-caveate exines the upper margin of the endexine is usually smooth and indicates an even or slightly undulating boundary with the foot layer. The lower margins are variable: in non-caveate pollen they appear smooth but in caveate pollen they are usually uneven and disrupted. Some of this irregularity has been interpreted to result from processing methods (Southworth, 1966; Heslop-Harrison, 1969; Faegri and Iversen, 1975).

Exine patterns at ultrastructural level

Diversity of exine units as outlined in the preceding section, permits the recognition of several structural patterns. These are outlined below.

Helianthoid pattern. This pattern, established in earlier work (Skvarla and Turner, 1966a), is characteristic of caveate exines with internal foramina. It is the only pattern we have observed for the Heliantheae, Eupatorieae, Astereae, Helenieae and Calenduleae. It is common also in the Inuleae and Senecioneae, and less so in the Anthemideae. In the Senecioneae and Anthemideae the Helianthoid pattern appears to occur in taxa of questionable systematic position (to be discussed under tribal relationships).

Future studies at the generic and specific levels may indicate that the Helianthoid pattern is too inclusive and that "modified" Helianthoid patterns should be recognized. These might be established for exines with columellar bases that are smooth, irregular or with "subcolumellae". By drawing attention to the internal foramina still other patterns might consider size and shape (elongate, circular, etc.). The scope of the present work, which is a summary-review of the family does not allow this fragmentation and our intention is only to note that modifications in the basic Helianthoid structure may provide important systematic data. At this time it is more useful provisionally to consider all caveate exines with internal foramina as Helianthoid. Then, as in-depth generic and specific investigations are forthcoming, the Helianthoid pattern accordingly can be modified.

Senecioid pattern. The Senecioid pattern, also established in previous work (Skvarla and Turner, 1966a), is distinguished from the Helianthoid pattern only by the absence of internal foramina. It is most common in the Senecioneae (Skvarla and Turner, 1966a, b) and Ambrosieae (Payne and Skvarla, 1970) and is observed frequently in the Inuleae (Plate 5E and unpublished). Work currently in progress indicates that it may exist also in some Mutisieae and Arctotideae.

There are several factors which make the Senecioid pattern of special interest, the most prominent being that by which it is distinguished from the Helianthoid pattern, i.e. the absence of internal foramina. Examination of electron micrographs having the Senecioid pattern frequently indicates that at intermediate and high magnification the tectum and columellae display minute holes. This has been noted for exines in the Senecioneae, Ambrosieae and Inuleae. Quantification is difficult as the holes usually are obscured, partly because of the heavy metal stains employed to contrast the exines. However, we have also noted these holes in exines which did not receive any staining treatment. These holes, as indicated by Payne and Skvarla (1970), are subject to several interpretations. They may be (1) free spaces between coiled or ramified columellae, (2) artifacts of a physical, chemical or preparatory nature, or (3) internal foramina of a vestigial nature. If they are interpreted as vestigial internal foramina then the exine pattern would have to be regarded as Helianthoid. The incongruity is that the Helianthoid pattern then includes exines such as *Arnica* (Plate 13B) and *Neurolaena* (Plate 16D) with very prominent internal foramina as well as exines of *Tetradymia* (Plate 14A), *Gynoxys* (Plate 14B), *Cacalia* (Plate 14E), and *Petasites* (Plate 14I) with vestigial internal foramina only visible under high magnification. Furthermore, when visible they are appreciably less clearly outlined than the internal foramina of the "normal" Helianthoid pattern. In view of this incongruity it might be worthwhile to place a magnification limit on recognizing internal foramina. Other alternatives are classification according to size or number per unit area. We realize that such systems are purely artificial and mention them for pragmatic reasons.

To summarize, it is obvious that careful quantitative and qualitative electron microscopy must be carried out before the internal foramina can be used satisfactorily for systematic purposes. Our feeling is that no exine will be shown to be totally devoid of holes, but their relative frequency and shapes might provide data for a more detailed classification than currently exists.

Another aspect of the Senecioid pattern is that it does not occur without exception in a given genus. In a current study of the genus *Senecio* we have noted that most species, as expected, possess the Senecioid pattern. However, *S. heritieri* (Plate 13D) is decidedly Helianthoid. Similarly, earlier work (Skvarla and Turner, 1966a) showed a Helianthoid pattern for *S. glabellus*. Such specific exceptions will become clear only with more intensive investigation.

Arctotoid pattern. This pattern, not previously recognized by us, is actually a modification of the Senecioid pattern. It is distinctive in having (1) a single layer of internal tecta which almost equally divides the solid columellae, (2) highly bifurcating (disjunct) columellae bases, and (3) a prominent spine-channel outlined by columellar bases with extensions hanging freely into the cavus or attached to the foot layer. The latter character apparently was used by Stix (1960) to distinguish *Arctotis* and *Berkheya* pollen-types. The Arctotoid pattern is noted in the genera *Arctotis* (Plate 20B), *Cymbonotus* (Plate 20C), *Arctotheca* (Plate 20D), and *Haplocarpha* (Plate 20E). It does not, however, characterize *Cullumia* (Plate 19D, E) and *Berkheya* (Plate 19G–I). The latter two exceptions will be discussed elsewhere (tribal discussion).

Anthemoid pattern. This pattern, earlier recognized in the Anthemideae (Skvarla and Turner, 1966a), is characteristic of non-caveate exines. Essentially, it consists of a thick, long series of basal columellae which support shorter levels of columellae alternating with internal tecta. In the present report the Anthemoid pattern is extended to include the tribe Cynareae (Plate 21B–D) and *Glossarion rhodanthum* (Plate 23D), *Trixis angustifolia* (Plate 22B), *Onoseris odorata* (Plate 22C), and *Perezia wrightii* (Plate 22D) of the Mutisieae. Earlier work attempted to quantify the various levels of internal tecta and columellae (Skvarla and Turner, 1966b). We have not extended these observations here but believe that such refinement may be of value at the species level.

A modified Anthemoid pattern (i.e. Liabioid) is present in *Liabum*, *Cacosmia* and in the Vernonieae. It is distinguished by having much broader columellae than noted for the Anthemoid pattern. In *Liabum* the lateral branches of the distal ends of the columellae form a uniform internal tectum. From this internal tectum a short, uniform set of columellae arise which in turn are capped by a perforate tectum. Commonly, the tectum areas between the large columellae show gentle undulations. In the Vernonieae the major distinction from *Liabum* is that the lateral branches are frequently (but not always) a thick network of solid rods or tubules with complex anastomosing patterns. This modified Anthemoid pattern is of interest in that it includes echinate (*Liabum*) and lophate (Vernonieae) exines and seemingly covers the five pollen-types established for *Liabum* and the Vernonieae by Stix (1960).

Another modified Anthemoid pattern would appear to exist for *Doniophyton patagonicum* (Plate 23E). As mentioned elsewhere, the structure of the columellae in this taxon resembles the upper complex columellar structures in other Mutisieae and, pending more detailed work, we will refer to this pattern as Anthemoid.

To summarize this section, echinate pollen possesses Helianthoid, Senecioid, Arctotoid and Anthemoid patterns. These are determined by the internal structure of the exine and as such provide a means of distinguishing groups of exines which are otherwise, at least superficially, similar. The terms, as stressed previously, are artificial designations and can be applied to the pollen of any tribe of Compositae. It is possible to distinguish yet other ultrastructural patterns, particularly in lophate exines (Vernononieae, Cichorieae, members of the Arctotideae and Mutisieae). Examination of plates showing lophate exines indicates that the internal morphology may include "caveate" and "noncaveate" patterns. The former appears to parallel the lacunae and consists of solid, single level columellae resembling Senecioid pollen. The "noncaveate" pattern appears to parallel the ridges and indicates some attachment with the foot layer. Exine structure is more complex than the "caveate" pattern and superficially resembles Anthemoid pollen. In view of this, as well as the paucity of ultrastructural information on lophate pollen, we will not propose new exine patterns. As pollen studies continue, new or modified patterns will surely emerge and they should take into consideration these exomorphic subtleties of the lophate character.

Taxa without designated patterns. The above patterns have been proposed in an attempt to group those taxa which possess morphologic similarities at the fine structure level. We reemphasize that they are primarily intended for descriptive purposes; taken alone, the patterns are not meant to imply meaningful relationships. However, they may prove valuable in assessing taxonomic positions and in our discussion of the individual tribes we will make some effort to point this out. Still, a number of taxa included in this report cannot be classified according to exine pattern in any of the above groups. Therefore it seems appropriate to mention briefly those taxa which are at present considered as "patternless".

1. *Dasyphyllum-Schlechtendalia.* In these Mutisieae genera each interapertural area is characterized by a large depression which lies between and roughly parallels the three aperture systems (Plate 23A). This feature was first recognized by Wodehouse (1928b) and he coined the term intercolpar concavities. Thin sections parallel to the pollen grain equator and including the intercolpar concavity show the exine to be composed of columellae which are short, straight, solid rods directly attached to the foot layer (i.e. they are non-caveate) (Plate 23B, C). The areas bounding the concavities are different. In *Dasyphyllum* the exine is caveate with the columellae showing complex bases (Plate 23B). In *Schlechtendalia* the exine is non-caveate with the columellae showing two distinct levels of internal tecta (Plate 23C).

2. *Cullumia-Barnadesia. Cullumia* (Plate 19C) in the Arctotideae and *Barnadesia* (Plate 22E) in the Mutisieae are psilolophate, differing exomorphologically from most lophate pollen grains by the absence of spines. In thin section these exines appear to be unusual for several reasons. In *Barnadesia* the crests formed where the ridges intersect are capped by a narrow but distinct layer of sporopollenin (Plate 22F); the columellae are not clearly defined and resemble ektexine columellae of *Doniophyton* (Plate 23E); and the foot layer is granular (like the columellae and tectum) and appears to be separated from the endexine, resulting in a cavus (Plate 22F and inset) which differs from those described for Helianthoid and Senecioid patterns in that the columellar bases are not split, so that they occur on each side of the cavus. In essence, the endexine is completely free of a foot layer. In contrast, *Cullumia* (Plate 19D) exhibits a different internal morphology (described in tribal discussion). Clearly, these and related taxa are in need of additional study. For example, preliminary observations of the genus *Gazania* (Arctotideae) indicate a pollen exomorphology resembling *Cullumia* and *Barnadesia.*

Summary

An alternate system to ours is that of Stix (1960). Stix surveyed 225 Compositae taxa at the LM level and established 45 pollen-types. Of these, 11 were representative of one taxon, while the remaining 34 types, like our own terminological system, included taxa from a number of tribes. Although we are in agreement with many of these categories we have not attempted to integrate our data into her pollen-types for several reasons: (1) detection is dependent upon different instrumental approaches LM (by Stix) and TEM (by us); (2) internal foramina were not recognized by Stix: the addition of this character eliminates several of her pollen types; and (3) many of her types are based on various kinds of structure in the exine layers, and our work has not confirmed all of these variations (e.g. absence of the foot layer in certain taxa).

Our major disenchantment with Stix's pollen types centers on the unwieldiness of her system. The Liabinae, where Stix established three types, illustrates this point very well in that rapid perusal of these types does not provide the reader with any idea as to where the *Liabum* genera might be placed systematically. In comparison, our designation of Senecioid or Senecioid-like for *Liabum ovatum* and Anthemoid (modified, i.e. as occurs in the Vernonieae) for the remaining genera makes taxonomic assessments somewhat more readily comprehended. We are not criticizing the approach of Stix, as her work is classic; instead, we are offering an alternative which may be more useful in practice.

Finally, it is accepted that some workers may complain that our terminological system is oversimplified. This may be so but we feel that this is its strongest virtue: making possible the ready introduction of pollen information into the accruing stream of anatomical, morphological, taxonomic, cytological and chemical data for the Compositae, without subjecting non pollen morphologists to a plethora of palynological descriptions and terminology. These inherent dangers have been pointed out by Davis and Heywood (1963).

TRIBAL CONSIDERATIONS

The sections describing the general ultrastructural pollen morphology of Compositae should form background material for the discussion that follows. A previous review article cautioned against formulating tribal circumscriptions on the paucity of data available at the time (Skvarla and Turner, 1966b). That was 10 years ago and the message should be reiterated. However, with such a large group there has to be a point, short of examining the entire family, when palynological correlatives become obvious. With some exceptions, we believe that this report will contribute to this end. We have followed the system of Bentham (1873) which consists of 13 tribes beginning with the Vernonieae and ending with the Cichorieae.

The discussion below is intended to review previous studies, summarize and update the ultrastructural work and discuss, when relevant, the application of pollen morphology to current problems in the family. Our review of the literature is inevitably incomplete and abbreviated, and we have focused upon those papers which treat Compositae pollen within a taxonomic framework. Although we hope to have cited those studies having direct bearing on the present paper, it is inevitable that some papers must be overlooked, both intentionally and unintentionally. For a more exhaustive bibliography, attention should be given to Erdtman, 1971, Thanikaimoni (see Appendix, this chapter), *Review of Palaeobotany and Palynology* **12**, (1971) and Tralau (*Bibliography and Index to Palaeobotany and Palynology*, 1950-70).

Vernonieae

Early studies by Wodehouse (1928c, 1935) showed that pollen of this tribe was highly distinctive because of its lophate exomorphology. These pollen grains were distinguished from the lophate pollen grains of the Cichorieae by the usually greater number of lacunae (Wodehouse, 1935). It was believed that "... the nature of the lophate character when present is of the highest phylogenetic value..." (Wodehouse, 1935, p. 469). The genus *Vernonia*, in particular, was singled out as having diversity in ridge and lacunae arrangements (Wodehouse, 1928c). Other workers also have concentrated on *Vernonia*. Stix (1960) established four pollen-types on the basis of the lophate character; Jones (1970) and Kingham (1976) using fine series of scanning electron micrographs, made phylogenetic and taxonomic deductions, and Smith (1969) used lophate morphology with pollen grain size to make sectional groupings. Although the lophate condition characterizes the Vernonieae, echinate pollen, the typical condition for the *family*, also occurs. We have noted echinate morphology in *Vernonia* (unpublished work), Wodehouse (1928c) found it throughout the tribe, and Stix (1960) reported it for *Albertinia* (= *Vanillosmopsis*). The latter taxon, like *Vernonia*, contains highly variable pollen exomorphology.

Our scanning electron micrographs (Plate 1A-E) reveal a marked diversity in the lophate pattern. As depicted, the most striking morphological character is that of lacunar-ridge changes in relation to the pore-shaped germinal aperture. Contrasting with this great variability of exomorphology is the endomorphology. *Vernonia amygdalina* (Plate 1F), *Harleya oxylepis* (Plate 1G) and *V. pacchensis* (Plate 1H) correspond to our Anthemoid (Liabioid) pattern. We have not attempted to integrate this pattern with the three pollen-types—*Vernonia, Lychnophora, Elephantopus*—recognized by Stix (1960). There are two reasons for this: (1) the taxa examined by

Stix are not the same ones that have been examined by us; (2) it appears that both *Harleya* and *Vernonia* could be included in several of her types.

In summary, pollen of the Vernonieae is distinctive for the Compositae. Surface morphology has its counterparts in the Cichorieae and a few taxa in the Mutisieae (*Barnadesia*, Plate 22E) and Arctotideae (*Cullumia*, Plate 19C; *Gazania*; and *Berkheya*, Plate 19F). Internal morphology resembles the Anthemoid pattern and can be considered a modification of it (i.e. Liabioid). Furthermore, most species of *Liabum* (to be discussed under the Senecioneae) appear to have an internal morphology similar to the Vernonieae, a point of interest in the genus considered by some to be near the Vernonieae (Turner and Powell, Chapter 25). Clearly, pollen morphology is of considerable utility to the taxonomy of the tribe (Cabrera, 1944; Ueno, 1972; and above references) and expanded studies should continue to prove enlightening. The multiplicity of surface configurations is particularly amenable to examination by LM and SEM; this, along with more detailed study of the internal structure, should do much to clarify pollen relationships within the tribe.

Eupatorieae

Pollen morphology in this tribe has been studied less than in most other Compositae tribes. While several surveys of selected groups have been made (i.e. Dalmau, 1961; Heusser, 1971; Thanikaimoni, Appendix, this chapter) it is probably best known through the work of King and Robinson (1967; 1968; 1970) whereby the very natural genus *Stevia* was shown to express many pollen forms.

The taxa represented on Plates 2 and 3 are our first attempts at ultrastructural studies in the tribe. It is obvious from the scanning electron micrographs of *Carminatia tenuiflora* (Plate 2B), *Kanimia purpurescens* (Plate 2C) and *Liatris aspera* (Plate 3B) that the surface morphology is primarily echinate. However, as noted in *Decachaeta haenkana* (Plate 3C), subechinate pollen also occurs. The colpi enclosing the germinal apertures are distinctive. As is particularly evident in *Carminatia* (Plate 2B) and *Decachaeta* (Plate 3C) they are long and deeply incised. Additionally, the colpi are outlined with a thin granular exine layer. This is evident in *Kanimia* (Plate 2C) and *Liatris* (Plate 3B). The extent to which this furrow lining occurs in Compositae pollen is at present unknown. The internal structure is clearly of the Helianthoid pattern (Plate 2A, 2D, 2E, 3A, D, E). Several of these taxa are of special interest: *Carphephorus* (Plate 3E) has huge internal foramina which are commonly seen in the Heliantheae and Helenieae, and *Mikania* (Plate 2E) indicates elongate rather than circular internal foramina such as seen in the Astereae and Inuleae. Stix (1960) examined 12 members of the tribe and placed them all in her *Eupatorium* pollen-type. This latter type, according to Stix, is distinguished from all other similar types by possession of a large opening directly beneath a flattened spine base (Stix, 1960, Fig. 12). As noted earlier, our interpretation of the significance of this character differs from hers.

Astereae

This tribe has been characterized as being consistently echinate but also as having widespread abnormalities in size and colpus number (Wodehouse, 1935). Our SEM examination of a limited number of Astereae pollen grains supports this contention. *Calotis echinacea* is a good example of variation in colpus number: Plate 4A shows two apertures which appear to encircle the pollen grain. Observations of other pollen grains from the same collection indicate considerable variations in this character. These variations are presumably the result of chromosome abnormalities during maturation divisions, polyploidy, etc. TEM, as exemplified by *Calotis* (Plate 4B) *Aphanostephus skirrhobasis* (Plate 4C), *Chaetopappa asteroides* (Plate 4D) and *Boltonia diffusa* (Plate 4E), indicates the Helianthoid pattern. It is notable that the internal foramina suggest an elongate outline. Although this report and that of Skvarla and Turner (1966b) include only a small sampling, we believe that the internal morphology of the tribe is basically Helianthoid. Stix (1960) also noted a consistency of morphology and established the *Baccharis* pollen type which was distinguished from her *Eupatorium*-type by columellae beneath spine bases.

Inuleae

Pollen of the Inuleae has been the subject of extensive investigation. Wodehouse (1928c) used the rounded spine tips of *Inula helenium* to strengthen his case for the adaptation of the entire tribe to insect pollination. In more recent times light microscope work by Liens (1968, 1971) and Besold (1971) report studies on more than 800 taxa. Pollen morphology was used in conjunction with other characters, such as chromosome numbers and styler morphology to assess tribal groupings. Three subtribes were established: Inulinae, Gnaphalinae and Athrixiinae (Liens, 1971 and Chapter 21). All pollen was considered to be echinate; however, internal features were shown to differ: the Inulinae was characterized as possessing a single layer of columellae and spine-bases with "dome-like thickenings" beneath, whereas the Gnaphalinae and Athrixiinae were shown to consist of two columellae layers.

In order to integrate the work of Liens (1968, 1971) and Besold (1971) with that of our own it is necessary first to review briefly the contributions of these investigators and then to interpret it with the data from TEM. In examining the pollen diagrams presented for the Inulinae, Gnaphaliinae and Athrixiinae, it is apparent that the pollen of all subtribes have a double columellae (bacula) layer, at least in part. The Inulinae, the subtribe considered by these investigators (see Merxmüller *et al.*, Chapter 21) to have a single columellar layer, shows this for interspinal areas. Beneath the spines, however, the columellae are frequently interrupted by an internal tectum which results in two columellar layers. Unquestionably the reference to "dome-like thickenings" alludes to this feature (Chapter 21). Examination of several drawings of Inulinae members by Liens (1971) indicates that *Tessaria integrifolia* (Fig. 6), *Porphyrostemma grantii* (Fig. 7), *Epaltes tatei* (Fig. 10), *Sachsia polycephala* (Fig. 12) and *Anisopappus* (Fig. 29-31) have only single columellae supporting the spines. (The same has been noted for *Polycline* by Besold, 1971.) However, examination of *Inula eupatorioides* (Fig. 15) and *I. graveolens* (Fig. 16) suggest that an additional columellar layer may be present. *Stenachaenium campestre* (Fig. 11) is even of greater interest. This exine is depicted by Liens as having two columellar layers in regions lacking spines and a multiplicity of columellae layers beneath the spines. Therefore, it appears that exines in the Inulinae, while similar in many taxa, are also highly diverse.

Pollen in the subtribes Gnaphaliinae and Athrixiinae are interpreted by us, from the sketches of Liens (1971) and Besold (1971), to consist of a thick, slightly perforated, basal layer from which one level of prominent columellae arise. Spine-bases are shown to be supported by straight columellae at a single level.

TEM provides additional insights into pollen of this tribe. Grains of the subtribe Inulinae have the Helianthoid pattern (*Polycline proteiformis*, Plate 5D) as well as the Senecioid pattern (*Blumea mollis*. Plate 5E; *Allogopappus dichotomus*, Plate 6C, D). In reviewing the genera included in the Inulinae by Merxmüller *et al.* (Chapter 21) our electron microscopy of many of these same taxa (unpublished work) substantiates the single columellar layer. The addition of the character of internal foramina (shape, presence or absence) should provide further clarification of the Inulinae. Apart from the already discussed differences in *Polycline* and *Blumea* other examples can be cited. *Inula* (Skvarla and Turner, 1966b), *Allagopappus*, *Laggera* and *Blepharispermum* demonstrate trends of internal foramina. The first two genera are without them, *Laggera* apparently has vestiges and *Blepharispermum* has them in great abundance. Because of the complexity of organization of the columellar (bacular) bases we have not attempted to substantiate relationships to the spines. As noted in Plates 5C, 6A, B the columellar bases can be interpreted as being highly ramified or as constituting an exine layer distinct from the above extending columellae. We accept the former (i.e. ramified bases with the spines supported by one layer of columellae). It is our belief that the columellae are more complex and less clearly defined than the drawings of Liens and Besold indicate. This is not to imply that Inulinae are without a double columellar layer;

rather, that the elucidation of spine–columellae relationships require more sophisticated analyses.

Electron microscopy of exines in the Gnaphaliinae (*Pithocarpha corymbosa*, Plate 5B, C; *Helichrysum davenportii*, Plate 6B) and Athrixiinae (*Stoebe capitata*, Plate 6A) presents a striking contrast to those of the Inulinae. As is especially evident in *Stoebe* (Plate 6A) and *Helichrysum* (Plate 6B), bases of the columellae ramify so greatly that they appear as an irregular smaller set of columellae uniformly alternating with the longer columellae. This irregularity has been discussed earlier (see section on Pollen Wall Units) as "sub-columellae" and we are convinced that these are equivalent to the basal sexine layer described by Merxmüller *et al.* in Chapter 21. While in a strict morphogenetic sense such an equivalence cannot be substantiated, this presents no difficulty for purely descriptive purposes. However, interpretative problems do arise because the sub-columellae appear to be less regular than characterized by these investigators. In some exines the difference between sub-columellae bases of the Gnaphaliinae and Athrixiinae, as compared to the less complex Inulinae, are indistinguishable. For example, *Callilepsis* (unpublished work) and *Polycline* (Plate 5D) appear to have very similar columellae bases (they can be interpreted either as sub-columellae or simply as columellae with ramifying bases), yet the former is in the Athrixiinae and the latter in the Inulinae. The point we wish to emphasize is that a wide spectrum of columellar bases are present in the tribe and it is extremely difficult to be objective in classifying exines as having or not having sub-columellae (or a basal sexine layer). A case could be made for establishing an Inuloid pattern for those exines with prominent sub-columellae, and it is surprising that Stix (1960) did not do this. The pattern is so recognizable, as reference to Liens (1968, 1971), Besold (1971) and the present report shows, and Stix's work was so thorough, that we must conclude that she did not consider it as a sufficiently important character. Stix did sanction two pollen types—*Gnaphalium* and *Inula*—but a bilayered sexine is not evident in her drawings. Although we may be guided by a certain prejudice, her *Gnaphalium*-type seems to approach our interpretation of highly ramifying columellar bases. Our present reticence in designating the pattern as Inuloid is based mainly on the previous comments regarding transitional forms. Examination of exines of *Bebbia juncea* (Skvarla and Larson, 1965b), *Argyroxiphium virescens* (Skvarla and Larson, 1965a), *Craspedia richea*, *Marshallia caespitosa*, *Amblyopappus pusillus* (Skvarla and Turner, 1966b), *Alepidocline annua* (Plate 7D), *Baileya multiradiata* (Plate 9D), *Whitneya dealbata* (Plate 10C), *Amblyolepis setigera* (Plate 11C), *Ceratogyne obionoides* (Plate 12H), *Gynoxys parvifolia* (Plate 14B) and *Tussilago farfara* (Plate 14G) emphasize this reticence since they, to various degrees, could be interpreted as having two sexine layers in deference to our regarding them as having single columellae with complex bases. Still, a case can be made for establishing an Inuloid pattern when the internal foramina are taken into consideration. As noted in *Pithocarpha* (Plate 5C), as well as in unpublished electron micrographs of *Gilruthia, Helichrysum, Helipterum,* and *Athrixia,* the internal foramina are extremely elongate, without, as far as we are presently aware, counterparts in other Compositae tribes. Incidentally, these elongate internal foramina are not equivalent to the openings in the lower sexine layer as seen in the work of Liens (1968, 1971) and Besold (1971); the latter openings are simply empty spaces resulting from complexly anastomosing columellae and were so interpreted by these investigators. The Inuloid case is weakened, however, when circular or reduced internal foramina are noted in *Stoebe* (Plate 6A) and *Raoulia* (Plate 6E, F) and unpublished micrographs of *Antennaria, Stuartina, Leptorynchus* and *Syncephalum.* The internal foramina of *Callilepis* (unpublished work) further complicate the situation because of their markedly reduced or vestigial nature. In view of all this it seems best to reserve the term "Inuloid pattern" until more exhaustive work is forthcoming in the tribe. As concerns our interpretation of the relationship of spines to columellae in these two subtribes we are in agreement with that of Merxmüller *et al.*

To summarize, it is clear that extended pollen studies on the Inuleae will be of considerable value in constructing taxonomic and phyletic relationships. The pioneering work of Liens and Besold has served as a foundation upon which future investigations should develop. Our TEM studies have refined, rather than refuted, the interpretations of these investigators.

Apart from the above discussion the Inuleae possess a number of genera whose position within the tribe is open to question. Since we have examined the pollen of many of these genera we will comment on them below.

Tarchonantheae. The genera *Tarchonanthus, Brachylaena* and *Synchodendron* (=*Brachylaena*), which encompass approximately 30 species, was considered to be a subtribe of the Inuleae by Bentham (1873). *Tarchonanthus minor* (Plate 5G, H) and most other members of the group (unpublished work) exhibit an Anthemoid morphology that is unlike anything else in the tribe. In short, the Tarchonantheae occupy an anomalous position within the Inuleae.

Liens (1971) feels that the subtribe should be excluded from the Inuleae, suggesting a relationship with the Mutisieae. We tend to agree with this, which is in line with observations by Carlquist (1957b), as well as the more recent extensive work of Parra and Marticorena (1972). However, comparable pollen-types also can be found in Anthemideae, Vernonieae, and various sections of other tribes of the Compositae. Particularly striking are comparisons with the Anthemideae. Our scanning electron micrographs (Plate 5G and unpublished work) compare very favorably with those shown by Praglowski (1971) for the Anthemideae, and transmission electron micrographs (Skvarla and Larson, 1965a, b; Skvarla and Turner, 1966a, b, 1971; Plate 5H and unpublished work) reinforce this association. If brought to a decision at this time we would position Tarchonantheae as a subtribe in the Anthemideae (or a distinct tribe next to this), relating it specifically to the South African, shrubby genus, *Eriocephalus* (see Plate 12F). *Tarchonanthus* itself is a woody African group and cursory examination of the flowers of the two genera make credible a relationship which, while not striking, seems natural. In fact, it would not be surprising to learn that other synantherologists have made this observation, so superficially similar do they appear to be. The only difficulty in this disposition is that voiced by Bentham (1873), for he reckons *Eriocephalus*, while positioned as the first genus in the Anthemideae, to be of uncertain affinity in that tribe. However, the case for relationship of the Tarchonantheae with the Anthemideae is strengthened by the work of Bremer (1972) on *Osmitopsis* (to be discussed below), for he places this genus (as did Cassini in 1817!) as unequivocally in the latter tribe. *Osmitopsis* is also a somewhat shrubby, African genus possessed of pappus and style branches similar to that of *Tarchonanthus*; indeed, however crude the chemistry, they are both reportedly strongly possessed of camphor smells. We conclude, therefore, on palynological, megamorphic and geographical evidence, that the woody Tarchonantheae can be readily positioned in the Anthemideae as a first subtribe, accommodating the four genera *Tarchonanthus*, *Brachylaena*, *Eriocephalus* and *Osmitopsis*. Clearly, additional studies are needed, both within the tribe and among yet other tribes considered to be closely associated, before any definitive conclusions can be attained.

Adenocaulon. The pollen ultrastructure of *Adenocaulon* (Plate 6I) also displays an Anthemoid pattern. It has been described under light microscopy by several investigators (Ikuse, 1956; Stix, 1960; Heusser, 1971). Liens (1968) made superficial reference to it and cited the work of Stix (1960) as support for placement within the Anthemideae. Although *Adenocaulon* does indeed exhibit Anthemoid pollen characters (see Plate 12, also Skvarla and Larson, 1965a; Skvarla and Turner, 1966a, b, for transmission electron micrographs; and Praglowski, 1971, for scanning micrographs), placement of the genus within the Anthemideae on palynological criteria alone must be met with a certain degree of caution in view of previous work on the genus. For example, Stebbins is cited in personal communication (Ornduff *et al.*, 1963, 1967) as stating that *Adenocaulon* shows pollen affinities with the Mutisieae. Megamorphic evidence also suggests a non-Anthemideae position for *Adenocaulon*: Bentham (1873) related it to the Heliantheae; Cronquist (1955a) and Wagenitz (1964) favored a position within the Senecioneae; and Hoffmann (1894) referred it to the Inuleae (subtribe Inulineae). Whatever tribal position *Adenocaulon* occupies (see additional comments below) its complex ultrastructural pollen morphology appears significant in view of the remarks by Ornduff *et al.* (1967) who feel that the chromosome data and distribution (eastern Asia, North, Central and South America) suggest that the genus is a relatively remote one with obscure tribal relationships. Nevertheless, one of us (B.L.T.) feels that its white rays, floral morphology, habit and general habitat predilections mark it as close to the Anthemideae and with its Anthemoid pollen might be positioned with better ease in that tribe.

Gymnarrhena. Pollen of *G. micrantha* is of the Anthemoid pattern but differs from Tarchonantheae and *Adenocaulon* in that the exine above the broadened basal columellae consists of a complex mesh-like network (Plate 6J).

Gymnarrhena was placed in the Inuleae by both Bentham (1873) and Hoffmann (1894), although the former put it in the subtribe Buphthalminae while the latter favored the Filagineae. Bentham's disposition was based on its assumed relationship to *Geigeria*, although it lacked most of the essential characters of the group. Similarly, Cronquist (1955b) also indicated an Inuloid (*Geigeria*) relationship. The pollen evidence does not support Buphthalminae (including *Geigeria*), Filagineae, or the tribe Astereae (Bentham's alternate choice for the genus). Small (1917–19) considered *Gymnarrhena* to be in the Buphthalminae and to be nearly identical to *Centaurea* of the Cynareae, suggesting therefore a direct origin of the Cynareae out of the Inuleae. The numerous LM studies on the Cynareae (Wagenitz, 1955; Schtepa, 1958, 1967; Stix, 1960; Avetisian, 1964; Parra, 1969–70) and the present EM (Plate 21) all indicate at least a superficial resemblance to *Gymnarrhena*. Dimon (1971) also examined pollen of *Gymnarrhena* and found it to be exceptional for the Inuleae. We are in complete agreement with her statements that the taxon does not possess pollen similar to *Evax* or *Micropus*, genera with which classical taxonomists might relate the group (Boissier, 1875). *Gymnarrhena* pollen, like that of the Tarchonantheae and *Adenocaulon*, is difficult to place. Megamorphically, its clasping leaves, xeric proclivity and geographic distribution suggest a closer relationship with the Anthemideae, perhaps near *Cotula* (see Plate 12B) since both *Gymnarrhena* and *Cotula* possess species with base chromosome numbers of $x=5$ or 10 (Moore, 1972). In short, neither of these genera is clearly in the Inuleae on megamorphic grounds, and consequently one must depend somewhat more on pollen characters which are decidedly Anthemoid. It should be remembered however, that pollen of this type also occurs in otherwise typical Senecioid elements and in yet other tribes.

Osmites and *Osmitopsis.* These closely related genera have recently been combined into *Osmitopsis* (Bremer, 1972). As indicated by electron microscopy (Plate 6G, H) they have an Anthemoid pattern and the exines very closely resemble taxa in the Anthemideae (Plate 12). These genera were originally placed in the Inuleae subtribe Buphthalminae by Bentham and Hooker (1862–83) and Hoffmann (1894). In Bremer's (1972) revision of the genus, pollen morphology was considered along with other characters (viz. odour, involucral bracts, style shape, reduced pappus, etc.) as support for placement within the Anthemideae subtribe Anthemidinae. He did this with some reluctance but felt it a somewhat better alternative than positioning it within the only other subtribe, the Chrysantheminae. It would appear that this reluctance is well grounded. Although the Anthemideae have not been exhaustively studied with TEM, enough evidence is available to indicate that the two subtribes are not readily circumscribed by palynological characters. For example TEM has revealed that exines of *Anthemis* and *Leucanthemum* (Anthemidinae) show a single level of internal tecta; however, other taxa within this subtribe such as *Achillea* (Fig. 21, Skvarla and Larson, 1965a) tend to show greater complexity of internal tecta. Within the Chrysantheminae, both single level tecta (*Chrysanthemum*) and multiple levels (*Crossostephium* and *Matricaria*) are found (Skvarla and Larson, 1965a; Skvarla and Turner, 1966b). To date, we have not attempted to catalogue the levels of internal tecta into pollen types. Rather, we consider this spectrum of internal tecta levels to reflect an Anthemoid pattern.

Eriachaenium. This monotypic genus has been placed in several tribes: Calenduleae (Bentham, 1873), Inuleae (next to *Adenocaulon*: Cabrera, 1971) and Mutisieae (Robinson and Brettell, 1973a). Our pollen studies of *Eriachaenium magellanicum* (in preparation) indicate an Anthemoid morphology very similar to *Adenocaulon*. Thus, Cabrera's position for the genus is supported but as already discussed *Adenocaulon* appears out of place in the Inuleae. However, the two genera might be grouped together as a distinct subtribe in the Anthemideae.

Eremothamnus. This monotypic genus has had several tribal positions suggested for it: Senecioneae (Hoffmann, 1894; Merxmuller, 1954), Inuleae (Moore, 1929), Arctotideae (Liens, 1970) and a new tribe, the Eremothamneae (Robinson and Brettell, 1973d). The pollen morphology has not yet been examined by us but analyses of the diagrams of Liens (1970) and Stix (1960) appear to be contradictory. Liens' drawing (of a spine) suggests an Arctotoid morphology while Stix considers the genus to possess a *Lychnophora* pollen-type which is characteristic of the Vernonieae. Her drawing, also through a spine, is strikingly different from that of Liens.

Craspedia. Besold (1971) refers to the work of Skvarla and Turner (1966b) as removing the *Craspedia* from the Inuleae in favor of a position in the Heliantheae. This is incorrect as these latter authors merely stated that the ultrastructural morphology of *Craspedia* was of the "Helianthoid" pattern; they do not suggest that the genus belongs to the Heliantheae nor was it their intention to imply such a position.

Dimeresia. This genus was considered to be better placed in the Helenieae (Besold, 1971). The ultrastructural evidence of Skvarla and Turner (1966b) were cited by Besold (1971) to support this transfer. As seen in Plate 5F the exine of *Dimeresia howellii* exhibits the Helianthoid pattern, but, as discussed earlier, this pattern occurs in a number of tribes, including the Inuleae, so transfer on the basis of pollen morphology alone is subject to question.

Heterolepis. The exine morphology of *Heterolepis decipiens* was diagrammatically represented by Besold (1971) as like that of the tribe Arctotideae. We have not examined this species but the lucid sketch and pollen description by Besold (1971) leave no doubt of an Arctotoid placement.

Heliantheae

Pollen in the Heliantheae has received considerable investigation but this work has been concentrated primarily on *Ambrosia* and closely allied genera. *Ambrosia* and related groups have been variously considered as a subtribe of the Heliantheae, as a distinct tribe and as a separate family (Payne, 1963). Thus, much of our knowledge of the Heliantheae has a decidedly "atypical" or Ambrosioid slant; pollen of the Heliantheae, in general, has received little study. Wodehouse (1931) examined the pollen of *Dahlia*, concentrating upon the six germinal furrow pattern which frequently occurs in the genus. Stix (1960) examined six genera (*Bidens, Flourensia, Helianthus, Polymnia, Rumfordia* and *Zinnia*) and referred them to her Helianthus pollen-type. Felippe and Labouriau (1964) extended this work to include nine genera and 18 species. While agreeing with the *Helianthus* type of Stix the latter workers recognized three distinct groups based upon holes in the spine apices: presence and number (as well as absence). This subject has been discussed previously (under General Pollen Wall Morphology). The most incisive description of Heliantheae pollen comes from the SEM studies of *Cosmos bipinnatus* by Heslop-Harrison (1969): whole pollen grains as well as fractured internal surfaces were diagnosed and earlier interpretations based on light and TEM were clarified. In many respects analysis of the caveate exine of *Cosmos* serves as a structural model for all such Compositae pollen. In two studies of the genus *Polymnia* (Fisher and Wells, 1962; Wells, 1971), great diversity in size, shape and surface sculpturing was indicated for pollen examined from individual plants. Finally, investigation by TEM of a number of

Helianthoid genera (Skvarla and Larson, 1965a, b; Skvarla and Turner, 1966a, b, 1969) revealed a common structural pattern which has been termed Helianthoid.

SEM of *Guizotia abyssinica* (Plate 7A) and *Eryngiophyllum pinnatisectum* (Plate 7F) illustrates the typical echinate surface morphology for the tribe. TEM of *G. abyssinica* (Plate 7B), *Enhydra sessilis* (Plate 7C), *Alepidocline annua* (Plate 7D), *Eryngiophyllum pinnatisectum* (Plate 7E) and *Helianthus giganteus* (Plate 7G) all display typical Helianthoid patterns, as presumably do nearly all genera of this tribe. However, as discussed earlier, investigations at the specific level will undoubtedly lead to various refinements on this basic pattern.

Fitchia has been singled out for specific comment because of the detailed studies of the wood anatomy and pollen by Carlquist (1957a, 1963). The genus has been placed in several tribes (Cronquist, 1955b; Carlquist, 1957a) but is more commonly treated as a subtribe, Fitchiinae, within the Heliantheae (Stuessy, Chapter 23). SEM of *F. cuneata* (Plate 8B) and *F. speciosa* (Plate 8C) indicates a basic echinate morphology but variation is evident. *Fitchia cuneata* has spines with blunt tips (Plate 8A, B) while in *F. speciosa* the spine tips are markedly pointed (Plate 8C, D). Additionally, the three germinal furrows in *F. cuneata* are considerably shorter than in *F. speciosa*. Polar regions in pollen of the latter suggest that the furrows are continuous, a morphological condition termed syncolpate (Faegri and Iversen, 1975). The internal morphology of *Fitchia* (Plate 8A, D) shows a Helianthoid pattern and the columellae and tectum (but not the foot layer) are characterized by large internal foramina. Because of the variation already detected and that noted by Carlquist, examination of all eight species of the genus should prove highly instructive.

Ambrosia. This and related genera have received considerable study by Wodehouse (1928a, 1935). He accurately schematized the internal wall-structure and established a phyletic scheme which considered long spines and long germinal furrows, as exemplified by *Oxytenia*, as primitive; the almost total absence of spines and furrows such as seen in *Ambrosia* and *Xanthium* was thought to suggest advancement.

TEM studies of the Ambrosieae show considerable heterogeneity of pollen structure (Skvarla and Larson, 1965a; Skvarla and Turner, 1966b; Payne and Skvarla, 1970). While these references should be consulted for in-depth descriptions, the basic morphological structures will be briefly presented here. The genus *Iva* was indicated to possess both Helianthoid and Senecioid structural patterns as well as having members displaying internal tecta (Skvarla and Larson, 1965a; Skvarla and Turner, 1966b). Pollen patterns of the other genera of this group as recognized by Payne (1963) are as follows: *Euphrosyne* has the Helianthoid pattern; *Dicoria*, *Hymenoclea* and *Ambrosia* (including *Franseria*) have the Senecioid pattern; *Xanthium*, like several species of *Iva*, contains a distinct internal tectum above the cavus. It is noteworthy that this internal tectum, particularly as illustrated by *Xanthium spinosum* (Fig. 37 of Skvarla and Larson, 1965a) bears a certain likeness to the internal morphology of certain Arctotideae taxa included in the present investigation (see Plate 20B–E).

In reviewing our previous work an observation has arisen which, with our increased experience, is in need of correction. This concerns our interpretation in electron micrographs of sections passing through sub-polar or subequatorial regions (Skvarla and Larson, 1965a; Payne and Skvarla, 1970). These sections are now more accurately considered to pass through regions of the colpi which show columellar union with the foot layer. Although designations such as "subpolar" etc. may have technical merit, they can be misleading since the impression can be conveyed of a cavus being absent at the poles. With this clarification it is apparent that the additional minute columellae layer noted in *Ambrosia* (Payne and Skvarla, 1970) is located in the colpal rather than the polar areas. This is not to vitiate, however, such columellae as taxonomic characters.

Finally, it should be noted that the so-called "micro-Anthemoid" morphology found for *Ambrosia* pollen (Payne and Skvarla, 1970) has not been studied in depth but the examination of additional taxa of the Ambrosieae should be pursued since it has pollen which is somewhat anomalous in the Heliantheae.

Helenieae

Pollen of the Helenieae was considered by Stix (1960) to be similar to the Heliantheae. This was based on examination of *Tagetes minuta*, *Schkuhria pinnata*, *Helenium autumnale* and *H. lanatum*. TEM studies confirm these observations (Skvarla and Turner, 1966b).

In the present study the internal morphology is reported for *Tagetes elongata* (Plate 9C), *Baileya multiradiata* (Plate 9D), *Espejoa mexicana* (Plate 10A), *Jaumea peduncularis* (Plate 10B), *Whitneya dealbata* (Plate 10C), *Flaveria anomala* (Plate 10D), *Orochaenactis thysanocarpha* (Plate 10E), *Chaenactis glabriuscula* (Plate 10F, G), *Pericome caudata* (Plate 10H), *Oxypappus scaber* (Plate 10I), *Hymenothrix wislizenii* (Plate 10J), *Eriophyllum caespitosum* (Plate 10K), *Cacosmia rugosa* var. *arachnoidea* (Plate 11A), *Amblyopappus pusillus* (Plate 11B), *Amblyolepis setigera* (Plate 11C) and *Dyssodia anthemidifolia* (Plate 11D). With the exceptions of *Cacosmia* and *Amblyopappus* the structural pattern is unquestionably Helianthoid. In contrast, *Cacosmia* pollen resembles that of *Liabum* while *Amblyopappus* possesses the Senecioid structural pattern as noted by Skvarla and Turner (1966b).

The validity of the Helenieae as a tribe is vigorously challenged by Turner and Powell in Chapter 25. Their study suggests that the genera within it can be positioned in the Heliantheae, Senecioneae, Astereae and

Eupatorieae. Reposition in these several tribes cannot be refuted on palynological grounds since all possess Helianthoid pollen types as already noted. The two genera with anomalous pollen structure, *Cacosmia* and *Amblyopappus*, received individual placement. *Cacosmia* was placed in the Liabinae, close to or within the Vernonieae, while *Amblyopappus* was considered to be an isolated member of the Eriophyllinae of the Senecioneae. The pollen evidence supports these tribal placements, except that *Amblyopappus* does not readily relate to Eriophyllinae, as pointed out by Turner and Powell (Chapter 25).

Anthemideae

This tribe was one of the first reported upon by Wodehouse and much of his initial impression of pollen structure in the Compositae was undoubtedly colored by this fact. For example, Wodehouse (1926, 1935) felt that the relationship of spine-length, number, arrangement and size were constant for a given species. He also felt that the tribe could be divided into two major groups: (1) those containing spines, an adaptation to insect pollination and, (2) those without spines, an adaptation to wind pollination. In addition, as shown by his diagram of *Tanacetum camphoratum* (Wodehouse, 1935, p. 509), he correctly interpreted the internal structure of the pollen grain wall in this tribe.

Subsequently, a number of pollen studies have been conducted in the Anthemideae using both LM and EM (Thanikaimoni, Appendix, this chapter). Many of these studies (e.g. Praglowski, 1971) have been concerned with the identification of *Artemisia* pollen and a few have been of a physiological nature (Brewer and Henstra, 1970; Southworth and Branton, 1971). Stix (1960) was the first to make a comprehensive survey of pollen in the tribe, from which she recognized two basic pollen-types, *Anthemis* and *Artemisia*, distinguishing them by spine presence in the former and absence (or great reduction) in the latter. It is notable that Stix observed nexine-2 (i.e. endexine) only in the aperture regions.

Pollen of the Anthemideae was first examined by TEM by Skvarla and Larson (1965a) and Skvarla and Turner (1966a, b, 1971). On the basis of cavus absence and the complexity of internal wall layers, an Anthemoid pattern was recognized by these workers.

Scanning electron micrographs of *Anthemis cotula* (Plate 12A), in addition to serving as an example of the echinate members of the tribe, shows a highly perforate surface. This surface texturing is what we believe Wodehouse (1926, 1935) described as "granular", considering it as one of the most distinctive morphological characters for the tribe. Transmission electron micrographs of *Cotula coronopifolia* (Plate 12B), *Soliva stolonifera* (Plate 12C), *Artemisia alaskana* (Plate 12G) and *Anthemis ruthenica* (Plate 12I) emphasize the complexity of levels of internal tecta: in *Cotula* and

Anthemis there appears to be just one, while in *Soliva* and *Artemisia* there appear to be two levels. Notable in all these genera is the presence of an endexine. While an endexine is consistently present in pollen of the Anthemideae, it does vary in thickness among taxa. *Eriocephalus hoffmannianus* in Plate 12F is represented by a pollen grain that had been fractured and then examined with SEM. It clearly shows the long individual columellae which bifurcate distally into a massive spongy layer. As is known from TEM the spongy layer consists of complex levels of internal tecta and columellae. Plate 12F is of further interest in that the columellae can be seen to be supported by a solid basal layer, the nexine of Erdtman (1971) and Stix (1960). This layer is actually composed of two units, foot-layer and endexine.

A number of Anthemideae genera appear to be out of place on the basis of megamorphic data (see Heywood and Humphries, Chapter 31). TEM studies of some of these groups prove interesting. *Centipeda* (Plate 12D) displays an internal pollen morphology which is not characteristically Anthemoid. This is also true of its chemistry (Sørenson, Chapter 13). Our interpretation of the internal morphology is that the highly ramifying columellae (above the cavus), which appear to be "sub-columellae" as discussed previously, suggest a relationship with the Inuleae. *Elachanthus* (Plate 12E), *Ceratogyne* (Plate 12H), *Isoetopsis* and *Abrontanella* (the latter two genera not illustrated in this study), all possess the Helianthoid pattern of wall morphology. Turner (1970) suggests that both *Ceratogyne* and *Isoetopsis* might be better placed in the Inuleae while Robinson and Brettell (1973b, c) place *Isoetopsis* in the Astereae. Other anomalous genera listed by Heywood and Humphries (Chapter 31) have yet to be examined. These include *Ischnea*, *Plagiocheilus* and *Polygyne*. The genus *Ursinia*, included by some workers in the Anthemideae or as a new tribe (Robinson and Brettell, 1973c), will be discussed under the Arctotideae. Similarly, the genus *Osmites* (=*Osmitopsis*), which is now considered in the Anthemideae (Bremer, 1972) has been discussed previously (see Inuleae).

Senecioneae

Stix (1960) and Skvarla and Turner (1966a, b) have made the only detailed studies of internal pollen structure for this tribe. Stix recognized two pollen types, the *Senecio*-type and the *Arnica*-type. The primary distinction between the two was the occurrence of small "infrategillar bacula" (=internal columellae) which occasionally bridged the cavus areas in pollen of the *Senecio*-type. TEM shows the infrategillar bacula to be part of the columellae bases; as such they are common throughout the tribe (Skvarla and Turner, 1966b). Random extensions across the cavus and attachment with the foot layer, as indicated by Stix (1960), have not been substantiated, although exines with columellae bases that appear to be in

8. POLLEN MORPHOLOGY

superficial contact with the foot layer have been observed in many caveate exines in the Compositae in general. Not all of the taxa examined by Stix (1960) have been examined by us. However, it is interesting that *Neurolaena*, *Schistocarpha* and *Werneria* of her *Senecio*-type, as well as *Arnica* of her *Arnica*-type, all have a Helianthoid structural pattern.

In a TEM study of Senecioneae pollen a structural pattern termed Senecioid was established (Skvarla and Turner, 1966a). While this pattern was found in a number of genera in the Senecioneae, the tribe was also shown to possess taxa with the Helianthoid pattern. Altogether, 19 genera of the Senecioneae were classified as having either Senecioid or Helianthoid structural patterns (Skvarla and Turner, 1966b). Subsequently, the patterns for some of these genera have been more intensively investigated and this has resulted in certain reinterpretations. Thus, the genera *Gynoxys*, *Tetradymia*, *Cacalia*, *Crassocephalum*, *Cineraria* and *Lepidospartum* are now considered to have a Senecioid pattern while *Arnica*, *Haploesthes*, *Schistocarpha*, *Bartlettia* and *Senecio glabellus* have the Helianthoid pattern.

In addition to the Senecioid and Helianthoid patterns, the Senecioneae were indicated as having an "Anthemoid-like" pattern for *Sinclaria* and *Liabum* (Skvarla and Turner, 1966a, b). These taxa are of doubtful position within the Senecioneae (Turner and Powell, Chapter 25; Nordenstam, Chapter 29).

In the present electron microscope study we have examined 25 taxa, all of which were treated by Bentham (1873) as belonging to the Senecioneae. Some of these, as will be noted below, are no longer considered to be in this tribe or else are controversial. However, our primary intent is to illustrate exine patterns among the tribe as classically circumscribed. The echinate sculpturing, which is typical for this tribe, is shown by the SEM of *Arnica acaulis* (Plate 13A). With TEM, *Arnica chamissonis* (Plate 13B), *Doronicum altaicum* (Plate 16E), *Raillardella scaposa* (Plate 13C), *Senecio heritieri* (Plate 13D), *Schistocarpha platyphylla* (Plate 16A), *S. sinforosii* (Plate 16B, C), and *Neurolaena cobanensi* (Plate 16D), all show Helianthoid structural patterns. By contrast, the Senecioid pattern is evident in *Senecio lyallii* (Plate 13E), *Tetradymia canescens* (Plate 14A), *Gynoxys parvifolia* (Plate 14B), *Odontotrichum cervinum* (Plate 14C), *Erechtites valerianaefolia* (Plate 14D), *Cacalia goldsmithii* (Plate 14E), *Emilia coccinea* (Plate 14F), *Tussilago farfara* (Plate 14G), *Gamolepis brachypoda* (Plate 14H) and *Petasites hyperboreus* (Plate 14I).

The tribal position of some taxa is debatable, as mentioned above and elsewhere (Turner and Powell Chapter 25; Nordenstam, Chapter 29). However, one genus *Liabum*, is sufficiently distinct palynologically to deserve additional discussion here. As seen in SEM, *Liabum megacephalum* (Plate 15A) and *L. sagittatum* (Plate 15B) are prominently echinate. TEM of *L. glabrum* var. *hypoleucum* (Plate 15C), *L. caducifolium* (Plate 15D) and *L. kluttii* (Plate 15E) characteristically display an Anthemoid or "modified" Anthemoid pattern as mentioned earlier. A similar internal structure occurs in *Cacosmia* (of Bentham's Helenieae) (Skvarla and Turner, 1966b) and in the Vernonieae. Exomorphically, however, the lophate pollen of Vernonieae is quite different from the echinate pollen of *Liabum* and *Cacosmia*. *Liabum* and related genera have been variously treated, but most often as a tribe or subtribe related to the Senecioneae. However, Turner and Powell (Chapter 25) would include the taxon in or near the Vernonieae. Other workers might differ, at least as to inclusion *within* the Vernonieae (Robinson and Brettell, 1973e, 1974). Regardless of the exact positioning, the similarity of pollen within the group is unchallenged, that is, until the electron micrograph of *Liabum ovatum* (Plate 15F) is examined. Interpretation of this exine is difficult because, while there is a suggestion of a faint cavus, the basal portions of the columellae, which are laterally connected, may be attached to the foot layer. This is particularly evident in the region of the spine (see far right corner of Plate 15F). While the exine structure can be considered tentatively as possessing a "modified" Senecioid structure we have not encountered this morphology previously in Helianthoid or Senecioid patterns and can do no more than call attention to its unique structure. Stix (1960) also studied *Liabum ovatum*, erecting for this a separate pollen-type because of its strikingly different pattern as contrasted with other species of *Liabum*. Her diagram differs from our electron micrograph principally in that she indicates the columellae as being uniformly fused to the foot layer. However, she expressed some uncertainty about this interpretation. *Liabum ovatum* is of additional interest because it was originally described as *Paranephelius ovatus* Wedd. and is still so recognized in recent treatment of the group (Robinson and Brettell, 1974; Nordenstam, Chapter 29). *Paranephelius*, along with 13 other genera, are considered to constitute the Liabeae (Robinson and Brettell, 1974). In any case, the group as currently treated is palynologically quite heterogeneous. This heterogeneity is emphasized by Stix (1960) as she established two other *Liabum* pollen types (i.e. in addition to her *L. ovatum*-type). *Liabum* and related genera are clearly in need of additional study.

Calenduleae

Pollen of this small tribe is known primarily through the work of Stix (1960). She established four pollen-types, three of which were monogeneric: *Calendula-*, *Dimorphotheca-* and *Castalis*-types. The fourth type, *Osteospermum*, was also found in the genus *Gibbaria*. The *Calendula-* and *Castalis*-types were depicted as having infrategillar bacula "freely hanging" in the cavus region while the other two types lacked such infrategillar bacula. Holes in the spine-tips and spine-construction

(*Osteospermum*-type) were also used by Stix in distinguishing between types.

SEM of *Chrysanthemoides incana* (Plate 18A), *Tripteris* (= *Osteospermum*) *clandestinum* (Plate 18B), *Gibbaria ilicifolia* (Plate 17A), *Osteospermum vaillantii* (Plate 17B) and *Dimorphotheca pluvialis* (Plate 19A) all indicate a typical echinate sculpturing. Likewise, TEM (Plates 17C, and inset; 18C, D; 19B and inset) reveal a typical Helianthoid pattern, although the internal foramina have small diameters.

The Helianthoid pattern for Calenduleae is noteworthy in view of the suggested relationship of the tribe to the Heliantheae (Cassini, 1817). In contrast, it seems less related to the Senecioneae and even less to the Cynareae (Bentham, 1873). The one genus in question, *Eriachaenium* (discussed above under Inuleae), has a pollen structure distinct from the Calenduleae.

Arctotideae

Arctotideae pollen appears to be quite variable. *Berkheya* was given considerable attention by Wodehouse (1935) because of its lophate exine. Stix (1960) examined 44 species from 11 genera and established five pollen types: *Ursinia*-type (*Ursinia*), *Arctotis*-type (*Arctotis* and *Haplocarpha*), *Berkheya*-type (*Berkheya, Cullumia, Didelta, Heterorhachis*), *Gazania*-type (*Gazania*) and *Gorteria*-type (*Gorteria, Hirpicium*). Liens and Thyret (1971) examined the subtribe Gorteriinae, one of the two Arctotideae subtribes, in an attempt to correlate pollen morphology with floral data. Three phylogenetic lines were postulated for the Gorteriinae, each starting from an "irregular spiny" type pollen grain and ending with a lacunar type. Such observations were made earlier by Wodehouse (1928c, 1935) with the "irregular spiny" type designated as echinate or sub-echinate and the lacunar type as psilolophate.

SEM illustrates the diverse exomorphology of this tribe: *Cullumia setosa* (Plate 19C) is lophate; *Berkheya rhapontica* (Plate 19F) is subechinolophate and *Arctotis verbasifolia* (Plate 20A) is echinate. The prominent lophate character of *Cullumia* is found also in *Gazania* (unpublished work; Ueno, 1972). The genus *Berkheya* appears to possess numerous transitional lophate or echinolophate characters as shown by Liens and Thyret (1971) and Wodehouse (1935). The latter described *B. heterophylla* as containing 29 lacunae.

With TEM the major features of the echinate pollen are the cavus, prominent spine channels, columellae apparently connecting spine areas with the foot layer, extremely thin foot layer (i.e. that portion beneath the cavus) and centrally located internal tectum (Plate 20B, C, D, E). In some taxa the exine over the cavus appears depressed and to be in contact with the foot layer (Plate 20E; and Stix 1960, Plate 4C). Whether or not there is complete fusion with the foot layer is uncertain. Plate 20E is somewhat misleading because the exine has not been acetolyzed and the electron-dense material forming the "solid layer" between foot layer and columellae base may be destroyed by acetolysis. As indicated earlier, *Arctotis, Cymbonotus, Arctotheca,* and *Haplocarpha* possess what has been termed an "Arctotoid" pattern. The single species of *Berkheya* examined (Plate 19G–I) does not have an internal tectum and hence differs from the Arctotoid pattern. It should be emphasized that this distinction is provisional since, as mentioned earlier, *Berkheya* appears to be a highly pleomorphic genus and additional taxa need examination. Liens and Thyret (1971) indicate that *Berkheya* is quite diverse and they show species both with and without a central internal tectum. Stix (1960) shows an internal tectum for her *Berkheya*-type and, in contrast to Liens and Thyret (1971), indicates that "infrategillar bacula" (columellae) beneath the spines are continuous from foot layer to spine interior. This has been noted also by us (Plate 19G).

TEM of the lophate pollen grains of *Cullumia setosa* (Plate 19D, E) and *Gazania* (unpublished work) indicate a thin caveate single columellae layer along the lacunae (Plate 19D). This interpretation is also implicit in the drawings of Liens and Thyret (1971) and to a lesser extent in those of Stix (1960). Enigmatically, Stix includes *Cullumia* in her *Berkheya*-type rather than with *Gazania* in her *Gazania*-type. She reports the former as having an internal tectum while the latter is without one. In our electron microscope sections parallel to the ridges the exine is mesh-like and considerably thicker than shown by Stix. This mesh-like nature of the ridges has been noted also by Liens and Thyret (1971). A central internal tectum is suggested in Plate 19E but it has not yet been substantiated. *Cullumia* and *Gazania* are of interest also because, while they are not considered to possess spines, sections through the corners of the ridges outlining the lacunae are sharply pointed and contain channels, and possible connections, with the foot layer.

"Infrategillar bacula" (columellae) connecting the nexine (foot layer) with the spines appears to be an extremely important character and was used by Stix to distinguish *Berkheya* and *Arctotis* pollen types. While both types displayed infrategillar bacula, connection with the nexine was shown only for the *Berkheya*-type. It is our feeling that this character is in need of additional study; for example, in contrast to Stix's placement of *Haplocarpha scaposa* in the *Arctotis*-type because of infrategillar bacula not being united with the nexine, our electron micrograph (Plate 20E) indicates an opposite interpretation. The presence of infrategillar bacula connecting with the nexine would affect the definition of caveate exines since, as originally conceived, the cavus is considered to occupy interapertural areas.

There is no question but that the Arctotideae is in need of much additional study. Its pollen is quite variable both between and within individual taxa. Thus, our preliminary TEM with *Ursinia* shows the structural pattern to be Senecioid. The only disparity is that the foot layer is remarkably thickened and, up to the present time, without equal in the family. It is noteworthy that Stix (1960) examined six species of *Ursinia* and considered them to form her *Ursinia*-type. On the basis of her interpretation, which did not recognize a foot layer (i.e. nexine-1), she placed the genus with the Anthemideae, although Robinson and Brettell (1973c) treated it as a monogeneric tribe.

Cynareae

The pollen of this tribe was intensively investigated by Wagenitz (1955), primarily the genus *Centaurea*. Although this work is reviewed elsewhere (Dittrich, Chapter 36), it is noteworthy that Wagenitz established phyletic trends starting with what we consider as an Anthemoid pattern and ending with pollen having a great reduction of absence of the inner columellae layer, resembling what we consider as the Helianthoid pattern. Other significant studies of Cynareae pollen morphology are those of Schtepa (1958, 1962, 1965, 1966, 1967, 1973), Stix (1960), Avetisian (1964) and Parra (1969-70). Additional studies were directed at developmental aspects of the pollen wall of *Echinops banaticus* by polarization microscopy (Stix, 1964, 1970).

With SEM prominent echinate sculpturing is evident for *Cirsium americanum* (Plate 21A). This sculpturing is common for the tribe but, as review of the above mentioned references indicates, a smooth (psilate) sculpturing is not uncommon. TEM reveals a basic Anthemoid pattern but considerable variation is noted in the number of columellae and internal tectum levels: *Carthamus tinctorius* (Plate 21B) has two levels of internal tecta and three levels of columellae, a morphology apparently corresponding to the *Carlina* and *Cirsium* types of Stix (1960) and to *Centaurea americana* (Serratula-type) of Wagenitz (1955, Fig. 1). *Cirsium lanceolatum* has a single internal tectum level but it appears to be highly ramifying (Plate 21C) and does not correspond to any of the eight Cynareae types recognized by Stix (1960). *Arctium minus* (Plate 21D) is similar to *Cirsium* in having a highly ramified internal tectum.

Continued pollen studies in the Cynareae should prove highly interesting, especially in *Centaurea*. We have not yet examined this genus with electron microscopy but feel that those species which Wagenitz (1955) indicated as losing basal columellae should be closely investigated. As sketched by Wagenitz, these taxa (*Jacea*-type, *Scabiosa*-type) have a distinct cavus and a single level of columellae. It is the columellae which we are interested in examining. Similarly, our interpretation of the sketches of Avetisian (1964) for the genera *Tomanthea*, *Grossheimia*, *Cheirolepis*, *Chartolepis*, *Tetramorphaea*, and *Acroptilon* is that these exines also possess a cavus; since all of these genera are usually treated as belonging to *Centaurea*, in the light of Wagenitz's study this is not unexpected. What is surprising, however, is that the portion of the exine above the cavus, in contrast to the apparent single columellae of *Centaurea* (Wagenitz, 1955), appears to have an Anthemoid morphology. Additional study of *Centaurea* pollen might provide considerable new insight into relationships in the family. It seems obvious (Skvarla and Larson, 1965a; Skvarla and Turner, 1966b; this review) that the Anthemoid and Helianthoid patterns dominate the Compositae. While the Helianthoid pattern predominates in those tribes usually thought to be basal for the family, it is our thesis that the Helianthoid pattern is derived. The investigation by Wagenitz (1955) provides support for this concept. Examination of still other taxa also appear to hold much interest. For example, our current work with the genus *Galactites* suggests that the exine has a definite Helianthoid pattern. This is the first evidence of Helianthoid pollen in the Cynareae suggesting, unless the taxon is misplaced, that the Helianthoid pattern has arisen independently in several tribal lines. Of course, it could be argued that the pollen forms are relictual in those tribes where "it appears" anomalous, but it is to be hoped that comparative study of yet other characters will decide between these alternatives.

Mutisieae

Pollen in this tribe has perhaps been studied more extensively than in any other. Wodehouse (1928b, 1929a, b, 1935) performed an elegant series of investigations spanning nearly every major genus. Although the succeeding 46 years have shown him in error in certain morphologic and phyletic interpretations, he established a solid conceptual base for subsequent students of the tribe. It would not be feasible to review all of his work here and the aforementioned citations should be consulted for a thorough background.

Carlquist (1957b) reported on pollen of the subtribe Gochnatinae, Gerberinae and the monotypic *Glossarion*. Although numerous variations in the pollen morphology of these taxa were shown, all exhibited what we have termed an Anthemoid pattern. Carlquist felt that the pollen morphology represented "... a basic pattern upon which variations have taken place" (p. 451). This was supported by his studies of floral anatomy. He further emphasized that pollen of the Heliantheae differed considerably from that of the Mutisieae, noting that these represented quite different phyletic lines. Portions of Carlquist's work have recently been reaffirmed and extended by pollen SEM of Barroso and Maguire (1973). Stix (1960) established eight Mutisieae pollen-types: *Onoseris, Saussurea, Berardia, Dicoma, Erythrocephalum, Mutisia, Ameghinoa, Oxyphyllum* and *Trixis*; each type being confined to its own genus.

8. POLLEN MORPHOLOGY

The first TEM was done by Southworth (1966) on *Gerbera jamesonii*. She indicated that an internal tectum which was termed a "non-homogeneous layer" interrupted a single layer of bacula [columellae] which ". . . may branch at least four times (Fig. 15) and finally join the outer tectum" (p. 327). Presumably, at least one set of these branches formed the "non-homogeneous layer". Southworth also indicated that the endexine layer in young pollen contained coarse-branched lamellae which disappeared at pollen maturity. Our electron micrograph of *Mutisia campanulata* (Skvarla and Turner, 1966b) also indicates similar lamellae in pollen we have believed to be mature. The morphogenetic significance of this apparent discrepancy deserves further study. Parra and Marticorena (1972) presented diagrams of pollen of 27 genera representing 160 species. All taxa except *Doniophyton*, *Chuquiraga* and *Dasyphyllum* resembled highly diverse Anthemoid patterns. These latter genera were depicted as having long, slender columellae directly attached to a foot layer and without an endexine. These same authors (1975) indicate that *Hesperomannia* and *Moquinia* have yet other pollen patterns.

Our EM survey (Plates 22 and 23) is of pollen from the four subtribes as recognized by Cabrera (1966): Barnadesiinae, Gochnatiinae, Mutisiinae and Nassauviinae. SEM of *Trixis angustifolia* (Plate 22A) reveals long, wide germinal furrows which terminate in round thick polar caps; globular particles are common in the furrows; and exine surface is studded with numerous small granules. Internal morphology resembles the Anthemoid pattern, characterized by thickened and long basal columellae supporting a delicate network of internal tectal membranes which in turn supports a finer level of columellae (Plate 22B). This description closely parallels that of Wodehouse (1929b) for the six species of the genus he examined. He considered the absence of spines and the extremely long germinal furrows, as typified by *Trixis*, to distinguish his Nassauvinae from other Mutisioid groups. Additionally, the globular particles within the germinal furrows were interpreted as being significant to the expansion–contraction mechanism of the pollen grain. The presence or absence of these characters enabled him to divide the tribe into groups which he believed natural.

TEM of *Perezia* pollen (Plate 22D) is similar to that of *Trixis*, the former possessing a slightly longer upper columellae layer. Wodehouse (1929b) believed that the two genera had a similar morphology, differing only by the absence or poor development of polar caps in *Perezia*. Our unpublished SEM studies of *P. wrightii* confirm this observation.

TEM studies of both *Onoseris odorata* (Plate 22C) and *Glossarion rhodanthum* (Plate 23D) reveal an Anthemoid pattern; they differ from *Trixis* and *Perezia* in possessing numerous fine columellae which form the outer part of the exine. Carlquist (1957b) indicated the presence of a third ektexine layer in *Glossarion*, presumably located between the two levels of disproportionate columellae. This served to distinguish the genus from other Guayana Mutisieae. We have not been able to substantiate this layer in our TEM work, other than noting a fine internal layer of the tectum which separates the two levels of columellae.

SEM and TEM of *Barnadesia* (Plate 22E–F) clearly indicate its strikingly different morphology which distinguishes it from all other Mutisioid genera considered in this report. The pollen is prominently lophate (Plate 22E). The internal morphology, as shown for *B. lehmanii* (Plate 22F) and *B. horrida* (inset to Plate 22F) has already been described in detail (section entitled Exine patterns, *Cullumia-Barnadesia* Pattern) and is not recounted here. Wodehouse (1928b) intensively studied the lophate character in *Barnadesia* and used it to identify species. He concluded that *Barnadesia* was closely allied to the Vernonieae and only distantly so to the Cichorieae, the only other tribes in which he found this character. As mentioned earlier, it is now known that *Gazania* and *Cullumia* of the Arctotideae also possess the lophate condition. Because of our limited knowledge of the distribution of lophate pollen within the Compositae generally, questions as to the palynological affinities of *Barnadesia* seem premature at this time.

The fine structure of *Dasyphyllum excelsum* (Plate 23A, B) and *Schlechtendalia luzulaefolia* (Plate 23C) has been discussed previously (see Exine Patterns). Wodehouse (1928b) considered *Schlechtendalia* to be quite distinct in its possession of concavities between the furrow regions (i.e. intercolpar depressions). His observation of similar concavities in some species of *Chuquiraga* can be discounted in that Cabrera (1959) placed those taxa with intercolpar concavities in *Dasyphyllum*. The occurrence of intercolpar concavities has recently been noted for *Dasyphyllum diacanthoides* and *D. excelsum* by Parra and Marticorena (1972). Wodehouse (1928b) attached considerable phyletic significance to this character and stated (p. 461) "... the exine of the grains of *Schlechtendalia*, suggests it as a possible link between the form of the grain of *Barnadesia* and that of the Feroces of *Chuquiraga*" [=*Dasyphyllum*]. We have not examined all species of *Dasyphyllum* but preliminary SEM observations suggest that intercolpar concavities may not occur throughout the genus. Additionally, other Mutisieae genera may possibly have these concavities. The difficulty in evaluating intercolpar concavities by SEM is that artifacts caused by drying and vacuum can result in collapsed areas of the exine which resemble intercolpar concavities. Therefore, specialized preparation techniques such as critical-point drying must be employed to identify these features correctly.

The unusual structural morphology of *Doniophyton patagonicum* (Plate 23E) was considered earlier to be a "modified" Anthemoid pattern (see section, Exine Patterns), being given that designation solely on the basis of having columellae undivided by a cavus. While this analysis is highly tentative, a point of considerable interest is that the foot layer is extremely reduced (Plate 23E); indeed, it often appears absent and must rank as the most reduced foot layer so far observed in the Compositae. However, Parra and Marticorena (1972) have represented diagrammatically similar reduction in the foot layer for *Chuquiraga* as well as *Doniophyton*.

To summarize, it is quite obvious that the Mutisieae has a most diverse assemblage of pollen grains, both at the exo- and endomorphic levels. Wodehouse (1929b) concluded that this diversity reinforced the polyphyletic nature of the tribe and suggested a possible relationship with the Cynareae. The latter view is also held by Cronquist (1955b). Pollen of some Mutisieae genera (*Trixis, Onoseris, Perezia*) does indicate similarity to the Cynareae, at least within the limits of this study. Although the tribe contains a wide variety of pollen structures, it is notable that definite Helianthoid, and with the exception of *Dasyphyllum*, Senecioid patterns have yet to be observed. The occurrence of such patterns, coupled with the other diversified structures, could be a significant factor toward understanding the evolutionary development of the tribe.

Cichorieae

It is surprising that the remarkably sculptured pollen grains of the Cichorieae (Lactuceae) have received comparatively little attention. Wodehouse (1928c, 1935) examined a large number of taxa primarily in an effort to help formulate a phylogeny for the Compositae. He was interested in finding pollen grains intermediate in sculpturing between the primitive echinate types and advanced psilolophate types. His studies also led to the characterization of the basic echinolophate patterns common to the Cichorieae. Stix (1960) examined *Taraxacum officinale* and *Lactuca sativa* but apparently did not designate additional pollen types since she felt, like Pausinger-Frankenburg (1951), that the pollen of the Cichorieae was easily distinguished from all other tribes in the Compositae. Ueno (1969) examined surface replicas of *Taraxacum* with TEM and made comparisons with *Ambrosia*. Other studies of Cichorieae pollen are mainly as adjuncts to systematic investigations (Stebbins, 1940, 1953; Saad, 1961; Davis and Raven, 1962; Pons and Boulos, 1972; Tomb, 1972a, b, 1973). The bibliographic index of Thanikaimoni (Appendix, this chapter) should be consulted for a complete listing of pollen studies in this tribe.

SEM and TEM have been used to examine a substantial portion of the Cichorieae. Tomb *et al.* (1974) investigated pollen of the Stephanomeriinae and indicated the subtribe to be almost exclusively echinate. *Lygodesmia* was the only member to be echinolophate. Feuer (1974) examined the Microseridinae, which, in contrast to the Stephanomeriinae, was predominantly echinolophate. The genus *Picrosia* was shown to have echinate as well as echinolophate pollen. Of particular interest was the occurrence of a second endexine layer in some species of *Agoseris* and *Apargidium*

(Feuer, 1974). This layer was clearly defined and was considerably less dense to electrons than the overlying first endexine layer. Tomb (1976) concluded a survey of the entire tribe and presented data showing highly modified echinolophate patterns. *Scorzonera* contained both echinate and echinolophate pollen. His TEM studies of *Scolymus grandiflorus* and *Catananche arenaria* deserve special note: pollen sections of *Scolymus* are strikingly like those of *Liabum*, *Cacosmia* and the tribe Vernonieae, and *Catananche* is of interest because of the prominent internal foramina which occur throughout the columellae and tectum. This will be discussed below.

It has already been suggested that lophate exines may consist of a structural pattern paralleling the ridges and another which parallels the lacunae (see Exine Patterns). This concept is supported in the echinolophate taxa shown here by TEM and SEM (*Prenanthes*, Plate 24A, B; *Picris*, Plate 24I; *Scorzonella*, Plate 25B). Sections of *Prenanthes trifoliata* (Plate 24C), *Leontodon nudicaulis* (Plate 24D, E), *Rafinesquia californica* (Plate 24F), *Hedypnois polymorpha* (Plate 24G), *Lapsana communis* (Plate 24H), *Scorzonella* (=*Microseris*) *nutans* (Plate 25D) and *Scorzonera divericata* (Plate 25E) include both ridge and lacunar areas. The ridges appear to be at least partially connected to the foot layer while the lacunae appear to have highly reduced "cavus like areas". We have used the term cavus in a qualified sense because it was originally described as occurring between apertural areas (Skvarla and Larson, 1965a). The above cited transmission electron micrographs are deceptive because while a "true" cavus may occur (i.e. between apertures), short columellae appear to extend from the more prominent columellae bases and periodically interrupt the cavus to join with the foot layer. These short columellae are not always observed in thin section because of their size and distribution. However, such connections are evident in areas where ridges and spines are accentuated (Plate 24E at extreme right; Plate 24F). Furthermore, broken portions of these columellae can be seen in the other included electron micrographs. Feuer (1974) has provided good evidence for these columellae connections. Pollen grains were fractured and then examined with SEM. The connections were clearly shown to be cylindrical where they attach to the columellae base and then taper into thread-like strands across the cavus. Apparently, these columellae were recognized also in other studies. Stix (1960), as noted earlier, described them in a number of tribes as "infrategillar bacula" and Saad (1961) described them in *Sonchus* as "endosexine". Analogies to the "two" exine patterns in the echinolophate grains can be made to structural patterns described in echinate pollen. The structure paralleling the ridges is not too far from that of the Vernonieae while that paralleling the lacunae is reminiscent of the Senecioid pattern.

Catananche has an echinate surface morphology (Plate 25A) and an internal structure identical to the Helianthoid pattern (Plate 25C), with the notable exception of broad columellae periodically unifying the columellae bases with the foot layer. This exine structure is at present unknown elsewhere in the Compositae and was first noted by Tomb (1976). Although internal foramina have not been recognized in other members of the tribe it must be emphasized that the echinate or subechinate pollen grains have not yet been comprehensively investigated with TEM (the only other echinate taxon included in this report is *Rafinesquia*, Plate 24F). Indeed, Tomb (1976) implies that internal foramina may be present in the echinolophate pollen of *Tragopogon*. *Catananche* was regarded by Wodehouse as primitive and served as a link between echinate and echinolophate pollen. Tomb (1976) also considers the genus as primitive but presents evidence indicating that not all echinate exines are primitive in the tribe.

It is evident that the Cichorieae is in need of additional study, primarily at the internal rather than exomorphic level.

RELATED FAMILIES

Background review and rationale

In an effort to gain some insight into the possible relationships between the Compositae and other angiosperm groups we have conducted pollen-wall surveys among a number of select families. The latter were chosen from systems of classification proposed by Cronquist (1968), Thorne (1976), Takhtajan (1969) and Hutchinson (1969). From these systems three orders emerge as candidates for housing the Compositae: Campanulales, Dipsacacales and Rubiales. However, the various families treated within these orders are not the same in each of the systems. Families such as the Brunoniaceae, Goodeniaceae, Valerianaceae, and Calyceraceae are included variously in the Dipsacacales, Campanulales, or as separate orders. Furthermore, the order Rubiales is subject to varying interpretations. Cronquist (1955b, 1968) considers the Rubiales as comprised of only one family, the Rubiaceae, whereas others (e.g. Thorne, 1968; Takhtajan, 1969; etc.) place it in the Gentianales. The diverse interpretations of familial and ordinal classification schemes of these four authorities have been summarized in tabular form by Becker (1973). In order to provide a rationale for our approach we have summarized below the principal ideas on Compositae ancestry given by these same investigators.

Cronquist system

Cronquist (1955b, 1968) makes the suggestion that the Rubiales are ancestral to the Compositae. However, no specific rubialian taxa or groups are indicated as showing direct linkages. He does not regard the Dipsacales and Campanulales as being directly related to the Compositae. The Dipsacales are considered to be too advanced to be ancestral, and the family Dipsacaceae is believed to have undergone parallel changes much like that which led to the origin of the Compositae, both families arising from some Rubiales-Dipsacales ancestral complex. The Campanulales are completely discounted as showing affinities with the Compositae, and any similarities which exist are regarded as insufficient to demonstrate strong ancestral linkages. For example, similar pollen presentation mechanisms which occur in both groups are shown to occur also in the Rubiaceae. The relatively small family Calyceraceae is acknowledged as being closest to the Compositae in floral morphology and inflorescence, but this relationship is thought to be merely collateral, and ovule differences are cited as negating close affinity (apical for Calyceraceae, basal for Compositae).

Takhtajan system

The views of Takhtajan (1969), while appearing to differ significantly from Cronquist (1968), are actually quite similar. The Asterales (Compositae), Calycerales (treated as a family within the Dipsacales by Cronquist), and Campanulales are considered to be derived from the Gentianales. The seeming disparity is reconciled when it is realized that Takhtajan includes Rubiaceae within the Gentianales. There is agreement with Cronquist in that the general floral morphology of the Calycerales shows the closest affinities to the Compositae. Additionally, similar pollen morphology links the Calycerales and Goodeniaceae (Campanulales). The basis for this pollen relationship is presumably Erdtman (1952). In contrast, the Dipsacales are considered to be more remote and to have a common origin, with the Cornales, from the Saxifragales.

Hutchinson system

Hutchinson (1969) indicated that parallelisms exist between the Compositae and Dipsacales (specifically, the families Valerianaceae and Dipsacaceae) but that they do not imply direct affinity. His belief was that the Dipsacales have a phylogeny distinct from the Compositae and supposed affinities (e.g. the inflorescence) are due merely to convergent evolution. The Compositae and Rubiales were described as showing a similar trend in inflorescence evolution, but Hutchinson did not feel that this justified phyletic derivations. Contrary to Cronquist and Takhtajan, he regarded the Campanulales as ancestral to the Compositae but brought up the extremely interesting point that the latter are probably polyphyletic in part. Of the two families constituting his Campanulales, the more primitive Campanulaceae were suggested as ancestral to the Compositae tribe Heliantheae (more specifically the subtribe Verbesininae), whereas the

Lobeliaceae were regarded as advanced and similar to the more highly evolved tribes of the Compositae.

Thorne system

Thorne (1976) believes that the Dipsacales are closest to the Compositae. The Campanulales are thought to be quite removed. Although the Rubiales are placed nearer to the Compositae than the Campanulales, Thorne feels that other orders, such as the Cornales and Lamiales, are closer than even the Rubiales.

REVIEW OF POLLEN MORPHOLOGY

Although the above systems are based upon a wealth of morphologic data, they are but a sampling of ideas bearing on the ancestry of the Compositae. For example, Philipson (1953) presented an excellent account of the family, including views as to its ancestry. Nearly all of these systems have lacked any significant input from palynology. This is emphasized by Cronquist's statement: "It should be noted that although the classical taxonomic characters have been fairly well observed, the more recondite characters such as the structure of the pollen . . . are still inadequately known in the Rubiaceae and most other families. Examination of more species is likely to bring surprises" (1968, p. 304). Additional emphasis is provided by Muller (1970, p. 442), who comments, "Future studies should focus on the possible ancestors of Compositae by studying the pollen characters of their primitive members . . .".

In the pollen literature there are studies which exist for several of the families mentioned above but they are mainly taxonomic in scope and, with few exceptions, do not make comparison with pollen of the Compositae (Campanulaceae: Avetisian, 1948; Tarnavschi and Radulescu, 1959; Chapman, 1966; Dunbar, 1973, 1975a, b; Badre et al. 1975; Dipsacaceae: Van der Spoel-Walvius and De Vries, 1964; Goodeniaceae: Duigan, 1961; Valerianaceae: Wagenitz, 1956; Rubiaceae: Bahadur, 1964; Baker, 1956, Lewis, 1964, 1966; Moens, 1962). To the present time Erdtman's textbook (1971) is still the best reference for a synthesis of the palynology.

Three pollen studies appear to be highly significant when put in the context of association with the Compositae. In a systematic investigation of Middle European Valerianaceae, Wagenitz (1956) described pollen of *Valeriana* and *Valerianella* and showed a plate of light photomicrographs bearing a great deal of superficial resemblance to Compositae. This resemblance was presumably recognized since a generalized comparison was made with *Centaurea* of the Compositae.

Duigan (1961) described pollen of the Goodeniaceae and Brunoniaceae. This work was aimed at identifying Australian plants by their pollen and intended to serve as a basis for paleobotanical investigations. Genera examined were *Goodenia, Selliera, Scaevola, Velleia, Brunonia*, and *Dampiera*. Pollen descriptions and light photomicrographs of these taxa appear to bear considerable resemblance to the Compositae.

Van der Spoel-Walvius and De Vries (1964) described and illustrated the pollen of the genus *Dipsacus* (Dipsacaceae) comparing this with *Cephalaria* (Dipsacaceae). The significance of this study is that Stix (1960) first showed *Cephalaria* pollen to be similar to that of the Compositae. Therefore, another genus in the Dipsacaceae having Compositae pollen structure would appear to strengthen the evidence for a Compositae–Dipsacaceae relationship.

Noteworthy in the reports of Wagenitz (1956), Duigan (1961), and Van der Spoel-Walvius and De Vries (1964) is that the pollen morphology of those taxa which seem Compositae-like are all related to one Compositae morphologic pattern, the Anthemoid pattern. These studies suggest that the Anthemoid pattern is ancestral for the family. Although they have not, to our knowledge, been further extended to a consideration of pollen prototypes, the studies have provided us with initial points of reference for our EM work with potentially allied Compositae groups.

REVIEW OF PHYTOCHEMISTRY

Of the three major groups thought to be related to the Compositae by traditional taxonomy, chemical evidence somewhat favors the Campanulales. Hegnauer (1964) indicates that both the Compositae and Campanulaceae share many groups of similar chemical constituents, namely alkaloids, polyacetylenes, terpenes, and inulin. Inulin, which is notable in the Compositae, has also been detected in other Campanulalean associated families; the Lobeliaceae, Goodeniaceae and Stylidiaceae (Hegnauer, 1964). Cronquist (1968) cites phenolic evidence for a Rubiaceae-Compositae relationship. This is somewhat discounted because similar classes of flavonoids have been found in the Rubiales, Dipsacales and Campanulales (Bate-Smith and Metcalfe, 1957; Bate-Smith, 1962). In addition to these groups, chemical data suggest relationships with the Rutaceae, Boraginaceae, and the Umbelliferae (Hegnauer, 1964). The latter relationship appears to be gaining in popularity (Hegnauer, 1969, 1971; Bohlmann, 1971; Boulter, 1973; Herout, 1973; Harborne, 1973) and receives some support from pollen morphology (to be discussed later).

ELECTRON MICROSCOPY OF RELATED FAMILIES

Using SEM and TEM we have examined the pollen morphology of taxa from a variety of families including most of those that have been suggested as related to the Compositae. These will be discussed on an individual basis.

Calyceraceae

The Calyceraceae is a South American family of four genera and approximately 40 species. Pollen was examined of *Boopis alpina*, *B. spathulatus*, *Calycera pulvinata*, *Moschopsis monocephala*, *Acicarpha procumbens* and *A. tribuloides*. SEM of these taxa (Plate 26A, B, E, Plate 27A–D) shows a smooth exine surface covered with highly reduced spines not unlike members of the tribe Anthemideae. Of particular significance are *Acicarpha procumbens* (Plate 27C–D), *Boopis spathulatus* (Plate 26B) and *Calycera pulvinata* (Plate 26E). In these taxa regions between the three colpi possess flattened or depressed areas resembling the intercolpar concavities of Wodehouse (1928b). These are not peculiar to the Calyceraceae, as they have been noted in other angiosperm families (e.g. Boraginaceae by Nowicke and Skvarla, 1974; Dipterocarpaceae by Maury, Muller, and Lugardon, 1975); their relevance to the present study is that the Mutisioid taxa, *Dasyphyllum excelsum* (Plate 23A) and *Schlechtendalia luzulaefolia*, also possess them. However, the similarity of *Dasyphyllum* and *Schlechtendalia* to these genera of the Calyceraceae is somewhat discounted when comparisons are made with their respective aperture systems. *Boopis spathulatus* (Plate 26B), *Calycera pulvinata* (Plate 26E), and to a lesser extent, *Acicarpha procumbens* (Plate 27C–D), have a distinct layer of granular sporopollenin outlining each colpus region. Furthermore, these taxa have their aperture openings covered by pouch-like protrusions of the exine. Such protrusions are noted in other angiosperm families (viz. Boraginaceae—Nowicke and Skvarla, 1974). Although the pouch-like protrusions and granular sporopollenin lining of the aperture region presently are not well understood in the Compositae, SEM studies strongly suggest that a granular lining may be common to Compositae pollen (e.g. Plates 2B–C, 3C, 9B, 15B, 17B, 23A).

TEM of those Calyceraceae taxa possessing intercolpar concavities indicate a twofold morphology. Plates 26D, F and 27E, G display a single layer of columellae attached to a thick foot layer which in turn is attached to a prominent endexine. This morphology is similar to comparable sections of *Dasyphyllum* and *Schlechtendalia* (Plate 23B–C). In contrast, the adjacent raised areas (Plate 26D, G) suggest an Anthemoid pattern, unlike that found in the latter two genera. As discussed earlier, the raised areas of *Dasyphyllum* suggest a Senecioid exine pattern while those of *Schlechtendalia* show a "modified" Anthemoid pattern.

Calyceraceae taxa without intercolpar concavities (i.e. *Moschopsis monocephala*, Plate 27F; *Boopis alpina*, Plate 26C) possess patterns which we would unhesitatingly refer to as Anthemoid.

To summarize, it seems clear that the Calyceraceae possess a pollen morphology which is strikingly similar to the Compositae. This similarity extends to the specialized intercolpar concavities already discussed. However, differences are evident in apertural structure, but as indicated, they may not be as marked as presently thought. We intend to examine pollen from all available species of the Calyceraceae in order to obtain a better understanding of its variability since it now appears that the family has a variety of morphologic patterns.

Valerianaceae

This family consists of about 15 genera and approximately 400 species. SEM of *Plectritis* (Plate 28A), *Nardostachys* (Plate 28D), *Patrinia* (Plate 28E), and *Valeriana* indicate various degrees of exomorphic similarity to the Compositae. *Plectritis* has highly reduced spines, *Nardostachys* has short, thickened spines, and *Patrinia* and *Valeriana* have spines supported by swollen bases. *Plectritis* resembles Anthemideae pollen, and *Nardostachys* resembles *Ambrosia*. In all taxa there is a lining circumscribing the apertures. Unlike the apertural lining of Calyceraceae and some Compositae, in Valerianaceae it consists of prominent spines.

The endomorphology, observed by TEM (Plate 28B, F, G), with exception of *Nardostachys* (Plate 28C), suggests only a distant relationship with the Compositae. In all taxa a cavus is absent and the columellae are branched distally, but internal tecta are not clearly in evidence. Furthermore, a well-defined endexine has not been observed. *Nardostachys* differs from the other taxa in that a weakly defined internal tectum is present, and the exine does bear some resemblance to the Anthemoid pattern. However, as discussed above, the presence of an endexine is problematical.

In view of the preceeding analyses the Valerianaceae appear less similar to the Compositae than the Calyceraceae. However, enough similarity does exist to link Valerianaceae with Compositae even though such a linkage is remote.

Dipsacaceae

In the present study only *Dipsacus pilosus* (Plate 29G, H) and *Cephalaria transylvanica* (Plate 29I) were examined. Endomorphology is similar to the Valerianaceae (cf. *Valeriana edulis*, Plate 28G); exomorphology differs principally in having larger pollen and highly reduced aperture systems. The two families seem clearly related but much additional sampling among the Dipsacaceae will be needed before conclusive statements can be made.

Campanulaceae

Electron microscopy of *Downingia elegans* (Plate 29D), *D. portarella* (Plate 29F), and *Lobelia puberula* (Plate 29E), members of the large family

Campanulaceae (approximately 60–70 genera and 2000 species) does not indicate any morphologic relationship with the Compositae. *Lobelia* was selected for study since Small (1917–19) suggested that the Compositae, at least as represented by *Senecio*, was related to the segregate family, Lobeliaceae. *Lobelia* itself is a large genus (approximately 200 species) so our present sampling can hardly be considered as sufficient. *Jasione montana*, because of the suggestion by Cronquist (1955b) that it showed certain resemblances (connate anthers, milky juice, vegetative characters) to advanced Compositae, was also examined (unpublished observations). Its pollen, like that of *Downingia* and *Lobelia*, is quite different from the Compositae, a conclusion also reached by Dunbar (1973).

Brunoniaceae

This is a monotypic Australian family, consisting of *Brunonia australis*. In some taxonomic treatments it has been placed in the family Goodeniaceae. SEM indicates a surface covered with minute spines like the Compositae but also with an aperture system more complex than noted in the Compositae (Plate 30B). TEM (Plate 30A) seemingly reveals an Anthemoid pattern, but the upper areas of the columellae, while markedly bifurcate (i.e. digitate), do not form the internal tecta-columellae structure typical of Anthemoid pollen. Therefore, palynological data remotely link the family with the Compositae, although such linkage is not a strong one.

Goodeniaceae

The family consists of about 14 genera and 300 species and is restricted chiefly to Australia. Surface morphology as observed with SEM (Plate 30H–J) shows a wide spectrum of variability from short spines in *Scaevola* and *Goodenia* (Plate 30F, I) to prominent striae in *Dampiera* (Plate 30J) and *Anthotium* to tetrad pollen grains in *Leschenaultia*. With TEM (Plate 30C–G) all these taxa (except the tetrads of *Leschenaultia*) appear nearly similar and are characterized by thickened columellae which bifurcate distally. This morphology is similar to that of the Brunoniaceae (Plate 30A) and supports Duigan's (1961) inclusion of Brunoniaceae in the Goodeniaceae. Pollen of the latter is not especially close to the Compositae. However, an Anthemoid type of morphology is approached and the Goodeniaceae, like the Brunoniaceae, are felt to be distantly associated with the Compositae.

Caprifoliaceae

Pollen of the Caprifoliaceae has been included because it is considered as the most primitive family in the Dipsacales as well as close to the Rubiales (Cronquist, 1968). Electron microscopy of the pollen of *Symphoricarpos occidentalis* (Plate 29A, B) and *Kolkwitzia amabilis* (Plate 29C) does indicate some similarity to *Dipsacus* and *Cephalaria* (Plate 29G-I). In *Symphoricarpos* this similarity is primarily exomorphic; in *Kolkwitzia* it is endomorphic. The latter appears to approach an Anthemoid pattern. As indicated by Punt *et al.* (1974), the family is highly diverse. Consequently, much additional sampling will be needed before meaningful results can be expected.

Rubiaceae

Pollen of only two genera of Rubiaceae have been examined, *Morinda longiflora* (Plate 31F, G) and *Pentanisia schweinfurthii* (Plate 31H, I). Electron microscopy does not suggest a relationship with Compositae. Rather, it more closely parallels some Campanulaceae. A review of the literature (references given earlier) supports our meagre pollen sampling. The Rubiaceae is a very large, heterogeneous family and our current work can therefore not be considered as representative.

Umbelliferae

Although this family is treated as belonging to the Umbellales (Cronquist, 1968; Hutchinson, 1969) or Cornales (Thorne, 1968; Takhtajan, 1969), it was selected for study here because of phytochemical evidence which suggests a direct linkage to the Compositae (Bohlmann, 1971; Chapter 41; Hegnauer, 1971). Earlier, Merxmüller (1954), in reviewing the pollen work of Wagenitz (1955) on the Compositae genus *Centaurea*, suggested an umbelliferoid relationship. It is particularly evident from the work of Cerceau-Larrival (1971; see Thanikaimoni, 1972, 1973, for complete references) that a number of genera such as *Heracleum, Cachrys, Echinophora, Pycnocycla, Orlaya, Trachymene, Lisaea,* and *Turgenia* have a pollen morphology resembling some advanced Compositae groups. In order to gain further insight into the Umbelliferae, pollen of *Trachymene arfakensis, T. anisocarpa* and *Angelica venenosa* was examined by TEM and SEM. As seen in Plate 31, the ultrastructure of these taxa is highly diverse. *Trachymene* (Plate 31C, D) possesses broadened columellae which bifurcate distally. The resultant branches of the columellae do not appear to fuse laterally but rather form a loose net or reticulum. Because of this we do not interpret *Trachymene* to possess an internal tectum like that which occurs in the Compositae. However, the pattern is "Anthemoid" in the sense that it is superficially similar to patterns found in the Vernonieae, including *Liabum*. This similarity to the Compositae is not as striking as that of the Calyceraceae and is further diminished by the totally different surface morphology, as illustrated in the scanning electron micrographs

(Plates 31A, B). The ultrastructural morphology of *Angelica* (Plate 31E) is quite different from *Trachymene* and is also quite different from the Compositae. It consists of simple columellae with a thick tectum. Nevertheless, the morphology of Umbelliferae pollen, or at least certain sections of the family, is close enough to the Compositae to warrant additional intensive comparative studies.

Summary of related families

It is obvious that our investigation of the above-mentioned families is vastly incomplete and of restricted scope; nevertheless, since this is the first attempt of which we are aware to make direct palynological comparisons to the Compositae, we believe that tentative suggestions as to the phyletic import of these data is in order.

Of the nine families investigated, the Calyceraceae exhibit ultrastructural exine features seemingly identical to the Compositae. Next in line are the Valerianaceae, Brunoniaceae, Goodeniaceae, Dipsacaceae, and Umbelliferae. These five families show an exine morphology somewhat apart from the Compositae. However, enough similarity exists to suggest distant linkages. The Campanulaceae, Rubiaceae, and Caprifoliaceae appear to bear little or no resemblance to the Compositae and are not considered to possess pollen affinities. It should be reiterated, however, that our sampling is quite small and study of more taxa (e.g. the Caprifoliaceae) will improve our understanding of these pollen relationships.

Considering these interfamilial comparisons, what can be said about the primitive versus advanced pollen structure for the Compositae as a whole? The Calyceraceae have a pollen morphology like the Compositae, yet this morphology represents only one of the major pollen types elucidated for the family: the Anthemoid pattern. It is an interesting fact that the other families examined also show, to some extent, an Anthemoid pattern. Why, then, do we not find the Helianthoid pattern among these various families if indeed this tribe (i.e. Heliantheae) is thought to be primitive or at least basal for the family? There is no doubt that *exomorphic* similarities to Compositae are encountered in other, completely unrelated angiosperm groups. Examples are evident in Juglandaceae and Rhoipteleaceae (Stone and Broome, 1971). This however, is not thought to be of any significance because the foundation upon which the exine surface is constructed is totally different. In perusing the pollen literature and in our various studies we have not yet found, other than the families discussed above, an internal morphology similar to the Helianthoid pattern and we feel, at least at this time, that it is unique to the family. Similar feelings are expressed for the other Compositae exine patterns we have recognized. Should any of these exine patterns emerge in other groups we would conjecture that these have arisen independently unless they are correlated with yet other characters of the pollen and the plant as a whole. For example, internal foramina in the exine of *Stellaria* (Erdtman, 1968) and *Cerastium* (Skvarla and Nowicke, 1976), members of the Calyophyllaceae, approach that found in the Helianthoid pattern, but, the exine itself is more Anthemoid in structure. Of course, the pollen exomorphology of these two genera is quite different and the Caryophyllaceae by all other criteria is so remote from the Compositae that such comparisons are idle exercises. The Anthemoid pattern, as already discussed, appears to be the only pollen-type which links the Compositae with other families. To our knowledge, the only family in which an Anthemoid pattern may exist, other than those already discussed, is the Sapindaceae. According to George and Erdtman (1969) *Diplopeltis huegelii* has "umbelliferoid" pollen, and while these authors stressed the distinctiveness of this species from others of the genus, it is of interest to note that the Sapindales have been suggested as being related to the Umbellales, on morphological and chemical grounds (Cronquist, 1968; Hegnauer, 1964).

Finally, so as to provide some perspective with what has been said above, and in view of the chemical data (Mabry and Bohlman, Chapter 41; Boulter, 1973) which suggest an umbelloid ancestry for the Compositae, it should be emphasized that the studies of Cerceau-Larrival (references given earlier) and our photographs of Umbelliferae pollen clearly reveal that the latter family also has pollen which is both like and unlike that of the Compositae. This would suggest that a direct evaluation of pollen morphology and phytochemistry are necessary to achieve a more informative presentation.

CONCLUDING REMARKS

The present conspectus of Compositae pollen morphology indicates that four basic structural patterns occur in the family: Anthemoid, Helianthoid, Senecioid, and Arctotoid. Although the present and previous reports have not included every genus of the family we feel reasonably confident that the pollen morphology has now been documented to the point of not yielding new *major* structural patterns.

An extremely interesting aspect of the present study has been that of the relationship(s) of the Compositae with yet other angiosperm groups. This question has occupied the thinking of synantherologists for well over a hundred years but yet it has not been given a great deal of investigation. Part of the reason for this is probably the size and diversification of the family and its phyletic history. Turner (Chapter 2) indicates that the Compositae has descended from a very old phyletic line, the family itself being present as far back as the Cretaceous. With this in mind (i.e. great modern diversity and old lineage) the difficulty in tracing Compositae affinities is compounded because many of the intermediate or interlinking groups simply do not exist. This imposes considerable limitation on anatomical and morphological studies. Still, it does not mean that such studies should be ignored; rather, that they should be intensified, that is, our investigations must eventually become more sophisticated. In addition, there must be an attempt at synthesis of these studies with those from yet other disciplines, especially the fledgling field of chemosystematics.

ACKNOWLEDGEMENTS

This study was supported by grants from the National Science Foundation to J. J. Skvarla, B. L. Turner and A. S. Tomb. We are grateful to R. Wibel (University of Illinois at Chicago Circle) for preliminary scanning electron microscopy. Sincerest appreciation is extended to W. F. Chissoe, B. Richey and C. Pyle for their technical assistance.

REFERENCES

AVETISIAN, E. M. (1948). Palynologia caucasia. 3. Campanulaceae. *Trudy bot Inst., Erevan* **5**, 199–206.

AVETISIAN, E. M., 1964. Palynosystématique de la tribu des Centaureinae des Asteraceae. *Tr. Bot. Inst. Akad. Nauk Sci. Armen.*, **14**, 31–47.

BADRÉ, F., CADET, T., CUSSET, G. and HIDEUX, M. (1975). Position systématique, étude morphologique et palynologique du genre *Berenice*. *Adansonia* **15**, 139–145.

BAHADUR, B. (1964). Pollen exine dimorphism in heterostyled *Oldenlandia umbellata* (Rubiaceae). *Rhodora* **66**, 56–60.

BAKER, H. G. (1956). Pollen dimorphism in Rubiaceae. *Evolution* **10**, 23–31.

BARROSO, G. M. and MAGUIRE, B. (1973). A review of the genus *Wundelichia* (*Mutisieae*, Compositae). *Revta brasil. Biol.* **33**, 379–406.

BATE-SMITH, E. C. (1962). The phenolic constituents of plants and their taxonomic significance. I. Dicotyledons. *J. Linn. Soc. (Bot.)* **58**, 95–173.

BATE-SMITH, E. C. and METCALFE, C. R. (1957). Leucoanthocyanins. 3. The nature and systematic distribution of tannins in dicotyledonous plants. *J. Linn. Soc. (Bot.)* **55**, 669–705.

BECKER, K. M. (1973). A comparison of angiosperm classification systems. *Taxon* **22**, 19–50.

BENTHAM, G. (1873). Notes on the classification, history and geographical distribution of Compositae. *J. Linn. Soc. (Bot.)* **13**, 335–557.

BENTHAM, G. and HOOKER, J. D. (1862–1883). "Genera Plantarum". 3 volumes.

BESOLD, B. (1971). Pollenmorphologische untersuchungen in Inuleen. (Angianthinae, Relhaniinae, Athrixiinae). *Dissertationes Botanicae* **14**, 1–72.

BOHLMANN, F. (1971). Acetylenic compounds in the Umbelliferae. *In* "The Biology and Chemistry of the Umbelliferae" (V. Heywood, ed.), pp. 279–291. Academic Press, London and New York.

BOISSIER, E. (1875). "Flora orientalis, Composées", **3**, 151–883.

BOULTER, D. (1973). The use of amino acid sequence data in the classification of higher plants. *In* "Chemistry in Botanical Classification" (G. Benz and J. Santesson, eds), pp. 211–216. Academic Press.

BREMER, K. (1972). The genus *Osmitopsis* (Compositae). *Bot. Notiser* **125**, 9–48.

BREWER, J. G. and HENSTRA, S. (1970). A membrane investing mature individual pollen grains of *Pyrethrum* (*Chrysanthemum cinerariaefolium*, Vis). *Euphytica* **19**, 121–124.
CABRERA, A. L. (1944). Vernonieas Argentinas (Compositae). *Darwiniana* **6**, 265–379.
CABRERA, A. L. (1959). Revision del género *Dasyphyllum* (Compositae). *Revta Mus. La Plata, Bot.* (N.S.) **9**, 21–100.
CABRERA, A. L. (1966). Revision del genero *Mutisia* (Compositae). *Op. lilloana* **13**, 1–127.
CABRERA, A. L. (1971). Flora Patagonica. Parte VII. Compositae, pp. 1–451.
CARLQUIST, S. (1957a). The genus *Fitchia* (Compositae). *Univ. Calif. Publs Bot.* **29**, 1–143.
CARLQUIST, S. (1957b). Anatomy of Guayana Mutisieae. *Mem. N.Y. bot. Gdn* **9**, 441–476.
CARLQUIST, S. (1963). Studies in *Fitchia* (Compositae): Novelties from the Society Islands; Anatomical studies. *Pacif. Sci.* **17**, 282–298.
CASSINI, H. (1817). Apercu des genres noaveaux formés par M. Henri Cassini dans la famille des Synanthérées. *Bull. Sci. Soc. philom. Paris* **6**, 153–154.
CERCEAU-LARRIVAL, M.-TH. (1971). Morphológie pollinique et corrélations phylogénétiques chez les Ombelliféres. *In* "The Biology and Chemistry of the Umbelliferae" (V. Heywood, ed.), pp. 109–155. Academic Press, London and New York.
CHAPMAN, J. (1966). Comparative palynology in Campanulaceae. *Trans. Kansas Acad. Sci.* **69**, 197–200.
CRONQUIST, A. (1955a). Compositae. *In* "Vascular Plants of the Pacific Northwest" (C. L. Hitchcock, A. Cronquist, M. Ownbey, and J. W. Thompson, eds). *Univ. Wash. Publs Biol.* **17**, 1–343.
CRONQUIST, A. (1955b). Phylogeny and taxonomy of the Compositae. *Am. Midl. Nat.* **53**, 478–551.
CRONQUIST, A. (1968). "The Evolution and Classification of Flowering Plants". Houghton-Mifflin, Boston.
DALMAU, J. (1961). Polen. Estructura y características de los granos de polen.—Precisiones morfológicas sobre el polen de especies recolectadas en el N.E. de Espana.—Polinización y aeropalinología.
DAVIS, P. H. and HEYWOOD, V. H. (1963). "Principles of Angiosperm Taxonomy". Van Nostrand, Princeton, N.J.
DAVIS, W. S. and RAVEN, P. H. (1962). Three new species related to *Malacothrix clevelandii*. *Madroño* **16**, 258–266.
DIMON, M.-T. (1971). Problémes généraux soulevés par l'etude pollinque de composées méditerranéennes. *Naturalia monspeliensia* **22**, 129–144.
DUIGAN, S. L. (1961). Studies of the pollen grains native to Victoria, Australia. 1. Goodeniaceae (including *Brunonia*). *Proc. R. Soc. Vict.* **74**, 89–109.
DUNBAR, A. (1973). A short report on the fine structure of some Campanulaceae pollen. *Grana* **13**, 25–28.
DUNBAR, A. (1975a). On pollen of Campanulaceae and related families with special reference to the surface ultrastructure. I. Campanulaceae Subfam. Campanuloidae. *Bot. Notiser* **128**, 73–101.
DUNBAR, A. (1975b). On pollen of Campanulaceae and related families with

special reference to the surface ultrastructure. II. Campanulaceae Subfam. Cyphioidae and Subfam. Lobelioidae; Goodeniaceae; Sphenocleaceae. *Bot. Notiser* **128**, 102-118.

ERDTMAN, G. (1952). "Pollen morphology and plant taxonomy. I. Angiosperms". Almquist and Wiksell, Stockholm.

ERDTMAN, G. (1960). The acetolysis method. A revised description. *Svensk bot. Tidskr.* **54**, 561-564.

ERDTMAN, G. (1963). Palynology. *In* "Vistas in Botany" (W. B. Turrill, ed.), pp. 23-54. Pergamon Press, New York.

ERDTMAN, G. (1966). Sporoderm morphology and morphogenesis. A collocation of data and suppositions. *Grana Palynol.* **6**, 317-323.

ERDTMAN, G. (1968). On the exine in *Stellaria crassipes* Hult. *Grana Palynol.* **8**, 271-276.

ERDTMAN, G. (1969). "Handbook of Palynology". Munksgaard, Copenhagen.

ERDTMAN, G. (1970). Topography and non-topography in exine organization. *Grana* **10**, 243-245.

ERDTMAN, G. (1971). "Pollen Morphology and Plant Taxonomy. Angiosperms" (Corrected reprint of the edition of 1952 with a new addendum). Hafner, New York.

FAEGRI, K. (1956). Recent trends in palynology. *Bot. Rev.* **22**, 639-664.

FAEGRI, K. and IVERSEN, J. (1975). "Textbook of Modern Pollen Analysis". Hafner, New York.

FELIPPE, G. M. and LABORIAU, M. L. S. (1964). Pollen grains of plants of the "Cerrado". VI. Compositae—Tribe Heliantheae. Acad. Brasileira de Ciencias **36**, 85-101.

FEUER, S. M. (1974). "Pollen Morphology and Ultrastructure in the Subtribe Microseridinae (Tribe Lactuceae: Family Asteraceae)". M.S. thesis, University of Illinois at Chicago Circle.

FISCHER, H. (1890). "Beitrage zur vergleichenden Morphologic der Pollenkorner". Berlin.

FISHER, T. R. and WELLS, J. R. (1962). Heteromorphic pollen grains in *Polymnia. Rhodora* **64**, 336-340.

GEORGE, A. S. and ERDTMAN, G. (1969). A revision of the genus *Diplopeltis* Endl. *Grana Palynol.* **9**, 92-109.

HARBORNE, J. B. (1973). Flavonoids as systematic markers in angiosperms. *In* "Chemistry in Botanical Classification" (eds G. Benz and J. Santesson), pp.103-116. Academic Press.

HEGNAUER, R. (1964). "Chemotaxonomic der Pflanzen. III. Dicotyledoneae: Acanthaceae-Cyrrillaceae". Birkhauser Verlag, Basel and Stuttgart.

HEGNAUER, R. (1969). Chemical evidence for the classification of some plant taxa. *In* "Perspectives in Phytochemistry" (J. B. Harborne and T. Swain, eds), pp. 121-138. Academic Press, London and New York.

HEGNAUER, R. (1971). Chemical patterns and relationships of Umbelliferae. *In* "The Biology and Chemistry of the Umbelliferae" (V. H. Heywood, ed.), pp. 267-277. Academic Press, London and New York.

HEROUT, V. (1973). A chemical compound as taxonomic character. *In* "Chemistry in Botanical Classification", pp. 55-62. Benz, G. and Santesson, J. (eds). Academic Press, London and New York.

HESLOP-HARRISON, J. (1969). Scanning electron microscope observations on the wall of the pollen grain of *Cosmos bipinnatus* Compositae. *In* Proc. 2nd Stereoscan Colloquium, pp. 89–96. Engis Equipment Co., Merton Grove, Ill.

HEUSSER, C. J. (1971). "Pollen and Spores of Chile". University of Arizona Press, Tucson.

HOFFMANN, O. (1894). Compositae. *In* "Die natürliche Pflanzenfamilien" (Engler and Prantl, eds) **4** (5), 87–387.

HUTCHINSON, J. (1969). "Evolution and Phylogeny of the Flowering Plants". Academic Press, London and New York.

IKUSE, M. (1956). "Pollen Grains of Japan". Hirokawa Publishing Co., Tokyo.

JONES, S. B. (1970). Scanning electron microscopy of pollen as an aid to the systematics of *Vernonia* (Compositae). *Bull. Torrey bot. Club* **97**, 325–335.

KING, R. M. and ROBINSON, H. (1967). Multiple pollen forms in two species of the genus *Stevia* (Compositae). *Sida* **3**, 165–169.

KING, R. M. and ROBINSON, H. (1968). Studies in the Compositae—Eupatorieae VIII. Observations on the microstructure of *Stevia*. *Sida* **3**, 257–269.

KING, R. M. and ROBINSON, H. (1970). Studies in the Eupatorieae (Compositae). XIV. Another example of dimorphic pollen? *Phytologia* **19**, 301–302.

KINGHAM, D. L. (1976). A study of the pollen morphology of tropical African and certain other Vernonieae (Compositae). *Kew Bull.* **31**, 9–26.

KNOX, R. B. and HESLOP-HARRISON, J. (1969). Cytochemical localization of enzymes in the wall of the pollen grain. *Nature, Lond.* **223**, 92–94.

KNOX, R. B. and HESLOP-HARRISON, J. (1970). Pollen-wall proteins: Localization and enzymic activity. *J. Cell Sci.* **6**, 1–27.

LEWIS, W. H. (1964). *Oldenlandia corymbosa* (Rubiaceae). *Grana palynol.* **5**, 330–341.

LEWIS, W. H. (1966). The Asian Genus *Neanotis* nomen novum (*Anotis*) and allied taxa in the Americas (Rubiaceae). *Ann. Mo. bot. Gdn* **53**, 32–46.

LIENS, P. (1968). Versuch einer Gliederung der Inulinae und Buphthalminae nach den Pollenkorntypen. *Ber. dt. bot. Ges.* **81**, 498–504.

LIENS, P. (1970). Die Pollenkorner und Verwandtschaftsbeziehungen der Gattung *Eremothamnus* (Asteraceae). *Mitt. bot. St Samml., Münch.* **7**, 369–376.

LIENS, P. (1971). Pollensystematische studien an Inuleen. I. Tarchonanthinae, Plucheinae, Inulinae, Buphthalminae. *Bot. Jb.* **91**, 91–146.

LIENS, P. and THYRET, G. (1971). Pollen phylogeny and taxonomy exemplified by an African Asteraceae group. *Mitt. bot. St Samml., Münch.* **10**, 280–286.

MANTEN, A. A. (1970). Ultra-violet and electron microscopy and their application in palynology. *Rev. Palaeobot. Palynol.* **10**, 5–37.

MARTICORENA, C. and PARRA, O. (1975). Morfologia de los granos de polen de *Hesperomannia* Gray y *Moquinia* DC. (Compositae-Mutisieae). Estudio comparativo con generos afines. *Gayana* **29**, 3–22.

MAURY, G., MULLER, J. and LUGARDON, B. (1975). Notes on the morphology and fine structure of the exine of some pollen types in Dipterocarpaceae. *Rev. Palaeobot. Palynol.* **19**, 241–289.

MERXMÜLLER, H. (1954). Beitrage zur Taxonomie der Compositen. *Ber. dt. bot. Ges.* **67**, 23–24.

MOENS, P. (1962). Observations sur le pollen de quelques especes du genre *Coffea* et de certains genres voisins (Rubiacees). *Pollen et Spores* **4**, 47–64.

Moore, S. (1929). Alabastra diversa. Part 36. 2. Notes on African Compositae. **67**, 273-276.
Moore, R. J. (1972). Index to plant chromosome numbers for 1970. *Regnum Vegetabile* **84**, 1-138.
Muller, J. (1970). Palynological evidence on early differentiation of angiosperms. *Biol. Rev.* **45**, 417-450.
Nowicke, J. W. and Skvarla, J. J. (1974). A palynological investigation of the genus *Tournefortia* (Boraginaceae). *Am. J. Bot.* **61**, 1021-1036.
Ornduff, R., Raven, P. H., Kyhos, D. W. and Kruckeberg, A. R. (1963). Chromosome numbers in Compositae. III. Senecioneae. *Am. J. Bot.* **50**, 131-139.
Ornduff, R., Mosquin, T., Kyhos, D. W. and Raven, P. H. (1967). Chromosome numbers in Compositae. VI. Senecioneae. II. *Am. J. Bot.* **54**, 205-213.
Parra, O. (1969-1970). Morfologia de los granos de polen de las Campuestas Cynareas Chilenos. *Bol. Soc. biol. Concepcion* **42**, 89-96.
Parra, O. and Marticorena, C. (1972). Granos de polen de plantas Chilenas. II. Compositae-Mutisieae. *Gayana* **21**, 1-107.
Pausinger-Frankenburg, F. (1951). Vom Blutenstaub der Wegwarten. (Die Pollengestaltung der Cichorieae.) *Corinthia* **II**, 3-47.
Payne, W. W. (1963). The morphology of the inflorescence of ragweeds (*Ambrosia-Franseria*: Compositae). *Am. J. Bot.* **50**, 872-880.
Payne, W. W. (1972). Observations of harmomegathy in pollen of Anthophyta. *Grana* **12**, 93-98.
Payne, W. W. and Skvarla, J. J. (1970). Electron microscope study of *Ambrosia* pollen (Compositae: Ambrosieae). *Grana* **10**, 89-100.
Philipson, W. R. (1953). The relationships of the Compositae particularly as illustrated by the morphology of the inflorescence in the Rubiales and the Campanulatae. *Phytomorphology* **3**, 391-404.
Pons, A. and Boulos, L. (1972). Révision systématique du genre *Sonchus* L. s.l. III. Etude palynologique. *Bot. Notiser* **125**, 310-319.
Praglowski, J. (1971). The pollen morphology of the Scandinavian species of *Artemisia* L. *Pollen et Spores* **13**, 381-404.
Punt, W., Reitsma, Tj. and Reuvers, A. A. M. L. (1974). Caprifoliaceae. *Rev. Palaeobot. Palynol.* **17**, 5-29.
Robinson, H. and Brettell, R. D. (1973a). Tribal revisions in the Asteraceae. VI. The relationship of *Eriachaenium*. *Phytologia* **26**, 71-72.
Robinson, H. and Brettell, R. D. (1973b). Tribal revisions in the Asteraceae. VII. The relationship of *Isoetopsis*. *Phytologia* **26**, 73-75.
Robinson, H. and Brettell, R. D. (1973c). Tribal revisions in the Asteraceae. VIII. A new tribe, Ursinieae. *Phytologia* **26**, 76-85.
Robinson, H. and Brettell, R. D. (1973d). Tribal revisions in the Asteraceae. XI. A new tribe, Eremothamnae. *Phytologia* **26**, 163-166.
Robinson, H. and Brettell, R. D. (1973e). Tribal revisions in the Asteraceae. III. A new tribe, Liabeae. *Phytologia* **25**, 404-407.
Robinson, H. and Brettell, R. D. (1974). Studies in the Liabeae (Asteraceae). II. Preliminary survey of the genera. *Phytologia* **28**, 43-63.
Rowley, J. R. and Nilsson, S. (1972). Structural stabilization for electron microscopy of pollen from herbarium specimens. *Grana palynol.* **12**, 23-30.

SAAD, S. I. (1961). Pollen morphology in the genus *Sonchus*. *Pollen et Spores* **3**, 247–260.

SCHTEPA, I. S. (1958). Ad cognitionem pollinis morphologiae generum nonnulorum tribus Cynareae familiae Compositae. *Notulae systematiceae Instituti Botanici Thbilissiensis* **20**, 53–62.

SCHTEPA, I. S. (1962). Palynological data for the systematics of the genus *Cousinia* Cass. *Pollen et Spores* **4**, 375.

SCHTEPA, I. S. (1965). Materies ad studium pollinis characterum generis *Cirsium* Mill. *Notulae systematiceae Institute Botanici Thbilissensis* **25**, 69–82.

SCHTEPA, I. S. (1966). On the problem of affinity between the genera *Arctium* L. and *Cousinia* Cass. of the family Compositae. *In* "The Importance of Palynological Analysis for the Stratigraphic and Paleofloristic Investigations". Acad. Sci. USSR, Moscow.

SCHTEPA, I. S. (1967). Commentationes de palinologia et systematica specierum caucasicarum generis *Couisinia* Cass. *Notulae Systematiceae Instituti Botanici Thbilissiensis* **26**, 57–62.

SCHTEPA, I. S. (1973). On the natural boundaries between the genera *Cousinia* and *Arctium* as suggested by palynological evidence. *In* "Pollen and Spore Morphology of the Recent Plants", pp. 37–40. Proc. 3rd Int. Palyn. Conf., Acad. Sci. USSR.

SHAW, G. (1971). The Chemistry of sporopollenin. *In* Sporopollenin (J. Brooks, M. Muir, P. van Gijzel, and G. Shaw, eds), pp. 305–350. Academic Press, London and New York.

SKVARLA, J. J. (1966). Techniques of pollen and electron microscopy. Part I. Staining, dehydration and embedding. *Okla. geol. Notes* **26**, 179–186.

SKVARLA, J. J. (1973). Pollen. *In* "Encyclopedia of Microscopy and Microtechnique" (P. Gray, ed.), pp. 456–459. Van Nostrand Reinhold, New York.

SKVARLA, J. J. and LARSON, D. A. (1965a). An electron microscopic study of pollen morphology in the Compositae with special reference to the Ambrosiinae. *Grana palynol.* **6**, 210–269.

SKVARLA, J. J. and LARSON, D. A. (1965b). Interbedded exine components in some Compositae. *Southwest Nat.* **10**, 65–68.

SKVARLA, J. J. and TURNER, B. L. (1966a). Pollen wall ultrastructure and its bearing on the systematic position of Blennosperma and Crocidium (Compositae). *Am. J. Bot.* **53**, 555–563.

SKVARLA, J. J. and TURNER, B. L. (1966b). Systematic implications from electron microscopic studies of Compositae pollen—a review. *Ann. Mo. bot. Gdn* **53**, 200–256.

SKVARLA, J. J. and TURNER, B. L. (1969). Fine structure of Petrobinae (Compositae-Heliantheae) pollen walls. *Am. J. Bot.* **56**, 418–491.

SKVARLA, J. J. and TURNER, B. L. (1971). Fine structure of the pollen of *Anthemis nobilis* L. (Anthemideae-Compositae). *Okla. Acad. Sci.* **51**, 61–62.

SKVARLA, J. J. and NOWICKE, J. W. (1976). The structure of the exine in the order Centrospermae. *Plant Syst. Evol.* **126**, 55–78.

SMALL, J. (1917–1919). The origin and development of the Compositae. *New Phytol.* **16**, 157–177; 198–221; 253–276. **17**, 13–40; 69–94; 114–142; 200–230. **18**, 1–35; 65–89; 129–176; 201–234.

SMITH, C. E. (1969). Pollen characteristics of African species of *Vernonia*. *J. Arnold Arbor.* **50**, 469–477.
SOUTHWORTH, D. (1966). Ultrastructure of *Gerbera jamesonii* pollen. *Grana palynol.* **6**, 324–337.
SOUTHWORTH, D. and BRANTON, D. (1971). Freeze-etched pollen walls of *Artemisia pycnocephala* and *Lilium humboldtii*. *J. Cell Sci.* **9**, 193–207.
STEBBINS, G. L. (1940). Studies in the Cichorieae: *Dubyaea* and *Soroseris*, endemics of the Sino-Himalayan region. *Mem. Torrey bot. Club* **19**, 1–76.
STEBBINS, G. L. (1953). A new classification of the tribe Cichorieae, family Compositae. *Madroño* **12**, 65–81.
STIX, E. (1960). Pollenmorphologische untersuchungen and Compositae. *Grana palynol.* **2**, 41–114.
STIX, E. (1964). Polarisationsmikroskopische untersuchungen am sporoderm von *Echinops banaticus*. *Grana palynol.* **5**, 289–297.
STIX, E. (1970). Beitrag zur morphogenese der pollenkerner von *Echinops banaticus*. *Grana* **10**, 240–242.
STONE, D. E. and BROOME, C. R. (1971). Pollen ultrastructure: evidence for relationship of the Juglandaceae and the Rhoipteleaceae. *Pollen et Spores* **13**, 5–14.
TAKHTAJAN, A. (1969). "Flowering Plants: Origin and Dispersal" (Translated from the Russian by C. Jeffrey). Oliver and Boyd, Edinburgh.
TARNAVSCHI, I. T. and RADULESCU, D. (1959). Untersuchungen uber die morphologie des pollens der Campanulaceae aus der flora der Rumanischen Volkarepublik. *Rev. biol. Acad. Rep. pop. Roum.* **4**, 5–17.
THANIKAIMONI, G. (1972). "Index bibliographique sur la Morphologie des Pollens d'Angiosperms". Institut Francaís de Pondichéry. All India Press.
THANIKAIMONI, G. (1973). "Index bibliographique sur la Morphologie des Pollens d'Angiosperms". Supplément-1. Institut Francaís de Pondichéry. All India Press.
THORNE, R. (1976). A phylogenetic classification of the Angiospermae. *Evol. Biol.* **9**, 35–106.
TOMB, A. S. (1972a). The systematic significance of pollen morphology in the family Compositae. Tribe Cichorieae. *Brittonia* **24**, 129.
TOMB, A. S. (1972b). Re-establishment of the genus *Prenanthella* Rydb. (Compositae: Cichorieae). *Brittonia* **24**, 223–228.
TOMB, A. S. (1973). *Shinnersoseris* gen. nov. (Compositae: Cichorieae). *Sida* **5**, 183–189.
TOMB, A. S. (1976). A preliminary survey of pollen morphology and detailed structure in tribe Lactuceae, family Compositae. *Grana* (in press).
TOMB, A. S., LARSON, D. A. and SKVARLA, J. J. (1974). Pollen morphology and detailed structure of family Compositae. Tribe Cichorieae. I. Subtribe Stephanomeriinae. *Am. J. Bot.* **61**, 486–498.
TURNER, B. L. (1970). Chromosome numbers in the Compositae. XII. Australian species. *Am. J. Bot.* **57**, 382–389.
UENO, J. (1969). The fine structure of pollen surface. I. *Taraxacum* and *Ambrosia*. *Rep. Fac. Sci., Shizuoka Univ.* **4**, 67–74.
UENO, J. (1972). The fine structure of pollen surface. III. *Gazania* and *Stokesia*. *Rep. Fac. Sci. Shizuoka Univ.* **7**, 103–116.

VAN CAMPO, M., BRONCKERS, F. and GUINET, PH. (1967). Electron microscopy's contribution to the knowledge of the structure of acetolysed pollen grains (1). *Palynol. Bull.* **3-4**. Supplement.

VAN DER SPOEL-WALVIUS, M. R. and DE VRIES, R. J. (1964). Description of *Dipsacus fullonum* L. pollen. *Acta bot. neerl.* **13**, 422-431.

WAGENITZ, G. (1955). Pollenmorphologie und systematik in der gattung *Centaurea* L. *Flora, Jena* **142**, 213-279.

WAGENITZ, G. (1956). Pollenmorphologie der mitteleuropaischen Valerianaceen. *Flora, Jena* **143**, 473-485.

WAGENITZ, G. (1964). Compositae. *In* Englers Syllabus II, **12**. Aufl. Berlin.

WALKER, J. W. (1974). Evolution of exine structure in the pollen of primitive angiosperms. *Am. J. Bot.* **61**, 891-902.

WELLS, J. R. (1971). Variations in *Polymnia* pollen. *Am. J. Bot.* **58**, 124-130.

WITTMANN, G. and WALKER, D. (1965). Towards simplification in sporoderm description. *Pollen et Spores* **7**, 443-456.

WODEHOUSE, R. P. (1926). Pollen grain morphology in the classification of the Anthemideae. *Bull. Torrey bot. Club* **53**, 479-485.

WODEHOUSE, R. P. (1928a). Pollen grains in the identification and classification of plants. I. The Ambrosiaceae. *Bull. Torrey bot. Club* **55**, 181-198.

WODEHOUSE, R. P. (1928b). Pollen grains in the identification and classification of plants. II. *Barnadesia*. *Bull. Torrey bot. Club* **55**, 449-462.

WODEHOUSE, R. P. (1928c). The phylogenetic value of pollen grain characters. *Ann. Bot.* **42**, 891-934.

WODEHOUSE, R. P. (1929a). Pollen grains in the identification and classification of plants. III. The Nassauvinae. *Bull. Torrey bot. Club* **56**, 123-138.

WODEHOUSE, R. P. (1929b). Pollen grains in the identification and classification of plants. IV. The Mutisieae. *Am. J. Bot.* **16**, 297-313.

WODEHOUSE, R. P. (1929c). The origin of symmetry patterns of pollen grains. *Bull Torrey bot. Club* **56**, 339-350.

WODEHOUSE, R. P. (1930). Pollen grains in the identification and classification of plants. V. *Haplopappus* and other Astereae: the origin of their furrow configurations. *Bull. Torrey bot. Club* **57**, 21-46.

WODEHOUSE, R. P. (1931). The origin of the six-furrowed configuration of *Dahlia* pollen grains. *Bull Torrey bot. Club* **57**, 371-380.

WODEHOUSE, R. P. (1935). "Pollen Grains". McGraw-Hill, New York.

TABLE I. Pollen examined and collection data

Taxon[a]	Locality	Collector	Herbarium	Plate
COMPOSITAE (Tribe VERNONIEAE)				
Elephantopus carolinianus Willd.	Georgia	Cronquist 4687	US	1B
Harleya oxylepis (Benth.) Blake	British Honduras	Bartlett 12042	US	1G
Vernonia amygdalina Del.	Tanganyika	Richards 25767B	K	1F
V. capreaefolia (Sch.-Bip.) Gleason	Mexico	Cronquist 9705	US	1E
V. pacchensis Benth.	Peru	Wurdack 715	TEX	1H
V. patula (Dryland) Merrill var. *patula*	Thailand	King 5563	TEX	1A
V. subulata Baker	Brazil	Irwin et al. 15553	TEX	1C
V. venosissima Sch.-Bip.	Brazil	Irwin et al. 15642	TEX	1D
COMPOSITAE (Tribe EUPATORIEAE)				
Carminata tenuiflora DC.	Texas	Correll 33677	TEX	2A, B
Carphephorus bellidifolius (Michx.) T. & G.	North Carolina	Radford 28891	TEX	3E
Decachaeta haenkeana DC.	Mexico	Cronquist & Fay 10901	TEX	3C, D
Kanimia purpurescens Baker	Brazil	Irwin et al. 13054	TEX	2C, D
Liatris aspera Michx.	Texas	Duval 257	TEX	3A, B
Mikania cordata (Burm.) B.L. Rob.	Liberia	Barker 1083	K	2E
COMPOSITAE (Tribe ASTEREAE)				
Aphanostephus skirrhobasis (DC.) Trel.	Oklahoma	Perino 500	OKL	4C
Boltonia diffusa Ell.	Oklahoma	Taylor & Taylor 3463	OKL	4E
Calotis erinacea Steetz	Australia	Turner 5202	TEX	4A, B
Chaetopappa asteroides DC.	Oklahoma	Perino 487	OKL	4D

[a] Taxa names and authorities taken from herbarium labels.

Taxon	Locality	Collector	Herbarium	Plate
COMPOSITAE (Tribe INULEAE)				
Acomis rutidosis F. Muell.	Australia	Everist s.n. (in 1955)	US 2243730	5A
Adenocaulon bicolor Hook.	Michigan	Hiltunen 438	MSU	6I
Allogopappus dichotomus Cass.	Canary Islands	Murray (in 1892)	MO 2329870	6C, D
Blumea mollis (Don) Merr.	Indo-China	Clemens & Clemens 3676	MSU	5E
Dimeresia howellii A. Gray	California	Sharsmith 4490	US	5F
Gymnarrhena micrantha Desf.	Algeria	Balsana s.n.	MO 1592062	6J
Helichrysum davenportii F. Muell.	Australia	Turner 5441	TEX	6B
Osmites parvifolia DC.	South Africa	Parker 4598	MO	6H
Osmitopsis asteriscoides (L.) Cass.	South Africa	Wall	S	6G
Pithocarpha corymbosa Lindl.	Australia	George 2350	MO	5B, C
Polycline proteiformis Humbert	Madagascar	Humbert & Swingle 5709	US	5D
Raoulia monroi Hook. f.	New Zealand	Sledge 379	US	6E, F
Stoebe capitata Berg.	South Africa	Pillans 10747	MO	6A
Tarchonanthus minor Less.	South Africa	Sidey 3465	NY	5G, H
COMPOSITAE (Tribe HELIANTHEAE)				
Alepidocline annua Blake	Guatemala	Skutch 722	TEX	7D
Enhydra sessilis (SW.) DC.	Jamaica	Proctor 33128	TEX	7C
Eryngiophyllum pinnatisectum P. G. Wilson	Mexico	Hinton et al. 8020	TEX	7E, F
Fitchia cuneata Moore ssp. *tahaaensis* Grant & Carlquist	Tahiti	Grant 5161	RSA	8A, B
F. speciosa Cheeseman	Hawaii	Carlquist 1684	RSA	8C, D
Guizotia abyssinica (L.f.) Cass.	Ethiopia	Albers 62168	TEX	7A, B
Helianthus giganteus L.	Oklahoma	Bebb 6079a	OKL	7G

8. POLLEN MORPHOLOGY 203

			MO 174411[b]	
COMPOSITAE (Tribe HELENIEAE)				
Amblyolepis setigera DC.	—	Chile	MO	11C
Amblyopappus pusillus H. & A.	Worth & Morrison 16173	Texas	MO	11B
Baileya multiradiata Torr.	Goodman 7656	Ecuador	OKL	9B
Cacosmia rugosa var. *arachnoidea* Hieron.	Wiggins 10810	California	TEX	11A
Chaenactis glabriuscula var. *lanosa* (DC.) Hall	Wolf 6966		OKL	10F, G
Dyssodia anthemidifolia Benth.	Wiggins 15975	Mexico	TEX	11D
Eriophyllum caespitosum Dougl.	Cusick 2605	Oregon	MO	10K
Espejoa mexicana DC.	Johnston 5979	Mexico	TEX	10A
Flaveria anomala B.L. Rob.	Waterfall 15735	Mexico	OKL	10D
Hymenothrix wislizenii A. Gray	Clark 11093	Arizona	OKL	10J
Jaumea peduncularis (H. & A.) Oliv. & Hieron.	Cronquist 9818	Mexico	TEX	10B
Orochaenactis thysanocarpha (A. Gray) Cov.	Alexander & Kellogg 3345	California	MO	10E
Oxypappus scaber Benth.	Cronquist 9777	Mexico	MO	10I
Pericome caudata A. Gray	Goodman 7569	New Mexico	OKL	10H
Tagetes elongata Willd.	King 2248	Mexico	TEX	9C
T. zypaquirensis H. & B.	Camp E-4005	Ecuador	NY	9A
Whitneya dealbata A. Gray	Heller 10824	California	MO	10C
COMPOSITAE (Tribe ANTHEMIDEAE)				
Anthemis cotula L.	Taylor & Taylor 3984	Oklahoma	OKL	12A
A. ruthenica M. & B.	Legerstrom s.n.	Sweden	TEX	12I
Artemisia alaskana Rydb.	Taylor 2787	Alaska	OKL	12G
Centipeda cunninghamii (DC)	Constable 18730	New South Wales	MO	12D
Ceratogyne obionoides Turcz.	Turner 5302	Australia	TEX	12H

[b] Identified by herbarium accession number.

Taxon	Locality	Collector	Herbarium	Plate
COMPOSITAE (Tribe ANTHEMIDEAE) cont'd.				
Cotula coronopifolia L.	Australia	Turner 5454	TEX	12B
Elachanthus pusillus F. Muell.	Australia	Wilson 2168	AD	12E
Eriocephalus hoffmannianus Schl.	S. W. Africa	Schlechter 10888	MO	12F
Soliva stolonifera (Brat.) Landon	Alabama	Harper 3351	MO	12C
COMPOSITAE (Tribe SENECIONEAE)				
Arnica acaulis Britt.	South Carolina	Bozeman 9152	TEX	13A
A. chamissonis Less.	Colorado	Porter 9154	TEX	13B
Cacalia goldsmithii B.L. Rob.	Mexico	McVaugh 15460	TEX	14E
Doronicum altaicum Pall.	Tibet	Chapman 449	K	16E
Emilia coccinea (Sims) Sweet	Costa Rica	King 5342	TEX	14F
Erechtites valerianaefolia DC.	Brazil	Irwin et al. 9640	TEX	14D
Gamolepis brachypoda DC.	South Africa	Rogers 28687	K	14H
Gynoxys parvifolia Cuatr.	Peru	Wurdack 1702	TEX	14B
Liabum caducifolium B.L. Rob. & Bartl.	Mexico	Cronquist 9764	TEX	15D
L. glabrum var. *hypoleucum* Greenm.	Mexico	Cronquist 9765	MO	15C
L. kluttii B.L. Rob. & Greenm.	Mexico	King 2499	TEX	15E
L. megacephalum Sch.-Bip.	Peru	Hutchinson, Wright & Straw 5934	MO	15A
L. ovatum (Wedd.) J. Ball	Peru	Hutchinson & Wright 4406	MO	15F
L. sagittatum Sch.-Bip.	Columbia	Archer 1207	TEX	15B
Neurolaena cobanensis Greenm.	Guatemala	Tuerckheim 8414	K	16D
Odontotrichum cervinum Rydb.	Mexico	McVaugh 15313	TEX	14C
Petasites hyperboreus Rydb.	Alaska	Anderson 6423	TEX	14I
Raillardella scaposa A. Gray	California	Raven 9800	TEX	13C

Schistocarpha platyphylla Greenm.	Guatemala	King 3147	TEX	16A
S. sinforosii Cuatr.	Peru	Wurdack 796	TEX	16B, C
Senecio heritieri DC.	Canary Islands	Asplund 971	K	13D
S. lyallii Klatt	New Zealand	Mailbrung 1859	K	13E
Tetradymia canescens DC.	Oregon	Cronquist 8427	TEX	14A
Tussilago farfara L.	Czechoslovakia	Deyl 97	TEX	14G
COMPOSITAE (Tribe CALENDULEAE)				
Chrysanthemoides incana (Burm. f.) T. Norl.	Africa	Norlindh	S	18A
C. monolifera (L.) T. Norl.	Australia	Turner 5197	TEX	18C
Dimorphotheca pluvialis (L.) Moench.	Africa	Hutchinson 550	K	19A, B (inset)
Gibbaria ilicifolia (L.) T. Norl.	Africa	Norlindh 5595	S	17A, C
Osteospermum vaillantii (Decne) T. Norl.	Africa	Eggeling 2809	K	17B, C (inset)
Tripteris clandestinum Less.	Australia	Turner 5280	TEX	18B, D
COMPOSITAE (Tribe ARCTOTIDEAE)				
Arctotheca calendula (L.) Levyns	Australia	Symon 23362	K	20D
Arctotis aspera L.	Africa	Norlindh 5939	S	20B
A. verbasifolia Harv.	Africa	Esterhuysen 31619	S	20A
Berkheya rhapontica (DC.) Hutch. & Davy	Africa	Mogg 6896	K	19F–I
Cullumia setosa (L.) R. Br.	Africa	Ryder (in 1931)	K	19C–E
Cymbonotus lawsonianus Gaud.	Australia	McKee (in 1962)	K	20C
Haplocarpha scaposa Harv.	Africa	Norlindh 5769	S	20E
COMPOSITAE (Tribe CYNAREAE)				
Arctium minus (Hill) Bernh.	Oklahoma	Olney 74	OKL	21D
Carthamus tinctorius L.	Ethiopia	Albers 62175	TEX	21B

Taxon	Locality	Collector	Herbarium	Plate
COMPOSITAE (Tribe CYNAREAE) cont'd.				
Cirsium americanum (A. Gray) Daniels	—	—	TEX, 81722[b]	21A
C. lanceolatum (L.) Hill		Deane	TEX	21C
COMPOSITAE (Tribe MUTISIEAE)				
Barnadesia horrida Muschl.	Peru	Ferreyra 9843	CONC	22F (inset)
B. lehmanii Hieron.	Peru	Hutchinson & Wright 6684	CONC	22E, F
Dasyphyllum excelsum (Don) Cabr.	Chile	Garaventa s.n.	CONC	23A, B
Doniophyton patagonicum (Phil.) Hieron.	Argentina	Correa & Nicora 3712	CONC	23E
Glossarion rhodanthum Maguire & Wurdack	Venezuela	Maguire, Wurdack & Bunting 37126	US	23D
Onoseris odorata (D. Don) H. & A.	Peru	Ferreyra 6353	OKL	22C
Perezia wrightii A. Gray	Texas	Whitson s.n.	OKL	22D
Schlechtendalia luzulaefolia Less.	Uruguay	Cabrera 21873	CONC	23C
Trixis angustifolia DC.	Mexico	Rinehart 7019	OKL	22A, B
COMPOSITAE (Tribe CICHORIEAE)				
Catananche arenaria Cos. & Dr.	Morocco	P. & J. Davis 48787	NY	25A
C. caespitosa Desf.	Algeria	St. Lager s.n.	NY	25C
Hedypnois polymorpha DC.		Hopplinger s.n.	TEX 184168	24G
Lapsana communis L.	Czechoslovakia	Nitka 85	TEX	24H
Leontodon nudicaulis (L.) Banks	Oregon	Rafei s.n.	TEX 162955	24D, E
Microseris nutans (Hook.) Sch. Bip.	California	Rose s.n.	TEX 163478	25B, D
Picris hierocioides var. *kamtschatica* (Led.) Boivin.	Alaska	York 44505	TEX	24I
Prenanthes trifoliata (Cass.) Fern.	Pennsylvania	Schaeffer 5510	TEX	24A–C
Rafinesquia californica Nutt.	California	Wolf 10119	TEX	24F
Scorzonera divaricata Turcz.	Inner Mongolia	Roerich 162	US	25E

8. POLLEN MORPHOLOGY

CALYCERACEAE				
Acicarpha procumbens Less.	Argentina	Killip 39507	US	27C, D, G
A. tribuloides Juss.	Argentina	Petersen 9215	US	27E
Boopis alpina Poepp. & Endl.	Chile	Pennell 12406	US	26A, C
B. spathulatus Phil.	Chile		US	26B, D
Calycera pulvinata Remy	Argentina	Balb 5995	US	26E-G
Moscopsis monocephala (Phil.) Reiche	Peru	Welurbauer 7357	US	27A, B, F
VALERIANACEAE				
Nardostachys jatamansii DC.	China	Rock 5224	US	28C, D
Patrinia heterophylla Bunge	China	Cheo 63	US	28F
P. palmata Maxim.	Japan	Yatabe	US	28E
Plectritis aphanoptera (A. Gray) Suksd.	Washington State	Meyer 1493	US	28A, B
Valeriana edulis Nutt.	New Mexico	Hess 1294	OKL	28G
CAPRIFOLIACEAE				
Kolkwitzia amabilis Graebn.	Oklahoma	Felton 2	OKL	29C
Symphoricarpos occidentalis Hook.	Oklahoma	Waterfall 7945	OKL	29A, B
CAMPANULACEAE				
Downingia elegans (Lindl.) T. & G.	Nevada	Mason 12486	OKL	29D
D. portarella (Dougl.) Torr.	Oregon	Spellenberg & Hitchcock 1784	OKL	29F
Lobelia puberula Michx.	Oklahoma	Taylor & Taylor 2036	OKL	29E
DIPSACACEAE				
Cephalaria transylvanica Schrad.	Bohemia	Krist 915	K	29I
Dipsacus pilosus L.	Turkey	Davis 20768	K	29G, H

Taxon	Locality	Collector	Herbarium	Plate
BRUNONIACEAE				
Brunonia australis J. Sm.	Australia	Earle, Warnock & Nash (in 1934)	OKL	30A, B
GOODENIACEAE				
Anthotium humile R. Br.	W. Australia	10264	K	30D
Dampiera lanceolata DC.	Australia	Pedley 1043	K	30J
D. linearis R. Br.	Australia	det. R. C. Carolin (in 1973)	K	30G
Goodenia caerula R. Br.	Australia	Mann & George 200	K	30I
Scaevola mollis H. & A.	Hawaii	Fosberg 14146	OKL	30C
Velleia glabrata Sm.	Australia	Carolin 6314	K	30H
V. montana Hook. f.	Australia	Hoagland 10061	K	30E
RUBIACEAE				
Morinda longiflora G. Don.	Africa	Small 671	K	31F, G
Pentanisia schweinfurthii Hiern.	Tanganyika	Richards 20530	K	31H, I
UMBELLIFERAE				
Angelica venenosa (Greenway) Fern.	Missouri	Deane (in 1897)	OKL	31E
Trachymene anisocarpa (Turcz.) B. L. Burtt	Australia	Cullimore 182	K	31A–C
T. arfakensis (Gibbs) Bow.	New Guinea	Kostermans 2140	K	31D

LEGENDS TO PLATES

PLATE 1. A–H. Electron micrographs of pollen walls of the Vernonieae. Unless otherwise indicated, lines on individual figures equal 1 μm.
A. *Vernonia patula*. SEM. × 1500.
B. *Elephantopus carolinianus*. SEM. View of aperture enclosed by ridges with spines. × 4500.
C. *Vernonia subulata*. SEM. Ridges enclosing aperture are open at two positions by interlacunar gaps. × 1400.
D. *Vernonia venosissima*. SEM. × 950.
E. *Vernonia capreaefolia*. SEM. × 1400.
F. *Vernonia amygdalina*. TEM. × 1900.
G. *Harleya oxylepis*. TEM. × 7300.
H. *Vernonia pacchensis*. TEM. × 4600.

PLATE 2. A–E. Electron micrographs of pollen walls of the Eupatorieae. Lines on individual figures equal 1 μm.
A–B. *Carminatia tenuiflora*
    A. TEM. Mesocolpal view. Large internal foramina are common. × 8700.
    B. SEM. Polar view. Apertures have deep margins. × 2800.
C–D. *Kanimia purpurescens*.
    C. SEM. Equatorial view. × 2400.
    D. TEM. × 13 100.
  E. *Mikania cordata*. TEM. × 11 900.

PLATE 3. Electron micrographs of pollen walls of the Eupatorieae. Lines on individual figures equal 1 μm.
A–B. *Liatris aspera*.
    A. TEM. × 11 600.
    B. SEM. Polar view. × 2400.
C–D. *Decachaeta haenkana*.
    C. SEM. Both polar and equatorial views are evident. × 2300.
    D. TEM. × 23 200.
  E. *Carphephorus bellidifolius*. TEM. × 8300.

PLATE 4. Electron micrographs of pollen walls of the Astereae. Lines on individual figures equal 1 μm.
A–B. *Calotis erinacea*.
    A. SEM. Two furrows (colpi) encircle the pollen. Pore is evident in furrow at top right. × 2600.
    B. TEM. × 2700.
C. *Aphanostephus skirrhobasis*. TEM. × 18 800.
D. *Chaetopappa asteroides*. TEM. × 16 200.
E. *Boltonia diffusa*. TEM. × 19 000.

PLATE 5. A–H. Electron micrographs of pollen walls of the Inuleae. Unless otherwise indicated, lines on individual figures equal 1 μm.
  A. *Acomis rutidosis.* SEM. Equatorial view. × 2500.
B–C. *Pithocarpha corymbosa.*
  B. Polar view. × 3000.
  C. TEM. The basal region of the columellae (immediately above cavus) is highly ramified and appears as a "layer" distinct from the above columellae. × 18 700.
  D. *Polycline proteiformis.* TEM. Columellae bases are less complex than in Fig. 5C.
  E. *Blumea mollis.* TEM. Internal foramina are absent. × 18 700.
  F. *Dimeresia howellii.* TEM. × 15 000.
G–H. *Tarchonanthus minor.*
  G. SEM. Polar view with pore in colpus (at left). × 1500.
  H. TEM. The internal morphology is of the Anthemoid pattern. × 11 200.

PLATE 6. A–J. Electron micrographs of pollen walls of the Inuleae. Lines on individual figures equal 1 μm.
  A. *Stoebe capitata.* TEM. Internal morphology similar to *Pithocarpha corymbosa* (Plate 5C). × 14 300.
  B. *Helichrysum davenportii.* TEM. Internal morphology similar to 6A. × 6100.
C–D. *Allogopappus dichotomus.* TEM sections through a single spine indicating variability of hole in the tip.
  C. × 11 700.
  D. × 10 400.
E–F. *Raoulia monroi.* TEM sections through a single spine. In E the tip is solid while in F a hole is present. × 12 400.
  G. *Osmitopsis asteriscoides.* TEM. The structural pattern is Anthemoid. Note thin, uniform endexine. × 4600.
  H. *Osmites parvifolia.* TEM. Morphology is identical to 6G. × 5800.
  I. *Adenocaulon bicolor.* TEM. The structural pattern is Anthemoid but differs from Plate 6G, H in having a thick, delicate, highly anastomosing internal tectum. × 8700.
  J. *Gymnarrhena micrantha.* TEM. The Anthemoid pattern differs from Plate 6G–I in that the region above thickened columellae is a complex anastomosing net. × 9500.

PLATE 7. A–G. Electron micrographs of pollen walls of the Heliantheae. Unless otherwise indicated, lines on individual figures equal 1 μm.
A–B. *Guizotia abyssinica.*
  A. SEM. Polar view. × 1900.
  B. TEM. × 7000.
  C. *Enhydra sessilis.* TEM. × 8700.
  D. *Alepidocline annua.* TEM. × 12 200.
E–F. *Eryngiophyllum pinnatisectum.*
  E. TEM. × 4200.
  F. SEM. Polar view. × 1900.
  G. *Helianthus giganteus.* TEM. × 7600.

PLATE 8. A–D. Electron micrographs of pollen walls of *Fitchia* (Heliantheae). Unless otherwise indicated, lines on individual figures equal 1 μm.

8. POLLEN MORPHOLOGY

A–B. *Fitchia cuneata* subsp. *tahaaensis*.
    A. TEM. ×6000.
    B. SEM. Polar view. ×1200.
C–D. *Fitchia speciosa*.
    C. SEM. The polar views contrast with Plate 8B in that the apertures suggest continuity at poles (i.e. syncolpate). ×525.
    D. TEM. Morphology similar to Plate 8A except that spines are acute rather than flattened and blunt. The contrasting spine morphology is supported by the scanning electron micrographs (Plate 8B, C). ×5900.

PLATE 9. A–D. Electron micrographs of pollen walls of the Helenieae. Unless otherwise indicated, lines on individual figures equal 1 μm.
A. *Tagetes zypaquirensis*. SEM. Polar view. ×2600.
B. *Baileya multiradiata*. SEM. Equatorial view. ×1900.
C. *Tagetes elongata*. TEM. ×11 000.
D. *Baileya multiradiata*. TEM. Although appearing somewhat different in structure than Plate 9C, *B. multiradiata* is similar but magnification (×25 400) is more than doubled.

PLATE 10. A–K. Transmission electron micrographs of pollen walls of the Helenieae. Lines on individual figures equal 1 μm.
    A. *Espejoa mexicana*. ×6400.
    B. *Jaumea peduncularis*. ×12 400.
    C. *Whitneya dealbata*. ×7500.
    D. *Flaveria anomala*. ×14 300.
    E. *Orochaenactis thysanocarpha*. ×14 300.
F–G. *Chaenactis glabriuscula* var. *lanosa*. F and G are different sectional views illustrating wide cavity beneath spine tip in F and near absence of cavity in G.
    F. ×3800.
    G. ×5400.
    H. *Pericome caudata*. ×5400.
    I. *Oxypappus scaber*. ×8500.
    J. *Hymenothrix wislizenii*. ×17 600.
    K. *Eriophyllum caespitosum*. ×13 700.

PLATE 11. Transmission electron micrographs of pollen walls of the Helenieae. Lines on individual figures equal 1 μm.
A. *Cacosmia rugosa* var. *arachnoidea*. The structural pattern is a "modified" Anthemoid and is similar to Vernonieae pollen (see Plate 1 F–H). ×6600.
B. *Amblyopappus pusillus*. Internal foramina are absent. ×8600.
C. *Amblyolepis setigera*. ×9900.
D. *Dyssodia anthemidifolia*. ×15 200.

PLATE 12. Electron micrographs of pollen walls of the Anthemideae. Lines on individual figures equal 1 μm.
A. *Anthemis cotula*. SEM. Equatorial view. ×2000.
B. *Cotula coronopifolia*. TEM. ×2300.
C. *Soliva stolonifera*. TEM. ×9500.
D. *Centipeda cunninghamii*. TEM. The internal morphology is difficult to characterize. A narrow cavus is present in the mesocolpal regions. Immediately above cavus a thick complex layer is evident which extends upward into distinct columellae which in turn form the tectum. Similarities are evident with some Inuleae (see Plates 5C, 6A, B). ×2700.

E. *Elachanthus pusillus*. TEM. Note typical Helianthoid morphology. × 15 700.
F. *Eriocephalus hoffmannianus*. This is a SEM of a pollen wall that has been fractured. It demonstrates the true columnar nature of the ektexine. The spongy appearing layer above the thick, forked columellae is interpreted by TEM to consist of complex, anastomosing tecta and columellae. Foot layer is not differentiated from endexine in this fractured sample. × 7300.
G. *Artemisia alaskana*. TEM. × 19 600.
H. *Ceratogyne obionoides*. TEM. The internal morphology is Helianthoid. × 13 300.
I. *Anthemis ruthenica*. TEM. × 8700.

PLATE 13. A–E. Electron micrographs of pollen walls of the Senecioneae. Unless otherwise indicated, lines on individual figures equal 1 μm.
A. *Arnica acaulis*. The SEM includes both polar and equatorial views. × 6300.
B. *Arnica chamissonis*. TEM. Helianthoid pattern. × 14 100.
C. *Raillardella scaposa*. TEM. Helianthoid pattern. × 6500
D. *Senecio heritieri*. TEM. Helianthoid pattern. × 14 300.
E. *Senecio lyallii*. TEM. Senecioid pattern. × 4700.

PLATE 14. Transmission electron micrographs of pollen walls of the Senecioneae. Lines on individual figures equal 1 μm.
A. *Tetradymia canescens*. × 3400.
B. *Gynoxys parvifolia*. The columellae bases are slightly thickened and resemble somewhat exines in Inuleae (Plates 5C, 6A, B) and Anthemideae (Plate 12D). × 6400.
C. *Odontotrichum cervinum*. × 6700.
D. *Erechtites valerianaefolia*. × 7400.
E. *Cacalia goldsmithii*. × 4300.
F. *Emilia coccinea*. × 6700.
G. *Tussilago farfara*. Note similarity to 14B. × 8400.
H. *Gamolepis brachypoda*. × 9100.
I. *Petasites hyperboreus*. × 5600.

PLATE 15. A–F. Electron micrographs of pollen walls of the genus *Liabum* (Senecioneae). Unless otherwise indicated, lines on individual figures equal 1 μm.
A. *L. megacephalum*. SEM. Polar view. × 1600.
B. *L. sagittatum*. SEM. Equatorial view. × 1500.
C. *L. glabrum* var. *hypoleucum*. TEM. × 2600. Inset shows longitudinal section of spine. × 2500.
D. *L. caducifolium*. TEM. × 1800.
E. *L. kluttii*. TEM. × 2700. The internal morphology of 15C–E is similar to Cacosmia (Plate 11A) and Vernonieae (Plate 1F–H).
F. *L. ovatum* (= *Paranephelius Ovatus*). TEM. Unlike the structural morphology of 15C–E, *L. ovatum* appears "Senecioid-like". × 10 700.

PLATE 16. A–E. Transmission electron micrographs of pollen walls of the Senecioneae. Lines on individual figures equal 1 μm.
A. *Schistocarpha platyphylla*. TEM. × 11 500.
B–C. *Schistocarpha sinforosii*.
    B. TEM. × 7600.
    C. TEM. × 7600.
    The internal foramina in all three electron micrographs of *Schistocarpha* are filled with electron-opaque material. The micrographs illustrate how

different sections through spines can have various interpretations. In 16A the spine tip indicates an elongate channel as well as a prominent downward extension of the spine base. In 16B the spine bases are more regular (an elongate opening is evident in each spine tip). In 16C, which is simply a different section of 16B, large cavities are evident beneath the two spine bases (in 16B columellae bases are evident beneath the spine bases).
- D. *Neurolaena cobanensis.* × 12 900.
- E. *Doronicum altaicum.* × 13 100.

PLATE 17. A–C. Electron micrographs of pollen walls of the Calenduleae. Lines on individual figures equal 1 μm.
- A. *Gibbaria ilicifolia.* SEM. Polar view. × 2300.
- B. *Osteospermum vaillantii.* SEM. Equatorial view. × 2200.
- C. *Gibbaria ilicifolia.* TEM. Minute internal foramina are evident throughout ektexine. × 5400. The inset is a TEM of *Osteospermum vaillantii*. Although the internal morphology is Helianthoid, note apparent irregular extensions of columellae bases. × 14 800.

PLATE 18. A–D. Electron micrographs of pollen walls of the Calenduleae. Unless otherwise indicated, lines on individual figures equal 1 μm.
- A. *Chrysanthemoides incana.* SEM. Equatorial view. × 2700.
- B. *Tripteris clandestinum.* SEM. Equatorial view. × 2200.
- C. *Chrysanthemoides monolifera.* TEM. In contrast to *Osteospermum vaillantii* (inset to 17C) the extensions of the columellae bases show less contact with the foot layer. × 10 300.
- D. *Tripteris clandestinum.* TEM. × 7900.

PLATE 19. A–I. Electron micrographs of pollen walls of the Calenduleae (19A, B, inset to B) and Arctotideae. Unless otherwise indicated, lines on individual figures equal 1 μm.
- A–B. *Dimorphotheca pluvialis.*
  - A. SEM. Equatorial view. × 1900.
  - B. TEM. × 4900. Inset shows minute internal foramina. × 15 700.
- C–E. *Cullumia setosa.*
  - C. SEM. Polar view demonstrating lophate morphology. Note that at ridge corners the exine appears to have "spines". × 1600.
  - D. TEM of section along a lacuna. × 5200.
  - E. TEM of section along a ridge. × 5200.
- F–I. *Berkheya rhapontica.*
  - F. SEM. Polar view demonstrating echinolophate morphology. × 1600.
  - G. TEM showing connection of spine columellae with foot layer. × 4900.
  - H. TEM of section along a ridge. × 8500.
  - I. TEM showing possible contact of columellae base with foot layer. × 1800.

PLATE 20. A–E. Electron micrographs of pollen walls of the Arctotideae. Lines on individual figures equal 1 μm.
- A. *Arctotis verbasifolia.* SEM. Equatorial view. × 2700.
- B. *Arctotis aspera.* TEM. Note delicate internal tectum which nearly equally divides columellae. A thin foot layer, while present, is not clearly evident in this micrograph. × 6200.
- C. *Cymbonotus lawsonianus.* TEM. × 14 100.

D. *Arctotheca calendula.* TEM. Note deep spine channel characteristic of many exines of the Arctotideae. × 10 700.

E. *Haplocarpha scaposa.* TEM. This pollen grain has been rehydrated according to the method of Rowley and Nilsson (1972). Note columella connection with foot layer (beneath spine at left). Between the two spines the exine appears to be in contact with foot layer but this image is partially influenced by the processing method. × 15 400.

PLATE 21. A–D. Electron micrographs of pollen walls of the Cynareae. Unless otherwise indicated, lines on individual figures equal 1 μm.
A. *Cirsium americanum.* SEM. Polar view. × 1500.
B. *Carthamus tinctorius.* TEM. × 6600.
C. *Cirsium lanceolatum.* TEM. × 10 500.
D. *Arctium minus.* TEM. × 15 900.

PLATE 22. A–F. Electron micrographs of pollen walls of the Mutisieae. Unless otherwise indicated, lines on individual figures equal 1 μm.
A–B. *Trixis angustifolia.*
    A. SEM. Equatorial view. Note smooth polar caps and globular fragments in the wide colpus. × 1600.
    B. TEM. × 5500.
C. *Onoseris odorata.* TEM. × 1900.
D. *Perezia wrightii.* TEM. × 16 700.
E–F. *Barnadesia lehmanii.*
    E. SEM. Note similarity to *Cullumia setosa* (Plate 19C). × 1100.
    F. TEM. The tops of the two ridges are capped with a thin, uniform layer of sporopollenin. × 2300. Inset to 22F is TEM of *Barnadesia horrida* indicating cavus-like nature of the ektexine. × 1400.

PLATE 23. A–E. Electron micrographs of pollen walls of the Mutisieae. Lines on individual figures equal 1 μm.
A–B. *Dasyphyllum excelsum.*
    A. SEM including polar and equatorial-lateral views. Notable in both pollen grains are the depressions between the colpi, termed "intercolpar concavities" (Wodehouse, 1928b). × 2400.
    B. TEM along colpus margin is caveate. Above cavus the ektexine is similar to that in some Inuleae (Plates 5C, 6A, B), Anthemideae (Plate 12D) Senecioneae (Plate 14B, G). In the intercolpar concavity a cavus is absent and reduced columellae are capped by a thick tectum. × 8900.
C. *Schlechtendalia luzulaefolia.* TEM comparable to 23B. However, a cavus is not present in the colpus margin: the ektexine indicates delicate levels of internal tecta. × 6200.
D. *Glossarion rhodanthum.* TEM. Note that the upper half of the ektexine is similar to the ektexine of the colpus margin of 23C. × 3500.
E. *Doniophyton patagonicum.* TEM. The foot is barely perceptible while the columellae are similar to the colpus margin of 23C and the upper half of the ektexine in 23D. × 7400.

PLATE 24. A–I. Electron mircographs of pollen walls of the Cichorieae. Unless otherwise indicated, lines on individual figures equal 1 μm.
A–C. *Prenanthes trifoliata.*
    A. SEM. Polar view. × 1800.
    B. SEM showing echinolophate morphology. × 3200.
    C. TEM. Section includes lacuna between two ridges. × 3200.

8. POLLEN MORPHOLOGY

D–E. *Leontodon nudicaulis.*
    D. TEM along ridge. × 3600.
    E. TEM approximating a lacuna. × 3800.
    Note that in both D and E large channels are beneath spines (similar to some Arctotideae in Plates 19D, E, 20B, D, E).
F. *Rafinesquia californica.* TEM. Note thin columellae of spine attached to foot layer similar to that of *Leontodon* (Plate 23E), *Berkheya* (Plate 19G) and *Haplocarpha* (Plate 20E). × 9700.
G. *Hedypnois polymorpha.* TEM. × 3600.
H. *Lapsana communis.* TEM. × 7800.
I. *Picris hieracioides.* SEM. Equatorial-lateral view. × 2000.

PLATE 25. A–E. Electron micrographs of pollen walls of the Cichorieae. Unless otherwise indicated, lines on individual figures equal 1 μm.
A. *Catananche arenaria.* SEM. Polar view. × 1160.
B. *Microseris nutans.* SEM. Polar-lateral view. × 1900.
C. *Catananche caespitosa.* TEM. The sectional view in this electron micrograph is misleading because the enlarged columellae which rise from the foot layer are not shown to be attached to the *overlying ektexine*. The latter has a typical Helianthoid pattern and the entire exine (i.e. enlarged columellae and overlying ektexine) appears transitional between Anthemoid and Helianthoid patterns. × 7600.
D. *Microseris nutans.* TEM. Although the surface is echinolophate (Plate 25B) the internal morphology appears to have a Senecioid pattern. × 9700.
E. *Scorzonera divaricata.* TEM. Note small columellae rising from foot layer to interrupt cavus. × 22 400.

PLATE 26. A–G. Electron micrographs of pollen walls of the family Calyceraceae. Lines on individual figures equal 1 μm.
A. *Boopis alpina.* SEM. Equatorial view. × 2000.
B. *Boopis spathulatus.* SEM. Equatorial view. The colpus is lined with a low inner ridge. Intercolpar concavities, similar to the SEM of *Dasyphyllum excelsum* (Plate 23A) are noteworthy. × 2400.
C. *Boopis alpina.* TEM. The internal morphology resembles the Anthemoid structural pattern. × 15 600.
D. *Boopis spathulatus.* TEM section plane identical to that described for *Dasyphyllum excelsum* (Plate 23B) and *Schlechtendalia luzulaefolia* (Plate 23C). However the colpus margin is clearly an Anthemoid pattern while the structure of the intercolpar depression is identical to *D. excelsum* and *S. luzulaefolia.* × 10 500.
E–G. *Calycera pulvinata.*
    E. SEM. Equatorial view. Identical to *Boopis spathulatus* (Plate 26B). × 2000.
    F. TEM of intercolpar concavity. × 14 300.
    G. TEM (section not passing through intercolpar concavity). The internal morphology is characteristically Anthemoid. × 7700.

PLATE 27. A–G. Electron micrographs of pollen walls of the family Calyceraceae. Lines on individual figures equal 1 μm.
A–B. *Moschopsis monocephala.*
    A. SEM. Equatorial view. × 2000.
    B. SEM. Polar view. × 2000.
    These surfaces are similar to *Boopis alpina* (Plate 26A).

C–D. *Acicarpha procumbens.*
    C. SEM. Equatorial view. × 3700.
    D. Polar view. × 3700. These surface views are similar to *Boopis spathulatus* (Plate 26B), *Calycera pulvinata* (Plate 26E) and *Dasyphyllum excelsum* (Plate 23A).
  E. *Acicarpha tribuloides.* TEM showing intercolpar concavities, Anthemoid-like colpus margins and colpi. × 4600. Inset shows characteristic Anthemoid pattern of colpus margin. × 14 100.
  F. *Moschopsis monocephala.* TEM showing Anthemoid pattern. × 6500.
  G. *Acicarpha procumbens.* TEM. Internal morphology is similar to Plate 27E. × 6300.

PLATE 28. A–G. Electron micrographs of pollen walls of the family Valerianaceae. Unless otherwise indicated, lines on individual figures equal 1 μm.
A–B. *Plectritis aphanoptera.*
    A. SEM. Polar view. Note thickened spine flecks outlining the three colpi. × 900.
    B. TEM showing columellae which become distended to form the tectum. A thin, problematic endexine is present at extreme lower portion of exine. × 12 200.
C–D. *Nardostachys jatamansii.*
    C. TEM. The internal morphology closely resembles the Anthemoid pattern. × 9300.
    D. SEM. Polar view. Short spines line the colpi. × 900.
  E. *Patrinia palmata.* SEM. Spines arise from inflated bases. Colpi are lined with short spine flecks. × 900.
  F. *Patrinia heterophylla.* TEM includes part of aperture region (at left). × 7400.
  G. *Valeriana edulis.* TEM. × 11 600.

PLATE 29. A–I. Electron micrographs of pollen walls of the Caprifoliaceae, Campanulaceae, and Dipsacaceae. Unless otherwise indicated, lines on individual figures equal 1 μm.
A–B. *Symphoricarpos occidentalis* (Caprifoliaceae).
    A. SEM. Equatorial view. × 1400.
    B. TEM. × 10 300. These electron micrographs do not suggest any Compositae similarity.
  C. *Kolkwitzia amabilis* (Caprifoliaceae). TEM. The internal morphology has a superficial Anthemoid pattern. Of all the taxa of Caprifoliaceae examined to the present time only *K. amabilis* indicates a Compositae-like morphology. × 6200.
  D. *Downingia elegans* (Campanulaceae). SEM. Polar view. × 2000.
  E. *Lobelia puberula* (Campanulaceae). TEM. × 2700.
  F. *Downingia portarella* (Campanulaceae). TEM. × 7900.
G–H. *Dipsacus pilosus* (Dipsacaceae).
    G. SEM. Equatorial view. × 900.
    H. TEM suggests a distant Anthemoid pattern. × 3900.
  I. *Cephalaria transylvanica.* TEM. × 6100.

PLATE 30. A–J. Electron micrographs of pollen walls of the Brunoniaceae and Goodeniaceae. Unless otherwise indicated, lines on individual figures equal 1 μm.

8. POLLEN MORPHOLOGY 217

A–B. *Brunonia australis* (Brunoniaceae).
    A. TEM indicates a superficial Anthemoid morphology. × 8700.
    B. SEM. Polar view. Note additional ridge-like layer of exine in colpus region. × 1800.
  C. *Scaevola mollis* (Goodeniaceae). TEM. × 5900.
  D. *Anthotium humile* (Goodeniaceae), TEM. × 16 900.
  E. *Velleia montana* (Goodeniaceae). TEM. × 7900.
  F. *Goodenia caerula* (Goodeniaceae). TEM. × 9900.
  G. *Dampiera linearis* (Goodeniaceae). TEM. × 9900.
  H. *Velleia glabrata* (Goodeniaceae). SEM. Polar view. × 2400.
  I. *Goodenia caerula* (Goodeniaceae). SEM. Equatorial view. × 1400.
  J. *Dampiera lanceolata* (Goodeniaceae). SEM. Equatorial view. × 3200.

PLATE 31. A–I. Electron micrographs of pollen walls of the Umbelliferae and Rubiaceae. Unless otherwise indicated, lines on individual figures equal 1 μm.
A–C. *Trachymene anisocarpa* (Umbelliferae).
    A. SEM. Polar view. × 2800.
    B. SEM. Equatorial view. × 1600.
    C. TEM suggests a remote Anthemoid pattern. × 8300.
  D. *Trachymeme arfakensis* (Umbelliferae). TEM. × 10 600.
  E. *Angelica venenosa* (Umbelliferae). TEM does not indicate Compositae morphology. × 3800.
F–G. *Morinda longiflora* (Rubiaceae).
    F. TEM. × 9900.
    G. SEM. Equatorial view. × 1300.
H–I. *Pentanisia schweinfurthii* (Rubiaceae).
    H. TEM. × 6100.
    I. SEM. Polar view. × 1500.

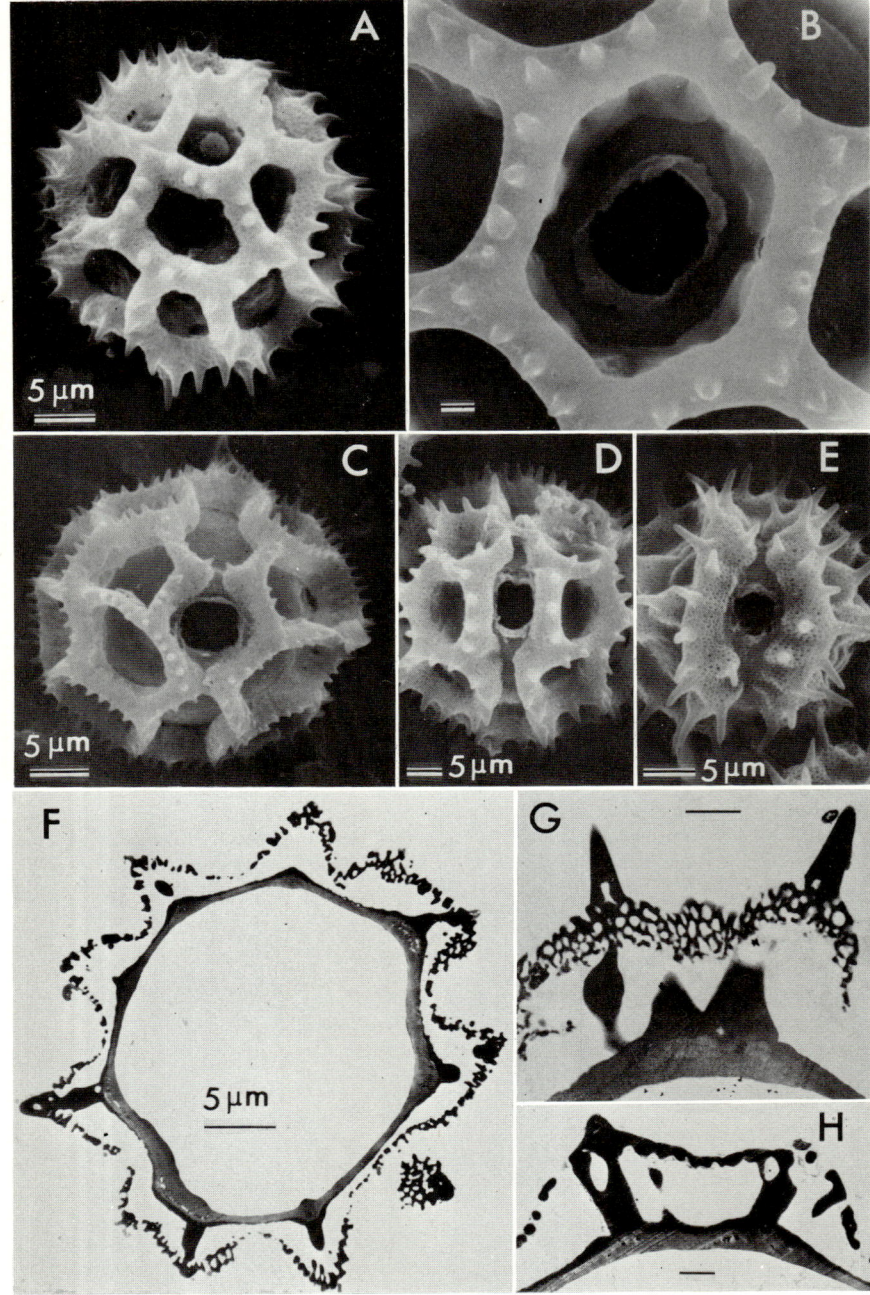

Plate 1.

8. POLLEN MORPHOLOGY

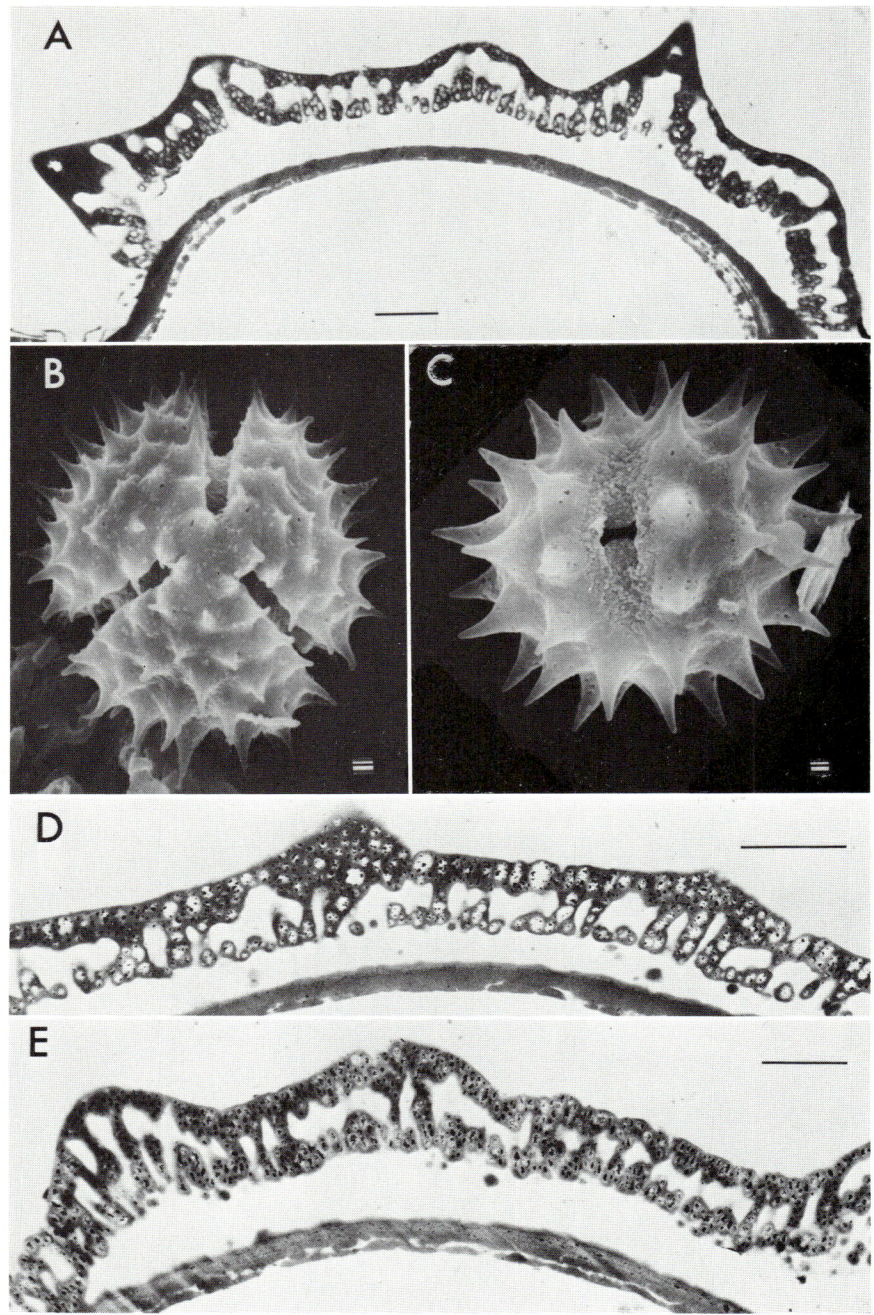

Plate 2.

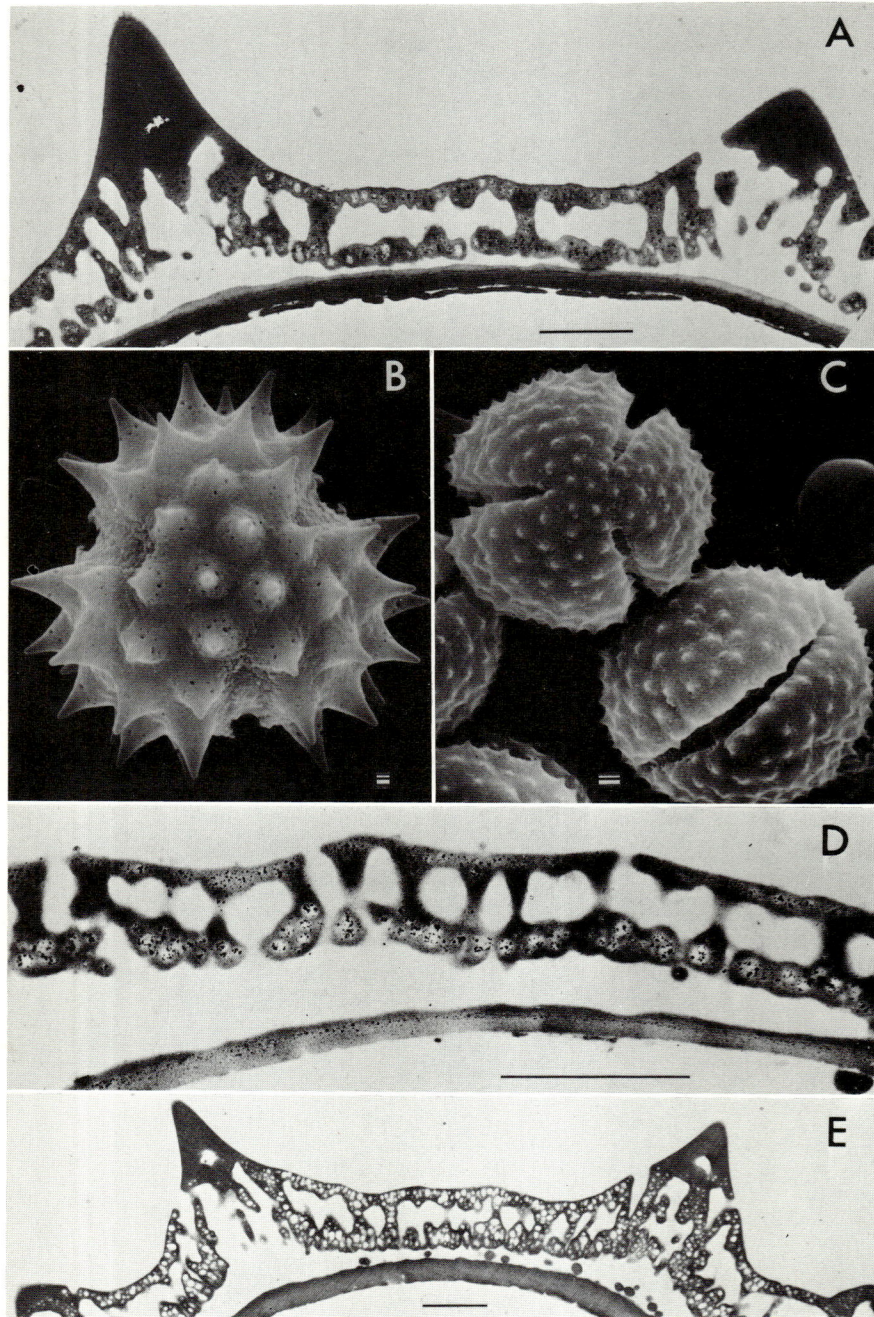

PLATE 3.

8. POLLEN MORPHOLOGY 221

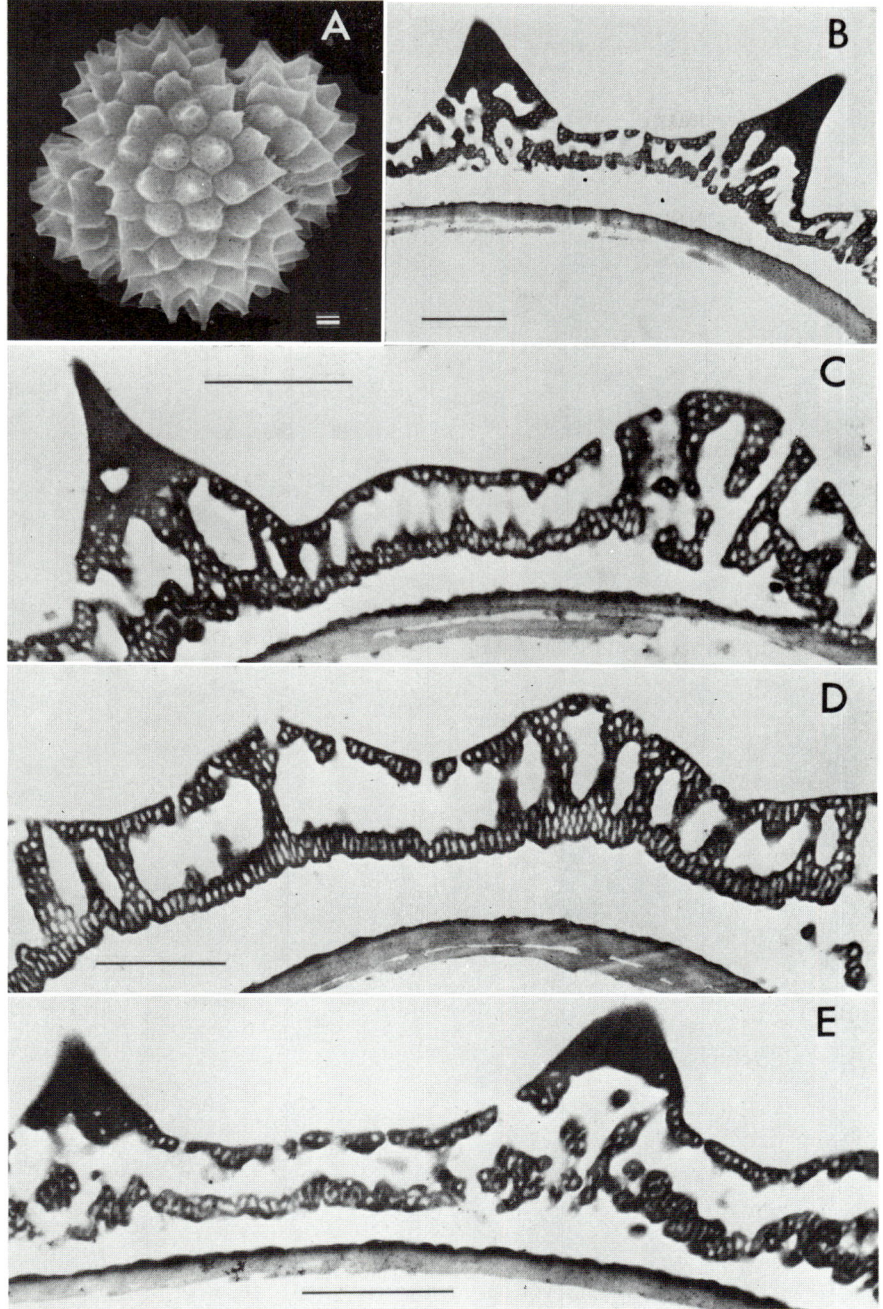

Plate 4.

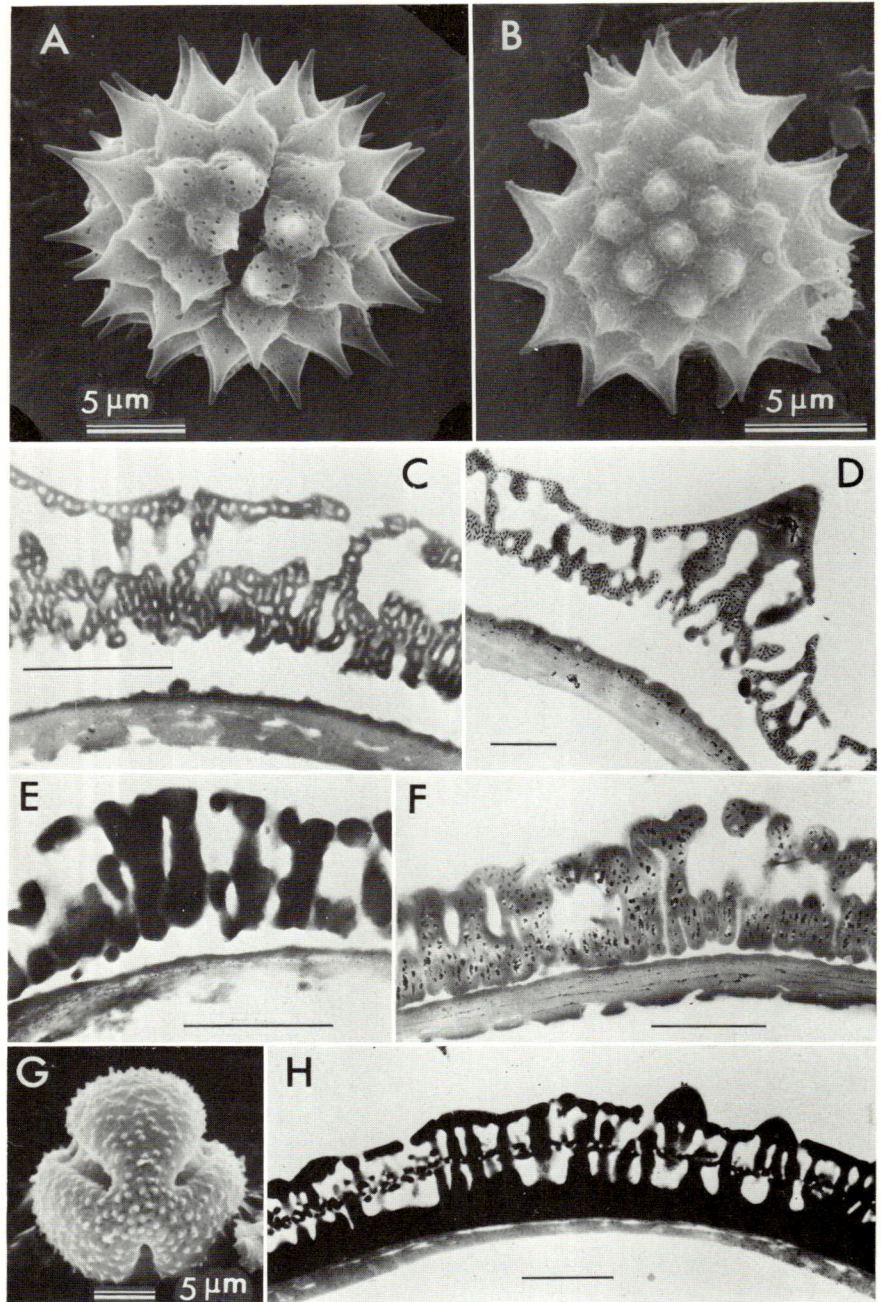

Plate 5.

8. POLLEN MORPHOLOGY

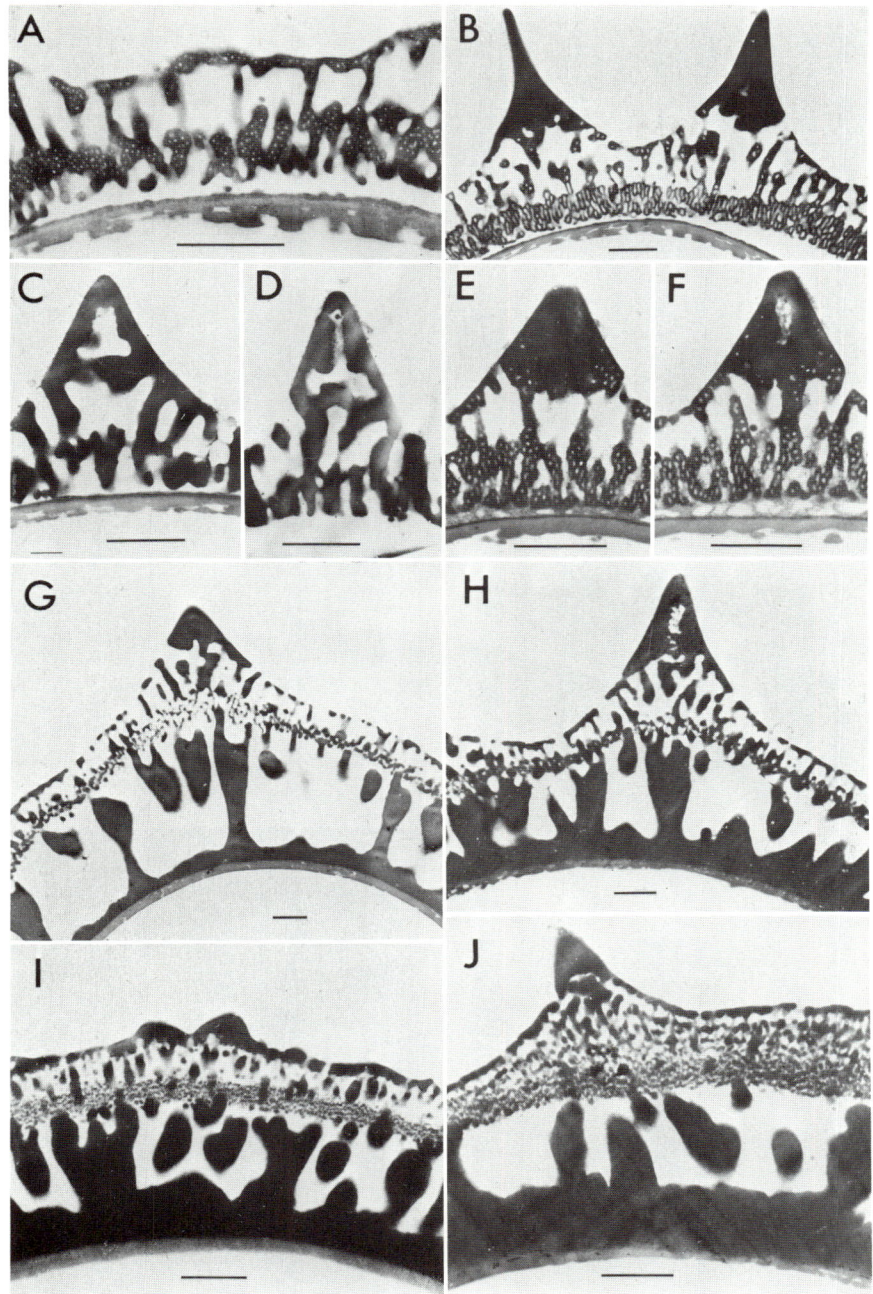

Plate 6.

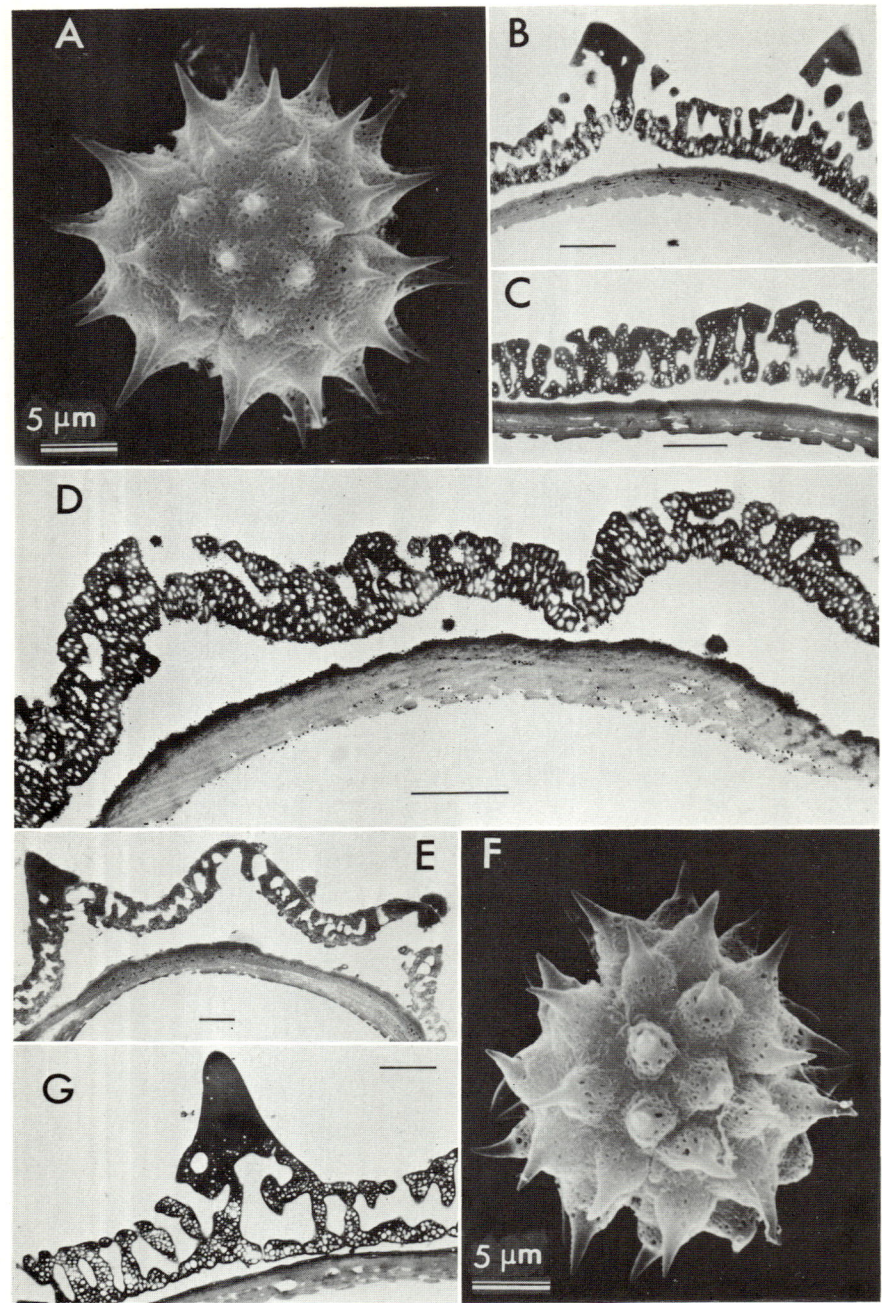

PLATE 7.

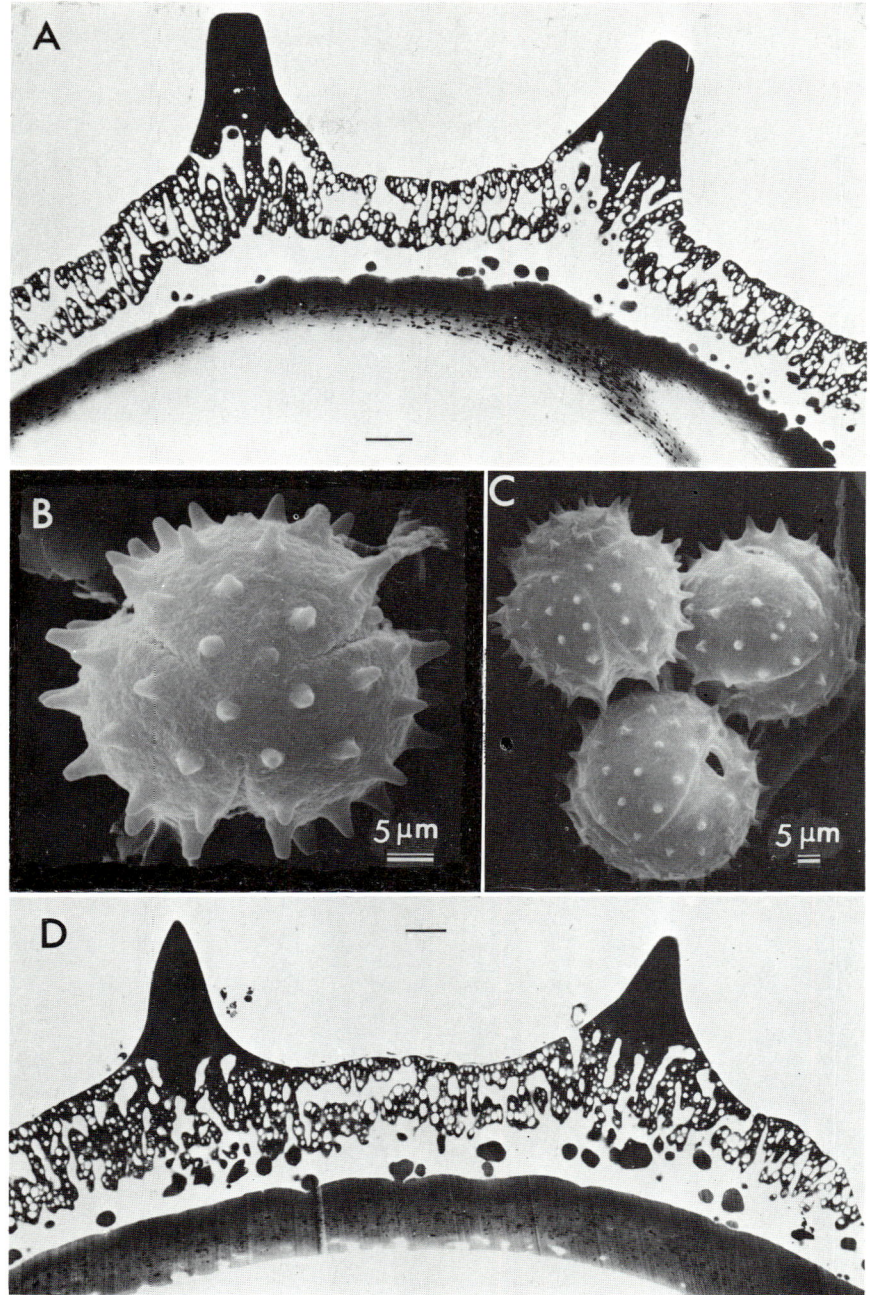

PLATE 8.

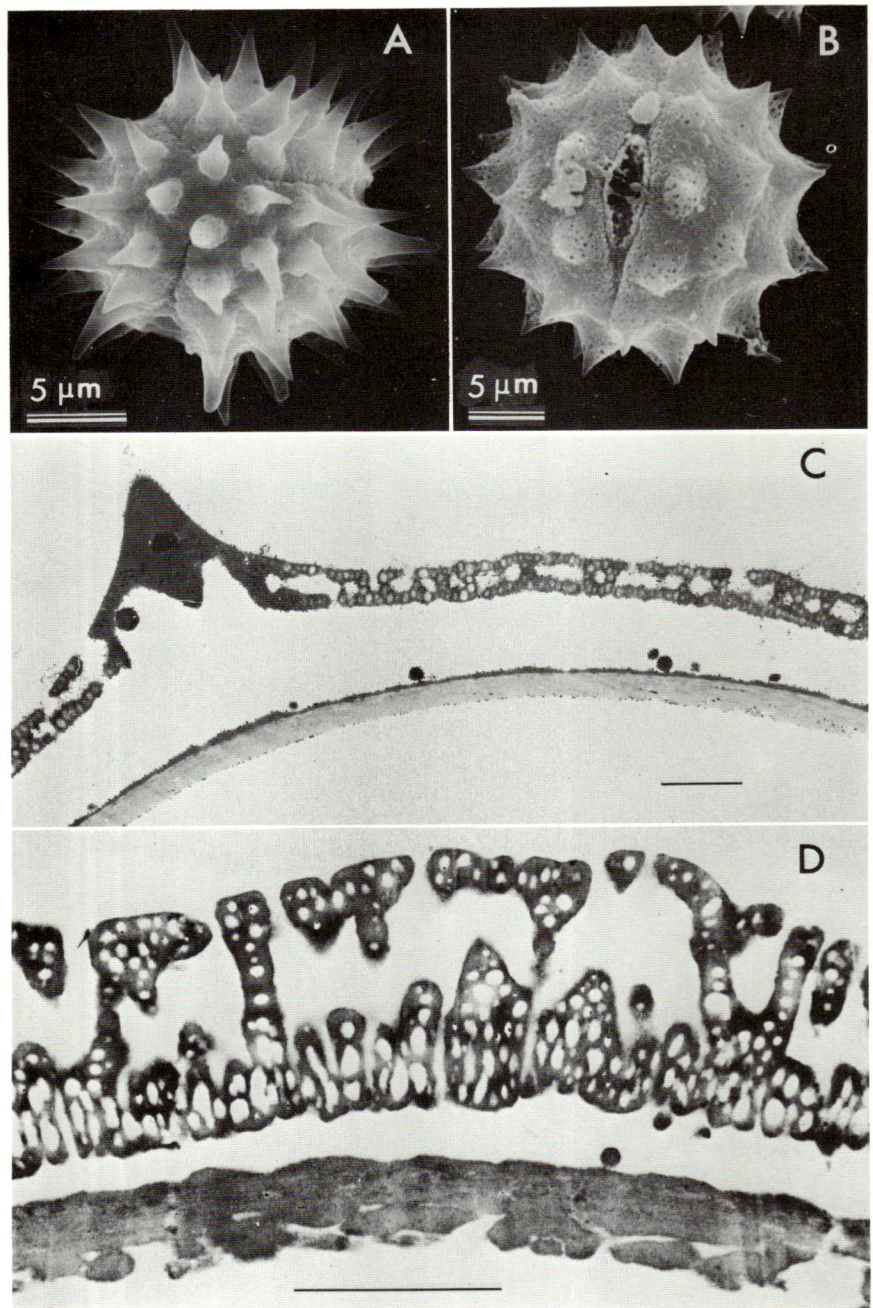

Plate 9.

8. POLLEN MORPHOLOGY

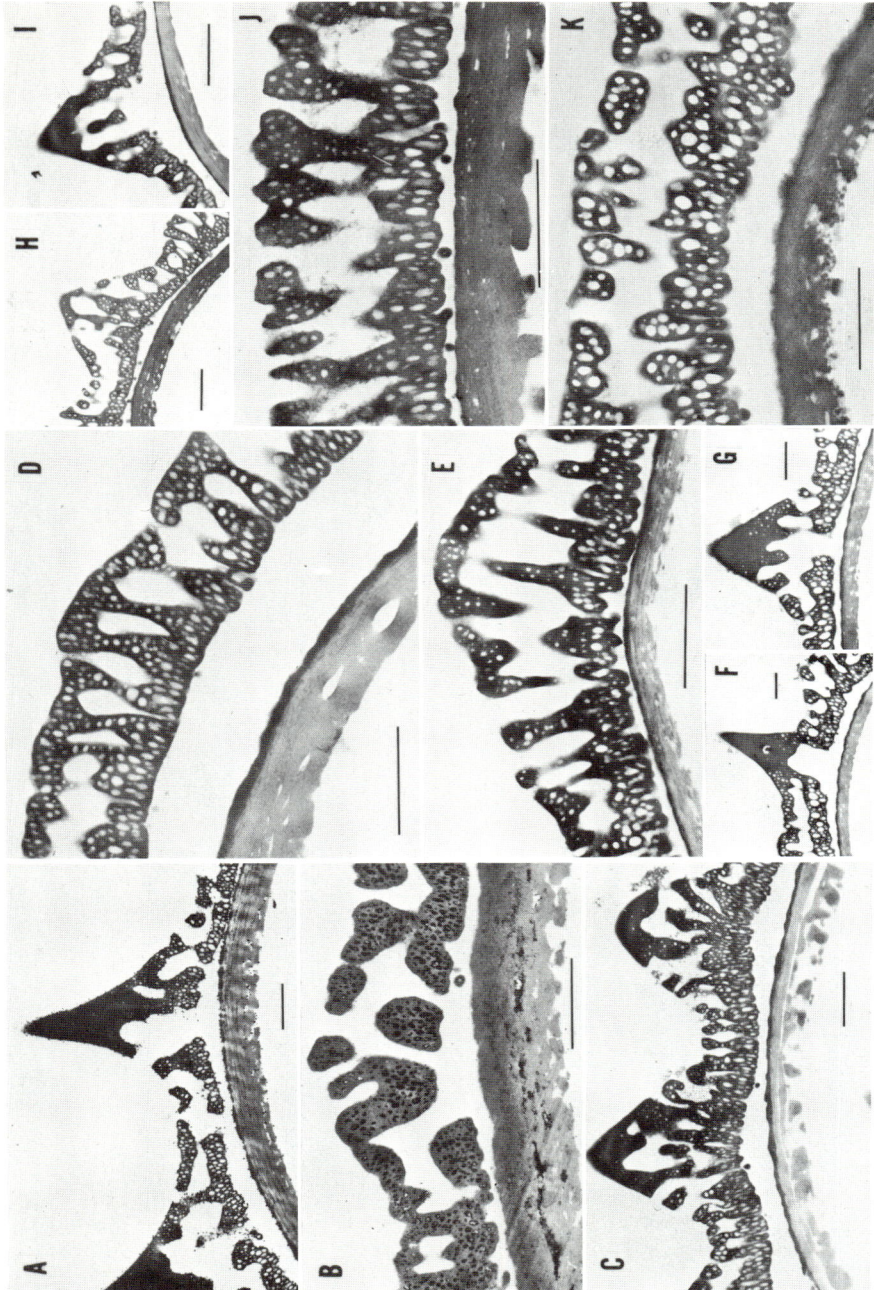

Plate 10.

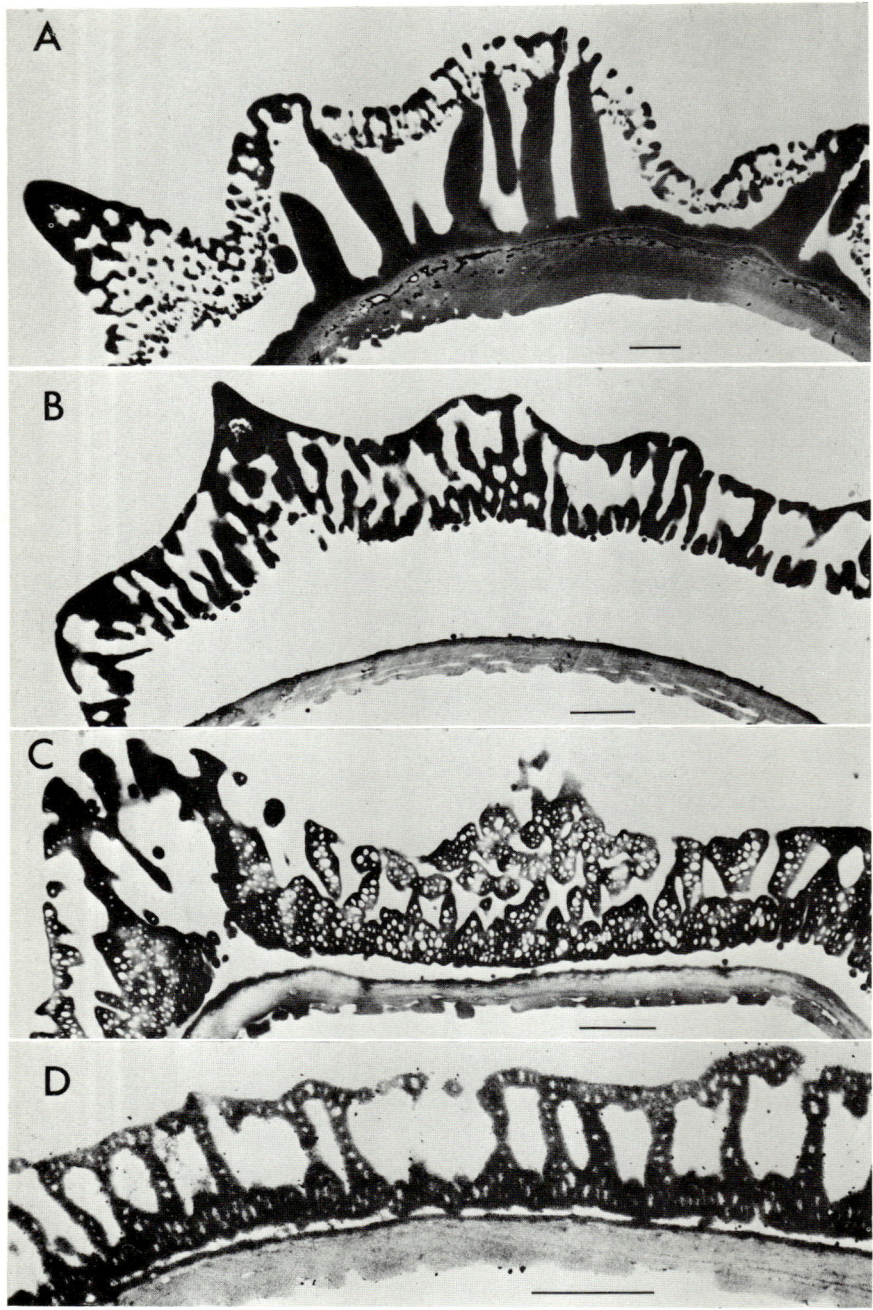

Plate 11.

8. POLLEN MORPHOLOGY

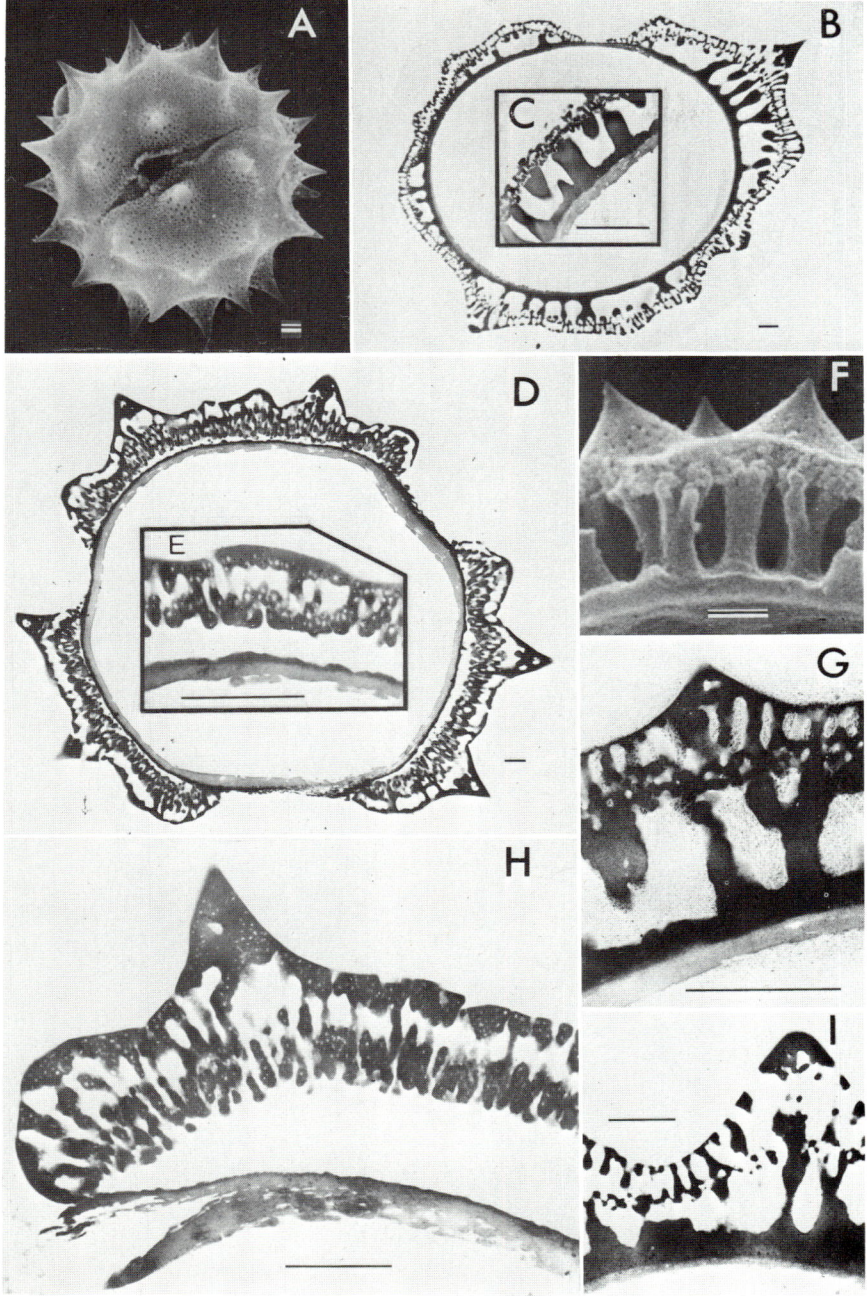

PLATE 12.

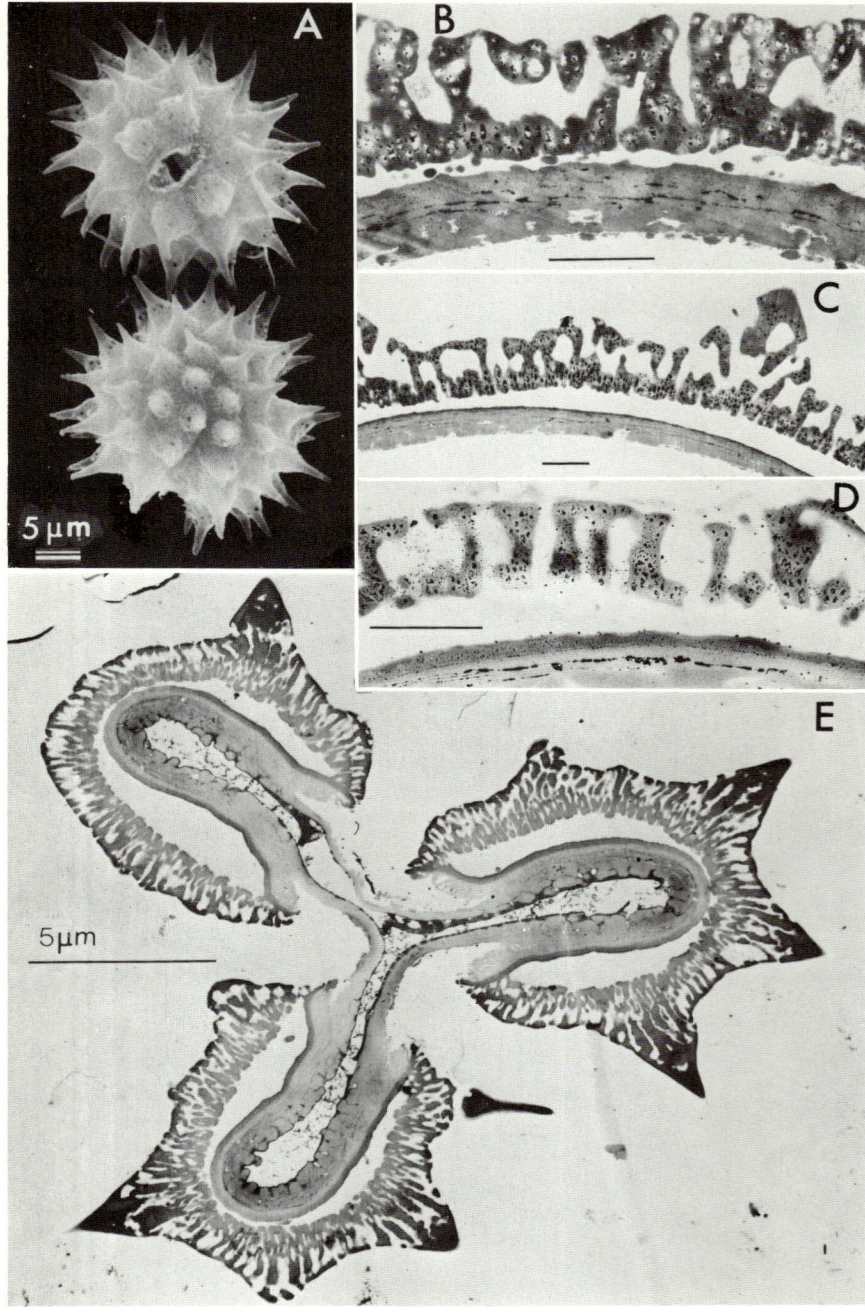

Plate 13.

8. POLLEN MORPHOLOGY

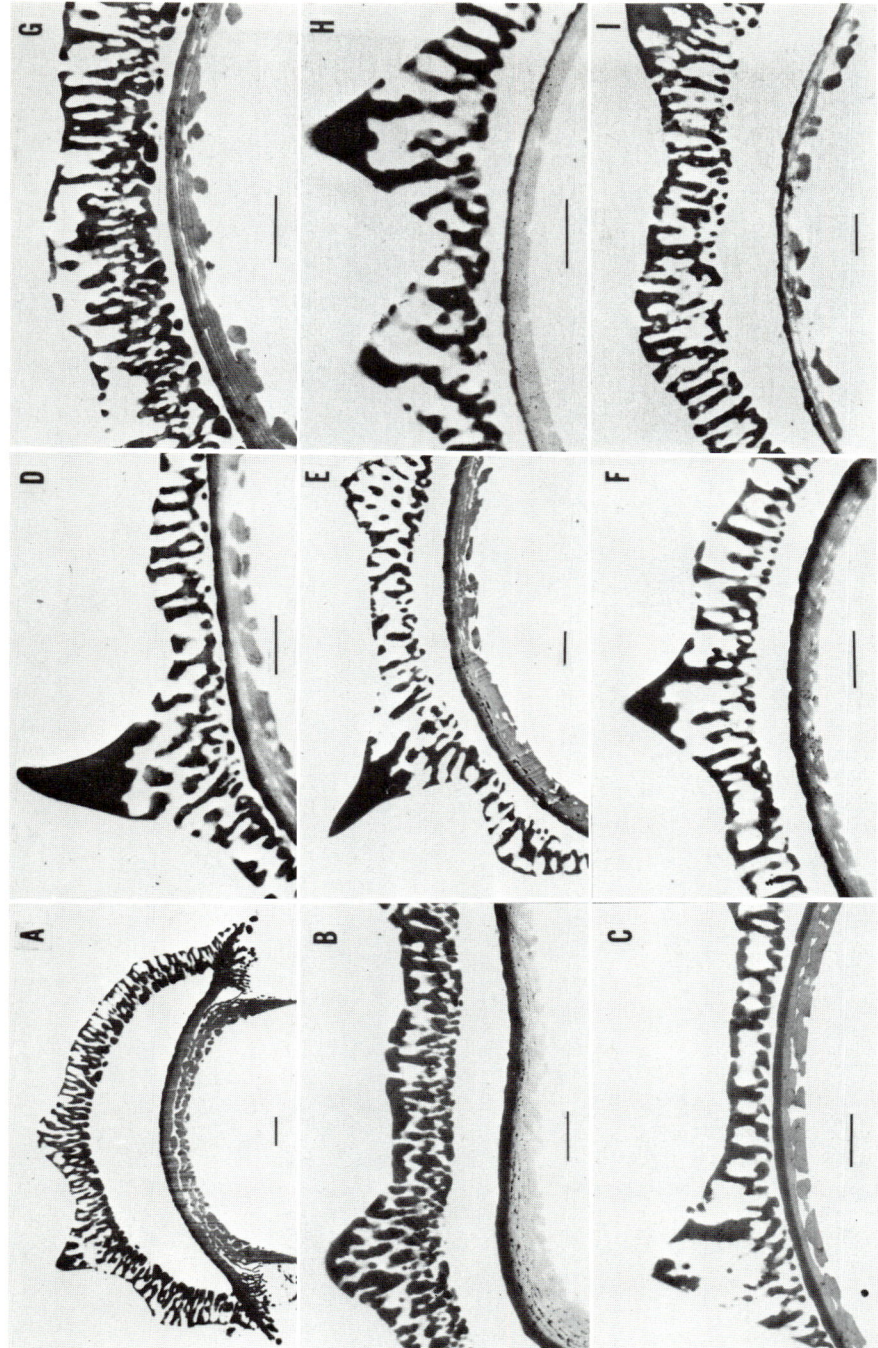

Plate 14.

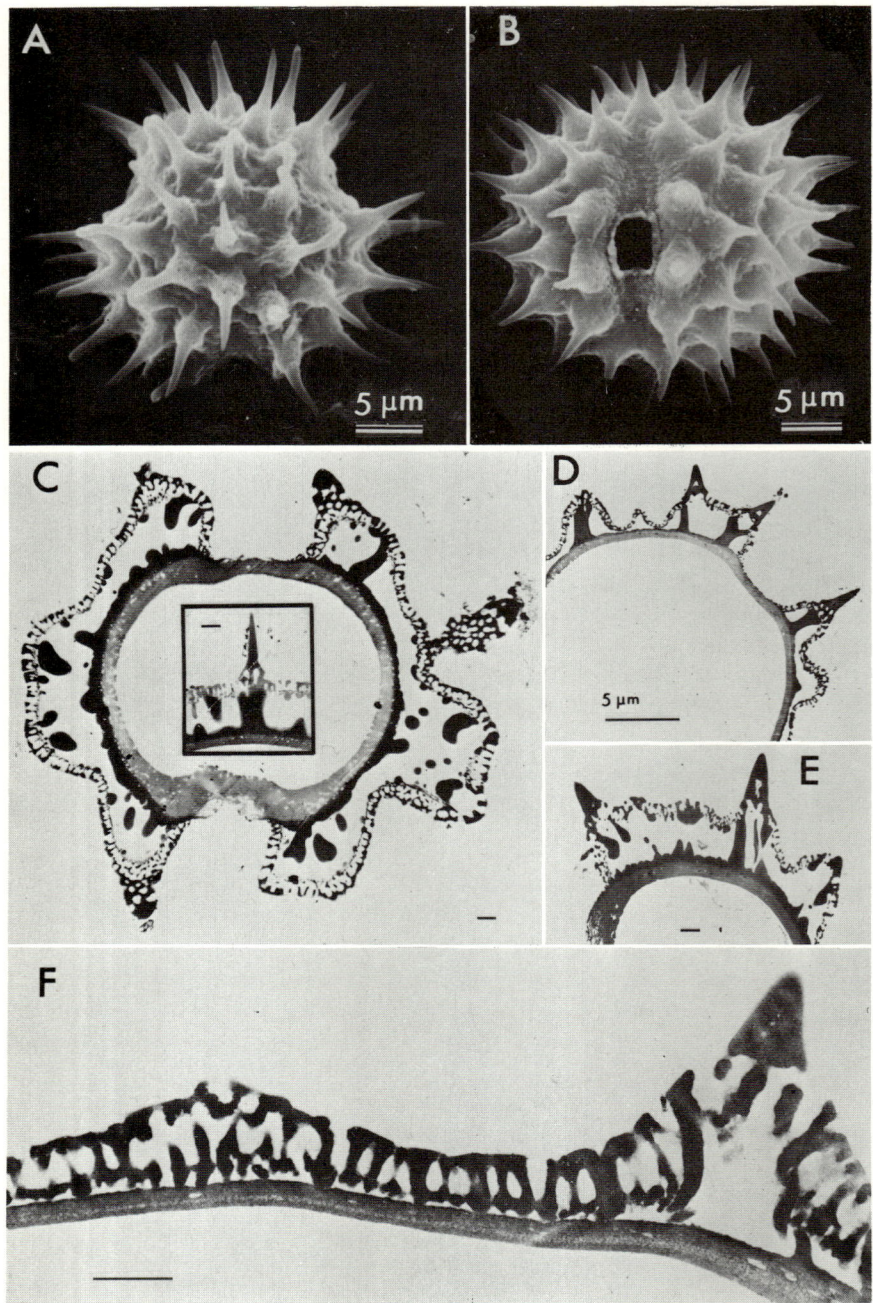

Plate 15.

8. POLLEN MORPHOLOGY

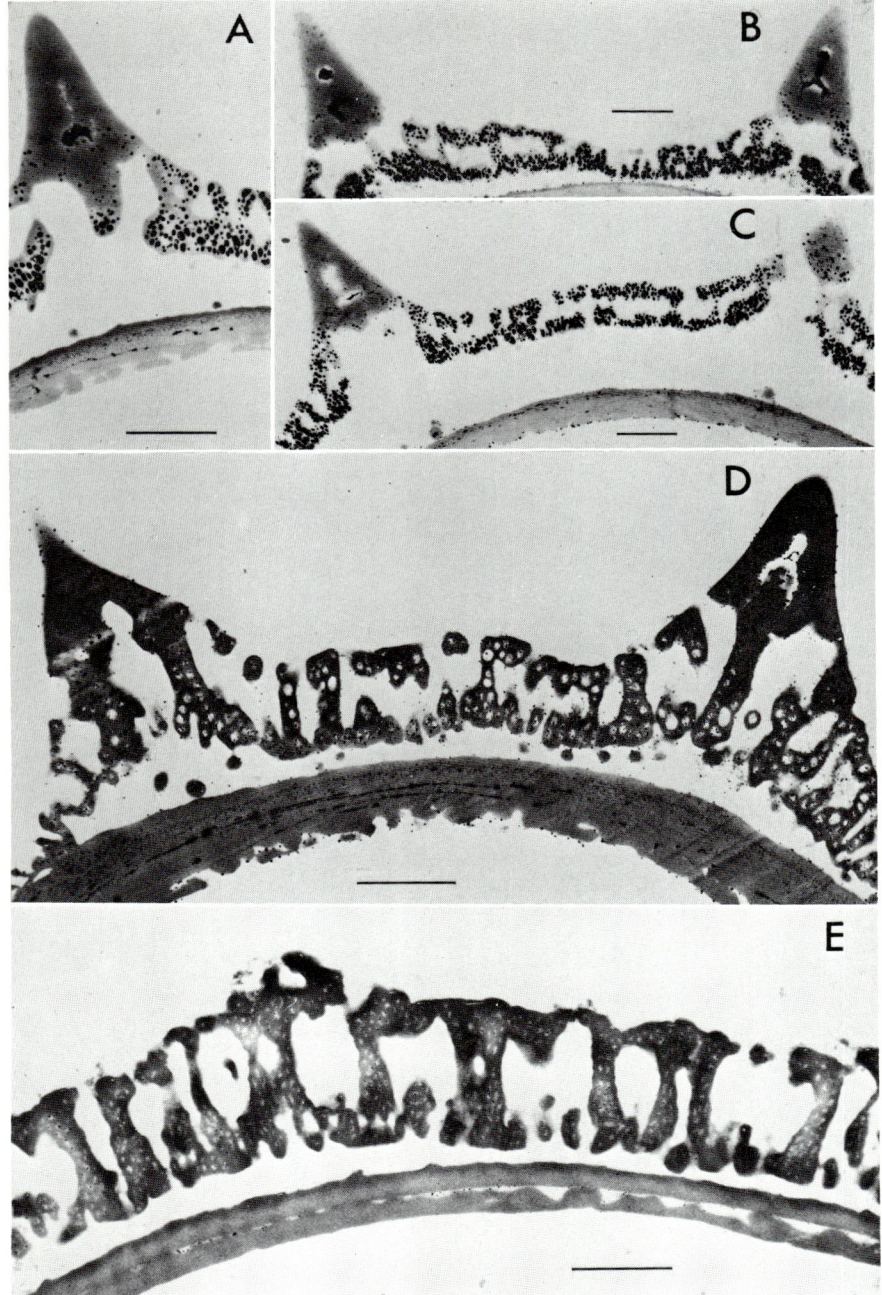

PLATE 16.

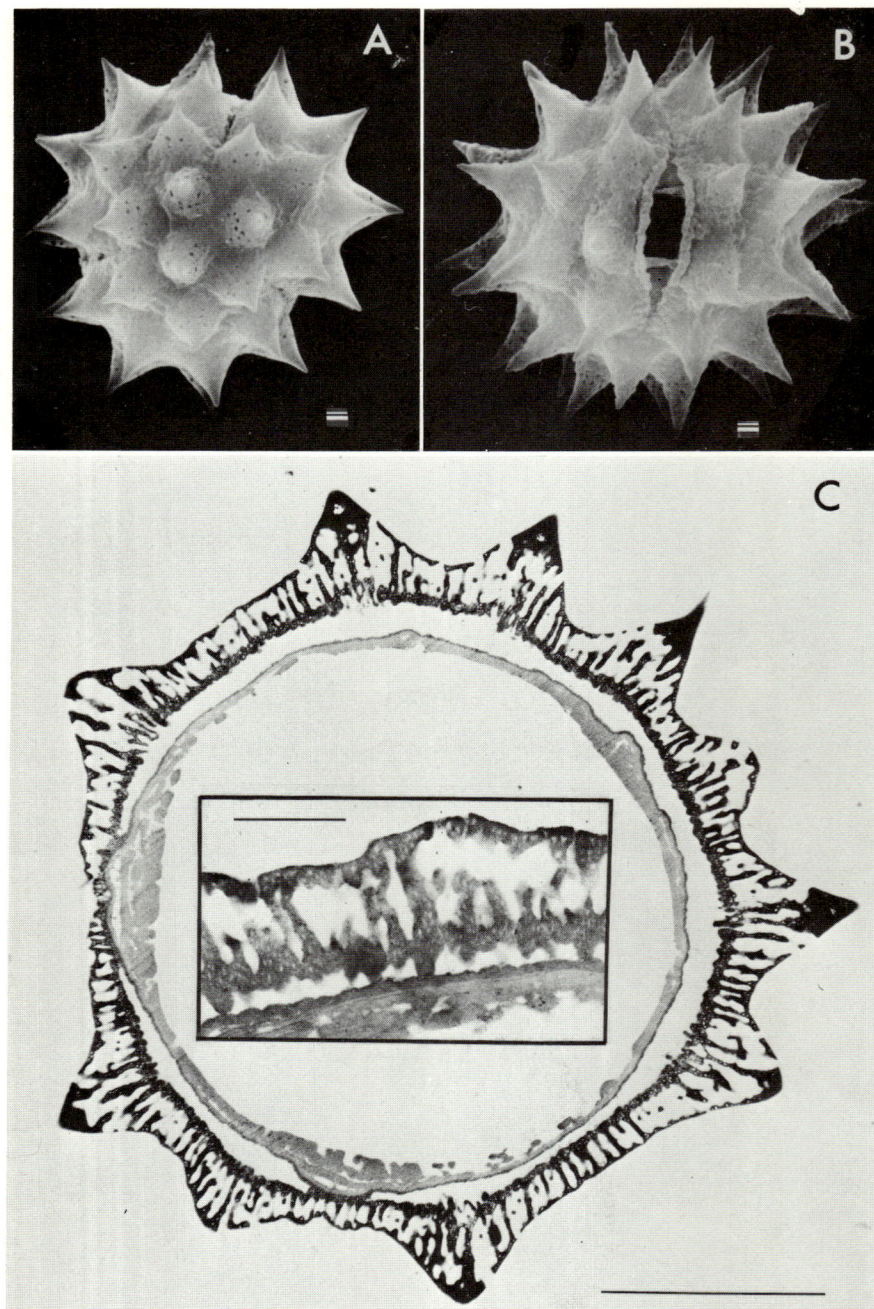

Plate 17.

8. POLLEN MORPHOLOGY

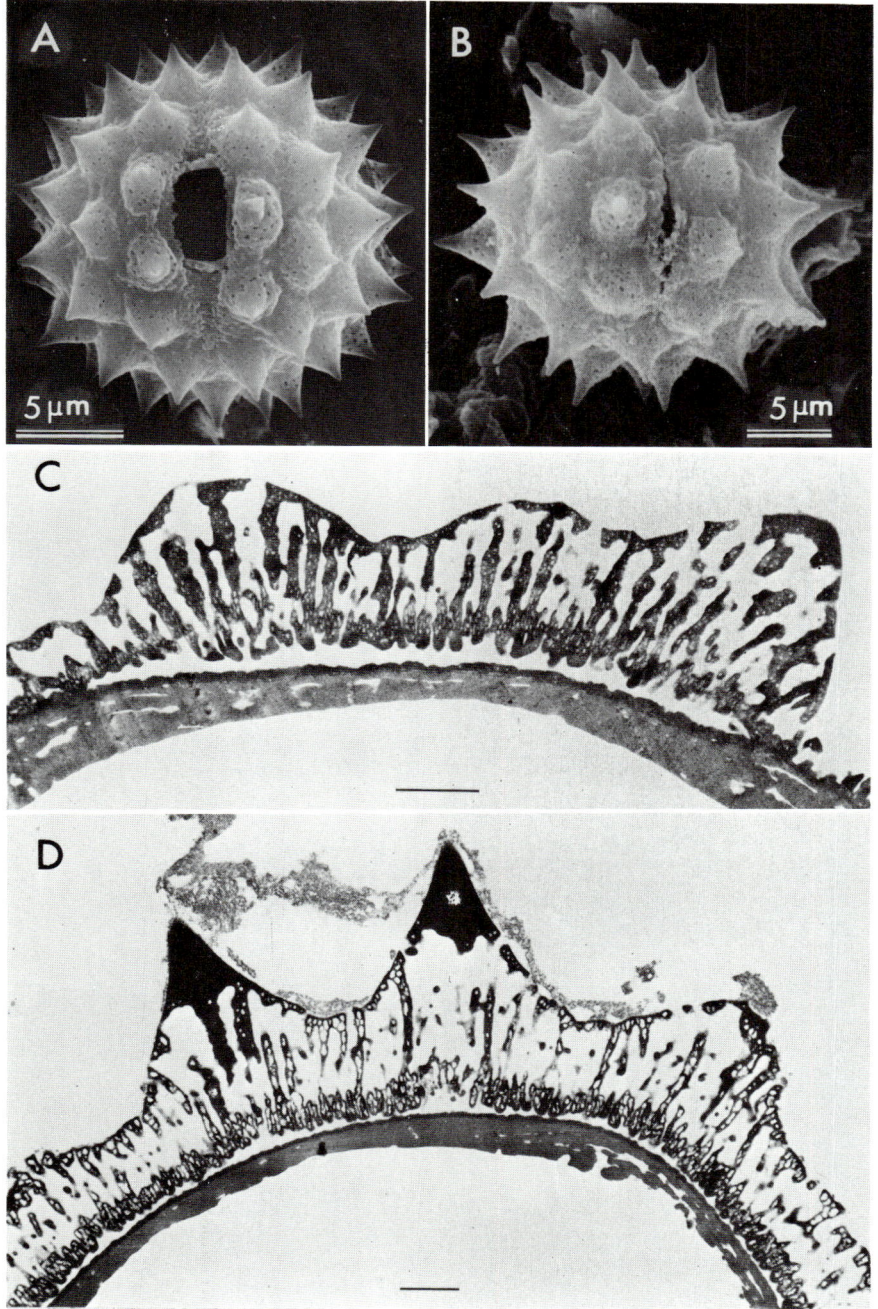

Plate 18.

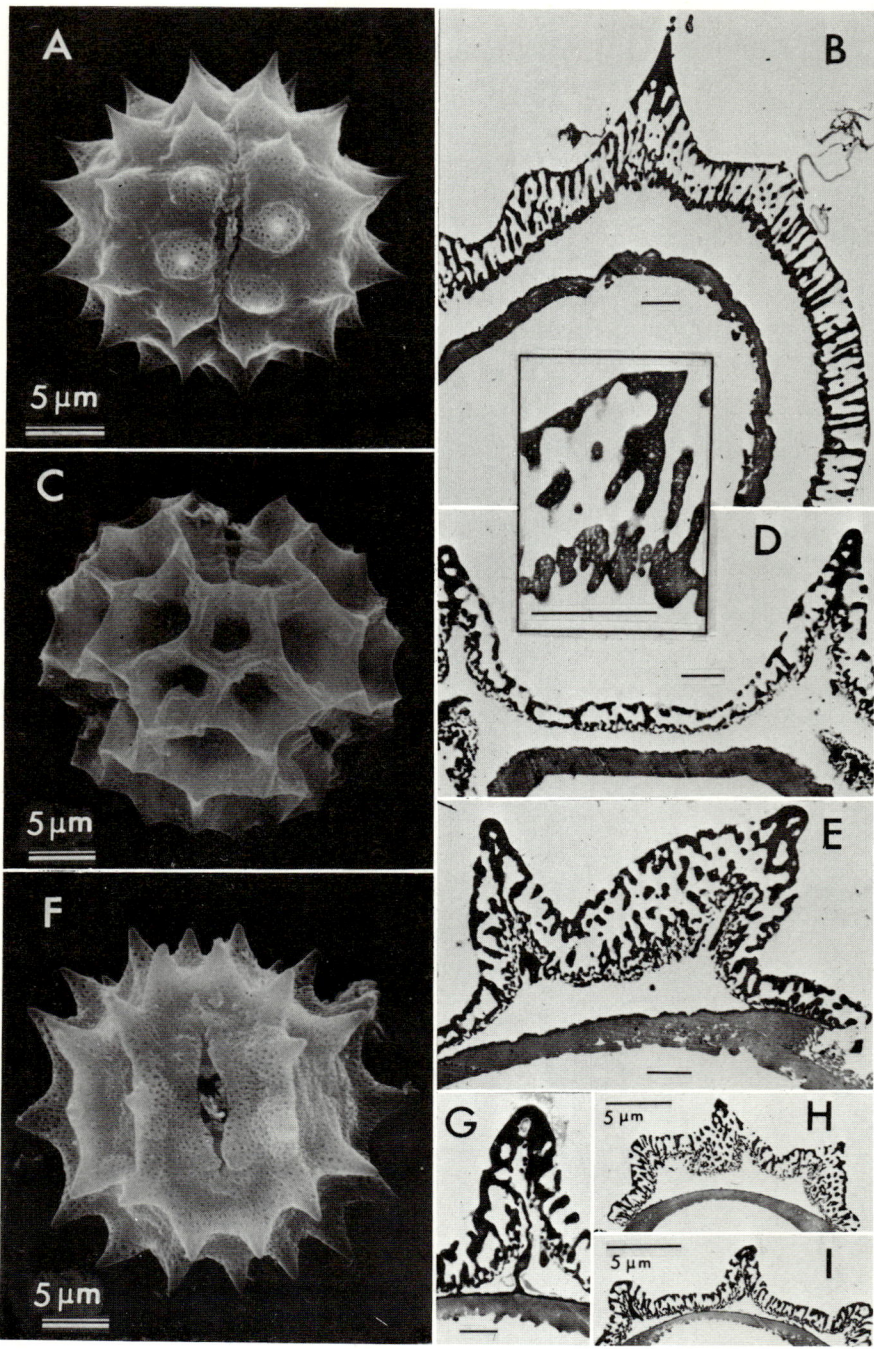

Plate 19.

8. POLLEN MORPHOLOGY

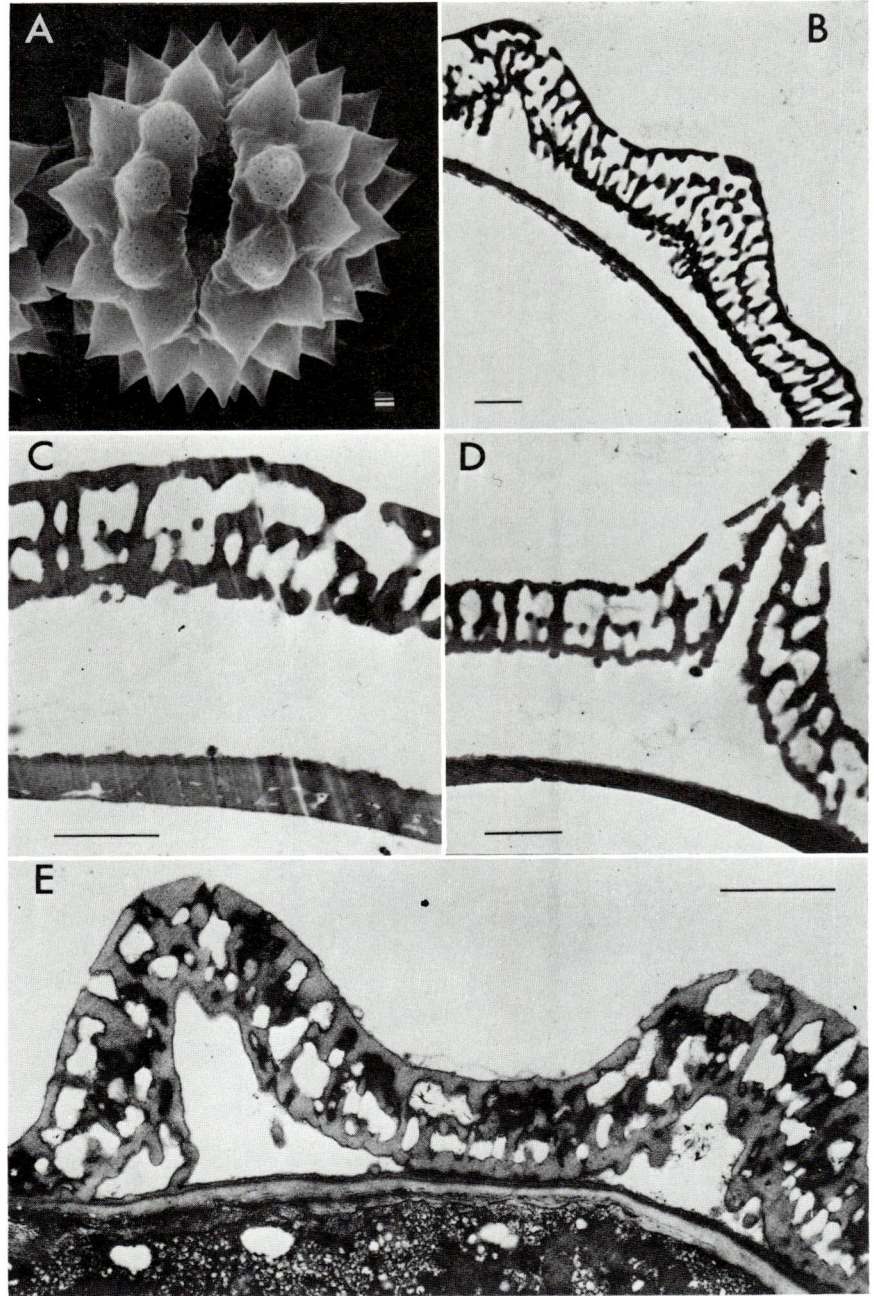

Plate 20.

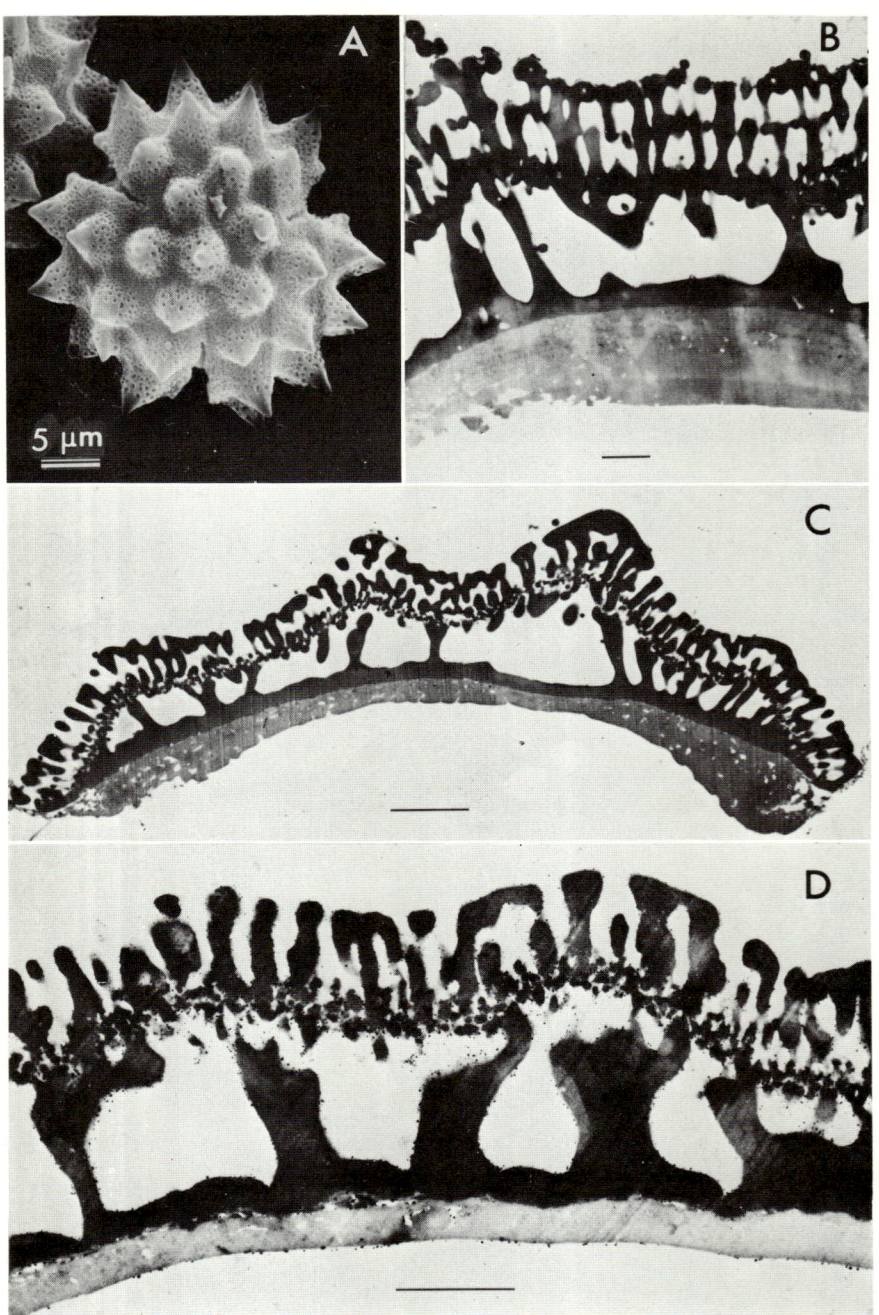

Plate 21.

8. POLLEN MORPHOLOGY

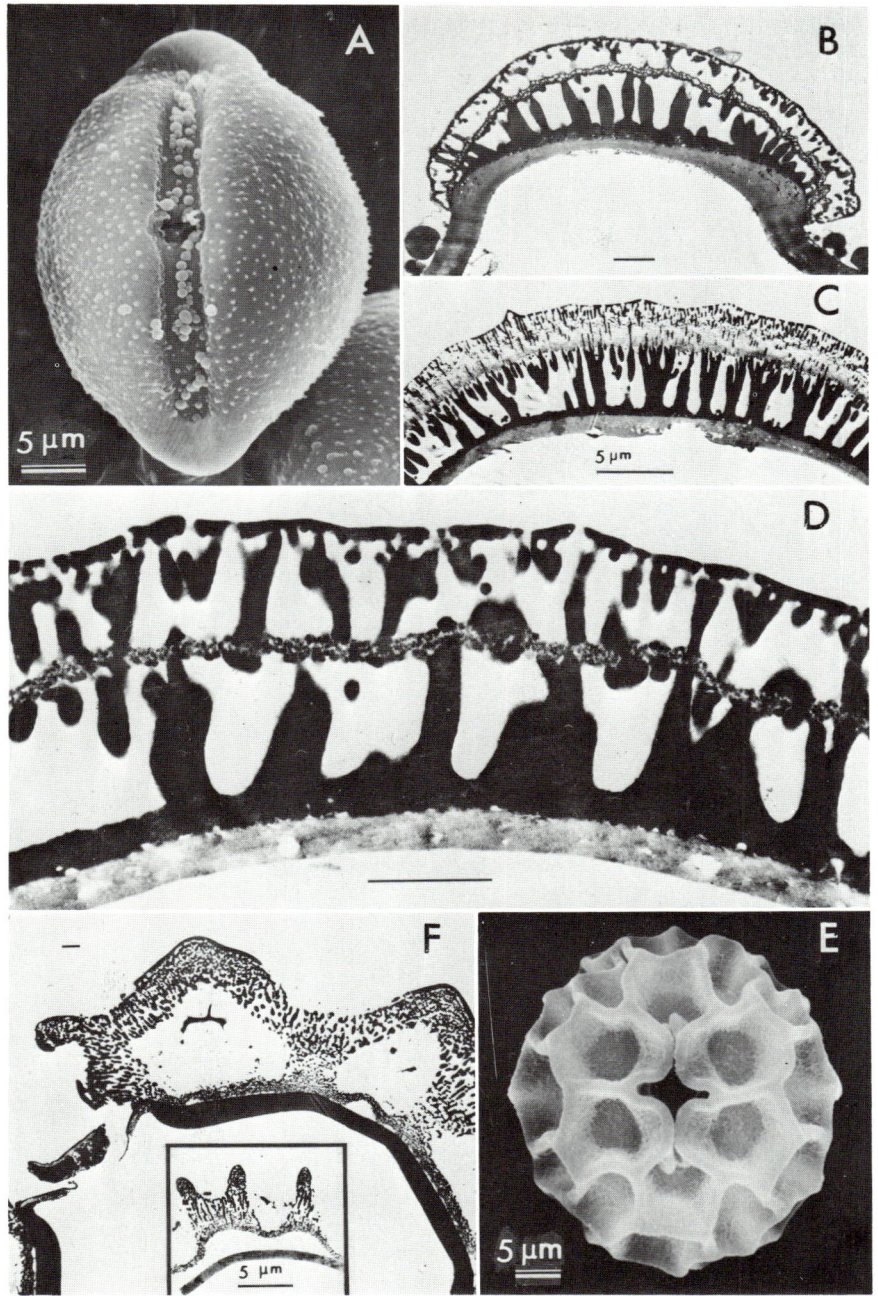

Plate 22.

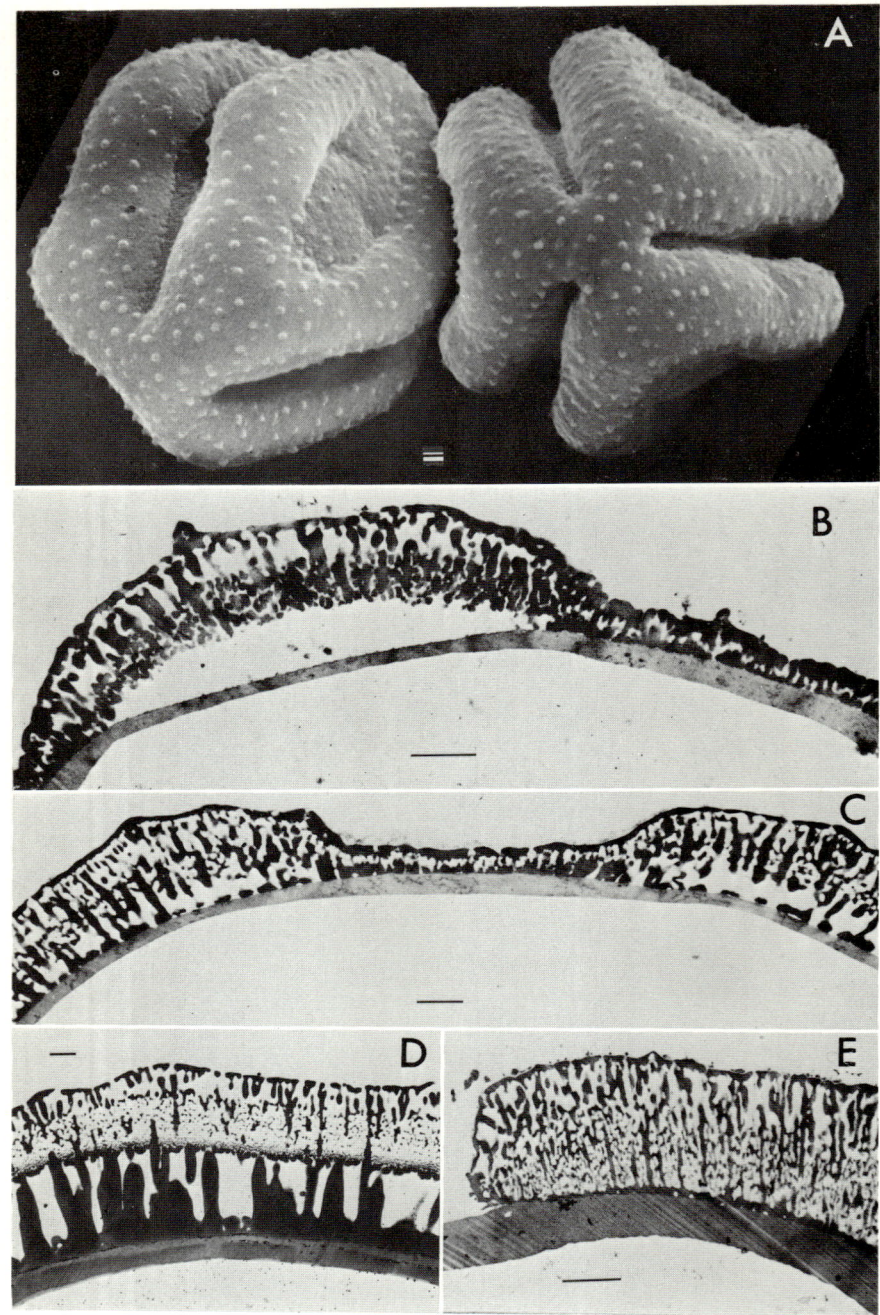

Plate 23.

8. POLLEN MORPHOLOGY

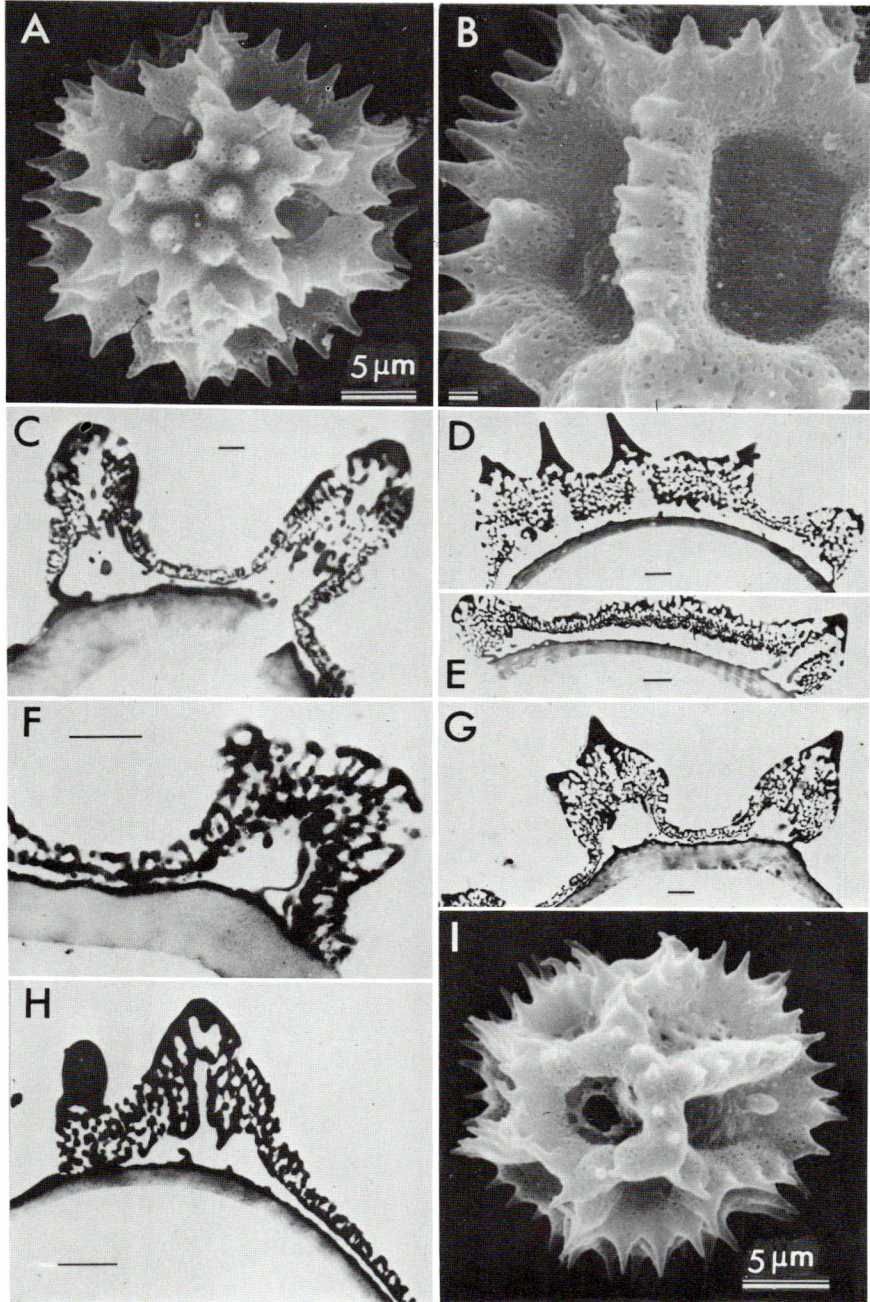

Plate 24.

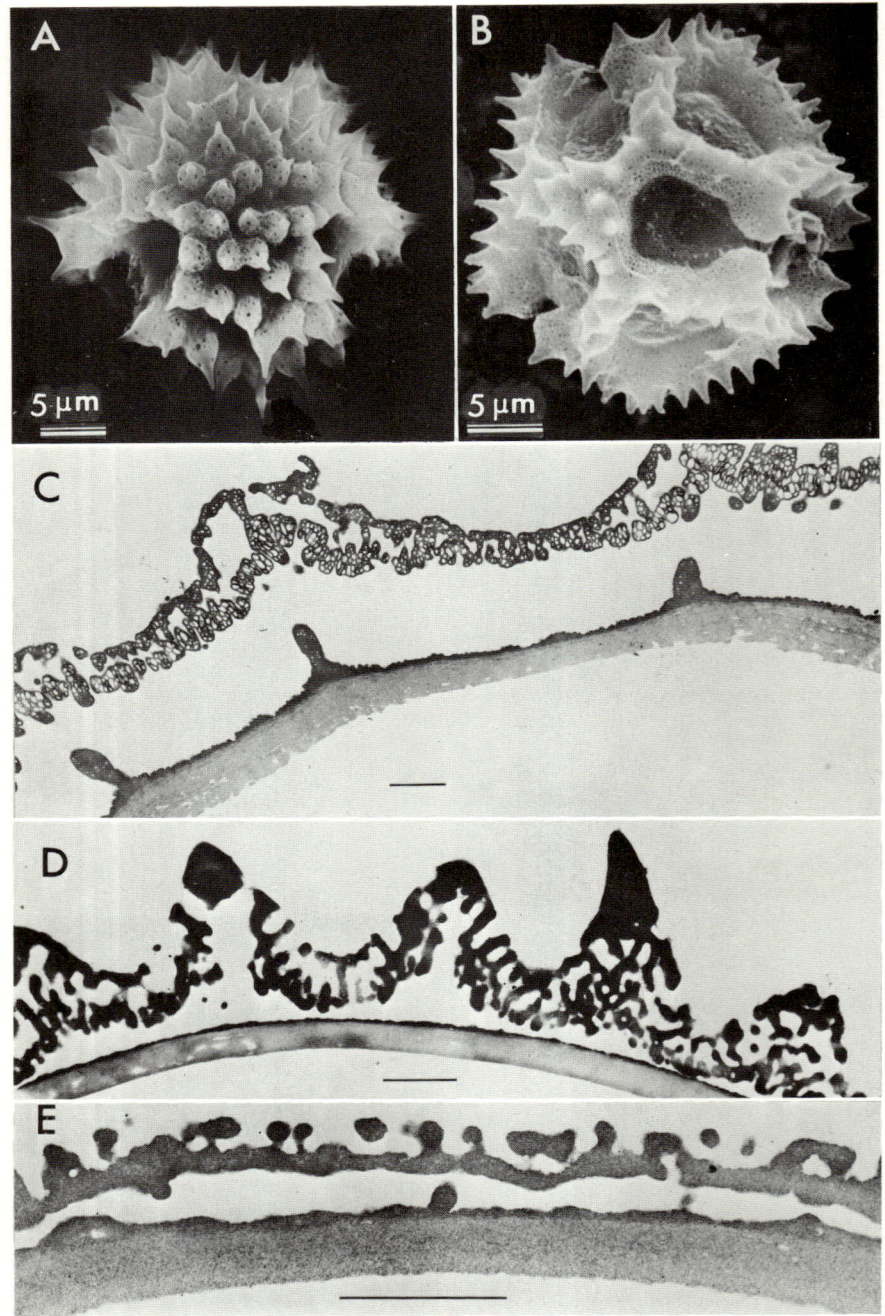

PLATE 25.

8. POLLEN MORPHOLOGY

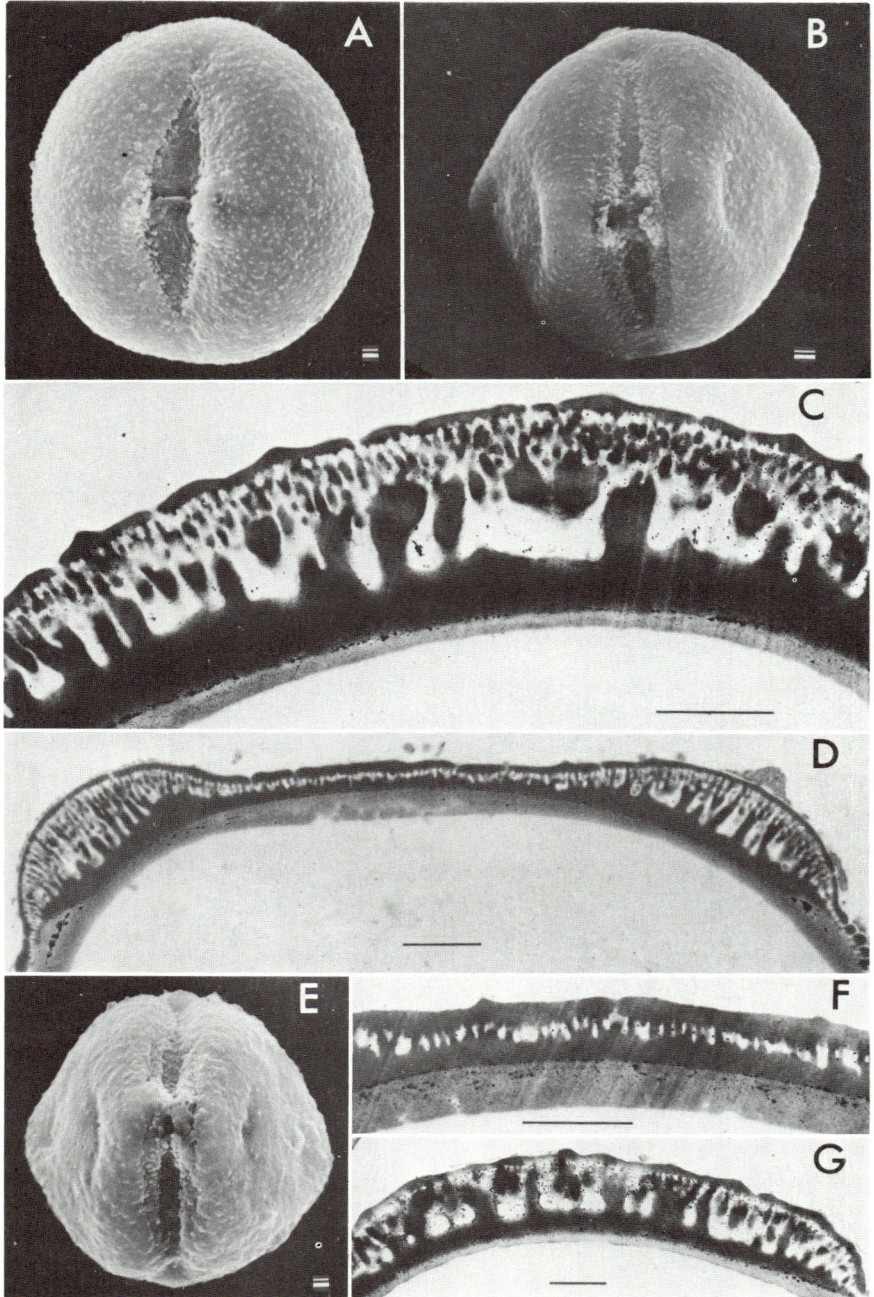

Plate 26.

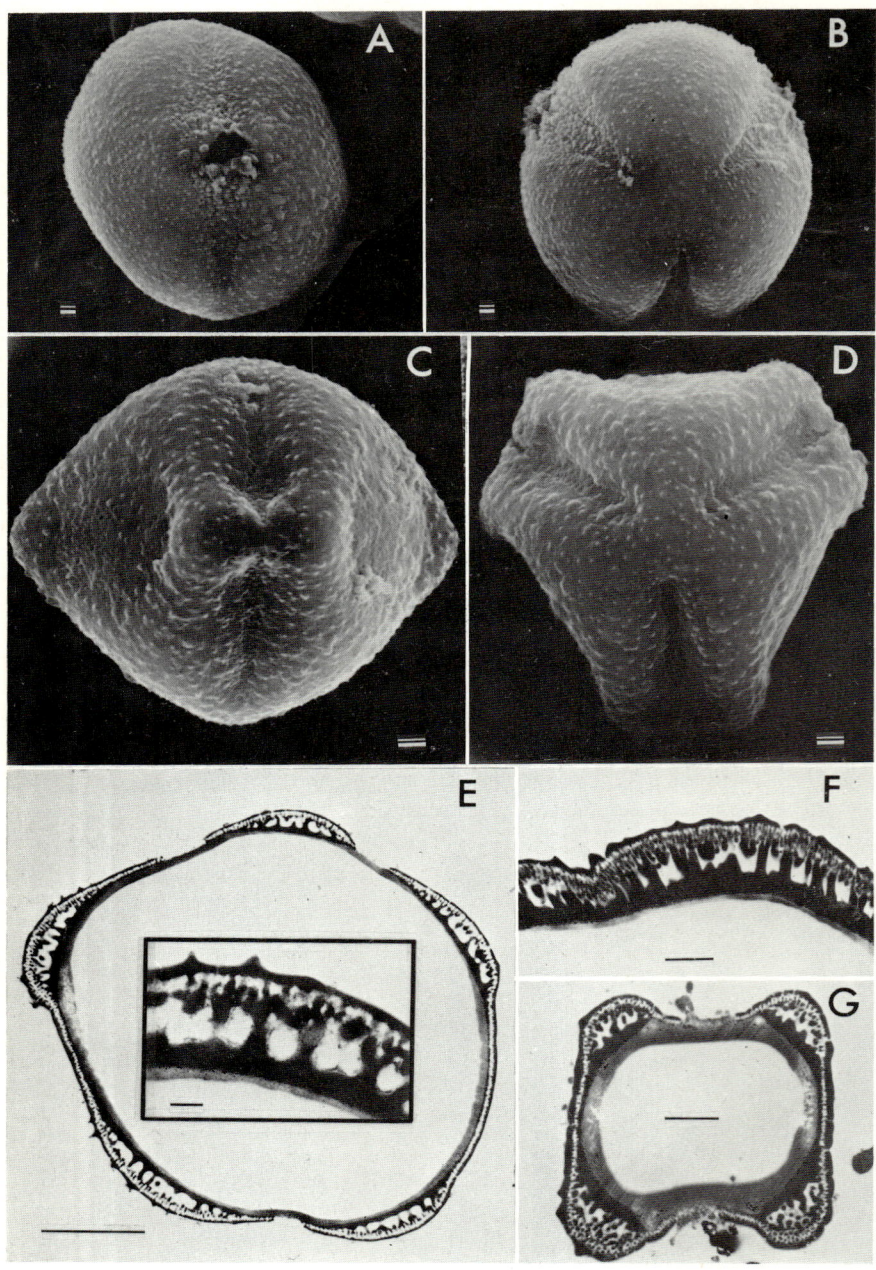

Plate 27.

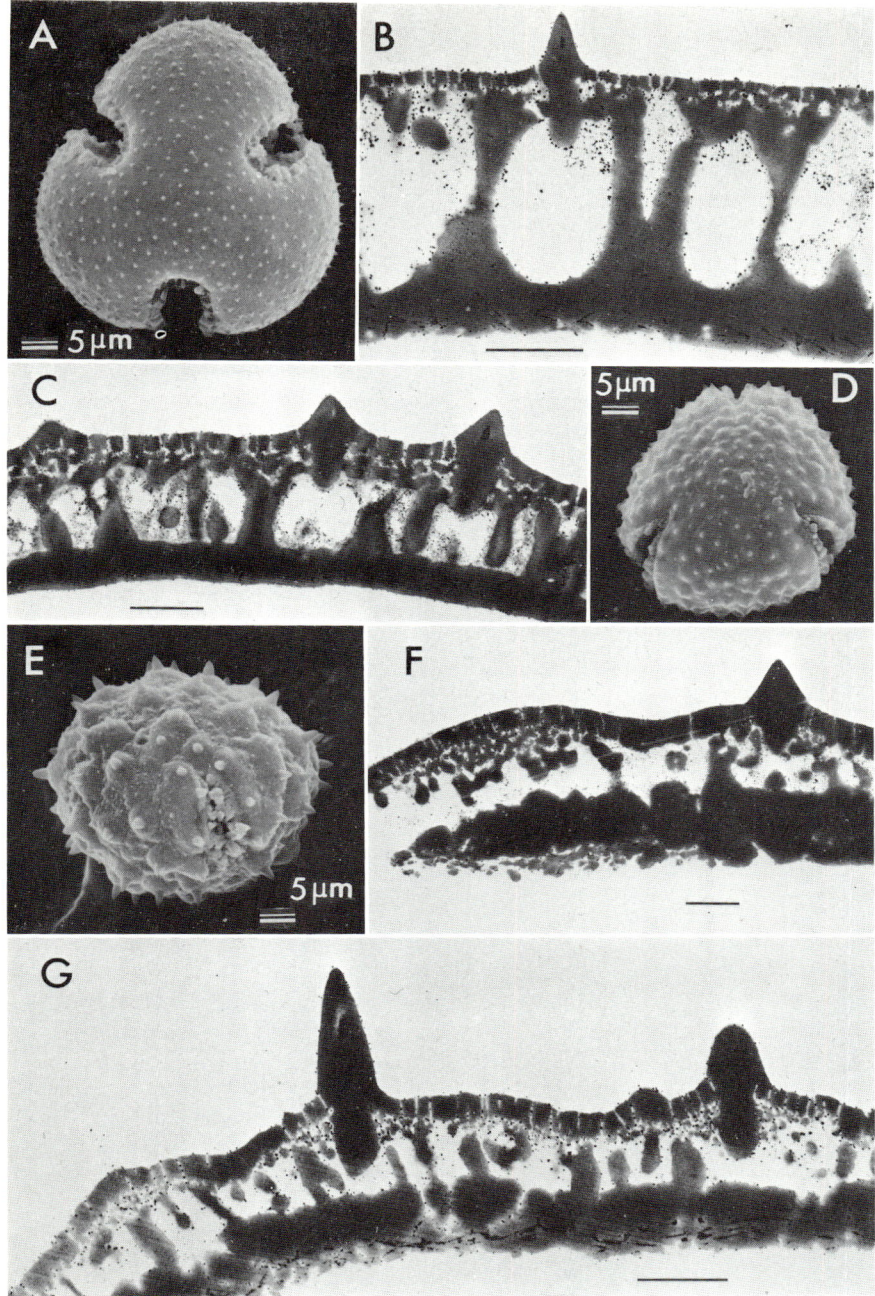

PLATE 28.

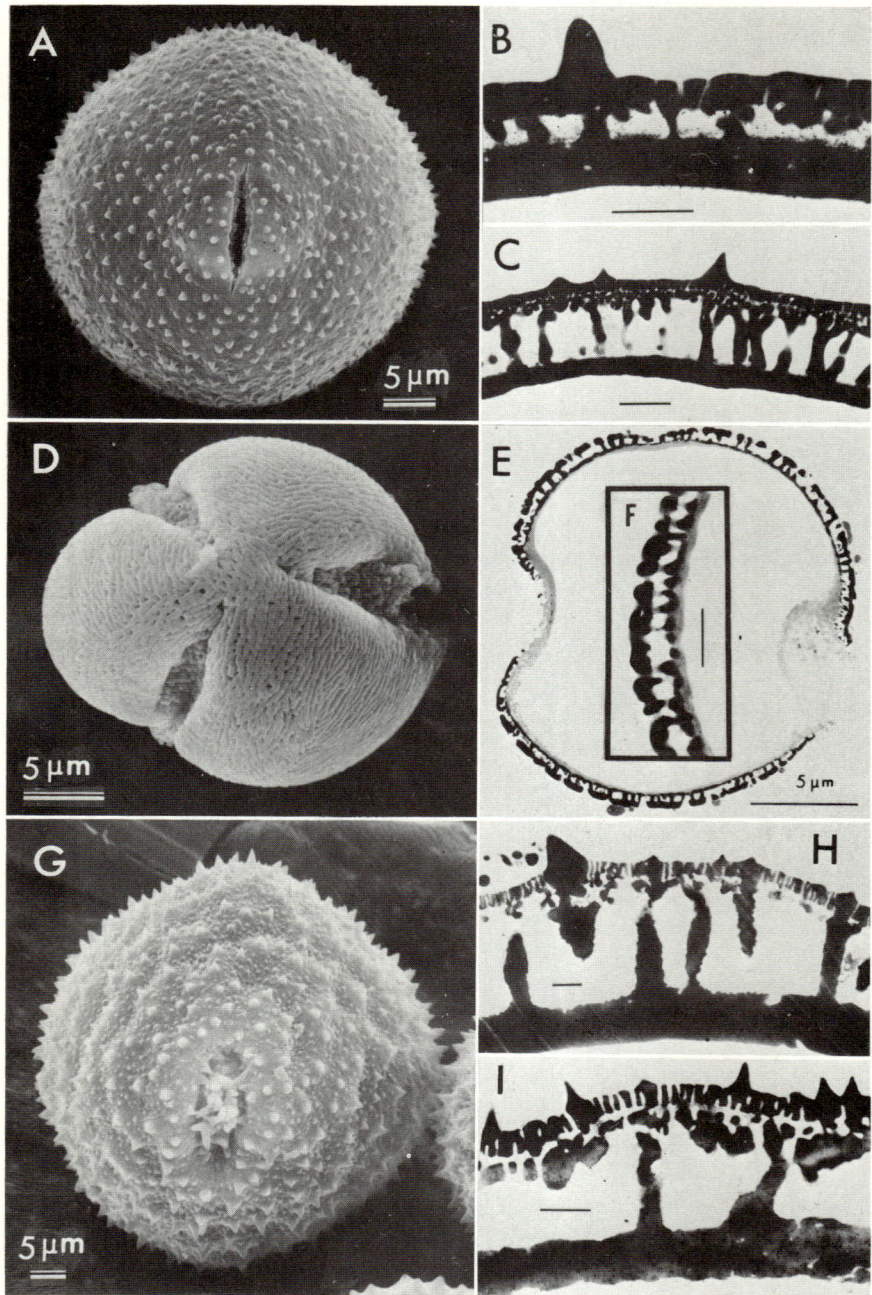

PLATE 29.

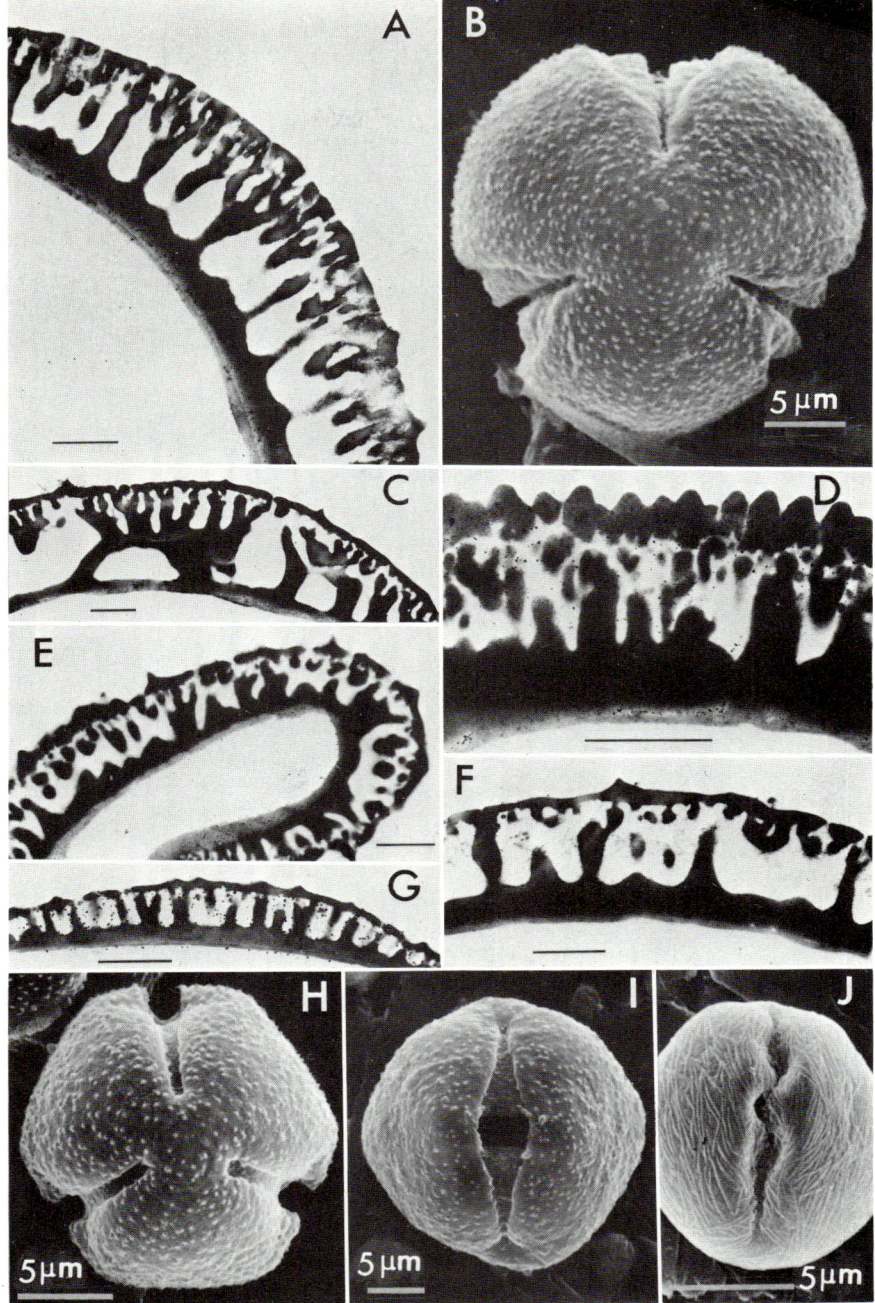

Plate 30.

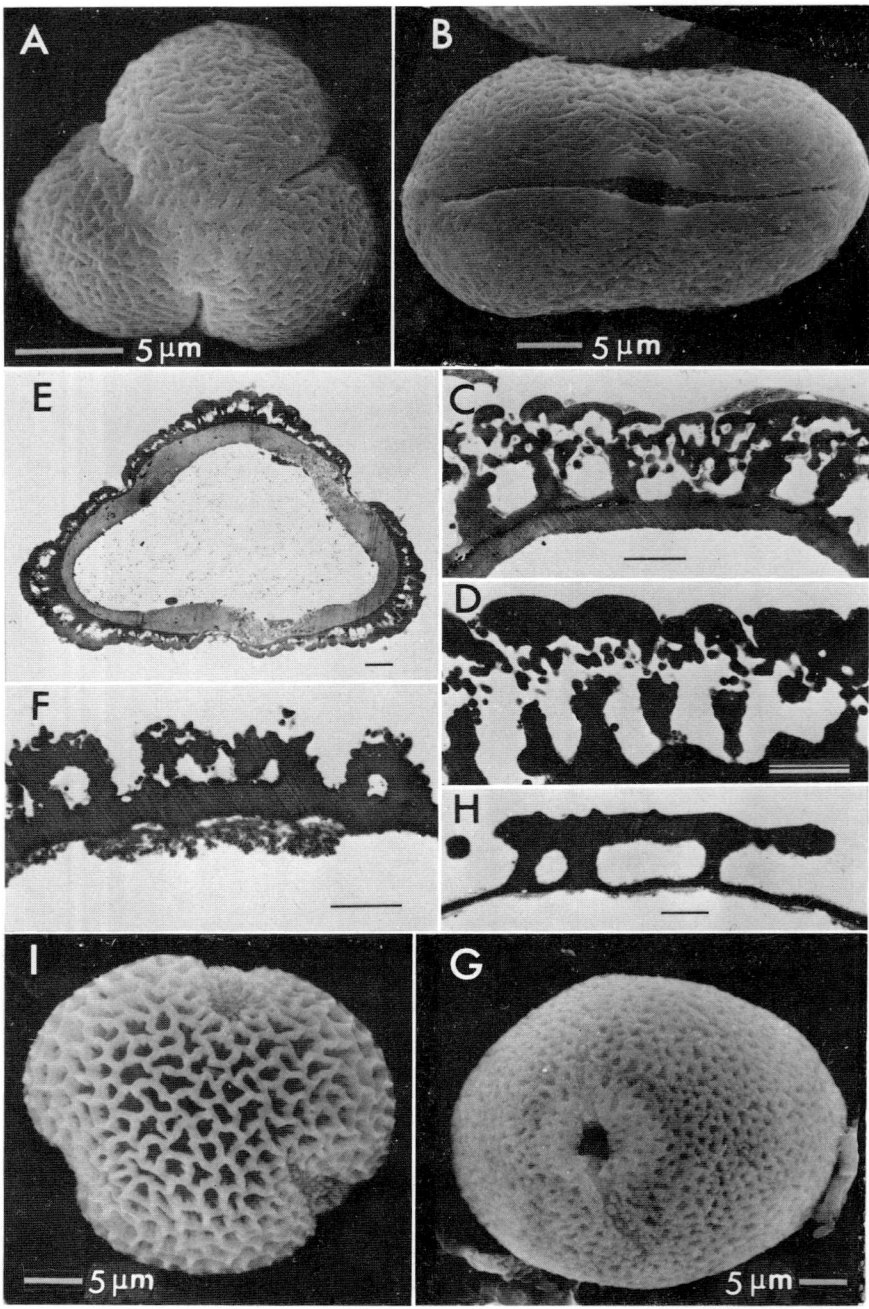

PLATE 31.

Appendix

Principal works on the pollen morphology of the Compositae

G. THANIKAIMONI
French Institute, Pondicherry, India

A list of 228 references, and index to 608 genera with indications regarding Replica (R), Scanning (S) and Transmission (T) electron microscope studies.

REFERENCES

1. AGASHE, S. N. and VINAY, P. (1975). Aeropalynological studies of Bangalore city. Part I. Pollen morphology of *Parthenium hysterophorus* L. *Curr. Sci., India* **44**, 216–217.
2. ANDERSON, C. (1972). A monograph of the Mexican and Central American species of *Trixis* (Compositae). *Mem. N.Y. Bot. Gdns* **22**(3), 1–68.
3. ANON. (1973). Pollen development. In Review of the work of the Royal Botanic Gardens, Kew, from June 1971 to December 1972. *Kew Bull.* **28**, 275, pl. 42 (T).
4. ARACHI, J. X. (1968). Pictorial presentation of Indian flora special study: Courtallam. St. Xavier's College, Palayamkottah, S. India. 193 pp.
5. ASKEROVA, R. K. (1970a). [A contribution to the palynological characterization of the tribe *Cichorieae of Compositae*.] *Bot. J. Moscow* **55**, 660–668.
6. ASKEROVA, R. K. (1970b). [La morphologie du pollen du genre *Scorzonera* L.] (en russe). Izvest. Akad. Nauk azerbajdzh. S.S.R., Biol. Nauk, Baku, 4: 43–46.
7. ASKEROVA, R. K. (1970c). [Données palynologiques sur la systématique et la phylogénie du genre *Scorzonera* L.]. *Izvest. Akad. Nauk azerbajdzh. S.S.R., Biol. Nauk, Baku* **6**, 11–15.
8. ASKEROVA, R. K. (1971). [On the pollen of certain genera of the tribe *Cichorieae (Compositae)*]. *Bot. J. Moscow* **56**(7), 971–978, pl. 1.
9. ASKEROVA, R. K. (1973). [Palynology of the tribe *Cichorieae (Compositae)*.]— In "Pollen and Spore Morphology of the Recent Plants". *Proc. 3rd Int. Palyn. Conf. Acad. Sci. USSR*, 33–37, 2 pl.
10. AVETISIAN, E. M. (1964). Palynosystématique de la tribu des *Centaureinae* des Asteraceae. *Bot. Inst. bot. Acad. Sci. Arménie* **14**, 31–47.
11. BAAGØE, J. (1974). The genus *Guizotia* (Compositae). A taxonomic revision. *Bot. Tidsskr.* **69**, 1–39. (S).
12. BANERJI, I. (1942). A contribution to the life history of *Blumea laciniata* L. *J. Indian bot. Soc.* **21**, 295–307.

13. BASSET, I. J., MULLIGAN, G. A. and FRANKTON, C. (1962). Poverty weed, *Iva axillaris* in Canada and the United States. *Can. J. Bot.* **40**, 1243-1249.
13a. BATALLA, M. A. (1940). Estudio morphologico de los granos de polen de las plantas vulgares del valle de Mexico. *An. Inst. Biol. Univ. Nacl. Mex.* **11**, 129-161.
14. BESOLD, B. (1971). Pollenmorphologische Untersuchungen an *Inuleen* (*Angianthinae, Relhaniinae, Athrixiinae*). *Dissertationes Botanicae* **14**, 1-72.
15. BHATTACHARYYA, U. C. (1963). *Soliva anthemifolia* R. Br. (Compositae)—a new record for India. *Bull. bot. Surv. India* **5**, 375-376.
16. BOULOS, L. (1973). Révision systématique du genre *Sonchus* L. s.l. IV. Sous-genre 1. *Sonchus. Bot. Notiser* **126**, 155-196.
17. BOULOS, L. (1974). Révision systematique du genre *Sonchus* L. s.l. *Bot. Notiser* **127**, 402-451.
18. BREMER, K. (1972). The genus *Osmitopsis* (Compositae). *Bot. Notiser* **125**, 9-48.
19. BREWER, J. G. and HENSTRA, S. (1970). A membrane investing mature individual pollen grains of Pyrethrum (*Chrysanthemum cinerariaefolium* Vis.). *Euphytica* **19**, 121-124.
20. CARLQUIST, S. (1957a). Anatomy of Guayana *Mutisieae*: pollen. *Mem. N.Y. Bot. Gdn* **9**(3), 441-476.
21. CARLQUIST, S. (1957b). The genus *Fitchia* (Compositae). *Univ. Calif. Publs Bot.* **29**, 1-144.
22. CARLQUIST, S. (1958). Anatomy and systematic position of *Centaurodendron* and *Yunquea* (Compositae). *Brittonia* **10**, 78-93.
23. CARLQUIST, S. (1959). Studies on *Madinae*: anatomy, cytology and evolutionary relationships. *Aliso*, p. 171-236, 113 fig.
24. CARLQUIST, S. (1967). Anatomy and systematics of *Dendroseris* (*sensu lato*). *Brittonia* **19**(2), 99-121.
25. CARLQUIST, S. and GRANT, M. L. (1963). Studies in *Fitchia* (Compositae): novelties from the Society Islands; anatomical studies. *Pacific Science* **XVII**, 282-298.
26. CHAUBAL, P. D. and DEODIKAR, G. B. (1965). Pollen morphotypes in the family Compositae from parts of Western Ghats (India). *Palynol. Bull.* **I**, 56-58, 4 fig.
27. CHOPRA, S. (1964). Illustrations of Indian plants.—*Taraxacum*. I. *Bull. natn. bot. Gdns*, n° 91.
28. CHOPRA, S. (1965). Illustrations of Indian Plants.—*Taraxacum*. II.—*Bull. nat. bot. Gdns Lucknow*, n° 114.
29. CHOPRA, S. and NAIR, P. K. K. (1965). Palynological studies on Indian Taraxaca I. *Proc. Indian Acad. Sci.*, Sect. B **61**, 214-221, 26 fig.
29a. CRANWELL, L. M. (1942). New Zealand Pollen Studies. I. *Rec. Auckland Inst. Mus.* **2**, 280-308.
30. CRISCI, J. V. (1971). Sobre una especie de *Leuceria* (*Compositae*) de Chile. *Darwiniana* **16**, 627-633.
31. CRISCI, J. V. (1974a). *Marticorenia*: a new genus of Mutisieae (Compositae). *J. Arnold Arbor.* **55**, 38-45. (S).
32. CRISCI, J. V. (1974b). A numerical taxonomic study of the sub-tribe Nassuvinae (Compositae, Mutisieae). *J. Arnold Arbor.* **55**, 568-610 (pollen p. 592-594).
33. DABROWSKA, J. (1971). La taille des cellules de garde des stomates et des grains de pollen de quatre espèces d'*Achillea* (en polonais). *Herba polon. Pologne* **17** (1-2), 13-30.

34. DAKSHINI, K. M. M. and SINGH, P. (1973). Contribution to the palynology of Compositae. (Asteraceae). *Palynol. Bull.* **VI**(2), 99–107; 1970.
34a. DAVIS, G. L. (1962). Embryological studies in the Compositae, I and II. *Aust. J. Bot.* **10**, 1–12 (pollen fig. 19), 65–75 (pollen fig. 16).
35. DESHPANDE, P. K. (1962). Contribution to the embryology of *Caesulia axillaris* Roxb. *J. Indian bot. Soc.* **41**, 540–550.
36. DESHPANDE, P. K. and KOTHARE, M. M. (1973). A contribution to the life history of *Chrysanthellum indicum* DC. *Proc. 60th Ind. Sc. Cong.* Part III, Abstr. 327.
37. DIMON, M. TH. (1971a). Problèmes généraux soulevés par l'étude pollinique de Composées Mediterranéennes. *Naturalia Monspeliensia* **22**, 129–144; pl. I–V. (S).
38. DIMON, M. TH. (1971b). "Etude des Types polliniques des *Composées* échinulées du Bassin Mediterraneen Occidental". Thèse de specialité. Univ. Sci. et Tech. Montpelier (S).
38a. EDGEWORTH, M. P. (1877). "Pollen". Hardwicke & Bogue, London.
39. ERDTMAN, G. (1952). "Pollen Morphology and Plant Taxonomy. Angiosperms. Almqvist & Wiksell, Stockholm. (Russian edition translated by L. A. Kozjar in 1956).
40. FELIPPE, G. M. and LABOURIAU, M. L. S. (1964). Pollen grains of plants of the "Cerrado". VI—Compositae. Tribus Helianthae. *Anals Acad. bras. Cienc*, 36, n° 1, p. 85–101, 32 fig.
41. FEUER S. M. (1974). Pollen morphology and ultrastructure in the subtribe Microseridinae (Tribe Lactuceae, family Asteraceae). *Am. J. Bot.* **61** (suppl. 5), 43.
41a. FISCHER, H. (1890). "Beiträge zur vergleichenden Morphologie der Pollenkörner". Thesis, Breslau.
42. FISHER, T. R. and WELLS, J. R. (1962). Heteromorphic pollen grains in *Polymnia. Rhodora* **64**, 336–340.
43. GHOSH, R. B. (1962). A contribution to the life history of *Wedelia calendulaceae* Less. *J. Indian bot. Soc.* **41**, 196–206.
43a. GONZALEZ QUINTERO, L. (1969). Morphologia polinica. *Palaeoecologia, Mexico* **3**, 1–185.
44. GORODKOW, B. N. (1952). [Examination of the possibility to distinguish species of *Artemisia* by pollen.] *Bot. J. Moscow* **37**, 659–660.
45. GRASHOFF, J. L. (1974). Novelties in *Stevia* (Compositae: Eupatorieae). *Brittonia* **26**, 347–384.
46. GRASHOFF, J. L. and BEAMAN, J. H. (1970). Studies in *Eupatorium* (Compositae), III. Apparent wind pollination. *Brittonia* **22**, 77–84. (S).
47. GREUTER, W. (1973). Monographie der Gattung *Ptilostemon* (Compositae). *Boissiera* **22**, 1–215; pl. I–VIII (pollen: p. 51, 53–55, 76, pl. II).
48. GREUTER, W. and DITTRICH, M. (1973). Neuer Beitrag zur Kenntnis der Gattung *Lamyropsis* (Compositae): die Identität von *Cirsium microcephalum* Moris. *Ann. Mus. Goulandris* **1**, 85–98.
49. GRIERSON, A. J. C. (1972). A new species of *Anaphalis* (Compositae) from Mexico. *Notes R. bot. Gdn. Edinb.* **31**, 389–392. (T).
50. GUINET, PH. and MALEY, J. (1974). Compositae. *Trav. Doc. Geogr. Trop. CEGET, Talence.* **16**, 96–97, 102–105.
51. GUSTAFSSON, M. and SNOGERUP, S. (1972). *Scorzonera scyria*, a new chasmophytic species from Greece. *Bot. Notiser* **125**, 323–328.
51a. HASSALL, A. H. (1842). Observations on the structure of the pollen granule. *Ann. Mag. nat. Hist.* **8**, 92–108, **9**, 544–573.

52. HENDERSON, R. J. F. (1969). *Podolepis monticola* a new species of *Compositae* from Queensland. *Contr. Qd Herb.* **2**, 1–9.
53. HERNANDEZ, P. J. (1966). Notas palinologicas del norte Argentino, provincia Altoandia. I. Compositae. *Ameghiniana, Argent.* **4**(9), 305–308.
54. HESLOP-HARRISON, J. (1969a). The origin of surface features of the pollen wall of *Tagetes patula* as observed by scanning electron microscopy. *Cytobios* **1**(2), 177–186.
54a. HESLOP-HARRISON, J. (1969b). Scanning electron microscopic observations on the wall of pollen grains of *Cosmos bipinnatus* (Compositae). *Proc. Engis Stereoscan Colloquium Wis.*, 89–96.
54b. HESLOP-HARRISON, J. (1969c). An acetolysis-resistant membrane investing tapetum and sporogenous tissue in the anthers of certain Compositae. *Can. J. Bot.* **47**(4), 541–542. (S).
55. HEUSSER, C. J. (1971). "Pollen and Spores of Chile". University Arizona Press, Tucson.
55a. HUANG, T. C. (1972). "Pollen Flora of Taiwan". National Taiwan University Botany Department Press.
55b. IKUSE, M. (1956). "Pollen grains of Japan". Hirokawa Publishing Co., Tokyo.
56. IKUSE, M. (1962). [On pollen grains of the Compositae collected by the Kyoto University scientific expedition to the Karakoram and Hindukush 1955.] *Acta phytotax. Geocot.* **20**, 112–119.
57. JAIN, R. K. and NANDA, S. (1966). Pollen morphology of some desert plants of Pilani, Rajasthan. *Palynol. Bull.* **II** & **III**: 56–59.
58. JEFFREY, C. (1968). Notes on Compositae—III. *Kew Bull.* **22**, 107–140.
59. JONES, S. B. (1970). Scanning electron microscopy of pollen as an aid to the systematics of *Vernonia* (Compositae). *Bull. Torrey bot. Club* **97**, 325–335. (S).
60. JONES, S. B. Jr. (1973). Revision of *Vernonia* section *Eremosis* (Compositae) in North America. *Brittonia* **25**, 86–115. (S).
61. KAPIL, R. N. and SETHI, S. B. (1962). Gametogenesis and seed development in *Ainsliaea aptera* DC. *Phytomorphology* **12**, 222–234.
62. KAUL, V. (1973). Embryology and development of fruit in *Cichorieae*. III. *Youngia japonica* DC. Proc. 60th Indian Science Congress, Part III: Abstr. 320.
63. KESSLER, L. G. and LARSON, D. A. (1969). Effects of polyploidy on pollen grain diameter and other exomorphic exine features in *Tridex coronifolia*. *Pollen et Spores* **11**, 203–221.
64. KHIDIR, M. O. (1970). A note on the inheritance of pollen colour in safflower (*Carthamus* L.). *Can. J. Genet. Cytol.* **12**, 360–361.
65. KING, R. M. and ROBINSON, H. (1967). Multiple pollen forms in two species of the genus *Stevia* (Compositae). *Sida* **3**, 165–169, 2 fig., 2 tabl.
66. KING, R. M. and ROBINSON, H. (1970). Studies in Eupatorieae (Compositae) XIV. Another example of dimorphic pollen? *Phytologia* **19**, 301–302.
67. KING, R. M. and ROBINSON, H. (1972). Studies in the Eupatorieae (Asteraceae). LXVI. The genus *Pachythamnus*. *Phytologia* **23**, 153–154.
68. KING, R. M. and ROBINSON, H. (1972a). Studies in the Eupatorieae (Asteraceae). LXXII. *Phytologia* **23**, 395–396.
69. KING, R. M. and ROBINSON, H. (1972b). Studies in the Eupatorieae (Asteraceae). LXXIV. *Phytologia* **23**, 405–408.
70. KING, R. M. and ROBINSON, H. (1972c). Studies in the Eupatorieae (Asteraceae). LXXVI. *Phytologia* **24**, 57–59.

71. KING, R. M. and ROBINSON, H. (1972d). Studies in the Eupatorieae (Asteraceae). LXXVII. *Phytologia* **24**, 60–62.
72. KING, R. M. and ROBINSON, H. (1972e). Studies in the Eupatorieae (Asteraceae). LXXX. *Phytologia* **24**, 67–69.
73. KING, R. M. and ROBINSON, H. (1972f). Studies in the Eupatorieae (Asteraceae). LXXXIII. *Phytologia* **24**, 74–76.
74. KING, R. M. and ROBINSON, H. (1972g). Studies in the Eupatorieae (Asteraceae). LXXXV. *Phytologia* **24**, 79–104.
75. KING, R. M. and ROBINSON, H. (1972h). Studies in the Eupatorieae (Asteraceae). LXXXVI. *Phytologia* **24**, 105–107.
76. KING, R. M. and ROBINSON, H. (1972i). Studies in the Eupatorieae (Asteraceae). LXXXVII. *Phytologia* **24**, 108–111.
77. KING, R. M. and ROBINSON, H. (1972j). Studies in the Eupatorieae (Asteraceae). LXXXVIII. *Phytologia* **24**, 112–117.
78. KING, R. M. and ROBINSON, H. (1972k). Studies in the Eupatorieae (Asteraceae). XCI. *Phytologia* **24**, 173–175.
79. KING, R. M. and ROBINSON, H. (1972l). Studies in the Eupatorieae (Asteraceae). XCVI. *Phytologia* **24**, 185–186.
80. KING, R. M. and ROBINSON, H. (1972m). Studies in the Eupatorieae (Asteraceae). XCVII. *Phytologia* **24**, 187–191.
81. KING, R. M. and ROBINSON, H. (1972n). Studies in the Eupatorieae (Asteraceae). XCVIII. *Phytologia* **24**, 192–194.
82. KING, R. M. and ROBINSON, H. (1972o). Studies in the Eupatorieae (Asteraceae). CI. *Phytologia* **24**, 281.
83. KING, R. M. and ROBINSON, H. (1972p). Studies in the Eupatorieae (Asteraceae). CIII. *Phytologia* **24**, 382–386.
84. KING, R. M. and ROBINSON, H. (1972q). Studies in the Eupatorieae (Asteraceae). CVIII. *Phytologia* **24**, 397–400.
85. KING, R. M. and ROBINSON, H. (1972r). Studies in the Eupatorieae (Asteraceae). CIX. *Phytologia* **24**, 401–403.
86. KING, R. M. and ROBINSON, H. (1972s). Studies in the Eupatorieae (Asteraceae). CX. *Phytologia* **24**, 404–406.
87. KIRPICHNIKOV, M. E. and KUPRIANOVA, L. A. (1950). [Morphological, geographical and palynological contributions to the understanding of the subtribe Gnaphaliinae.] *Acta Inst. Bot. Acad. Sci. URSS. Ser. 1* **9**, 7–37.
87a. KNOX, R. B. and HESLOP-HARRISON, J. (1971). Pollen wall proteins; localization of antigenic and allergenic proteins in the pollen grain walls of *Ambrosia* spp. (ragweeds). *Cytobios* **4**, 49–54.
88. KUPRIANOVA, L. A. and ALYOSHINA, L. A. (1972). [Pollen and spores of plants from the flora of European part of USSR.], Vol 1. Akad. Sci. USSR Komorov Bot. Inst. 170 p.
88a. LABOURIAU, M. L. S. (1973). "Contribuçao à Palinologia dos Cerrados". Acad. Bras. Ciencias, Rio de Janeiro.
89. LARSON, D. A., SKVARLA, J. J. and LEWIS, C. W. (1962). An electron microscope study of exine stratification and fine structure. *Pollen et Spores* **4**, 233–246(T).
90. LEINS, P. (1968). Versuch einer Gliederung der *Inulinae* und *Buphtalminae* nach den Pollenkorntypen. *Ber. dt. bot. Ges.* **81**(11), 498–504, 5 fig.
91. LEINS, P. (1970). Die Pollenkörner und Verwandschafts-beziehungen der Gattung *Eremothamnus* (Asteraceae). *Mitt. Bot. München* **7**, 369–376.
92. LEINS, P. (1971). Pollensystematische Studien an Inuleen. I. Tarchonanthinae, Plucheinae, Inulinae, Buphthalminae. *Bot. Jb.* **91**(1), 91–146.

93. Leins, P. and Thyret, G. (1971). Pollen phylogeny and taxonomy exemplified by an African Asteraceae group. *Mitt. bot. Staatssamml. München* **10**, 280–286.
93a. Macko, S. (1957). Lower Miocene pollen flora. *Trav. Soc. Sci. Lettres Wroclaw, Ser. B* **96**, 1–178.
93b. Maguire et al. (1967). The Botany of the Guayana Highland, Part VII. *Mem. N.Y. bot. Gdn* **17**, 1–439.
94. Mallea, M. and Soler, M. (1974). Compositae. *Trav. Doc. Geogr. trop. CEGET, Talence* **16**, 98–101.
95. Manum, S. (1955). Some remarks on the pollen grains of *Gomphrena globosa* and *Chrysanthemum cariantum*. *Blyttia* **13**, 90–95; 1 pl.
96. Marticorena, C. and Crisci, J. V. (1972). Sobre *Haplopappus scrobiculatus* (Compositae) de chile y Argentina y su Sinonima. *Darwiniana* **17**, 467–472.
96a. Martin, P. S. and Drew, C. M. (1969). Scanning electron photomicrographs of Southwestern pollen grains. *J. Arizona Acad. Sci.* **5**, 147–176. (S).
96b. Martin, P. S. and Drew, C. M. (1970). Additional scanning electron photomicrographs of Southwestern pollen grains. *J. Arizona Acad. Sci.* **6**, 140–161 (S).
97. Maurizio, A. and Louveaux, J. (1963). Pollens de plantes mellifères d'Europe IV. *Pollen et Spores* **5**, 213–232.
98. Misra, S. (1957). Floral morphology of the family Compositae. *J. Indian bot. Soc.* **36**, 503–512.
99. Mitra, J. (1947). A contribution to the embryology of some Compositae. *J. Indian bot. Soc.* **26**, 105–123.
100. Mitsuoka, S. (1968). [Abnormal pollen grain formation in some hibrids of *Matricaria* and *Anthemis*.] *Jap. J. Palynol.* **2**, 6–7.
100a. Mohl, H. (1835). Sur la structure et les formes des graines de pollen. *Ann. Sci. nat., Ser. 2* **3**, 148–180, 220–236, 304–346.
101. Monoszon, M. C. (1948). [Morphology of *Artemisia* pollen grains.] *Konferencija po sporovo-pyl'cevomu analizu, Tezisy dokl., Moskva*; pp. 37–39.
102. Monoszon, M. C. (1950a). [Morphology of *Artemisia* pollen grains.] *Tr. Konferencii po Sporovo-pyl'cevomu analizu 1948*, pp. 251–259.
103. Monoszon, M. C. (1950b). [Description of pollen grains of *Artemisia* spp. of the USSR.] *Trav. Inst. Geog. Akad. Nauk URSS.* **46**, 271–360.
104. Moreira, A. X. (1958). Contribuiçao ao estudo da familia *Compositae*. IV. Consideraçoes sobre a morfologia e descriçao do polen de *Haplopappus velutinus* Remy. *Bolm. Mus. nac. bot., Bras.* **19**, 1–6; 6 fig.
105. Moreira, A. X. (1969). Catalogo de polens do estado da Guanabera e Arredores. Rio de Jeneiro, G.B., Mus. nacion.
106. Moreira, A. X. (1971). Contribuçao ao estudo da familia *Compositae*. V. *Bol. Mus. nac. (n.s.) Rio de Janeiro, Bot.* **39**, 1–7.
107. Movsesjan, S. N. and Oganesjan, R. A. (1969). La microsporogenèse et le dévelopment du gamétophyte mâle chez *Rudbeckia triloba*. *Biol. Zh. Armenii, Erevan* **22**(5), 56–66, 3 pl.h.t.
108. Nair, P. K. K. (1961). Pollen morphology of some Indian medicinal plants. *J. sci. ind. Res. C* **20**(2), 45–50.
109. Nair P. K. K. (1965). "Pollen Grains of Western Himalayan Plants". Asia Monographs n° 5, India, VIII + 102 pp.
110. Nair, P. K. K. (1966). "Essentials of Palynology". Asia Publishing House, Lucknow.
111. Nair, P. K. K. (1969). Significance of pollen and spores in investigations of Medicinal plants. *Palynol. Bull.* **V**, 119–122.

8. POLLEN MORPHOLOGY

112. NORDENSTAM, B. (1968). The genus *Euryops*. II. Aspects of morphology and cytology. *Bot. Notiser* **121**, 209–232.
113. OSCAR PARRA, B. (1970). Morfologia de los granos de polen de las compuestas cynareas chilenas. *Bol. Soc. biol. Concepcion* **42**, 89–96, 2 fig.
114. OSWIECIMSKA, M. and GAWLOWSKA, M. (1967). Auxiliary methods of taxon determination within the collective species *Achillea millefolium*. I. Correlation between ploidy, pollen grain diameter and chemical composition. *Herba polon.* **13** (1–2), 3–11.
115. PARRA, O. and MARTICORENA, C. (1972). Granos de polen de plantas chilenas. II. Compositae Mutisieae. *Gayana* **21**, 1–107.
116. PAYNE, W. W. (1962). Sexine structure as an indicator of the relationships of the Ambrosieae (Compositae). International Conference on Palynology, Tucson (Abstract) *Pollen et Spores* **4** 369–370.
117. PAYNE, W. W. (1964). A re-evaluation of the genus *Ambrosia*. *J. Arnold Arbor.* **45**, 401–438.
118. PAYNE, W. W. and SKVARLA, J. J. (1970). Electron microscope study of *Ambrosia* pollen (Compositae: Ambrosieae). *Grana* **10**, 89–100. (T).
119. PETERSON, K. M. and PAYNE, W. W. (1973). Genus *Hymenoclea* (Compositae Ambrosieae). *Brittonia* **25**, 243–256.
120. PLA DALMAU, J. M. (1961). "Polen". Talleres graficos D.C.P., Gerona.
121. PODDUBNAJA-ARNOLDI, V. and DIANOWA V. (1934). Eine zytoembryologische Untersuchung einiger Arten der Gattung *Taraxacum*. *Planta* **23**(2), 19–46.
122. PONS, A. and BOULOS, L. (1972). Révision systématique du genre *Sonchus* L. *s.l.* III. Etude palynologique. *Bot. Notiser* **125**, 310–319.
123. PRAGLOWSKI, J. (1971). The pollen morphology of the Scandinavian species of *Artemisia* L. *Pollen et Spores* **XIII**, 381–404. (S, T).
124. RANDERIA, A. J. (1960). The Compositae genus *Blumea*, a taxonomic revision. *Blumea* **10**, 176–317.
125. RAO, A. N. and ONG, E. T. (1971). Pollen dimorphism in *Wedelia biflora* DC. *Curr. Sci. India* **40**, 44–45.
126. ROBINSON, H. (1972). Studies in the Heliantheae (Asteraceae). I. *Phytologia* **24**, 209–210.
127. ROBINSON, H. and BRETTELL, R. D. (1972). Studies in the Heliantheae (Asteraceae) II. *Phytologia* **24**, 361–377.
128. ROBINSON, H. and BRETTELL, R. B. (1973). Tribal revisions in the Asteraceae II. *Phytologia* **24**, 259–261.
129. SAAD, S. I. (1961). Pollen morphology in the genus *Sonchus*. *Pollen et Spores* **3**, 247–260.
130. SAYEEDUD-DIN, M. SALAM, M. A. and SUXENA, M. R. (1942). A comparative study of the structure of pollen grains in some of the families of Angiosperms. *J. Osmania Univ.* **10**, 12–25; 12 fig.
131. SCHTEPA, I. S. (1958). Ad congnitionem pollinis morphologiae generum nonnulorum, tribus Cynareae, familiae Compositae. *Notulae Syst. Geogr. Inst. Bot. Thbilissiensis* **20**, 53–62, 3 tabl.
132. SCHTEPA, I. S. (1962). Palynological data for the systematics of the genus *Cousinia* Cass. *Pollen et Spores* **4**, 375.
133. SCHTEPA, I. S. (1965). Materies ad studium Pollinis characterum generis *Cirsium* Mill. *Notulae Syst. Geogr. Inst. Thbilissiensis* **25**, 69–82.
134. SCHTEPA, I. S. (1966). [On the problem of affinity between the genera *Arctium* L. and *Cousinia* Cass. of the family Compositae.] (English summary). In "The importance of Palynological analysis for the Stratigraphic and Paleofloristic Investigation". Acad. Sci. USSR, Moscow, 1966. For the 2nd Internat. Palyn. Conf., Utrecht, 1966, pp. 35–36.

135. SCHTEPA, I. S. (1967). Commentationes de Palinologia et Systematica specierum Caucasicarum generis *Cousinia* Cass. (en russe, résumé en gëorgien). *Notulae Syst. Geogr. Inst. Bot. Thbilissiensis* **26**, 57–62, 2 tables, 1 fig.
136. SCHTEPA, I. S. (1973). [On the natural boundaries between the genera *Cousinia* and *Arctium* as suggested by palynological evidence.] *In* "Pollen and Spore Morphology of the Recent Plants". Proc. 3rd Int. Palyn. Conf., Acad. Sci. USSR. 37–40.
137. SELLING, O. H. (1947). Studies in Hawaiian Pollen Statistics. Part II. The Pollens of the Hawaiian Phanerogams". Bernice P. Bishop Museum Special Publication No. 38; 430 p; 58 pl. Honolulu.
138. SHERIFF, E. E. (1937). The genus *Bidens. Field Mus. Nat. Hist. (Bot.)* **XVI**, 1–709.
139. SINGH, G. and JOSHI, R. D. (1969). Pollen morphology of some Eurasian species of *Artemisia. Grana Palynol.* **9**, 50–62.
140. SKVARLA, J. J. and LARSON, D. A. (1965a). An electron microscopic study of pollen morphology in the Compositae with special reference to the Ambrosiinae. *Grana Palynol.* **6**, 210–269 (T).
141. SKVARLA, J. J. and LARSON, D. A. (1965b). Interbedded exine components in some Compositae. *Southwest. Nat.* **10**, 65–68 (T).
142. SKVARLA, J. J. and TURNER, B. L. (1966a). Pollen wall ultrastructure and its bearing on the systematic position of *Blennosperma* and *Crocidum* (Compositae). *Am. J. Bot.* **53**(6:1), 555–563 (T).
143. SKVARLA, J. J. and TURNER, B. L. (1966b). Systematic implication from electron microscopic studies of Compositae pollen. A review. *Ann. Mo. bot. Gdn* **53**, 220–244 (T).
144. SKVARLA, J. J. and TURNER, B. L. (1969). Fine structure of Petrobinae (Compositae-Heliantheae) pollen walls. *Am. J. Bot.* **56**, 418–419 (T).
145. SKVARLA, J. J. and TURNER, B. L. (1971). Fine structure of the pollen of *Anthemis nobilis* L. (Anthemideae-Compositae) *Proc. Okla. Acad. Sci.* **51**, 61–62 (T).
146. SMITH, C. E. (1969). Pollen characteristics of African species of *Vernonia. J. Arnold Arbor.* **50**, 469–477.
147. SMITH, C. E. Jr. (1971). "Observations on Stengelioid Species of *Vernonia*". Agricultural Handbook N° 396. U.S.D.A., Washington. 87 p.
148. SMOLJANINOVA, L. A. and GOLUBKOVA, V. F. (1955). [On microtome sections of Compositae pollen.]. *Isv. Akad. Nauk Beloruss. S.S.S.R.* **2**, 127–129.
149. SOLBRIG, O. T. (1960). The South American sections of *Erigeron* and their relation to *Celmisia. Contr. Gray Herb.* 188, 65–86.
150. SOLBRIG, O. T. (1964). Infraspecific variation in the *Gutierrezia sarothrae* complex (Compositae-Astereae). *Contr. Gray Herb.* 193, 67–115.
150a. SOLOMON, A. M., KING, J. E., MARTIN, P. S. and THOMAS, J. (1973). Further scanning electron photomicrographs of Southwestern pollen grains. *Arizona Acad. Sci.* **8**, 135–157.
151. SOUTHWORTH, D. (1966). Ultrastructure of *Gerbera jamesonii* Pollen. *Grana palynol.* **6**, 324–337 (T).
152. SOUTHWORTH, D. (1969a). Development of *Gerbera jamesonii* pollen. *XI intern. bot. Congr. Seattle*, Abstr., p. 206.
153. SOUTHWORTH, D. (1969b). Ultraviolet absorption spectra of pollen and spore walls. *Grana palynol.* **9**, 5–15.
154. SOUTHWORTH, D. (1971). Incorporation of radioactive precursors into developing pollen walls. *In* "Pollen: Development and Physiology", pp. 115–120. Butterworth, London (reprinted in 1973).

8. POLLEN MORPHOLOGY

155. SOUTHWORTH, D. (1973). Cytochemical reactivity of pollen walls. *J. Histochem.* **21**(1), 73–80. (T).
156. SOUTHWORTH, D. and BRANTON D. (1971). Freeze-etched pollen walls of *Artemisia pycnocephala* and *Lilium humboldtii. J. Cell Sci.* **9**, 193–207. (T).
157. STIX, E. (1960). Pollenmorphologische Untersuchungen an Compositen. *Grana palynol.* **2**(2), 41–104; 49 fig., 21 tabl.
158. STIX, E. (1964). Polarisationsmikroskopische Untersuchungen am Sporoderm von *Echinops banaticus. Grana palynol.* **5**(3), 289–297, 5 fig.
159. STIX, E. (1970). Beitrag zur Morphogenese der Pollenkörner von *Echinops banaticus. Grana* **10**, 240–242.
160. STRAKA, H. (1952). Zur Feinmorphologie des Pollens von *Salix* und von *Artemisia. Svensk bot. Tidskr.* **46**, 204–227.
161. STROTHER, J. L. (1974). Taxonomy of *Tetradymia* (Compositae: Senecioneae). *Brittonia* **26**, 177–202.
162. SUNDARA RAJAN, S. (1968). Embryological studies in Compositae. Floral morphology, sporogenesis and gametogenesis in *Emilia sonchifolia* (Linn.) DC. *Curr. Sci.* **37**, 26–28.
163. TARNAVSCHI, I. T. and MITROIU, N. (1959). [Recherches sur la morphologie du pollen des *Composées* de la flore Roumainie.] *Stud. Cerc. Biol., Séer. Biol. Vég. Romin* **11**(3), 213–271, 14 pl.
164. TOMB, A. S. (1971). Karyotypes, pollen grains and systematics in the genus *Lygodesmia* (Compositae: Cichorieae). *Am. J. Bot.* **58** (5, part 2), 467.
165. TOMB, A. S. (1972a). The systematic significance of pollen morphology in the family Compositae. Tribe Cichorieae. *Brittonia* **24**, 129.
166. TOMB, A. S. (1972b). Re-establishment of the genus *Prenanthella* Rydb. (Compositae: Cichorieae). *Brittonia* **24**, 223–228. (S).
167. TOMB, A. S., LARSON, D. A. and SKVARLA, J. (1974). Pollen morphology and detailed structure in the Compositae, tribe Cichorieae. I. Subtribe *Stephanomeriinae. Am. J. Bot.* **61**, 486–498.
168. UENO, J. (1969). The fine structure of pollen surface. I. *Taraxacum* and *Ambrosia. Rep. Fac. Sci. Schizuoka Univ.* **4**, 67–74. (S, T).
169. UENO, J. (1970). Plants of pollinosis in Japan. *Jap. J. Palynol.* **6**, 22–36.
170. UENO, J. (1971). The fine structure of pollen surface. II. *Dahlia. Rep. Fac. Sci. Shizuoka Univ.* **6**, 149–164. (R).
171. UENO, J. (1972). The fine structure of pollen surface III. *Gazania* and *Stokesia. Rep. Fac. Sci., Shizuoka Univ.* **7**, 103–116. (R, S).
172. VARGHESE, T. M. (1964). Study of pollen grains of some members of Compositae. *Agra. Univ. J. Res. Sc.* **13**(1), 79–84.
173. VASANTHY, G. (1975). Structure et nomenclature de la paroi sporopollinique: Asteraceae. *In* "Structure et Terminologie de la Paroi Sporo-pollinique", pp. 44–48. A.P.L.F.
174. VASUDEVAN, R. (1966). *Struchium sparganophorum* (L.) O. Kuntze. *Bull. bot. Surv. India* **8**, 202–203.
175. VENKATESWARLU, J. and DEVI, H. M. (1955). Embryological studies in Compositae. II. Helenieae. *Proc. natn. Inst. Sci., India.* **21**, 149–161.
176. VISSET, L. (1974). Pollens de Compositae-Asteroideae observés au microscope Eléctronique à Balayage. *Beitr. Biol. Pfl.* **50**, 137–161. (S).
177. VUILLEMIER, B. S. (1973). The genera of Lactuceae (Compositae) in the southeastern United States. *J. Arnold Arbor.* **54**, 42–93.
178. WAGENITZ, G. (1955). Pollenmorphologie und Systematik in der Gattung *Centaurea* L. s.l. *Flora, Jena* **142**, 213–279.
179. WAGENITZ, G. (1958). Die Gattung *Myopordon* Boiss. (Compositae-Cynaraceae). *Ber. dt. bot. Ges.* **71**(7), 271–277.

179a. WANG, F. H. (1960). [Pollen Grains of China].
180. WELLS, J. R. (1971). Variations in *Polymnia* pollen. *Am. J. Bot.* **58**(2), 124–130. (S).
181. WODEHOUSE, R. P. (1926). Pollen grain morphology in the classification of the Anthemideae. *Bull. Torrey bot. Club* **53**, 479–485.
182. WODEHOUSE, R. P. (1928a). Pollen grains in the identification and classification of plants. I. The Ambrosiaceae. *Bull. Torrey bot. Club* **55**, 181–198.
183. WODEHOUSE, R. P. (1928b). Pollen grains in the identification and classification of plants, II. *Barnadesia*. *Bull. Torrey bot. Club* **55**, 449–462.
184. WODEHOUSE, R. P. (1928c). The phylogenetic value of pollen grain characters. *Ann. Bot.* **XLII**, 891–934.
185. WODEHOUSE, R. P. (1929a). The origin of symmetry patterns of pollen grains. *Bull. Torrey Bot. Club* **56**(7), 339–350.
186. WODEHOUSE, R. P. (1929b). Pollen grains in the identification and classification of plants. III. The Nassuvinae. *Bull. Torrey bot. Club* **56**, 123–138.
187. WODEHOUSE, R. P. (1929c). Pollen grains in the identification and classification of plants. IV. The Mutisieae. *Am. J. Bot.* **16**, 297–313.
188. WODEHOUSE, R. P. (1930). Pollen grains in the identification and classification of plants—V. *Haplopappus* and other Astereae: the origin of their furrow configurations. *Bull. Torrey bot. Club* **57**, 21–46.
189. WODEHOUSE, R. P. (1931). The origin of the six-furrowed configuration of *Dahlia* pollen grains. *Bull. Torrey bot. Club* **57**, 371–380.
190. WODEHOUSE, R. P. (1935). (reprinted in 1959). "Pollen Grains". Hafner, New York.
191. WODEHOUSE, R. P. (1945). "Hayfever Plants". Waltham, Mass.

* * *

REFERENCES NOT AVAILABLE FOR CONSULTATION

AREVSHATYAN, I. G. (1973). Nombre de chromosomes et dimensions des grains de pollen des espèces du genre *Taraxacum* Weber vivant en Armenie. *Biol. Zh. Armenii SSSR.* **26**(3), 38–44.
BREWER, J. G. (1970). The fine structure of Pyrethrum pollen (*Chrysanthemum cinerariaefolium* Vis.)—a note. *Pyrethrum Post, London.* **10**(3), 3–6.
BREWER, J. G. and HENSTRA, S. (1974). Pollen of Pyrethrum (*Chrysanthemum cinerariaefolium* Vis.). Fine structure and recognition reaction. *Euphytica* **23**, 657–663 (S).
CABRERA, A. L. (1956). Un nuevo genero de Eupatorieas-Compositae de Bolivia. *Bol. Soc. Arg. Bot.* **6**(2), 91–93.
CABRERA, A. L. (1959). "Revision del genero *Dasyphyllum* (Compositae). *Rev. Mus. La Plata Bot.*, n.s. **9**, 21–100.
CRISCI, J. V. (1971). Sobre una especie de *Jungia* (Compositae) del Peru. *Bol. Soc. Arg. Bot.* **13**, 341–346.
KING, R. M. and ROBINSON, H. (1968). Studies in the Compositae-Eupatorieae. VIII. Observations on the microstructure of *Stevia*. *Sida* **3**(4), 257–269.
LEPPIK, E. E. (1970). Evolutionary differentiation of the flower head of the Compositae 2. *Ann. bot. fenn.* **7**, 325–352.
LIPSHITS, S. Y. (1956). Neuer Untertribus'heue Gattung und Art der Familie Compositae aus Zentral Asien. *Akad. V. N. Sukachevu K. Izd. A.N. SSSR. Leningrad* **75**, 354–362.
MAHESHWARI, H. M. (1957). Embryological studies in Compositae. 3. *Gerbera jamesonii* Bolus. *Proc. Indian Acad. Sci.* **46**(B), 68–74.

8. POLLEN MORPHOLOGY

PAUSINGER-FRANKENBURG, F. (1951). Vom Blütenstaub der Wegwarten. (Die Pollengestaltung der *Cichorieae*) XIII. *Sonderh. Carintha* **II**.

PINKAVA, D. J. (1967). Biosystematic study of *Berlandiera*-Compositae. *Brittonia* **19**, 285–298.

SCHTEPA, I. S. (1973). [Les types voisins de la famille des Compositae.] *Tbilisskij Bot. Inst. Zamet. Sistemat. Geogr. Rasten. SSSR* **30**, 44–50.

SMOLINA, M. C. (1950). [Description of *Artemisia* pollen.] *Tr. Geogr. Inst. Akad. Nauk SSSR.* **46**.

SWEENEY, C. R. (1967). Biosystematic studies in the genus *Silphium* L. (*Compositae*). *Silphium compositum* Michaux. *Diss. Abstr. USA.* **27**(7), 2258b.

WAGENITZ, G. (1970). Compositen Korbblueter. II. Teil. Illustr. Flora Mitteleuropa Hegi ed. C. Hanser. Verlag. Muenchen. **63**(1), 80, 34.

WITTENBACH, A. J. and SKVARLA, J. J. (1969). Light and electron microscope studies of Inuleae (Compositae) pollen. XI *In. bot. Congr. Seattle*, Abstr. **241**, 11.

INDEX TO GENERA

Numbers refer to the references in the preceding section

A

Aaronsohnia (T), 142, 143
Abrotanella, 29a
Acanthambrosia, 182, 190
Acanthospermum, 26
Achillea (S, T), 33, 38, 55b, 56, 109, 114, 120, 140, 143, 157, 163, 176, 179a
Achnopogon, 20, 93b
Achyrachaena, 23
Achyrocline, 87
Acritopappus, 85
Acroptilon, 10, 179a
Adenocaulon, 55, 55b, 90, 179a
Adenoön, 26, 38
Adenostemma, 26, 55a, 55b, 137, 179a
Adenostyles (S), 39, 157, 163, 176
Adenothamnus, 23
Ageratina, 74
Ageratum, 34, 38, 39, 55a, 77, 99, 137, 157, 163, 179a
Agoseris, 41
Ainsliaea, 55a, 55b, 61, 179a, 187
Alfredia, 41a
Allagopappus, 92
Alomia, 76
Amberboa (T), 10, 34, 37, 38, 56, 57
Amblyocarpum, 92
Amblyopappus (T), 142, 143
Amblyopogon, 10
Ambrosia (S, T), 13a, 39, 43a, 46, 55, 55b, 87a, 89, 96a, 96b, 116, 117, 118, 120, 140, 143, 153, 155, 163, 168, 169, 182, 184, 190, 191
Ameghinoa, 32, 157
Amellus, 41a
Ammobium, 34a, 55b
Amphidoxa, 157

Amphiglossa, 14
Anacyclus, 37, 38, 120
Anaglypha, 14, 92
Anaphalioïdes, 87
Anaphalis (T), 49, 55a, 55b, 87, 109, 143, 179a
Anastraphia, 187
Andryala (T), 120, 143
Angianthus, 14, 157
Anisochaeta, 90, 92
Anisocoma (S), 167
Anisopappus, 90, 92, 179a
Antennaria, 87, 88, 163
Anthemis (S, T), 37, 38, 55b, 56, 100, 140, 143, 145, 157, 163, 176, 190
Antiphiona, 92
Antithrixia, 14
Anvillea, 37, 38, 90, 92
Aphanostephus (T), 43a, 140, 143
Arctium, 55b, 88, 134, 136, 163, 179a, 191
Arctotheca, 38a
Arctotis (T), 39, 55b, 143, 157
× *Argyrautia*, 137
Argyroxiphium (T), 23, 137, 140, 141, 143
Arnica (T), 55b, 143, 157, 163
Arnoseris, 5, 39
Arrowsmithia, 14
Artemisia, (S, T), 26, 34, 37, 38, 39, 43a, 44, 55, 55a, 55b, 56, 88, 96a, 101, 102, 103, 109, 110, 123, 137, 139, 140, 143, 153, 155, 156, 157, 160, 163, 169, 179a, 181, 184, 190, 191
Artemisiastrum, 181, 190
Aspilia, 40, 88a, 140, 143
Aster (R, S), 43a, 55a, 55b, 56, 88, 97, 109, 137, 163, 176, 179a, 190
Asteriscus, 37, 38, 92
Asteromoea, 179a
Asteropterus, 14, 37, 38

Athrixia, 14
Athroisma, 92
Atractylis, 37, 38
Atractylodes, 55b, 179a
Atrichoseris (S), 167
Ayapanopsis, 83

B

Babcockia, 17, 122
Baccharis (S), 39, 43a, 55, 55b, 88a, 96b, 143, 157
Baeria, 143
Bahia, 55, 142, 143
Baileya (S), 96b
Balduina (T), 140, 143
Balsamorhiza (T), 140, 143
Baltimora (T), 140, 143
Barnadesia, 39, 171, 183, 184, 190, 191
Bartlettia (T), 143
Bebbia (S, T), 140, 141, 143, 150a
Bellis (S), 37, 38, 88, 120, 143, 163, 176
Bellium, 37, 38
Berardia, 157
Berkheya, 91, 93, 157, 171, 190
Berkheyopsis (T), 143
Berlandiera, 182
Bidens (T), 13a, 26, 34, 37, 38, 40, 55a, 55b, 56, 88a, 94, 109, 137, 138, 140, 143, 157, 163, 179a, 182
Blainvillea, 26, 34, 179a
Blennosperma (T), 142, 143
Blepharipappus, 23
Blepharispermum, 26, 92
Blepharizonia, 23
Blumea, 12, 26, 34, 39, 55a, 92, 124, 172, 179a
Blumeopsis, 92
Boltonia, 55a
Brachyactis, 56, 179a
Brachyclados, 115
Brachycome, 29a
Brachyglottis, 29a

8. POLLEN MORPHOLOGY

Brachylaena, 39, 92
Brickellia, 43a, 157
Bryomorphe, 14
Buphthalmum, 92, 157

C

Cacalia (T), 55a, 55b, 143, 179a
Cacosmia (T), 142, 143
Caesulia, 14, 26, 35, 92, 109
Calea (T), 40, 88a, 140, 143
Calendula (S), 34, 37, 38, 55b, 108, 111, 157, 163, 176
Callilepis, 90, 92
Callistephus, 190
Calostephane, 92
Calycoseris (S), 96b, 167
Campuloclinium (S), 86
Cardopatium, 37, 38
Carduncellus, 37, 38
Carduus (S, T), 37, 38, 39, 55b, 56, 120, 132, 157, 163, 176, 179a
Carlina, 37, 38, 157, 163
Carpesium, 55a, 55b, 90, 92, 163, 179a
Carterothamnus, 66
Carthamus, 26, 34, 37, 38, 55b, 56, 64, 108, 109, 111, 120, 163
Cassinia, 14
Castalis, 157
Catananche (S), 5, 38, 171, 184, 190
Celmisia, 149
Centaurea (S, T), 10, 26, 37, 38, 39, 50, 55, 56, 58, 88, 97, 113, 120, 150a, 157, 163, 176, 178
Centaurodendron, 22, 113
Centipeda, 55b
Centratherum, 26, 39, 173
Cephalopappus, 32
Chaenactis (S), 96a, 96b
Chaetadelpha (S, T), 167
Chaetanthera, 115, 187
Chamartemisia, 181, 190
Chamomilla, 190
Chaptalia, 105, 115, 187
Charieis, 41a
Chimantaea, 20, 93b
Chionolaena, 87, 105
Chondrilla, 56, 88, 163
Chorisiva, 182, 190
Chrysactinia, 43a

Chrysanthellum, 36
Chrysanthemoïdes, 55, 157
Chrysanthemum (S, T), 19, 39, 55a, 55b, 56, 95, 108, 111, 120, 140, 143, 157, 163, 172, 176, 179a, 190, 191
Chrysocoma, 100a
Chrysogonum, 51a
Chrysophthalmum, 92
Chrysothamnus (S), 150a
Chuquiraga, 115, 184, 187
Cicerbita, 9, 56, 163
Cichorium, 5, 55b, 56, 88, 109, 120, 148, 163, 177, 184, 190
Cineraria, 163
Cirsium (S), 37, 38, 39, 48, 55a, 55b, 56, 88, 96a, 97, 120, 131, 132, 133, 157, 163, 176, 179a
Cladanthus, 37, 38
Cladochaeta, 87
Cleanthes, 186
Clibadium (T), 39, 140, 143
Cnicus, 37, 38, 55, 109, 120, 131, 163, 191
Codonocephalum, 92
Coleocoma, 92
Condylopodium, 84
Conyza, 26, 34, 39, 43a, 55a, 56, 109, 120, 157
Coreopsis (T), 26, 43a, 55a, 55b, 140, 143, 179a, 182
Corymbium, 100a
Cosmos (S, T), 3, 34, 54a, 55a, 55b, 140, 143, 163, 182
Cotula, 34a, 55, 56, 109, 157, 179a
Cousinia, 56, 132, 134, 135, 136
Craspedia (T), 14, 143
Cremanthodium, 179a
Crepidiastrum, 55a, 55b
Crepis, 9, 56, 88, 109, 120, 163, 177, 179a, 190
Critonia, 69
Crocidium (T), 142, 143
Crossostephium (T), 55a, 143, 181, 184, 190
Crupina, 163
Cullumia, 93, 157
Cuspidia, 93
Cyathocline, 26
Cyclachaena (T), 140, 182, 190, 191
Cyclolepis, 187

Cylindrocline, 92
Cynara (S), 37, 38, 48, 120
Cynthia, 190

D

Dahlia (R, T), 55a, 55b, 140, 143, 163, 170, 182, 185, 189
Dasycondylus, 80
Dasyphyllum, 55, 105, 106, 115
Delamerea, 92
Demidium, 87
Dendroseris, 24, 39, 171, 190
Denekia, 92
Diacranthera, 81
Dichrocephala, 26, 38, 55a
Dicoma, 26, 57, 157, 187
Dicoria (T), 140, 143, 182, 190
Didelta (T), 93, 143, 157
Dimeresia (T), 14, 143
Dimorphotheca, 55b, 157
Disparago, 14
Doniophyton, 115
Doronicum (S), 120, 163, 176, 179a
Dubautia, 23, 137
Dubyaea, 165
Dugesia, 13a
Duidaea, 20, 93b
Dyssodia, 13a, 26, 34

E

Echinacea (T), 140, 143
Echinops (S), 34, 37, 38, 39, 56, 88, 148, 157, 158, 159, 163, 179a, 191
Eclipta, 26, 34, 55a, 55b, 143, 172, 179a
Elephantopus, 26, 39, 55a, 88a, 157, 179a
Elytropappus, 14, 157
Embergeria, 17, 122
Emilia, 26, 55a, 142, 143, 162, 179a
Encelia, 55
Engelmannia (T), 140, 143
Enydra, 99
Epaltes, 55a, 92
Epilasia, 8, 9
Erechtites, 55a, 55b, 163
Eremanthus, 88a
Eremopappus, 10
Eremothamnus, 91

Erigeron (T), 26, 34, 39, 55a, 55b, 56, 88, 109, 143, 149, 163, 179a, 190
Eriocephalus, 38a
Eriophyllum (S), 96b
Erythroecephalum, 157
Ethulia, 38
Eupatorium (S), 26, 34, 38, 43a, 46, 55, 55a, 55b, 88, 88a, 105, 120, 157, 163, 176, 179a
Euphrosyne (T), 140, 143, 182, 190
Eurydochus, 93b
Euryops, 112, 142, 143
Evax, 37, 38

F

Faberia, 179a
Farfugium, 55a, 55b, 179a
Felicia, 41a
Filago (S), 38, 56, 157, 163, 176
Filifolium, 179a
Fitchia, 21, 25, 39
Flaveria, 26, 43a, 55, 98
Fleischmannia, 69, 82
Florestina, 43a
Flourensia, 43a, 157
Flyriella, 72
Fontquera, 92
Franseria (S, T), 55, 96b, 140, 143, 182, 190

G

Gaillardia (S), 34, 143, 150a, 163, 175
Galactites (S), 37, 38, 120
Galatella, 41a
Galinsoga (T), 13a, 39, 55b, 140, 141, 143, 163, 179a, 182
Garberia, 93a
Garhadiolus, 56
Gazania (R, S), 38, 93, 120, 157, 171
Geigeria, 37, 38, 90, 92, 157
Gerea (T), 140, 143
Gerbera (S, T), 26, 39, 151, 152, 154, 155, 179a, 187
Gibbaria, 157
Glossarion, 20, 93b
Glossocardia, 26, 34, 179a
Glossogyne, 55a
Glyptopleura (S, T), 167
Gnaphaliothamnus, 87

Gnaphalium (S, T), 26, 34, 49, 55, 55a, 55b, 56, 87, 88, 88a, 96b, 109, 120, 137, 157, 163, 176, 179a
Gochnatia, 32, 43a, 105, 115, 187
Gongylolepis, 20, 93b
Goniocaulon, 26
Gorteria, 39, 93, 157
Grangea, 26, 34, 99, 157
Grantia, 92
Grindelia, 41a
Grossheimia, 10
Guaicaia, 93b
Guardiola (T), 140, 143, 182
Guizotia (S), 11, 26, 38
Gundelia, 157
Gutenbergia, 157
Gutierrezia, 55, 150
Gymnarrhena, 37, 38
Gynoxys (T), 142, 143
Gynura, 26, 55a, 55b, 142, 143, 179a
Gypothamnium, 115

H

Haplocarpha, 14, 157
Haploesthes (T), 142, 143
Haplopappus (S), 43a, 55, 96b, 104, 188
Hebeclinium, 69
Hecastocleïs, 187
Helenium, 43a, 157
Helianthus, (S, T), 21, 26, 34, 39, 55b, 96b, 97, 110, 130, 140, 143, 157, 163, 170, 172, 179a, 182, 191
Helichrysopsis, 87
Helichrysum (S), 37, 38, 39, 55b, 56, 87, 88, 120, 163, 172, 176
Heliopsis (T), 140, 143
Helipterum, 55b
Hemizonia (T), 23, 140, 142, 143, 182,
Hertia, 56
Hesperomannia, 39, 137
Heterolepis, 14
Heteropappus, 55a
Heterorhachis, 93, 157
Heterothalamus, 157
Hidalgoa (T), 140, 143
Hieracium, 55b, 88, 163, 190
Hirpicium, 93, 157
Hirschia, 92

Hispidella, 5
Homognaphalium, 87
Homogyne, 163
Hulsea, 142, 143
Humea, 14
Hyalis, 115
Hyaloseris, 187
Hymenoclea (T), 119, 140, 143, 182, 190
Hymenonema, 5, 9
Hymenopappus (T), 140, 142, 143
Hyoseris, 5, 163
Hypelichrysum, 87
Hypochoeris, 53, 55, 105, 163, 179a, 190

I

Ichtyothere, 40, 88a
Ifloga, 37, 38, 109
Inula (S, T), 38, 55b, 56, 90, 92, 109, 120, 143, 157, 163, 176, 179a, 184
Iphiona, 90, 92
Isostigma, 40, 88a
Iva, 13, 140, 143, 150a, 163, 182, 190, 191
Ixeris, 55a, 55b, 179a

J

Jaegeria, 143
Jasonia, 37, 38, 92
Jaumea, 142, 143
Jungia, 31, 32, 186
Jurinea, 37, 38, 56, 88, 109, 157, 163, 179a

K

Kleinia, 100a
Koanophyllon, 68
Koelpinia, 5
Krigia, 190
Kuhnia, 157
Kyrsteniopsis, 70

L

Lachnospermum, 14
Lactuca, 9, 26, 55a, 55b, 56, 88, 109, 129, 163, 179a, 190
Lagascea (T), 26, 140, 143
Lagenifera, 55a, 137
Laggera, 55a, 92, 179a
Lagophylla, 23
Lagoseris, 163
Lamprachaenium, 26
Lamyropsis, 48
Lancisia, 41a

8. POLLEN MORPHOLOGY

Lapsana, 55b, 163
Lasiolaena, 79
Lasiopogon, 87
Lasthenia (T), 142, 143
Launaea, 26, 34, 56, 57, 109, 129, 172
Layia (S), 23, 143, 150a
Leibnitzia, 55b
Leontodon, 120, 163
Leontopodium, 55a, 55b, 87, 163, 179a
Leucanthemum (S, T), 37, 38, 120, 140, 143, 176, 190
Leucheria, 30, 31, 32, 115, 186
Leuciva, 182, 190
Leucogenes, 29a
Leunisia, 32, 115
Leuzea, 38, 163
Liabum (T), 142, 143, 157
Liatris, 157
Lifago, 38
Ligularia, 55a, 55b, 157, 163, 179a
Lindheimera, 41a
Lipochaeta, 137
Lonas, 37, 38
Lophopappus, 32, 115
Lucilia, 87
Lychnophora, 157
Lycoseris, 187
Lygodesmia (S, T), 164, 166, 167, 177

M

Macowania, 14
Macrachaenium, 32, 115
Macropodina, 78
Madia, 23, 55, 182
Malacothrix (S, T), 150a, 167
Mantisalca, 37, 38
Marshallia (T), 140, 143
Marticorenia (S), 31, 32
Maruta, 190
Matricaria (S, T), 37, 38, 55b, 56, 100, 120, 140, 143, 157, 163, 176
Melampodium (T), 43a, 140, 143, 182
Metalasia, 14
Micropus, 37, 38
Microseris, 55
Mikania, 55a, 157, 179a
Milleria (T), 140, 143
Miricacalia, 55b
Modestia, 131

Mollera, 92
Monarrhenus, 92
Monolopia, 142, 143
Montanoa, 43a
Moquinia (T), 143, 187
Moscharia, 32, 115, 186
Mutisia (T), 39, 55, 115, 143, 157, 187
Mycelis, 163
Myopordon, 179
Myriactis, 55a, 56, 179a

N

Nablonium, 90, 92
Nannoglottis, 179a
Nanothamnus, 26, 92
Nardosmia, 88
Nassauvia, 32, 115, 186
Neblinaea, 20, 93b
Neocuatrecasia, 75
Neomirandea, 82
Nestlera, 14
Neurolaena, 157
Nicolasia, 92
Nidorella, 157
Notonia, 26, 130
Notoptera (T), 140, 143

O

Oaxacania, 66
Odontospermum, 157
Oldenburgia, 39
Olearia, 29a, 38a, 41a
Olgaea, 179a
Oliganthemum, 92
Ondetia, 92
Onopordum, 37, 38, 56, 163
Onoseris, 32, 157, 187
Oparanthus, 21
Ophryosporus, 55
Ormenis, 37, 38
Orthopappus, 88a
Osmites, 90, 157
Osmitopsis, 18, 90, 157
Otanthus, 37, 38
Othonna, 163
Oxyphyllum, 32, 115, 157
Oxytenia (T), 140, 143, 182, 190, 191

P

Pachylaena, 115, 187
Pachystegia, 29a
Pachythamnus, 67
Pacourina, 38, 39, 171, 184
Palafoxia (T), 142, 143
Pallenis, 92
Pamphalea, 32, 186

Parthenice (T), 140, 143, 182
Parthenium (T), 1, 13a, 26, 34, 140, 143, 182
Pechuel-Loeschea, 92
Pegolettia, 37, 38, 90, 92, 157
Pelucha, 92
Pentatrichia, 90, 92
Pentzia, 37, 38
Perezia (S), 32, 53, 96a, 96b, 115, 186
Pericome, 142, 143
Perityle (S), 150a
Perotriche, 14
Perralderia, 37, 38, 92
Pertya, 55b, 56, 179a, 187
Perymenium, 43a
Petasites, 55a, 55b, 88, 142, 143, 163
Peucephyllum, 142, 143
Phagnalon, 37, 38, 56
Phania, 39
Philyrophyllum, 90, 92
Picris, 9, 55a, 55b, 56, 163
Picrosia, 41
Pinaropappus (S, T), 9, 167
Piptocarpha, 88a
Piqueria, 39, 191
Plagiobasis, 56
Platycarpha, 157
Plazia, 115, 187
Pleocarphus, 31, 32, 115
Pluchea, 34, 37, 38, 55a, 92, 179a
Podanthus (T), 55, 144
Podolepis, 14, 52
Polyachyrus, 32, 115, 186
Polycline, 14
Polymnia, (S, T), 42, 140, 143, 157, 180, 182
Porophyllum (S), 96b
Porphyrostemma, 92
Postia, 92
Prenanthella (S), 166, 167
Prenanthes, 55b, 109, 163, 177, 179a, 190
Printzia, 90, 92
Proustia, 32, 55, 115, 186
Psathyrotes, 142, 143
Pseudobrichkellia, 73
Pseudoclappia, 143
Pseudognaphalium, 87
Psilostrophe (S), 96b, 142, 143
Pterocaulon, 92
Pterothrix, 14
Ptilostemon 47

Pulicaria (S), 26, 34, 38, 56, 57, 92, 109, 120, 163, 176
Pyrrhopappus (S), 140, 143

Q
Quelchia, 20, 93b

R
Rafinesquia (S), 96a, 96b, 167
Raillardella, 23
Raillardia, 137
Raoulia, 87
Reichardia, 120
Relhania, 14, 157
Remya, 39, 137
Rhagadiolus, 41a
Rhanterium, 92
Rhaponticum, 39, 179a
Rhynchospermum, 109, 179a
Rhysolepis, 126
Riencourtia, 40, 88a
Robinsonia, 39
Rolandra, 39, 157
Rosenia, 14
Rothmaleria, 190
Rudbeckia (T), 55b, 107, 140, 143, 163, 182
Rumfordia, 157

S
Sabazia, 43a
Sachsia, 92
Salmea (T), 140, 143
Santolina, 37, 38, 120
Sanvitalia, 43a
Sartwellia, 43
Saussurea, 55a, 55b, 56, 109, 131, 157, 163, 179a
Schistocarpha, 142, 143, 157
Schkuhria, 43a, 157
Schlechtendalia, 187
Sclerocarpus, 26, 109
Scolymus, 9, 163, 171, 185, 190
Scorzonera, 6, 7, 8, 9, 51, 56, 88, 163, 171, 179a, 190
Senecio (S, T), 13a, 26, 37, 38, 43a, 53, 55, 55a, 55b, 56, 88a, 96a, 96b, 109, 120, 137, 142, 143, 157, 163, 176, 179a
Seris, 187
Serratula (S), 38, 55b, 88, 157, 163, 176, 179a

Sigesbeckia, 26, 55a, 163, 179a
Silphium (T), 140, 143, 163, 182, 184
Silybum, 37, 38, 55b, 120, 132
Simsia, 127
Solidago (S, T), 55a, 55b, 56, 88, 97, 120, 140, 143, 157, 163, 169, 176, 179a, 190, 191
Soliva, 15
Sonchus, (S, T), 13a, 16, 17, 26, 55a, 55b, 56, 57, 88, 109, 120, 122, 129, 143, 163, 171, 172, 177, 179, 190
Soroseris, 165, 179a
Sosnovskya, 10
Sphaeranthus, 26, 92
Spilanthes (T), 14, 26, 39, 40, 55a, 55b, 88a, 108, 111, 140, 143, 179a, 182
Spiracantha, 157
Staehelina, 38
Stenachaenium, 39
Stenopadus, 20, 93b
Stephanomeria, (S, T), 9, 96b, 167
Stevia, 43a, 45, 65
Steviopsis, 71
Stifftia, 20, 39
Stoebe, 14
Stokesia (S), 55b, 171, 184
Stomatochaeta, 20, 93b
Struchium, 157, 174, 184
Sventenia, 39, 171
Synchaeta, 87
Synedrella, 26, 94
Syneilesis, 55b
Synurus, 55b, 179a

T
Taekholmia, 17, 122
Tagetes (S), 54, 54b, 55a, 108, 111, 157, 163, 175, 179a
Tanacetum (T), 88, 120, 140, 143, 181, 184, 190
Taraxacum (S), 13a, 27, 28, 29, 55a, 55b, 88, 97, 109, 110, 120, 121, 163, 168, 171, 177, 179a, 190, 191
Tarchonanthus, 39, 92
Telekia, 92, 163
Tessaria, 55, 92
Tetradymia (T), 143, 161

Tetramolopium, 137
Thamnoseris, 39
Thelesperma (T), 140, 143
Thespidium, 92
Tithonia, 13a
Tolpis, 5
Tomanthea, 10
Tourneuxia, 8
Tragopogon, 8, 9, 55b, 56, 88, 120, 163, 171, 190
Trichocline, 115, 187
Tricholepis, 26, 39, 56
Trichospira, 128
Tridax (T), 26, 34, 43a, 55a, 57, 63, 140, 143
Tripleurospermum, 41a
Triplocephalum, 92
Triptilion, 32, 115, 186
Trixis (S), 2, 32, 43a, 88a, 96a, 96b, 105, 115, 157, 186
Tussilago (S), 163, 176, 179a

U
Urmenetia, 115
Urospermum (S), 38, 120
Ursinia, 157

V
Vanillosmopsis, 88a, 157
Varilla, 143
Varthemia, 56, 92
Venegasia, 142, 143
Venidium, 41a
Verbesina (S), 13a, 34, 43a, 88a, 96a, 96b, 140, 143
Vernonia (S, T), 4, 26, 34, 38, 50, 55, 55a, 55b, 59, 60, 94, 105, 108, 109, 143, 146, 147, 157, 171, 173, 179a, 184, 190
Vicoa, 26, 34, 172
Vieraea, 92
Viguiera (T), 40, 88a, 140, 143
Vittadinia, 29a, 41a

W
Warionia, 37, 38
Wedelia (T), 40, 43, 55a, 88a, 109, 125, 140, 143, 179a
Werneria, 53, 55, 142, 143, 157
Wilkesia, 23

Wunderlichia, 20, 105, 187
Wyethia (T), 21, 140, 143

X

Xanthium (S, T), 26, 34, 55b, 120, 130, 140, 143, 150a, 163, 179a, 182, 184, 190, 191

Xanthopappus, 179a
Xeranthemum, 37, 38, 56, 163

Y

Youngia, 55a, 55b, 62, 179a
Yunquea, 113

Z

Zaluzania, 43a
Zemisne, 137
Zinnia (S, T), 43a, 55b, 96a, 96b, 140, 143, 157, 163, 179a
Zoëgea, 100a
Zoutpansbergia, 14, 92

Chapter 9

Chromosomal cytology and evolution in the family Compositae

OTTO T. SOLBRIG
*Department of Biology, Harvard University,
Cambridge, Massachusetts, U.S.A.*

Abstract. The chromosome cytology of the Compositae is fairly well known. Over 7000 of the approximately 20 000 species in the family have been counted. The chromosome morphology is less well known, and least of all is our knowledge of meiotic behaviour and embryogenesis. To gain an overview of the chromosome cytology of the family, the tribe, subtribe, habit, geographical distribution, habitat and breeding system (as far as it is known) of each species were recorded on computer cards and the appropriate statistics obtained. These results are presented here, together with a resumé of the most notable cytological phenomena in the family.

CONTENTS

Introduction	267
Chromosome numbers in the family	268
Chromosome number and habit	271
Habit and low chromosome numbers	271
Habit and polyploidy	274
Chromosome number in trees and shrubs	274
Mechanisms of chromosome number change	275
Chromosome morphology and DNA content	277
Chromosome number and breeding system	278
References	279

INTRODUCTION

The use of chromosomes in solving systematic problems was introduced early in this century and its importance has grown gradually as the nature of chromosomes became understood and as faster and better methods for the study of chromosomes were developed. As a modern discipline, cytology arose with the development of good quality compound microscopes in the second half of the nineteenth century, but it was not until after the rediscovery of the Mendelian laws that it was possible for W. S. Sutton to be able to point out the similarity of the phenomena that occur in the nucleus during cell division and the predicted behavior of the genes. When Winge in 1917 showed experimentally how new species could arise by allopolyploidy, chromosome cytology as an important adjunct to systematics and evolution had come of age.

By far the majority of chromosomal investigations are restricted to ascertaining chromosome numbers. Such information can be very useful in verifying the integrity of a species and, in those instances when species have different chromosome numbers, for providing a reliable character to separate species. This is especially so when the chromosomal difference is correlated with morphological variation. Chromosome numbers are relatively easy to obtain, requiring only access to appropriate living material in the field or greenhouse.

Useful as the chromosome number is, it provides insufficient information if genetic and evolutionary inferences are sought. For the latter, more detailed information regarding the size and shape of the chromosomes and their behavior during cell and gamete formation is necessary. Such studies are not only more time-consuming, but they almost always require ready access to living material. This requirement usually necessitates cultivation of the species under study. For these reasons far less is known regarding these more dynamic aspects of cytology than is known about chromosome numbers.

In order to summarize the cytological work done to date on Compositae and to draw whatever general conclusions are warranted, I have been preparing an Index to the chromosome numbers of the Compositae. In addition to chromosome number, the Index contains the tribal and subtribal classification of each species following the system of Hoffmann (1894), the habit of the species (whether annual, perennial herb, shrub or tree), and its natural distribution by continent. Other pertinent data recorded, when available, are reference to breeding system, habitat and chromosome size. The Index has served as the source of information for the survey that follows. At present there are 7900 entries in the Index. A total of 1596 articles and books have been consulted so far, but, only a few of these references will be cited here. It is hoped eventually to publish the Index.

In this paper I wish to review briefly the distribution of chromosome numbers in the family, the correlation of chromosome number with plant habit, and the mechanism by which chromosomes increase and decrease in numbers. Finally I wish to mention briefly the correlation of some aspects of the breeding behavior of Compositae with the chromosome number.

CHROMOSOME NUMBERS IN THE FAMILY

The Compositae shows a great array of chromosome numbers, varying from a low of $n=2$ in *Haplopappus gracilis* (Nutt.) Gray and *Brachycome lineariloba* (DC.) Drace to highs of $n=110-120$ in *Melanthera aspera*; $n=106$ in *Werneria nebigena* Kunth and $n=103$ in *Werneria apiculata* Sch. Bip. However, of the 7900 species for which the chromosome number has been established, 6171 or 78% of the total have a chromosome number between $n=4$ and $n=18$. High polyploids then, are relatively infrequent, as are plants with very low numbers such as $n=2$ (2 species) and $n=3$ (33 species). The most common chromosome number in the family is $n=9$ found in 1634 species of roughly 21% of all taxa examined. Species with $n=10$, 12, 17, and 18 each constitute more than 5% of the total counted and species with the n numbers 9, 10, 12, 17 or 18 comprise 50% of all species for which the chromosome number is known (Figs 1, & 2).

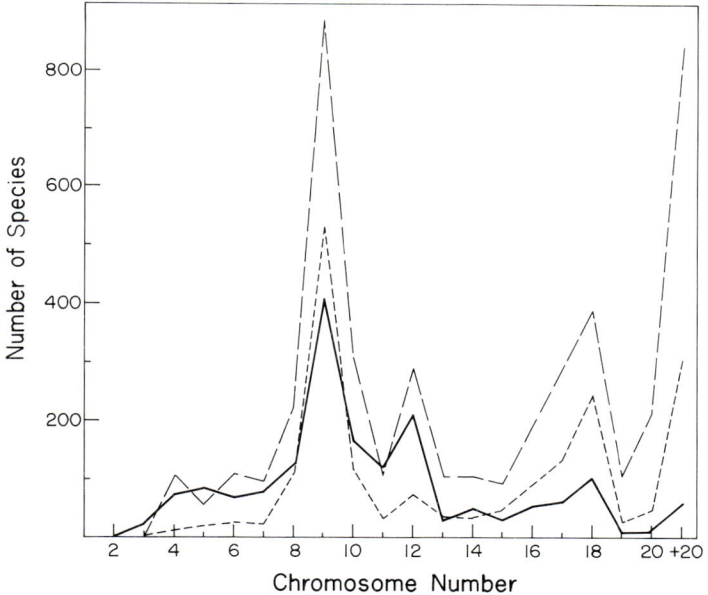

FIG. 1. Number of species with different chromosome numbers by habit. Solid line, annuals; long dashes, herbaceous perennial; short dashes, shrubs and trees.

Current data suggest that $n=9$ is the modal number for the family. If it is assumed that species with $n=18$ and higher multiples (27, 36, 54) are polyploid derivatives of species with $n=9$, then more than 30% of all species are based on $x=9$. Whether 9 is also the primitive number in the

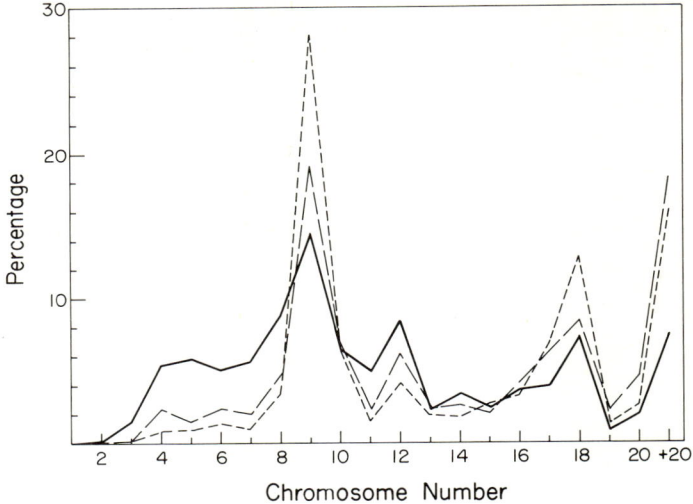

Fig. 2. Percentage of species with a given habit and chromosome number. Solid line, annuals; long dashes, herbaceous perennial; short dashes, shrubs and trees.

family requires detailed studies of what characteristics are primitive in the family or presumed to be so and their correlation with a chromosome number of $x=9$.

Although species with 9 are common in the family, they are not evenly distributed throughout the various tribes and subtribes. Only in the Arctodideae, Astereae, Anthemideae, and Lactuceae do species with $n=9$ constitute the modal group. In the Inuleae, Heliantheae, and Cardueae, species with $n=9$ (or $n=18$) are common, but species with another number in each case form the modal group of the tribe. In other tribes, species with a number other than 9 are much more common, as follows: Vernonieae: $n=17$; Eupatorieae: $n=10$; Helenieae: $n=17$; Senecioneae: $n=10$ and $n=20$; and Mutisieae: $n=12$. In the Mutisieae only one species with $n=9$ has been recorded. The samples for Calenduleae and Arctotideae are too small to make any valuable statement. Consequently when the modal class at the tribal level is not 9 it is either $n=10$ or $n=17$ with the sole exception of Mutisieae where the chromosome numbers of only 85 species are known.

At the subtribal level there is much more variation. So, for example, even though $n=9$ is a very common number in the Heliantheae, subtribe Helianthineae has a modal number of $n=17$, the Madineae $n=7$ and the Ambrosineae $n=18$. Likewise in the Lactuceae where species with $n=9$ prevail, in subtribe Hypochoerineae species with $n=6$ are the most abundant. When the next lower taxonomic level, the genus, is investigated no correlation between $n=9$ and number of species within a genus can be observed. In many genera (*Aster, Haplopappus, Artemisia*) there are species with more than one chromosome number, and in these cases it is occasionally observed that species with $n=9$ are more abundant (*Aster, Erigeron*) but the majority of the genera have all their species based on the same base chromosome number. Nevertheless, in tribes where $n=9$ is the modal number a majority of the genera will also have at least a species with $n=9$, and likewise for the entire family, genera based on $x=9$ constitute the modal class.

In any event, the observed pattern is consistent with the predictions of a model that postulates 9 as the primitive number in the family. If 9 is the primitive number and we assume a model of gradual radiation and evolution from a primitive stock, so that the lineages we recognize today as "tribes" radiated first, those we recognize today as "subtribes" second and genera and species last, then we would expect a majority of species with $n=9$ but not necessarily in each phyletic line. Nevertheless, suggestive though this correlation is, it does not constitute definite evidence. Correlation of $n=9$ with "primitive" characters in the family is required. Syneran-therologists have, however, not yet arrived at a consensus of what characters are primitive in the family.

CHROMOSOME NUMBER AND HABIT

Habit and low chromosome numbers

Stebbins (1950, 1958) has pointed out a correlation between the short-lived annual habit and low chromosome number. Short-lived plants are adapted to exploiting habitats with ephemeral resources. They repeatedly go through cycles of fast population growth, and are subjected to what is known as r-selection (MacArthur and Wilson, 1968; Gadgil and Solbrig, 1972). Under those circumstances, near genetic constancy resulting from lowered recombination may be favored (Stebbins, 1950, 1959; Grant, 1950, 1975). One way to decrease recombination is by lowering the chromosome number. Since Babcock (1974) and Clausen (1951) had verified experimentally that in the genus *Crepis* and in the subtribe Madineae there is a decrease of chromosome number associated with the annual habit (Fig. 3), a special effort was made to see how general that phenomenon is in the family.

Only 16% of all species of Compositae have a chromosome number lower than 9 (1251 spp.). Of 1447 species of annual Compositae for which the chromosome number is known, 467 or 32% of them have a chromosome number lower than 9, while among perennial herbs 632 out of 4600, or 14% have a chromosome number less than 9. Finally, among arboreal and shrubby Compositae only 152 out of 1853 species counted, or 8% of

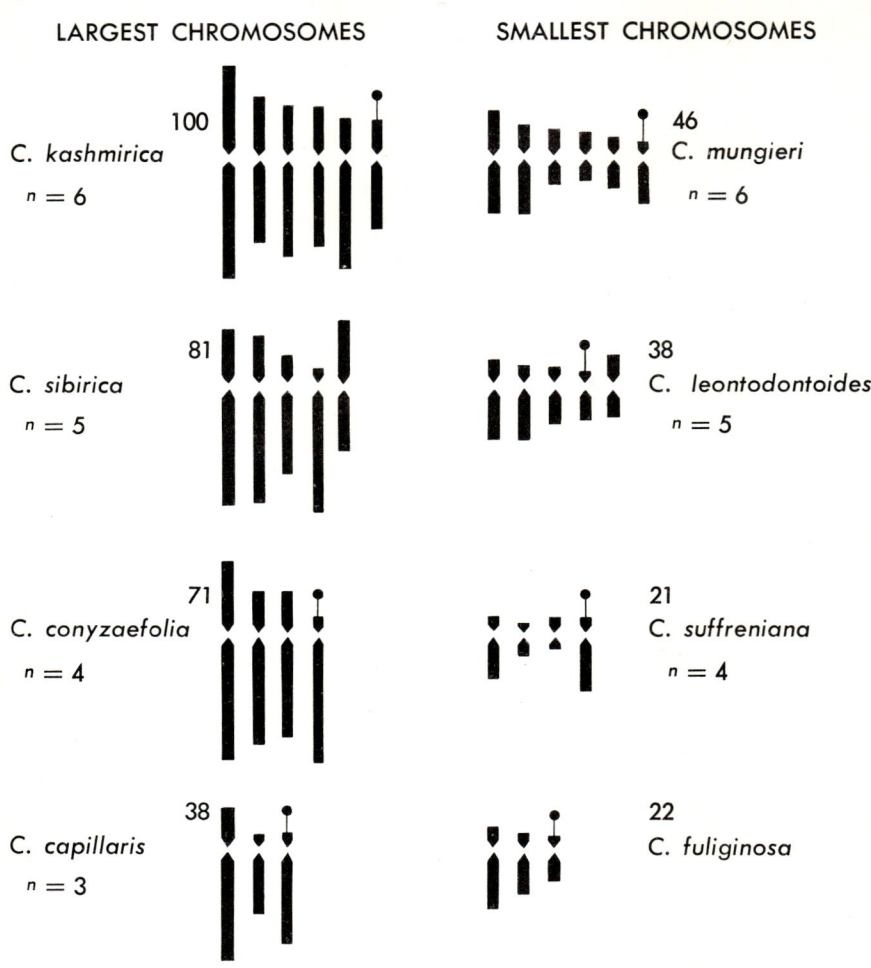

FIG. 3. Reduction in karyotype length in two phyletic lines of *Crepis*, according to Babcock (1947).

trees and shrubs, have chromosome numbers less than 9. If only the very lowest chromosome numbers are taken into account ($n = 2$, 3, 4, and 5) then 7% of all annuals, but only 4% of all perennial herbs and 2% of all trees and shrubs have such low numbers. Of those species with chromosome numbers less than 9, 37% are annuals, 51% are perennial herbs, and 23% trees and shrubs. This compares with 18% annuals, 58% perennial herbs, and 23% trees and shrubs overall in the family. Of species with the lower chromosome numbers, ($n = 2$, 3, 4, and 5) 44% are annuals, 47% are perennial herbs, and 9% are trees or shrubs. It can be seen that while the proportion of annual species increases with a decrease in the chromosome number, the proportion of perennial herbs and particularly that of trees and shrubs decreases. However, except among species with $n=2$ and $n=3$, perennial species always predominate, but not as much as among species with numbers higher than 9 (see Table I).

TABLE I. Chromosome numbers in species of Compositae, by habit.

Chromosome number	ANNUAL			PERENNIAL			SHRUB			Totals
	No.	%	% Total	No.	%	% Total	No.	%	% Total	
2	2	0.14	0.03	—	—	—	—	—	—	2
3	22	1.52	0.28	8	0.17	0.10	3	0.16	0.04	33
4	78	5.39	0.98	112	2.40	1.40	16	0.85	0.20	206
5	85	5.88	1.06	79	1.70	0.99	21	1.12	0.26	185
6	71	4.91	0.89	112	2.40	1.40	28	1.49	0.35	211
7	82	5.67	1.03	95	2.04	1.19	20	1.06	0.25	197
8	127	8.78	1.59	226	4.85	2.83	64	3.41	0.80	417
9	213	14.73	2.67	892	19.15	11.17	529	28.16	6.63	1634
10	93	6.43	1.16	316	6.78	3.96	120	6.39	1.50	529
11	71	4.91	0.89	112	2.40	1.40	31	1.65	0.39	214
12	122	8.44	1.53	297	6.37	3.72	77	4.10	0.96	496
13	32	2.21	0.40	106	2.28	1.33	37	1.97	0.46	175
14	50	3.46	0.63	108	2.32	1.35	36	1.92	0.45	194
15	32	2.21	0.40	94	2.02	1.18	48	2.56	0.60	174
16	54	3.73	0.68	196	4.21	2.46	63	3.35	0.79	313
17	60	4.15	0.75	290	6.23	3.63	133	7.08	1.67	483
18	106	7.33	1.33	392	8.41	4.91	245	13.04	3.07	743
19	12	0.83	0.15	102	2.19	1.28	26	1.38	0.33	140
20	28	1.94	0.35	218	4.68	2.73	49	2.61	0.61	295
+20	107	7.40	1.34	845	18.14	10.58	307	16.35	3.85	1259

The picture varies somewhat from tribe to tribe: 5% of the Anthemideae with known chromosome number are annuals, but only seven species of annual Anthemideae (0.05%) have a chromosome number lower than 9, while 24 perennial herbs (1.8%) and 31 shrubs (2.3%) have numbers lower than 9. Also among Heliantheae where 37% of all species in our sample are annuals, only 20% of them have chromosome numbers lower than 9. On the other hand, in the Astereae 201 species out of 1134 of known chromosome number in the tribe are annuals (18%). Of these, 87, or 43%, have chromosome numbers lower than 9 as compared to only 17% of the perennial herbs in the tribe (130 out of 755) and 13% of the trees and shrubs (24 out of 178). Likewise in the Lactuceae, where 255 species with known chromosome number out of 1219 species are annuals, 60% have chromosome numbers lower than 9 against 34% among herbaceous perennials and 40% among trees and shrubs.

It is very clear that, statistically speaking, there is a correlation between the annual habit and a low chromosome number as predicted by the presently accepted model of the genetic system (Stebbins, 1950; Grant, 1958). That the correlation is not absolute is not surprising: chromosome number is only one aspect of the genetic system. Breeding system, crossover rate and degree of heterochromatization are other aspects, of which more further on.

Habit and polyploidy

1259 species out of 7900, or 16% of the sample have a chromosome number higher than $n=20$. Unless the very unlikely thesis that the ancestral Compositae had more than 20 chromosomes is brought forward, all of these species have originated by polyploidy. That this is probably so has been verified experimentally in a great many instances (Ehrendorfer, 1953; Babcock and Stebbins 1938; Solbrig 1960, 1964). Furthermore, if the thesis is accepted that the ancestral chromosome number of the Compositae is $x=9$, then 5015 species, or 63% of the total have originated by eu- or aneupolyploidization.

Considering only species with known chromosome numbers above 20, only 107 are annuals, roughly 7% of all annuals; 845 are perennial herbs, or 18% of all plants with that habit, and 307 are trees or shrubs, that is 16% of all trees and shrubs. This distribution is again consistent with the prediction made by Stebbins (1950) regarding the ease with which polyploidy can occur. According to this model, perennial herbs with alternative means of vegetative reproduction have a greater probability of surviving when polyploidized (particularly in cases of segmental allopolyploidy), than either ephemeral annuals, or trees and shrubs that rely exclusively on sexual reproduction.

Chromosome number in trees and shrubs

The final correlation between habit and chromosome number worth pointing out is the high frequency of species with $x=9$ among trees and shrubs. In the Compositae, 42% of all trees and shrubs have either $n=9$ (29%) or $n=18$ (13%); 7% have $n=10$ ($n+1$) and 7% have $n=17$ ($2n-1$), while 16% are high polyploids mostly multiples of 9. Only 25% of all trees and shrubs have other numbers. If it is assumed that $n=9$ is indeed the ancestral number, the strong correlation between the woody habit and a base number of $x=9$ could be interpreted as evidence for the ancestral Compositae having been woody, a not implausible possibility.

9. CHROMOSOMAL CYTOLOGY AND EVOLUTION

MECHANISMS OF CHROMOSOME NUMBER CHANGE

The vast array of chromosome numbers in the family with at least one species with each number between 2 and 40 and a fair number of species with chromosome numbers higher than that indicates that chromosome number changes must have taken place repeatedly in the family. The documented mechanisms by which these changes have occurred are polyploidy, aneuploid reduction, loss of chromosomes and chromosome splitting. It is not possible to present all cases of chromosomal changes that have been verified experimentally, and therefore only a few selected examples will be discussed.

Probably the best known mechanism of chromosome number change in the Compositae is polyploidy, be it either allo- or autopolyploidy. Usually polyploidy is inferred from the chromosome numbers without experimental verification. However, there is a sufficiently large number of studies demonstrating the occurrence of polyploidy.

One of the best examples of the origin of species by allopolyploidy in the family is that of *Tragopogon miscellus* Ownbey and *T. mirrus* Ownbey. These two species, which are restricted in their distribution to a few localities in the western United States, originated by hybridization and subsequent doubling of the chromosomes of three European species of *Tragopogon*, introduced by man into the New World (Ownbey, 1950). *Tragopogon miscellus* ($n=12$) is an allopolyploid between *T. pratensis* ($n=6$) and *T. dubius* ($n=6$); *T. mirrus* ($n=12$) is an allopolyploid between *T. dubius* and *T. porrifolius* ($n=12$). From distributional and other data (Ownbey, 1950; Brehm and Ownbey, 1965) it is inferred that these allopolyploids originated within the last 50 years.

Not in all cases are the phylogenetic relations as clear as in *Tragopogon*. Often the polyploid and diploid populations are so close in their morphology that they are classified as the same species, or infraspecific category of the same species. One such example is found in the genus *Gutierrezia* (tribe Astereae), studied by Solbrig (1960, 1964, 1965, 1970). In this small genus of 12 species, taxa with $n=4$, 8, 12, 16, 24, and 28 are found. Studies by Rüdenberg and Solbrig (1963) indicate that all the perennial species of the genus which have (with one exception) multiples of 4 have arisen by polyploidy, while the two annual species (which are diploids with $n=4$) have originated by chromosomal changes. One of the shrubby species, *G. sarothrae*, was found to have diploid ($n=4$) and tetraploid populations ($n=8$). The latter constituted approximately 20% of the total, were scattered throughout the range of the species, and could not be separated from the diploid populations on the basis of any morphological character. Detailed field studies, and a numerical study (Solbrig, 1964, 1970) led to the hypothesis that polyploid populations in *G. sarothrae* had originated more than once, and that the polyploids are either autopolyploids or segmental allopolyploids resulting from the hybridization of plants from different diploid populations of the species. It is believed that this model of polyploid origin is of general applicability in the family, where species with polyploid populations are relatively frequent.

Aneuploid changes appear to be common in the family, judging from the variety of chromosome numbers that are found. However, there are few good documentations of such changes. One of the most elegant demonstrations of aneuploid reductions known to this author is that of Kyhos (1965) in *Chaenactis*. By means of crosses and subsequent analysis of the hybrids, Kyhos was able to show that *Chaenactis fremontii* ($n=5$) and *C. stevioides* ($n=5$) had both originated independently from *C. glabiuscula* ($n=6$) by aneuploid reduction involving at least two reciprocal translocations with chromosomal fusion and loss of a centromere. Another case where aneuploid reduction has been demonstrated is in the *Haplopappus ravenii* ($n=4$)–*Haplopappus gracilis* ($n=2$) complex (Jackson, 1962). Also the pioneering studies on the genus *Crepis* should be mentioned. Togby (1943) demonstrated that aneuploid reduction probably accounted for the origin of *Crepis fuliginosa* ($n=3$) from *C. neglecta* ($n=4$), and similarly Sherman (1946) showed that *Crepis kotschyana* ($n=4$) was probably derived from an ancestor with five pairs of chromosomes, similar or identical to *C. foetida* (Fig. 3).

Changes of chromosome number by loss of chromosomes was also shown to occur in *Taraxacum officinale* by Sörensen and Gundjonsson (1946). They were able to show in garden cultures the repeated appearance of the same sports in two related microspecies, sports that were known from nature. Upon investigation, these sports proved to be monosomics ($3n-1$). Eight such types were identified, corresponding to each of the eight basic chromosomes of the genome of this triploid species complex.

Polyploidy, aneuploid reduction and chromosome loss are believed to be the basic mechanisms of chromosome number change within the family, as in other angiosperm families. There is, however, one species, *Xanthisma texana*, that has an unusual chromosome situation. In this species, the root tissue always possesses eight chromosomes ($x=4$), while the shoot tissue has usually 10 chromosomes ($x=5$). Studies by Berger *et al.* (1956) indicate that, as expected (since the shoot has 10 chromosomes), the gametes have five and the zygote 10. One pair of chromosomes is, however, lost early in the embryogenesis of those cells that give rise to the radicle.

Finally we may mention the studies by Baker (1967, 1974) on the *Eupatorium microstemon* species aggregate. He showed how from an ancestral taxon with $n=10$, species with $n=4$ might have arisen by aneuploid reduction. Low chromosome number appears to be correlated with "weediness" in this aggregate.

CHROMOSOME MORPHOLOGY AND DNA CONTENT

The accepted way to demonstrate chromosome homology, or lack thereof, in the karyotypes of two species is by meiotic analysis of the hybrids. However, in many instances the species under study cannot be hybridized; consequently the chromosomal morphology of mitotic chromosomes is often used to infer homology. One such study is that of Huziwara (1957a, b; 1958a, b, 1959, 1962) on *Aster*. Using data such as chromosome number, position of the centromere, the relative length of each chromosome as well as the entire genome, he proposed a phylogeny for the *Aster ageratoides* polyploid complex. It was concluded that species of

TABLE II. Chromosome number and DNA content per cell for various species of Astereae (from Stucky and Jackson, 1975)

Species	Chromosome number	DNA per cell
Machaeranthera boltoniae	4	22.3 ± 0.38
M. tenuis	4	14.8 ± 0.36
M. parviflora	5	27.2 ± 0.47
Aster oblongifolius	5	22.0 ± 0.56
A. riparius	5	36.4 ± 0.66
Machaeranthera brevilingulata	9	10.1 ± 0.37
Aster hydrophyllus	9	19.4 ± 0.29

Aster with $x=5$ and $x=8$ were aneuploid derivatives of species with $x=9$. The problem is of interest because in the Astereae there are a large number of species with $n=4$ and $n=5$, but relatively few species with numbers of $n=6$, 7, or 8, especially $n=7$ (Solbrig et al., 1964), so that the alternative hypothesis, namely that species with $x=9$ are polyploid derivatives of species with $n=4$ and $n=5$, is plausible (Turner et al. 1961, 1964).

A new technique that can aid in the solution of problems of karyotype evolution is the study of the DNA content of the nucleus. In effect, if two species have markedly different chromosome numbers (such as 4 and 9) their DNA content should (1) be different if the species with the low number has arisen by aneuploid reduction from one with the high number, or (2) if the one with the high number is a polyploid derivative of the species with low chromosome numbers. In the first case the DNA content should be similar in both species; in the second case the high chromosome number species should have twice the DNA of its presumed diploid ancestors. This prediction holds only if no further DNA changes take place after the chromosome number changes take place. Stucky and Jackson (1975) applied this technique to the problem of origin of species with $x=4$, 5, and 9 in the genus *Machaeranthera*. Although their sample is too small to make definite conclusion (only two species each with $n=4$ and $n=9$ and three species with $n=5$ were analyzed) and the variation in DNA content was quite large, the results are in better agreement with the hypothesis of aneuploid reduction in the Astereae (Solbrig et al., 1964) than polyploid origin (Turner et al., 1961) (Table II).

Nagl and Ehrendorfer (1974) have applied these same techniques in a study of annual and perennial species in the Anthemideae. As was pointed out before, in the Anthemideae the correlation between low chromosome number and the annual habit is not observed. They selected pairs of closely related annuals and perennials in seven genera of the tribe and measured the nuclear volume, DNA content and karyogram length of each species. They found that, as was to be expected, karyogram length and nuclear volume are correlated with DNA content. However they could find no correlation of these factors with plant habit. They were, however, able to show a negative correlation between total euchromatin and the annual habit. That is, annuals had either a smaller genome than their close perennial relatives, or if they had a genome similar or larger than the perennial, a portion of it was heterochromatized, so that the euchromatic portion was less in the annual than in the perennial. Nagl and Ehrendorfer (1974) also showed a faster growth rate in the annual species under study compared to their perennial relatives and a higher mitotic index. They attributed both these factors in part to the possession by annuals of a smaller effective functional genome.

CHROMOSOME NUMBER AND BREEDING SYSTEM

In this paper I have shown that there is a correlation between chromosome number and habit in the Compositae: decreases in chromosome number are proportionally more frequent in annuals, whereas increases are more frequent in perennial herbs. The selective forces associated with these changes are of two types. First, there is selection in ephemeral plants toward a faster growth rate (Stebbins, 1950, 1958; Nagl and Ehrendorfer, 1974). Lower amounts of DNA, especially euchromatin, apparently allow a higher rate of cell division. Secondly, a lower chromosome number decreases the rate of recombination and lowers the release of variability. Stebbins (1950, 1958) and Grant (1959, 1975) have argued that ephemerals that occupy more unstable habits than their ancestors are subjected to much greater fluctuations in population sizes. Under those circumstances, a lowering of recombination may be favored because it decreases the possible number of inviable or poorly adapted recombinants, even though such a change may also decrease the number of possibly superior genotypes that may arise by recombination. The recombination index can be lowered in a number of ways: reduction of crossover frequency, increasing the asymmetry of the karyotype, reduction of the chromosome number and changes in the breeding system, especially changes from obligatory outbreeding to systems involving some kind of inbreeding. Since all these mechanisms operate in the same direction, it can be expected that they are to some extent interchangeable, and that annual species with a very low chromosome number may have a higher crossover rate than annual species with a higher chromosome number growing in the same habitat. Unfortunately, the available information is insufficient to draw specific conclusions regarding these phenomena in the Compositae as a whole. Stebbins (1959) has presented a detailed analysis of the relationships between longevity, habit, and the release of genetic variability in the tribe Lactuceae. He showed that in the Lactuceae species that on other criteria are considered primitive are preferentially perennials, they grow usually in a stable environment, and have a basic chromosome number of $x=9$. They also are sexual and obligatorily outcrossing. Species in more unstable habitats tend to have a lower chromosome number and/or a lower recombination index, and/or a breeding system favoring inbreeding, be it either self-fertilization or apomixis. The situation in the Lactuceae conforms to the proposed model. It is likely that on further investigation it will apply to the entire family.

REFERENCES

Babcock, E. B. (1947). The Genus *Crepis*, I and II. *Univ. Calif. Publs Bot.* **21** and **22**.

Babcock, E. B. and Stebbins, G. L. (1938). "The American species of *Crepis*: their Relationship and Distributions as affected by Polypoidy and Apomixis". Carnegie Institution, Washington, Publication **504**.

Baker, H. G. (1967). The Evolution of weedy taxa in the *Eupatorium microstemon* species aggregate. *Taxon* **16**, 293–300.

Baker, H. G. (1974). The evolution of weeds. *A. Rev. Ecol. Syst.* **5**, 1–24.

Berger, C. A., Feeley, E. J. and Witkus, E. R. (1956). The cytology of *Xanthisma texanum* DC. IV. Megasporogenesis and embryo sac formation, pollen mitosis, and embryo formation. *Bull. Torrey bot. Club* **83**, 428–434.

Brehm, B. G. and Ownbey, M. (1965). Variation in Chromatographic patterns in the *Tragopogon dubius–pratensis–porrifolius* complex (Compositae). *Am. J. Bot.* **52**, 811–818.

Clausen, J. (1951). "Stages in the Evolution of Plant Species". Cornell University Press, Ithaca, N.Y.

Ehrendorfer, F. (1953). Systematische und zytogenetische Untersuchungen an europäischen Rassen des *Achillea millefolium*-complexes. *Öst. bot. Z.* **100**, 583–592.

Gadgil, M. D. and Solbrig, O. T. (1972). The concept of "r" and "K" selection: evidence from widlflowers and theoretical considerations. *Am. Nat.* **106**, 14–31.

Grant, V. (1958). The regulation of recombination in plants. *Cold Spring Harb. Symp. quant. Biol.* **23**, 337–363.

GRANT, V. (1975). "Genetics of Flowering Plants". Columbia University Press, New York.
HOFFMAN, O. (1894). Compositae. In "Die natürlichen pflanzenfamilien" Eugler and R. Prautl, eds, vol. IV. W. Eugelmann, Leipzig.
HUZIWARA, Y. (1957a). Karotype analysis in some genera of Compositae. II. The karyotype of Japanese *Aster* species. *Cytologia* **22**, 96-112.
HUZIWARA, Y. (1957b). Karyotype analysis in some genera of Compositae III. The karyotype of the *Aster ageratoides* group. *Am. J. Bot.* **74**, 783-790.
HUZIWARA, Y. (1958a). Karyotype analysis in some genera of Compositae. IV. The karyotypes within the genera *Gymnaster*, *Kalimeris*, and *Heteropappus*. *Cytologia* **23**, 33-45.
HUZIWARA, Y. (1958b). Karyotype analysis in some genera of Compositae. V. The chromosomes of American *Aster* species. *Jap. J. Genet.* **33**, 129-137.
HUZIWARA, Y. (1959). Chromosome evolution in the subtribe Asterinae. *Evolution* **13**, 188-193.
HUZIWARA, Y. (1962). Karyotype analysis in some genera of Compositae. IX. Chromosomes of European species of *Aster*. *Bot. Mag., Tokyo* **75**, 143-149.
JACKSON, R. C. (1962). Interspecific hybridization in *Haplopappus* and its bearing on chromosome evolution in the *Blepharodon* section. *Am. J. Bot.* **49**, 119-132.
KYHOS, D. W. (1965). The independent aneuploid origin of two species of *Chaenactis* (Compositae) from a common ancestor. *Evolution* **19**, 26-43.
MACARTHUR, R. H. and WILSON, E. O. (1968). "The Theory of Island Biogeography'. Princeton University Press, New Jersey.
NAGL. W. and EHRENDORFER, F. (1974). DNA content, heterochromatin, mitotic index, and growth in perennial and annual Anthemideae (Asteraceae). *Plant Syst. Evol.* **23**, 35-54.
OWNBEY, M. (1950). Natural hybridization and amphiploidy in the genus *Tragopogon*. *Am. J. Bot.* **37**, 487-499.
RÜDENBERG, L. and SOLBRIG, O. T. (1963). Chromosome number and phylogeny in the genus *Gutierrezia* (Compositae-Astereae). *Phyton, B. Aires* **20**, 199-204.
SHERMAN, M. (1946). Karyotype evolution: a cytogenetic study of seven species and six interspecific hybrids of *Crepis*. *Univ. Calif. Publs Bot.* **18**, 369-408.
SOLBRIG, O. T. (1960). Cytotaxonomic and evolutionary studies in the North American species of *Gutierrezia* (Compositae). *Contr. Gray Herb.* **188**, 1-63.
SOLBRIG, O. T. (1964). Infraspecific variation in the *Gutierrezia sarothrae* complex (Compositae-Astereae). *Contr. Gray Herb.* **193**, 67-115.
SOLBRIG, O. T. (1965). The California species of *Gutierrezia* (Compositae-Astereae). *Madroño* **18**, 75-84.
SOLBRIG, O. T. (1970). The phylogeny of *Gutierrezia*: an eclectic approach. *Brittonia* **22**, 217-229.
SOLBRIG, O. T., ANDERSON, L. C., KYHOS, D. W., RAVEN, P. H. and RÜDENBERG, L. (1964). Chromosome numbers in Compositae. V. Astereae II. *Am. J. Bot.* **51**, 513-519.
SÖRENSEN, T. and GUNDJONSSON, G. (1946). Spontaneous chromosome aberrants in apomictic Taraxaca. *K. danske Vidensk. Selsk. Skr.* **4**(2), 1-48.
STEBBINS, G. L. (1950). "Variation and Evolution in Plants". Columbia University Press, New York.

STEBBINS, G. L. (1958). Longevity, habitat, and the release of genetic variability in the higher plants. *Cold Spring Harb. Symp. quant. Biol.* **23**, 365–378.
STUCKY, J. and JACKSON, R. C. (1975). DNA content of seven species of Astereae and its significance to theories of chromosome evolution in the tribe. *Am. J. Bot.* **62**, 509–518.
TOGBY, H. A. (1943). A cytological study of *Crepis fulginosa, C. neglecta* and their F_1 hybrid, and its bearing on the mechanism of phylogenetic reduction in chromosome number. *J. Genet.* **45**, 67–111.
TURNER, B. L., ELLISON, W. L. and KING, R. M. (1961). Chromosome numbers in the Compositae. IV. North American species, with phylogenetic interpretations. *Am. J. Bot.* **48**, 216–223.
TURNER, B. L., POWELL, M. and KING, R. M. (1964). Chromosome numbers in the Compositae. VI. Additional Mexican and Guatemalan species. *Rhodora* **64**, 251–271.

Chapter 10

The chemistry of the Compositae

R. HEGNAUER

*Laboratorium voor Experimentele Plantensystematiek,
Leiden, Netherlands*

Abstract. Secondary metabolites of Compositae are reviewed and their taxonomic significance is discussed. Sesquiterpene lactones, triterpene monols and diols, acetylenic compounds, methylated flavonols and flavones, inulin-type fructans, the cyclitols L-inositol and scyllitol and fatty oils in seeds are common constituents of many species; they probably occur in all tribes and form the chemical make-up of the family. Essential oils and diterpenoids are widely distributed also; they seem to be lacking, however, in Cichorieae. Alkaloids, cyanogenic glycosides, amides, coumarins and several types of phenolic constituents are of much more limited distribution but they might become interesting characters at infrafamilial taxonomic ranks. The same is true for those compound classes which occur characteristically throughout the family. It is concluded that the picture sketched in 1964 for the chemical affinities of composites is still valid. The presently available chemical data, however, not only provide a means of discerning relationships between families; many of the chemical characters will almost certainly become valuable aids to infrafamilial taxonomy in the near future.

CONTENTS

Introduction	284
Ubiquitous classes of constituents in the Compositae	284
Sesquiterpene lactones	285
Triterpenes	286
Acetylenic compounds	289
Flavonoids	290
Carbohydrates and cyclitols	292
Essential oils	293
Diterpenes	294
Amides	299
Alkaloids and alkaloid-like compounds	301
Cyanogenic glycosides	305
Coumarins	306
Some phenolic constituents	309
Simple phenolics	310
Phenolic acids	310

Phenylpropanoid compounds	311
Lignans	312
Chromenes and benzofurans	312
Some unclassified phenolics	315
Seed oils	316
Concluding remarks	320
References	321

INTRODUCTION

More than 10 years ago (Hegnauer, 1964) the present author summarized the chemistry of Compositae and discussed its systematic implications. Three conclusions were reached. These were as follows:

(1) The family is chemically very distinct. Inulin-type fructans, seed oils sometimes containing characteristic fatty acids, bitter sesquiterpene lactones, pentacyclic triterpene alcohols, accumulation of large amounts of derivatives of caffeic acid, of flavones and of methylated flavonols and a total lack of true tannins and of iridoid glycosides were especially mentioned. Acetylenic compounds, not reported from Senecioneae and Cichorieae, and essential oils, not accumulated by latex-bearing Cichorieae, were likewise considered to belong to the chemical make-up of the family Compositae.

(2) Additional classes of compounds as well as patterns of constituents belonging to the chemical classes already mentioned were expected to become very useful in future at infrafamilial taxonomic levels. Patterns of sesquiterpene lactones, acetylenic and flavonoid compounds and distribution and structures of cyclitols, diterpenes, alkaloids and cyanogenic glycosides were particularly selected as being important in this respect.

(3) By comparing the combinations of primary and secondary metabolites characteristic of Compositae with chemical patterns known from members of other families, an evolutionary line: Magnoliales → Rutales → Umbellales → Asterales was esteemed worthwhile of consideration by taxonomists. A relationship between Polemoniales (Boraginaceae) and Asterales, however, also seemed possible on chemical grounds (Fig. 1).

Since the summary mentioned above was written, our phytochemical knowledge of Compositae has more than doubled! It is my intention to evaluate the systematic meaning of the chemical characters in the light of the new evidence. Most attention will be paid to those classes of compounds which will probably be treated less thoroughly by other contributors. In this chapter, the classification of the family by Wagenitz (1964) will be used.

UBIQUITOUS CLASSES OF CONSTITUENTS IN THE COMPOSITAE

The combined occurrence of sesquiterpene lactones, acetylenic compounds and inulin-type fructans is almost as characteristic of the Compositae as are their headlike inflorescences (capitula). Triterpenes and

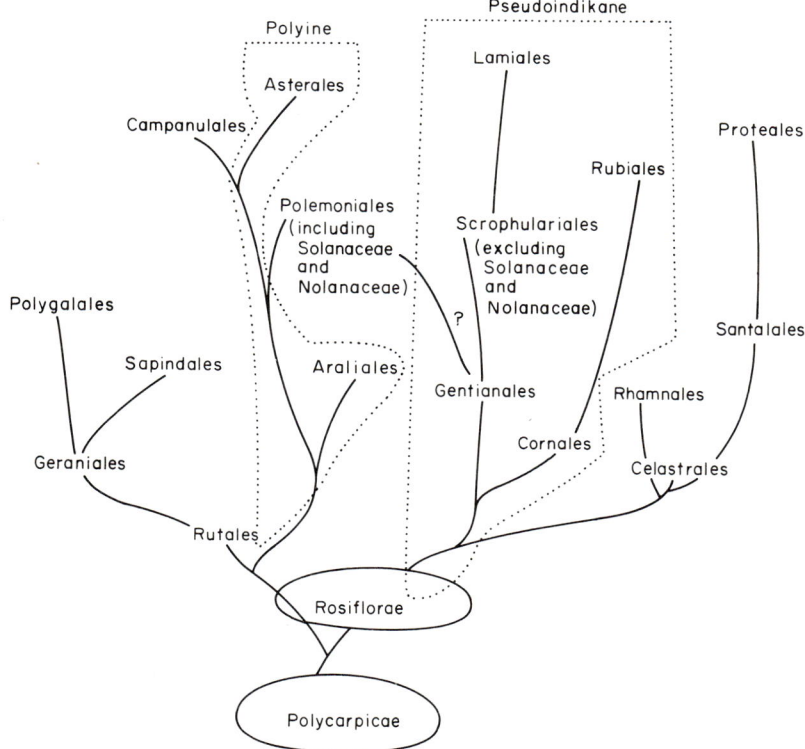

FIG. 1. Chemical (and phylogenetic?) relationships between some families of Angiosperms (from Hegnauer 1964, p. 544; Pseudoindikane = iridoid constituents).

flavonoids are present in every member of the family and some of their structural features suggest family-specific trends with regard to triterpene and flavonoid synthesis. Therefore some remarks about these five classes of constituents must be included in a general survey of the chemistry of Compositae in relation to taxonomy.

Sesquiterpene lactones
The bitter taste of *Cnicus benedictus* L. and the blue volatile oil of *Matricaria chamomilla* L., two medicinal plants of long standing, drew the attention of many chemists to these plants. The blue component of oil of chamomile was shown to be chamazulene. The latter was later proved to be an artefact which arises during steam distillation from the bitter guaianolide matricin (**1**). Matricin is one of the first chemically defined representatives of the now very large and structurally diverse group of

sesquiterpene lactones of Compositae. These bitter principles are deposited in latices in members of the Cichorieae. Members of other tribes excrete them in subcuticular cavities of glandular hairs (Weber and Deufel, 1951; Stahl, 1952, 1957) or extrude them by means of the glandular hairs (Shields, 1952). From the botanical point of view the bitter lactones may be interpreted as components of essential oils (see next section, "Essential oils") which have been rendered non-volatile by oxidation. Within a relatively short period of time an intensive study of bitter principles of Compositae yielded a picture of distribution of the sesquiterpene lactones (Hegnauer, 1964: pp. 459–460, 463–478, 654–656) which is still essentially valid. Such lactones occur in all tribes. Some of the most striking facts are the concentration of pseudoguaianolides (=ambrosanolides) in Inuleae and Heliantheae (Ambrosiinae included) and the production of eremophilanolides, such as the petasolides (**2**), and of furanoelemophilanes and related compounds with the intact or modified eremophilane skeleton by Senecioneae. The pseudoguaianolide-producing genus *Arnica* of the Senecioneae, however, is exceptional in an otherwise uniform tribe. Guaianolides also occur in members of the Umbelliferae; their presence strengthens the biochemical affinities between this family and the Compositae. However, the occurrence of lactones in the Magnoliaceae (germacranolides) and Hepaticae (eudesmanolides) demonstrates that metabolic convergence may also take place. Sesquiterpene lactones have been reviewed several times (e.g. Steelink and Spitzer, 1966; Herout and Šorm, 1969; Herout, 1971; Herz, 1968, 1974). Many species and genera are surprisingly versatile with regard to lactone production. The observed variation within species and genera renders these constituents suitable for taxonomic studies at specific and sectional levels (e.g. Geissman and Irwin, 1970) and for the study of speciation and plant migration (e.g. Geissman and Matsueda, 1968; Geissman *et al.*, 1969; Mabry, 1970; Potter and Mabry, 1972). Most of the lactones are intensely bitter and several of them are toxic. These properties suggest that sesquiterpene lactones are involved in defence against herbivores and plant parasites. It is, however, extremely difficult to provide irrefutable evidence for such an ecological function (Burnett *et al.*, 1974).

Triterpenes
Compositae are triterpene accumulators. Monols and diols of the oleanol (**3**), ursanol (**4**) and lupeol (**5**) type are most characteristic of the family. They occur free or, more frequently, esterified with acetic acid or fatty acids in the lipid fractions of roots, stems, flowers and fruits and, in Cichorieae, in latices (Hegnauer, 1964: pp. 483–488). Esters of triterpene monols may account largely for the often large amounts of unsaponifiable matter in "seed" oils of Compositae; they are, however, derived from the pericarp part of the fruits (Mikolajczak and Smith, 1967a). β-Amyrin

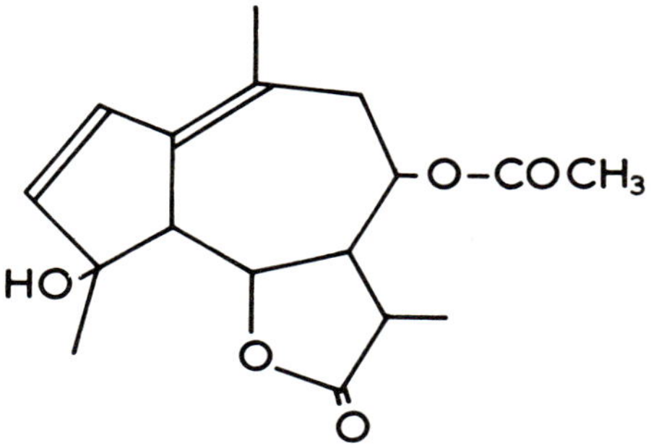

Fig. 2. Sesquiterpene lactones of Compositae, exemplified by the guaianolide matricin and the eremophilanolide petasolide.

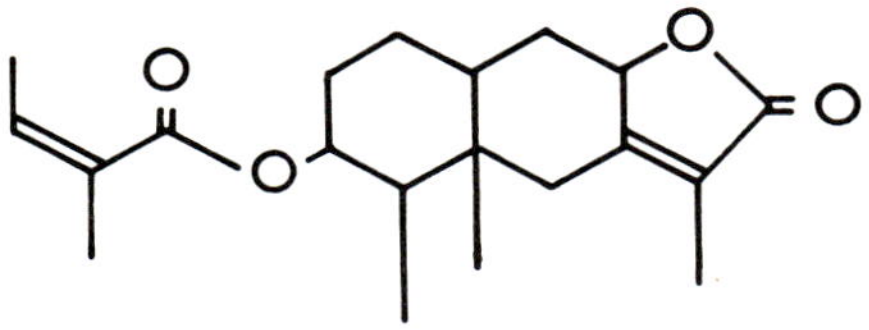

(**3**; $\Delta^{12,13}$), taraxerol (**3**; $\Delta^{14,15}$), germanicol (**3**; $\Delta^{18,19}$), α-amyrin (**4**; $\Delta^{12,13}$), taraxasterol (**4**; $\Delta^{20,30}$, 19βH), pseudotaraxasterol (**4**; $\Delta^{20,24}$, 19βH) and lupeol (**5**) have been isolated from many members of the family. Taraxerol, germanicol, taraxasterol and pseudotaraxasterol (for a review see Dutta *et al.*, 1972) are somewhat typical constituents of composites. In many yellow-flowering taxa triterpene monols are accompanied in flowers by the diols faradiol (**4**; $\Delta^{20,21}$, 19βH, 16βOH), arnidiol (**4**; $\Delta^{20,30}$, 19βH, 16βOH) and calenduladiol (**5**; 12βOH) (Kasprzyk and Kozierowska, 1966; Kasprzyk *et al.*, 1970; Pyrek and Baranowska, 1973). Other diols isolated recently from Compositae are erythrodiol (**3**; $\Delta^{12,13}$, 28OH), sophoradiol (**3**; $\Delta^{12,13}$, 22βOH) (Cambie and Parnell, 1969), uvaol (**4**; $\Delta^{12,13}$, 28OH), ursadiol (**4**; $\Delta^{12,13}$, 21OH) (Śliwowski *et al.*, 1973), brein (**4**; $\Delta^{12,13}$, 16βOH) and betulin (**5**; 28OH). Oleanolic acid (**3**; $\Delta^{12,13}$, 28 COOH) and ursolic acid (**4**; $\Delta^{12,13}$, 28COOH) are present also in many composites; they occur free and as 3-acetyl derivatives and seem to replace triterpene diols in flowers coloured by anthocyanins. Oleanolic acid and some of its derivatives (e.g. echinocystic acid) are the main sapogenins of saponin-accumulating taxa of the family (e.g. *Calendula, Helianthus, Solidago* and many others). The results of some recent investigations serve to illustrate the versatility of composites with regard to triterpene synthesis. Dammaradienyl acetate is present in appreciable amounts in *Inula helenium* L. (Yosioka and Yamada, 1963), *Eupatorium cannabinum* L. (Talapatra *et al.*, 1974) and *Olearia paniculata* (J. R. & G. Forst.) Druce (Corbett *et al.*, 1964) and α-euphorbol has been isolated from *Acroptilon picris* Pall. (Anyehchi and Eshaghzadeh, 1974); both are tetracyclic triterpenes. *Artemisia vulgaris* L. produces minor amounts of the "primitive" triterpene fernenol (Kundu *et al.*, 1966). Some members of Astereae synthesize shionone (6; *Aster baccharoides* Steetz, *A. tataricus* L.f.) or the closely related baccharis oxide (*Baccharis halimifolia* L.) (Takahashi *et al.*, 1967; Hui *et al.*, 1971; Mo *et al.*, 1972; Tachibana and Takahashi, 1975).

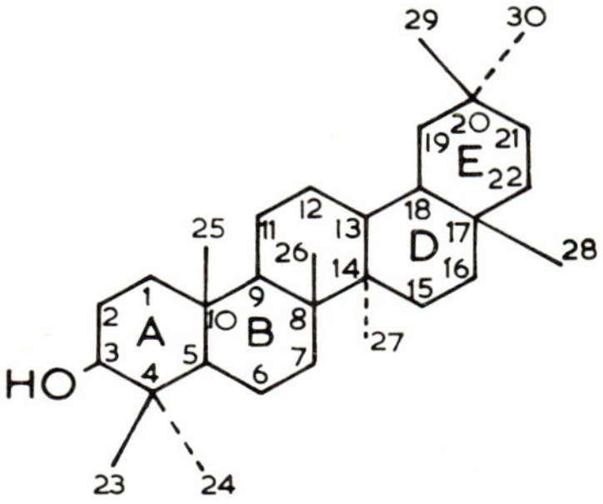

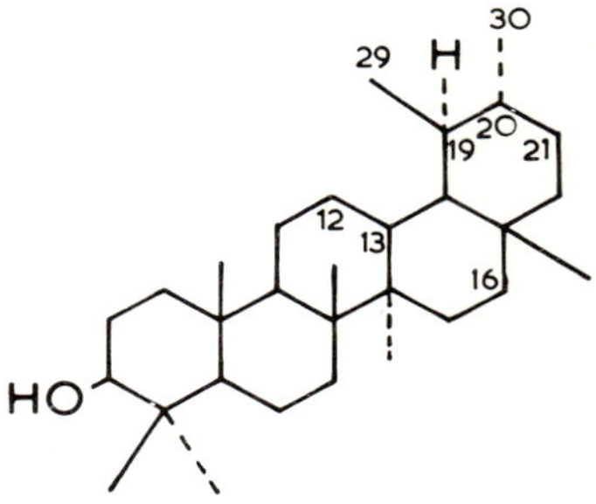

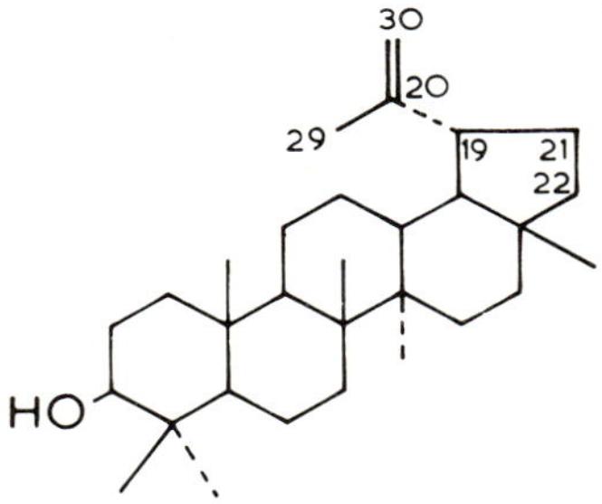

FIG. 3. Some triterpenes of Compositae.

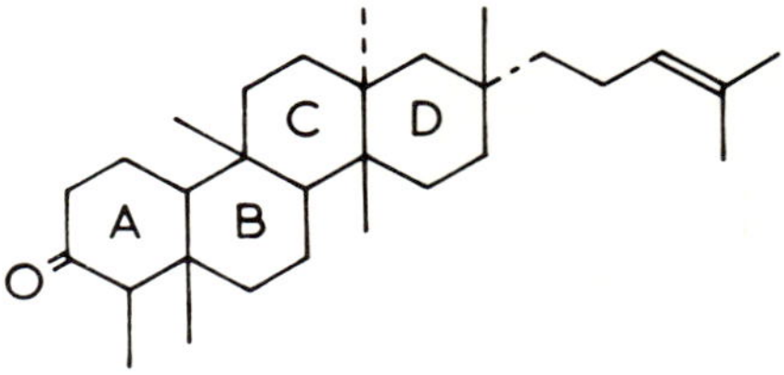

In the light of present evidence the co-occurrence of the monols and diols mentioned represents a metabolic trend of the family as a whole. The synthesis of triterpene acids, saponins and rare triterpenoids, such as shionone, may become taxonomically useful in future at the generic and tribal levels.

Acetylenic compounds
Earlier, in reviewing the distribution of acetylenes in Compositae (Hegnauer, 1964: pp. 490–507; Fig. 28, p. 505) their presence in all tribes, except the Senecioneae and Cichorieae, was stressed. At the same time a biogenetic relationship between the so-called phytomelanins of the fruits of many members of Eupatorieae, Heliantheae and Helenieae and of the involucra of some members of the family and polyacetylenes, such as the penta-yne-ene (**7**), was suggested. De Vries (1948) had shown with *Helianthus* and *Tagetes* that phytomelanin production in fruits starts with the excretion of colourless, oily substances in intercellular spaces; gradually these deposits solidify and become brown to black (Fig. 4, A and B); the process resembles the polymerization of highly unsaturated acetylenes. Similar observations concerning black-coloured hairs of *Erigeron humile* Grah. (= *E. unalaschkense* [DC]. Vierh.) have been reported by Sörensen (1968); in this instance, cumulenes seem to be the precursors of the dark colouring matter. Recent reviews by two scientists who have contributed much to our chemical knowledge of polyacetylenes in the Compositae have become available (Sörensen, 1968; Bohlmann *et al.*, 1973a). Both authors mention the absence of acetylenes from Senecioneae-Senecioninae, if *Arnica* and *Doronicum* are excluded, and the rareness of these constituents in the Cichorieae. In the latter tribe, however, crepenynic acid is often present in seed oils (see pp. 294–299), and species of *Lactuca* contain appreciable amounts of acetylenes in roots. It is worth mentioning in this context that Lam and Drake (1973) showed recently that *Senecio jacobaea* L. has small amounts of several C_{17} acetylenes in the capitula, but not in the roots and leaves. Thus it seems that Senecioninae have not totally lost the capacity to accumulate acetylenic compounds. Bohlmann *et al.* (1973a, p. 502) have outlined chemical relationships between the 13 tribes of Compositae based on their acetylenic constituents.

Just as with other classes of secondary metabolites, a hierarchy can be discerned in acetylene production. Synthesis of acetylenes is a character of the family as a whole and distinct patterns may be attributes of tribes and lower systematic categories. These general trends, however, are often considerably upset by certain deviating taxa. The causes of this variation and versatility in secondary metabolism are generally unknown, but are most probably the consequences of selection. An ecological function of many polyacetylenes is suggested by the nematicidal action of compounds such as (**7**) and (**8**) (Gommers, 1973), the antibiotic properties of carlina oxide and the alexin-like behaviour of the safflower acetylenes. The fast-acting poison ichthyothereol (**9**) of *Ichthyothere terminalis* (Spreng.) Malme (Chin *et al.*, 1965) and *Clibadium silvestre* (Aubl.) Baill. (Gorinsky *et al.*, 1973) may be toxic for many organisms other than fishes.

From the systematic point of view, the fairly general occurrence of composite polyacetylenes in the Campanulaceae and Umbelliferae is worth

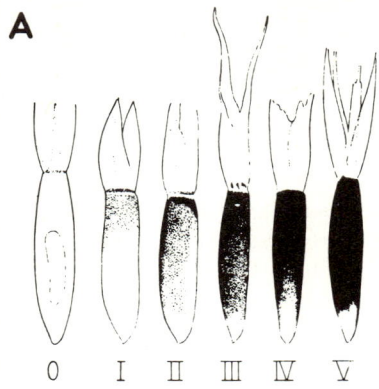

$$H_3C-(C\equiv C)_5-CH=CH_2 \qquad (7)$$

$$\text{(bithienyl)}-C\equiv C-CH=CH_2 \qquad (8)$$

$$H_3C-(C\equiv C)_3-CH=CH-\text{(tetrahydropyranyl-OH)} \qquad (9)$$

FIG. 4. Acetylenes and phytomelanins of Compositae.
A and B from de Vries (1948).
A. Different stages of fruit maturation (*Tagetes patula*). Note the gradual development of phytomelanins (V = ripe fruit).
B. Solidified dendritic masses of phytomelanins in the pericarp of ripe fruits.

mentioning. As with the sesquiterpene lactones, acetylenes form a biochemical link between composites and umbellifers.

Flavonoids

Flavonoids are universal in green land plants. As pointed out by Harborne (e.g. 1967, 1974) and other scientists, distinct patterns of flavonoid constituents often represent taxonomically valuable characters. In composites the common flavonols kaempferol and quercetin and the common flavones apigenin and luteolin are widely distributed; they are represented by several of their glycosides (Hegnauer, 1964: pp. 526–532; Harborne, 1967: pp. 226–232). The flavonol myricetin with a trihydroxylated B-ring is absent. At the flavone level, however, trihydroxylation of the B-ring occurs, tricin 5-glucoside being one of the main flavonoids of *Cirsium arvense* L. (Wallace, 1974). Some rarer flavonoids also occur in many composites and account for a somewhat distinctive flavonoid chemistry for the family. These features are:

(1) Frequent occurrence of 6-hydroxylated compounds such as quercetagetin (6-hydroxyquercetin), scutellarein (6-hydroxyapigenin) and 6-hydroxyluteolin.

(2) Methylation of flavonoids. Methylated flavonoids occur free or as glycosides. Several of these compounds have cytotoxic properties (e.g. Kupchan *et al.*, 1969; Dobberstein *et al.*, 1974). Examples of methylated flavonoids are artemitin (**10**; $R_1 = R_3 = R_4 = OCH_3$, $R_2 = CH_3$), jaceidin (**10**; $R_1 = R_4 = OCH_3$, $R_2 = H$, $R_3 = OH$), centaureidin (**10**; $R_1 = R_3 = OCH_3$, $R_2 = H$, $R_4 = OH$) (Wagner *et al.*, 1973), eupatilin (**10**; $R_1 = R_2 = H$, $R_3 = R_4 = OCH_3$), eupatorin (**10**; $R_1 = H$, $R_2 = CH_3$, $R_3 = OH$, $R_4 = OCH_3$), eupafolin (**10**; $R_1 = R_2 = H$, $R_3 = R_4 = OH$) and the tetramethoxyflavone (**10**; $R_1 = H$, $R_2 = CH_3$, $R_3 = R_4 = OCH_3$).

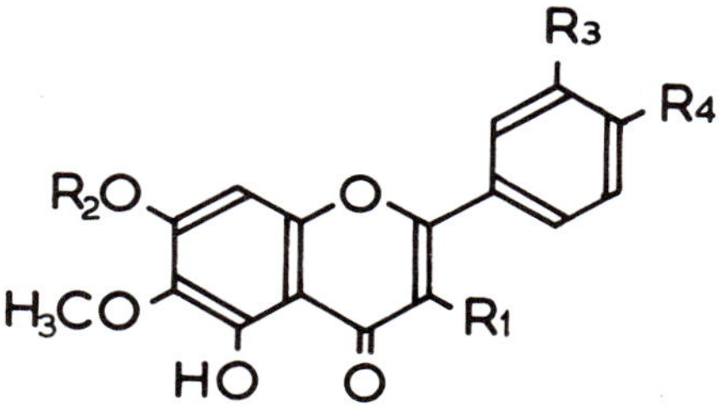

(10)

FIG. 5. Some methylated flavonoids of Compositae.

(3) The relatively frequent occurrence of chalcone glycosides. They are flower pigments in the genera *Baeria*, *Bidens*, *Carthamus*, *Coreopsis*, *Dahlia*, *Helichrysum* and *Viguiera* where they are often accompanied by the corresponding aurones (Farkas and Pallos, 1967) or flavanones. A recent addition to the Compositae chalcones is dehydro-*para*-asebotin from flowers of *Helichrysum affine* D.Don (Aritomi and Kawasaki, 1974). Chalcones and the isomeric flavanones are not restricted to flowers, however. Leaves and twigs of *Baccharis salicifolia* (R. & P.) Pers. contain sakuranetin and the 7,3'-dimethyl ether of butin which was formerly isolated from the rutaceous plant *Melicope sarcococca* (Kavak *et al.*, 1973).

Leaves of *Eupatorium odoratum* L. yield isosakuranetin and the 2-hydroxy-4′, 5′, 6′, 4-tetramethoxychalcone odoratin (Bose *et al.*, 1973). Flavanonols are rare but not totally absent. Aromadendrin (dihydrokaempferol) and its 7-methyl-ether have been isolated from leaves of *Psiadia altissima* Benth. & Hook. (Canonica *et al.*, 1969).

Unusual flavonoids in the family include the flavonolignans of *Silybum marianum* (Fig. 11; **66**), the *Helichrysum* flower pigment (Fig. 13; **96**) and the isoflavonoids wedelolactone (*Eclipta, Wedelia*), demethylwedelolactone and its 7-glucoside (*Wedelia*; Bhargava *et al.*, 1972) and rotenone (*Balduinia angustifolia* [Pursh] Robins; Lee *et al.*, 1972).

The strong tendency in Compositae to methylate flavonoids is shared by many sympetalous families and by the Rutaceae.

Carbohydrates and cyclitols

Inulin-type fructans probably occur in all composites except a few annual species (Hegnauer, 1964: pp. 536–537). For a long time fructan accumulation has been considered an indication of the close affinities between Campanulaceae and Compositae. Plouvier (1963) reviewed the distribution of readily isolable amounts of cyclitols. He noted the occurrence of the cyclohexenetetrol leucanthemitol in some species of *Chrysanthemum* and the presence of L-viburnitol, a cyclohexanepentol, in some related genera of Anthemideae (*Achillea, Chrysanthemum, Tanacetum*). The cyclohexanehexol L-inositol is known from 12 species belonging to 11 genera and seven tribes. Synthesis and accumulation of L-inositol is believed to be one of the metabolic features of the family. Scyllitol is known from *Vernonia altissima* Nutt. only. L-Quebrachitol, 2-O-methyl-L-inositol, seems to be universally present in the genus *Artemisia*; in one species, *A. dracunculus* L., it is replaced by 3-O-methyl-L-inositol (L-pinitol). Later Plouvier (1964b, 1970) confirmed the wide but sporadic occurrence of L-inositol and, incidentally, observed the general presence of appreciable amounts of potassium nitrate. Scyllitol, too, was found by Plouvier (1971a, b, 1972) to occur widely in Compositae; he isolated it from 61 species belonging to genera representing all tribes of the family. As a rule the concentrations of L-inositol or scyllitol are rather low. As L-inositol and scyllitol originate by direct epimerization of *myo*-inositol (Kindl and Hoffmann-Ostenhof, 1966, 1967), *myo*-inositol-epimerizing enzymes seem to be rather characteristic of the family; they are lacking in Campanulaceae which are able, however, to acetylate *myo*-inositol (Plouvier, 1970). During their investigations of the biogenesis of L-viburnitol in *Chrysanthemum leucanthemum* L., Kindl and Hoffmann-Ostenhof (1967) found *myo*-inositol and L-viburnitol to be the main cyclitols of this plant; they are accompanied by smaller amounts of L-inositol, L-(-)-quercitol and L-leucanthemitol. From the capitulae of the same species, Plouvier (1973) reported the isolation of D-quercitol which is strange, because optical antipodes of the same cyclitol do not generally occur together in the one plant.

Summarizing, it may be stated that inulin-type fructans and accumulation of the isomeric cyclitols L-inositol and scyllitol together form a very distinct feature of the family.

ESSENTIAL OILS

In a technical sense, volatile or essential oils are complex mixtures of steam-volatile plant constituents. Botanists connect the term essential oils with excretion. Essential oils are the volatile products deposited in dead cells (oil idioblasts), in oil cavities and ducts or in subcuticular spaces of glandular hairs. Members of the Compositae, with the exception of the Cichorieae, have secreting glandular hairs and schizogenous ducts. Many of them are aromatic plants and yield appreciable amounts of essential oil. Volatile constituents other than monoterpenoids and sesquiterpenoids are mentioned elsewhere notwithstanding the fact that some of them may be major components of a given essential oil (especially some acetylenes, phenylpropanoids and chromenes). During the last decade, essential oils of Compositae have received much chemical attention. The present discussion is restricted to some apparently taxon-linked trends in essential oil biosynthesis (Fig. 6).

The phenolic monoterpenoid thymol (**11**) and a large number of thymol derivatives (e.g. **12-14**), both free and esterified, are common constituents of certain taxa, especially of species of *Eupatorium* (Eupatorieae), *Inula* (Inuleae) and *Helenium* and *Gaillardia* (Helenieae) (Bohlmann et al., 1969a; Anthonsen and Kjösen, 1971; Shtacher and Kashman, 1971; Romo de Vivar et al., 1971; Gommers, 1973). p-Menth-3-ene-3, 6-diol isolated from *Eupatorium macrocephalum* Less. (Gonzalez et al., 1972) could be a precursor of thymol and/or p-thymohydroquinone. Compound (**13**), a thymol-derived dihydrobenzofuranoid of species of *Helenium*, has strong nematicidal properties (Gommers, 1973). Other nematicidal metabolites of Compositae such as (**7**) and (**8**) (Gommers, 1973) belong to the polyacetylenes. Species of *Arnica* (Rinn, 1970; Bohlmann and Zdero, 1972d) and *Doronicum* (Bohlmann and Zdero, 1970) produce thymol derivatives (**14**) together with polyacetylenes; senecionine-type alkaloids are lacking and pseudoguaianolide-type sesquiterpene lactones are present. These genera are out of place in Senecioneae for many reasons. Irregular monoterpenoids having the artemisyl, santolinyl or chrysanthemyl skeleton seem to be common in Anthemideae, but have also been reported from Astereae (*Cyathocline*) (Epstein and Poulter, 1973).

Furanosesquiterpenes, such as 12-acetoxy-10, 11-dehydrongaione (**15**), have been isolated from species of the South African genera *Asaemia*, *Athanasia*, *Eumorphia*, *Lasiospermum*, *Phymaspermum*, *Stilpnophytum* and *Ursinia*, all classified in Astereae (Bornowski 1971; Bohlmann and Zdero

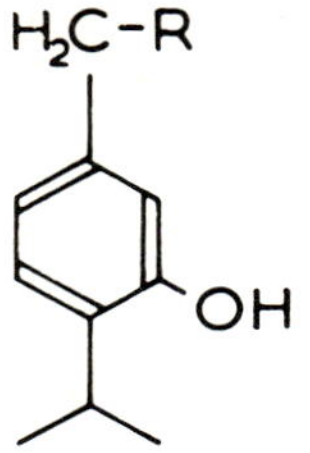

Fig. 6. Some components of essential oils of Compositae.

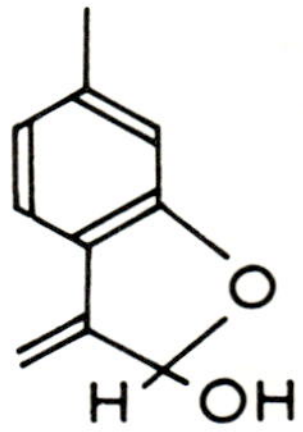

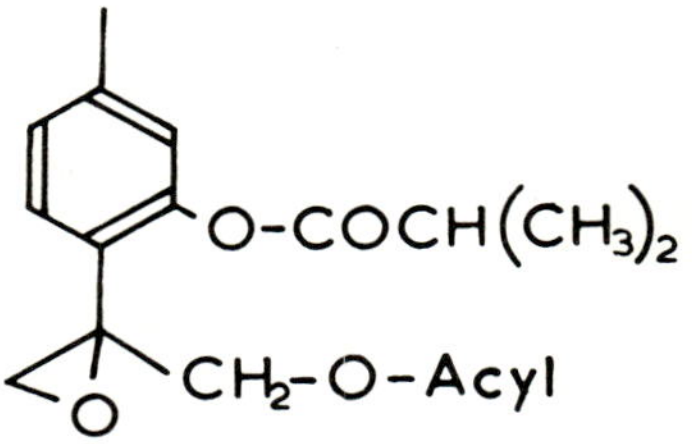

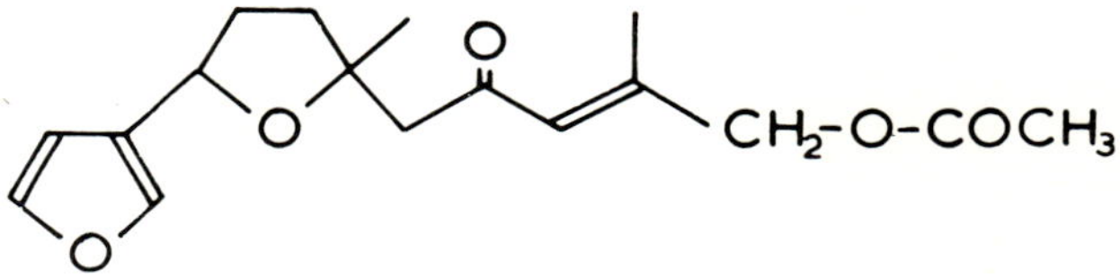

1974b). The *para*-quinonoid sesquiterpenoids perezone and hydroxyperezone and the pipitzols should also be mentioned; they occur in the related genera *Perezia* and *Trixis* (Mutisieae) (Thomson, 1971). An irregular phenolic sesquiterpene is elvirol from *Elvira biflora* DC. (Heliantheae) (Bohlmann and Körnig, 1974). It is thus apparent that biosynthetic trends with regard to essential oil production often represent valuable characters at subgeneric, generic, subtribal or tribal levels.

DITERPENES

"Diterpene sind in der Familie vermutlich verbreitet; sie wurden bisher jedoch verhältnismässig wenig bearbeitet" (Hegnauer 1964: p. 478). Royleanone and lycoctonine (*Inula royleana* DC.), the grindelic acid group (*Grindelia* spp.), darutigenol (*Siegesbeckia orientalis* L.), stevioside (*Stevia rebaudiana* Bertoni) and atractyloside (*Atractylis gummifera* L.)

were the only well characterized diterpenoids to be reported some 10 years ago. They are dicyclic (grindelic acid group), tricyclic (royleanone, darutigenol) or tetracyclic (steviol, atractyligenine, lycoctonine) compounds representing different classes and groups or series (Hanson, 1968; Ourisson, 1974) of diterpenoids.

In the meantime much work has been performed with diterpene-accumulating composites. The presently known diterpenoids of the family belong to four main classes, some of them being represented by two groups. This grouping is based on the carbon skeletons only and ignores structural and stereochemical details (Fig. 7).

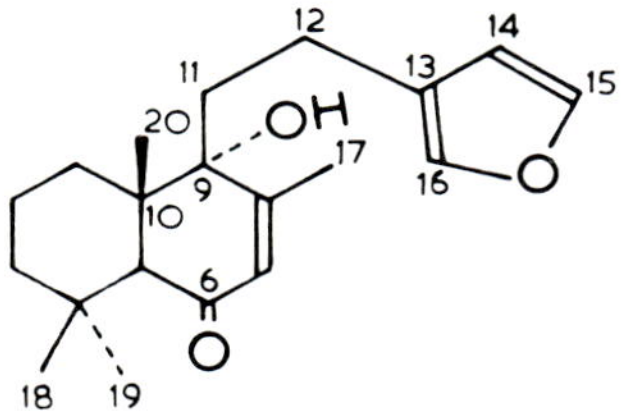

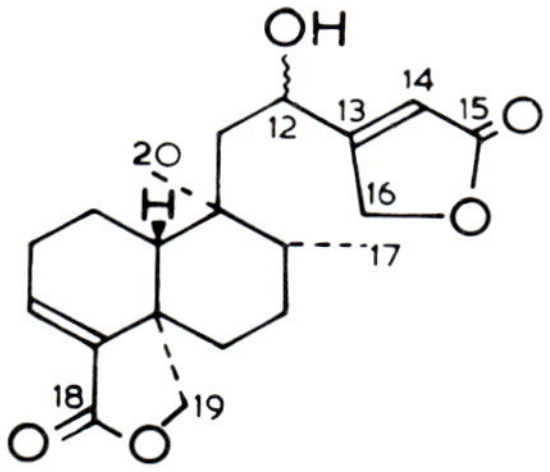

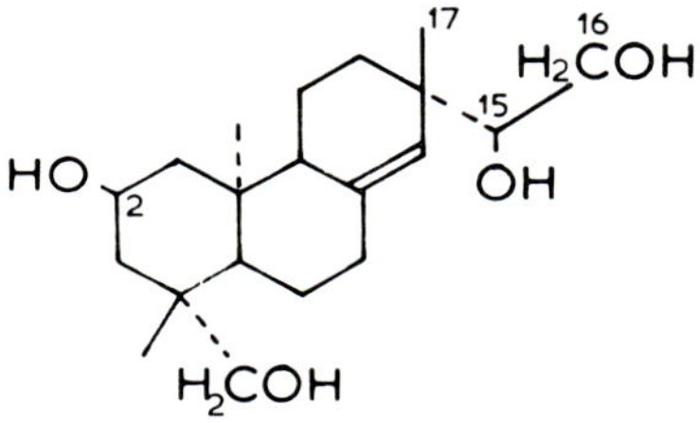

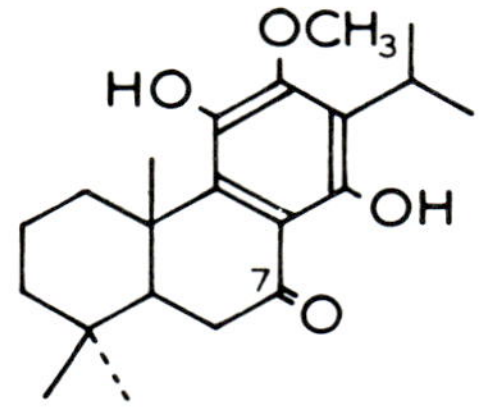

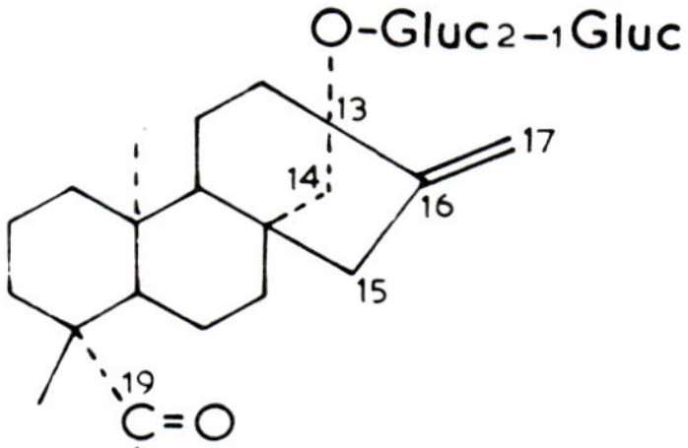

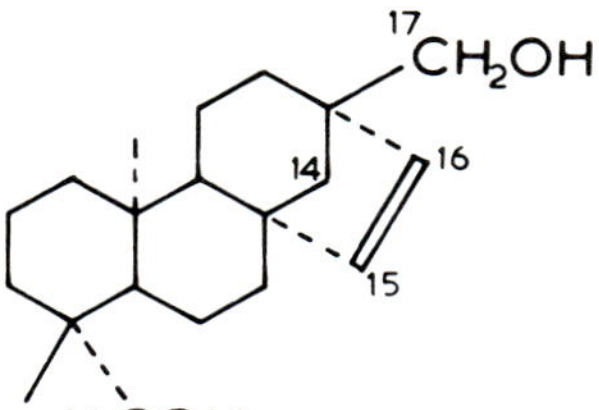

FIG. 7. Some diterpenoids of Compositae.

I. *Bicyclic diterpenoids*
(a) *Regular*: The labdane-manooloxide type (**16**); represented in the family by the "normal" (**16**) and *ent*(enantio) series.
(b) *Rearranged*: Clerodane-type (**17**); *cis*- and *trans* (**17**)-clerodane derivatives occur in composites.

II. *Tricyclic diterpenoids*
(a) Pimarane-type (**18**); occurring in composites as derivatives of sandracopimarene ("normal") and of (−)-pimarene (*ent*-series) (**18**).
(b) Abietane-type (**19**).

III. *Tetracyclic diterpenoids*
(a) *Kaurane-type* (**20**); represented only by (−)-kaurene derivatives (*ent*-series).
(b) *Stachane-type* (**21**); represented only by (+)-stachene derivatives (*ent*-series).

IV. *Pentacyclic diterpenoids*
Trachylobane-type. The combinations of diterpenoids belonging to one or several classes and groups mentioned with special structural features, such as the furanoid (**16**), butenolide (**17**), lactone (**17**) or phenolic (**19**) systems, often seem to represent taxon-characteristic trends within the family. This is illustrated by the following review which by no means pretends to be exhaustive.

EUPATORIEAE. Stevioside (**20**), an extremely sweet-tasting glycoside of *Stevia rebaudiana*, still seems to be the sole well-characterized diterpenoid of this tribe. It is present in the plant in large amounts (Mitsuhashi *et al.*, 1975) and has been shown to have contraceptive properties (Mazzei Planas and Kuć, 1968).

ASTEREAE. Accumulation of diterpenoids seems to be frequent in this tribe. *Solidago* is the genus of composites most intensively investigated for diterpene constituents.
Subtribe Solidagininae. *Grindelia robusta* Nutt. produces a resin covering the young anthodia which contains grindelic acid and several derivatives; grindelic acid is a labdane-type diterpenoid with an oxygen bridge between C_9 and C_{13}. *Gymnosperma glutinosum* (Spr.) Less. produces a labdanetriol called gymnospermin (Miyakado *et al.*, 1974) and *Gutierrezia dracunculoides* (DC.) Blake yields the dilactone gutierolide; one of the lactonoid systems is a butenolide ring; gutierolide is a 3-chloro-*cis*-clerodane derivative (Cruse *et al.*, 1971). The taxonomically notoriously difficult genus *Haplopappus* is the source of haplopappic acid, a *cis*-clerodane dicarboxylic acid (Silva and Sammes, 1973: *H. foliosus* DC. and

H. angustifolius?) and of juslimtetrol and norjuslimdiolon (a C_{19} compound), two pimarane-type diterpenoids (Dominguez and Jiménez, 1973: *H. divaricatus* [Nutt.] Gray = *Croptilon divaricatum* [Nutt.] Raf.). *Solidago altissima* L. contains several *trans*-clerodane-type diterpenoids like kolavenol, kolavenic acid, 6-oxokolavenic acid (= altissimic acid) and the angelate of 6-hydroxykolavenic acid in its roots; the main bitter principle of the roots, solidagonic acid, is 7-acetoxykolavenic acid (Kusumoto *et al.*, 1968, 1969; Ohsuka *et al.*, 1972). *S. arguta* Ait. has *cis*- and *trans*-clerodane-type furanoid diterpenoids in roots (Anderson *et al.*, 1974; Ferguson *et al.*, 1975). *S. canadensis* L. yields solidagenone (**16**) and a mixture of epimeric 9-13-oxygen-bridged dihydrofuranoid solidagenone derivatives (Anthonsen *et al.*, 1967, 1969). Roots of *S. elongata* Nutt., a species closely related to *S. canadensis*, has a series of *trans*-clerodane-type diterpenoids, such as kolavenol, kolavenic acid, 6-acetoxykolavenic acid, 6-angeloyloxykolavenic acid and the elongatolides A–E which contain a butenolide ring (Anthonsen and McCrindle, 1969). *S. gigantea* Ait. var. *serotina* (Kuntze) Cronq. (= *S. serotina* Ait.) has a very complex mixture of acidic and non-acidic labdane-type diterpenoids, many of which possess a furan ring; solidagoic acid B has an angeloyloxy group at C_{18} (Anthonsen *et al.*, 1968; Henderson *et al.*, 1973a). *S. juncea* Ait. contains the abietene-type junceanols X, Y and Z, the *trans*-clerodane-type diterpenoids, junceic acid, its epoxide and hardwickic acid and *ent*-16-kauren-19-oic acid (Henderson *et al.*, 1973b). *S. missouriensis* Nutt. has a mixture of *ent*-labdane-type compounds (13-epi-*enantio*-manooloxide and its 3-oxo-derivative) and the missourienols which are derivatives of abiet-9(14)-en-13α-ol (Anthonsen and Bergland, 1970, 1973). On chemical examination, *S. rigida* L. yielded only *enantio*-kaur-16-en-19-oic acid and *S. shortii* Torr. & Gray had two elongatolide-type diterpenoids in its roots. No diterpenoids, but only acetylenic compounds, could be detected in roots of *S. flexicaulis* L. and *S. virgaurea* L. (Anthonsen and Bergland, 1971). These results indicate that the genus *Solidago* is very versatile in its diterpene metabolism.

Subtribe *Asterinae*. Three furanoid *trans*-clerodane-type diterpenoids occur in aerial parts of *Hinterhubera imbricata* Cuatr. (Bohlmann *et al.*, 1973b). Olearin (**17**), a dilactone of the *trans*-clerodane series, is present in *Olearia heterocarpa* S. T. Blake; one of its five membered lactone rings is α, β-unsaturated (= butenolide ring) (Pinhey *et al.*, 1971). A *trans*-clerodane-type dihydroxy acid having a furan ring has been isolated from *O. muelleri* (Sond.) Benth. (Jefferies *et al.*, 1974).

Subtribe *Conyzinae*. Furanoid *enantio*-labdane-type diterpenoids occur free and as fatty acid esters in leaves of *Psiadia altissima* Benth. & Hook. (Canonica *et al.*, 1967, 1969) and *Conyza ivaefolia* Less. yields hautriwaic acid and its lactone, two *trans*-clerodane-type diterpenoids with a furan ring (Bohlmann and Grenz, 1972).

Subtribe *Baccharidinae*. Bacchofertine, a furanoid diterpene of the *trans*-

clerodane series, was isolated from *Baccharis conferta* H.B.K. (Guerro and de Vivar, 1973).

INULEAE. Inuroyleanol (**19**) is a recent addition to the diterpenoids of *Inula royleana* DC. which belong to the abietane group (Bhat *et al.*, 1975). 16α,17-Dihydroxy-16β-(−)-kauran-19-oic acid was isolated from *Helichrysum diosmifolium* (Vent.) Sweet by Lassak and Pinhey (1968) and *H. dendroideum* N. H. Wakefield contains a series of kaurene-type and stachene-type diols; 17-hydroxymonogynol (**21**) is one of the stachene derivatives occurring in this plant (Lloyd and Fales, 1967).

HELIANTHEAE. Diterpenoids of this tribe seem to belong mainly to the (−)-kaurene series; they may be accompanied by pentacyclic trachylobane-type compounds or by (−)-pimarene derivatives.

Subtribe Melampodinae. (−)-Kaurene derivatives occur widely in the genus *Espeletia*. Grandiflorolic acid, grandiflorenic acid, kaurene, kaurenol, kaurenal and kaurenoic acid were isolated from *E. grandiflora* Humb. & Bonpl. (Piozzi *et al.*, 1968, 1971). Roots (Bohlmann and Rao, 1973) and aerial parts (Brieskorn and Pöhlmann, 1968; Kloss, 1969) of *E. schultzii* Wedd. yielded a series of similar kaurene derivatives. Subterranean parts of *E. tenorae* Aristeg. (Usubillaga and Morales Méndez, 1970) and of *E. weddellii* Sch. Bip. (Morales Méndez *et al.*, 1973) yielded similar kaurenoid compounds. Cunabic acid isolated from *Ichthyothere* (=*Icthyothere*) *cunabi* Mart. was shown to be (−)-kaur-16-en-19-oic acid by Leonico d'Albuquerque *et al.* (1969).

Subtribe Verbesininae. Kaurene-type diterpenoids were reported from *Enhydra fluctuans* Lour. (Pakrashi *et al.*, 1970), *Helianthus annuus* L. (Pyrek, 1970; trachyloban-19-oic acid is present also), *H. ciliaris* DC. (Bjeldanes and Geissman, 1972; 7-hydroxytrachylobanoic acid [=ciliaric acid] also), *Montanoa tomentosa* Cerv. (Caballero and Walls, 1970), *Siegesbeckia pubescens* Makino (Canonica *et al.*, 1969; Murakami *et al.*, 1973) and *Viguiera stenoloba* Blake (Cuevas *et al.*, 1972). In the genus *Siegesbeckia* kaurenoids are accompanied by (−)-pimarene derivatives, such as darutigenol of *S. orientalis* and kirenol (**18**) of *S. pubescens* (Murakami *et al.*, 1973).

ANTHEMIDEAE. Two kaurene derivatives were found in *Abrotanella nivigena* F. Muell., but *A. forsterioides* (Hook. f.) Benth. did not yield any diterpenoids (Anthonsen and Chantharasakul, 1971).

CALENDULEAE. Roots of *Garuleum bipinnatum* Less., *G. pinnatifidum* DC., *Osteospermum fruticosum* (L.) Norl., *O. junceum* Berg and *O. oppositifolium* Ait. yielded sandaracopimar-15-en-8β-ol and some derivatives of this diterpenoid of the "normal" pimarane series H (Bohlmann *et al.*, 1973c).

CARDUEAE. *Atractylis gummifera* L. is still the only member of this tribe with well characterized diterpenoids. The strange toxic glycoside atractyloside (a nor-kaurene derivative) is accompanied by its C_{20} progenitor in the plant (Danieli *et al.*, 1972).

The function of diterpenoids is still unknown. The growth-regulating gibberellins, of course, are essential to land plants and occur universally; they are present, however, in trace amounts only. Diterpenoids which are major constituents most probably fulfil different ecological functions, depending on their structure and localization. Such functions have been demonstrated convincingly for the labdane-type sclareols of *Nicotiana glutinosa* which prevent rust diseases (Bailey *et al.*, 1975) and for a series of bitter furanoid clerodane-type compounds which possess strong anti-feeding properties against insects (Hosozawa *et al.*, 1974).

Summarizing the known facts about diterpenoids of Compositae, some trends which may become valuable to plant taxonomy in the future may be mentioned.

(1) Diterpenoids are more or less common in some tribes (Astereae, Calenduleae) and extremely rare or totally lacking in others (e.g. Cynareae, Cichorieae).

(2) (−)-Kaurene derivatives predominate in Heliantheae and sandaracopimarene derivatives in Calenduleae. Astereae on the other hand are astonishingly versatile in their diterpenoid metabolism. Within this tribe diterpenoids are often characters at specific and generic levels.

(3) Diterpenoids occur predominantly free, esterified or glycosylated, depending on the taxa concerned and on the plant parts where accumulation takes place.

(4) The strong tendency of some Astereae taxa to produce furanoid and lactonoid diterpenoids may cause misidentifications during plant screening for pharmacologically active constituents. Lactones of the butenolide type could be confused with cardenolides; this has possibly happened with those species of *Vernonia* which were reported as containing cardenolides (Patel and Rowson, 1964).

AMIDES (FIG. 8)

In 1964 isobutyl amides of long-chain fatty acids with characteristic olefinic and acetylenic unsaturation patterns were known from members of Anthemideae and Heliantheae. Many of these constituents have a pungent taste and possess insecticidal properties. Examples are pellitorin from roots, leaves and anthodia of *Anacyclus pyrethrum* DC., spilanthol (=affinin) from anthodia of *Spilanthes oleracea* Jacq. and from roots of *Heliopsis longipes* (A. Gray) Blake, scabrin from roots of *Heliopsis scabra* Dunal, heliopsin from roots of *H. scabra* and echinacein from roots of *Echinacea angustifolia* DC. and *E. pallida* (Nutt.) Britton. Echinacein (**23**+**27**) was shown to be identical with the insecticidal amide neoherculin (=α-sanshool) from members of the rutaceous *Fagara-Zanthoxylum* complex. These insecticidal amides are generally present in plants as complex mixtures of isomers and homologues. They may be accompanied by physiologically more or less inert but closely related amides. Therefore some discrepancies concerning structures of individual compounds still exist in the literature (Jacobson and Crosby, 1971; Bohlmann et al., 1973a). The active isobutylamides have C_{10}, C_{11}, C_{12}, C_{14} or C_{18} polyunsaturated fatty acids. A *trans*-double bond in position 2 and a dimethylene interruption between the two systems of unsaturation seems to be essential for pungency and insecticidal activity. An example of a relatively inactive compound is anacyclin (**22 + 27**) which occurs together with pellitorin in roots of *Anacyclus pyrethrum*. According to Bohlmann et al., (1973a) it is isobutyltetradeca-2,4-diene-8,10-diynamide; partial reduction of the two acetylene linkages results in a highly active compound. Burden and Crombie (1969) showed that isobutylamides are accompanied in *Anacyclus* roots by the corresponding tyramine amides of *trans*-2, *trans*-4–dienoic acids and Bohlmann and collaborators (Bohlmann et al., 1973a, 1974b; Rente et al., 1972) have detected many new amides during the past 5 years. They showed that such amides are restricted to some related genera of Heliantheae and Anthemideae and that phenylethylamine, piperidine and piperideine also occur as amides in these taxa. The amines seem to be derived from the amino acids valine (isobutylamine), tyrosine (tyramine), phenylalanine (phenylethylamine), lysine (piperideine) and pipecolic acid (piperidine). The biosynthesis of the polyunsaturated acids seems to be connected with the production of the acetylenes which are highly characteristic constituents of Compositae. Oleic, linoleic and crepenynic acid are the overall precursors. By β-oxidation the carbon chain is shortened starting with the carboxyl end. Another type of chain shortening can take place at the methyl end. It is preceded by the introduction of an additional acetylene linkage in position 14 of crepenynic acid. An undecapentadienoic acid containing a thiophen ring is otanthic acid (**25**) which occurs as the isobutylamide in *Otanthus maritimus* (L.) Hoffmsgg. & Link (= *Diotis maritima* Smith) and *Anacyclus tomentosus* DC. A corresponding decatetraenoic acid (**26**) is present in the insecticidal isobutylamide fraction of some species of *Chrysanthemum* which lack cinerins and pyrethrins (Romo de Vivar et al., 1974).

Amides of the types mentioned are so far known to occur in the following taxa.

HELIANTHEAE: *Echinacea* (2–3 taxa); *Heliopsis* (3–4 taxa); *Spilanthes* (3–4 taxa).

ANTHEMIDEAE: *Achillea* (6 taxa); *Chrysanthemum* (4 species); *Anacyclus* (3 taxa); *Anthemis* (2 species of section Ormenis); *Otanthus maritimus*; *Cladanthus arabicus* Cass.

10. THE CHEMISTRY OF THE COMPOSITAE 301

(22)  C 14

(23)  C 12

(24)  C 11

(25) 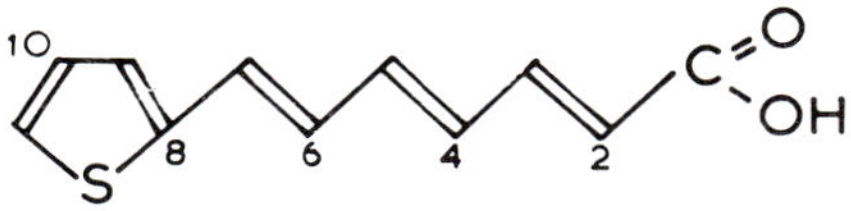 C 11

(26) 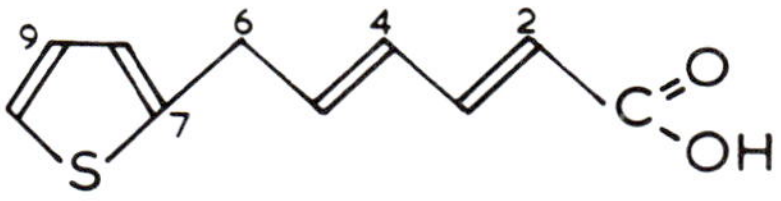 C 10

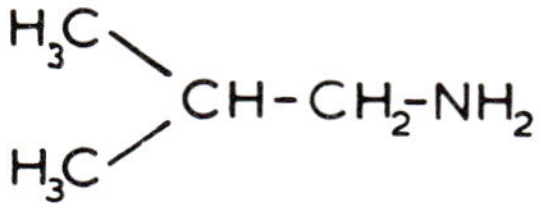

(27)

FIG. 8. Some unsaturated acids and a common amine occurring in insecticidal amides of Compositae.

These amides form a biochemical link between Heliantheae and Anthemideae. It seems worth mentioning that similar constituents occur in Rutaceae (*Fagara-Zanthoxylum*). Isobutylamides of *trans*-2, *trans*-4-dienoic fatty acids, however, are also present in some species of *Piper* (Piperaceae).

An amide was recently isolated from *Senecio rivularis* DC. and shown to be 3-methyl-penta-3-ene amide (Klásek *et al.*, 1969). Biogenetically this amide seems to be connected with the metabolism of necic acids (see next section) rather than with the amides of Heliantheae and Anthemideae.

ALKALOIDS AND ALKALOID-LIKE COMPOUNDS (FIG. 9)

Alkaloid-bearing plants have long been known in the Compositae but structural investigations have been restricted mainly to some notoriously toxic or medicinally useful plants. Alkaloid chemistry of Compositae was summarized by Hegnauer (1964, 1966). Senecioneae (*Cacalia, Erechtites, Kleinia, Nardosmia* [=*Petasites* p.p.], *Senecio*) producing pyrrolizidine-type ester alkaloids, Heliantheae (*Eclipta, Zinnia*) synthesizing nicotine, nornicotine and anabasine and the echinopsine-producing genus *Echinops* (Cardueae) can all be mentioned as taxa with well characterized alkaloids. From other taxa alkaloid-like substances such as the diterpene alkaloids from *Inula royleana* DC. (Inuleae) and betaines (betaine, trigonelline, chrysanthemine [=stachydrine], achillein [=betonicine] and moschatine [=betaine of pipecolic acid]) have been reported. A hypotensive pseudo-alkaloid from *Carduus acanthoides* L. was described in 1962 by Frydman and Deulofeu and the first observation of the occurrence of pyrrolizidine alkaloids in a composite not classified in Senecioneae dates back to 1963; *Eupatorium maculatum* L., a species toxic to cattle, was shown to contain echinatine and trachelanthimidine, two alkaloids characteristic of Boraginaceae rather than of Senecioneae. In general, the Compositae was considered as a group in which alkaloids are not rare. At the same time available evidence suggested that alkaloid patterns are characteristic of species, genera or tribes rather than of the family as a whole.

More recently, new occurrences of alkaloids have been reported in the family (Willaman and Li, 1970), but in most instances structural investigations have not yet been performed. In fact, from the taxonomic point of view, no progress has been made. None of the many species of Vernonieae, Astereae, Inuleae (except *Inula royleana* and *Gaillardia pulchella*), Heliantheae (except *Eclipta alba* Hassk. and *Zinnia elegans* Jacq.), Cardueae (except *Carduus acanthoides* and the genus *Echinops*) and Cichorieae reported as alkaloid-bearing has been studied thoroughly. Most attention has been paid to the hepatotoxic and carcinogenic pyrrolizidine alkaloids of Senecioneae and Eupatorieae and to the quinoline alkaloids of the genus *Echinops*. The work on pyrrolizidine alkaloids has been summarized by Bull *et al.* (1968). At present such alkaloids are known to occur also in *Brachyglottis repanda* J. R. & G. Forst. (Mortimer and White, 1967), *Cacalia hastata* L. subsp. *orientalis* Kitamura (Hayashi *et al.*, 1972), *C. floridana* A. Gray (=*Gymnostyles floridana* Greene) (Cava *et al.*, 1968), *Emilia sonchifolia* DC., *E. flammea* Cass. (Kohlmünzer *et al.*, 1969, 1971), *Farfugium japonicum* Kitamura (Furuya *et al.*, 1971), several species of *Ligularia* (Klásek *et al.*, 1971), *Petasites hybridus* (L.) G.M. Sch. (Bull *et al.*, 1968), *Syneilesis palmata* Maxim. (Hikichi and Furuya, 1974) and *Tussilago farfara* (Bull *et al.*, 1968). The more widespread occurrence of pyrrolizidine alkaloids in the genus *Eupatorium* has been demonstrated also. *Eupatorium serotinum* Michx. contains supinine (**29**) and rinderine (**30**) (Locock *et al.*, 1966) and *E. fortunei* Turcz. (=*E. stoechadorum* Hance), a species occurring in Japan and China, contains supinine and lindelofine (Furuya and Hikichi, 1973). Pyrrolizidine alkaloids seem to be more or less universal in Senecioneae and may be considered as one of their outstanding biochemical features. Senecionine (**28**) appears to be the most common alkaloid of this tribe; it is a diester of retronecine with senecic acid (a C_{10} necic acid). Pyrrolizidine alkaloids form a biochemical link between

Senecioneae and Eupatorieae. It is worthwhile, however, to mention that all alkaloids isolated hitherto from species of *Eupatorium* are characteristic of Boraginaceae rather than of Senecioneae; they are monoesters of C_7 necic acids with the necines retronecine, heliotridine, isoretronecanol, trachelanthamidine or supinidine.

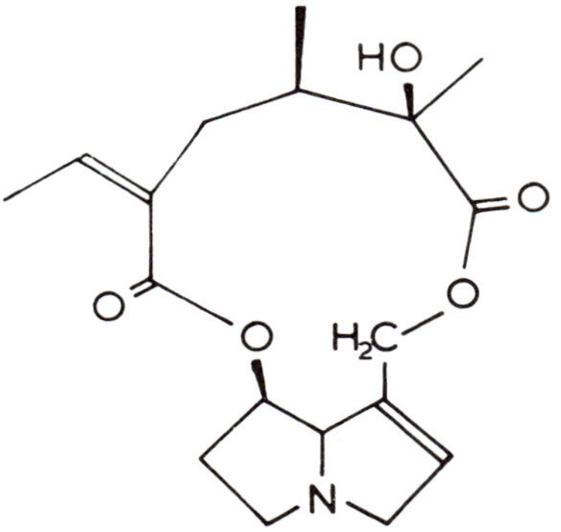

(28)

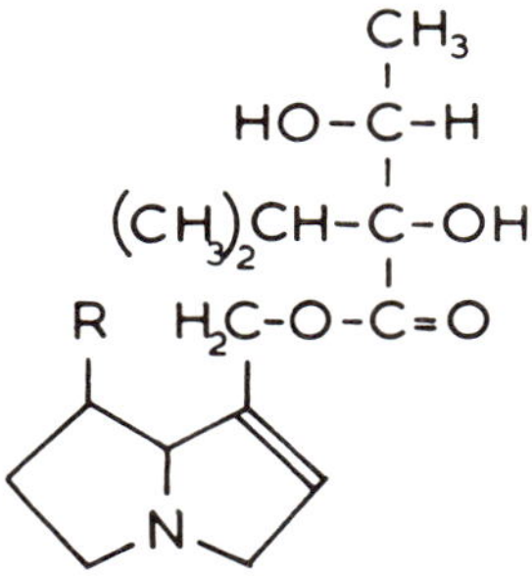

R=H (29)
R=OH (30)

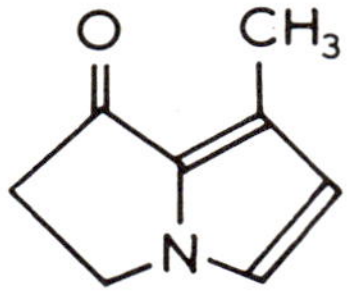

(31)

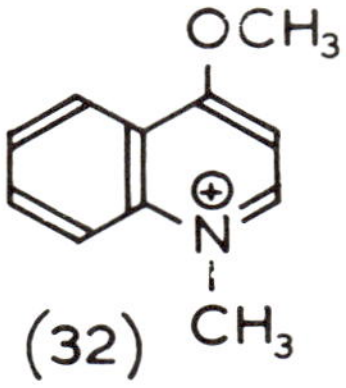

(32)

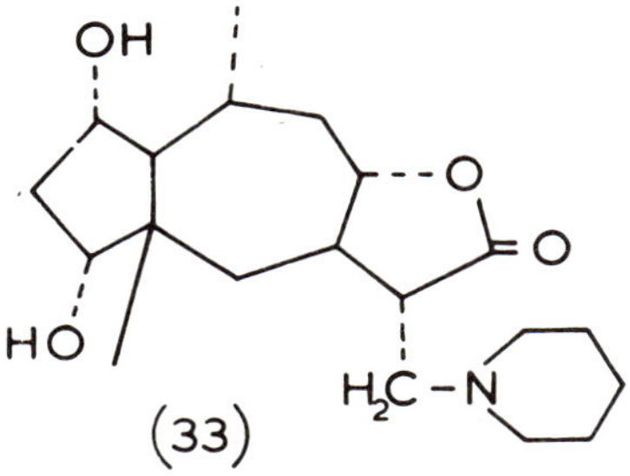

(33)

FIG. 9. Some alkaloids and alkaloid-like constituents of Compositae.

Progress with other alkaloids and alkaloid-like substances of Compositae is rather scanty. The alkaloids of *Echinops*, echinopsine and echinopsidine were shown to be artefacts; the genuine alkaloid is echinorine (**32**). Its biosynthesis has been studied: like the quinoline alkaloids of Rutaceae, its skeleton seems to be derived from anthranilic acid and acetate (Luckner, 1969). A dihydroquinoline alkaloid, echinine, is present also in seeds of *Echinops ritro* L. (Döpke and Fritsch, 1969). A new alkaloid was isolated from *Gaillardia pulchella* Foug. (Helenieae) and identified as a derivative of the pseudoguaianolide pulchelline (Yamagita *et al.*, 1969, 1970). Like the alkaloids of *Inula royleana*, pulchellidine (**33**) is a terpenoid rather than a true alkaloid. Similar nitrogen-containing sesquiterpenes and diterpenes may perhaps occur more frequently in Compositae than is indicated by the presently known facts; some of them might be artefacts of isolation. An "alkaloid" isolated from *Pluchea lanceolata* L. (Inuleae) and named pluchine was later shown to be betaine (Dasgupta *et al.*, 1968). It is highly probable that several of the still uncharacterized "alkaloids" of Compositae in future will prove to be betaines.

A series of observations with danaid butterflies (Edgar and Culvenor, 1974, 1975; Edgar *et al.*, 1974) merit mention in this context. Many danaid species have evolved very peculiar behaviour patterns. Caterpillars feed on cardenolide-containing apocynaceous and asclepiadaceous plants, and the heart poisons of the host plants are accumulated and transmitted to the imagines during metamorphosis. Caterpillars and butterflies use cardenolides in defence against predators. The male butterflies of many of these species produce courtship pheromones containing dihydropyrrolizine bases (**31**). For pheromone production they depend on plants producing pyrrolizidine alkaloids; adult feeding of males on Boraginaceae, *Senecio*, *Eupatorium* and *Crotalaria* has been observed. This strange dependence of many danaid species on several apparently unrelated plant groups was interpreted in evolutionary terms. The original food plants of these butterflies are assumed to have been the ancestors of present-day Apocynaceae and to have produced both cardenolides and pyrrolizidine alkaloids. Severe predation by insects which had evolved a dependence on both these types of secondary metabolites resulted in divergent evolution of the plant group concerned. In some of the evolutionary lines, cardenolide production was preserved and in others the ability to synthesize pyrrolizidine alkaloids was retained. The present-day behaviour of danaid butterflies becomes understandable by assuming that they preserved their defensive device by continuing larval feeding on the cardenolide-line of the original food plants and evolved adult feeding in males for obtaining the active constituents of their courtship pheromones. All presently known facts seem to fit in with this view. This hypothesis also assumes a common origin of present-day Boraginaceae and Compositae from apocynacean-like ancestors. The search for a plant still producing both cardenolides and pyrrolizidine alkaloids resulted in the detection of the boraginaceous alkaloids lycopsamine and intermedine (or indicine) (**30**) in two Australian species of *Parsonsia*, a genus belonging to the cardenolide-producing subfamily Echitoideae of Apocynaceae! Insect behaviour, in this instance, seems to provide a clue for plant relationships; it suggests a more or less direct evolutionary link between apocynacean-like ancestors and present-day Boraginaceae and Compositae. It should be stressed, however, that primary and secondary metabolism of composites still favours close affinities with Umbelliferae rather than with Apocynaceae and Boraginaceae. Conformities with the latter are still restricted to fructan production, necine alkaloids and mass accumulation of caffeic acid derivatives (Hegnauer, 1964: p. 543).

CYANOGENIC GLYCOSIDES

Cyanogenic glycosides are erratically distributed among vascular plants. Most of the presently known glycosides are biosynthetically derived from one or other of the protein amino acids (Conn, 1973; Seigler, 1975). On the basis of these amino acids, several biogenetic pathways may be discerned; the latter are more meaningful as taxonomic characters than is the mere release of hydrocyanic acid (Hegnauer, 1973a).

Cyanogenic plants are known from six tribes of Compositae but only from three of them is evidence available concerning the structures of the cyanophoric constituents.

ASTEREAE: *Calotis scapigera* Hook. in Mitch.; immature plants are cyanophoric (Hurst, 1942).

HELIANTHEAE: *Acanthospermum australe* O. Kuntze and *A. hispidum* DC., two introduced weeds of South Africa, are strongly cyanogenic (Walt, 1944).

HELENIEAE: In addition to *Florestina pedata* Cass. and *Hymenoxys tweedii* Hook. reported long ago as cyanogenic taxa, *Picradeniopsis oppositifolia* (Nutt.) Rydb. (Deem *et al.*, 1939) and *Florestina tripteris* DC. (Boughton and Hardy, 1939) were identified as cyanophoric plants which can cause the intoxification of cattle.

ANTHEMIDEAE: Fruits of species of *Anacyclus* and *Anthemis* and the aerial parts of *Achillea millefolium* L., *A. pseudo-pectinata* Janka and *Chrysanthemum leucanthemum* L. have been known for a long time to be very weakly to strongly cyanophoric. In several instances the release of benzaldehyde during glycoside hydrolysis was demonstrated; this indicates the presence of prunasin-type glycosides. Greshoff isolated an amygdalin-like glycoside from the fruits of two species of *Anthemis*. Two of the taxa mentioned, *Achillea millefolium* and *Chrysanthemum leucanthemum*, are in fact polyploid complexes. In our experience the *A. millefolium* aggregate is polytypic and polymorphic with respect to cyanogenesis. We have never found any cyanophoric plants in its component taxa *A. roseo-alba* Ehrend. (diploid) and *A. stricta* Schleicher (hexaploid). On the other hand, *A. millefolium* L. sensu stricto (hexaploid) and *A. collina* Becker (tetraploid) are more or less cyanophoric in some regions and acyanophoric in others. Leaves of *Achillea macrophylla* L., which does not belong to the *A. millefolium* aggregate, are rather strongly cyanophoric in our experience; they contain prunasin.

CALENDULEAE: Most composites which are strongly cyanogenic belong to this predominantly Southern African tribe. Formerly all taxa concerned were included in *Dimorphotheca*, but are now classified in several genera: *Castalis, Dimorphotheca sensu stricto, Osteospermum*. Linamarin was isolated from *Castalis spectabilis* T. Norl., *Dimorphotheca cuneata* Less., *D. zeyheri* Sond., *Osteospermum ecklonis* T. Norl. and *O. fruticosum* T. Norl. Other taxa of this aggregate were reported to be toxic and to release large amounts of hydrocyanic acid. Evidently the Calenduleae use another pathway to produce cyanogenic glycosides than that used by the Anthemideae or the Cardueae. Butler (1965) showed that *Dimorphotheca barberiae* Harv. and *Osteospermum jucundum* T. Norl. also contain much linamarin and that the latter is accompanied only by trace amounts of lotaustralin in these taxa. In this respect Calenduleae seem to differ from most other linamarin-producing taxa (e.g. species of *Linum*; species of *Trifolium* and *Lotus*).

CARDUEAE: *Centaurea aspera* L. (aerial parts), *C. crocodylium* L. (aerial parts), *Chardinia orientalis* (Mill.) O. Kuntze (= *Ch. xeranthemoides* Desf.) (fruits), *Xeranthemum annuum* L. (fruits) and *X. foetidum* Moench (= *X. cylindraceum* Sibth. & Sm.) (fruits) are known to be strongly cyanophoric and to release benzaldehyde together with hydrocyanic acid. Like members of the Anthemideae, these plants contain prunasin-type glycosides. Some other species of the same tribe (*Centaurea montana* L., *C. solstitialis* L., *Cirsium arvense* L., *Saussurea candicans* DC. and *Xeranthemum inapertum* [L.] Willd. [aerial parts]) were reported in the literature to be weakly cyanophoric. Some of these reports could not be confirmed in later work. We have had some personal experience with representatives of the *Centaurea scabiosa* L. aggregate and with species of *Xeranthemum*. In Switzerland some *C. scabiosa* plants are cyanophoric and some plants are acyanophoric. Hence polymorphism and polytypism with regard to cyanogenesis may be the cause of the above discrepancies. *Xeranthemum* species have strongly cyanophoric fruits; young plants, however, are acyanophoric according to our observations; in this instance, discrepancies may be due to the localization of the cyanogenic glycosides in certain tissues. Available evidence indicates that Anthemideae and Cynareae strongly resemble each other with regard to cyanogenesis and cyanogenic constituents.

COUMARINS (FIG. 10)

In 1964 only coumarin (**34**) itself (from 12 species belonging to the genera *Ageratum, Eupatorium, Liatris, Trilisa* [Eupatorieae], *Chrysanthemum*,

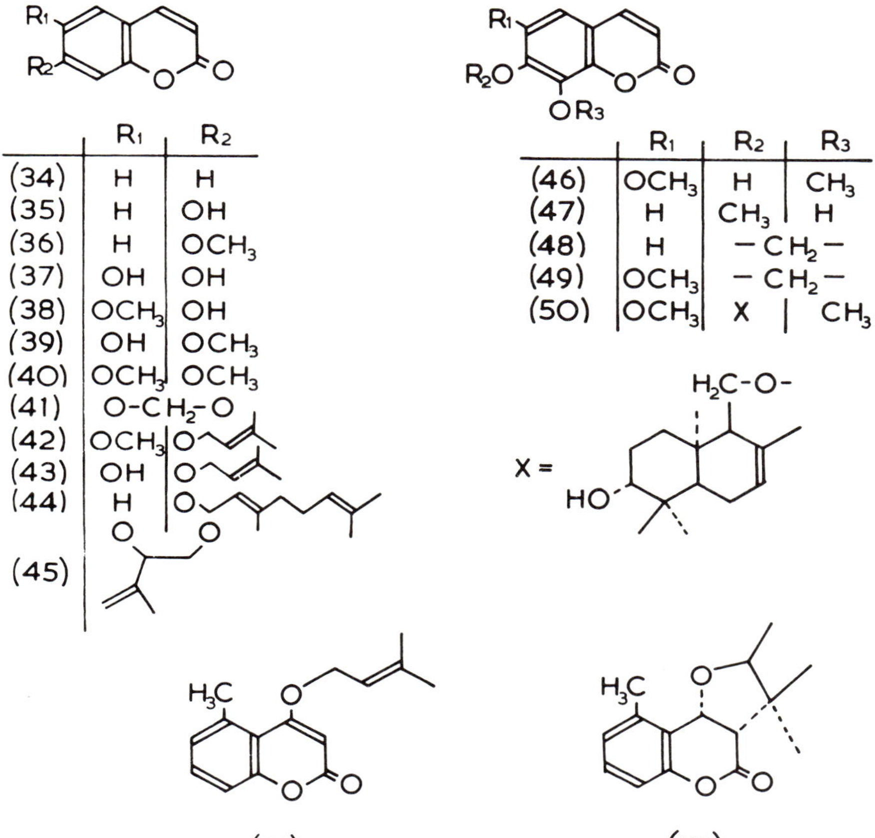

10. THE CHEMISTRY OF THE COMPOSITAE

	R_1	R_2
(34)	H	H
(35)	H	OH
(36)	H	OCH_3
(37)	OH	OH
(38)	OCH_3	OH
(39)	OH	OCH_3
(40)	OCH_3	OCH_3
(41)	$O-CH_2-O$	
(42)	OCH_3	$O\sim$
(43)	OH	$O\sim$
(44)	H	$O\sim$
(45)		

	R_1	R_2	R_3
(46)	OCH_3	H	CH_3
(47)	H	CH_3	H
(48)	H	$-CH_2-$	
(49)	OCH_3	$-CH_2-$	
(50)	OCH_3	X	CH_3

FIG. 10. Some coumarins of Compositae.

Matricaria [Anthemideae] and *Rudbeckia* [Heliantheae]), umbelliferone (**35**), herniarin (**36**), esculetin (**37**), scopoletin (**38**), scoparone (**40**) and ayapin (**41**) and some of their glucosides could be reported (Hegnauer, 1964) as constituents of a number of species representing several tribes. Since then many new coumarins and coumarin-accumulating Compositae have been detected. *Artemisia* (Anthemideae) is still the genus having by far the largest number of known coumarin-producing species. Geissman and Irwin (1970) reported herniarin, scopoletin and scoparone as common constituents of members of section *Dracunculus*. Schmersahl (1966) added isofraxidin (**46**) to the coumarins known to occur in *A. abrotanum* L. Isofraxidin is also present in *A. camphorata* Vill. (Danielak and Borkowski, 1970) and *A. afra* Jacq. (Bohlmann and Zdero, 1972b). Herz *et al.* (1970) isolated scopoletin, scoparone, daphnetin-7-methyl ether (**47**), 7, 8-

methylenedioxycoumarin (**48**) (i.e. the methylene ether of daphnetin), 7, 8-methylenedioxy-6-methoxycoumarin (**49**) and scopoletin-γ,γ-dimethylallyl ether (**42**) from *A. dracunculoides* Pursh. *A. carruthii* Wood also contains (**49**) (Geissman *et al.*, 1971). Shafizadeh and Melnikoff (1970) found isoscopoletin (**39**), mainly as its glucoside methylesculin, to be present in *A. tridentata* Nutt. subsp. *vaseyana* (Rydb.) Beetle and some other American *Artemisia* taxa. *A. tridentata* subsp. *vaseyana* collected in Wyoming yielded 10 coumarins: The glucosides 7-methylesculin, esculin, skimmin (umbelliferone glucoside), cichoriin, the corresponding aglucones isoscopoletin, esculetin and umbelliferone and scoparone and artelin (5,6,7,8-tetramethoxycoumarin) (Brown *et al.*, 1975). Finally Bohlmann *et al.* 1974a) isolated an ether of isofraxidin with a new sesquiterpenediol (3,11-dihydroxydrymene) (**50**) from roots of *A. pontica* L.; this compound closely resembles conferol, a coumarin ether of species of *Ferula* (Umbelliferae). Coumarins like scopoletin, isoscopoletin and herniarin have been isolated from many more species of *Artemisia*, especially by Russian authors. Arscotin, 7-hydroxy-5,6-dimethoxycoumarin, is present in *Artemisia scotina* Nevski (Yusupov and Sidyakin, 1973).

Some further new coumarins and coumarin-producing taxa may serve to illustrate the more or less general tendency in Compositae to produce appreciable amounts of coumarins.

EUPATORIEAE: *Alomia fastigiata* Benth. contains ayapin (Pozetti and Ferreira, 1967). Haskins *et al.* (1972) investigated the origin of coumarin in cured leaf of *Carphephorus odoratissimus* (J.F.Gmel.) Hebert (= *Trilisa odoratissima* [J. F. Gmelin] Cass. = *Liatris odoratissima* J. F. Gmel.); fresh leaves contain large amounts of the glucosides of *trans-* and *cis-o-*hydroxycinnamic acid; during curing coumarin arises by glucoside hydrolysis and spontaneous lactonization of the aglucone.

ASTEREAE: *Haplopappus baylahuen* Remy contains 1% of prenyletin (**43**), a coumarin already known from *Ptaeroxylon obliquum* (Thunb.) Radlk. (Schwenker *et al.*, 1967). Roots of *Aster yunnanensis* Franch. yielded auraptene (**44**) and five biogenetically related umbelliferone ethers; the same coumarins were detected in 19 other species of *Aster* (Bohlmann *et al.*, 1968).

INULEAE: Roots of *Phaenocoma prolifera* (L.) D. Don yielded obliquin (**45**), another coumarin already known from *Ptaeroxylon obliquum* (Bohlmann and Franke, 1973) and *Pterocaulon sphacelatum* Benth. & Hook. f. contains scoparone (Johns *et al.*, 1968).

HELENIEAE: *Amblyolepis* (*Helenium*) *setigera* DC. yielded appreciable amounts of coumarin (Herz and Bhat, 1970).

ANTHEMIDEAE (genera other than *Artemisia*): *Anthemis nobilis* L. has scopolin (Hérisset *et al.*, 1970; anthodia), *A. pseudocotula* Boiss. herniarin and scopoletin (Saleh and Rizk, 1974; whole plant), *Chrysanthemum uliginosum* Pers. calycanthoside (glucoside of isofraxidin) (Plouvier, 1968; stems) and *Matricaria matricarioides* (Less.) Porter has herniarin (Jain and Karchesy, 1971).

CARDUEAE: Simple coumarins were isolated from roots of *Atractylodes* (*Atractylis*) *ovata* (Thunb.) DC. (Studennikova and Khaletsky, 1965), anthodia of *Centaurea depressa* M.B. (Khalmatov and Aliev, 1967) and whole plants of *Onopordon acanthium* L. (Bogs and Bogs, 1965: esculin).

MUTISIEAE: Bohlmann *et al.* (1973e) isolated gerberacoumarin (**51**) and isogerbera-coumarin (**52**) from roots of *Gerbera crocea* Kuntze? and suggested that these constituents are biogenetically derived from 6-methyl-salicylic acid. If this turns out to be the truth, the *Gerbera* compounds cannot be viewed as true coumarins, but rather as acetogenins or perhaps as derivatives of acetylenes (cf. capillarin-type isocoumarins).

CICHORIEAE: Cichoriin, the 7-glucoside of esculetin, is a well known constituent of *Cichorium intybus* L. and umbelliferone of *Hieracium pilosella* L. Gorecki and Mrugasiewicz (1974) showed that flowering tops of *Cichorium intybus* contain up to $2 \cdot 4\%$ of a mixture of coumarins consisting mainly of cichoriin, esculin (the 6-glucoside of esculetin) and esculetin. Bramwell and Dakshini (1971) detected scopoletin in 20 Canary Island species of *Sonchus*; in 15 species it was accompanied by cichoriin and in some by esculin. These authors also mention the presence of cichoriin in *Launaea arborescens* (Batt.) Murb. Bate-Smith *et al.* (1968) found that umbelliferone is not present in leaves of species of *Hieracium sensu stricto*; it seems to be restricted to a group of related species of *Pilosella* Hill.

Hegnauer (1973b, pp. 626–629) pointed out that coumarins such as auraptene (*Aster*) tend to biochemically link Compositae with Umbelliferae (and Rutaceae). This link is strengthened by the detection of the *Artemisia pontica* coumarin and of the *Ptaeroxylon* coumarins prenyletin and obliquin. It is true that *Ptaeroxylon* itself is a taxon *incertae sedis* (Hegnauer, 1969: pp. 68–69, 426–428); its morphology and chemistry, however, agree with the generally accepted rutalean-sapindalean affinity. In several instances (*Artemisia, Hieracium*), coumarins seem to be useful characters at the generic level.

SOME PHENOLIC CONSTITUENTS

Phenolics represent a very heterogeneous class of plant constituents. Attention will be paid in this section to some simple glucosides, to some phenolic acids, including pyromeconic acid which is difficult to place elsewhere, to phenylpropanoids and lignans and to benzofurans and chromenes. Flavonoids and phenolic monoterpenes, sesquiterpenes and diterpenes have already been mentioned. It should be noted that some of the constituents treated here are steam volatile and are often included under essential oils, and that the phenolic monoterpene thymol may give rise to benzofuranoid compounds (see Fig. 6 [**13**]).

Simple phenolics (Fig. 11)
Arbutin (**54**) was isolated from *Serratula bracteifolia* (Iljin) Stankov, *S. isophylla* Claus, *S. xeranthemoides* Bieb. (Yatsyuk *et al.*, 1968) and *S. sogdiana* Bunge (Zatsny *et al.*, 1973). Hydroquinone (**53**), not arbutin, is present in appreciable amounts in seeds and seedlings of species of *Xanthium*; it seems to be the cause of cattle poisoning after consumption of plants in the seedling stage (Keeler, 1975). An isoprenylated *p*-benzoquinone (**55**) occurs in *Phagnalon saxatile* Cass.; it is accompanied by the corresponding hydroquinone derivative (Bohlmann and Kleine, 1966). The recent isolation of picein (**68**; Fig. 12) from *Homogyne alpina* (L.) Cass. (Schwendimann *et al.*, 1974) is of considerable interest because most of the benzofurans and chromones of Compositae seem to be derived from *p*-hydroxyacetophenone (**67**).

Phenolic acids (Fig. 11)
Caffeic acid is one of the major phenolics of Compositae. It is usually present as a mixture of chlorogenic acids including cynarin and sometimes as cichoric acid (Hegnauer, 1964; Zane and Wender, 1966; Kahl *et al.*, 1969). 2,3-Dihydroxycinnamic acid (anthenobilic acid) accompanies caffeic acid in *Anthemis nobilis* L. (Hérisset *et al.*, 1970). Fukinolic acid (**56**; R=caffeyl) and fukiic acid (**56**; R=H) (=hydroxypiscidic acid), new *o*-dihydroxy aromatic acids, have been isolated recently from *Petasites japonicus* (S. & Z.) Maxim. (Sakamura *et al.*, 1973). The two catechol derivatives of *Chrysanthemum morifolium* Ram. (Chang *et al.*, 1975) may be metabolites of fukiic acid. Dimethoxyhydroxybenzyl benzoate (**58**) is a constituent of *Aster ptarmicoides* Torr. & Gray; it is accompanied by methyl and benzyl 6-methoxysalicylate (Bohlmann *et al.*, 1969c); (**58**) occurs also in *Solidago rigida* L. (Anthonsen and Bergland, 1971). Benzyl 2,6-dimethoxybenzoate is present in *Solidago virgaurea* L. and *Aster ptarmicoides* Torr. & Gray (Anthonsen and Bergland, 1971). *Erigeron canadense* L. produces *o*-benzylbenzoic acid (Aziz-ur Rahman and Gatica, 1969), a non-phenolic compound of course.

Pyromeconic acid (**57**) is only included for convenience here; it was shown by Plouvier (1964a) to occur as a glucoside (=erigeroside) in many species of *Erigeron* (Astereae). The free acid is abundantly present in anthodia and leaves and stems of *Parthenium integrifolium* L. (Heliantheae) (Herz and Subramanian, 1971). Maltol (2-methylpyromeconic acid) is a constituent of *Helichrysum ramosissimum* Hook. (Lassak and Pinhey, 1968).

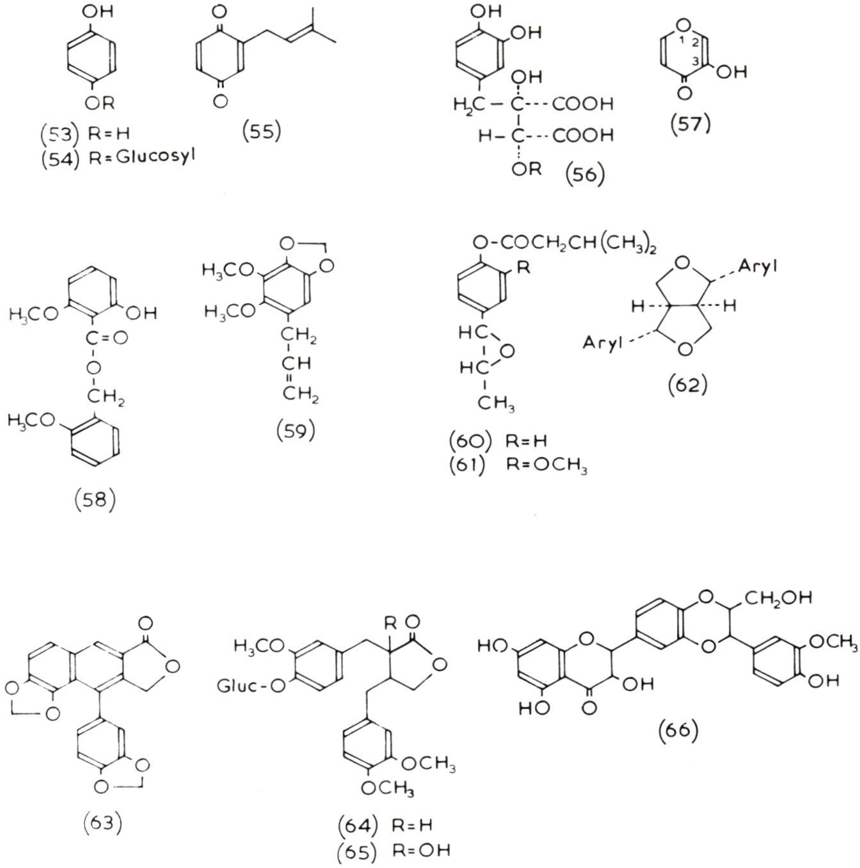

FIG. 11. Some simple phenolics, phenylpropanoids and lignans of Compositae.

Phenylpropanoid compounds (Fig. 11)
Phenylpropanoids are major components of essential oils of a limited number of Compositae (Hegnauer, 1964). Some recent additions are the following. *Caesulia axillaris* Roxb.: 2,4,5-Trimethoxyallylbenzene (γ-asarone) (Devgan and Bokadia, 1968). *Erigeron pappochroma* Labill.: Dillapiole (**59**) (Sörensen and Sörensen, 1969; in some populations of the species only). *Tagetes filifolia* Lag.: Esdragole (Bohrmann and Youngken, 1968). Less volatile and probably taxonomically more valuable phenylpropanoids have been detected recently in members of this family. The presence of similar compounds in Umbelliferae (Hegnauer, 1974: pp. 623–624, 627) is another fact which relates the Compositae to the Umbelliferae. Roots of *Blumea lacera* DC. (Inuleae) contain coniferyl diangelicate (Bohlmann and Zdero, 1969b). *Coreopsis gigantea* (Kell.) H. M. Hall

(Heliantheae) has anol isovalerate (Sörensen and Sörensen, 1966) and its epoxide (**60**) and a corresponding isoeugenol epoxide (**61**) in roots (Bohlmann and Zdero, 1969a). From *Cotula australis* (Sieb. ex Spreng.) Hook. f. Sörensen *et al.* (1968) isolated the 3,3-dimethylallyl ether of *p*-hydroxy methylcinnamate and *Lepidophorum grisleyi* Samp. yielded a new derivative of isoeugenol, lepidophorone (Bohlmann and Zdero, 1973a); both these plants belong to Anthemideae.

Lignans (Fig. 11)
Ten years ago, lignans had been reported only from a few taxa of Compositae (Hegnauer, 1964). Today lignans are known from members of several tribes. Most probably, they are rather common constituents in the family, or of some tribes at least.

EUPATORIEAE: Eudesmin (**62**) and epieudesmin from *Carphephorus odoratissimus* (Wahlberg *et al.*, 1972).

HELIANTHEAE: Helianthoidin and helioxanthin (**63**) in *Heliopsis scabra* Dunal (Burden *et al.*, 1969).

ANTHEMIDEAE: Sesamin (**62**) was isolated from anthodia of *Chrysanthemum cinerarifolium* (Trev.) Vis. (Doskotch and El-Feraly, 1969) and sesamin-type lignans (**62**) from *Artemisia fragrans* Willd. (Bohlmann *et al.*, 1973d).

CARDUEAE: The bitter principle of the seeds of Centaureinae and Carduinae are arctiin-like lignan glycosides (**64**) (Hänsel *et al.*, 1964). *Carthamus tinctorius* L. has matairesinol glucoside (bitter) and tracheloside (**65**) (tasteless; laxative) in its seeds (Palter *et al.*, 1971, 1972). Seeds of *Silybum marianum* (L.) Gaertn. and *S. eburneum* Coss. & Dur. contain dehydrodiconiferylalcohol (a dimer) (Weinges *et al.*, 1970) and the flavonolignans silybin (**66**), silydianin and silychristin (Hänsel *et al.*, 1972; Koch, 1975).

Available evidence suggests that there are differences in lignan synthesis between members of different tribes.

Chromenes and benzofurans (Fig. 12)
The presence of ageratochromene (**79**) in the essential oils of species of *Ageratum*, of euparin (**69**) in some species of *Eupatorium* and of a toxic mixture of benzofuranoid compounds called "tremetol" in *Eupatorium rugosum* Houtt. (= *E. urticaefolium* Reich.; white snake-root) and species of *Haplopappus* (rayless goldenrod; e.g. *H. heterophyllus* [Gray] Blake) have been known for 20 years (Hegnauer, 1964). Tremetol-containing plants cause "trembles" in cattle and "milk sickness" in humans (Christensen, 1965; Keeler, 1975). Later "tremetols" from *Eupatorium* and

Haplopappus were shown to be mixtures of several compounds, such as dehydrotremetone (**70**), hydroxytremetone (**71**), tremetone (**72**) and toxol (**73**).

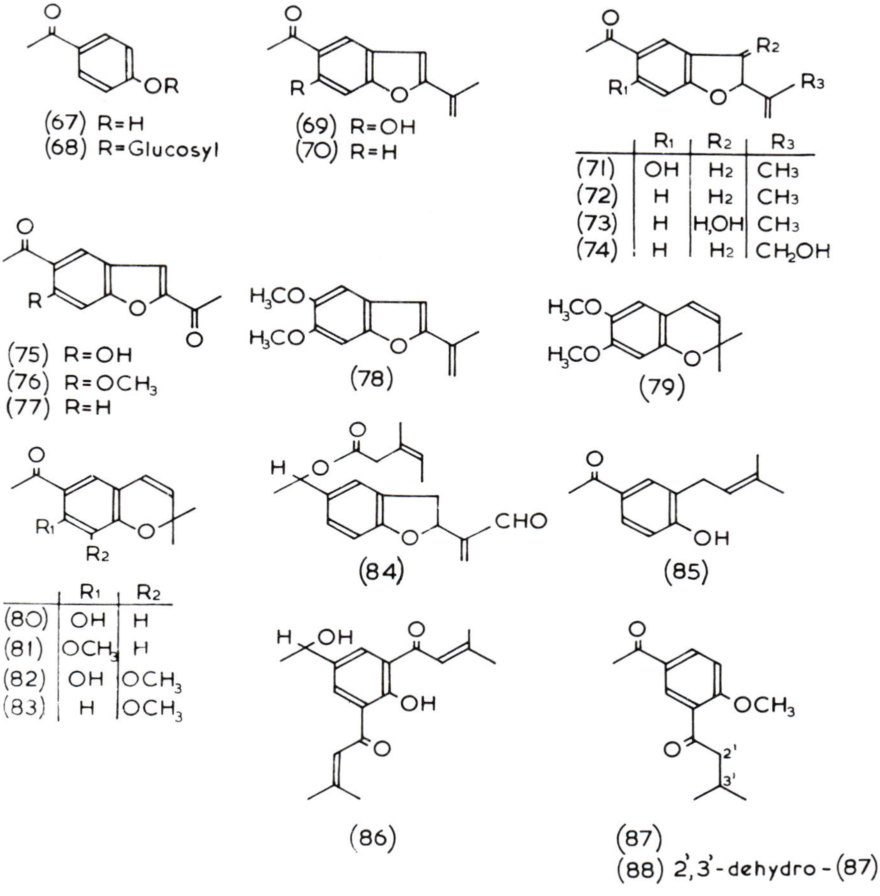

(67) R=H
(68) R=Glucosyl

(69) R=OH
(70) R=H

	R_1	R_2	R_3
(71)	OH	H_2	CH_3
(72)	H	H_2	CH_3
(73)	H	H,OH	CH_3
(74)	H	H_2	CH_2OH

(75) R=OH
(76) R=OCH_3
(77) R=H

(78)

(79)

	R_1	R_2
(80)	OH	H
(81)	OCH_3	H
(82)	OH	OCH_3
(83)	H	OCH_3

(84)

(85)

(86)

(87)
(88) 2',3'-dehydro-(87)

FIG. 12. *Para*-hydroxyacetophenones, benzofurans and chromenes of Compositae.

Today an extremely varied array of chromenes and benzofurans is known to occur in Compositae, especially in some tribes.

EUPATORIEAE: 6-Demethoxyageratochromene and a dimeric ageratochromene (Kasturi *et al.*, 1973: *Ageratum conyzoides* L.). Ageratone and dihydroageratone, a (**74**)-type dihydrobenzofuran (Anthonsen and Chantharasakul, 1970: *A. houstonianum* Mill.). A toxol (**73**) derivative with a terminal epoxide group from *Liatris provincialis* Godfrey (Herz and Wahlberg, 1973). Euparin (**69**) from *Eupatorium quasi-tripartitum* Hayata (Sasaki *et al.*, 1966) and from roots of *E. fortunei* Turcz. and *E. lindleyanum* DC. (Yoshizaki *et al.*, 1974). Anthonsen (1969) reported that eupatoriochromene (**80**) is present in many species of *Eupatorium*; it is accompanied in Australian samples of *E. riparium* Regel by the ripariochromenes A (**82**), B and C and by the methyl ether of (**82**). Samples of the same species from Jamaica yielded the methyl ether of (**82**) and acetovanillochromene (**83**) (Taylor and Wright, 1971). *E. glandulosum* H.B. K. (Australia) only had the methyl ether of eupatoriochromene (**81**) (Anthonsen, 1969). *Carelia cistifolia* Less. yielded euparin (**69**), 6-hydroxy-(**73**) angelicate and 6-methoxy-(**73**) angelicate from roots (Bohlmann and Zdero, 1971).

ASTEREAE: The presence of "tremetol" in members of *Haplopappus* (=*Aplopappus*) has already been mentioned. *Haplopappus heterophyllus* has been reinvestigated; the absolute configuration of toxol at C-3 was elucidated (Zalkov *et al.*, 1972) and tremetone (**72**), and desoxyeuparone (**77**) were isolated in addition to the previously isolated constituents (**70**) and (**73**) (Zalkov *et al.*, 1968).

INULEAE: *Helichrysum stoechas* L. contains (**85**) and corresponding benzofuranoid and dimethylchromanoid constituents (Quesada *et al.*, 1972). Leysseralangelicate (**84**) was reported from *Leyssera gnaphaloides* L. and *L. tenella* DC. (Bohlmann and Zdero, 1972a) and stoebenone (**86**) and dehydrostoebenone from *Stoebe plumosa* Thunb. (Bohlmann and Zdero, 1972b).

HELIANTHEAE: *Encelia californica* Nutt. yielded encecalin (**81**), euparin (**69**) and euparone methyl ether (**76**), but no sesquiterpene lactones (Bjeldanes and Geissman, 1969). *Helianthella uniflora* Torr. & Gray contains (**72**), (**80**), 7-methoxy-(**72**) and several isopentenyl *p*-hydroxyacetophenone (**85**) derivatives (Bohlmann & Grenz, 1970). Zexmeniol, a ripariochromene (**82**)-type compound, was detected in *Zexmenia brevifolia* A. Gray (Ortega and Romo, 1974). Espeletone (**87**) and dehydroespeletone (**88**) were detected in *Espeletia schultzii* Wedd. (Bohlmann and Rao, 1973).

HELENIEAE: *Lasthenia glabrata* Lindl. contains small amounts of euparin (**69**) and larger amounts of (**78**) (Alertsen *et al.*, 1971).

ANTHEMIDEAE: *Abrotanella nivigena* F. Muell. and *A. forsterioides* (Hook. f.) Benth., two Australian species, contain euparin (**69**); in the latter (**69**) is accompanied by hydroxytremetone (**71**) (Anthonsen and Chantharasakul, 1971).

SENECIONEAE: *Doronicum austriacum* Jacq. has several isobutyric and isovaleric esters of (**74**)-type compounds (Bohlmann and Zdero, 1970). The euparinoid compound (**78**) with the substitution pattern of ageratochromene has also been isolated from *Ligularia stenocephala* Matsum. & Koidz. (Murae *et al.*, 1968).

MUTISIEAE: Roots of *Gerbera crocea* Kuntze(?) and *G. asplenifolia* Spr. contain dehydrodidesoxostoebenone ([**86**]-type compound) (Bohlmann *et al.*, 1973e).

Anthonsen (1969), Bjeldanes and Geissman (1969) and Bohlmann and Zdero (1972: leysseral angelicate) suggest that in future benzofuranoid and chromenoid constituents may become taxonomically useful markers within the Compositae. This seems to be true indeed. These compounds seem to be fairly common constituents of Eupatorieae, Inuleae and Heliantheae; they have never been found in Cardueae or Cichorieae.

Ageratochromene (**79**) and its 6-demethoxy derivative were detected recently in the essential oil of the rutaceous plant *Boenninghausenia albiflora* Reichb. (Suga *et al.*, 1975). Ripariochromene A-type (**82**) chromenes also occur in several rutaceous genera (Hegnauer, 1973b: evodional p. 178; acronylin pp. 203–204). Most chromenes of Rutaceae, however, are derivatives of phloracetophenones. The co-occurrence of xanthoxylin (=brevifolin, the 4,6-dimethyl ether of phloracetophenone) and of the chromene precursor acronylin (2-O-methyl-3-isopentenylphloracetophenone) suggests that they are totally acetogenic. Brevifolin, however, does also occur in Compositae (*Artemisia brevifolia* Wall. and *Blumea balsamifera* Blume [Hegnauer, 1964]; *Artemisia santolina* Schrenk and *A. scotina* Nevski [Akyev *et al.*, 1973]). Perhaps two pathways leading to acetophenone derivatives are present in Compositae: one starting with phenylalanine (or tyrosine) resulting in derivatives of *p*-hydroxyacetophenone and a polyacetate pathway giving rise to phloracetophenone derivatives. A sound taxonomic interpretation of the presence of similar chromenes in Rutaceae and Compositae is impossible without information derived from biosynthetic studies. Finally it may be mentioned that evodionol-type compounds occur also in the monocotyledonous family Cyperaceae (Allan *et al.*, 1969, 1970) and that euparone (**75**) has been isolated from roots of *Ruscus aculeatus* L. (Liliaceae) (El Sohly *et al.*, 1975).

Some unclassified phenolics (Fig. 13)
Phthalides and α-pyrones occur in species of *Helichrysum* and *Gnaphalium* (Inuleae) and a phthalide (**92**) was detected also in roots of *Othonna cylindrica* DC. (Senecioneae: Bohlmann and Grenz, 1974). The flavanone obtusifolin (**96**) was isolated from *Gnaphalium obtusifolium* L. (Hänsel *et al.*, 1970; Narayanan *et al.*, 1970), the phthalides (**90**), (**91**) and helipyrone (**93**) occur in *Helichrysum italicum* G. Don (Opitz and Hänsel, 1970, 1971)

and (**89**) and (**90**) occur together with arenol (**94**) and homoarenol (**95**) in *Helichrysum arenarium* (L.) Moench (Vrkoč *et al.*, 1971, 1973).

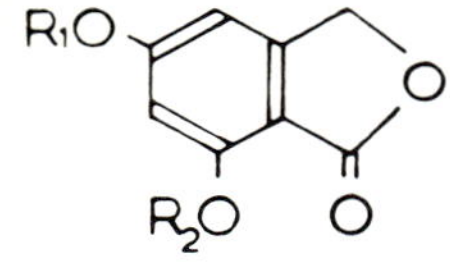

(89) $R_1 = R_2 = H$
(90) $R_1 = CH_3, R_2 = H$
(91) $R_1 = R_2 = CH_3$

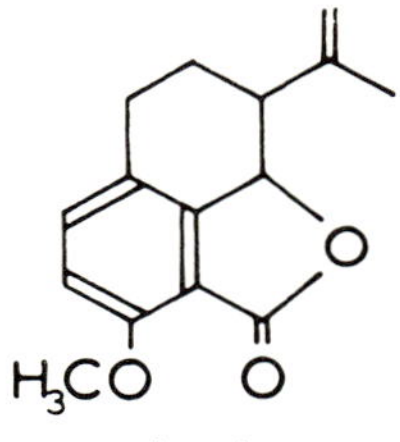

(92)

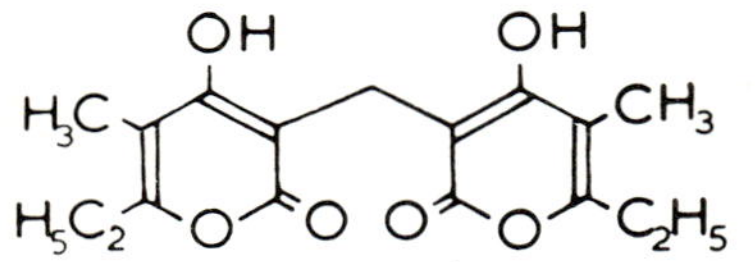

(93)

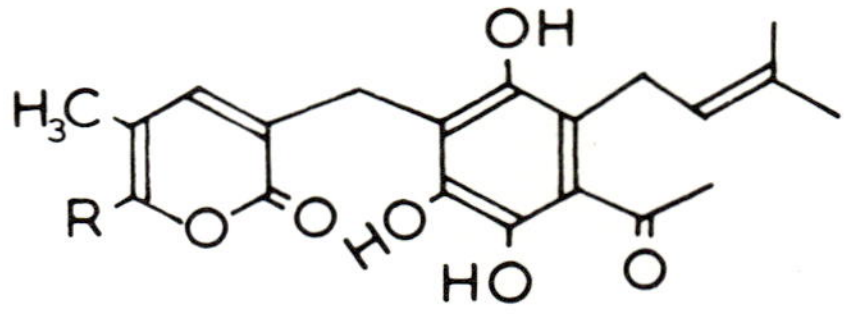

(94) $R = CH_3$
(95) $R = C_2H_5$

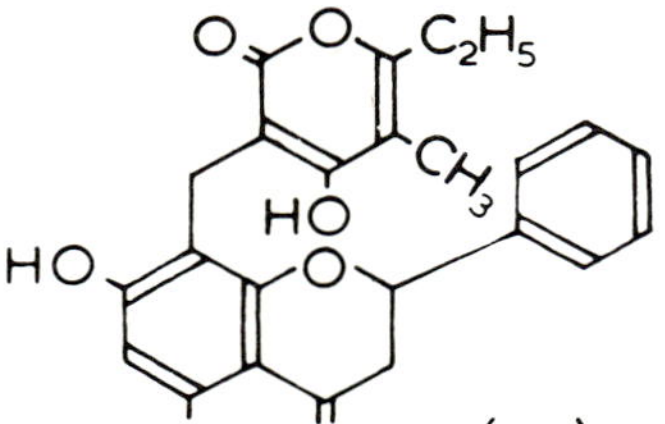

(96)

FIG. 13. Some unusual phenolic constituents of Compositae.

SEED OILS

Plants of the Compositae store mainly proteins and oils in their seeds. As a rule, the seed oils are rich in linoleic acid and contain lesser amounts of oleic and palmitic acid; linolenic and stearic acid are minor fatty acids (group Ib of Shorland, 1963). On the whole, the family may be designated as a group with the linoleic-rich variant of seed oils of the usual type. There is, however, a rapidly increasing number of exceptions to this rule. Shorland (1963), Hilditch and Williams (1964) and Hegnauer (1964) already mentioned some taxa with unusual C_{18} unsaturated acids as major (i.e. > 10% of total) fatty acids in seed oils. Since then, much attention has been paid to seed oils of Compositae. Jones and Earle (1966) and Barclay and Earle (1974) reported on a large number of species, many of which produce oils with unusual fatty acids. Wolff (1966) discussed, in his review of seed lipids, a number of unusual fatty acids which more or less replace linoleic (or oleic) acid in seed oils of certain Compositae taxa; more such examples have been detected since 1966.

The usual type of seed oils of Spermatophytes is represented by methylene-interrupted *cis*-unsaturation in C_{18} fatty acids. Deviations from this pattern imply additional biosynthetic mechanisms. This is illustrated by a survey of structural features of C_{18} fatty acids known from seed oils of Compositae (Table I).

TABLE I. C_{18} unsaturated acids known to occur as major fatty acids in seed oils of Compositae.

Common name	Position and stereo-chemistry of double bonds	Position of acetylene linkage	Oxygen functions
USUAL TYPE:			
Oleic acid	9-*cis*	—	—
Linoleic acid	9-*cis*,12-*cis*	—	—
Linolenic acid[a]	9-*cis*,12-*cis*,15-*cis*	—	—
UNUSUAL TYPE:			
—	*trans*-3	—	—
—	*trans*-3,*cis*-9	—	—
—	*trans*-3,*cis*-9,*cis*-12	—	—
—	*cis*-5,*cis*-9,*cis*-12	—	—
Calendic acid	*trans*-8,*trans*-10,*cis*-12	—	—
Epoxystearic acid	—	—	9,10-epoxy
Vernolic acid	*cis*-9	—	12,13-epoxy
Coronaric acid	*cis*-12	—	9,10-epoxy
Dimorphecolic acid[b]	10-*trans*,12-*trans*	—	9-hydroxy
—[b]	10-*trans*,12-*cis*	—	9-hydroxy
—[b]	9-*cis*,11-*trans*	—	13-hydroxy
Phloionolic acid	—	—	9,10,18-trihydroxy
—	12-*cis*	—	9,10,18-trihydroxy
Crepenynic acid	9-*cis*	12	—
Helenynolic acid	10-*trans*	12	9-hydroxy

[a] $H_3\overset{18}{C}-\overset{17}{C}H_2-\overset{16}{C}H=\overset{c\ 15}{C}H-\overset{14}{C}H_2-\overset{13}{C}H=\overset{c\ 12}{C}H-\overset{11}{C}H_2-\overset{10}{C}H=\overset{c\ 9}{C}H-[CH_2]_7-\overset{1}{C}OOH$

[b] α-Hydroxy conjugated dienoic acids (α with regard to conjugated unsaturation); e.g. dimorphecolic acid:

$H_3C-[CH_2]_4-CH\overset{t\ 12}{=}CH-CH\overset{t\ 10}{=}CH-\overset{9}{C}HOH-[CH_2]_7-COOH$

Table I shows that several additional biosynthetic mechanisms must be active during seed oil production in some Compositae taxa. These are: (a) Introduction of double bonds controlled by the carboxyl ēnd: *cis*- or *trans*-unsaturation in positions 3 or 5 (some species also have *trans*-3-hexadecenoic acid); (b) Desaturation of linoleic acid to give crepenynic acid; (c) Epoxidation of oleic or linoleic acid resulting in epoxystearic, vernolic and coronaric acid; (d) Hydroxylation of linoleic (or crepenynic) acid involving conjugation of double bonds (α-hydroxy conjugated dienoic acids) and sometimes followed by dehydration (conjugated trienoic acids like calendic acid); and (e) Hydroxylation of the terminal methyl group (phloionolic acid).

The known systematic distribution of well characterized unusual seed oils suggests that in future the chemistry of oils may become taxonomically valuable at infrafamiliar levels.

VERNONIEAE: Vernolic acid is a main oil constituent of most species of *Vernonia* (8–80%) (Hegnauer, 1964; White *et al.*, 1971). Vernolic acid is also major fatty acid in seed oils of *Erlangea tomentosa* S. Moore (Phillips *et al.*, 1969) and *Stokesia laevis* (Hill) Greene, a taxon whose classification in Vernonieae is still in doubt (Gunn and White, 1974).

ASTEREAE: *Grindelia oxylepis* Greene has two monoenoic acids with *trans*-3 unsaturation (Kleiman *et al.*, 1966; 14% 16:1 and 2% 18:1). An oil containing several acids with *trans*-3 unsaturation is also produced by *Aster alpinus* L. (Morris *et al.*, 1968).

INULEAE: *Helichrysum bracteatum* (Vent.) Andr. has an oil with 36% linoleic acid and the same amount of a mixture of unusual fatty acids of which crepenynic, helenynolic, coronaric and two isomeric α-hydroxy conjugated dienoic acids were identified (Powell *et al.*, 1965). *Stenachaenium macrocephalum* (DC.) Benth. & Hook. f. has *trans*-3, *cis*-9, *cis*-12-octadecatrienoic acid as a major fatty acid; after long storage several oxygenated acids were also present (Kleiman *et al.*, 1971).

HELIANTHEAE: *trans*-3,*cis*-9,*cis*-12-Octadecatrienoic acid is a major fatty acid of the seed oil of *Calea urticaefolia* (Mill.) DC. (Bagby *et al.*, 1965) and a species of *Heliopsis* was reported to have 13% of an epoxy monoenoic acid in its seed oil (Hilditch and Williams, 1964).

HELENIEAE: *Helenium bigelovii* A. Gray, not *H. hoopesii* A. Gray, has about 10% of *trans*-3-hexadecenoic acid in its seed oil (Hopkins and Chisholm, 1964) and the oil of *Tagetes erecta* L. contains two isomeric α-hydroxy conjugated dienoic acids (Hopkins and Chisholm, 1965).

ANTHEMIDEAE: *Artemisia absinthium* L. and *Chrysanthemum coronarium* L. seed oils contain substantial amounts of epoxy monoenoic acids and of α-hydroxy conjugated dienoic acids; coronaric acid was isolated from *Ch. coronarium* for the first time (Hilditch and Williams, 1964). The seed oil of *Chrysanthemum viscidum* Thell. contains 15% of crepenynic acid (White *et al.*, 1971).

CALENDULEAE: Since 1960 when dimorphecolic acid was detected in species of *Dimorphotheca*, much work has been performed with representatives of this tribe. Earle *et al.* (1964) and Barclay and Earle (1965) showed that dimorphecolic acid is major fatty acid in all species of *Dimorphotheca*, in *Castalis nudicaulis* DC. and in five species belonging to section *Blaxium* of the genus *Osteospermum*. The other species of this tribe investigated hitherto have seed oils with 14–60% of the biogenetically related calendic acid: 14 species of *Osteospermum* (including *Tripteris hyposeroides* DC. = *O. hyposeroides* [DC.] Norl.; Conacher *et al.*, 1970), *Calendula arvensis* L. (including *C. stellata* Cav. = *C. arvensis* L. var. *stellata* [Cav.] Maire; Chisholm and Hopkins, 1966) and *C. officinalis* L. and two species of *Chrysanthemoides*.

ARCTOTEAE: *Arctotis grandis* Thunb. was reported to contain 7% of an epoxy monoenoic acid in its seed oil (Hilditch and Williams, 1964).

CARDUEAE: *Cynara cardunculus* L. has 12% of epoxy monoenoic acids in its seed oil (Hilditch and Williams, 1964). The major fatty acids of the seed oils of *Carlina acaulis* L. and *C. corymbosa* L. are *cis*-5-octadecenoic acid (21–24%) and linoleic acid (50–52%) (Spencer *et al.*, 1969). *Arctium minus* L. produces an oil with 74% of linoleic acid and 10% of *trans*-3, *cis*-9,*cis*-12-octadecatrienoic acid (Morris *et al.*, 1968). Oils of *Chamaepeuce afra* (Jacq.) DC. and *Ch. hispida* DC. contain approximately 60% of linoleic acid and appreciable amounts of two trihydroxy acids, 9% of phloionolic acid and 14% of the corresponding *cis*-12-octadecenoic acid (Mikolajczak and Smith, 1967b). In *Xeranthemum annuum* L. seed oil linoleic acid is accompanied by appreciable amounts (25%) of a mixture of 5-*cis*,9-*cis*,12-*cis*-octadecatrienoic acid, epoxystearic acid, coronaric acid, vernolic acid and two unsaturated hydroxy acids (Powell *et al.*, 1967). Crepenynic acid (13–36%) is present in the seed oils of *Centaurea glaberrima* Tausch, *Jurinea anatolica* Boiss., *J. carduiformis* (Jaub & Spach) Boiss. and *Saussurea candicans* C. B. Clarke (White *et al.*, 1971).

CICHORIEAE: Since the detection of crepenynic acid in *Crepis foetida* L. in 1964, seed oils of more species of *Crepis* have been investigated (Earle *et al.*, 1966; Tallent *et al.*, 1966; Conacher *et al.*, 1970). Three species of section *Hostia* (*C. foetida*, *C. rubra* L. and *C. thomsonii* Babc.) appeared to have 36–65% crepenynic acid and *C. biennis* L., *C. aurea* (L.) Cass., *C. vesicaria* L., all belonging to other sections of the genus, had vernolic acid (47–68%) as one of their major fatty acids; a third category of seed oils was observed in representatives of section *Psilochaenia* (*C. intermedia* A. Gray; *C. occidentalis* Nutt.); these species produce oils with approximately 10% of crepenynic acid and 20–35% of vernolic acid. Later (White et al., 1971) *Crepis alpina* L., *Lampsana communis* L., *L. grandiflora* Bieb., *Picris comosa* (Boiss.) Benth. & Hook. f., *P. echioides* L. and *P. hieracioides* L. were announced to have 15–75% of crepenynic acid in their seed oils. *Crepis alpina* (also belonging to section *Hostia*) with 70–75% crepenynic acid is investigated for its crop potentialities (White et al., 1973). *Crepis* cf. *aspera* L. was also mentioned by White et al. (1971) as having an oil with 59% crepenynic acid; *C. aspera* is the only species of section *Nemauchenes* investigated so far. *Tragopogon porrifolius* L. has minor amounts of epoxystearic acid and of α-hydroxy conjugated dienoic acids in its seed oil (Shorland, 1963).

It is now evident that seed oils of many Compositae contain unusual fatty acids besides or replacing linoleic acid and that the chemistry of seed oils is often characteristic of taxa (e.g. Calenduleae; sections of the genus *Crepis*). Since linoleic acid and crepenynic acid are the precursors of acetylenic constituents and of the C_{14}–C_{10} polyunsaturated acids present in the insecticidal amides (see pp. 277–279), it is to be expected that seed oils often contain crepenynic acid; trace amounts of this acid may be present in every member of the family.

The commercially available safflower oil is the seed oil of *Carthamus tinctorius* L., a crop plant of warmer regions. Safflower oil normally has approximately 75% of linoleic acid and 15% of oleic acid. When Indian strains, probably of Eastern Pakistan origin, were analysed, it was found that they produce an oil with 80% of oleic and only 12% of linoleic acid. Using these strains it was possible to show that the composition of seed oils is under genetic control. The factor inducing predominance of oleic acid is a recessive allele (Knowles and Mutwakil, 1963; Knowles, 1965).

CONCLUDING REMARKS

Phytochemical and chemotaxonomic research of the past decade has produced an overwhelming number of new facts. Plant taxonomists can and should make use of the systematically important part of the new chemical evidence. From a chemosystematic point of view, it is highly satisfying that a chemotaxonomic evaluation of Compositae performed more than 10 years ago stands the test made possible by the availability of a very large number of new facts. Moreover the new facts definitely show that some of the chemical characters are extremely useful at infrafamiliar taxonomic levels. It is the conviction of the present author that intensive and well-planned collaboration between a team of specialized botanists and specialized phytochemists should settle some of the many still controversial points (e.g. classification of genera like *Arnica* and *Doronicum*; interrelationships of genera, subtribes and tribes) of Compositae classification within a relatively short period of time.

REFERENCES

AKYEV, B., YUSUPOV, M. I., KASIMOV, SH. L. and SIDYAKIN, G. P. (1973). [Xanthoxylin from *Artemisia santolina* and *A. scotina.*] *Khim. Prir. Soedin* **1973**, 422–423.

ALERTSEN, A. R., ANTHONSEN, T., RAKUES, E. and SÖRENSEN, N. A. (1971). Benzofuran derivatives from *Lasthenia glabrata*. *Acta chem. scand.* **25**, 1919–1920.

ALLAN, R. D., CORRELL, R. L. and WELLS, R. J. (1969). A new class of quinones from certain members of the family Cyperaceae. *Tetrahedron Letters* **1969**, 4669–4672. Two new phenolic ketones from *Remirea maritima* (Cyperaceae). *Tetrahedron Letters* **1969**, 4673–4674.

ALLAN, R. D., WELLS, R. J. and MCLEOD, J. K. (1970). Further phenolic ketones from *Remirea maritima* Aubl. *Tetrahedron Letters* **1970**, 3945–3946.

ANDERSON, A. B., MCCRINDLE, R. and NAKAMURA, E. (1974). Diterpenoids of *Solidago arguta* Ait. The stereochemistry of *cis*-clerodanes. *J. chem. Soc. chem. Commun.* **1974**, 453–454.

ANTHONSEN, T. (1969). New chromenes from *Eupatorium* species. *Acta chem. scand.* **23**, 3605–3607.

ANTHONSEN, T. and BERGLAND, G. (1970). The diterpenoids of *Solidago missouriensis* Nutt. *Acta chem. scand.* **24**, 1860–1861.

ANTHONSEN, T. and BERGLAND, G. (1971). The diterpenoids of some *Solidago* species. *Acta chem. scand.* **25**, 1924–1925.

ANTHONSEN, T. and BERGLAND, G. (1973). Constitution and stereochemistry of diterpenoids from *Solidago missouriensis*. *Acta chem. scand.* **27**, 1073–1082.

ANTHONSEN, T. and CHANTHARASAKUL, S. (1970). Ageratone and dihydroageratone, new benzofuran derivatives from *Ageratum houstonianum*. *Acta chem. scand.* **24**, 721–722.

ANTHONSEN, T. and CHANTHARASAKUL, S. (1971). Isolation of *ent*-16-kauren-19-oic acid and *ent*-16-kauren-19-ol from *Abrotanella nivigena* Muell. *Acta chem. scand.* **25**, 1925–1927.

ANTHONSEN, T. and KJÖSEN, B. (1971). New thymol derivatives from *Inula salicina*. *Acta chem. scand.* **25**, 390–392.

ANTHONSEN, T. and MCCRINDLE, R. (1969). The constitution of diterpenoids from *Solidago elongata*. *Acta chem. scand.* **23**, 1068–1070.

ANTHONSEN, T., MCCABE, P. H., MCCRINDLE, R. and MURRAY, R. D. H. (1967). The constitution and stereochemistry of solidagenone. *Acta chem. scand.* **21**, 2289.

ANTHONSEN, T., HENDERSON, M. S., MARTIN, A., MCCRINDLE, R. and MURRAY, R. D. H. (1968). Furan-containing diterpenoids from *Solidago serotina* Ait. *Acta chem. scand.* **22**, 351–352.

ANTHONSEN, T., MCCABE, P. H., MCCRINDLE, R. and MURRAY, R. D. H. (1969). The constitution and stereochemistry of diterpenoids from *Solidago canadensis* L. *Tetrahedron* **25**, 2233–2239; see also *Tetrahedron* **26**, 3091–3097 (1970).

ANTHONSEN, T., HENDERSON, M. S., MARTIN, A., MURRAY, R. D. H., MCCRINDLE, R. and MCMASTER, D. (1973). Solidagoic acid A and B, diterpenoids from *Solidago gigantea* var. *serotina*. *Can. J. Chem.* **51**, 1332–1345.

ANYEHCHI, Y. and ESHAGHZADEH, S. (1974). Terpenoids and hydrocarbons of *Acroptilon picris*. *Phytochemistry* **13**, 2000–2001.

ARITOMI, M. and KAWASAKI, T. (1974). Dehydro-*para*-asebotin, a new chalcone glucoside in the flowers of *Helichrysum affine* D.Don. *Chem. Pharm. Bull. (Tokyo)* **22**, 1800–1805.

AZIZ-UR RAHMAN and GATICA, H. S. E. (1969). Isolation of *o*-benzylbenzoic acid from *Erigeron canadense*. *Rec. Trav. Chim. Pays-Bas* **88**, 1332–1344.

BAGBY, M. O., SIEGL, W. O. and WOLFF, I. A. (1965). A new acid from *Calea urticaefolia* seed oil: *trans*-3, *cis*-9, *cis*-12-octadecatrienoic acid. *J. Am. Oil Chemists' Soc.* **42**, 50–53.

BAILEY, J. A., CARTER, G. A., BURDEN, R. S. and WAIN, R. L. (1975). Control of rust diseases by diterpenes from *Nicotiana glutinosa*. *Nature, Lond.* **255**, 328–329.

BARCLAY, A. S. and EARLE, F. R. (1965). The search for new industrial crops V. The South African Calenduleae (Compositae) as a source of new seed oils. *Econ. Bot.* **19**, 33–43.

BARCLAY, A. S. and EARLE, F. R. (1974). Chemical analysis of seeds III: Oil and protein content of 1253 species. *Econ. Bot.* **28**, 178–236.

BATE-SMITH, E. C., SELL, P. D. and WEST, C. (1968). Chemistry and taxonomy of *Hieracium* L. and *Pilosella* Hill. *Phytochemistry* **7**, 1165–1169.

BHARGAVA, K. K., KRISHNASWAMY, N. R. and SESHADRI, T. R. (1972). Desmethylwedelolactone from *Eclipta alba* leaves. *Indian J. Chem.* **10**, 810–811.

BHAT, S. V., KALYANARAMAN, P. S., KOHL, H. and SOUZA, DE, N. J. (1975). Inuroyleanol and 7-ketoroyleanol, two novel diterpenoids of *Inula royleana* DC. *Tetrahedron* **31**, 1001–1004.

BJELDANES, L. F. and GEISSMAN, T. A. (1969). Euparinoid constituents of *Encelia californica*. *Phytochemistry* **8**, 1293–1296.

BJELDANES, L. F. and GEISSMAN, T. A. (1972). Constituents of *Helianthus ciliaris*. *Phytochemistry* **11**, 327–332.

BOGS, H.-U. and BOGS, U. (1965). Über Inhaltsstoffe von *Onopordon acanthinum* 1. Cumarine und Flavone. *Pharmazie* **20**, 706–709.

BOHLMANN, F. and FRANKE, H. (1973). Isolation of obliquin from *Phaenocoma prolifera*. *Phytochemistry* **12**, 726–727.

BOHLMANN, F. and GRENZ, M. (1970). Neue Isopentenylacetophenon-Derivate aus *Helianthella uniflora*. *Chem. Ber.* **103**, 90–96.

BOHLMANN, F. and GRENZ, M. (1972). Notiz über die Isolierung von Hautriwa-Säure aus *Conyza ivaefolium* Less. *Chem. Ber.* **105**, 3123–3125.

BOHLMANN, F. and GRENZ, M. (1974). Ein neues Methylsalicylsäurederivat aus *Othonna cylindrica*. *Tetrahedron Letters* **1974**, 1681–1682.

BOHLMANN, F. and KLEINE, K.-M. (1966). Über ein neues Chinon aus höheren Pflanzen. *Chem. Ber.* **99**, 885–888.

BOHLMANN, F. and KÖRNIG, D. (1974). Synthese des Sesquiterpens aus *Elvira biflora* DC. *Chem. Ber.* **107**, 1777–1779.

BOHLMANN, F. and RAO, N. (1973). Neue Hydroxyacetophenon-Derivate aus *Espeletia schultzii*. *Chem. Ber.* **106**, 3035–3038.
BOHLMANN, F. and ZDERO, CH. (1969a). Über Inhaltsstoffe von *Coreopsis gigantea*. *Chem. Ber.* **102**, 1691–1697.
BOHLMANN, F. and ZDERO, CH. (1969b). Über ein neues Coniferylalkohol-Derivat aus *Blumea lacera*. *Tetrahedron Letters*, 69–70.
BOHLMANN, F. and ZDERO, CH. (1970). Neue Benzofuran-Derivate aus *Doronicum austriacum*. *Tetrahedron Letters* **1970**, 3575–3576.
BOHLMANN, F. and ZDERO, CH. (1971). Notiz über die Inhaltsstoffe von *Carelia cistifolia* Less. *Chem. Ber.* **104**, 964–966.
BOHLMANN, F. and ZDERO, CH. (1972a). Leysseral-angelicat, ein neuartiges Benzofuran-Derivat. *Chem. Ber.* **105**, 2534–2538.
BOHLMANN, F. and ZDERO, CH. (1972b). Weitere Inhaltsstoffe der Tribus Inuleae. *Chem. Ber.* **105**, 2604–2606.
BOHLMANN, F. and ZDERO, CH. (1972c). Constituents of *Artemisia afra* Jacq. *Phytochemistry* **11**, 2339–2340.
BOHLMANN, F. and ZDERO, CH. (1972d). Neue Thymol-Derivate aus *Arnica amplexicaulis*. *Tetrahedron Letters*, 2827–2828.
BOHLMANN, F. and ZDERO, CH. (1973a). Über ein neues Isoeugenol-Derivat aus *Lepidophorum grisleyi* Samp. *Chem. Ber.* **106**, 379–381.
BOHLMANN, F. and ZDERO, CH. (1973b). Neue Inhaltsstoffe aus *Achillea*-Arten. *Chem. Ber.* **106**, 1328–1336.
BOHLMANN, F. and ZDERO, CH. (1974a). Neue Acetylenverbindungen aus südafrikanischen Vertretern der Tribus Anthemideae. *Chem. Ber.* **107**, 1044–1048.
BOHLMANN, F. and ZDERO, CH. (1974b). Über ein neues Furansesquiterpen aus *Stilpnophytum linifolium* (Thunb.) Less. *Chem. Ber.* **107**, 1071–1073 (also review of furanosesquiterpenes in Compositae).
BOHLMANN, F., ZDERO, CH. and KAPTEYN, H. (1968). Über Cumarin-Derivate der Gattung *Aster*. *Liebigs Ann. Chem.* **717**, 186–192.
BOHLMANN, F., NIEDBALLA, U. and SCHULZ, J. (1969a). Thymolderivate aus *Gaillardia*- und *Helenium*-Arten. *Chem. Ber.* **102**, 864–871.
BOHLMANN, F., SCHULZ, J. and BÜHMANN, U. (1969b). Struktur und Synthese eines Thymolderivates aus *Helenium*-Arten. *Tetrahedron Letters*, 4703–4704.
BOHLMANN, F., ZDERO, CH. and KAPTEYN, H. (1969c). Über die Acetylenverbindungen der Astereae. *Chem. Ber.* **102**, 1682–1690.
BOHLMANN, F., BURKHARDT, T. and ZDERO, CH., eds (1973a). "Naturally Occurring Acetylenes". Academic Press, London and New York.
BOHLMANN, F., GRENZ, M. and SCHWARZ, H. (1973b). Neue Diterpene aus *Hinterhubera imbricata*. *Chem. Ber.* **106**, 2479–2484.
BOHLMANN, F., WEICKGENANNT, G. and ZDERO, CH. (1973c). Neue Diterpene aus der Tribus der Calenduleae. *Chem. Ber.* **106**, 826–840.
BOHLMANN, F., ZDERO, CH. and FAASS, U. (1973d). Über die Inhaltsstoffe von *Artemisia fragrans* Willd. *Chem. Ber.* **106**, 2904–2909.
BOHLMANN, F., ZDERO, CH. and FRANKE, H. (1973e). Über Inhaltsstoffe der Gattung *Gerbera*. *Chem. Ber.* **106**, 382–387.
BOHLMANN, F., SCHUMANN, D. and ZDERO, CH. (1974a). Über ein neues Sesquiterpen-Derivat aus *Artemisia pontica* L. *Chem. Ber.* **107**, 644–649.

BOHLMANN, F., ZDERO, CH. and SUWITA, A. (1974b). Weitere Amide aus der Tribus Anthemideae. *Chem. Ber.* **107**, 1038–1043.

BOHRMANN, H. and YOUNGKEN, JR., H. W. (1968). Esdragole, the main compound in the volatile oil of *Tagetes filifolia* (Compositae). *Phytochemistry* **7**, 1415–1416.

BORNOWSKI, H. (1971). Die Struktur des Lasiospermans. Ein neuer Typ von Furanosesquiterpenen. *Tetrahedron* **27**, 4101–4108.

BOSE, P. K., CHAKRABARTI, P., CHAKRAVARTI, S., DUTTA, S. P. and BARUA, A. K. (1973). Flavonoid constituents of *Eupatorium odoratum*. *Phytochemistry* **12**, 667–668.

BOUGHTON, I. B. and HARDY, W. T. (1939). Feeding trials of suspected plants. *Texas agric. Exp. Stn A. Rep.* **52**, 239.

BRAMWELL, D. and DAKSHINI, K. M. M. (1971). Luteolin 7-glucoside and hydroxycoumarins in Canary Island *Sonchus* species. *Phytochemistry* **10**, 2245–2246.

BRIESKORN, C.-H. and PÖHLMANN, E. (1968). Diterpene vom Kaurantyp aus der Composite *Espeletia schultzii*. *Tetrahedron Letters*, 5661–5664. *Chem. Ber.* **102**, 2621–2628.

BROWN, D., ASPLUND, R. O. and MCMAHON, V. A. (1975). Phenolic constituents of *Artemisia tridentata* ssp. *vaseyana*. *Phytochemistry* **14**, 1083–1084.

BULL, L. B., CULVENOR, C. C. J. and DICK, A. T. (1968). "The pyrrolizidine Alkaloids". North-Holland, Amsterdam.

BURDEN, R. S. and CROMBIE, L. (1969). A new series of alka-2,4-dienoic tyramine amides from *Anacyclus pyrethrum* DC. *J. chem. Soc.* (C), 2477–2481.

BURDEN, R. S., CROMBIE, L. and WHITING, D. A. (1969). Extraction of *Heliopsis scabra*: Constitution of two lignans. *J. chem. Soc.* (C), 693–701.

BURNETT, JR., W. C., JONES, JR., S. B., MABRY, T. J. and PADOLINA, W. G. (1974). Sesquiterpene lactones—insect feeding deterrants. *Biochem. Systematics and Ecology* **2**, 25–29.

BUTLER, G. W. (1965). The distribution of the cyanoglucosides linamarin and lotaustralin in higher plants. *Phytochemistry* **4**, 127–131.

CABALLERO, Y. and WALLS, F. (1970). Natural products of *Montanoa tomentosa*. *Bol. Inst. quim. Univ. Nac. Anton. Mexico* **22**, 79–102. Ex *Chem. Abstr.* **74**, 136398 (1971).

CAMBIE, R. C. and PARNELL, J. C. (1969). A New Zealand phytochemical survey. 7. Constituents of some dicotyledons. *N.Z. J. Sci.* **12**, 453–466.

CANONICA, L., RINDONE, B., SCOLASTICO, C., FERRARI, G. and CASAGRANDE, C. (1967). Constituenti estrattivi della *Psiadia altissima* Benth. et Hook. *Gazz. chim. ital.* **99**, 260–275. See also *Tetrahedron Letters* (1967), 2639–2642: Structure and stereochemistry of psiadiol.

CANONICA, L., RINDONE, B., SCOLASTICO, C., KOO DONG HAN and JAE HOON KIM. (1969). New diterpene with pimarane skeleton. *Tetrahedron Letters*, 4801–4804.

CAVA, M. P., RAO, K. V., WEISBACH, J. A., RAFAUFF, R. F. and DOUGLAS, B. (1968). Alkaloids of *Cacalia floridana*. *J. org. Chem.* **33**, 3570–3573.

CHANG, CH.-F., KAMIYA, Y., NAGASAWA, H. and TAMURA, S. (1975). Occurrence of two catechol derivatives in *Chrysanthemum morifolium*. *Agr. biol. Chem., Tokyo* **39**, 573–574.

CHIN, C., JONES, E. R. H., THALOR, V., APLIN, R. T., DURHAM, J. C., CASCON, S. C., MORS, W. R. and TURSCH, B. M. (1965). A toxic C_{14} polyacetylenic tetrahydropyranol alcohol from Compositae. *Chem. Commun.*, 152–154.

CHISHOLM, M. J. and HOPKINS, C. Y. (1966). Kamlolenic acid and other conjugated fatty acids in some seed oils. *J. Am. Oil Chemists' Soc.* **43**, 390–392.

CHRISTENSEN, W. I. (1965). Milk sickness: A review of the literature. *Econ. Bot.* **19**, 293–300.

CONACHER, H. B. S., GUSTONE, F. B., HORNBY, G. M. and PADLEY, F. B. (1970). Glyceride studies IX: Intraglyceride distribution of vernolic acid and of five conjugated octadecatrienoic acids in seed glycerides. *Lipids* **5**, 434–441.

CONN, E. E. (1973). Biosynthesis of cyanogenic glycosides. *Biochem. Soc. Symp.* **38**, 277–302.

CORBETT, R. E., YOUNG, H. and WILSON, R. S. (1964). Extractives from the bark of *Olearia paniculata*. *Aust. J. Chem.* **17**, 712–714.

CRUSE, W. B. T., JAMES, N. M. G., AL-SHAMMA, A. A., BEAL, J. K. and DOSKOTCH, R. W. (1971). The molecular structure of gutierolide, a novel chloro-diterpenoid lactone. *J. chem. Soc. chem. Commun.* **1971**, 1278–1280.

CUEVAS, L. A., GAREIA GIMENEZ, F. and ROMO DE VIVAR, A. (1972). Estructura de la estenolobina. *Rev. latinoamer. quim.* **3**, 22–27.

DANIELAK, R. and BORKOWSKI, B. (1970). Isolation of isofraxidin from herb of *Artemisia abrotanum* L. and search for it in some other *Artemisia* species. *Dissert. Pharm. Pharmacol. polon.* **22**, 231–235.

DANIELI, B., BOMBARDELLI, E., BONATI, A. and GAMBETTA, B. (1972). Structure of the diterpenoid carboxyatractyloside. *Phytochemistry* **11**, 3501–3504.

DASGUPTA, B., BASU, K. and DASGUPTA, S. (1968). Chemical investigation of *Pluchea lanceolata* II. Identity of pluchine with betaine hydrochloride. *Experientia* **24**, 882.

DEEM, A. W., THORP, JR., F. and DURRELL, L. W. (1939). Range plants newly found to be poisonous. *Science*, N.Y. **89**, 435.

DEVGAN, O. N. and BOKADIA, M. M. (1968). Isolation of 2,4,5-trimethoxyallyl-benzene from *Caesulia axillaris* oil. *Aust. J. Chem.* **21**, 3001–3003.

DOBBERSTEIN, R. H., TIN-WA, N., FONG, H. H. S., CRANE, F. A. and FARNSWORTH, N. R. (1974). Cytotoxic flavones from *Eupatorium altissimum* L. *Lloydia* **37**, 640.

DOMINGUEZ, X. A. and JIMÉNEZ JIMÉNEZ, S. (1973). Aislamento y estructura del juslimtetrol, *nor*-juslimdiolona, y la isocumambramina, metabolitos secudarios del *Croptylon divaricatum* (Compuesta). *Rev. Latinoamer. Quim.* **3**, 177–182.

DÖPKE, W. and FRITSCH, G. (1969). Echinin, ein Dihydrochinolin-Alkaloid aus den Samen von *Echinops ritro* L. *Pharmazie* **24**, 782.

DOSKOTCH, R. W. and EL-FERALY, F. S. (1969). Isolation and characterization of (+)-sesamin and β-cyclopyrethrosin from *Pyrethrum* flowers. *Can. J. Chem.* **47**, 1139–1142.

DUTTA, C. P., RAY, L. P. K. and ROY, D. N. (1972). Taraxasterol and its derivatives from *Cirsium arvense*. *Phytochemistry* **11**, 2267–2269.

EARLE, F. R., MIKOLAJCZAK, K. L. and WOLFF, I. A. (1964). Search for new industrial oils X. Calenduleae. *J. Am. Oil Chemists' Soc.* **41**, 345–347.

EARLE, F. R., BARCLAY, A. S. and WOLFF, I. A. (1966). Compositional variation in seed oils of the *Crepis* genus. *Lipids* **1**, 325–327.

EDGAR, J. A. and CULVENOR, C. C. J. (1974). Pyrrolizidine ester alkaloid in danaid butterflies. *Nature, Lond.* **248**, 614–615.

EDGAR, J. A. and CULVENOR, C. C. J. (1975). Pyrrolizidine alkaloids in *Parsonsia* species (family Apocynaceae) which attract Danaid butterflies. *Experientia* **31**, 393–394.

EDGAR, J. A., CULVENOR, C. C. J. and PLISKE, T. E. (1974). Coevolution of Danaid butterflies with their host plants. *Nature, Lond.* **250**, 646–648.

EL SOHLY, L., KNAPP, J. E., SLATKIN, D. J. and SCHIFF, JR., P. L. (1975). Constituents of *Ruscus aculeatus*. *Lloydia* **38**, 106–108.

EPSTEIN, W. W. and POULTER, C. D. (1973). A survey of some irregular monoterpenes and their biogenetic analogies to presqualene alcohol. *Phytochemistry* **12**, 737–747.

FARKAS, L. and PALLOS, L. (1967). Natürlich vorkommende Auronglykoside. *Fortschr. Chem. Org. Naturstoffe* **25**, 150–174.

FERGUSON, G., MARSH, W. C., MCCRINDLE, R. and NAKAMURA, E. (1975). Stereochemistry of clerodanes. X-Ray structure of a key diterpenoid from *Solidago arguta* Ait. *J.C.S. chem. Commun.* **1975**, 299.

FRYDMAN, B. and DEULOFEU, V. (1962). Alkaloids from *Carduus acanthoides* L. Structure of acanthoine and acanthoidine and synthesis of racemic acanthoidine. *Tetrahedron* **18**, 1063–1072.

FURUYA, T. and HIKICHI, M. (1973). Lindelofine and supinine: pyrrolizidine alkaloids from *Eupatorium stoechadosum*. *Phytochemistry* **12**, 225.

FURUYA, T., MURAKAMI, K. and HIKICHI, M. (1971). Senkirkine, a pyrrolizidine alkaloid from *Farfugium japonicum*. *Phytochemistry* **10**, 3306–3307.

GEISSMAN, T. A. and IRWIN, M. A. (1970). Chemical contributions to taxonomy and phylogeny in the genus *Artemisia*. *Pure appl. Chem.* **21**, 167–180.

GEISSMAN, T. A. and IRWIN, M. A. (1974). Chemical constitution and botanical affinity in *Artemisia*. In "Chemistry in Botanical Classification". (G. Bendz and J. Santesson, eds). pp. 135–143. Nobel Symposium **25**, (1973). Almquist-Wiksell, Uppsala.

GEISSMAN, T. A. and MATSUEDA, S. (1968). Sesquiterpene lactones. Constituents of diploid and polyploid *Ambrosia dumosa* Gray. *Phytochemistry* **7**, 1613–1622.

GEISSMAN, T. A., GRIFFIN, S., WADDELL, T. G. and CHEN, H. H. (1969). Sesquiterpene lactones. Some new constituents of *Ambrosia* species: *A. psilostachya* and *A. acanthicarpa*. *Phytochemistry* **8**, 145–150.

GEISSMAN, T. A., LEE, K. and MITCHELL, R. E. (1971). 6-Methoxy-7,8-methylenedioxycoumarin from *Artemisia carruthii* Wood. *Phytochemistry* **10**, 902–903.

GOMMERS, F. J. (1973). Nematicidal principles in Compositae. *Mededel. Landbouwhoogesch. Wageningen* 73–17, 71 pp.

GONZALEZ, A., BERMEJO BERRARA, J. BERMEJO BERRARA, J. L. and MASSANET, G. M. (1972). Chemistry of natural compounds XI. *Eupatorium macrocephalum*. *An. Quim.* (*Madrid*) **68**, 319–323.

GORECKI, P. and MRUGASIEWICZ, K. (1974). [Attempts to utilize the herb of chicory for isolation of esculetin.] *Herba polon.* **20**, 339–343. (In Polish with English summary).

GORINSKY, C., TEMPLETON, W. and ZAIDI, S. A. H. (1973). Isolation of ichthyothereol and its acetate from *Clibadium sylvestre*. *Lloydia* **36**, 352–353.

GUERRO, C. and VIVAR, DE, R. (1973). Estructura y estereoquimica de la bacchofertina, diterpene aislado de *Baccharis conferta*. *Rev. latinoamer. quim.* **4**, 178–184.

GUNN, CH. R. and WHITE, G. A. (1974). Stokesia laevis: Taxonomy and economic value. *Econ. Bot.* **28**, 130–135.

HÄNSEL, R., SCHULZ, H. and LEUCKERT, CH. (1964). Das Lignanglykosid Arctiin als chemotaxonomisches Merkmal in der Familie der Compositae. *Z. Naturf.* **19b**, 727–734.

HÄNSEL, R., OHLENDORF, D. and PELTER, A. (1970). Obtusifolin, ein Flavanon mit einem biogenetisch unüblichen C_9-Baustein. *Z. Naturf.* **25b**, 989–994.

HÄNSEL, R., SCHULZ, J. and PELTER, A. (1972). Structure of Silybin. *J.C.S. chem. Commun.* **1972**, 195–196.

HANSON, J. R. (1968). "The Tetracyclic Diterpenes". Pergamon Press, Oxford.

HARBORNE, J. B., ed. (1967). "Comparative Biochemistry of the Flavonoids". Academic Press, London and New York.

HARBORNE, J. B. (1974). Flavonoids as systematic markers in the angiosperms. *In* "Chemistry in Botanical Classification" (G. Bendz and J. Santesson, eds). Nobel Symposium 25, pp. 103–115. Almquist-Wiksell, Uppsala.

HASKINS, F. A., GORZ, H. J. and LEFFEL, R. C. (1972). Form and level of coumarin in Deer's Tongue, *Trilisa odoratissima*. *Econ. Bot.* **26**, 44–48.

HAYASHI, K., NATORIGAWA, A. and MITSUHASHI, A. (1972). Integerrimine from *Cacalia hastata* L. subsp. *orientalis* Kitamura. *Chem. pharm. Bull.* (*Tokyo*) **20**, 201–202.

HEGNAUER, R. (1964). "Chemotaxonomie der Pflanzen", Band 3. Birkhäuser Verlag, Basel.

HEGNAUER, R. (1966). Comparative phytochemistry of alkaloids. *In* "Comparative Phytochemistry" (T. Swain, ed.), pp. 211–230. Academic Press, London and New York.

HEGNAUER, R. (1969). "Chemotaxonomie der Pflanzen", Band 5. Birkhäuser Verlag, Basel.

HEGNAUER, R. (1973a). Die cyanogenen Verbindungen der Liliatae und Magnoliatae-Magnoliidae: Zur systematischen Bedeutung des Merkmals der Cyanogenese. *Biochemical Systematics* **1**, 191–197.

HEGNAUER, R. (1973b). "Chemotaxonomie der Pflanzen", Band 6. Birkhäuser Verlag, Basel.

HENDERSON, M. S., MCCRINDLE, R. and MCMASTER, D. (1973a). Non-acidic diterpenoids from *Solidago gigantea* var. *serotina*. *Can. J. Chem.* **51**, 1346–1358.

HENDERSON, M. S., MURRAY, R. D. H., MCCRINDLE, R. and MCMASTER, D. (1973b). The constitution of diterpenoids from *Solidago juncea* Ait. *Can. J. Chem.* **51**, 1322–1331.

HÉRISSET, A., CHAUMONT, J.-P. and PARIS, R. (1970). Les polyphénols de la chamomille romaine. *Plantes Méd. Phytothérapie* **4**, 189–200.

HEROUT, V. (1971). Chemotaxonomy of the family Compositae (Asteraceae). *In* "Pharmacognosy and Phytochemistry" (E. Wagner and L. Hörhammer, eds), pp. 93–110. Springer-Verlag, Berlin, Heidelberg and New York.

HEROUT, V. and ŠORM, F. (1969). Chemotaxonomy of the sesquiterpenoids of Compositae. In "Perspectives in Phytochemistry" (J. B. Harborne and T. Swain, eds), pp. 139-165. Academic Press, London and New York.

HERZ, W. (1968). Pseudoguaianolides in Compositae. *Recent Advances in Phytochemistry* **1**, 229-269. North-Holland, Amsterdam.

HERZ, W. (1974). Pseudoguaianolides in Compositae. In "Chemistry in Botanical Classification" (G. Bendz and J. Santesson, eds). Nobel Symposium 25 (1973), pp. 153-172. Almquist-Wiksell, Uppsala.

HERZ, W. and BHAT, S. V. (1970). Coumarin in *Amblyolepis setigera*. *Phytochemistry* **9**, 817.

HERZ, W. and SUBRAMANIAN, P. S. (1971). Pyromeconic acid in *Parthenium integrifolium*. *Phytochemistry* **10**, 1689-1690.

HERZ, W. and WAHLBERG, I. (1973). A new dihydrobenzofuran from *Liatris provincialis*. *Phytochemistry* **12**, 429-432.

HERZ, W., BHAT, S. V. and SANTHANAM, P. S. (1970). Coumarins of *Artemisia dracunculoides* and 3',6-dimethoxy-4',5,7-trihydroxyflavone in *A. arctica*. *Phytochemistry* **9**, 891-894.

HIKICHI, M. and FURUYA, T. (1974). Syneilesine, a new pyrrolizidine alkaloid from *Syneilesis palmata*. *Tetrahedron Letters* **1974**, 3657-3660.

HILDITCH, T. P. and WILLIAMS, P. N. (1964). "The Chemical Constitution of Natural Fats" (4th edition). Chapman and Hall, London.

HOPKINS, C. Y. and CHISHOLM, MARY J. (1964). Occurrence of *trans*-3-hexadecenoic acid in a seed oil. *Can. J. Chem.* **42**, 2224-2227.

HOPKINS, C. Y. and CHISHOLM, MARY J. (1965). Occurrence in seed oils of some fatty acids with conjugated unsaturation. *Can. J. Chem.* **43**, 3160-3164.

HOSOZAWA, S., KATO, N. and MUNAKATO, K. (1974). Antifeeding active substances for insect in *Caryopteris divaricata* Maxim. *Agric. biol. Chem. (Tokyo)* **38**, 823-826.

HUI, W. H., LAM, W. K. and TYE, S. M. (1971). Triterpenoid and steroid constituents of *Aster baccharoides*. *Phytochemistry* **10**, 903-904.

HURST, E. (1942). "The Poison Plants of New South Wales". Snelling Printing Works, Sydney.

JACOBSON, M. and CROSBY, D. G., eds (1971). "Naturally Occurring Insecticides", pp. 137-176. Marcel Dekker, New York.

JAIN, C. and KARCHESY, J. J. (1971). Concerning the chemical constituents of *Matricaria matricarioides*. *Phytochemistry* **10**, 2825-2826.

JEFFERIES, P. R., KNOX, J. R., PRICE, K. R. and SCAF, B. (1974). Constituents of the tumor-inhibitory extract of *Olearia muelleri*. *Aust. J. Chem.* **27**, 221-225.

JOHNS, S. R., LAMBERTON, J. A., PRICE, J. R. and SIOUMIS, A. A. (1968). Identification of coumarins from *Lepiniopsis ternatensis* (Apocynaceae), *Pterocaulon sphacelatum* (Compositae) and *Melicope melanophloia* (Rutaceae). *Aust. J. Chem.* **21**, 3079-3080.

JONES, Q. and EARLE, F. R. (1966). Chemical analysis of seeds II: Oil and protein content of 759 species. *Econ. Bot.* **20**, 127-155.

KAHL, W., GRODZINSKA-ZACHWIEJA, Z. and HOLIK, Z. (1969). Isolation of isomers of chlorogenic acid from *Cichorium intybus*. *Diss. pharm. pharmacol. polon.* **21**, 449-455.

KASPRZYK, Z. and KOZIEROWSKA, T. (1966). Distribution of sterols and triterpenic alcohols in plants of the Compositae family. *Bull. Acad. polon. Sci.* Cl. II **14**, 645–649.

KASPRZYK, Z., PYREK, J. ST., JOLAND, S. D. and STEELINK, C. (1970). The identity of calenduladiol and thurberin: A lupenediol found in marigold flowers and organpipe cactus. *Phytochemistry* **9**, 2065–2066.

KASTURI, T. R., MANI, THOMAS and ABRAHAM, E. M. (1973). Essential oil of *Ageratum conyzoides*: Isolation and structure of two new constituents. *Indian J. Chem.* **11**, 91–95.

KAVAK, J., GUERREIRO, E., GIORDANO, O. S. and ROMO, J. (1973). Flavanoides del *Baccharis salicifolia* (R. et P.) Persoon. *Rev. latinoamer. quim.* **4**, 101–104.

KEELER, R. F. (1975). Toxins and teratogens of higher plants. *Lloydia* **38**, 56–86.

KHALMATOV, KH. KH. and ALIEV, KH. (1967). *Centaurea depressa*. *Chem. Abstr.* **67**, 71123.

KINDL, H. and HOFFMAN-OSTENHOF, O. (1967). Die Bildung von L-Viburnit in wechsel und Vorkommen. *Fortschr. Chem. Org. Naturstoffe* **24**, 149–205.

KINDL, H. and HOFFMAN-OSTENHOF, O. (1967) Die Bildung von L-Viburnit in *Chrysanthemum vulgaris*. *Phytochemistry* **6**, 77–83.

KLÁSEK, A., NEUNER JEHLE, N. and ŠANTAVY, F. (1969). Structure of the amide $C_6H_{11}NO$ from *Senecio rivularis*. *Chemy Ind.*, 987–988.

KLÁSEK, A., SEDMERA, P. and ŠANTAVY, F. (1971). Alkaloids from some plants of the genus *Ligularia*. *Coll. Czech. Chem. Commun.* **36**, 2205–2215.

KLEIMAN, R., EARLE, F. R. and WOLFF, I. A. (1966). The *trans*-3-enoic acids of *Grindelia oxylepis* seed oil. *Lipids* **1**, 301–304.

KLEIMAN, R., SPENCER, G. F., TJARKS, L. W. and EARLE, F. R. (1971). Oxygenated *trans*-3-olefinic acids in a *Stenachaenium* seed oil. *Lipids* **6**, 617–622.

KLOSS, P. (1969). Inhaltsstoffe von *Espeletia schultzii*. *Arch. Pharm.* **302**, 376–381.

KNOWLES, P. F. (1965). Variability in oleic and linoleic acid contents of safflower oil. *Econ. Bot.* **19**, 53–62.

KNOWLES, P. F. and MUTWAKIL, A. (1963). Inheritance of low iodine value of safflower selections from India. *Econ. Bot.* **17**, 139–145.

KOCH, H. (1975). Einfluss von Silymarin auf Keimung und embryonales Wachstum von Gartenkresse (*Lepidium sativum*). *Experientia* **31**, 281–283.

KOHLMÜNZER, S. and TOMCZYK, H. (1969). Pyrrolizidine alkaloids of *Emilia flammea* Cass. *Diss. pharm. pharmacol. polon.* **21**, 433–441.

KOHLMÜNZER, S., TOMCZYK, H. and SAINT-FIRMIN, A. (1971). Emiline, a new othonecine ester from *Emilia flammea* Cass. *Diss. pharm. pharmacol. polon.* **23**, 419–427.

KUNDU, S. K., CHATTERJEE, A. and RAO, R. S. (1966). Isolation of fernenol, a new pentacyclic alcohol from *Artemisia vulgaris*. *Tetrahedron Letters*, 1043–1045.

KUPCHAN, S. M., SIGEL, C. W., HEMINGWAY, R. J., KNOX, J. R. and UDAYMURTHY, M. S. (1969). Cytotoxic flavones from *Eupatorium* species. *Tetrahedron* **25**, 1603–1615.

KUSUMOTO, SH., OKAZAKI, T., OHSUKA, A. and KOTAKE, M. (1969). Structure and stereochemistry of solidagonic acid. *Bull. chem. Soc. Japan* **42**, 812–820. See also *Tetrahedron Letters*, 4325–4327.

LAM, J. and DRAKE, D. (1973). Polyacetylenes from *Senecio jacobaea*. *Phytochemistry* **12**, 149–151.
LASSAK, E. V. and PINHEY, J. T. (1968). The constituents of some species of *Helichrysum* (Compositae). *Aust. J. Chem.* **21**, 1927–1929.
LEE, K. H., ANUFORO, D. C., HUANG, E.-SH. and PIANTADOSI, L. (1972). Angustibalin, a new cytotoxic lactone from *Balduinia angustifolia* (Pursh) Robins. *J. pharm. Sci.* **61**, 626–628.
LEONICO D'ALBUQUERQUE, I., CORIO, E., DELLE MONACHE, F., TUCCI, A. P. and MARINI-BETTOLO, G. B. (1969). Occurrence of (-)-kaur-16-en-19-oic acid in *Ichthyothere cunabi*. *Ann. Ist. Super. Sanita* **5**, 557–558. Ex *Chem. Abstr.* **73**, 117128 (1970).
LLOYD, H. A. and FALES, H. M. (1967). Terpene alcohols of *Helichrysum dendroideum*. *Tetrahedron Letters*, 4891–4895.
LOCOCK, R. A., BEAL, J. L. and DOSKOTCH, R. W. (1966). Alkaloid constituents of *Eupatorium serotinum*. *Lloydia* **29**, 201–205.
LUCKNER, M. (1969). Chinoline. In "Biosynthese der Alkaloide" (K. Mothes and H. R. Schütte, eds), pp. 510–550. VEB Deutscher Verlag der Wissenschaften, Berlin.
MABRY, T. J. (1970). Infraspecific variation of sesquiterpene lactones in *Ambrosia* (Compositae): applications to evolutionary problems at the populational level. In "Phytochemical Phylogeny" (J. B. Harborne, ed.), pp. 269–300. Academic Press, London and New York.
MAZZEI PLANAS, G. and KUĆ, J. (1968). Contraceptive properties of *Stevia rebaudiana*. *Science, N.Y.* **162**, 1007.
MIKOLAJCZAK, K. L. and SMITH, JR., C. R. (1967a). Pentacyclic triterpenes of *Jurinea anatolica* Boiss. and *J. consanguinea* DC. fruit. *Lipids* **2**, 127–132.
MIKOLAJCZAK, K. L. and SMITH, JR., C. R. (1967b). Optically active trihydroxy acids of *Chamaepeuce* seed oils. *Lipids* **2**, 261–265.
MITSUHASHI, H., UNEO, J. and SUMITA, T. (1975). Studies on the cultivation of *Stevia rebaudiana*. Determination of stevioside. *J. pharm. Soc. Japan* **95**, 127–130.
MIYAKADO, M., OHNO, M., YOSHIOKA, H., MABRY, T. J. and WHIFFIN, T. (1974). Gymnospermin: a new labdantriol from *Gymnosperma glutinosa*. *Phytochemistry* **13**, 189–190.
MO, F., ANTHONSEN, T. and BRUUN, T. (1972). Revised structure of the triterpenoid *Baccharis* oxide. *Acta chem. scand.* **26**, 1287–1288.
MORALES MÉNDEZ, A., USUBILLAGA, A., BANERJEE, A. K. and NAKANO, T. (1973). Studies on the constituents of *Espeletia weddellii*. *Planta Medica* **24**, 243–248.
MORRIS, L. J., MARSHALL, M. O. and HAMMOND, E. W. (1968). *Trans*-3-enoic acids of *Aster alpinus* and *Arctium minus* seed oils. *Lipids* **3**, 91–95.
MORTIMER, P. H. and WHITE, E. P. (1967). Hepatotoxic substance in *Brachyglottis repanda*. *Nature, Lond.* **214**, 1255.
MURAE, T., TANAHASHI, Y. and TAKAHASHI, T. (1968). 5,6-Dimethoxy-2-isopropenyl benzofuran from *Ligularia stenocephala*. *Tetrahedron* **24**, 2177–2181.
MURAKAMI, T., ISA, T. and SATAKE, T. (1973). Eine Neuuntersuchung der Inhaltstoffe von *Siegesbeckia orientalis*. *Tetrahedron Letters*, 4991–4994.

NARAYANAN, P., ZECHMEISTER, K., RÖHRL, M. and HOPPE, W. (1970). The crystal structure analysis of obtusifolin, $C_{24}H_{22}O_7$, a flavanone. *Tetrahedron Letters* 3643–3644.

OHSUKA, A., KUSUMOTO, SH. and KOTAKE, M. (1972). Structure of the diterpene carboxylic acid from *Solidago altissima*. *Nippon Kagaku Kaishi* **1972**, 963–967. Ex *Chem. Abstr.* **77**, 45510 (1972).

OPITZ, L. and HÄNSEL, R. (1970). Helipyron, ein Methylen-bis-triacetsäurelacton aus *Helichrysum italicum*. *Tetrahedron Letters*, 3369–3370.

OPITZ, L. and HÄNSEL, R. (1971). Phthalide aus *Helichrysum italicum*. *Arch. Pharm.* **304**, 228–230.

ORTEGA, A. and ROMO, J. (1974). Zexmeniol, un acetilcromeno aislado de *Zexmenia brevifolia*. *Rev. Latinoamer. Quim.* **5**, 223–224.

OURISSON, G. (1974). Some aspects of the distribution of diterpenes in plants. *In* "Chemistry in Botanical Classification" (G. Bendz and J. Santesson, eds). Nobel Symposium 25, pp. 129–134. Nobel Foundation, Stockholm.

PAKRASHI, S. C., GOSH DASTIDAR, P. P. and GUPTA, S. K. (1970). Diterpenoids from *Enhydra fluctuans*. *Phytochemistry* **9**, 459.

PALTER, R., HADDON, W. F. and LUNDIN, R. E. (1971). The complete structure of matairesinol monoglucoside. *Phytochemistry* **10**, 1587–1589.

PALTER, R., LUNDIN, R. E. and HADDON, W. F. (1972). A cathartic lignan glycoside isolated from *Carthamus tinctorius*. *Phytochemistry* **11**, 2871–2874.

PATEL, M. B. and ROWSON, J. M. (1964). Investigation of certain Nigerian medicinal plants I. Preliminary pharmacological and phytochemical screenings for cardiac activity. *Planta Medica* **12**, 33–42.

PHILLIPS, B. E., SMITH, JR., C. R. and HAGEMANN, J. W. (1969). New source of vernolic acid. *Lipids* **4**, 473–477.

PINHEY, J. T., SIMPSON, R. F. and BATEY, I. L. (1971). The constituents of *Olearia heterocarpa*. The structure of olearin, a diterpene dilactone of the cascarillin group. *Aust. J. Chem.* **24**, 2621–2637.

PIOZZI, F., PASSANNANTI, S., PATERNOSTRO, M. P. and SPRIO, V. (1971). Kaurenoid diterpenes from *Espeletia grandiflora*. *Phytochemistry* **10**, 1164–1166. See also *Gazz. Chim. Ital.* **98**, 907–910 (1968): Grandiflorolic acid.

PLOUVIER, V. (1963). Distribution of aliphatic polyols and cyclitols. *In* "Chemical Plant Taxonomy" (T. Swain, ed.), pp. 313–336. Academic Press, London and New York.

PLOUVIER, V. (1964a). Sur deux hétérosides nouveaux, l'érigéroside isolé des *Erigeron* et le dianthoside isolé des *Dianthus*. *C. r. hebd. Séanc. Acad. Sci., Paris* **258**, 1099–1102.

PLOUVIER, V. (1964b). Recherches des L-inositol, L-quebrachitol et D-pinitol dans quelques groupes botaniques. *C. r. hebd. Séanc. Acad. Sci., Paris* **258**, 2921–2924.

PLOUVIER, V. (1968). Recherche du fraxoside et hétérosides coumariniques voisins dans quelques groupes botaniques. *C. r. hebd. Séanc. Acad. Sci., Paris* **267D**, 1883–1885.

PLOUVIER, V. (1970). Présence d'un mono-acetate de *myo*-inositol dans les campanules. Recherches des L-inositol, scyllitol, mannitol et sorbitol dans quelques groupes botaniques. *C. r. hebd. Séanc. Acad. Sci., Paris* **270D**, 560–563.
PLOUVIER, V. (1971a). Sur la recherche du scyllitol, du *myo*-inositol et du dulcitol dans quelques groupes botaniques. *C. r. hebd. Séanc. Acad. Sci., Paris* **272D**, 141–144.
PLOUVIER, V. (1971b). Nouvelle recherche du scyllitol dans quelques groupes botaniques: sa large répartition chez les Composées. *C. r. hebd. Séanc. Acad. Sci., Paris* **273D**, 1625–1628.
PLOUVIER, V. (1972). Compléments sur la recherche du scyllitol chez les plantes supérieures. *C. r. hebd. Séanc. Acad. Sci., Paris* **275D**, 2993–2996.
PLOUVIER, V. (1973). Recherche de cyclitol dans quelques groupes botaniques: Cryptogames, Polygonacées, Apocynacées etc. *C. r. hebd. Séanc. Acad. Sci., Paris* **277D**, 1945–1948.
POTTER, J. L. and MABRY, T. J. (1972). Origin of Texas Gulf populations of *Ambrosia psylostachya*. *Phytochemistry* **11**, 715–723.
POWELL, R. G., SMITH, JR., C. R., GLASS, C. A. and WOLFF, I. A. (1965). *Helichrysum* seed oil I. and II. *J. Am. Oil Chemists' Soc.* **42**, 165–169; *J. org. Chem.* **30**, 610–615.
POWELL, R. G., SMITH, JR., C. R. and WOLFF, I. A. (1967). *cis*-5,*cis*-9,*cis*-12-Octadecatrienoic acid and some unusual oxygenated acids in *Xeranthemum annuum* seed oil. *Lipids*, **2**, 172–177.
POZETTI, G. L. and FERREIRA, P. C. (1967). *Rev. Fac. Farm. Bioquim. São Paulo* **5**, 253–255. Ex *Chem. Abstr.* **68**, 104895 (1968).
PYREK, J. ST. (1970). New pentacyclic diterpene acid. Trachyloban-19-oic acid from sunflower. *Tetrahedron* **26**, 5029–5032.
PYREK, J. ST. and BARANOWSKA, E. (1973). Faradiol and arnidiol—Revision of the structure. *Tetrahedron Letters* **1973**, 809–810.
QUESADA, DE, T. G., RODRIGUEZ, B. and VALVERDE, S. (1972). The constituents of *Helichrysum stoechas*. *Phytochemistry* **11**, 446–449.
RENTE, RUTH, BONNET, P.-H. and BOHLMANN, F. (1972). Über Inhaltsstoffe von *Anacyclus pyrethrum* DC. *Chem. Ber.* **105**, 1694–1700.
RINN, W. (1970). Isobuttersäurethymylester-Hauptbestandteil des ätherischen Öles der Rhizome und Wurzeln von *Arnica chamissonis*. *Planta Medica* **18**, 147–149.
ROMO DE VIVAR, A., CUEVAS, L. A. and GUERRERO, C. (1971). Euglabrin from *Eupatorium glabratum* H.B. et K. *Rev. Latinoamer. Quim.* **2**, 32–34. Ex *Chem. Abstr.* **75**, 45625 (1971).
ROMO DE VIVAR, A., MONTIEL, F. and DIAZ, E. (1974). Estructura e estereoquimica de las amidas sulfuradas aisladas di differentes especies de *Chrysanthemum*. *Rev. Latinoamer. Quim.* **5**, 32–40.
SAKAMURA, S., YOSHIHARA, T. and TOYODA, K. (1973). The constituents of *Petasites japonicus*. Structures of fukuiic acid and fukinolic acid. *Agric. biol. Chem. (Tokyo)* **37**, 1915–1921.
SALEH, M. M. and RIZK, A. M. (1974). Flavonoids and coumarins of *Anthemis pseudocotula*. *Planta Medica* **25**, 60–62.

SASAKI, S., CHIANG, H. C., HABAGUCHI, K., YAMADA, T., NAKANISHI, K., MAT-SUEDA, S., HSÜ, K.-Y. and WU, W.-N. (1966). Studies on the constituents of medicinal plants of Taiwan. *J. pharm. Soc. Japan* **86**, 869–870.

SCHMERSAHL, P. (1966). Über das Vorkommen von Cumarin-Derivaten im Kraut von *Artemisia abrotanum*. *Planta Medica* **14**, 179–183.

SCHWENDIMANN, J.-M., TABACCHI, R. and JACOT-GUILLARMOD, A. (1974). Identification des composés phénoliques dans les feuilles de *Homogyne alpina*, *Petasites albus*, *P. hybridus* et *Adenostyles alliariae*. *Helv. chim. Acta* **57**, 552–557.

SCHWENKER, G., KLOSS, P. and ENGELS, W. (1967). Über die Isolierung von Prenyletin aus *Haplopappus baylahuen*. *Pharmazie* **22**, 724–725.

SEIGLER, D. S. (1975). Isolation and characterization of naturally occurring cyanogenic compounds. *Phytochemistry* **14**, 9–29.

SHAFIZADEH, F. and MELNIKOFF, A. B. (1970). Coumarins of *Artemisia tridentata*. *Phytochemistry* **9**, 1311–1316.

SHIELDS, L. M. (1952). Distribution of the bitter principle in the shoot of *Helenium tenuifolium*. *Bot. Gazz.* **113**, 471–475.

SHORLAND, F. B. (1963). The distribution of fatty acids in plant lipids. In "Chemical Plant Taxonomy" (T. Swain, ed.), pp. 253–311. Academic Press, London and New York.

SHTACHER, G. and KASHMAN, Y. (1971). Chemical investigation of volatile constituents of *Inula viscosa*. *Tetrahedron* **27**, 1343–1349.

SILVA, M. and SAMMES, P. G. (1973). A new diterpenic acid and other constituents of *Haplopappus foliosus* and *H. angustifolius*. *Phytochemistry* **12**, 1755–1758.

ŚLIWOWSKI, J., DZIEWANOWSKA, K. and KASPRZYK, Z. (1973). Ursadiol, a new triterpene diol from *Calendula officinalis* flowers. *Phytochemistry* **12**, 157–160.

SÖRENSEN, N. A. (1968). The taxonomic significance of acetylenic compounds. *Recent Advances in Phytochemistry* **1**, 187–227. North-Holland, Amsterdam.

SÖRENSEN, J. S. and SÖRENSEN, N. A. (1966). A preliminary investigation of *Coreopsis gigantea* (Kell.) H. M. Hall. *Acta Chem. Scand.* **20**, 992–1002.

SÖRENSEN, J. S. and SÖRENSEN, N. A. (1969). Investigation of *Erigeron* ssp. from the Australian mountains and Tasmania. *Aust. J. Chem.* **22**, 751–760.

SÖRENSEN, J. S., VE, B., ANTHONSEN, T. and SÖRENSEN, N. A. (1968). Studies related to naturally occurring acetylene compounds. XXXIV Some Australian members of the genus *Cotula* L. *Aust. J. Chem.* **21**, 2037–2051.

SPENCER, G. F., KLEIMAN, R., EARLE, F. R. and WOLFF, I. A. (1969). Cis-5-monoethenoid fatty acids of *Carlina* (Compositae) seed oils. *Lipids* **4**, 99–101.

STAHL, E. (1952). Nachweis der Vorstufe des Azulens in den Drüsenhaaren der Schafgarbe. *Naturwissenschaften* **39**, 551–552.

STAHL, E. (1957). Über Vorgänge in den Drüsenhaaren der Schafgarbe. *Z. Bot.* **45**, 297–315.

STEELINK, C. and SPITZER, J. C. (1966). Sesquiterpene lactones in chemotaxonomy. *Phytochemistry* **5**, 357–365.

STUDENNIKOVA, L. D. and KHALETSKY, A. M. (1965). Chemical composition of *Atractylodes ovata* (Thunb.) DC. *Aptechn. Delo* **14**, No 6, 23–26. Ex *Chem. Abstr.* **64**, 7037 (1966).

SUGA, T., SHISHIBORI, T., KOSELA, S. and SOOD, V. K. (1975). The neutral volatiles of *Boenninghausenia albiflora*. *Phytochemistry* **14**, 308–309.

TACHIBANA, K. and TAKAHASHI, T. (1975). The conversion of shionone into dihydrobaccharis oxide. *Tetrahedron Letters*, 1857–1858.

TAKAHASHI, T., MORIYAMA, Y., TANAHASHI, Y. and OURISSON, G. (1967). The structure of shionone. *Tetrahedron Letters*, 2991–2996.

TALAPATRA, S. K., BHAR, D. S. and TALAPATRA, B. (1974). Dammaradienyl acetate and taraxasterol from *Eupatorium cannabinum*. *Aust. J. Chem.* **27**, 1137–1142.

TALLENT, W. H., COPE, DIANA G., HAGEMANN, J. W., EARLE, F. R. and WOLFF, I. A. (1966). Identification and distribution of epoxyacyl groups in new, natural epoxy oils. *Lipids* **1**, 335–340.

TAYLOR, D. R. and WRIGHT, J. A. (1971). Chromenes from *Eupatorium riparium*. *Phytochemistry* **10**, 1665–1667.

THOMSON, R. H. (1971). "Naturally Occurring Quinones" (2nd edition). Academic Press, London and New York.

USUBILLAGA, A. and MORALES MÉNDEZ, A. (1970). Kaurene derivatives in *Espeletia tenorae*. *Rev. Latinoamer. Quim.* **1**, 128–131. Ex *Chem. Abstr.* **75**, 16119 (1971).

VRIES, M. A. DE (1948). "Over de vorming van phytomelaan bij *Tagetes patula* L. en enige andere Compositae". Thesis, University of Leiden.

VRKOČ, J., DOLEJŠ, J., SEDMERA, P., VAŠIČKOVÁ, S. and ŠORM, F. (1971). The structure of arenol and homoarenol, α-pyrone derivatives from *Helichrysum arenarium*. *Tetrahedron Letters*, 247–250.

VRKOČ, J., UBIK, K. and SEDMERA, P. (1973). Phenolic extractives from the achenes of *Helichrysum arenarium*. *Phytochemistry* **12**, 2062.

WAGENITZ, G. (1964). Campanulatae. In Engler's "Syllabus der Pflanzenfamilien" (H. Melchior, ed., 12th edition), vol. 2, pp. 478–497. Gebr. Borntraeger, Berlin.

WAGNER, H., HÖER, R., MURAKAMI, T. and FARKAS, L. (1973). Isolierung, Strukturaufklärung und Synthese von 4′,5,7-Trihydroxy-3′,6-dimethoxy-7-mono-β-D-glucopyranosid (Jaceosid), einem neuen Flavonglucosid aus den Wurzeln von *Centaurea jacea* L. *Chem. Ber.* **106**, 20–27. It seems that the names jaceidin and centaureidin are used rather inconsistently for the two trimethylated quercetagetin derivatives; compare: FARKAS, L., HÖRHAMMER, L., WAGNER, H., RÖSLER, H. and GURNIAK, R. (1964). Die Struktur des Jaceins und dessen Synthese aus dem Aglykon und Acetobromglucose. *Chem. Ber.* **97**, 610–615. Names interchanged.

WAHLBERG, I., KARLSSON, K. and ENZELL, C. R. (1972). Non-volatile constituents of Deertongue leaf. *Acta chem. scand.* **26**, 1383–1388.

WALLACE, J. W. (1974). Tricin-5-O-glucoside and other flavonoids of *Cirsium arvense*. *Phytochemistry* **13**, 2320–2321.

WALT, S. J. VAN DER (1944). Some aspects of the toxicology of hydrocyanic acid in ruminants. *Onderstepoort J. vet. Sci. Anim. Ind.* **19**, 79–153.

WEBER, U. and DEUFEL, J. (1951). Zur Cytologie der Drüsenhaare von *Achillea millefolium*. *Arch. Pharm.* **284**, 318–323.

WEINGES, K., MÜLLER, R., KLOSS, P. and JAGGY, H. (1970). Isolierung und Konstitutionsaufklärung eines optisch aktiven Dehydro-diconiferylalkohols aus den Samen der Mariendistel, *Silybum marianum.* *Liebigs Ann. Chem.* **736**, 170–172.

WHITE, G. A., WILLINGHAM, B. C., SKRDLA, W. H., MASSEY, J. H., HIGGINS, J. J., CALHOUN, W., DAVIS, A. W., DOLAN, D. D. and EARLE, F. R. (1971). Agronomic evaluation of prospective new crop species. *Econ. Bot.* **25**, 22–54.

WHITE, G. A., WILLINGHAM, B. C. and CALHOUN, W. (1973). Agronomic evaluation of prospective new crop species III. *Crepis alpina*—source of crepenynic acid. *Econ. Bot.* **27**, 320–322.

WILLAMAN, J. J. and LI, H.-L. (1970). "Alkaloid-bearing Plants and their Contained Alkaloids 1957–1968". *Lloydia* **33**, supplement September 1970, number 3A.

WOLFF, I. A. (1966). Seed lipids. *Science, N.Y.*, **154**, 1140–1149.

YAMAGITA, M., IHAYAMA, S., KWATAMA, T. and OKURA, T. (1969, 1970). Pulchellidine, a novel sesquiterpene alkaloid isolated from *Gaillardia pulchella* Foug. *Tetrahedron Letters*, 2073–2076; *Tetrahedron Letters*, 131–134.

YATSYUK, YA. K., LASHENKO, S. S. and BATYUK, V. S. (1968). Arbutin occurrence in some species of *Serratula*. *Khim. Prir. Soedin*, 54. Ex *Chem. Abstr.* **69**, 8878 (1968).

YOSHIZAKI, M., SUZUKI, H., SANO, K., KIMURA, K. and NAMBA, T. (1974). On the constituents of *Eupatorium* spp. *J. pharm. Soc. Japan* **94**, 338–342.

YOSIOKA, I. and YAMADA, Y. (1963). Isolation of dammaradienyl acetate from *Inula helenium* L. *J. pharm. Soc. Japan* **83**, 801–802.

YUSUPOV, M. I. and SIDYAKIN, G. P. (1973). [Oxycoumarin arscotin from *Artemisia scotina*.] *Khim. Prir. Soedin*, 430.

ZALKOV, L. H., CABOT, G. A., CHETTY, G. L., GHAL, M. and KEEN, G. (1968). Isolation of a new sterol: Stigmasta-8(14), 22-dien-3β-ol. *Tetrahedron Letters* 5727–5729.

ZALKOV, L. H., KEINAN, E., STEINDL, S., KALYANARAMAN, A. R. and BERTRAND, J. A. (1972). On the absolute configuration of toxol at C-3. *Tetrahedron Letters*, 2873–2876.

ZANE, A. and WENDER, S. H. (1966). Depsides in sunflower leaves. *Nature, Lond.* **209**, 80–81.

ZATSNY, I. L., GOROVITS, M. B. and ABUBAKIROV, N. K. (1973). [Arbutin from *Serratula sogdiana*.] *Khim. Prir. Soedin*, 437–438.

Chapter 11
Sesquiterpene lactones in the Compositae*

WERNER HERZ

*Department of Chemistry, The Florida State University,
Tallahassee, Florida, U.S.A.*

Abstract. Sesquiterpene lactones seem to be characteristic secondary metabolites of Compositae. The various classes of sesquiterpene lactones found in the family are surveyed and arranged in order of increasing "biogenetic complexity", emphasis being given to the idea that use of chemical characters in taxonomy must be based on knowledge of their biosynthesis. The distribution of the various lactone types in the tribes of Compositae is examined. Although the available data are still very limited, some apparent regularities have emerged to which attention is directed. Possible implications at the tribal and subtribal level are discussed.

CONTENTS

Introduction	337
Sesquiterpene lactones and "biogenetic complexity"	339
Distribution of sesquiterpene lactones in Compositae	345
Possible implications of lactone distribution	348
References	353

INTRODUCTION

As a family still undergoing dynamic evolution, the Compositae are rich in secondary metabolites with whose distribution in the various tribes the chemical surveys in this text are concerned. Two classes of secondary metabolites seem to have been selected for special consideration, namely the polyacetylenes and the sesquiterpene lactones. Selection of the former topic needs no justification, because the enormous amount of material which continues to accumulate needs to be evaluated at regular intervals. Appearance of the second topic is a more recent phenomenon (Herz, 1968, 1971, 1973; Herout and Sorm, 1969; Herout, 1971) caused largely by

* Financial support from the U.S. Public Health Service (CA-13121) through the National Cancer Institute and Hoffmann-LaRoche, Inc. is gratefully acknowledged.

developments in chemical instrumentation which have facilitated rapid acquisition of knowledge in what used to be a rather esoteric area of natural products chemistry. Thus, whereas by 1960 chemists had satisfactorily solved the structures of perhaps one dozen naturally occurring sesquiterpene lactones and were struggling with two dozen others, the number of known compounds of this class currently exceeds 600 and the pace of their discovery is quickening.

There can be little doubt that sesquiterpene lactones are characteristic constituents of Compositae or certain subdivisions thereof, although instances of their occurrence in other plant families are coming to light (Herz, 1973, refs 1–11; Holub and Samek, 1973a, b; Holub *et al.*, 1973; Holub, *et al.*, 1973; Serkerov, 1970, 1971, 1972a, b; Konovalova *et al.*, 1972, 1973; Talapatra *et al.*, 1973; Wiedhopf *et al.*, 1973; Gonzalez *et al.*, 1974). This is illustrated by Table I which shows the number of taxa known at the time to contain sesquiterpene lactones of reasonably well established structure this volume went to press (December 1976). Taxa which contain lactones of unkown structure were arbitrarily excluded.

TABLE I. Distribution of sesquiterpene lactones in the plant kingdom

A. Lactones formed by oxidation of "head" methyl group	
Compositae	c. 450 taxa
Umbelliferae	12
Lauraceae	1
Bursereae	1
Magnoliaceae	5
Hepaticae	4
B. Other lactones	
Amaranthaceae	1
Aristolochiaceae	2
Cannellaceae	2
Lauraceae	2

To a certain extent this apparent preference for the Compositae may stem from the intensity with which sesquiterpene lactones have been looked for in certain genera of this plant family, notably in *Artemisia*, *Ambrosia*, *Helenium* and *Vernonia*. Nevertheless, even on the basis of the sometimes rather arbitrary sampling which characterized much phytochemical research prior to the advent of general interest in chemotaxonomy, the incidence of sesquiterpene lactones in Compositae is unusually high; consequently their occurrence in other plant families must be attributed to parallelism. Moreover, the distribution of sesquiterpene lactones within

Compositae appears to harmonize at least partially with the divisions laid down by classical plant taxonomy, especially when attention is paid to those structural features that involve alterations in the carbon skeleton. A modernization of this thesis, as put forward by Herout (1971), will constitute the main goal of this presentation.

SESQUITERPENE LACTONES AND "BIOGENETIC COMPLEXITY"

Biogenetic theory assumes that the biosynthesis of sesquiterpenoids involves modification and/or cyclization of the pyrophosphate esters of *trans, trans*-farnesol, *cis, trans*-farnesol or nerolidol (Parker *et al.*, 1967), although evidence for this in higher plants is exceedingly meager. However, compared with the bewildering variety of sesquiterpenoid structures arising from such cyclizations (Mills and Money, 1974), the number of sesquiterpene lactone skeletal types so far encountered in higher plants is quite low.

By far the largest number, typical of Compositae, are γ-lactones whose formation on paper (Scheme I) involves oxidation of one of the two methyl

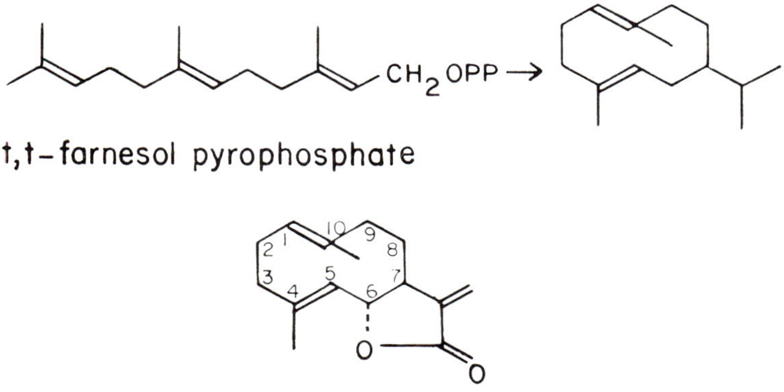

t,t-farnesol pyrophosphate

I costunolide

Scheme I

groups in the isopropyl "head" of the farnesol type precursor to a carboxyl, oxidation of an adjacent methylene group to a secondary alcohol function and eventual ring closure (cf. costunolide (1) which represents the most elementary cyclic sesquiterpene lactone since it retains two of the three double bonds of farnesol pyrophosphate in the *trans, trans* configuration). The details of this process are not known although two possible biogenetic schemes have been suggested. The first of these (Scheme II: Hikino *et al.*,

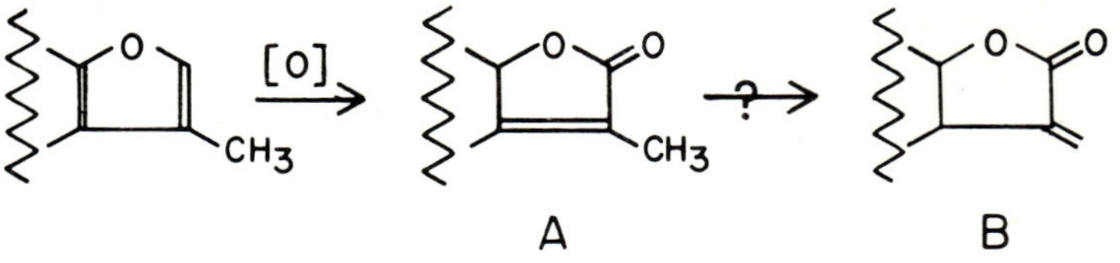

Scheme II

1962) may well be responsible for the occasional occurrence of lactones of type A in those plant groups (Lauraceae, Senecioneae) which normally elaborate furanoid sesquiterpenes, although its relevance to the formation of type B lactones so prevalent in the Compositae is less clear.

An alternative proposal (Scheme III: Geissman and Crout, 1969; Geissman, 1973) which involves oxidation of the isopropenyl side chain of sesquiterpenes followed by introduction of oxygen at C-6 or C-8 and

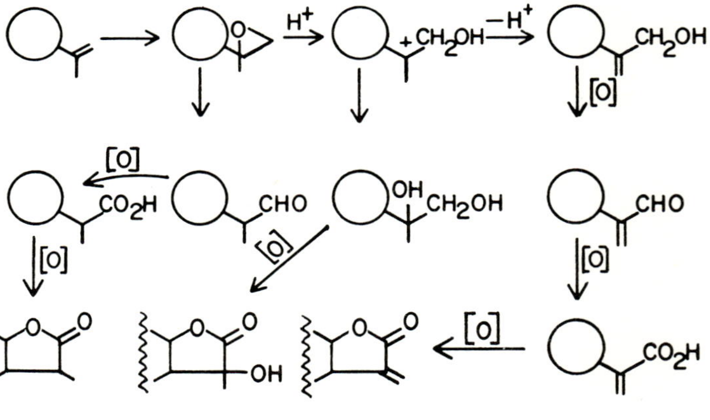

Scheme III

lactone ring closure appears more attractive as a general route to sesquiterpene lactones of the Compositae, since some of the postulated intermediates occasionally accompany the lactone end products and since it can be modified to lead to the furanosesquiterpenes of Senecioneae.

A second, much rarer type of γ-lactone results from oxidation of a non-terminal methyl group [e.g. aristolactone (**2**), from *Aristolochia* species (Martin-Smith *et al.*, 1964), drimenin (**3**), from *Drymis winteri* Forst. (Appel *et al.*, 1960) and (**4**) from *Ligularia hodgsonii* Hook. (Ishizaki *et al.*, 1969), the only substance of this kind so far found in Compositae] but compounds embodying both types of lactone rings are being encountered with increasing frequency in Compositae [e.g. elephantopin (**5**) (Kupchan et al., 1967) and miscandenin (**6**) (Cox et al., 1973)].

Names and formulae of the various lactone skeletal types so far found in Compositae are listed in Scheme IV which also adumbrates the presumed biogenetic relationships. For the sake of simplicity, only one mode of lactone ring closure, that toward C-6, has been depicted for the germacranolides and their most frequently encountered biological transformation products, the elemanolides, eudesmanolides and guaianolides, although lactone ring closure toward C-8 is common. In xanthanolides,

2 aristolactone   3 drimenin   4

5 elephantopin   6 miscandenin

lactone ring closure to C-6 and C-8 is known. Ambrosanolides and *seco*-ambrosanolides are invariably closed to C-6, whereas helenanolides and *seco*-helenanolides are generally closed to C-8. The few known eremophilanolides which, it will be recalled, are lactones of type A, Scheme II, (e.g. (**7**), Ishii et al., 1966) and their biological derivatives, the bakkenolides, are closed to C-8 also. Lactone ring closure in the so far very rare *seco*-eudesmanolides (two compounds, e.g. ivangulin (**8**), Herz et al., 1967), *seco*-germacranolides (one compound, pycnolide (**9**), Herz and Sharma, 1975), chrymoranolides (one compound, chlorochrymorin **10**, Osawa et al., 1973) and cadinanolides (one compound, arteannuin (**11**), Jeremić et al., 1973; Uskoković et al., 1974; Leppard et al., 1974) is depicted in the manner exhibited by the few examples known.

Lactones grouped together in the same vertical column of Scheme IV are produced, at least superficially, from the precursor farnesol pyrophosphate by the same number of changes in the carbon skeleton and thus might be said to exhibit the same degree of "biogenetic complexity", even

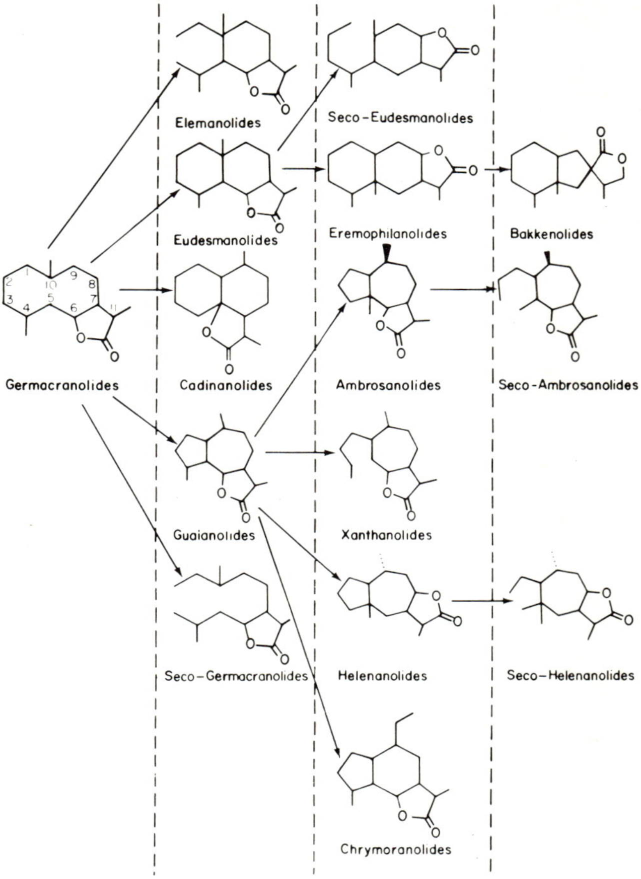
Elemanolides
Seco-Eudesmanolides
Eudesmanolides
Eremophilanolides
Bakkenolides
Germacranolides
Cadinanolides
Ambrosanolides
Seco-Ambrosanolides
Guaianolides
Xanthanolides
Seco-Germacranolides
Helenanolides
Seco-Helenanolides
Chrymoranolides

Scheme IV

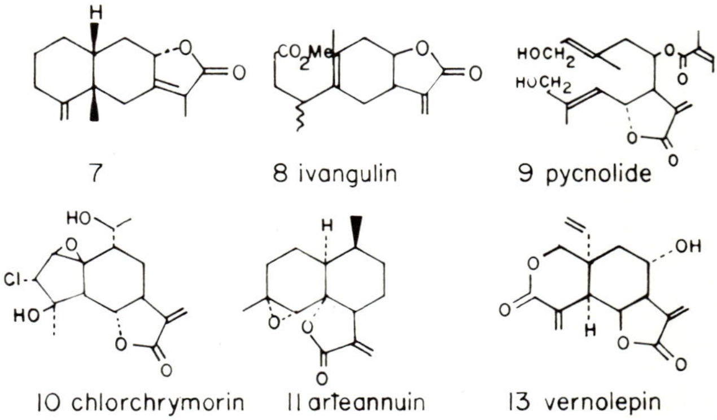
7
8 ivangulin
9 pycnolide
10 chlorchrymorin
11 arteannuin
13 vernolepin

though individual members of a particular class may differ widely in oxidation state at various sites within the molecule. Thus eudesmanolides, guaianolides and cadinanolides all appear to be derived by different modes of cyclization of 1(10),4,5-germacradienes or their epoxide equivalents, presumably under the influence of different enzyme systems (for details on this and the ensuing steps, see Herz, 1971, 1973; Geissman, 1973). Subsequent, formally similar methyl migrations in the eudesmanolide and the guaianolide class would give rise to eremophilanolides from the former and ambrosanolides and helenanolides from the latter. Oxidative cleavage of germacranolides, eudesmanolides, ambrosanolides and helenanolides is clearly responsible for the genesis of the respective seco-derivatives, while the nature of functionalization in the naturally occurring xanthanolides suggests the operation in the guaianolide class of another, different mode of five-membered ring scission. Lastly, enzymatically induced ring contraction of guaianolides, on the one hand, and eremophilanolides of type B, Scheme II, on the other, leads to chrymoranolides and bakkenolides, each of which type has been found in only one species so far.

Some caveats are in order lest the deceptive simplicity of this scheme lead to misinterpretation. For example, some elemanolides, e.g. saussurea lactone (**12**, Rao *et al.*, 1961), appear to be artifacts formed as the result of the Cope rearrangement of costunolide-type germacradienolides during work-up of certain plant extracts (Scheme V). On the other hand, the

Scheme V, Cope Rearrangement

dihydrocostunolide   12 saussurea lactone

discovery of more complex elemadienolides such as **6** (Cox *et al.*, 1973) and vernolepin (**13**, Kupchan *et al.*, 1968) clearly demonstrates the existence of a biological equivalent of the Cope rearrangement in other species. Hence the isolation of elemanolides is not, *per se*, a criterion of biosynthetic capacity.

Recent observations also suggest the possibility that all germacranolides and guaianolides do not lie on the same biogenetic pathway. Discovery of the *cis*-1(10),*trans*-4,5-germacradiene or "heliangolide" (e.g. heliangin (**14**), Nishikawa *et al.*, 1966; Neidle and Rogers, 1972) and *trans*-1(10), *cis*-4,5-germacradiene or "melampolide" (e.g. melampodin (**15**), Neidle and Rogers, 1972) subgroups raises the question whether their formation involves isomerization of a *trans*-1(10)-*trans*-4,5-germacradiene precursor or cyclization of a *trans*, *cis*- (or *cis*, *trans*) farnesol pyrophosphate (Schemes VI and VII). The additional presence of enzymes required for either process should be taken into account when considering sesquiterpene lactones as possible taxonomic markers.

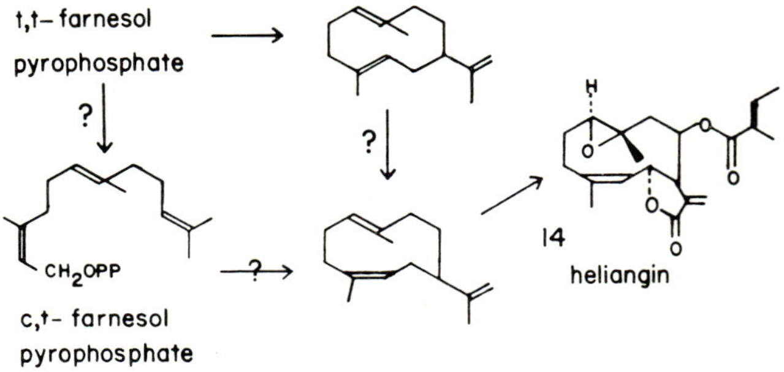

Scheme VI Possible Routes to Heliangolides

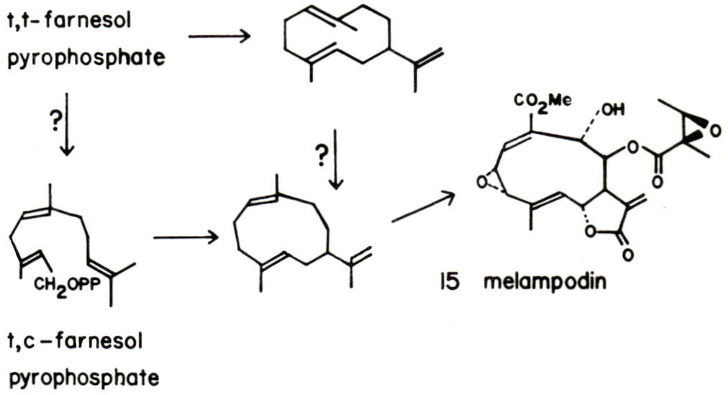

Scheme VII Possible Routes to Melampolides

Similarly, it is probable that a biosynthetic distinction must be made between the large and widely distributed class of *cis*-fused guaianolides, e.g. cynaropicrin (**16**, Samek *et al.*, 1971; Corbella *et al.*, 1972) and the very small group of transfused guaianolides of the gaillardin type (**17**, Dullforce *et al.*, 1969). The *cis*-fused compounds are derived from germacranolides of the costunolide type and are thought to be biological precursors of the ambrosanolides (Scheme VIII). On the other hand, the trans-fused guaianolides, presumably formed by cyclization of *cis*-1(10),*trans*-4,

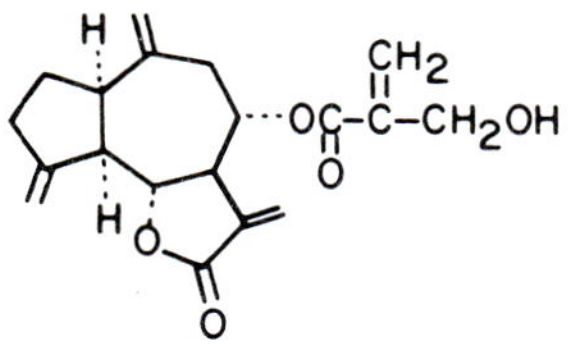

16 cynaropicrin

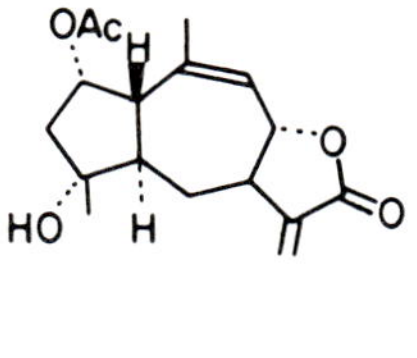

17 gaillardin

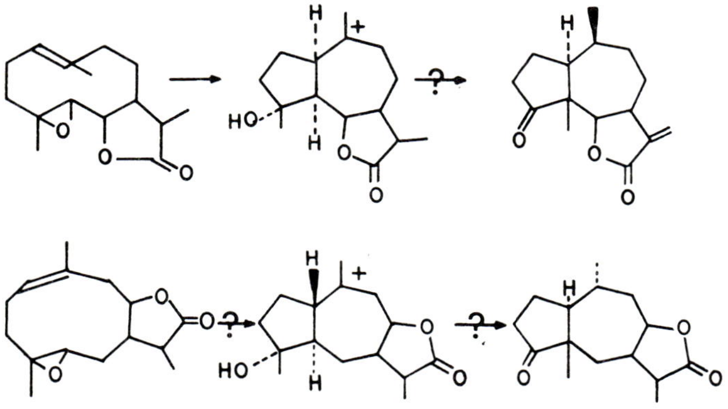

Scheme VIII Possible Routes to Ambrosanolides and Helenanolides

5-germacradienolides, are postulated as biosynthetic intermediates *en route* to the helenanolides (Parker et al., 1967; Herz, 1973). Thus knowledge of guaianolide stereochemistry as a possible manifestation of biosynthetic pathways would seem to be required before guaianolides can be used as a measure of close biological affinity.

DISTRIBUTION OF SESQUITERPENE LACTONES IN COMPOSITAE

In the preceding discussion, it was stressed that use of chemical characters in taxonomy must be based on knowledge of their biosynthesis; as regards sesquiterpene lactones in higher plants generally and in the Compositae specifically, the proposed biosynthetic routes to the various classes are based on inference rather than on actual experiments and possibly significant biosynthetic differences may reveal themselves in superficially rather minor structural alterations. If this be kept in mind, knowledge of sesquiterpene lactone distribution may be an aid in examining taxonomic problems at the tribal level or below.

Table II, which is an updated version of information previously presented by Herout (1971), shows the distribution of the various classes of

TABLE II. Distribution of sesquiterpene lactones in Compositae

	Genera	Taxa	GE[a]	SGE	EL	EU	SEU	ER	BA	GU	XA	AM	SAM	HE	SHE	CA	CH
Vernonieae	4	53	52[n]		3					4							
Eupatorieae	6	30	22[b]	1	2					9		1					
Astereae	1	1								1							
Inuleae	5	16	2		1	8				5[c]	5			2		1	
Heliantheae	24	53	27[d,e]		3	8		1		6	1	11[f]		2[g]			
Ambrosiinae	5	53	11[h]			7	1	1		5	16	24	12				
Helenieae	11	64	8[i]		1	3	1			8[j]				60			
Tageteae	—	—															
Senecioneae	4	8									2			3[l]			
Anthemideae	12	99	1[k]			56		4	2	46					1[m]	1	1
Arctoteae–Calenduleae	1	1								1							
Cynareae	11	47	26		3	1				18							
Mutisieae	1	1															
Cichorieae	7	10	1			2				8							

[a] GE = Germacranolides, SGE = Secogermacranolides, EL = Elemanolides, EU = Eudesmanolides, SEU = Secoeudesmanolides, ER = Eremophilanolides, BA = Bakkenolides, GU = Guaianolides, XA = Xanthanolides, AM = Ambrosanolides, SAM = Secoambrosanolides, HE = Helenanolides, SHE = Secohelenanolides, CA = Cadinanolides, CH = Chrymanolides

[b] Includes 9 taxa containing heliangolides
[c] Includes one known trans-fused guaianolide
[d] Includes 10 taxa with known heliangolides
[e] Includes 11 taxa with known melampolides
[f] All in Parthenium
[g] All in Balduina
[h] Includes 1 taxon with melampolide
[i] Includes 2 taxa with known heliangolides
[j] Includes 3 taxa with known trans-fused guaianolides
[k] Heliangolide
[l] All in Arnica
[m] Possible misidentification
[n] Includes 3 taxa with known heliangolides

sesquiterpene lactones in Compositae according to the tribal arrangements mostly suggested by Bentham. Subtribe Ambrosiinae of Heliantheae is listed separately because information on it is reasonably complete (53 of c. 85 species). The numbers in the various columns represent the taxa from which lactones of a particular type have been isolated, generally by extraction of the epigeal, mainly herbaceous, parts. Since some species, or different collections of the same species, produce more than one type of lactone, the sum of the numbers in a horizontal row generally exceeds the number of taxa in the third column. This number, while providing an indication of the thoroughness with which the various tribes have been investigated, is somewhat misleading since it represents only those taxa for which positive results have been reported. Negative results, i.e. absences of sesquiterpene lactones from particular extracts, are frequently not thought worth recording and are therefore not included in the Table, especially as they represent subjective judgments which depend on isolation, separation and identification techniques that have improved markedly within the last decade. Thus, until relatively recently, it was customary to report only those lactones that could be conveniently crystallized.

However, when negative results can be substantiated, they must be accorded attention and may be of considerable interest to the taxonomist. For example, the complete absence of sesquiterpene lactones from some members of a genus, ordinarily characterized by its proclivity toward elaboration of such compounds, has been verified by two groups of workers in *Ambrosia eriocentra* (Gray) Payne (Higo *et al.*, 1971; Herz *et al.*, 1973), *A. grayi* (Nels.) Shinners (Higo *et al.*, 1971; Herz *et al.*, 1975), *A. linearis* (Rydb.) Payne (Higo *et al.*, 1971), *A. tomentosa* Nutt. (Higo *et al.*, 1971; Herz *et al.*, 1975) and *A. trifida* L. (Herz and Högenauer, 1961; Higo *et al.*, 1971). Whether the taxa in question represent relatively primitive species within the genus or subgroups thereof which did not develop the biogenetic capacity for sesquiterpene lactones or whether the absence of lactones is due to mutations which have completely blocked relatively early stages of biosynthetic paths characteristic of the genus as a whole is a question that requires further study; it is possible that both reasons apply (Payne, 1964; Mabry, 1970; Payne *et al.*, 1972).

It was mentioned in a previous paragraph that extraction of one taxon may provide representatives of several classes of lactones, either of the same degree or of different degrees of "biogenetic complexity." A good example of the latter is *Baileya multiradiata* Harv. & Gray (Waddell and Geissman, 1969; Pettit *et al.*, 1975) which elaborates several helenanolides as well as a germacranolide and a guaianolide of unverified stereochemistry. This need occasion no surprise; the appearance of such a series of lactones could be attributed either to accumulation of biosynthetic intermediates *en route* to biogenetically more complex lactones or to partial diversion of precursors under the influence of other enzyme systems into products not susceptible to the action of enzymes catalyzing cyclizations, methyl migrations and the like.

A second situation of considerably greater interest to the taxonomist arises when lactones of dissimilar types are isolated from separate collections of the same taxon. In the absence of such obvious causes as misidentification, seasonal variations and differences in isolation procedures which may have resulted in genesis of artifacts or preferential separation of one class of product, such differences within the same taxon must be genetically controlled and deserve the same scrutiny as variations in morphological, anatomical and cytological characters.

A few such situations which have been related to other taxonomic criteria may be cited, e.g. infraspecific variations in the sesquiterpene lactone content of *Ambrosia psilostachya* DC. (Miller *et al.*, 1968; for results on a California collection see Geissman *et al.*, 1969 and Mabry, 1970), *A. chamissonis* (Less.) Greene (Payne *et al.*, 1973) and *A. confertiflora* DC. (Yoshioka *et al.*, 1970a; Mabry, 1970, Higo *et al.*, 1971) as well as in *Gaillardia pulchella* Foug. (Herz *et al.*, 1963; Herz and Inayama, 1964; Kupchan *et al.*, 1965; Herz *et al.*, 1967; Dullforce *et al.*, 1969; Aota *et al.*, 1970; Yoshioka *et al.*, 1970b; Yanagita *et al.*, 1969, 1970; Inayama *et al.*, 1973; W. P. Stoutamire, pers. comm.). Interesting examples of infraspecific variations which have not yet received attention from taxonomists include, *inter alia*, *Iva microcephala* Nutt. (Herz and Högenauer, 1962; Herz, *et al.*, 1964, 1965; Anderson *et al.*, 1973), and *Helenium autumnale* L. (Furukawa *et al.*, 1973; Herz *et al.*, 1969; Herz and Subramaniam, 1972; Hikino *et al.*, 1968; Kozuka *et al.*, 1975; Lee *et al.*, 1974a, b; Lucas *et al.*, 1964; McPhail 1973, 1975; Pettit *et al.*, 1975).

POSSIBLE IMPLICATIONS OF LACTONE DISTRIBUTION

Inspection of Table II reveals certain patterns of sesquiterpene lactone distribution within the family which invite further comment. Thus, while one should remember the comments on negative results mentioned in the previous section, the impression is given that sesquiterpene lactones are not characteristic secondary metabolites of the subtribe Astereae (see Dominguez and Jimenez, 1973, for the only report so far) and Mutisiae (one: Tomasini and Gilbert, 1972). Similarly there is only one definite record from the Arctoteae-Calenduleae (Grabarczyk and Makowska, 1973), although some references to lactones of unknown structure have appeared (Suchy and Herout, 1961; Grabarczyk, 1973). The apparent absence of lactones from the Tageteae (Herz, unpublished) tends to support exclusion of these genera from the classical Helenieae.

What chemical information is available on Vernonieae, Eupatorieae, Cynareae and Cichorieae suggests that members of these tribes may not have developed the biogenetic capacity to transform lactones beyond the

second (i.e. eudesmanolide, guaianolide, elemanolide or *seco*-germacranolide) stage, although further work may upset this bold generalization.*
However, some qualitative differences are discernible. In the Vernonieae, where the results are derived mainly from work with *Vernonia* and *Elephantopus* species, polyoxygenated germacranolides seem to predominate, the North American species of *Vernonia* exhibiting remarkable chemical uniformity (Abdel-Baset *et al.*, 1971; Mabry *et al.*, 1975). Furthermore, the elemanolides (such as **13**) which have been isolated from African *Vernonia* species are clearly not artifacts. In the Eupatorieae, *Mikania* and *Eupatorium* (*sensu stricto*, Robinson and King, Chapter 15), have yielded, with two guaianolide exceptions, complex germacranolides, while *Liatris* species produce a more even mix of germacranolides and guaianolides. The incidence of heliangolides in Eupatorieae so far also serves to distinguish this tribe from all but Heliantheae and Helenieae.

In Cynareae, the incidence of guaianolides in subtribe Centaurinae is somewhat higher than in Cynareinae, but not significantly so. The elemanolides which have been isolated from members of this tribe may well be artifacts. The apparent absence of eudesmanolides from Vernonieae, Eupatorieae and Cynareae could be meaningful; a very much smaller sampling of the Cichorieae, which strongly favor guaianolides, shows two taxa producing eudesmanolides and only one taxon where biosynthesis has stopped at the first germacranolide stage.

In contrast to their relatives, the Arctoteae, the Anthemideae (mainly *Artemisia* and *Chrysanthemum*) seem to be prolific producers of sesquiterpene lactones and divide their attention between eudesmanolides and guaianolides. This tribe also includes two examples of so far unique biosynthetic capacity, i.e. *Artemisia annua* L. in which cyclization of a germacradiene precursor to a cadinanolide (arteannuin **11**) has occurred and *Chrysanthemum morifolium* L. which elaborates the unusual class III lactone chlorchrymorin (**10**), presumably by rearrangement of a guaianolide precursor.

The large essentially or exclusively American tribes of the Heliantheae and Helenieae differ from the preceding tribes in the diversity of their lactone constituents. Table II indicates that double bond isomerization to heliangolides and melampolides is relatively common, particularly in Heliantheae, and that many species in Heliantheae and Helenieae are capable of effecting transformation of sesquiterpene lactone precursors beyond the guaianolide stage. Cyclization of germacranolide precursors to eudesmanolides seems to occur less frequently in these tribes and further modification of eudesmanolides is a rather rare phenomenon which has

---

* The isolation of an ambrosanolide from *Stevia rhombifolia* HBK (Eupatorieae) (Rios *et al.*, 1967) remains an unexplained anomaly.

been observed to date only in *Iva angustifolia* Nutt. (Heliantheae, Ambrosiinae) (Herz et al., 1967) and *Eriophyllum lanatum* Forbes (Helenieae) (Kupchan et al., 1973).

However, aside from the occurrence of heliangolides and melampolides this apparent increase in biosynthetic capacity is limited to relatively few genera. In Heliantheae these are four of the five genera in Ambrosiinae (*Ambrosia, Iva* section *Cyclochaena, Hymenoclea* and *Xanthium**), as well as *Parthenium* and *Parthenice* in Melampodiinae and *Balduina* in Galinsoginae. The situation in Helenieae will be discussed later.

The highly characteristic class III and class IV lactone constituents of *Ambrosia, Hymenoclea, Parthenium* and certain taxa in *Iva* section Cychlochaena are ambrosanolides such as ambrosin (**18**) and *seco*-ambrosanolides such as psilostachyin (**19**) which, as will be recalled, are probably rearrangement products of *cis*-fused guaianolide precursors. It should also be noted that infraspecific variations which appear to be under genetic control are relatively common in *Ambrosia* (*vide supra*). *Xanthium*, *Parthenice* and other members of *Iva* section *Cyclochaena* characteristically elaborate xanthanolides (Herz, 1971, 1973; McMillan et al., 1975) e.g. xanthumin (**20**) and parthemollin (**21**), by enzyme-catalyzed five-membered ring cleavage of guaianolides. This obvious chemical affinity between *Parthenium* and *Parthenice* on the one hand and Ambrosiinae on the other, has been adduced as an additional argument for removing *Parthenium* and *Parthenice* from the rather heterogeneous assembly of the Melampodiinae and placing them in the Ambrosiinae (Stuessy Chapter 23). The desirability of separate tribal status for the members of this group has been debated vigorously and could be defended on chemical grounds, i.e. the possession of advanced chemical characters not duplicated elsewhere in Heliantheae (Herout, 1971), but possibly not on others (Stuessy, Chapter 23).

The genus *Balduina* occupies a highly anomalous position in Heliantheae from the chemical point of view because all three of its taxa elaborate helenanolides (Herz, 1973). These helenanolides (e.g. helenalin, **22**) are identical with or slight variants of sesquiterpene lactones which are very characteristic secondary metabolites of subtribe Gaillardiinae (= Helenieae, of Chapter 25) (Herz, 1973). Rock (1957), Stuessy (Chapter 23) and Turner and Powell (Chapter 25), among others, position this subtribe in the Heliantheae. Such a repositioning would certainly be supported by the sesquiterpene lactone chemistry.

The situation within the classically constituted Helenieae is also not immediately obvious from Table II. Many systematists consider this

---

* The fifth, *Dicoria*, also seems to produce lactones, but no homogeneous lactone fractions have so far been isolated (Herz et al., 1973).

18 ambrosin   19 psilostachyin   20 xanthumin

21 parthemollin   22 helenalin

tribe to be an artificial assemblage of genera, although submergence of the entire tribe in Heliantheae advocated by Cronquist (1955) has been questioned (Turner and Powell, Chapter 25). However, the existence of some monophyletic groups within the tribe seems well established.

One such group seems to be that of the Gaillardiinae, mentioned above. The increase in biosynthetic capacity within Helenieae manifested in Table II is almost entirely due to genera within this subtribe, although it must be granted that genera in other subtribes have not been investigated systematically. Thus all *Helenium* and *Gaillardia* species, with but few exceptions (Herz, 1973), elaborate helenanolides (e.g. helenalin (**22**) and pulchellin (**23**)); the exceptions being due, in the main, to intraspecific variations in *Helenium autumnale* L. and *Gaillardia pulchella* Foug. These variations have also led to the discovery of some *trans*-fused guaianolides, e.g. gaillardin (**17**) (Dullforce et al., 1969; Lee et al., 1974) which have been postulated as biosynthetic intermediates *en route* to helenanolides (Herz, 1973). Taxa which depart from this generalization are *H. pinnatifidum* (Nutt.) Rydb., a eudesmanolide producer, *H. virginicum* Blake which yields a guaianolide of unspecified stereochemistry, possibly a helenanolide precursor, and *H. drummondii* Rock as well as *Amblyolepis setigera* DC., both of which appear to contain little or no lactones. *Dugaldia* and *Hymenoxys* species occasionally contain helenanolides, but generally display a capacity for their further transformation to *seco*-helenanolides such as hymenovin (**24**) and vermeerin (**25**) (Herz, 1973; Ivie et al., 1975, 1976).

This capacity for biogenesis of helenanolides and *seco*-helenanolides is also exhibited by *Baileya* and *Psilostrophe* (subtribe Psilostrophinae), whereas scattered representatives of Rydberg's (1914) helenioid subtribes Jaumeinae, Bahiinae, Chaenactinae and Eriophyllinae have yielded only

23 pulchellin   **24** hymenovin   **25** vermeerin

germacranolides, elemanolides or guaianolides. Thus on chemical grounds, a good case could be made for a close relationship among members of the Gaillardiinae and possibly Psilostrophinae, but our present knowledge of sesquiterpene lactone chemistry of Rydberg's other helenioid subtribes is not sufficiently advanced to contribute to the realignments proposed by Turner and Powell (Chapter 25) and Stuessy (Chapter 23).

The same can be said about Inuleae as only a few species of this large, cosmopolitan tribe have been examined, mainly in subtribe Inulinae. In general, biosynthesis seems to stop at the eudesmanolide and guaianolide stage; however, two *Carpesium* species additionally yielded the cyclopropanoid xanthanolide carabrone (**26**)* (Minato *et al.*, 1964; Konovalova *et al.*, 1972). *Geigeria aspera* Harv. (subtribe Buphthalminae) contains guaianolides and a helenanolide geigerinin (stereochemistry uncertain), while *G. africana* Gries. yielded xanthanolides of uncertain stereochemistry and a *seco*-helenanolide vermeerin (**25**) also found in some *Hymenoxys* species. Thus at least some genera in Inuleae exhibit enhanced biosynthetic capacity.

As regards Senecioneae, the lactone incidence is very low although many species have been studied chemically. Until 1969 the only lactones found were obvious oxidation products, such as (**7**), of the furanoid eremophilane sesquiterpene hydrocarbons, alcohols and ketones which are widely distributed in the tribe and, in the case of *Petasites japonicus* Maxim., the bakkenolides (e.g. **27**), which are products of further rearrangement of eremophilanes (Abe *et al.*, 1968a, b; Shirahata *et al.*, 1968; Naya *et al.*, 1968, 1972a, b). These observations, the occurrence of the necine alkaloids which are absent from other Compositae and the absence of acetylenes coincide with the view that the tribe represents a comparatively uniform section of an independent phylogenetic line.

Since then, chemical examination of two *Arnica* species (*A. foliosa* Nutt.

---

* The stereochemistry of carabrone at C-1 is unusual and suggests its possible derivation from a *trans*-fused guaianolide (cf. gaillardin **17**), a class of compounds which, as has been mentioned may also lie on the biosynthetic pathway to helenanolides. Hence it is of considerable interest that isolation of carabrone has also been reported from *Helenium quadridentatum* Labill. (Hernandez *et al.*, 1968) and a race of *Arnica foliosa* Nutt. (Holub, *et al.*, 1972). Both are members of genera that normally elaborate helenanolides.

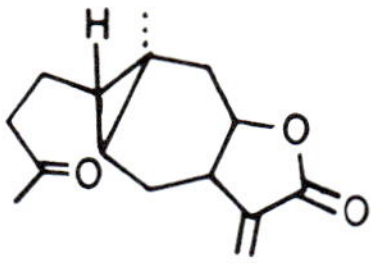

26 carabrone

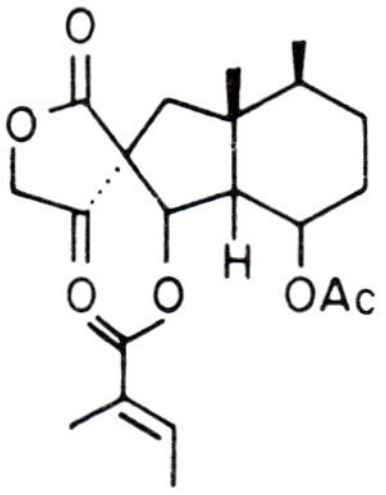

27 bakkenolide B

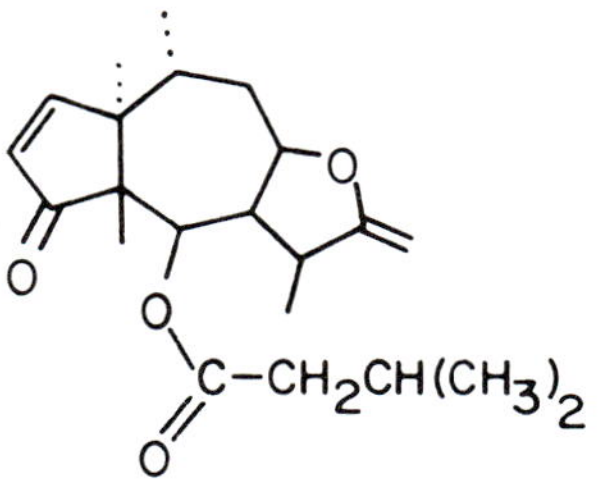

28 arnicolide B

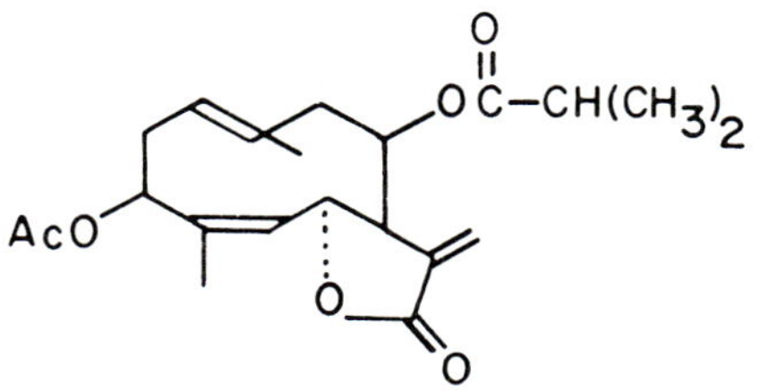

29

and *A. montana* L.) has provided a series of helenanolides such as (**28**), intimately related to the helenanolides of Helenieae (Evstratova *et al.*, 1969, 1971; Poplawski *et al.*, 1971). These observations, coupled with the presence of acetylenes in *Arnica* species (Bohlmann *et al.*, 1973) and the existence of significant morphological differences, have led to the suggestion (Herout, 1971) that *Arnica* should be removed from Senecioneae (cf. also Nordenstam, Chapter 29 and Turner and Powell, Chapter 25).

More recently, the discovery of a heliangolide (**29**) in *Peucephyllum schottii* Gray (Begley *et al.*, 1975) seems to reinforce earlier doubts about the location of *Peucephyllum* in Senecioneae (see Strother and Pilz 1975 and Nordenstam, Chapter 29).

REFERENCES

ABDEL-BASET, Z. A., SOUTHWICK, L., PADOLINA, W. G., YOSHIOKA, H. and MABRY, T. J. (1971). *Phytochemistry* **10**, 2201.
ABE, N., ONODA, R., SHIRAHATA, K., KATO, T., WOODS, M. C. and KITAHARA, Y. (1968a). *Tetrahedron Lett.* 369.
ABE, N., ONODA, R., SHIRAHATA, K., KATO, T., WOODS, M. C., KITAHARA, Y., RO, K. and KURIHARA, T. (1968b). *Tetrahedron Lett.* 1993.
ANDERSON, G. D., GITANY, R., MCEWEN, R. S. and HERZ, W. (1973). *Tetrahedron Lett.* 2409.

AOTA, K., CAUGHLAN, C. N., EMERSON, M. T., HERZ, W., INAYAMA, S. and MAZHAR-UL-HAQUE. (1970). *J. org. Chem.* **35**, 1448.

APPEL, H. H., CONNOLLY, J. D., OVERTON, K. H. and BOND, R. P. M. (1960). *J. chem. Soc.* 4685.

BEGLEY, M. J., PATTENDEN, G., MABRY, T. J., MIYAKADO, M. and YOSHIOKA, H. (1975). *Tetrahedron Lett.* 1105.

BOHLMANN, F., BURKHARDT, T. and ZDERO, C. (1963). "Naturally Occurring Acetylenes", p. 438. Academic Press, London and New York.

CORBELLA, A., GARIBOLDI, P., JOMMI, G., SAMEK, Z., HOLUB, M., DROZDZ, B. and BLOSZYK, E. (1972). *Chem. Commun.* 386.

COX, P. J., SIM, G. A., ROBERTS, J. S. and HERZ, W. (1973). *J. chem. Soc. Comm.* 428.

CRONQUIST, A. (1955). *Am. Midl. Nat.* **53**, 478.

DOMINGUEZ, X. A. and JIMENEZ, S. (1973). *Rev. latinoam. quim.* **3**, 179.

DULLFORCE, T. A., SIM, G. A., WHITE, D. N. J., KELSEY, J. E. and KUPCHAN, S. M. (1969). *Tetrahedron Lett.* 973.

EVSTRATOVA, R. I., BANKOVSKII, A. I., SHEICHENKO, V. I. and RYBALKO, K. S. (1971). *Khim. farm. Zh.* **7**, 270.

EVSTRATOVA, R. I., SHEICHENKO, V. I., RYBALKO, K. S. and BANKOVSKII, A. K. (1969). *Khim. farm. Zh.* **3**, 39.

FURUKAWA, H., LEE, K. H., SHINGU, T., MECK, R. and PIANTADOSI, C. (1973). *J. org. Chem.* **38**, 1722.

GEISSMAN, T. A. (1973). "Recent Advances in Phytochemistry", Vol. 6; p. 65. Academic Press, New York and London.

GEISSMAN, T. A. and CROUT, D. H. G. (1969). "Organic Chemistry of Secondary Plant Metabolism", p. 283. Freeman, Cooper and Co., San Francisco.

GEISSMAN, T. A., GRIFFIN, S., WADDELL, T. G. and CHEN, H. H. (1969). *Phytochemistry* **8**, 145.

GONZALEZ, A. G., BRETON, J. L., GALINDO, A. and RODRIGUEZ, A. (1974). *Anal. quim.* **69**, 1339.

GRABARCZYK, H. (1973). *Pol. J. Pharmacol. Pharm.* **25**, 469.

GRABARCZYK, H. and MAKOWSKA, B. (1973). *Pol. J. Pharmacol. Pharm.* **25**, 477.

HERNANDEZ, R., SANDOVAL, A., SETZER, A. and ROMO, J. (1968). *Bol. Inst. quim. Univ. nacl auton. Mex.* **20**, 81.

HEROUT, V. (1971). *In* "Pharmacognosy and Phytochemistry" (H. Wagner and L. Hörhammer, eds), p. 93. Springer Verlag, Berlin, Heidelberg and New York.

HEROUT, V. and SORM, F. (1969). *In* "Perspectives in Phytochemistry" (J. B. Harborne and T. Swain, eds), p. 139. Academic Press, London and New York.

HERZ, W. (1968). *In* "Recent Advances in Phytochemistry" (T. J. Mabry, R. E. Alston and V. C. Runeckles, eds), p. 229. Appleton-Century Crofts, New York.

HERZ, W. (1971). *In* "Pharmacognosy and Phytochemistry" (H. Wagner and L. Hörhammer, eds), p. 64. Springer Verlag, Berlin, Heidelberg and New York.

Herz, W. (1973). In "Chemistry in Botanical Classification". Nobel Symposium **25**, 153.
Herz, W., Anderson, G. D., Wagner, H., Maurer, G., Maurer, I., Flores, G. and Farkas, L. (1975). Tetrahedron **31**.
Herz, W., Fitzhenry, G. and Anderson, G. (1973). Phytochemistry **12**, 1181.
Herz, W. and Högenauer, G. (1961). J. org. Chem. **26**, 5011.
Herz, W. and Högenauer, G. (1962). J. org. Chem. **27**, 905.
Herz, W., Högenauer, G. and Romo de Vivar, A. (1964). J. org. Chem. **29**, 1700.
Herz, W. and Inayama, S. (1964). Tetrahedron **20**, 341.
Herz, W., Rajappa, S., Lakshmikantham, M. V., Raulais, D. and Schmid, J. J. (1967). J. org. Chem. **32**, 1042.
Herz, W., Raulais, D. and Anderson, G. D. (1973). Phytochemistry **12**, 1415.
Herz, W., Romo de Vivar, A. and Lakshmikantham, M. V. (1965). J. org. Chem. **30**, 118.
Herz, W. and Sharma, R. P. (1975). J. org. Chem. **40**, 392.
Herz, W. and Subramaniam, P. S. (1972). Phytochemistry **11**, 1101.
Herz, W., Ueda, K. and Inayama, S. (1963). Tetrahedron **19**, 483.
Herz, W., Subramaniam, P. S. and Dennis, N. (1969). J. org. Chem. **34**, 2915.
Herz, W., Sumi, Y., Sundarsanam, V. and Raulais, D. (1967). J. org. Chem. **32**, 3658.
Higo, A., Hammam, Z., Timmermann, B. N., Yoshioka, H., Lee, J., Mabry, T. J. and Payne, W. W. (1971). Phytochemistry **10**, 2241.
Hikino, H., Hikino, Y. and Yosioka, T. (1962). Chem. pharm. Bull. **10**, 641.
Hikino, H., Kuwano, D. and Takemoto, T. (1968). Chem. pharm. Bull. **16**, 1601.
Holub, M. and Samek, Z. (1973a). Coll. Czech. chem. Commun. **38**, 731.
Holub, M. and Samek, Z. (1973b). Coll. Czech. chem. Commun. **38**, 1428.
Holub, M., Samek, Z., de Groote, R., Herout, V. and Sorm, F. (1973). Coll. Czech. chem. Commun. **38**, 1551.
Holub, M., Samek, Z., Popa, D. P. and Herout, V. (1973). Coll. Czech. chem. Commun. **38**, 1804.
Holub, M., Samek, Z. and Toman, J. (1972). Phytochemistry **11**, 2627.
Inayama, S., Kawamata, T. and Yanagita, M. (1973). Phytochemistry **12**, 1741.
Ishii, H., Tozyo, T. and Minato, H. (1966). J. chem. Soc. 1545.
Ishizaki, Y., Tanahashi, Y., Takahashi, T. and Tori, K. (1969). Chem. Comm. 551.
Ivie, G. W., Witzel, D. A., Herz, W., Kannan, R., Norman, J. O., Rushing, D. D., Johnson, J. J., Rowe, L. D. and Veech, J. A. (1975). Agr. Food Chem. **23**, 841.
Ivie, G. W., Witzel, D. A., Herz, W., Sharma, R. P. and Johnson, A. E. (1976). Agr. Food Chem. **24**, 681.
Jeremić, D., Jokić, A., Behbud, A. and Stefanović, M. (1973). Tetrahedron Lett. 3039.
Konovalova, O. A., Bankovskii, A. I., Rybalko, K. S., Sheichenko, V. I., Zakhorov, P. I. and Pimenov, M. G. (1972). Khim. Prir. Soedin. **8**, 651.

Konovalova, O. A., Rybalko, K. S. and Kabanov, V. S. (1972). *Khim. Prir. Soedin.* **8**, 721.
Konovalova, O. A., Rybalko, K. S. and Pimenov, M. G. (1973). *Khim. Prir. Soedin.* **9**, 122.
Kozuka, M., Lee, K. H., McPhail, A. T. and Onan, K. D. (1975). *Chem. Pharm. Bull.* (Japan). **23**, 1895.
Kupchan, S. M., Baxter, R. L., Chiang, C. K., Gilmore, C. J. and Bryan, R. F. (1973). *J. chem. Soc. chem. Comm.* 842.
Kupchan, S. M., Cassady, J. M., Bailey, J. and Knox, J. R. (1965). *J. pharm. Sci.* **54**, 1703.
Kupchan, S. M., Hemingway, R. J., Werner, D., Karim, A., McPhail, A. T. and Sim, G. A. (1968). *J. Am. chem. Soc.* **90**, 3596.
Kupchan, S. M., Kelsey, J. E. and Sim, G. A. (1967). *Tetrahedron Lett.* 2863.
Lee, K. H., Ibuka, T., McPhail, A. T., Onan, K. D., Geissman, T. A. and Waddell, T. G. (1974a). *Tetrahedron Lett.* 1149.
Lee, K. H., Ibuka, T., Kozuka, M., McPhail, A. T. and Onan, K. D. (1974b). *Tetrahedron Lett.* 2287.
Leppard, D. G., Rey, M., Dreiding, A. S. and Grieb, R. (1974). *Helv. chim. Acta* **57**, 602.
Lucas, R. A., Smith, R. G. and Dorfman, L. (1964). *J. org. Chem.* **29**, 2101.
Mabry, T. J. (1970). *In* "Phytochemical Phylogeny" (J. B. Harborne, ed.), p. 269. Academic Press, London and New York.
Mabry, T. J., Abdel-Baset, Z., Padolina, W. G. and Jones, S. B. (1975). *Biochem. Syst. Ecol.* **2**, 185.
McMillan, C., Chavez, P. I., Plettmann, S. G. and Mabry, T. J. (1975). *Biochem. Syst. Ecol.* **2**, 181.
McPhail, A. T., Luhan, P. A., Lee, K. H., Furukawa, H., Meck, R., Piantadosi, C. and Shingu, T. (1973). *Tetrahedron Lett.* 4087.
McPhail, A. T., Onan, K. D., Furukawa, H. and Lee, K. H. (1975). *Tetrahedron Lett.* 1229.
Martin-Smith, M., De Mayo, P., Smith, S. J., Stenlake, J. B. and Williams, W. D. (1964). *Tetrahedron Lett.* 2391.
Miller, H. E., Mabry, T. J., Turner, B. L. and Payne, W. W. (1968). *Am. J. Bot.* **55**, 316.
Mills, R. W. and Money, T. (1974). *In* "Terpenoids and Steroids", Vol. 4, p. 77. The Chemical Society, London.
Minato, H., Nosaka, S. and Horibe, I. (1964). *J. chem. Soc.* 5503.
Naya, K., Hayashi, M., Takagi, I., Nakamura, S. and Kobayashi, M. (1972a). *Bull. chem. Soc. Japan* **45**, 3673.
Naya, K., Kawai, M., Naito, M. and Kasai, T. (1972b). *Chem. Lett.* 241.
Naya, K., Takagi, I., Hayashi, M., Nakamura, S., Kobayashi, M. and Katsumuura, S. (1968). *Chemy Ind.* 318.
Neidle, S. and Rodgers, D. (1972). *Chem. Commun.* 140.
Nishikawa, M., Kamiya, K., Takabatake, A. and Oshio, H. (1966). *Tetrahedron* **22**, 3601.
Osawa, T., Suzuki, A., Tamura, S., Ohashi, Y. and Sasada, Y. (1973). *Tetrahedron Lett.* 5135.

Parker, W., Roberts, J. S. and Ramage, R. (1967). *Q. Rev. Chem. Soc.* **21**, 311.
Payne, W. W. (1964). *J. Arnold Arbor.* **45**, 401.
Payne, W. W., Geissman, T. A., Lucas, A. J. and Saitoh, T. (1973). *Biochem. System.* **1**, 21.
Payne, W. W., Scora, R. W. and Kumamoto, J. (1972). *Brittonia* **24**, 189.
Pettit, G. R., Budzinski, J. C., Cragg, G. M., Brown, P. and Johnston, L. D. (1974). *J. med. Chem.* **17**, 1013.
Pettit, G. R., Herald, C. L., Judd, G. F., Bolliger, G. and Thayer, P. S. (1975). *J. pharm. Sci.* **64**, 2023.
Poplawski, J., Holub, M., Samek, Z. and Herout, V. (1971). *Coll. Czech. chem. Commun.* **36**, 2189.
Rao, A. S., Paul, A., Sadgopal and Bhattacharyya, S. C. (1961). *Tetrahedron* **13**, 318.
Rios, T., Romo de Vivar, A. and Romo, J. (1967). *Tetrahedron* **23**, 4265.
Rock, H. F. L. (1957). *Rhodora* **59**, 101.
Rydberg, P. A. (1914). Helenieae (in part). *N. Am. Fl.* **34**, 1.
Samek, Z., Holub, M., Drozdz, B., Jommi, G., Corbella, A. and Gariboldi, P. (1971). *Tetrahedron Lett.* 4775.
Serkerov, S. V. (1970). *Khim. Prir. Soedin.* **6**, 428.
Serkerov, S. V. (1971). *Khim. Prir. Soedin.* **7**, 667.
Serkerov, S. V. (1972a). *Khim. Prir. Soedin.* **8**, 63.
Serkerov, S. V. (1972b). *Khim. Prir. Soedin.* **8**, 173.
Shirahata, K., Abe, N., Kato, T. and Kitahara, Y. (1968). *Bull. chem. Soc. Japan* **41**, 1732.
Strother, J. L. and Pilz, G. (1975). *Madroño* **23**, 24.
Suchy, M. and Herout, V. (1961). *Coll. Czech. chem. Commun.* **26**, 890.
Tomasini, T. C. B. and Gilbert, B. (1972). *Phytochemistry* **11**, 1177.
Talapatra, S. K., Patra, A. and Talapatra, B. (1973). *Phytochemistry* **12**, 1827.
Uskoković, M., Williams, T. H. and Blount, J. F. (1974). *Helv. chim. Acta* **57**, 600.
Waddell, T. G. and Geissman, T. G. (1969). *Phytochemistry* **8**, 2371.
Wiedhopf, R. M., Young, M., Bianchi, E. and Cole, J. R. (1973). *J. pharm. Sci.* **62**, 345.
Yanagita, M., Inayama, S., Kawamata, T., Okura, T. and Herz, W. (1969). *Tetrahedron Lett.* 2073 and errata 4170.
Yanagita, M., Inayama, S. and Kawamata, T. (1970). *Tetrahedron Lett.* **131**, 3007.
Yoshioka, H., Renold, W., Fischer, N. H., Higo, A. and Mabry, T. J. (1970a). *Phytochemistry* **9**, 823.
Yoshioka, H., Dennis, N., Herz, W. and Mabry, T. J. (1970b). *J. org. Chem.* **35**, 627.

Chapter 12

Flavonoid profiles in the Compositae

JEFFREY B. HARBORNE

Department of Botany,
University of Reading, England

Abstract. Flavonoid pigments are widely, indeed universally, distributed in the Compositae and are potentially promising as taxonomic markers. The distribution of flavonoid types, however, seems to be relatively sporadic and there are only a few clear correlations with tribal or subtribal classification. The restriction of anthochlor pigments mainly to the Coreopsidinae is one example; the characteristic occurrence of flavonoids lacking B-ring hydroxyls in the Inuleae is another. Few anthocyanins have been identified in composites; the pattern, where it is known, is strangely a simple one. By contrast, flavones and flavonols occur in an almost bewildering array of derivatives with extra hydroxylation and/or methylation.

Undoubtedly it is at the lower levels of classification where flavonoid profiles are at present of most interest to the experimental taxonomist. Some examples are discussed where study of these profiles has contributed to our overall understanding of plant relationships between species and between genera.

CONTENTS

Introduction	359
Anthocyanins	361
Anthochlor pigments	365
Yellow flavonols	369
Other flower flavonoids	372
Leaf flavonoids	374
Flavonoids and classification at the tribal and generic levels in the Compositae	376
Flavonoids and classification at the species level in the Compositae	378
Conclusion	381
References	381

INTRODUCTION

Considering the vast size of the Compositae, our information on the flavonoids in the family is very sparse. Relatively few species have been examined—perhaps some 400 taxa representing 100 genera—and, in most cases, the flavonoid profile has been studied only superficially. Much of the data available has arisen accidentally as a by-product of investigations quite unrelated to plant taxonomy. Thus the chemical studies of A. G. Perkin (1902, 1913) on the flavonol quercetagetin of *Tagetes erecta* were undoubtedly initiated because the substance was a natural pigment and of potential value as a mordant dye. Similarly, more recent investigations of quercetagetin methyl ethers in *Eupatorium* species have arisen from a screening programme on plants for substances with anti-tumour activity (e.g. Kupchan *et al.*, 1969).

Very few deliberate flavonoid surveys have been attempted in the Compositae and those that there have been were restricted to relatively small taxonomic groups. Bate-Smith (1962), in a leaf survey of dicotyledonous angiosperms, looked at only 39 species from 28 genera of composites. No-one has yet attempted a representative survey throughout the family. There are technical difficulties here because of the presence of large amounts of interfering caffeic acid derivatives in many species, but these could be overcome by suitably modifying existing screening techniques.

From the limited data available, it is clear that the Compositae is significantly heterogeneous in flavonoid patterns. A considerable range of flavonoid types and of glycosidic or other combinations have been encountered in different tribes and subtribes of the family. In the leaf, the predominant pattern is not that expected in a highly evolved family, i.e. one based on the flavone luteolin. Instead, according to Bate-Smith (1962), flavonols are very common: he records quercetin in 64% and kaempferol in 56% of the species he surveyed. However, flavones such as luteolin and apigenin are undoubtedly widespread in the floral tissues of these plants (see Harborne, 1967).

There is no special flavonoid structure that one can associate immediately with the family. Indeed, hardly any of the compounds so far discovered in the Compositae are unique to the group. Although the aurones, a rare type of yellow flower pigment, were first discovered in the family, in *Coreopsis* (Geissman and Heaton, 1943), these substances have subsequently been found in several other families. Even the biosynthetically remarkable series of flavanolignans, such as silybin, which occur in the fruit of *Silybum marianum* (see Wagner, Chapter 14) are not entirely restricted to the family, since related structures do occur elsewhere.

The Compositae then can only be characterized by a syndrome of flavonoid structural features. These include: a simple anthocyanin pattern; simple flavone glycosides (in flowers); anthochlors (chalcones and/or aurones); flavones and flavonols with 6- and/or 8-hydroxylation; polymethylation of the hydroxyl groups in flavones and flavonols; and absence of 5-hydroxylation. Such a combination of features is not found in any family that has been placed anywhere near to the Compositae in taxonomic systems, e.g. the Caprifoliaceae, Rubiaceae or Campanulaceae (see

Cronquist, 1968). Indeed, the only other plant family where many (but not all) of the same features are found together is the Leguminosae. The association between these two families in terms of flavonoid chemistry has been remarked upon on more than one occasion. The occurrence of isoflavonoids, which are found abundantly in the plant kingdom only in the subfamily Lotoideae of the Leguminosae and, in two or three composite genera (*Balduina, Eclipta, Wedelia*), provides a further link between these two large and diverse plant groups.

In the present account, it is proposed to consider briefly the main classes of flavonoid found within the family and then to discuss the relevance of the presently available data to the solution of systematic and evolutionary problems within the group. Detailed accounts of the flavonoids as they occur in the different tribes are provided elsewhere in this volume; here only a few summary tables will be provided.

ANTHOCYANINS

In spite of the fact that the dominant flower colour in the Compositae is yellow, cyanic pigmentation is still a significant feature in the family. For example, anthocyanins are important contributors to flower colour in most ornamental composites, especially in *Dahlia variabilis, Dendranthema morifolium* and *Callistephus chinensis*. In these and other plants, breeding experiments have yielded a range of colour forms and the inheritance of anthocyanin production has been actively studied (for *Dahlia*, see Lawrence and Scott-Moncrieff, 1935). In wild species, red and magenta colours are common in the flowers of *Carduus, Carlina, Centaurea, Cirsium* and many other genera. Even the humble daisy, *Bellis perennis*, has red pigment coating the undersurface of its white flower rays. Blue flower colours are less frequent but occur in such plants as chicory, *Cichorium intybus*, blue fleabane, *Erigeron acer*, sea aster, *Aster tripolium*, and cornflower, *Centaurea cyanus*. Anthocyanin also occurs in plants with yellow flowers, in combination with yellow carotenoid in the flower (e.g. *Tolpis barbata, Coreopsis tinctoria*) or in the stem and leaf (many species).

The ability to produce anthocyanin is undoubtedly universal in the family. Even if it is not expressed under normal conditions of growth, it may be formed under physiological stress. Young seedlings of *Lactuca* are well known to produce red coloration on cold treatment. An example showing the potential for anthocyanin production in these plants is the weedy annual *Machaeranthera gracilis*, which does not normally exhibit anthocyanin in the intact plant, but produces a deep red cyanic colour when grown in callus culture (Harborne, 1964). That the anthocyanin so produced artificially is characteristic of the whole plant has been demonstrated by comparison of pigments formed in callus tissue with those present in petal and stem of another composite species, *Dimorphotheca sinuata*. The same two compounds—cyanidin and delphinidin 3-glucosides—were present throughout (Ball *et al.*, 1972).

The first composite anthocyanin to be identified was that of the blue cornflower, *Centaurea cyanus*. This was cyanidin 3,5-diglucoside, called cyanin after the plant source. This isolation, accomplished by Willstätter and Everest (1913), remains a landmark in the history of anthocyanin chemistry, because this was the first such pigment to be obtained in crystalline form from any plant. Since then, anthocyanins have been examined in a range of other composites and these pigments have been fully identified in 16 species from 14 genera (Table I). Anthocyanins have been partly characterized (i.e. as cyanidin or delphinidin glycosides in which the nature of the sugar attachment is unknown) in at least another 20 species from some further 12 genera (see e.g. Lawrence *et al.*, 1939; Forsyth and Simmonds, 1954).

On the basis of this small sample, it is apparent that the anthocyanin pattern of the Compositae is a remarkably simple one. Cyanidin is the major anthocyanidin with delphinidin occurring occasionally. Pelargonidin is presumably rare in the wild, since it has so far been detected mainly in mutant colour forms of cultivated species of the genera *Callistephus*, *Centaurea* and *Dahlia*. There are apparently only three main glycosidic types: 3-glucoside (in 8/16 spp.), 3,5-diglucoside (in 7/16 spp.) and 3-rutinoside (in 2/16 spp.). Methylation is absent, since no peonidin or malvidin derivatives have been reported. Acylation, another complex feature of anthocyanin production, is also rare or absent. An acylated pigment, a caffeyl ester of pelargonidin 3,5-diglucoside, has been reported once; this was not in a wild species but in a garden form of *Centaurea cyanus* (Asen, 1967).

It is surprising to find only simple anthocyanins in a family which is highly specialized in so many other features. From a comparative study of anthocyanin types in the angiosperms (Harborne, 1963, 1967), it is clear that cyanidin 3-glucoside, a common pigment in the Compositae, is a "primitive" structure and that modifications involving extra hydroxylation (to give delphinidin), complex glycosylation (attachment of sugars other than glucose), methylation and acylation are characteristic of more specialized angiosperms (see Fig. 1). For example, anthocyanins in the families of the Tubiflorae nearly all have several of these "extra" structural features. Acylation in particular has been correlated with specialization, since families which contain acylated anthocyanins all have a high Sporne advancement index (Harborne, 1977). The malvidin 3-(p-coumarylrutinoside)-5-glucoside isolated from petals of *Brunfelsia calycina* (Solanaceae) can be taken as a typical anthocyanin of such highly advanced angiosperm groups.

TABLE I. Anthocyanins identified in the Compositae

Plant	Organ	Anthocyanin(s)[a]	Reference
Callistephus hortensis Cass.	Petal	Pg and Cy 3-glucoside	Willstätter and Burdick, 1916
Centaurea cyanus L.	Petal	Cy 3,5-diglucoside	Willstätter and Everest, 1913
C. scabiosa L.	Petal	Cy 3-glucoside and Cy 3,5-diglucoside	J. B. Harborne, unpublished
Cichorium intybus L.[b]	Petal	Dp 3,5-diglucoside	
Coreopsis tinctoria Nutt.	Petal	Cy 3-glucoside	Shimokoriyama, 1957
Cosmos sulphureus Cav.	Petal	Cy 3-rutinoside	Hayashi, 1941
Dahlia variabilis Desf.	Petal	Pg and Cy 3,5-diglucosides	Harborne and Sherratt, 1957
Dendranthema indicum (L.) Desmoulins	Petal	Cy 3-glucoside	Willstätter and Bolton, 1916
D. morifolium (Ramat) Tzveler	Petal	Cy 3-glucoside	Kawase *et al.*, 1970
Dimorphotheca sinuata DC.	Petal, Stem, Callus	Cy and Dp 3-glucosides	Ball *et al.*, 1972
Helenium autumnale L.	Petal	Cy 3,5-diglucoside	Willstätter and Bolton, 1916
Machaeranthera gracilis (Nutt.) Shinners[c]	Callus	Cy 3-glucoside Cy 3-rutinoside	Harborne, 1964
Senecio formosus H.B. & K.	Petal	Cy 3,5-diglucoside	J. B. Harborne, unpublished
Solidago virgaurea L.	Petal	Cy 3-diglucoside	Bjorkman and Holmgren, 1958
Tanacetum coccineum Willd.	Petal	Cy 3-glucoside	Harborne *et al.*, 1970
Zinnia elegans Jacq.	Petal	Cy 3,5-diglucoside	Willstätter and Bolton, 1916

[a] Key: Cy = cyanidin, Dp = delphinidin

[b] The earlier suggestion based only on R_f values that delphinidin 3-rhamnoside-5-glucoside was present in chicory flowers (Harborne, 1967) has now been shown to be incorrect—a recent re-examination in this laboratory showed that the only sugar associated with the anthocyanin is glucose and that the pigment is delphin.

[c] Synonymous with *Haplopappus gracilis* (Nutt.) A. Gray

Anthocyanin structure	Type and source
UMBELLIFERAE 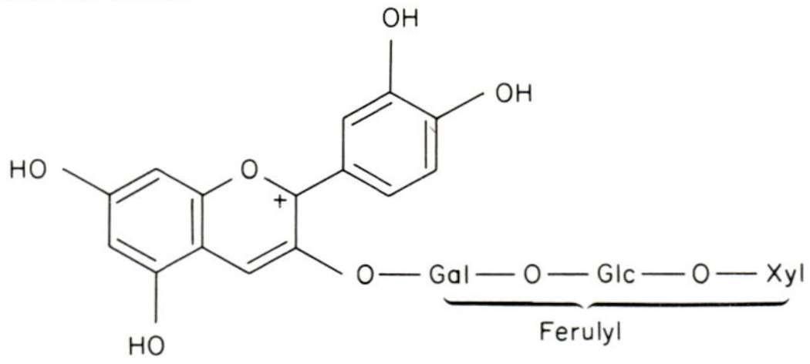 (cyanidin with O—Gal—O—Glc—O—Xyl at 3-position, Ferulyl on Glc–Xyl)	Acylated cyanidin triglycoside Typical source: *Daucus carota*
SOLANACEAE 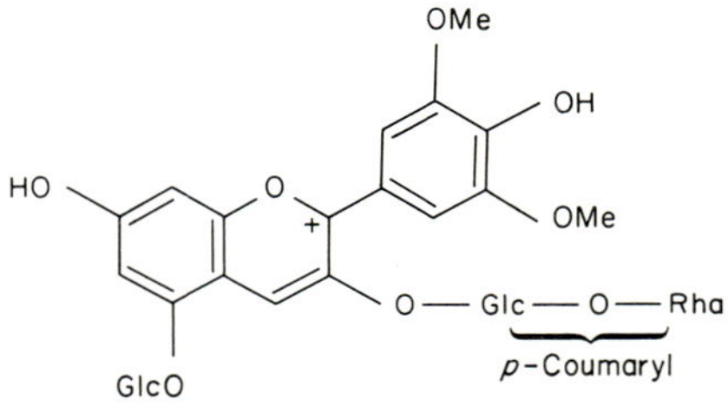 (malvidin with 5-OGlc and 3-O—Glc—O—Rha, p-Coumaryl on Glc–Rha)	Acylated malvidin triglycoside Typical source: *Brunfelsia calycina*
COMPOSITAE 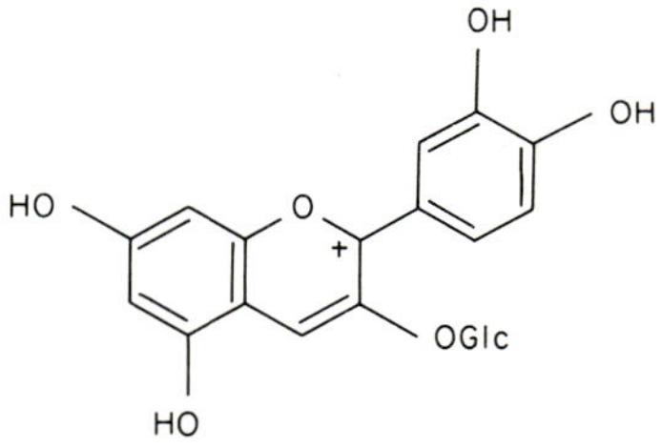 (cyanidin 3-OGlc)	Simple cyanidin monoside Typical source: *Dendranthema morifolium*

FIG. 1. Representative anthocyanin patterns in the Compositae, Solanaceae and Umbelliferae

The unusual simplicity of anthocyanin type in the Compositae is underlined when comparison is made with the Umbelliferae, a family which shows both close chemical (presence of polyacetylenes and sesquiterpene lactones) and at least distant morphological affinities with it. A representative survey (Harborne, 1976) of Umbelliferae pigments showed simplicity at the aglycone level—cyanidin was the only anthocyanidin detected—but complexity at the glycosidic level with presence of acylation. Indeed, attachment of a unique linear trisaccharide, a rare disaccharide, and acylation with p-coumaric, ferulic and sinapic acids were all involved.

Whether the simple anthocyanin pattern represents the retention of a primitive metabolic feature in the Compositae or whether it has been produced by "advancement through reduction" remains for future clarification. From what is argued elsewhere in this volume about the origins of the family and the status of other primitive features (e.g. woodiness, see Cronquist, Chapter 1), the former view seems the more likely. The conclusion that the anthocyanin profile is a simple one is based on a relatively small sampling and only future work will show whether this is representative for the family as a whole.

ANTHOCHLOR PIGMENTS

Yellow anthochlor pigments have long been of special interest to synantherologists, since they are amongst the most easily discerned of chemical characters. Their presence in yellow flowers can be determined in the field simply by fuming the petal with the alkaline vapour of a cigar, or of a bottle of ammonia (Gertz, 1938). An *in vivo* colour change to orange or red indicates that anthochlors are present; no colour change indicates that yellow colour is probably due to carotenoid. Simple colour tests of this type are of course not infallible and confirmation by chromatographic and spectral methods (see Harborne, 1967) is a necessary prerequisite for establishing the occurrence of these pigments in a particular species. Anthochlors may also occur in other tissues (e.g. in leaf of *Coreopsis mutica*, see Crawford, 1970) and in such cases chromatographic techniques are the only way of detecting their presence.

Chemically, anthochlors are of two interrelated structural types: chalcones and aurones. Chalcones are almost certainly intermediates in the synthesis of aurones and are probably converted to them by a specific enzymic oxidation. Pairs of structurally related chalcones and aurones occur together in many composites; some of the major structures identified variously in *Coreopsis* species are illustrated in Fig. 2. Anthochlors are relatively labile *in vitro* and undergo some hydrolysis and interconversion during extraction and isolation. The patterns of spots present on two-dimensional chromatograms or the mixtures isolated from flower tissues may therefore be more complex than the actual *in vivo* situation. Aglucones corresponding to the various natural glucosides nearly always appear in these mixtures, e.g. butein from the glucoside coreopsin etc. Complete details of the chemistry and identification of the natural anthochlors can be found in Geissman (1962), Harborne (1967) and Bohm (1975).

Yellow anthochlors are of limited distribution in composite flowers; indeed, they occur regularly in only one group of related taxa centred

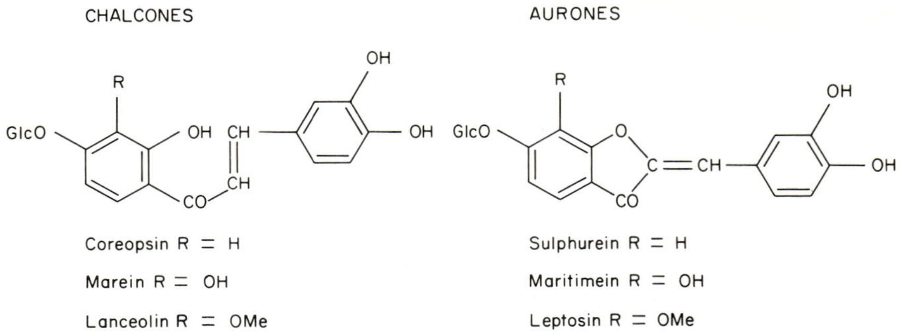

CHALCONES — AURONES

Coreopsin R = H
Marein R = OH
Lanceolin R = OMe

Sulphurein R = H
Maritimein R = OH
Leptosin R = OMe

FIG. 2. Anthochlor pigments of Coreopsis

about *Bidens, Coreopsis* and *Dahlia*. They are not the most useful of taxonomic markers, since their distribution encompasses at least three tribes of the family (Table II). Recent work (see Bohm, Chapter 26) has indicated that *Lasthenia* and *Syntrichopappus* (Helenieae) could well be accommodated on morphological grounds in the Coreopsidinae (Heliantheae) and the presence of anthochlors can be said to confirm this suggested alignment. Further, Turner and Powell (Chapter 25) propose tribal rank for this subtribe but without emphasis on anthochlors, which of course strengthens their case. At a lower level, the anthochlor character has been used in adjusting relationships in the genera *Baeria* and *Lasthenia* where these pigments are universally present. On this and other grounds, *Baeria* has been submerged into *Lasthenia*, being treated as a section of the latter (Ornduff, 1966).

Turning now to the occurrence of chalcones and aurones in the Cynareae and Inuleae, it is worth noting that the structures found here are different

TABLE II. Anthochlor pigments in the Compositae

Resorcinol-based Chalcones and Aurones
  Bidens, Cosmos, Coreopsis, Dahlia, etc. (Coreopsidinae) (HELIANTHEAE)
    Viguiera (Verbesininae) (HELIANTHEAE)
    Lasthenia (inc. *Baeria*), *Syntrichopappus* (HELENIEAE)
Phloroglucinol-based Chalcones and Aurones
    Carthamus (CYNAREAE)
    Gnaphalium, Helichrysum (INULEAE)

Outside the Compositae, chalcones occur in: Polypodiaceae, Pinaceae, Acanthaceae, Anacardiaceae, Caryophyllaceae, Gesneriaceae, Lauraceae, Leguminosae, Piperaceae, Plumbaginaceae, Ranunculaceae, Rutaceae, Salicaceae, Cyperaceae, Xanthorrhoeaceae, Zingiberaceae. Aurones occur in: Leguminosae, Plumbaginaceae, Gesneriaceae, Oxalidaceae, Scrophulariaceae, Cyperaceae (see Bohm, 1975).

from those of the Coreopsidinae. The pigments, as shown in Fig. 3, are all phloroglucinol-based (i.e. they have a hydroxyl at position 6'- or 4-) whilst those of the Coreopsidinae are resorcinol-based (there is no hydroxyl at 6'- or 4-). Bractein differs from any of the pigments of the Coreopsidinae in an additional feature, the presence of three instead of two hydroxyl groups in the B-ring.

Since flower pigments have not been extensively studied in the family, it is likely that anthochlors will be found elsewhere in the above tribes

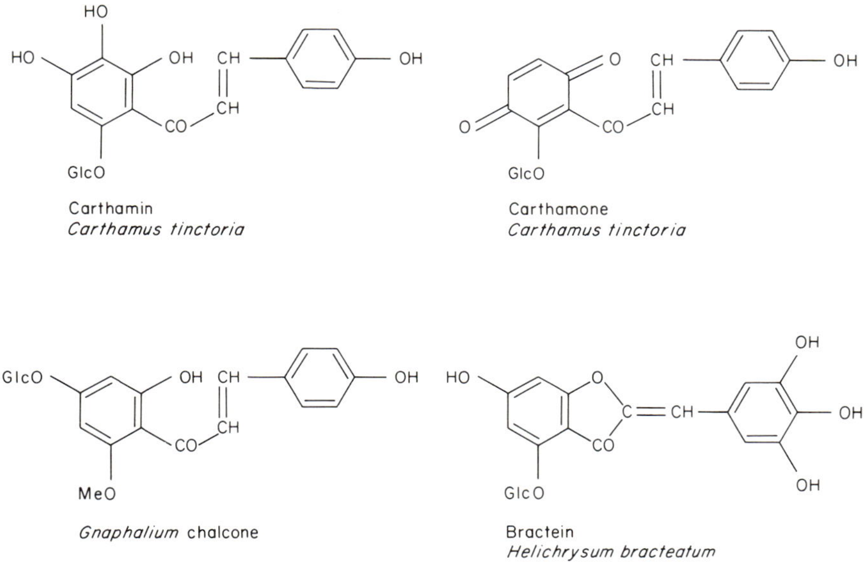

Fig. 3. Anthochlor pigment of Cynareae and Inuleae

and also in other tribes. Indeed, there is already a provisional report of anthochlors in the Cichorieae (in *Pyrrhopappus*; Northington, 1974). The main ligule pigment in four *Pyrrhopappus* species has subsequently been identified as coreopsin (see Fig. 2) (J. B. Harborne, unpublished results).

Since anthochlor pigments occur in a considerable number of unrelated angiosperm families (Table II), their presence in the Compositae cannot be regarded as having much phylogenetic significance. This is the more so because chalcones are, from the point of view of both biosynthesis and distribution, primitive pigments, while aurones represent advancement within the angiosperms. However, these pigments may jointly have some phylogenetic value at lower levels of classification. Bohm et al. (1974) have, for example, examined the distribution of anthochlors and other flavonoids in the various sections of *Lasthenia*. They have concluded that the ability to synthesize anthochlors is a primitive feature in the genus and that taxa which are more advanced on morphological and cytological grounds, have lost this dual character (see Fig. 4). They regard the direction of evolution as: *Baeria* and *Burrielia* → *Hologymne* → *Platycarpha* and *Ptilomeris*. This involves successive loss of okanin-maritimetin and butein-

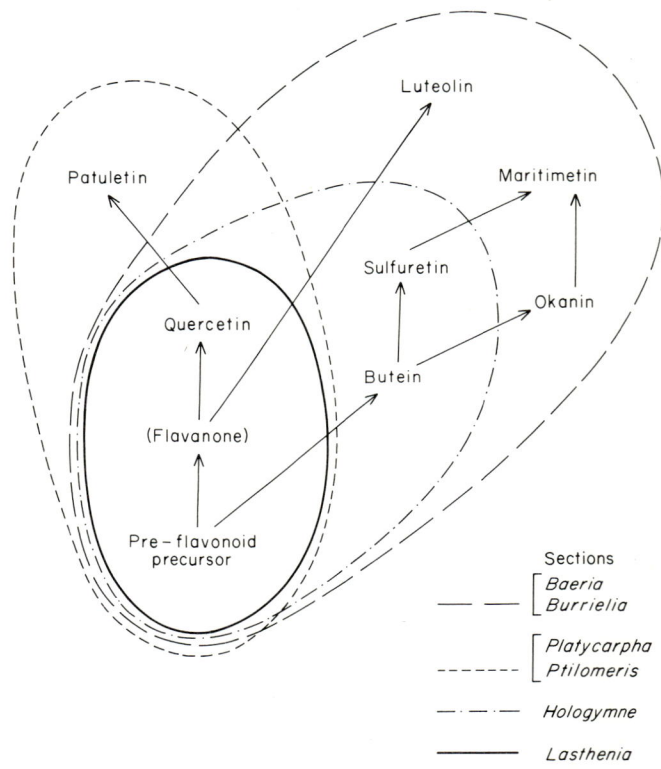

FIG. 4. Evolution of flavonoid pigments in the *Lasthenia* complex

sulfuretin and gain of patuletin. Thus one type of yellow phenolic pigment is gradually replaced by another, in this case patuletin, a yellow flavonol which is discussed in more detail below. Here, it is worth noting that yellow flavonols have been recognized recently to function in composite flowers as honey guides to bees and other insect visitors. Perhaps the most interesting unanswered question regarding the chalcones and aurones of the Coreopsidinae is their function. It is possible that pollinating insects can distinguish between yellow carotenoids, yellow flavonols and yellow anthochlors and that anthochlors function as honey guides in plants lacking yellow flavonols, but this has yet to be proved.

YELLOW FLAVONOLS

Yellow flavonols are, like the anthochlors, a class of flavonoid which significantly contributes to yellow flower colour in the Compositae. They are described as "yellow flavonols" (see Harborne, 1972) to distinguish them from the much more widespread common flavonols such as quercetin, which, if present in petals, provide pale cream or white colours. The two most important yellow flavonols are in fact simply derived from quercetin by the introduction of an extra hydroxyl group in the 8- or 6- position of the A-ring. There are thus two groups of pigment, those based on gossypetin where the extra hydroxyl is in the 8-position and those based on quercetagetin where the extra hydroxyl is at position 6 (see Fig. 5). The two groups have distinctly different distribution patterns within the angiosperms (Harborne, 1974). Gossypetin-based pigments are found mainly in the more primitive woody families, and especially in Ranunculaceae, Ericaceae, Primulaceae, Malvaceae and Sterculiaceae. By contrast quercetagetin pigments are almost entirely confined to the Compositae, with a few rare appearances in the Leguminosae and families of the Tubiflorae. The only plants where both types have been found to co-occur are in the Compositae; both quercetagetin and gossypetin occur in petals of *Chrysanthemum segetum* and *Leucanthemopsis flaveola*.

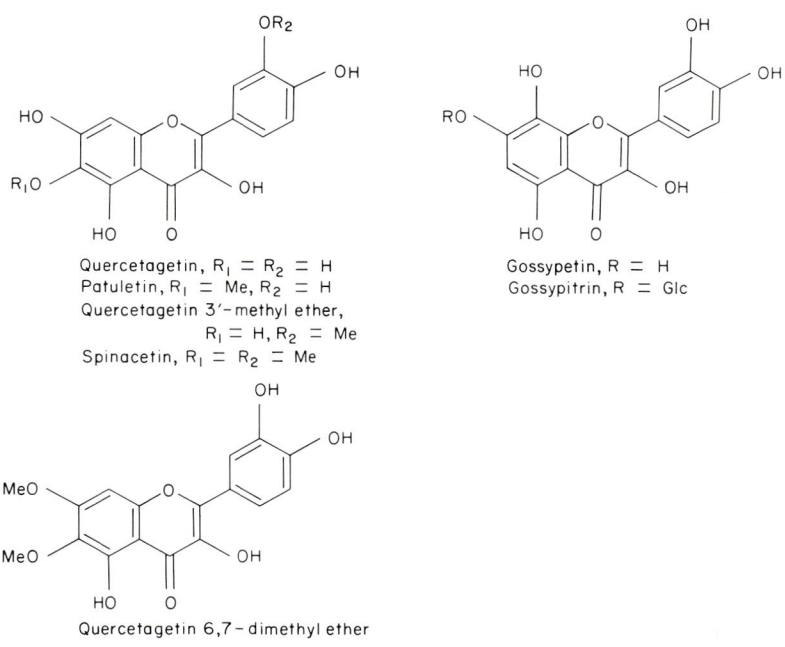

FIG. 5. Yellow flavonols of the Compositae

TABLE III. Evolution of yellow flavonols in the Chrysanthemum complex

Chemical Type:	Plants analysed:
Quercetagetin Gossypetin	*Chrysanthemum segetum* L. *Leucanthemopsis flaveola* (Hoffmanns. and Link) Heywood
Quercetagetin Patuletin (quercetagetin 6-methyl ether)	*Coleostephus myconis* (L.) Reichenb. f. *Heteranthemis viscidehirta* Schott
Patuletin alone	*Anthemis* spp. *Dendranthema arcticum* (L.) Tzvelev
Quercetagetin Quercetagetin 3'-methyl ether	*Chrysanthemum coronarium* L.
Quercetagetin Patuletin Quercetagetin 3'-methyl ether Quercetagetin 6,3'-dimethyl ether	*Lepidophorum repandum* (L.) DC.

Data from Harborne *et al.* (1970, 1976).

Like other classes of flavonoid, yellow flavonols are usually found *in vivo* in glycosidic combination. Gossypetin, for example, occurs in petals of *Chrysanthemum segetum* as the 7-glucoside, called gossypitrin, and quercetagetin as the 7-glucoside, quercetagitrin. These two yellow flavonols also occur in partly methylated form; patuletin, the 6-methyl ether of quercetagetin, first isolated from *Tagetes patula* (Rao and Seshadri, 1941), is particularly common. Other methyl ethers which have been isolated from composite flowers are the 3'-methyl ether (from *Chrysanthemum coronarium*), the 6,3'-dimethyl ether (from *Lepidophorum repandum*) and the 6,7-dimethyl ether (from *Rudbeckia hirta*). The structures of these various pigments are shown in Fig. 5.

The function of yellow flavonols is now known, at least as far as their occurrence in *Rudbeckia hirta* and other Heliantheae is concerned. In *Rudbeckia*, Thompson *et al.* (1972) have shown that they function as honey guides. They co-occur in the petals with carotenoids, but whereas the carotenoids are distributed over the whole petal, the yellow flavonols are found exclusively in the petal bases, which are thus u.v. absorbing.

They function here as nectar guides because, although invisible to the human eye, they are seen by pollinating insects which are drawn to the centre of the flower by this means. Observations in pressed flowers of other Heliantheae (namely *Bidens*, *Helianthus*, *Heliopsis* and *Viguiera* spp.) indicate that this phenomenon is relatively widespread in this tribe (Eisner *et al.*, 1973).

From the systematic viewpoint, yellow flavonols appear to have a relatively scattered distribution in the Compositae, but this is probably a false picture because few systematic surveys have been attempted. Their first isolation was from *Tagetes* (Latour and de la Source, 1877) and they are known to occur in petals of *T. erecta*, *T. patula* and *T. minuta* (see Chapter 28). As already mentioned above, they have also been obtained from *Rudbeckia hirta*, the petals of which contain quercetagetin and patuletin 7-glucosides and 6,7-dimethylquercetagetin 3-glucoside (Thompson *et al.*, 1972). However, their main occurrence to date is in the Anthemideae,

TABLE IV. Highly methylated quercetagetin ethers of the Compositae

Number of methyl groups and their position	Trivial name of compound	Composite source
Dimethyl Ethers		
3,6	axillarin	*Xanthium pennsylvanicum* Wallr.
3,7	—	*Parthenium tomentosum* DC.
6,7	—	*Eupatorium ligustrinum* DC.
6,3'	spinacetin	*Lepidophorum repandum* (L.) DC.
Trimethyl Ethers		
3,6,7	—	*Parthenium rollinsianum* Rzed.
3,6,3'	jaceidin	*Centaurea jacea* L.
3,6,4'	centaureidin	
3,7,3'	—	*Parthenium bipinnatifidum* (Ortega) Roll.
3,7,4'	oxyayanin B	*Pulicaria dysenterica* (L.) Bernh.
6,7,4'	eupatin	*Eupatorium semiserratum* DC.
Tetramethyl Ethers		
3,6,7,3'	chrysosplenetin	*Matricaria chamomilla* L.
5,6,7,4'	eupatoretin	*Eupatorium semiserratum* DC.
3,6,3',4'	—	*Bahia oppositifolia* DC.
6,7,3',4'	—	*Artemisia annua* L.
Pentamethyl Ether		
3,6,7,3',4'	artemetin	*Artemisia absinthium* L.

where they have been obtained from all yellow-flowered species that have been examined (Harborne et al., 1970, 1976). Their occurrence here seems to fall into an evolutionary pattern, with *Chrysanthemum segetum* containing "primitive" pigments (without methylation) and *Lepidophorum repandum* containing the more "advanced" pigments (with methylation at 6- and 3'-positions). Other related species can be fitted in between these two extremes (Table III). Assuming that increasing methylation is advantageous to the plant in stabilizing the pigment and that the pigments are useful in providing honey guides (as in the Heliantheae), then the various compounds can be placed in an evolutionary series, with gossypetin at one end and spinacetin at the other (Fig. 6). Clearly, the evolution of yellow

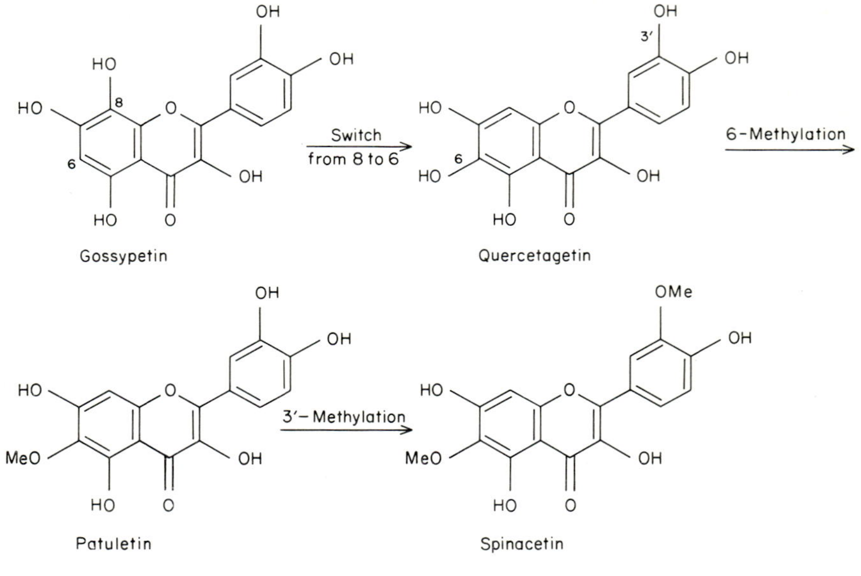

FIG. 6. Evolution of Yellow Flavonols in the Anthemideae

coloration in these plants is not necessarily related to morphological specialization, and indeed the positions shown in Table III are not directly correlated with taxonomic groupings.

OTHER FLOWER FLAVONOIDS

While the three classes of flavonoid pigment so far discussed are each of relatively limited distribution in flower tissues of Compositae, other kinds of flavonoid are more widespread. In particular, "colourless" flavone or flavonol glycosides are practically always present. In coloured flowers, they function as co-pigments to anthocyanins, sometimes having a blueing effect on flower colour (as in the blue cornflower) while in white or cream petals they provide pigments which, although not visible to the human eye, absorb strongly in the far ultraviolet and are attractive to certain kinds of insect pollinator.

The two most common of such flavonoids in composite flowers are undoubtedly apigenin and luteolin 7-glucosides, compounds which have been frequently isolated from many species (Harborne, 1967). For example, white rays of *Bellis perennis* and petals of *Achillea millifolium* contain apigenin 7-glucoside, while similar tissue of *Matricaria chamomilla* and of *Tripleurospermum* spp. contain luteolin 7-glucoside. White petals of *Tanacetum* spp. are slightly different in having luteolin 7-glucuronide (Harborne *et al.*, 1970).

Occasionally, rarer structures are encountered. One of the most distinctive compounds found in flowerheads of *Achyrocline*, *Gnaphalium* and *Helichrysum* spp. is 3,5-dihydroxy-6,7,8-trimethoxyflavone. This substance appears to be characteristic for the tribe Inuleae (see Chapter 22) but has been reported once in the Anthemideae, in *Artemisia klotzchiana* (Dominguez and Cardenas, 1975). Another chemical marker in flowers of the Inuleae is luteolin 4'-glucoside, found in *Antennaria dioica*, two *Gnaphalium* spp. and *Leontopodium alpinum*.

It is clear from the work that has been done so far on flavonoid patterns in the Compositae that it is usually worth investigating patterns in both leaf and flower, when carrying out systematic surveys. The flower pattern may be of taxonomic value and it may also differ significantly from that of the leaf. A comparison of flower and leaf patterns in one typical case, that of *Heywoodiella oligocephala*, is shown in Fig. 7.

In this plant, the spot pattern in leaf and flower looks superficially similar, with a group of five or six flavone constituents of low R_f in both

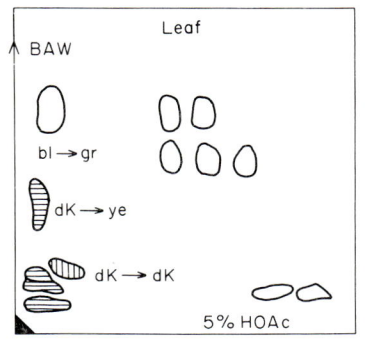
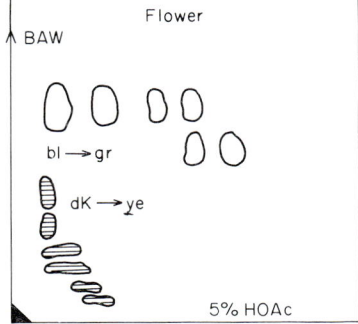

FIG. 7. Two dimensional chromatographic profiles of leaf and petal flavonoids of *Heywoodiella oligocephala*

solvents. However, detailed analysis has shown that there are clear differences. Six compounds occur in the petal; these are, in order of concentration: luteolin 7-glucoside, luteolin 7-glucuronide, luteolin 7-triglucoside, chrysoeriol 7-glucoside, an isoetin glucoside and an acylated apigenin glucoside. By contrast, in the leaf, the major component is isoetin 2'-xylosylarabinosylglucoside and there are lesser amounts of the related 2'-glucoside and 2'-galactoside and of the 7-diglucoside and 7-triglucoside of luteolin. Thus there are differences at both the aglycone and glycoside level. One of the aglycones present, isoetin (5,7,2',4',5'-pentahydroxyflavone) is a very rare substance and has previously only been found in two species of *Isoetes* of the Lycopsida (Voirin *et al.*, 1975). Isoetin is the first 2'-hydroxyflavone to be found in the Compositae. Besides occurring in *Heywoodiella*, it occurs in several related genera of the Cichorieae, namely in *Crepis* (in 2 of 9 spp. surveyed), *Hedypnois* (in 3 of 4 spp.), *Hypochaeris* (4/6 spp.), *Leontodon* (7/12), *Picris* (1/3) and *Reichardia* (1/6) (Harborne and King, 1976).

In some respects *Heywoodiella oligocephala*, which is a rare monotypic endemic of the Canary Islands, is unusually complex in having a range of differing flavonoids in both leaf and flower. In *Dendrosonchus* (also Cichorieae), the leaf pattern is uniformly simple (only luteolin 7-glucoside occurs) whereas floral tissue contains up to six or more flavone glycosides, mostly related to luteolin (Aldridge, 1975). In other parts of the Compositae, the reverse situation obtains. Thus in *Tripleurospermum* and *Matricaria* (Anthemideae) the petal constituents are few and the leaf constituents many (see Greger, 1975).

LEAF FLAVONOIDS

Much more is known of the flavonoid chemistry of the leaves than of the petals in the Compositae, presumably because leaves are more accessible and more generally studied. A large number of compounds have been isolated from leaves and it is impossible to mention here more than a few of the many interesting substances found. Sufficient data are available to discern several distinct patterns and also to predict with some confidence the kinds of compounds that are most likely to be found in future surveys of the family.

There are three common patterns of leaf constituents. These are, respectively: (1) common flavonols, i.e. kaempferol and quercetin, often as 3-glucoside or 3-rutinoside; (2) common flavones, i.e. apigenin and luteolin, usually as 7-glucoside or 7-rutinoside; and (3) *C*-glycosylflavones based on the two common flavones. These patterns may be superimposed one upon another and indeed, when *C*-glycosylflavones are present, they are often accompanied by either flavonol or flavone glycosides, as in, for example, the *Senecio radicans* complex (Glennie *et al.*, 1971). On the other hand, when C-glycosylflavones are absent, there is a tendency for genera to have either pattern 1 or pattern 2, but not both.

In addition to these common constituents, other substances may be encountered, with a frequency depending on the tribe under study. These rarer flavonoids can be considered under four headings: (1) simple methyl ethers; (2) 6-hydroxy compounds; (3) highly methylated derivatives; and (4) complex glycosides. Simple methyl ethers refer to monomethyl derivatives of the common flavones and flavonols. Substances such as isorhamnetin (quercetin 3'-methyl ether) and chrysoeriol (luteolin 3'-methyl ether) are not infrequent in the family. Methylation may also occur at the 3- or 7-positions and several such compounds have recently been isolated from leaves of *Vernonia* (Mabry et al., 1975).

6-Hydroxyflavonols have already been mentioned as occurring in flower petals (p. 351). The same substances also occur occasionally in the leaf; for example, in the Inuleae (see Chapter 22) and the Anthemideae (see Chapter 32). The more frequent leaf constituents are perhaps 6-hydroxyflavones and their simple methyl ethers. 6-Hydroxyapigenin (scutellarein) and its 6-monomethyl and 6,4'-dimethyl ethers are found regularly in leaves of *Centaurea* and *Cirsium* (Cynareae). The 6-monomethyl ether (hispidulin) and the 6,4'-dimethyl ether (pectolinaringenin) are both relatively common in leaf tissue of the Heliantheae. Finally, 6-hydroxyluteolin may be mentioned; it occurs in *Coreopsis* (Heliantheae) and has also been found in the Anthemideae, where it occurs in *Matricaria* but not usually in *Tripleurospermum* (see Greger, 1975).

A striking feature of the flavonoid patterns of advanced herbaceous angiosperm families—and this is especially true of the Compositae—is the presence in the leaf of lipid-soluble highly methylated derivatives. Because of their high lipid solubility and low water solubility, these compounds may often be missed when conventional extraction procedures are used, and they may therefore be much more widespread than at present appears. They normally occur without any sugars attached and it is possible that they are located either on the leaf surface or within the cytoplasm. Many such substances have been detected in different members of the Compositae. To illustrate the immense structural range of such derivatives, the various di-, tri-, tetra- and pentamethyl ethers of quercetagetin that have been obtained from the Compositae are listed in Table IV. It should be emphasized that the majority of these are unique to the family. Apart from the infrequency of 5-methylation, there seems to be no limit on the way that the different hydroxyls of the quercetagetin molecule are O-methylated. One can confidently predict that many of the missing isomers not yet reported for the family will be uncovered in the near future.

The last group of rarer leaf flavonoids are those which occur in unusual bound form. The very fact that much of the leaf flavonoid in many species is bound by methylation in lipid-soluble form means that much less is present in the cell vacuole as water-soluble glycoside. This could be one of the reasons why the glycosidic variation encountered in the Compositae is much less than in other comparable families. It is usually remarkably simple in both leaf and flower, involving two of the commonest patterns, combination with glucose or rutinose. However, other more complex patterns do occasionally appear. Flavones linked to glucuronic acid occur in several genera of the Anthemideae and one particular species, *Tanacetum corymbosum*, contains chrysoeriol linked to *p*-coumaric acid through a disaccharide containing both glucose and glucuronic acid (Harborne *et al.*, 1970). Another acylated flavonoid, found for the first time in the family, is quercetin 3-acetylglucoside which occurs in *Plummera* (= *Hymenoxys*) and *Helenium* (Wagner *et al.*, 1971). Glucosylation in a rare position—the 5-hydroxyl of flavonols—appears in leaves of several composite genera, notably in *Anthemis* and *Cotula*. A related flavone 5-glucoside, of luteolin, has also been found in *Cotula turbinata* (Glennie and Harborne, 1971).

FLAVONOIDS AND CLASSIFICATION AT THE TRIBAL AND GENERIC LEVELS IN THE COMPOSITAE

As yet, flavonoid data have had little impact on classification at the tribal or subtribal levels within this family, and are unlikely to have, until a much wider ascertainment has been achieved. Whilst it is possible to show that certain types of flavonoid (e.g. anthochlors) occur characteristically in some tribes, it is not yet clear that such types are definitely absent from others, since few efforts have been made to produce the necessary "negative" records. Particular tribes are apparently much richer in complex flavonoids than others, e.g. the Heliantheae, but this may be an accidental observation due to the fact that more chemical effort has been lavished on these tribes than on others. All that can be concluded from the present limited data is that there are indications of flavonoid heterogeneity at the tribal level which may well be worth pursuing.

Again, at the generic level, the situation is similar, largely because few genera have been properly and fully analysed for their flavonoid constituents. Some progress has been made in this respect in the classification of genera within the Anthemideae (Harborne *et al.*, 1970, 1976; Greger, 1969, 1975). One particular systematic problem has been the circumscription of the *Chrysanthemum* complex. Briefly, there are two main treatments of these plants: that of Hoffmann (1894) who adopted a very wide view of *Chrysanthemum* to include *Leucanthemum*, *Pyrethrum* and *Tanacetum*; and that of Briquet (Briquet and Vaillier, 1916) who maintained these and other genera as separate largely as a result of his carpological studies. Later taxonomic work (see Chapter 31) has largely reinforced Briquet's treatment.

In this situation, a chemical study was initiated of the flavonoids in leaf and flower of representative taxa of the *Chrysanthemum* complex. The results, some of which are summarized in Table V, unambiguously support the Briquet classification, since there are significant differences in flavonoid profiles within the group. These differences follow the division of the complex into smaller genera, along the lines of Briquet and, later, Heywood (see Chapter 31). In fact, each genus, so far as is known, can be separated on the basis of presence/absence of at least one chemical character. The characters detected include special glycosidic types (e.g.

TABLE V. Flavonoid profiles in leaf and flower of the *Chrysanthemum* complex

Genus and number of species studied	Flavones		Presence/absence of Flavonols		Glycosides[a]	
	Lu/Ap	Chr	Qu	Qut/Pat	7-Glur	7-α-Glc
Anthemis (3)	+	−	−	+	+	−
Chrysanthemum (5)	+	+	+	+	+	−
Dendranthema (1)	+	−	+	+	+	−
Leucanthemum (3)	+	−	−	−	+	−
Matricaria (1)	+	−	−	−	−	+
Tripleurospermum (4)	+	−	−	−	−	−
Tanacetum (4)	+	+	+	−	+	−

[a] Key: Lu = Luteolin, Ap = apigenin, Chr = chrysoeriol, Qu = quercetin, Qut = quercetagetin, Pat = patuletin, 7-Glur = 7-glucuronide, 7-α-Glc = 7-α-glucoside (as distinct from the more common 7-β-glucoside type). Data summarized from Harborne *et al.* (1970).

glucuronides), methylation of flavones (chrysoeriol formation) and 6-hydroxylation (production of quercetagetin). The same plants have been examined for polyacetylenes by Bohlmann *et al.* (1964) and their results agree closely with the flavonoid data in supporting the splitting of *Chrysanthemum* into a number of smaller genera.

More recent studies of petal pigments in these and related plants, mentioned earlier (p. 352), also support the re-evaluation of the taxonomic alignments in this group. Surveys of leaf flavonoids have also been helpful in providing support for the separation of *Tripleurospermum* from *Matricaria* (Greger, 1975). The *Chrysanthemum* flavonoid story is, however, far from complete and much more remains to be done in identifying constituents and determining distribution patterns, particularly at the populational level. Another example of the use of flavonoids at the generic level has been the work of M. W. Bierner (1973 and unpublished results) who has used such data very effectively to discriminate genera among the complex helenoid groups centring about *Helenium*, *Hymenoxys* and *Dugaldia*.

FLAVONOIDS AND CLASSIFICATION AT THE SPECIES LEVEL IN THE COMPOSITAE

Undoubtedly, the study of flavonoid profiles in the Compositae has been much more rewarding in its applications to systematic problems at the lower levels of classification than at the higher levels. Groups that have been studied in some depth include species of *Coreopsis, Lasthenia, Matricaria, Senecio, Tragopogon, Tripleurospermum* and *Vernonia*, to mention only a few. The results, although not always providing significant new characters for classificatory purposes, have been useful in reinforcing conclusions about plant populations derived purely from biological analyses. Because of the wealth of data available, only a few of the more prominent examples can be discussed here.

As an example of the impact of flavonoid data on infraspecific classification, the work of Crawford (1970) on *Coreopsis mutica* can be quoted. Following chromosomal, geographical, morphological, hybridization and flavonoid studies of these plants, this author recognized six subspecies. In drawing up his classification, he used four main flavonoid characters—presence/absence of common flavones, of common flavonols, of 6-hydroxyflavones and of anthochlors—and found them to correlate well with chromosome counts and geographical distribution (Table VI). Clearly, in such a variable plant species, it may be easier to identify with certainty different infraspecific taxa by flavonoid chromatography than by conventional methods.

A correlation between chemistry and chromosome number, as observed

TABLE VI. Flavonoid variation with subspecies of *Coreopsis mutica*

Name of Subspecies	Flavones	Flavonols	6-Hydroxy luteolin	Antho-chlors[a]	Chromosome number	Geographical site
microcephala	Lu	Qu	—	+	56	Guatemala
subvillosa	Lu/Ap	—	+	—	112	N.W. Oaxaco
carnosifolia	Lu/Ap	Qu	+	+	112	S.E. Oaxaco
multiligulata	—	Qu/Km	—	+	56	
leptomera	Lu/Ap	—	—	—	56	Central Mexico
mutica	Lu/Ap	Qu	+	+	112	

[a] Key: Lu = luteolin, Ap = apigenin, Qu = quercetin, Km = Kaempferol, anthochlors = butein + sulphuretin. Subsp *leptomera* also contains C-glycosylflavones (data from Crawford, 1970).

by Crawford (1970) in *Coreopsis*, may not always obtain and indeed a similar analysis of 44 clones from species within the *Senecio radicans* complex failed to show any relationship between flavonoids and chromosome numbers, the latter varying from $2n = 20$ to $2n = 180$. In this case, however, a correlation with geography was more apparent (Glennie *et al*., 1971). Thus, taxa from Madagascar, Canary Islands and Kenya markedly differed in leaf flavonoids from South and South West African plants. The former contained quercetin 3-glucoside, whereas the latter variously contained quercetin and kaempferol 3-rutinosides, kaempferol 3-glucoside, apigenin 7-glucoside, quercetin 3-methyl ether and a di-C-glycosylapigenin. Since the *Senecio radicans* complex undoubtedly originated in South West Africa, it is apparent in this case that geographical dispersal has produced populations depauperate in flavonoids.

A similar observation to the above has been made for species of the large genus *Vernonia* growing in the Americas. Here the South American taxa appear to provide the ancestral stock and they are rich in flavonoids, while the derived, more temperate, North American species are variously depauperate in these compounds (Mabry *et al*., 1975). In this case, differences in flavonoid content are also correlated with a changing pattern of sesquiterpene lactones (see Fig. 8).

Another group of plants in which geography apparently affects flavonoid content is *Tripleurospermum* (Greger, 1975). Here, Central Asian species

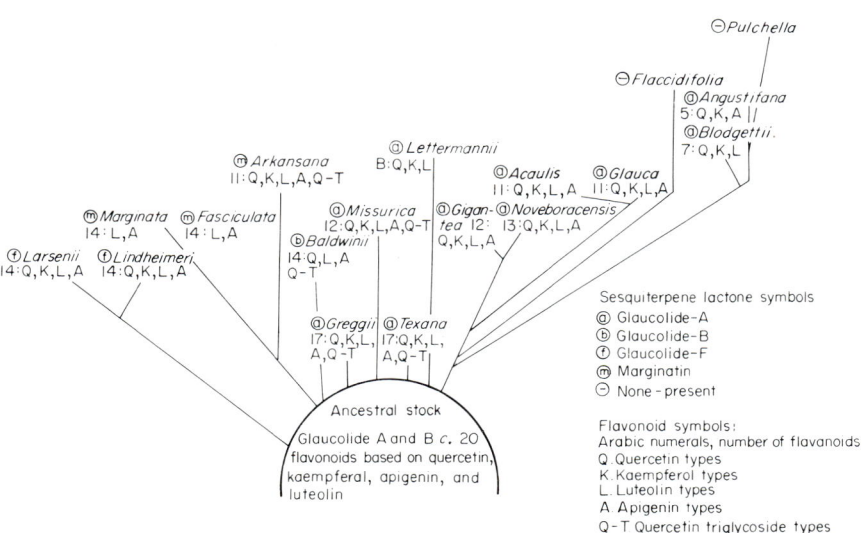

FIG. 8. Evolution of flavonoids and sesquiterpene lactones in North and South American *Vernonia* species

are distinguished from others by their ability to accumulate luteolin and apigenin 4'-glucosides and apigenin 7-glucoside. When the pattern in *Tripleurospermum* is compared with that of *Matricaria*, a line of evolutionary advancement is also apparent. The more primitive *Tripleurospermum* with the perennial habit contain 7-glucosides of quercetin and isorhamnetin, while in the more advanced *Matricaria*, these flavonols disappear and they are replaced by 7-glucosides of 6-hydroxyluteolin and of chrysoeriol.

Our final situation must be mentioned—the use of flavonoid chemistry in the detection of hybrids. The most important example here is the work of Brehm and Ownbey (1965) on the *Tragopogon dubius*, *T. pratensis* and *T. porrifolius* complex. These authors were able to distinguish hybrids of *T. dubius* × *T. porrifolius* and of *T. pratensis* × *T. porrifolius* simply by spot-pattern data of leaf flavonoids (Fig. 9). These authors were careful to measure both the relative intensities of the different flavonoid constituents and also the frequency of their distribution in plant populations. The chromatographic data were important in indicating that the transfer of genetic material among these diploid species was greater than was apparent simply from morphological studies. Although the flavonoids were not fully identified in this earlier work, subsequently identifications have been carried out on most of these compounds (see Kroschewsky *et al.*, 1969).

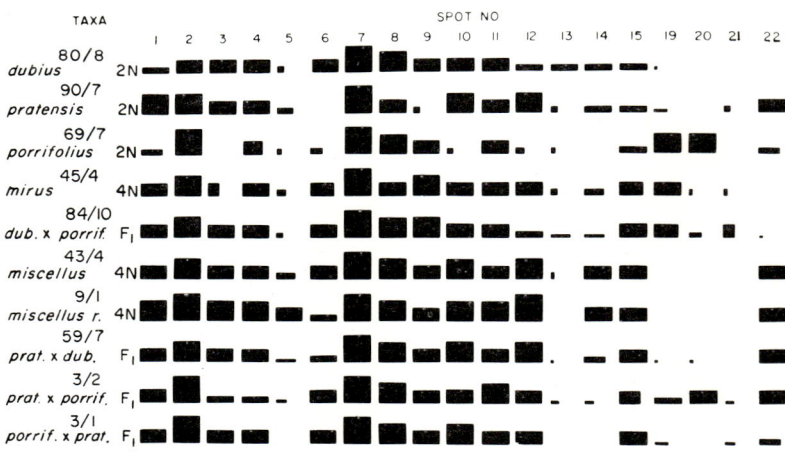

FIG. 9. Spot-pattern data of leaf flavonoids in species and their hybrids of *Tragopogon*. The figure shows two dimensional spot distribution, the horizontal dimension indicating population frequency of occurrence of the spot and the vertical dimension the average spot intensity (or concentration) (from Brehm and Ownbey, 1965).

CONCLUSION

To the organic chemist looking for new flavonoid structures in plants, the Compositae is one of the most promising family sources, since so much structural variation has already been revealed by past and present-day research. For example, a remarkable series of partly methylated 6-hydroxyflavonols have been uncovered in the leaves of these plants and clearly many more related structures remain to be found. The chemist should, however, be aware that there is a contradictory situation in the family in that some very simple structures are also widespread; this is especially true of the anthocyanins and flavones in floral tissues.

To the systematist interested in this family, the flavonoids represent a range of chemical characters, which may help him to solve taxonomic problems at the species and populational levels. There are some indications (e.g. with the anthochlor pigments) that flavonoids may be also of value at the subtribal and tribal levels. However, many more in-depth surveys will be needed before it will be possible to use flavonoids successfully at these higher levels of classification.

REFERENCES

ALDRIDGE, A. (1975). "Systematic and evolutionary studies in *Sonchus*, subgenus *Dendrosonchus*". Ph.D Thesis, University of Reading.

ASEN, S. (1967). Anthocyanins of cornflower cultivar 'Red Boy'. *Proc. Am. Soc. hort. Sci.* **91**, 653-657.

BALL, E. A., HARBORNE, J. B. and ARDITTI, J. (1972). Anthocyanins of *Dimorphotheca*: pigments in flowers, stems and callus tissues. *Am. J. Bot.* **59**, 924-930.

BATE-SMITH, E. C. (1962). The phenolic constituents of plants and their taxonomic significance—I—Dicotyledons. *J. Linn. Soc. (Bot.)* **58**, 95-173.

BEIRNER, M. W. (1973). Chemosystematic aspects of flavonoid distribution in twenty-two taxa of *Helenium Biochem. Syst.* **1**, 55-57.

BJORKMAN, O. and HOLMGREN, P. (1958). Anthocyanins and other flavonoids and respiration rates in different ecotypes of *Solidago virgaurea*. *Physiol. Plant.* **11**, 154-159.

BOHLMANN, F., ARNDT, C., BORNOWSKI, H., KLEINE, K. M. and HERBST, P. (1964). New acetylenic compounds from *Chrysanthemum* species. *Chem. Ber.* **97**, 1179-1192.

BOHM, B. A. (1975). Chalcones, aurones and dihydrochalcones. *In* "The Flavonoids" (J. B. Harborne, T. J. Mabry, and H. Mabry, eds), pp. 442-504. Chapman and Hall, London.

BOHM, B. A., SALEH, N. A. M. and ORNDUFF, R. (1974). Flavonoids of *Lasthenia*. *Am. J. Bot.* **61**, 551-561.

BREHM, B. G. and OWNBEY, M. (1965). Variation in chromatographic patterns in the *Tragopogon dubius–pratensis–porrifolius* complex. *Am. J. Bot.* **52**, 811–818.
BRIQUET, J. and VAILLIER, F. (1916). Compositae *In* "Flore des Alpes maritimes" (E. Burnet, ed.), **6**, 5–169.
CRAWFORD, D. J. (1970). *Coreopsis mutica*: flavonoid chemistry, chromosome numbers, morphology and hybridization. *Brittonia* **22**, 93–111.
CRONQUIST, A. (1968). "The Evolution and Classification of Flowering Plants". Nelson, London.
DOMINGUEZ, X. A. and CARDENAS, E. A. (1975). Achillin and desacetylmatricarin from two *Artemisia* species. *Phytochemistry* **14**, 2511–2512.
EISNER, T., EISNER, M., HYYPIO, P. A., ANESHANSLEY, D. and SILBERSGLIED, R. E. (1973). UV patterns visible in pressed herbarium specimens. *Science, N.Y.* **179**, 486.
FORSYTH, W. G. C. and SIMMONDS, N. W. (1954). A survey of the anthocyanins of some tropical plants. *Proc. R. Soc.* **B142**, 549–564.
GEISSMAN, T. A., ed. (1962). "Chemistry of Flavonoid Compounds". Pergamon Press, Oxford.
GEISSMAN, T. A. and HEATON, C. D. (1943). Anthochlor pigments. IV. The pigments of *Coreopsis grandiflora*. *J. Amer. Chem. Soc.* **65**, 677–680.
GERTZ, O. (1938). The distribution of anthochlors in the Compositae. *Kungl. Fysiog. Sallsk. Forh.* **8**, 62–70.
GLENNIE, C. W. and HARBORNE, J. B. (1971). Flavone and Flavonol 5-glucosides *Phytochemistry* **10**, 1325–1329.
GLENNIE, C. W., HARBORNE, J. B., ROWLEY, G. D. and MARCHANT, C. J. (1971). Correlations between flavonoid chemistry and plant geography in the *Senecio radicans* complex. *Phytochemistry* **10**, 2413–2417.
GREGER, H. (1969). Flavonoids and the systematics of the Anthemideae. *Naturwissenschaften* **56**, 467–468.
GREGER, H. (1975). Leaf flavonoids and systematics in *Matricaria* and *Tripleurospermum*. *Plant Syst. Evol.* **124**, 35–55.
HARBORNE, J. B. (1963). Distribution of anthocyanins in higher plants. *In* "Chemical Plant Taxonomy" (T. Swain, ed.), pp. 359–388. Academic Press, London and New York.
HARBORNE, J. B. (1964). Phenols in plant tissue cultures. *John Innes Inst. 54th Ann. Rept.* 45–46.
HARBORNE, J. B. (1967). "Comparative Biochemistry of the Flavonoids". Academic Press, London and New York.
HARBORNE, J. B. (1972). Evolution and function of flavonoids in plants. *Recent Adv. Phytochem.* **4**, 107–141.
HARBORNE, J. B. (1974). Flavonoids as systematic markers in the angiosperms. *In* "Chemistry in Botanical Classification" (G. Bendz and S. Santesson, eds), pp. 103–115. Nobel Foundation, Stockholm.
HARBORNE, J. B. (1976). Anthocyanin patterns in *Daucus carota* and other Umbelliferae. *Biochem. Syst. Ecol.*, **4**, 31–35.
HARBORNE, J. B. (1977). Flavonoids and the evolution of the angiosperms. *Biochem. system. Ecol.*, **5**, 251–269.

HARBORNE, J. B. and KING, L. (1976). Flavonoids in *Heywoodiella oligocephala*. *Phytochemistry*, in preparation.
HARBORNE, J. B. and SHERRATT, A. S. A. (1957). Variations in the glycoside pattern of anthocyanins. *Experientia* **13**, 486–490.
HARBORNE, J. B., HEYWOOD, V. H. and SALEH, N. A. M. (1970). Flavonoid patterns in the *Chrysanthemum* complex of the tribe Anthemideae. *Phytochemistry* **9**, 2011–2017.
HARBORNE, J. B., HEYWOOD, V. H. and KING, L. (1976). Evolution of yellow flavonols in *Lepidophorum repandum* and related Anthemideae. *Biochem. system. Ecol.*, **4**, 1–4.
HAYASHI, K. (1941). *Acta phytochim.*, *Japan* **12**, 65.
HOFFMAN, O. (1894). Compositae. *In* "Die natürlichen Pflanzenfamilien" (Engler and Prantl, eds) **4** (5), 267.
KAWASE, K., TSUKAMOTO, Y., SAITO, N. and OSAWA, Y. (1970). Flower colour in *Chrysanthemum morifolium*—anthocyanins. *P. Cell. Physiol.* **11**, 349.
KROSCHEWSKI, J. R., MABRY, R. J., MARKHAM, K. R. and ALSTON, R. E. (1969). Flavonoids from *Tragopogon*. *Phytochemistry* **8**, 1495–1498.
KUPCHAN, S. M., HEMINGWAY, R. J., KARIM, A. and WERNER, D. (1969). Tumor inhibitors. XLVII Vernodelin and Vernomygdin, two cytotoxic lactones from *Vernonia*. *J. org. Chem.* **34**, 3908–3911.
LATOUR and MAGNIER DE LA SOURCE (1877). *Bull. Soc. Chem. Paris* 228–337.
LAWRENCE, W. J. C. and SCOTT-MONCRIEFF, R. (1935). The genetics and chemistry of flower colour in *Dahlia*: a new theory of specific pigmentation. *J. Genet.* **30**, 155.
LAWRENCE, W. J. C., PRICE, J. R., ROBINSON, R. and ROBINSON, M. (1939). The distribution of anthocyanins in flowers, leaves and fruits. *Phil. Trans. R. Soc.* **230**, 149–197.
MABRY, T. J., BASET, Z. A., PADOLINA, W. G. and JONES, S. B. (1975). Systematic implications of flavonoids and sesquiterpene lactones in species of *Vernonia*. *Biochem. syst. Ecol.* **2**, 185–192.
NORTHINGTON, D. K. (1974). Chemosystematic studies of genus *Pyrrhopappus* (Cichorieae). *Spec. Publ. Mus. Tex. Tech. Univ.* **6**, 1–38.
ORNDUFF, R. (1966). A biosystematic survey of the gold-field genus *Lasthenia*. *Univ. Calif. Publ. Bot.* **40**, 1–92.
PERKIN, A. G. (1902). Quercetagetin and its sulphate potassium salt and acetyl compound. *Proc. chem. Soc.* **18**, 75.
PERKIN, A. G. (1913). Quercetagetin. *J. chem. Soc.* **103**, 209–219.
RAO, P. S. and SESHADRI, T. R. (1941). Isolation and constitution of quercetagitrin, a glucoside of quercetagetin. *Proc. Indian Acad. Sci.* **14A**, 289–296.
SHIMOKORIYAMA, M. (1957). Anthochlor pigments of *Coreopsis tinctoria*. *J. Am. chem. Soc.* **79**, 214–220.
THOMPSON, W. R., MEINWALD, J., ANESHANSLEY, D. and EISNER, T. (1972). Flavonols pigments responsible for UV absorption in nectar guide of flowers. *Science, N.Y.* **177**, 528.
VOIRIN, B., JAY, M. and HAUTEVILLE, M. (1975). Isoetin, a new flavone isolated from *Isoetes delilei* and *I. duriani*. *Phytochemistry* **14**, 257–259.

WAGNER, H., IYENGAR, M. A., MICHAHELLES, E. and HERZ, W. (1971). Quercetin 3-acetylglucoside in *Plummera floribunda* and *Helenium hoopesii*. *Phytochemistry* **10**, 2547.

WILLSTÄTTER, R. and BOLTON, E. K. (1916). The anthocyanin of winteraster, *Liebig's Ann. Chem.* **412**, 136–148.

WILLSTÄTTER, R. and BURDICK, C. L. (1916). Two anthocyanins in summer aster. *Liebig's Ann. Chem.* **412**, 149–164.

WILLSTÄTTER, R. and EVEREST, A. E. (1913). The colouring matter of cornflower. *Liebig's Ann. Chem.* **401**, 189–232.

Chapter 13

Polyacetylenes and conservatism of chemical characters in the Compositae

N. A. SØRENSEN
*Norges Tekniske Høgskole,
Trondheim, Norway*

Abstract. Acetylenic compounds have so far been isolated from 19 families of higher plants; among lower plants some genera of the Basidiomycete fungi are frequent producers of polyacetylenes. The dicotyledonous families most regularly synthesizing polyacetylenes—Araliaceae, Campanulaceae, Compositae, Santalaceae and Umbelliferae—all appear to use the same precursor as the Basidiomycetes, crepenynic acid.

The Compositae occupy, from the chemical point of view, a special position in so far as the acetylenic fatty acid derivatives to a very wide extent are transformed into cyclic compounds—aromatic or heterocyclic. With the exception of some lactones, which are formed spontaneously *in vitro* at neutral pH, such cyclic compounds are extremely rare amongst the Basidiomycete metabolites as well as amongst the acetylenic compounds from the remaining 18 families of higher plants.

Within the 13 tribes of the Compositae the occurrence of such cyclized acetylenes is very irregular, and a review is given of the occurrence of cyclized acetylenes in this family. For chemotaxonomic usefulness a regular occurrence of special types of compounds within sections, genera or tribes is necessary. Implicit in this is that the type of compound is constant within a species; in other words, chemovariants do not occur. The Compositae is well known for occurrence of chemovariants, as e.g. in sesquiterpenoid lactones. Systematic investigations of the occurrence of acetylenic compounds in this respect are lacking. However, scattered observations indicate that polyacetylenes do not belong to the secondary metabolites which give rise to chemovariants.

A number of polyacetylenic compounds from the Compositae are physiologically very active compounds against other organisms; their physiological importance inside the plants, however, is still unknown. What is known with certainty is that the type of compound varies between different parts of the plants—roots, leaves, stems—and in a number of cases with the season. In those cases where labelled polyacetylenes have been studied, they have been rapidly metabolized, with a half-life of 1–2 days.

For phylogenetic purposes the conservatism of a chemotaxonomic class of compounds is fundamental. Since the composites are mainly montane and xerophytic plants, paleontological evidence is mostly lacking. As with the family Santalaceae the occurrence is world-wide. Some indications of the conservatism of the polyacetylenes may be obtained from investigations of endemic representatives of genera or tribes from widely separated parts of the world, where geological facts indicate very ancient separations of parental stock. So far such investigations of members of the most regular producers of the two main types of polyacetylenes, the Compositae and the Santalaceae, confirm a far-reaching conservatism of polyacetylene synthesis.

CONTENTS

Introduction	386
The derivation of natural acetylenes	388
The degree of unsaturation among the acetylenes	389
The special characteristics of the polyacetylenes of the Compositae	391
The cyclic polyacetylenes of the Compositae	394
Conservatism of chemical characters in the Compositae	404
References	408

INTRODUCTION

Amongst the different classes of naturally occurring compounds the acetylenes occupy a somewhat special position, since organic chemists at one time did not believe they could exist in nature. It was at this time in history, when discussing the constitutional possibilities (Fig. 1) for "carlina

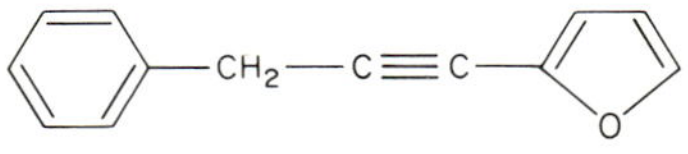

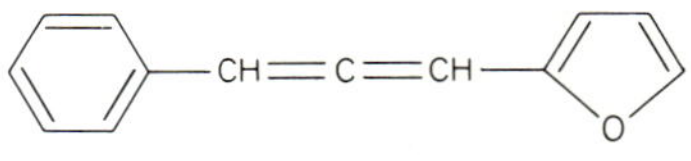

FIG. 1. Constitutional possibilities for "carlina oxide" in 1906.

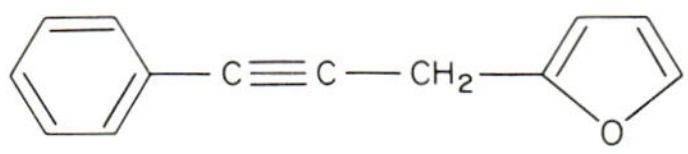

oxide", that L. F. Semmler (1906) stated: "Es ist von Hause aus sehr unwahrscheinlich, dass eine acetylenartige Verbindung vorliegt, sondern wir werden mit der Verbindung 2 zu tun haben." Looking back to the allenic derivatives known in 1906, it seems just as improbable that allenic compounds should exist regularly as natural products.

The most widespread occurrence of acetylenic compounds is closely connected to the dilemma of L. F. Semmler. Most of the structural variation encountered among the carotenoids is connected with the oxidation of the 5,6-bond and succeeding rearrangements. How far the two reaction

13. POLYACETYLENES AND CONSERVATISM OF CHEMICAL CHARACTERS

possibilities depicted in Fig. 2 are interconnected, is so far unknown; both lead to the allenic end group (4), found in all members of higher plants (neoxanthin). The transformation of the end group (4) into end group (5) leads to one group of acetylenic carotenoids. Acetylenic carotenoids are regular members of at least three algal families: the Chrysophyceae, the Bacillariophyceae (= diatoms) and the Xanthophyceae; but are not so far known from higher plants. The regular occurrence of type (5) carotenoids

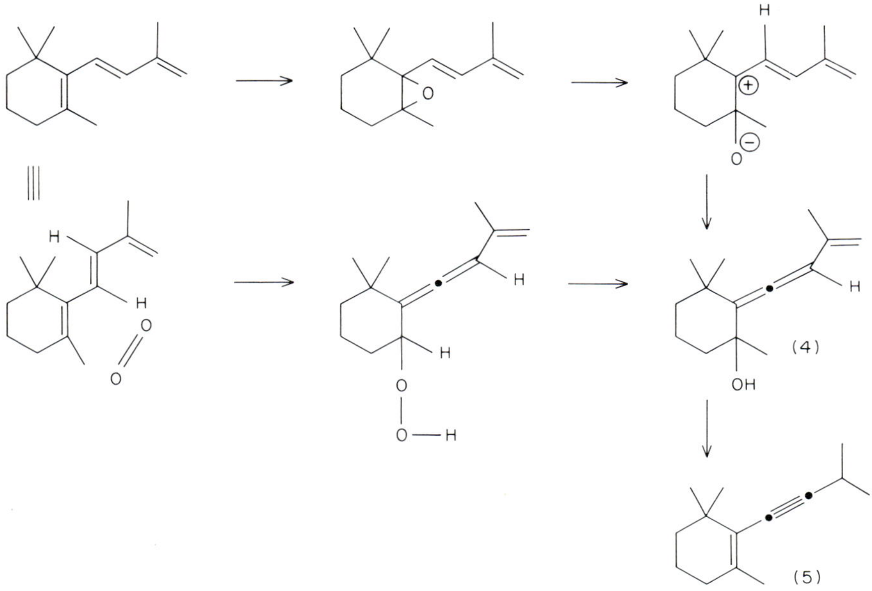

Fig. 2. Transformation of a normal carotenoid end-group into allenic and acetylenic end-group.

in these algal families probably outnumbers all other occurrences of acetylenic compounds. The presumption of Semmler has been answered as it deserved.

During the last 10–15 years, there have been isolated about a dozen monoacetylenic compounds belonging to very different classes of secondary plant products and from the most distant corners of the plant kingdom. They have one structural detail in common: they are all monosubstituted acetylenes. They co-occur to some extent with their vinyl analogues. Whether they are synthesized directly from the vinyl analogues by dehydrogenation or via hydration, is unknown (Fig. 3). To what extent these compounds are useful for chemotaxonomic purposes has not been determined.

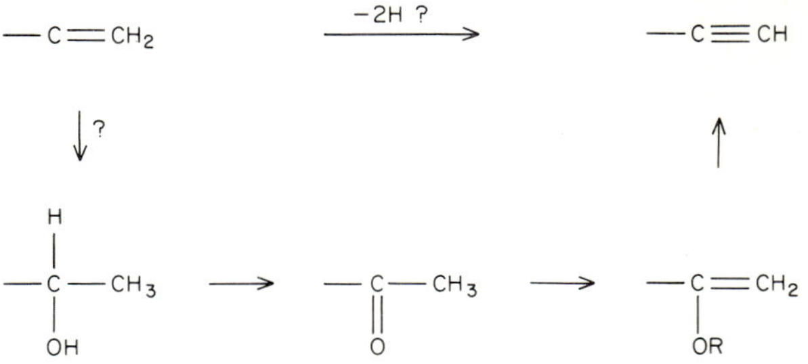

$$-C=CH_2 \xrightarrow{-2H\ ?} -C\equiv CH$$

$$\downarrow ?\qquad\qquad\qquad\qquad\qquad \uparrow$$

$$-\underset{OH}{\overset{H}{C}}-CH_3 \longrightarrow -\underset{O}{\overset{\|}{C}}-CH_3 \longrightarrow -\underset{OR}{C}=CH_2$$

FIG. 3. Pathways to monosubstituted acetylenic compounds.

THE DERIVATION OF NATURAL ACETYLENES

The isoprenoid skeleton of the carotenoids prohibits the formation of conjugated polyacetylenes. The only class of natural compounds suitable for a polyacetylenic element are the straight chain fatty acids. So far they may be divided into three groups (Fig. 4). Group 1 has so far only one member: tariric acid, the first acetylenic compound from nature whose structure was established as early as 1902 by Arnaud. Tariric acid is peculiar in so far as the unsaturation is in the same position as in petroselinic acid, one of the phytochemical markers of the seed oils from the Umbelliferae and Araliaceae. The real occurrence of monoacetylenic acids like tariric and stearolic acids and their analogues will not be known until a sensitive and nondestructive Raman spectrometer is commonplace in phytochemical laboratories.

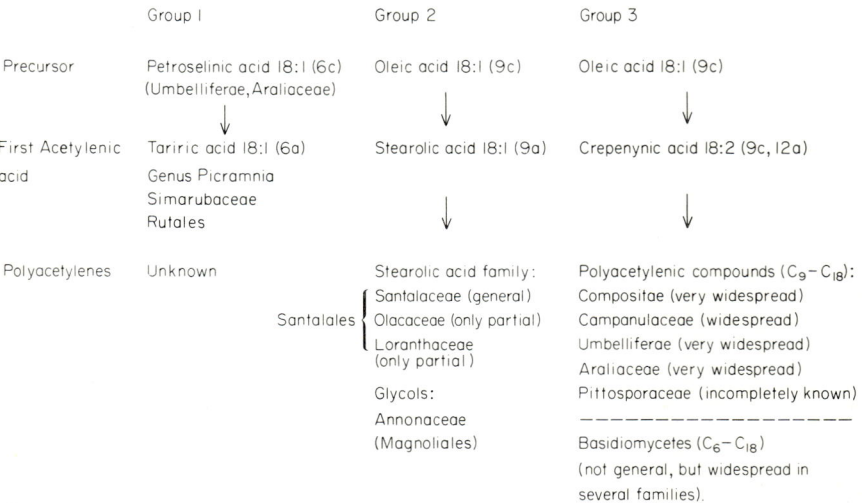

FIG. 4. Acetylenic compounds derived from fatty acids.

Group 2 comprises some 24 acetylenic fatty acids. Stearolic acid is transformed in four different ways (Fig. 5). These four transformations recur in Group 3, but within the order Santalales the transformations always stop at the acid level. Remarkably, when the seed kernels of *Ongokea klaineana* Pierre are allowed to rot, the acetylenic acids are split and a number of hydrocarbons with 10 carbon atoms and ene-diyne and ene-triyne unsaturation come into existence; compounds are formed with the characteristics of Group 3.

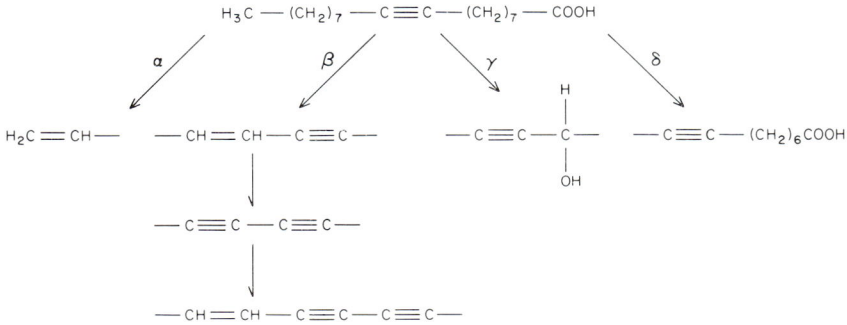

FIG. 5. Transformations of stearolic acid in group 2, Fig. 4.

From the chemotaxonomic point of view the results from the family Santalaceae are interesting: every species so far investigated contains members of the stearolic acid family. The family is world-wide in distribution and endemic genera confined to such distant regions as Alaska (*Geocaulon* Fernald) and Tasmania (*Leptomeria billardieri* R.Br.) have been investigated. The stearolic acid family is thus undoubtedly a very conservative phytochemical marker.

THE DEGREE OF UNSATURATION AMONG THE ACETYLENES

The main characteristic of the polyacetylenes of Group 3 is that analogues to the stearolic acid family of Group 2 are virtually absent. The 700 acetylenes so far known in Group 3 are dominated by short chains and are mostly neutral compounds: hydrocarbons, alcohols, aldehydes, ketones, nitriles, epoxides, amides and sulphur derivatives. Very common is the high degree of unsaturation; let us take two examples.

In the 1920s the Russians (cf. Anon, 1932) started a perfume industry based on the essential oil of *Lachnophyllum gossypinum* Bg. (Astereae). The main component was shown by Viljams *et al.* (1935) to be (**6**), which has the trivial name lachnophyllum ester (Fig. 6). This was the first diacetylenic compound with established constitution. Compounds (**7**) and (**8**) with extended unsaturation are still more common in the tribes Astereae and Anthemideae than (6); both as such and in the form of a large number of transformation products. These C_{10}-methyl esters have also been found in some fungi.

If compounds (6) to (8) occur at all outside the above two tribes, they are at least extremely scarce. One of the polyacetylenes most often found in most of the remaining composite tribes is the pentaacetylene (9), which is a pale yellow compound, very photosensitive and with a tendency to ignite between 10 and 40°C; (9) has a very characteristic u.v. spectrum, so that it will very seldom be overlooked during surveys. Does (9) represent the maximum of unsaturation found in plants?

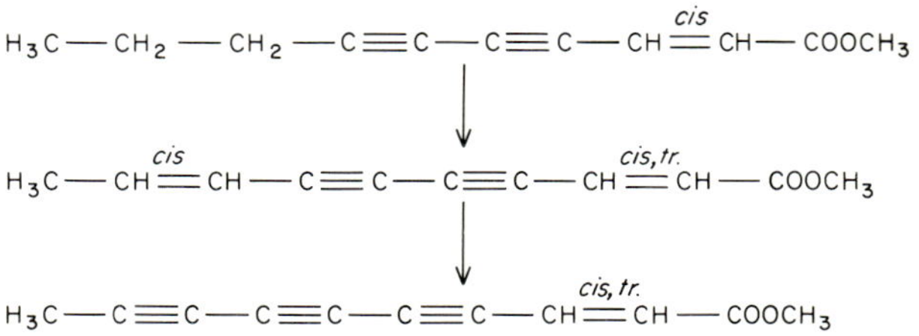

$$H_3C-CH_2-CH_2-C\equiv C-C\equiv C-CH\overset{cis}{=}CH-COOCH_3 \quad (6)$$

$$H_3C-CH\overset{cis}{=}CH-C\equiv C-C\equiv C-CH\overset{cis,tr.}{=}CH-COOCH_3 \quad (7)$$

$$H_3C-C\equiv C-C\equiv C-C\equiv C-CH\overset{cis,tr.}{=}CH-COOCH_3 \quad (8)$$

FIG. 6. Constitution of the most common C_{10}-methyl esters.

Glen et al. (1966) demonstrated that the main odoriferous principle of the root rot, *Fometopsis annosus* (Fr.) Karst., was triacetylene: $HC\equiv C-C\equiv C-C\equiv CH$. This explosive compound would normally escape detection since it decomposes during the normal isolation procedure, and the Glasgow group had to take all the precautions learnt while synthesizing this compound.

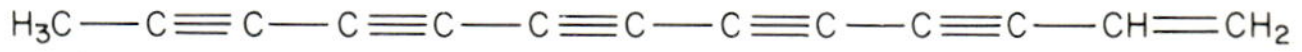

$$H_3C-C\equiv C-C\equiv C-C\equiv C-C\equiv C-C\equiv C-CH=CH_2 \quad (9)$$

$$H_3C-C\equiv C-C\equiv C-C\equiv C-C\equiv C-C\equiv C-C\equiv CH \quad (10)$$

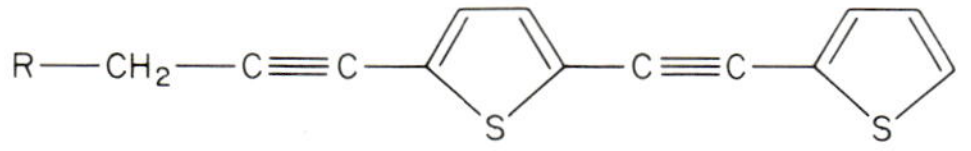

(11) R = H
(12) R = OH

Synthetic experience also tells us that (10) would not be isolated by the usual isolation techniques. It is reasonably well established that the thiophene derivatives of the Compositae—a very large group—originate from two acetylenic bonds and hydrogen sulphide. From *Berkheya heterophylla* (Thb.) O. Hoffm., Bohlmann (Bohlmann et al., 1973) has isolated the thiophenes (11) and (12), which are clear-cut analogues of (10). As in many other fields of natural product chemistry, the isolation techniques are not adequate for detecting either very volatile or extremely unstable compounds. This fact clearly restricts the usefulness of present survey techniques for chemotaxonomic purposes.

THE SPECIAL CHARACTERISTICS OF THE POLYACETYLENES OF THE COMPOSITAE

Except for the Pittosporaceae, from which family Professor Bohlmann and his collaborators in recent years have reported very typical polyacetylenes, the families of higher plants in Group 3 comprise families which in some phylogenetic schemes (cf. Bessey, 1915) are closely related. From a chemist's point of view, there is a large difference between them: the Basidiomycete acetylenes and the acetylenes from all but one of the families of higher plants are predominantly aliphatic; by contrast, the Compositae are crowded with "curled up" compounds, i.e. compounds which are aromatic, furanoid, thiophenic or spiroketal.

The easiest way is to consider the three exceptional occurrences of heterocycles outside the Compositae first. Since 1969, Sir Ewart Jones and his collaborators have isolated a number of acetylenic compounds from members of the Campanulaceae. They are mostly present in very low concentrations; a characteristic of their structures is the widespread occurrence of a tetrahydropyranyl end group. This end group, for which there so far seems to be no reason for it to have an acetylenic origin, has never been found in the Compositae.

$$HC\equiv C-\underset{\underset{O}{\|}}{C}-CH=C \overbrace{}^{O}{}_{H}-CH_2-CH_2-CH_2-CH_2-CH_3 \quad (13)$$

$$(14) \quad R= -C \begin{array}{c} H \\ \diagdown \\ O \end{array}$$

$$H_3C-C\equiv C-\underset{S}{\langle\rangle}-R$$

$$(15) \quad R= -\underset{\underset{O}{\|}}{C}-CHOH-CH_3$$

$$(16) \quad R= -\underset{\underset{O}{\|}}{C}-\underset{\underset{O}{\|}}{C}-CH_3$$

From the umbelliferous plant *Seseli hippomaranthum* Jacq., Bohlmann and Zdero (1971) have isolated the pyran, (**13**). Although (**13**) is unique, pyran formation by addition of an alcohol to an acetylenic bond is a common feature of the acetylenes of the Compositae.

As mentioned before, thiophenes are very common in the Compositae. The first compound, containing both an acetylene bond and a thiophene ring, junipal (**14**), however, was described by Birkinshaw and Chaplen (1955) from the fungus *Daedalea juniperina* Murr. This case is interesting since investigations with other strains of this fungus and other cultural conditions gave only aliphatic polyacetylenes. A reinvestigation by Curtis and Taylor (1969) confirmed that the strain used by Birkinshaw under his culture conditions produced (**14**), (**15**) and (**16**). These investigations emphasize how important it is to examine different strains under varying

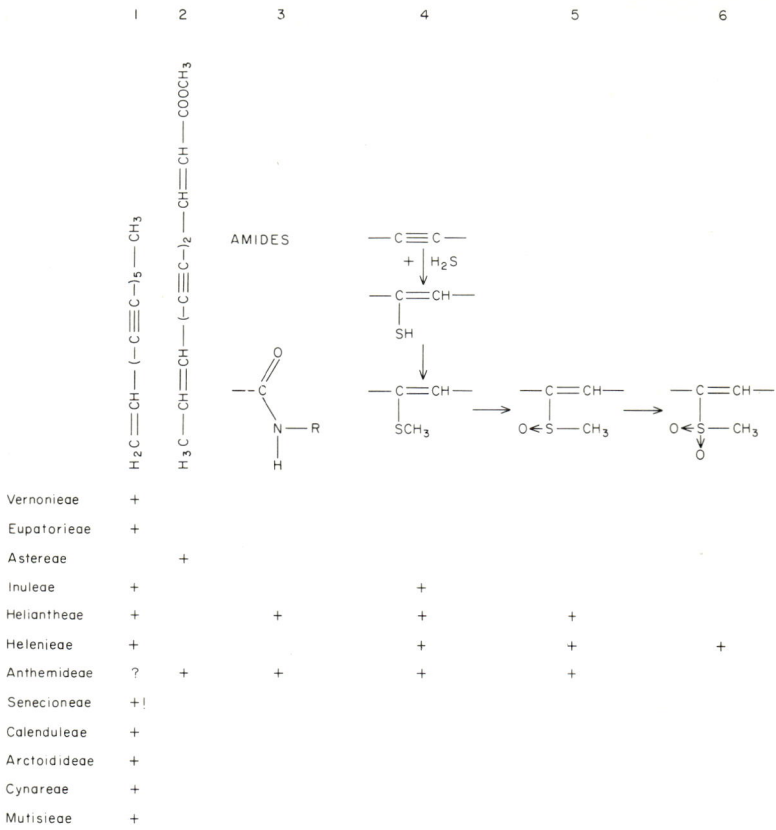

FIG. 7. Taxonomically important aliphatic polyacetylenes and their occurrence among the tribes.

culture conditions when conducting natural product chemistry on microorganisms.

Apart from three exceptions from other plant families, the Compositae is unique in containing some 26 cyclic or heterocyclic structural elements derived from acetylenic precursors and some four to five other heterocycles where formation from acetylenes is unlikely. This wealth of "curling up" at the ends is precisely the characteristic of the polyacetylenes of the Compositae. This does not mean that aliphatic polyacetylenes are not important in the family; two examples of aliphatic structures have already been given.

In Fig. 7 are collected these two examples and four further types of aliphatic derivative, the amides and three types of sulphur compounds. The figure contains a question and an exclamation mark. The question mark refers to the tridecapentaynene (9), which has been found in 10 out of the 13 tribes, and is due to a record in Anthemideae in the Australian *Centipeda*. This genus undoubtedly contain a compound with enepentayne chromophore as in (9) but the compound has yet to be isolated in sufficient quantity for complete characterization. The Bohlmann group has very extensively investigated a large number of representatives of the Anthemideae and never encountered (9). What has to be settled is that the methyl group is not exchanged with another alkyl residue in *Centipeda*, that is the only variation which would not show up in the u.v. spectrum. The absence of (9) and the very frequent occurrence of the matricaria ester—inclusive of analogues and derivatives—(column 2) in the Astereae and Anthemideae confirms the relationship between these two tribes, long ago realized by taxonomists.

The exclamation mark for the occurrence of (9) in the Senecioneae refers to the fact that polyacetylenes only occur in three genera of this tribe, namely: *Arnica*, *Doronicum* and *Gamolepis*; most of the other genera have been found to be practically negative. During the discussion at the Södergarn Meeting (Bendz and Santesson, 1974), I realized that many botanists would like at least to remove *Arnica* from this tribe.

In columns 4–6 of Fig. 7 are listed three types of noncyclic sulphur compounds which, Bohlmann and coworkers have shown, arise from acetylenes and are very numerous within the tribes Inuleae to Anthemideae. They are undoubtedly formed by the same reaction mechanism, present in these four tribes but seemingly absent from the other nine tribes; but this does not mean that the same methyl thioethers, sulphoxides and sulphones are found in all four tribes. The results are varied according to the type of acetylenic substrate available within each tribe; the enol thioether grouping is common, the individual compounds different.

Treating the Helenieae, Bohlmann states in his recent monograph (Bohlmann *et al.*, 1973): "No sign of the relationship to the tribe Anthemideae suggested by the botanical viewpoint can be observed. However, close relationships to the tribe Heliantheae are obvious, but the differences are relatively clear, especially for the better investigated genera" (p. 378). Two comments are pertinent at this point. What is most important from a chemotaxonomic point of view: identity in individual compounds, or identity in reaction mechanisms? Next a plus sign in these tables does not mean that these structural elements occur all through the members of a tribe that have been investigated. On the contrary, the acetylenic features are mostly restricted to some subtribes, some genera or some sections, so that the polyacetylenes may be useful at different levels below the tribe.

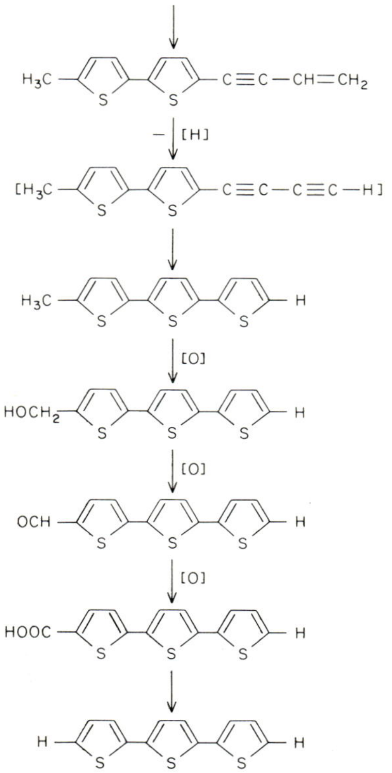

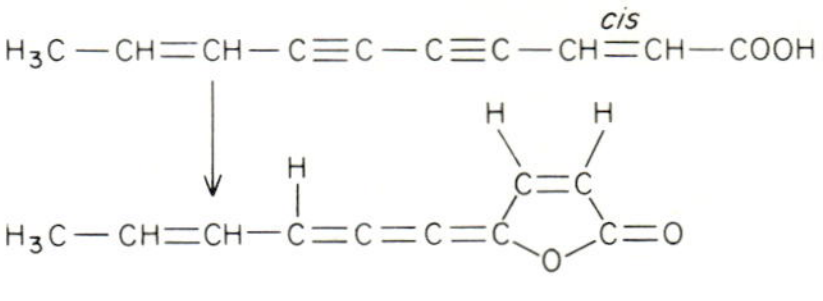

FIG. 8. Compounds defined as polyacetylenic on biosynthetic grounds.

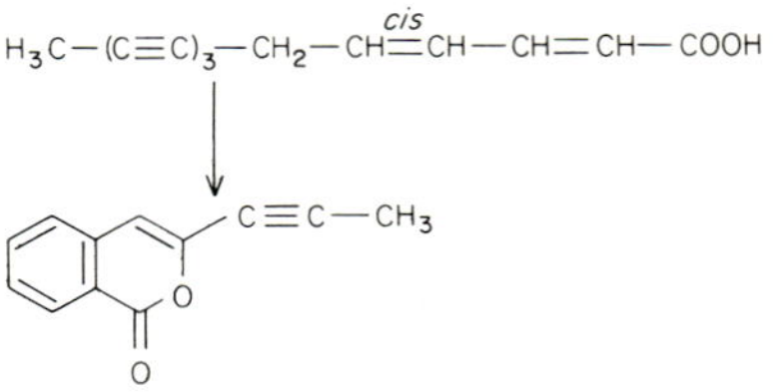

THE CYCLIC POLYACETYLENES OF THE COMPOSITAE

Turning now to the cyclic compounds characteristic of the Compositae, I must define what I mean by an "acetylenic compound." When Zechmeister and Sease (1947) isolated terthienyl from *Tagetes erecta*, no one would have classified it as an acetylenic compound. However, the biogenesis of terthienyl has now been worked out by Bohlmann and Zdero (1970) (Fig. 8) who have shown that it is of acetylenic origin. Thus all thiophenes which have "used up" the available diacetylenic groupings for thiophene synthesis are regarded as "polyacetylenic". In Fig. 8 is also shown the first natural butatriene (**17**), discovered by the Bohlmann group to occur in many Astereae; this has no acetylenic bonds but is still classified as "polyacetylenic".

The last example in Fig. 8—the formation of isocoumarins—has also been clarified by Bohlmann; in the Compositae I have classified this and

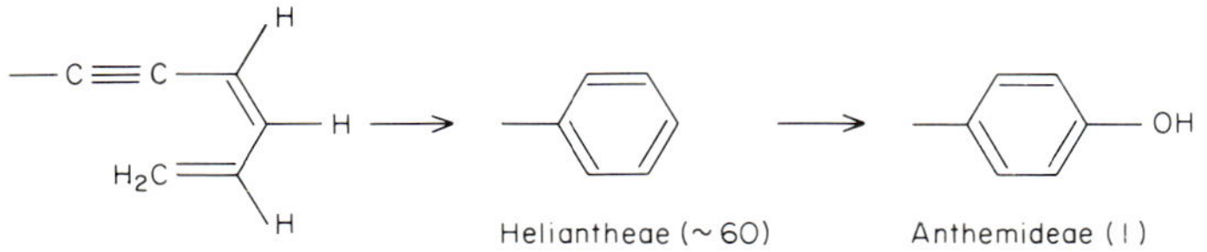

13. POLYACETYLENES AND CONSERVATISM OF CHEMICAL CHARACTERS 395

FIG. 9. The occurrence of phenyl derivatives in the Compositae.

Heliantheae (~60)
Anthemideae (36)
Cynareae (5)

Anthemideae (1)

some other aromatics as "polyacetylenic" even if there are no acetylenic bonds left in the side chain.

Using this definition, Semmler's classical "carlina oxide" is not a monoacetylenic compound, but a tetraacetylenic compound, two acetylenic bonds having been lost during the formation of the furan ring, and one for the benzene ring. It is somewhat disappointing from the chemotaxonomic point of view that a small group of *Carlina* thistles contain representatives of the aliphatic precursor, the monobenzenoid and the monofuranoid stages in addition to "carlina oxide".

Relatively few members of the Compositae possess the ability to synthesize phenyl rings from acetylenic precursors. As is clear from Fig. 9, they are restricted to the three tribes, Heliantheae, Anthemideae and Cynareae, but are common within the first two of the tribes.

I have indicated a direct cyclization from an dien-yn group (≡dien-allene-group), somewhat like the mechanism preferred by Raphael (see Eglinton *et al.*, 1960) for the "ψ-metatoluic acid" to metatoluic acid conversion. The reason is chemotaxonomic. Figure 10 gives the occurrence of acetylenic salicylic acids and their decarboxylation products; the biogenetic scheme, β-oxidation and Michael addition, is due to Bohlmann (1973 p. 102). The distribution is in the same three tribes, but they are so far less numerous (Fig. 10).

If one returns in the β-oxidation to the α,β-unsaturated acid, the process is again a "Raphael cyclization". As shown in Fig. 11 representatives are only known from the Anthemideae.

The most unusual of the aromatic acetylenes in the Compositae was isolated by the Bohlmann group (1963) from *Anthemis tinctoria* L. Tracer work has shown that methyl migration occurs during biosynthesis (Fig. 12) leading to a branched chain structure. The cyclization mechanism is again a "Raphael" cyclization, so that the Compositae must be able to operate at least two pathways from polyacetylenic precursors to aromatics.

The formation of five-membered lactones from suitably constructed acetylenic acids is a spontaneous process at neutral pH. The 2-*cis*-isomers of (6)–(8) on cautious saponification yield the lactone type shown in Fig. 13 on the left-hand side. That their occurrence is restricted to the same tribes as the methyl esters is therefore to be expected. Since most of these

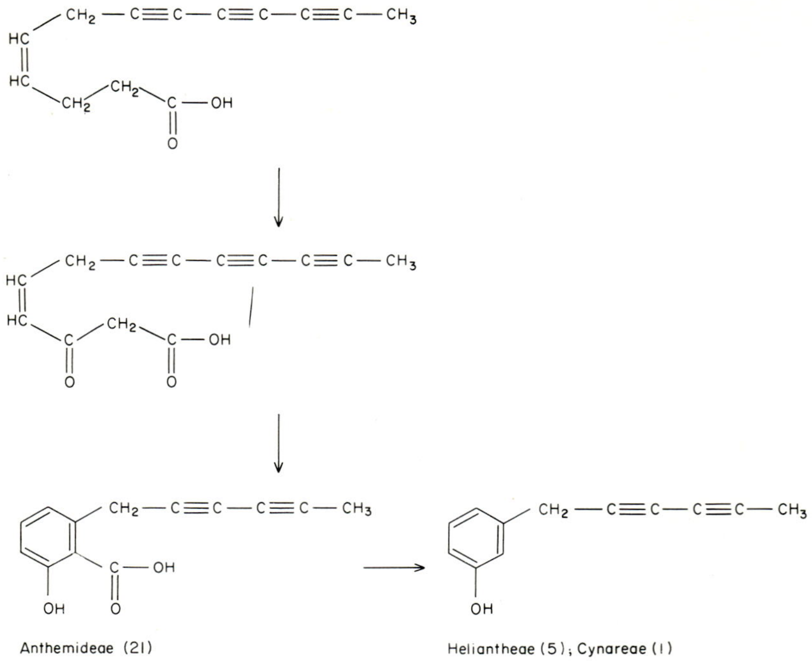

Fig. 10. The occurrence of salicylic acid derivatives in the Compositae.

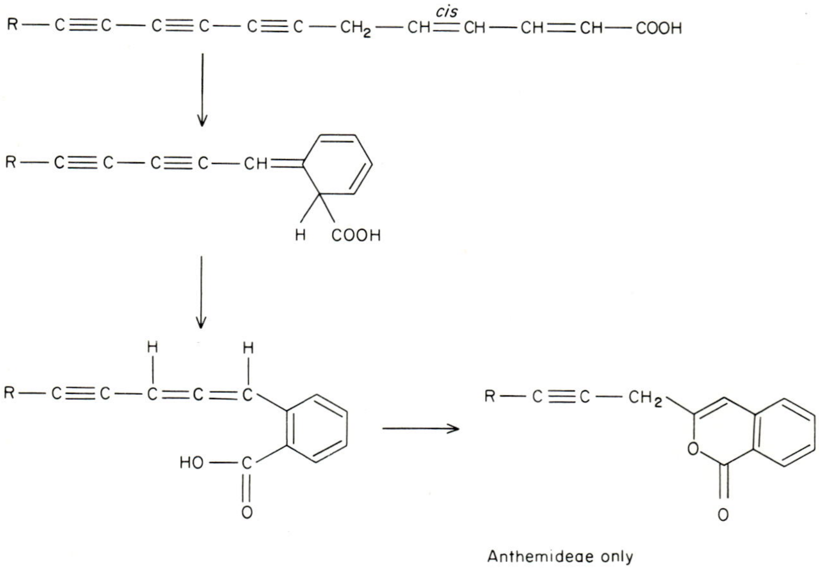

Fig. 11. Isocoumarins from polyacetylenes.

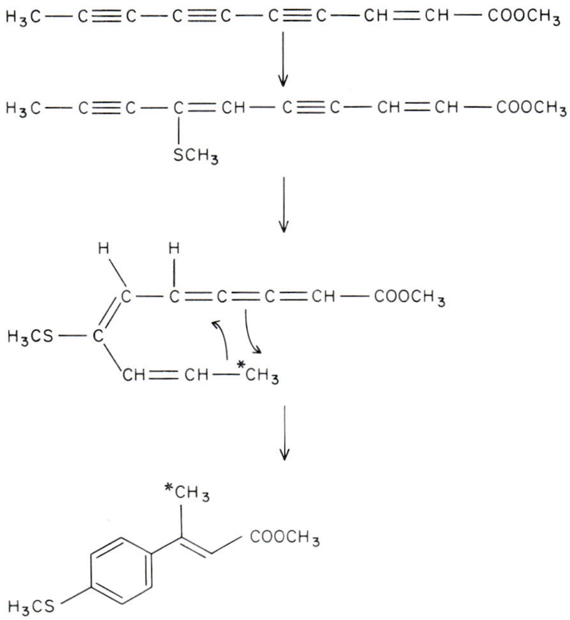

13. POLYACETYLENES AND CONSERVATISM OF CHEMICAL CHARACTERS 397

$H_3C-C\equiv C-C\equiv C-C\equiv C-CH=CH-COOCH_3$

↓

$H_3C-C\equiv C-C=CH-C\equiv C-CH=CH-COOCH_3$
             |
             SCH_3

↓

(structure with H_3CS group, vinyl, and $CH=CH-{}^*CH_3$ side chain, $COOCH_3$)

↓

(β-methylcinnamic acid methyl ester with H_3CS on para position and ${}^*CH_3$)

Anthemideae only

FIG. 12. Biogenetic scheme from dehydromatricaria ester to the β-methyl-cinnamic acid of *Anthemis tinctoria* according to Bohlmann et al. (1963).

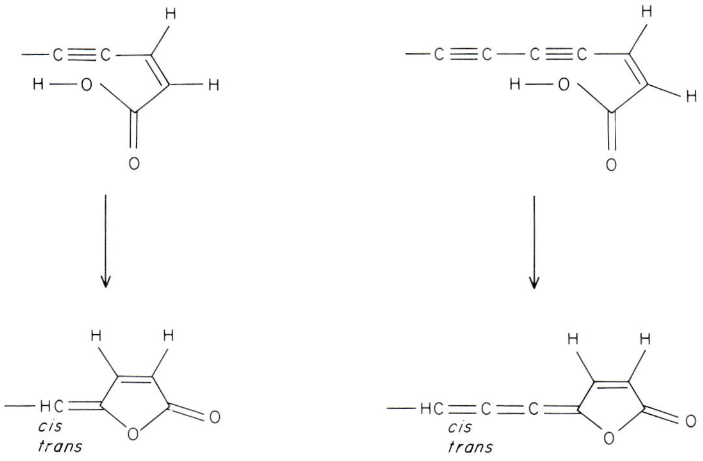

Astereae; Anthemideae; Cynareae   Astereae only

FIG. 13. Five-ring lactone formation from acetylenic acids.

methyl esters are conjugated diacetylenes, the formation of butatriene lactones is also possible. Bohlmann and co-workers have found such compounds many times in the Astereae, but never in the Anthemideae. The butatrienes, e.g. compound (**9**), are very reactive compounds, but they are certainly genuine natural products. A number of *Erigeron* species have glandular hairs. On contact with spectral hexane the fluid of the glandular hairs immediately goes into solution and one obtains the characteristic spectrum of (**17**). To isolate the pure substance is a formidable task.

The same acids may form six-membered-ring lactones, but this is not a spontaneous process. In the laboratory it has recently been achieved by strong acid treatment or by catalysis with mercury. The Bohlmann group have found the first representatives of these lactones in the Anthemideae (Fig. 14).

Although the addition of an alcohol group to the acetylene bond does not occur spontaneously, this is a very important process in the Compositae with many sorts of alcohol functions or their equivalents (enols, semiketals, epoxides) being involved. Two partial structures containing alcohol functions are shown in Fig. 15; compounds of this type have so far only been found in the Inuleae. Two other partial structures, one from the Inuleae and the other from the Cynareae are illustrated in Fig. 16.

Spiro-ketal groups contained in some of the Compositae acetylenes are collected together in Fig. 17. Such structures have only been found in the Anthemideae, and within this tribe spiro-ketals are limited to some subtribes and a few genera. Finally, three further partial structures formed by alcohol addition mechanisms are shown in Fig. 18. These types again are of very restricted distribution; one type has been found exclusively in the tribe Cichorieae, in the genus *Crepis*, which has given its name to the

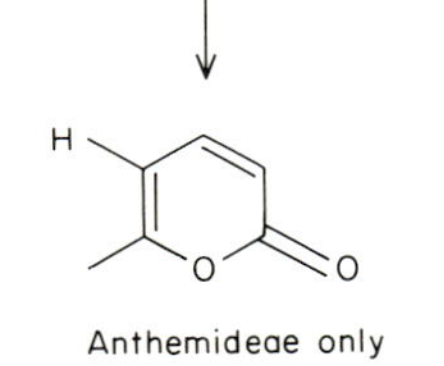

FIG. 14.   α-Pyrones from acetylenic acids.

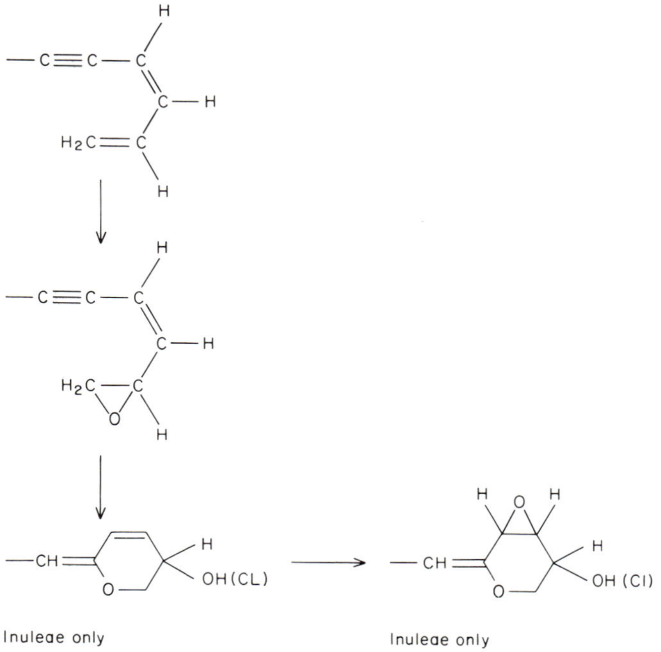

FIG. 15. Pyranoid end-group formed through reaction of epoxide with acetylenic bond (according to Bohlmann et al., 1973).

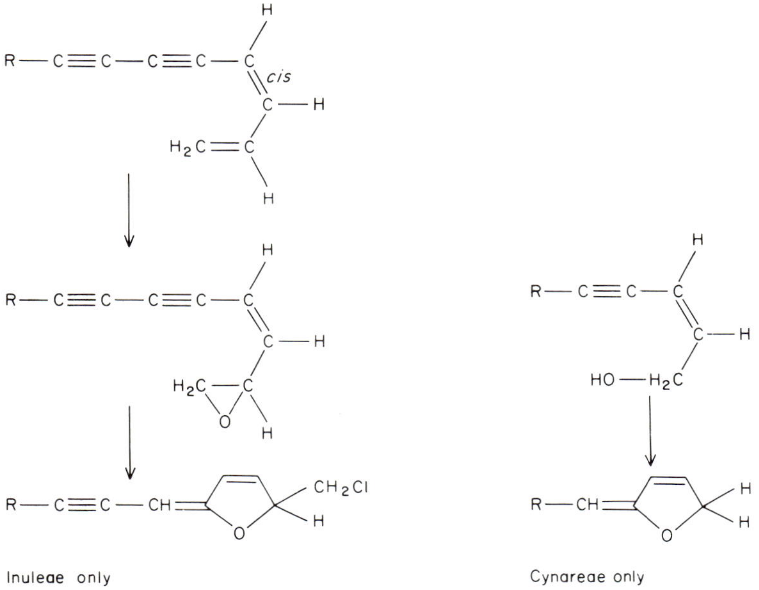

FIG. 16. Furanoid end-group formed from additions to acetylenic bond (according to Bohlmann et al., 1973).

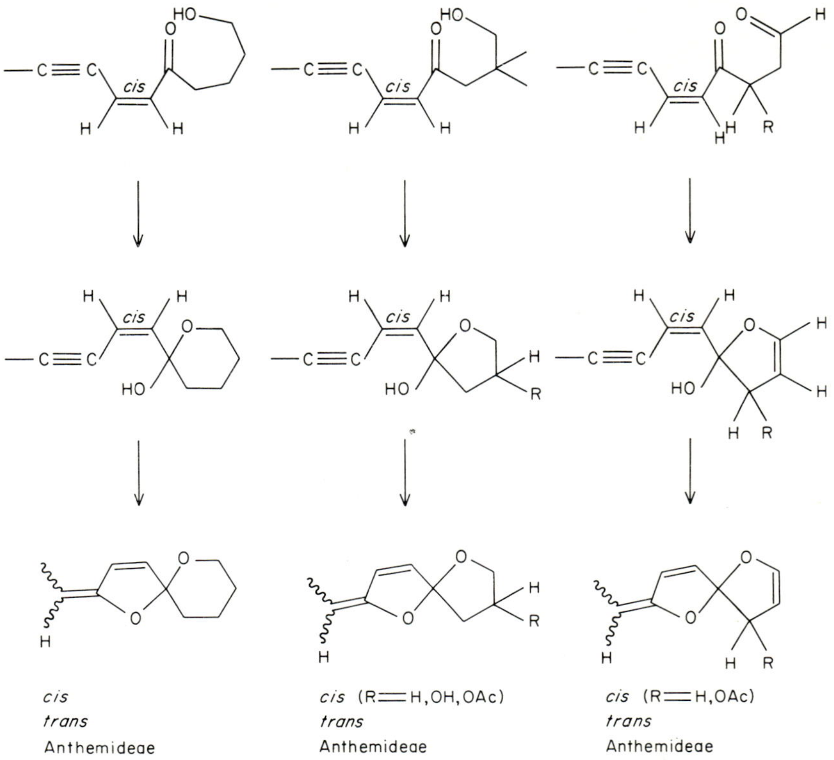

FIG. 17. The formation of spiroketal end-groups according to Bohlmann et al. (1973).

important acetylenic precursor, crepenynic acid. For a long time no polyacetylenes were found in this large tribe. The investigations of Sir Ewart Jones (see Bentley et al., 1969) completely changed the picture. Two facts explain this change. Firstly, the concentration of acetylenes is (as in the Campanulaceae) mostly very low. Secondly a large number of genera apparently have no acetylenes at all, and these happened to be the ones first investigated. Among the best polyacetylene producers in this tribe are members of *Lactuca*.

Let us now return to discuss the thiophenes of the Compositae. As is obvious from Fig. 19, these substances are very widely distributed in the family. Terthienyl has been found in four tribes, bithienyls in six tribes, monothiophene derivatives in no less than nine of the 13 tribes. To take just one example, the 3-hydroxythiophene (**19**) was found by Bohlmann et al. (1962) in *Artemisia arborescens* L. (Anthemideae) and by Atkinson and Curtis (1971) in *Liatris pycnostachya* Michx. (Eupatorieae). This is

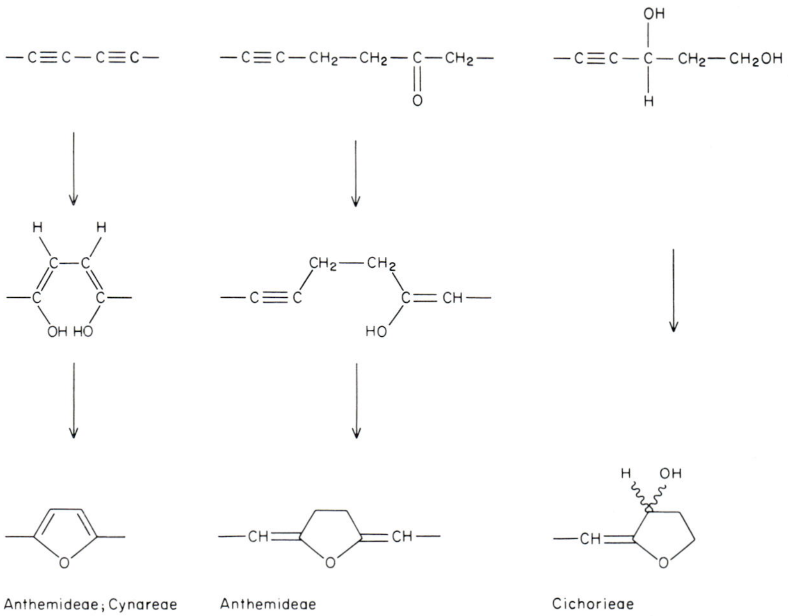

13. POLYACETYLENES AND CONSERVATISM OF CHEMICAL CHARACTERS 401

FIG. 18. Formation of three further furanoid compounds from acetylenes

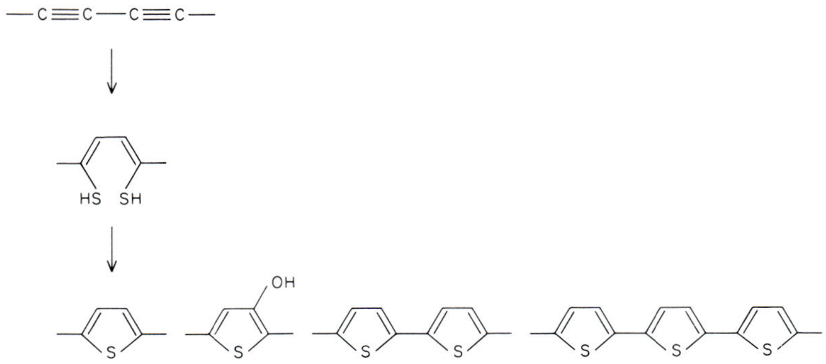

Vernonieae	+	−	−	−
Eupatoreae	+	+	−	−
Inuleae	+	−	+	−
Heliantheae	+	−	+	+
Helenieae	+	−	+	+
Anthemideae	+	+	−	−
Senecioneae	+	−	+	−
Arctodeae	+	−	+	+
Cynareae	+	−	+	+

FIG. 19. Occurrence of thiophenes within the Compositae.

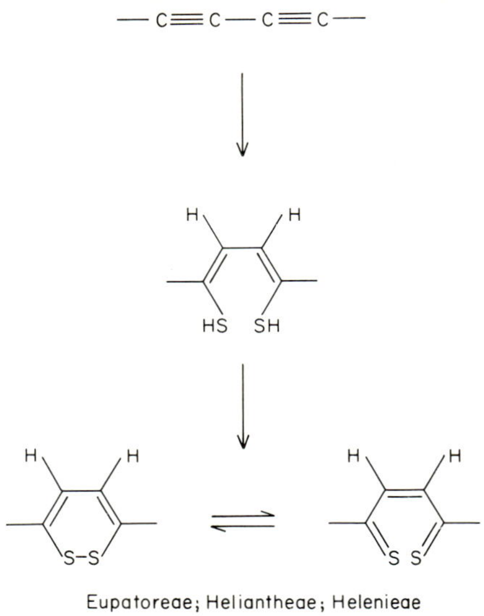

FIG. 20. The occurrence of dithiin compounds within the Compositae.

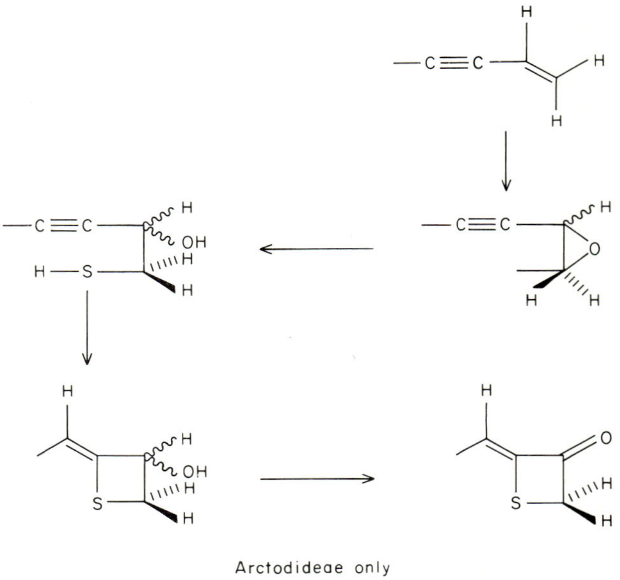

FIG. 21. The formation of thietanone derivatives from acetylenes according to Bohlmann.

13. POLYACETYLENES AND CONSERVATISM OF CHEMICAL CHARACTERS

an unusual case of the same compounds occurring in isolated species belonging to rather distant tribes.

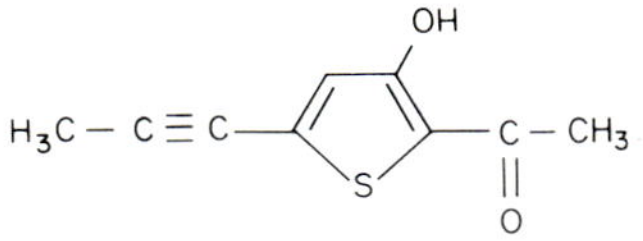

$$H_3C-C\equiv C-\underset{S}{\underset{|}{\diagdown}}\underset{}{\overset{OH}{\diagup}}-\underset{\underset{O}{||}}{C}-CH_3$$

The polyacetylenes of the Compositae are stored in the resin canals which are present characteristically in the 11 tribes of the subfamily Tubuliflorae. The canals are in some cases coloured red, owing to a structural element, unknown elsewhere in natural products—the 1,2-dithiin-ring (Fig. 20). These beautiful pigments should never be isolated in the crystalline state, since they are then destroyed by polymerization.

In recent years, the Bohlmann group have found another exceptional ring-sulphur structure in the Compositae, that of the thietanone (Fig. 21); so far, thietanones have been found to be restricted to the tribe Arctotideae, subtribe Gorterinae. Figures 22 and 23 show polyacetylene end

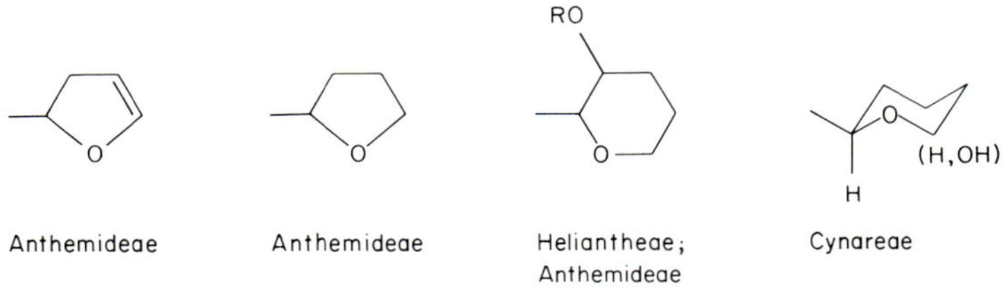

Anthemideae   Anthemideae   Heliantheae; Anthemideae   Cynareae

FIG. 22. Oxygen heterocyclic end-groups from the Compositae where acetylenic bonds do not take part.

Cynareae

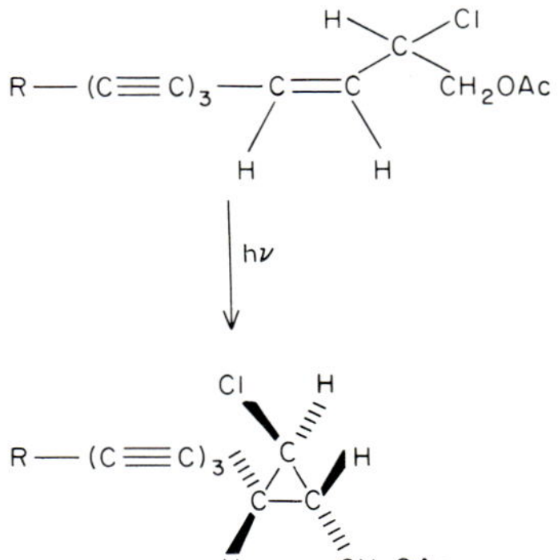

FIG. 23. Cyclopropane end-groups from the Compositae where acetylenic bonds do not take part.

groups also found in the Compositae, which do not appear to be derived directly by addition of functional groups to acetylenic bonds.

The present position regarding the natural distribution of polyacetylenes in the Compositae can be summarized by the following seven points. (a) Polyacetylenes in the widest sense have been found in all 13 tribes of the Compositae. (b) Polyacetylenes do not occur as phytochemical markers in every single member of the family, in the way that stearolic acid derivatives characterize all Santalaceae so far investigated. (c) In two of the largest tribes, the Senecioneae and the Cichorieae, polyacetylenes are very scarce; so far they are only known from a few subtribes or genera. (d) In the remaining 11 tribes the polyacetylenes are very frequent, but in most tribes some genera, sections or species seemingly lack polyacetylenes. (e) As distinct from the mainly aliphatic polyacetylenes of the related Umbelliferae, Araliaceae and Campanulaceae, the Compositae polyacetylenes are characterized by the presence of cyclic, aromatic or heterocyclic end groups; at present 26 such structural types have been discovered. (f) Some of these complex structures are restricted to one single tribe; no less than 10 types are confined to the Anthemideae and three other types are so far restricted to the Inuleae. (g) Some heterocyclic polyacetylenes, such as the thiophenes, have been found in the majority of the tribes; their occurrence is seemingly unrelated to morphological characters.

CONSERVATISM OF CHEMICAL CHARACTERS IN THE COMPOSITAE

In Chapter 2 of this book, Turner has drawn attention to George Bentham's classic paper of 1873 on the systematics of the family. In my opinion Bentham's treatise is very important to chemists working with the Compositae. Firstly, Bentham discusses in some detail most of the genera which in one way or another do not fit well into the schemes for subdivision of the family into tribes. Many of these critical genera have still not been investigated chemically. Secondly, Bentham drew very significant conclusions about the enormous age of the Compositae; this gives the natural product chemist the chance of realizing the possible conservative character of the various classes of naturally occurring compounds present in the family.

In his treatment of the Inuleae, Bentham remarks (p. 422): "The Helichryseae present one of those instances (such as Proteaceae, Restionaceae etc.) in which a large very natural group of plants had spread over two regions, South Africa and Australia, now quite isolated, but then possibly in connexion with each other, in times sufficiently remote for them to have diverged in each region into different forms, and have multiplied greatly in both, without having preserved a single species in common." The geological dating of the division of "Pangea" or "Gondwanaland" in

TABLE I. Australian members of the Inuleae studied for the occurrence of polyacetylenes; only *Podolepis canescens* Grah. is apparently negative.

A. Australian endemic species of genera represented outside Australia:
Pterocaulon sp. "1C-1343"
Helipterum albicans var. *aurea* (A. Cunn.) DC.
Helipterum jessenii F.v.M.
Helichrysum apiculatum (Labill.) DC.
Helichrysum obcordatum F.v.M.
Helichrysum scorpioides (Poir.) Labill.
Helichrysum semipapposum (Labill.) DC.

B. Species from endemic genera restricted to Australia:
Podosperma angustifolia Labill.
Craspedia chrysantha (Schlechtd.) Benth.
Craspedia uniflora Forst f.
Calocephalus citreus Less.
Myriocephalus guerinae F.v.M.
Ixodia achilleoides R.Br.
Humea elegans Sm.
Ammobium alatum R.Br.
Rutidosis leptorhynchoides F.v.M.
Cassinia longifolia R.Br.
Leptorhynchus squamatus (Labill.) Less.
Parantennaria uniceps Beauverd
Podolepis canescens Grah.

this connection is not very important; about 100 years after the publication of Bentham's treatise, the estimates seem to converge on a time of 150 million years.

What is of interest here is to look for polyacetylenic compounds in Australian Compositae and compare them with those isolated from plants, from the main centres of each tribe. Apart from the Senecioneae, three tribes are represented in Australia, first the Inuleae with about 35 endemic genera, then the Astereae and Anthemideae.

There has been no systematic study of the Australian Inuleae. However, Table I lists the representatives which have happened to have been studied by Bohlmann's or our own group. As to polyacetylene chemistry, the seven species of group A and the first 12 species of group B are similar to those species of the Inuleae examined from other regions. Compounds present are mostly aliphatic polyacetylenes, the tridecapentaynene (9) and its relatives, but also thiophenic compounds. Only the last species in group B is seemingly negative, so the picture is very normal.

The situation in the Astereae is much more uncertain (Table II). In

TABLE II. Australian members of the tribe Astereae so far studied for the occurrence of polyacetylenes.

A. Australian endemic species of genera represented outside Australia:
  Brachycome + in West-Austr. — East-Austr. (?)
  Conyza   +
  Erigeron +
  Vittadinia —
  Podocoma + (only investigated species South-American)
  Lagenophora + (?) ene-diynes

B. Species from endemic genera restricted to Australia:
  Erodiophyllum + specific acetylenes
  Calotis   + typical Astereae aliphatic acetylenes
  Solenogyne + (?) ene-diynes
  Minuria
  Olearia
  Celmisia

group A only *Vittadinia* is negative, although *Brachycome* needs some supplementary remarks. The first two species to be investigated were West Australian (*B. iberidifolia* Benth. and *B. lineariloba* (DC.) Druce); both were good producers of typical Astereae polyacetylenes. During the stay of my wife and myself with C.S.I.R.O., we obtained six species of *Brachycome* from east Australia; all were apparently devoid of acetylenes. Since the days of Bentham and Hooker, it has been well established that in typical Australian endemic genera, the species occurring in the west are usually completely different from those occurring in the east. The explanation is that back in the Cretaceous the Australian continent was in a horseshoe shape; a great bay occupied South Australia and Lake Erie. Since this "Cretaceous sea" closed up, the region enclosed is supposed to have been dominantly desert country. In this way south-western and south-eastern Australia have been separated nearly half as long as the division of "Pangea". Obviously the *Brachycome* case deserves a more thorough chemical study, but there is the drawback that most of the results are likely to be negative.

In group B (Table II) no less than half of the six genera investigated give negative results. These are large genera, two of them belonging to the antarctic flora element. I have placed *Calotis* in group B although one endemic species is stated to occur in Assam. The genus *Calotis* is rich in polyacetylenes, the yellow flowered *C. erinaceae* has a tremendous number of polyacetylenes in all parts of the plant, including the achenes.

As to the Anthemideae the material is scanty (see Table III). The arrangement given in this Table is in agreement with recent Australian Floras

TABLE III. Australian members of the tribe Anthemideae so far studied for the occurrence of polyacetylenes.

A. Australian endemic species of genera represented outside Australia:
Cotula + except *C. australis* (Less.) Hook.
Centipeda + but (9) which does not occur anywhere else in the tribe

B. Species from endemic genera restricted to Australia:
Dimorphocoma   not investigated
Abrotanella   —
Elachanthus   —
Isoetopsis   —
(*Ceratogyne obionoides* Turcz, extinct?)

which all are based on Hoffmann (1894). In the systematic survey of the Anthemideae by Heywood and Humphries (Chapter 31) these genera are mostly listed under Southern Hemisphere genera. However, in their Table I on accepted genera (pp. 855–858) *Abrotanella*, *Ceratogyne* and *Isoetopsis* are marked as doubtfully belonging to this tribe. If correct, these exclusions would appreciably reduce the anthemoid element in the Australian region.

Cotula has a number of endemic species in Australia and New Zealand; of the investigated members only *C. australis* is abnormal. Two *Centipeda* species have been investigated (*C. cunninghamii* F.v.M. and *C. minima* (L.) A.Br. & Asch.); both contain in the root an ene-pentayn compound. This is not an expected result for the Anthemideae. In their review of pollen fine structure, Skvarla *et al.* (Chapter 8) commented on the pollen wall structure of one species of *Centipeda* as not belonging to the otherwise rather clear "anthemoid" type. They note further that the occurrence of the tridecapentaynene (9) in *Centipeda* need not necessarily be a chemotaxonomic exception, if *Centipeda* is removed from the Anthemideae. This remark has led to further investigations by Dr J. Praglowski, The Palynological Laboratory, Stockholm, who in a preliminary report of October 3rd, 1975, mentions that after investigations of three species of *Centipeda* he entirely agrees with Professor Skvarla. The pentaynene (9) is undoubtedly the most common of all Compositae acetylenes, but although some 180 members of the Astereae and some 280 members of the Anthemideae have been investigated, (9) has never been encountered in these two tribes. If *Centipeda* is thus removed from the Anthemideae, the chemotaxonomic importance of (9) will be strengthened appreciably.

In group B, only four species have been investigated; all are apparently devoid of acetylenes.

Bentham emphasized that the flora of South Africa, so far as the Compositae is concerned, is one of the richest and most isolated in the world, most of the taxa being highly endemic. If one scans the table in Bohlmann's recent monograph for South African Astereae it is clear that the endemic genera: *Amellus*, *Mairia*, *Felicia* and *Chrysocoma* are all good producers of the classical Astereae-polyacetylenes. The only negative genus so far, *Charieis*, is somewhat parallel to that found in certain species of *Cosmos*; instead of acetylenes there are dehydrogenated terpenes.

Bohlmann's monograph makes similar scannings in the Compositae an easy task. Natural product chemistry can never approach the degree of completeness possible in classical taxonomy. At present less than 10% of the Compositae have been studied by chemists. But as to polyacetylenes, there is a clear outline: the ability to synthesize polyacetylenes must have been a property of the original stock; and the diversification of polyacetylenic pathways must be as old as the morphological diversification of the stock. The peculiarity of the polyacetylenes of the Compositae relative to other plant families producing acetylenic compounds is the further transformation of acetylene bonds, which partly follows some of the systems used by taxonomists, but very often operates independently of any taxonomic system.

REFERENCES

ANON. (1932). *Trudy Tadzhikskogo Botanicheskogo Sada.* **1**, 13.

ARNAUD, A. (1902). Sur la constitution de l'acide tarririque. *C.r. hebd. Séanc. Acad. Sci., Paris* **134**, 473.

ATKINSON, R. E. and CURTIS, R. F. (1971). A thiophene from *Liatris pycnostachia*. *Phytochemistry* **10**, 454.

BENDZ, G. and SANTESSON, J. (eds) (1974). "Chemistry in Botanical Classification". Academic Press, New York and London.

BENTLEY, R., JONES, E. R. H., LEEMING, P. R. and THALLER, V. (1969). Natural acetylenes, Part XXX. Polyacetylenes from *Lactuca* (Lettuce), Species of the Liguliflorae Subfamily of the Compositae. *J. chem. Soc.* (C), 1096.

BESSEY, C. E. (1915). The phylogenetic taxonomy of flowering plants. *Ann. Mo. bot. Gdn* **2**, 109.

BIRKINSHAW, J. H. and CHAPLEN, P. (1955). Biochemistry of the wood-rotting fungi 8. Volatile metabolic products of *Daedalia juniperina* Murr. *Biochem. J.* **60**, 255.

BOHLMANN, F. and ZDERO, C. (1971). Über die Inhaltsstoffe der Gattung *Seseli*. *Chem. Ber.* **104**, 2354.

BOHLMANN, F., KLEINE, K. H., and BORNOVSKI, H. (1962). Über zwei Thiophenketone aus *Artemisia arborescens* L. *Chem. Ber.* **95**, 2934.

BOHLMANN, F., ARNDT, C., BORNOWSKI, H. and KLEINE, K. M. (1963). Die Polyine der Gattung *Anthemis* L. *Chem. Ber.* **96**, 1485.

BOHLMANN, F. and ZDERO, C. (1970). Über die Inhaltsstoffe aus *Eclipta erecta* L. *Chem Ber.* **103**, 834.

BOHLMANN, F., BURKHARDT, T. and ZDERO, C. (1973). "Naturally Occurring Acetylenes". Academic Press, London and New York.

CURTIS, R. F. and TAYLOR, J. A. (1969). Naturally occurring thiophenes, Part V. Acetylenic Thiophenes from the Basidiomycete *Daedalea juniperina* Murr. *J. chem. Soc.* (C), 1813.

EGLINTON, G., RAPHAEL, R. A. and WILLIS, R. G. (1960). Rearrangement of Diacetylenes to Aromatic Compounds. *Proc. chem. Soc.*, 247.

GLEN, A. T., HUTCHINSON, S. A. and MCCORKINDALE, W. J. (1966). Hexa-1,3,5-Triyne—A Metabolite of *Fomes annosus*. *Tetrahedron Lett.*, 4223.

SEMMLER, L. F. (1906). Zusammensetzung des ätherischen Oels der Eberwurzel (*Carlina acaulis* L.). *Ber. dt. chem. Ges.* **39**, 730.

WILJAMS, W. W., SMIRNOV, V. S. and GOLJMOV, V. P. (1935). On the Properties of the crystalline compound from the essential oil of *Lachnophyllum gossypinum* Bge. *J. gen. Chem.* (U.S.S.R.) **5**, 1195.

ZECHMEISTER, L. and SEASE, J. (1947). A blue fluorescing compound, terthienyl, isolated from marigolds. *J. Am. chem. Soc.* **69**, 273.

Chapter 14

Pharmaceutical and economic uses of the Compositae

H. WAGNER
Institut für Pharmazeutische Arzneimittellehre
University of Munich, Germany

Abstract. The biological and therapeutical applications of the plants of the Compositae, with nearly 1000 genera and about 15 000 species, is more the result of systematically conducted chemical and pharmacological research than of tradition. In addition to drugs known since antiquity, from plants such as *Chamomilla*, *Cynara* and *Silybum*, there are over 25 species in the family which have found therapeutic application due to their antihepatotoxic, choleretic, spasmolytic, anthelmintic, antiphlogistic, antibiotic or antimicrobial activity. In food technology some drugs containing bitter principles have achieved industrial significance. The insecticide pyrethrins from the *Pyrethrum* species exemplify how plant products serve as models for the development of more active synthetic agents. The cytotoxic sesquiterpene lactones of certain species may act as pointers for future development of cancer drugs. There is a direct connection between the physico-chemical properties of the predominantly lipophilic compounds of the Compositae and their biological and pharmacological activity.

CONTENTS

Introduction	412
Pharmaceutical plants	412
Antiphlogistic and spasmolytic agents	412
Antihepatotoxic and cholerectic activity	414
Antibiotic, bacteriostatic, virostatic and fungistatic agents	417
Cytotoxic agents	418
Food industry	422
Sweet substances	422
Bitter substances	423
Spices and flavouring agents	424
Substances of general economic importance	424
Insecticides	424
Oil drugs	426
Polysaccharides	426
Caoutchouc	426

| Toxic substances | . | . | . | . | . | . | . | . | 427 |
| References | . | . | . | . | . | . | . | . | 428 |

INTRODUCTION

Despite the fact that the Compositae is one of the largest plant families, comprising 1000 genera and 15 000 species, it is the source of relatively few products of economic and medical importance. Only about 30 plant species are employed as crude drugs, and no more than about 20 well defined pure substances are commercially available, or used therapeutically. Only 16 drugs are found in the pharmacopeias.

Chamomilla, Arnica, Cynara, Echinacea and *Silybum* are some of the few drugs given to us by tradition and empiricism. The renewed importance of the family is the result of systematic, chemotaxonomic and pharmacological research, in which computer evaluation has played a great part. New screening methods and isolation techniques have made it possible to elucidate the mode of action of old drugs, and thereby reintroduce them into modern therapy. Many structures, discovered for the first time in this family, have served as models for the synthesis of biologically active compounds, and have promoted research into the activity of analogous structures. From our present state of knowledge, the scientific and medical potential of the Compositae family has yet to be fully exploited.

PHARMACEUTICAL PLANTS

Antiphlogistic and spasmolytic agents

The true camomile, *Matricaria chamomilla*, is one of our oldest pharmaceutical plants. Aqueous and alcoholic extracts have been used since antiquity, internally and externally, for their anti-inflammatory and wound-healing activity. For a long time the only known active principle was the blue azulene compound (**1**). It is produced from matricin (**2**) during steam distillation, via the equally unstable chamazulene carboxylic acid (**3**).

Matricin (2)   Chamazulene-carboxylic acid (3)   Chamazulene (1)

14. PHARMACEUTICAL AND ECONOMIC USES

More recently, as a result of systematic research, other active substances have been found, which possess even greater activity than azulene. The first compound was (−)-α-bisabolol (4), an unsaturated monocyclic sesquiterpene alcohol; the dextrorotatory form does not occur in camomile, but it was found in the leaf buds of poplar. α-Bisabolol can occur in the isopropylidine or in the isopropenyl form. Bisabolol is a more efficient antiphlogistic agent than guaiazulene when tested against the carragenin oedema of rat paw (Jakovlev and Schlichtegroll, 1969). It is interesting that laevorotatory bisabolol is a more powerful antiphlogistic and spasmolytic agent than the dextrorotatory form, or the racemate (Isaac, 1974). The chemistry of bisabolol has been fruitfully extended by the partial synthesis of bisabolol ethers and esters. Most of these synthetic compounds are more active, and usually have a lower toxicity than bisabolol (Thiele et al., 1969).

Two further compounds, bisabolol oxide (5, 6) and the spiroether (7), have also been found in oil of camomile.

(4) α-Bisabolol   (5) Bisabolol oxide − A   (6) Bisabolol oxide − B

(7) cis-Spiroether

Results from the pharmacological investigation of bisabolol oxides (5, 6) are not yet available, but it is known that the more abundant cis-spiroether (7) also has a better antiphlogistic activity than chamazulene (Breinlich and Scharnagel, 1968). In addition, the spasmolytic activity of the cis-spiroether is superior to that of papaverine (Breinlich and Scharnagel, 1968). It should also be mentioned that the drug contains a high proportion of spasmolytically active flavone glycosides (Hörhammer et al., 1963). Thus azulene no longer ranks as the sole substance responsible for the pharmacological properties of camomile. These latest discoveries must be taken in account in the design of methods for obtaining camomile preparations with optimal activity, and in methods for the determination of this activity. Camomile exists in many chemotypes in which either bisabolol or bisabolol oxide may be the predominant constituent. Thus the planned production of camomile preparations must start with the cultivation of the most appropriate chemovar.

Azulene and the en-in-dicycloether, which have just been mentioned, are not limited to camomile. As in the case of matricin or achillin (**8**) azulene is also found in the flowers of *Achillea millefolium* (Goldberg *et al.*, 1969; Smolenski *et al.*, 1967). Arnicolide (**9**), which is related to achillin, has been isolated from *Arnica montana*, which also has anti-inflammatory activity (Poplawski *et al.*, 1971).

(8) Achillin

(9) Arnicolide

Since the systematic studies of Breinlich (1970) have shown that various *Chrysanthemum* species also contain the en-in-dicycloether, it may be possible in the future to exploit these plants for pharmaceutical purposes.

Spasmolytic activity is found rather widely amongst the terpenes. As early as the Middle Ages, the leaves and roots of *Petasites hybridus* were used for their anticonvulsive activity in asthma and in disturbances of the alimentary canal; but it was not until 1956 that Stoll *et al.* and a few years later Aebi *et al.* (1958; Aebi and Waaler, 1959), succeeded in discovering the active principle in the terpene fraction of leaves and roots. It consists of the compounds petasin (**10**), isopetasin (**11**), S-petasin (**12**) and S-isopetasin (**13**), which belong to the group of eremophilane sesquiterpenes. They are esters of the C_{15}-alcohol petasol, or isopetasol, with angelic acid or methyl-mercaptoacrylic acid, respectively. Petasin is 14 times more active than papaverine; the esters of isopetasol are less active.

Antihepatotoxic and choleretic activity

An increase in the consumption of alcohol, over-feeding and other defects of civilization are responsible for the increasing frequency of liver diseases in the western world. The development of serviceable liver protection agents, or medicaments that increase the ability of the liver to regenerate, is therefore of pressing importance. This is the reason why, 10 years ago, several laboratories started chemical and pharmacological work on the

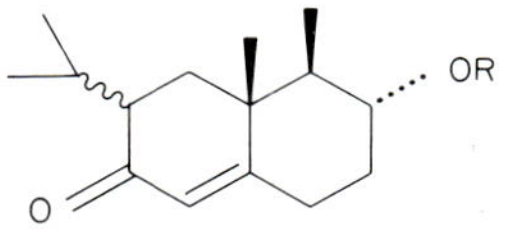

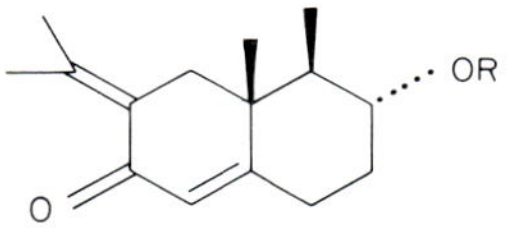

14. PHARMACEUTICAL AND ECONOMIC USES 415

(10) Petasin

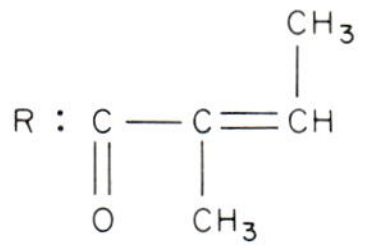

(11) Isopetasin

R : C(=O)—C(CH₃)=CH—CH₃

(12) S-Petasin

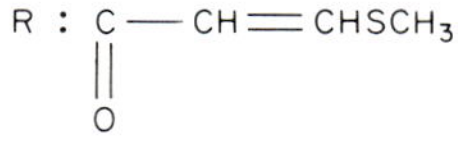

(13) S-Isopetasin

R : C(=O)—CH=CH—SCH₃

fruits of *Silybum marianum*. As a liver remedy, the fruits of this drug were listed in the "Materia medica" of Dioscorides, who lived in about A.D. 50. The first scientific confirmation of the protective action of this drug for the liver was obtained by Eichler and Hahn (1949) and by Mayer and Menge (1949). These authors reported that the tincture was able to protect the liver against the action of trinitrotoluene and carbon tetrachloride, and reported its successful use against hepatitis. With the aid of various liver tests, the active principle was found to be localized in the flavonoid fraction of the drug (Hahn *et al.*, 1968). Finally the compounds silybin (**14**), silydianin (**15**) and silychristin (**16**) were isolated and shown to be responsible for the activity, and their structures were elucidated (Wagner *et al.*, 1968; Pelter and Hänsel, 1968; Abraham *et al.*, 1970; Wagner *et al.*, 1971).

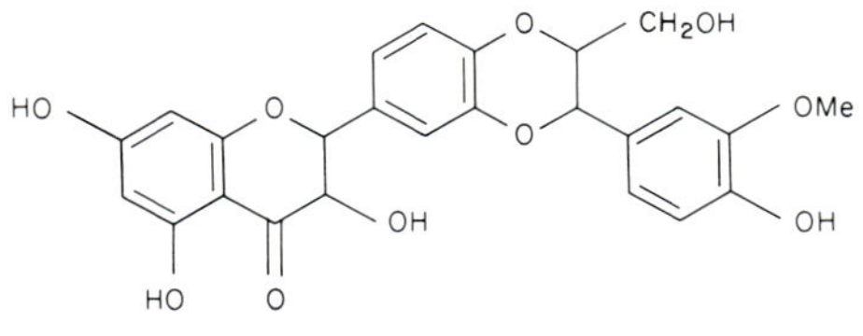

(14) Silybin

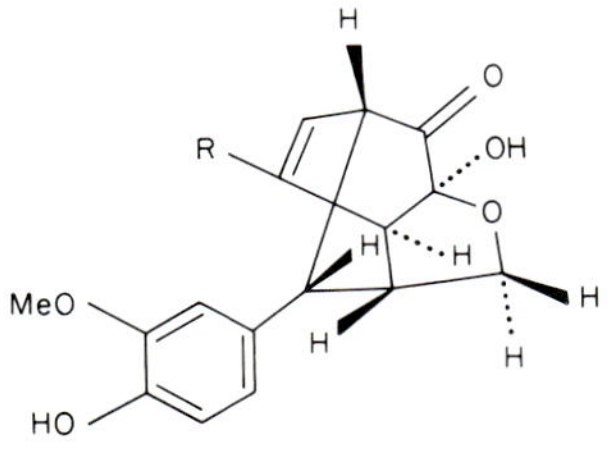

(15) Silydianin

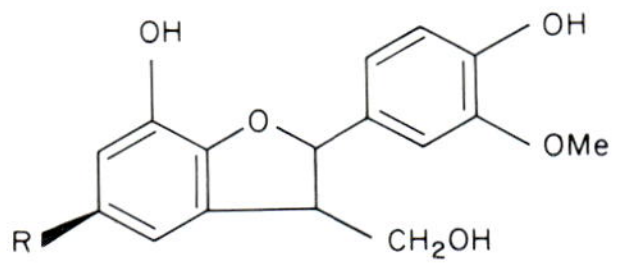

(16) Silychristin

The collective term silymarins embraces a new, hitherto undescribed group of natural substances, in which a flavanone molecule is coupled with coniferyl alcohol. They are called flavonolignans. Studies on the pharmacodynamics, the site of action and the mechanism of action have shown that one site of action of silymarin is the outer cell membrane of the liver (Vogel et al., 1975). Silybin exerts a dose-dependent, protective action against phalloidin, by blocking the attachment of the poison to specific membrane receptors (Frimmer and Kroker, 1975). It is, so far, the only compound known that can displace phalloidin after the latter has become bound to the liver cell (Petzinger et al., 1975). Recently, Sonnenbichler (1975) found that silybin strongly stimulates the synthesis of ribosomal RNA in the nuclei of hepatocytes. This stimulation results from an increase in the activity of polymerase A, which is required for the formation of rRNA. Probably the increase in the production of rRNA causes a stimulation of protein synthesis. The clinical activity of silymarin is no doubt due to its ability to "stabilize" the liver cell membranes and thereby render them impermeable to the entry of endogenous and exogenous toxins. Irreversibly damaged hepatocytes are not affected, but healthy or slightly damaged cells are protected from further damage (Vogel et al., 1975). The extremely low toxicity of silymarin is crucial to the therapy; 20 g per kg, administered orally to mice, shows no sign of incompatibility.

Drug preparations of the flavonoid-rich *Helichrysum arenarium* have been used in folk medicine also as anti-choleretic and as remedy of liver diseases. According to Prokopenko et al. (1972) and Szadowska (1962) the naringenin glycosides helichrysin and salipurposide, which are thought to be responsible for this action, possess about a third of the activity of dehydrocholic acid.

Since antiquity, the artichoke *Cynara scolymus* has also enjoyed a large reputation in folk medicine. The property of liver protection and choleretic activity were ascribed to extracts of the leaf. Extracts and tinctures were used therapeutically in jaundice. The chief active principle was isolated and its structure elucidated by Panizzi and Scarpati (1954a, b). The active principle cynarin was shown to be a 1,5-dicaffeylquinic acid (**17**) (Panizzi, 1965).

Cynarin increases the secretion of bile, which is thought to be primarily responsible for the described choleretic and cholagogic activity (Struppler

(17) Cynarin

and Rössler, 1957; Schreiber *et al.*, 1970). In recent work Maros *et al.* (1966, 1968) showed that aqueous leaf extracts of the artichoke cause a marked increase in the number of binucleate hepatocytes, and in the RNA concentration of the liver cells. Whether silybin has the same or a similar mechanism of action is still not known. An anticholeretic action is similarly ascribed to absinth (see bitter principles).

Antibiotic, bacteriostatic, virostatic and fungistatic agents

The structures of the polyacetylenes of the family Compositae have been elucidated in the last 10 years, especially by Bohlmann *et al.* (1973). Some of them possess remarkable bacteriostatic or fungistatic properties (Schulte and Rücker, 1970), and they probably participate in the pharmaceutical activity of some drugs. Known polyacetylenes are carlina oxide (**18**) from *Carlina acaulis* and capillin (**19**) from *Artemisia capillus*. At a dilution of 1:200 000, a *Carlina* extract still inhibits *Staphylococcus aureus*, typhus-paratyphus, and dysentery bacteria (Schmidt-Thomé, 1950). Capillin is a good antibiotic against dermal mycoses (Schulte *et al.*, 1967). Good bactericidal and fungicidal activity are also shown by the acetylenes trideca-1-monoene-3,5,7,9,11-pentayne (**20**) and trideca-1,11-diene-3,5,7,9-tetrayne (**21**), which have been isolated from the flowers of *Arnica montana*, the roots of *Arctium lappa*, and species of *Echinacea* and *Pulicaria* (Rücker, 1963; Schulte and Rücker, 1966; Schulte *et al.*, 1967). The various thymol ethers isolated are also implicated in the bactericidal and fungicidal activity of the *Arnica* root drug (Schulte *et al.*, 1963; Rücker, 1963).

(18) Carlina oxide (*Carlina acaulis*)

(19) Capillin (*Artemisia capillus*)

$$CH_3 - (C \equiv C)_5 - CH = CH_2$$

(20) Tridecene −1− pentayne − (3,5,7,9,11)

$$CH_3 - CH = CH - (C \equiv C)_4 - CH = CH_2$$

(21) Tridecadiene −(1,11)− tetrayne − (3,5,7,9)

(*Arnica* 5pp., *Arctium* 5pp., *Echinacea* species)

The first polyacetylene shows a fungicidal activity, which is about 10 orders of magnitude greater than that of capillin. Its activity against *Staphylococcus aureus* is about the same as that of capillin (Rücker, 1963).

Since the therapeutic availability of these compounds is limited by their high instability and considerable toxicity, the synthesis of more stable acetylene compounds has been undertaken (Reisch *et al.*, 1967). For example, acetylphenylacetylene and benzoylphenylacetylene were found to have greater fungistatic activity than capillin. Systematic microbiological investigations with the natural and synthetic polyacetylenes have provided useful information concerning the relationship between structure and activity. The active compounds had the following characteristics:

1. With a non-terminal triple bond, one substituent is an aromatic residue. The second carries a functional group adjacent to the triple bond; this can be an ester, thioamide, carbonyl, *sec.*-hydroxyl, aldehyde, halogen, ethylene or acetylene group.

2. With a terminal triple band, the substituent is an aromatic acyl residue, and compounds of weaker activity are obtained if this is replaced by an aliphatic residue. This residue can carry other functional groups, which are widely separated from the acetylene bond.

Fungistatic activity increases with the polarization of the triple bond and with the lipid solubility. The bactericidal activity increases with the hydrophilic nature of the compound (Reisch *et al.*, 1967).

Bacteriostatic activity is also exhibited by the many phenolic carboxylic acids that occur in the Compositae, e.g. caffeic acid and chlorogenic acid. Part of the bacteriostatic activity of *Echinacea* preparations, which are used pharmaceutically, is due to a complex depside, consisting of dihydroxyphenyl-ethanol, caffeic acid, 1 mol rhamnose and 2 mol glucose (Stoll *et al.*, 1950). It is not yet clear which compounds of the *Echinacea* total extract, as investigated by Orinda *et al.* (1973), are responsible for the antiviral activity of the drug.

Cytotoxic agents

The availability of modern, refined methods for testing anticarcinogenic agents has encouraged the systematic search for cancerostatic agents amongst natural products. Many of the sesquiterpenes in the Compositae, chiefly lactones from the germacranolide, guaianolide, pseudoguaianolide and elemanolide class, are especially active. The formulae of some representatives of the three classes of substances are shown below (formulae 22–26).

The first positive results were obtained with camomile extracts and chamazulene on Ehrlich-Ascites carcinoma and mouse chondroma respectively (Barton *et al.*, 1954; Kraul and Schmidt, 1957). Table I

14. PHARMACEUTICAL AND ECONOMIC USES

Cytotoxic active sesquiterpene lactones from Compositae

Germacranolide

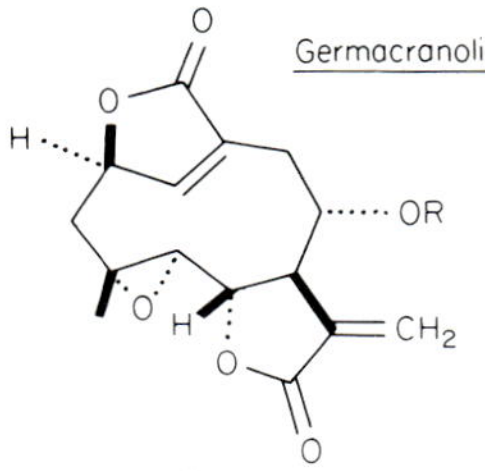

(22) R : CO—CH=C(CH₃)(CH₃) : Elephantin

(23) R : CO—C(CH₃)=CH₂ : Elephantopin

Guaianolide

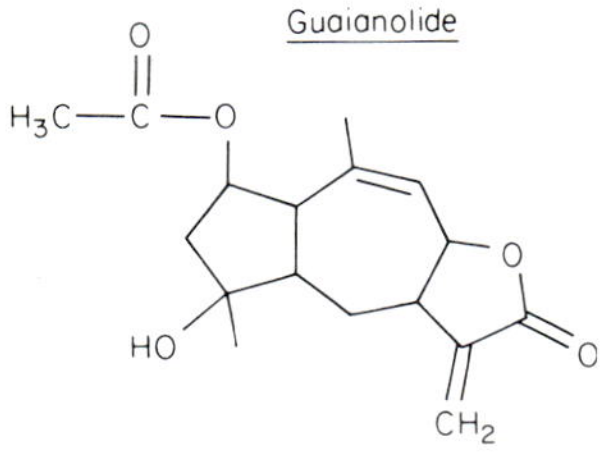

(24) Gaillardin

Elemanolide

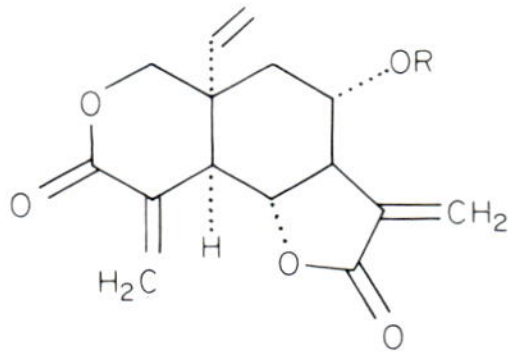

(25) R : H : Vernolepin
(26) R : CO—C(CH₂OH)=CH₂ : Vernodalin

shows the sesquiterpene lactones that have so far been isolated and shown to have inhibitory activity against tumours.

These compounds were active against Walker-carcinoma-sarcoma 256, P-388-leukemia of the mouse, against nasopharynx carcinoma KB cells, or against B-16-melanoma. According to the studies of Kupchan *et al.* (1967, 1969c), the antitumour activity is due to the presence of an α-methylene group in the γ-lactone ring, and an oxirane group or another double bond susceptible to nucleophilic attack. Since the growth inhibitor activity of vernolepin (**25**) can be blocked with sulphydryl compounds, and the double bond of the lactone ring reacts extremely easily with the SH-group of cysteine at pH 7·4 (Kupchan, 1970), the mechanism of action would appear to depend on an interaction of the lactone ring and SH-groups of enzymes.

The therapeutic use of these sesquiterpenes has hitherto been prevented by their relatively high toxicity. Attempts to increase the activity of the molecule by chemical modification have so far been unsuccessful (Pettit *et al.*, 1974).

TABLE I Sesquiterpene lactones with antitumour activity

Substance	Plant	References
	GERMACRANOLIDES	
Elephantin	*Elephantopus elatus*	Kupchan *et al.*, 1966
Elephantopin		Kupchan *et al.*, 1969a
		Kupchan and Banerschmidt, 1970
Phantomolin }		
Molephantin }	*Elephantopus mollis*	Lee *et al.*, 1973, 1975
Molephantinin }		McPhail *et al.*, 1974a
Costunolide	*Saussurea lappa*	Rao *et al.*, 1960
	Artemisia balchanorum	Herout and Sorm, 1959
	Cosmos sp.	Bohlmann *et al.*, 1964
Vernomygdin	*Vernonia amygdalina*	Kupchan *et al.*, 1969b
Provincialin	*Liatris provincialis*	Herz and Wahlberg, 1973
Ovatifolin	*Podanthus ovatifolius*	Gnecco *et al.*, 1973
Erioflorin and erioflorin esters		
Liatrin	*Liatris chapmanii*	Kupchan *et al.*, 1971a
Eupaserrin	*Eupatorium semiserratum*	Kupchan *et al.*, 1973a
Deacetyl eupaserrin		
Eupacunin and other lactones	*Eupatorium cuneifolium*	Kupchan *et al.*, 1971b
		Kupchan *et al.*, 1973b
Eupaformonin	*Eupatorium formosanum*	McPhail *et al.*, 1974b
Cnicin	*Cnicus benedictus*	Vanhaelen-Fastré, 1973
	GUAIANOLIDES AND PSEUDOGUAIANOLIDES	
Gaillardin	*Gaillardia pulchella*	Kupchan *et al.*, 1965a
Euparotin	*Eupatorium rotundifolium*	Kupchan *et al.*, 1967
Euparotin acetate		
Eupachlorin		
Eupachlorin acetate		
Damsin	*Ambrosia ambrosioides*	Doskotch and Hufford, 1969
Damsinic acid	*A. maritima*	Doskotch and Hufford, 1970
	A. hispida	
Helenalin	*Helenium autumnale*	Pettit *et al.*, 1974
Autumnolide	var. *montanum* and other *Helenium* species *Gaillardia* species *Balduina angustifolia* *Arnica montana*	
	ELEMANOLIDES	
Vernolepin	*Vernonia hymenoleposis*	Kupchan *et al.*, 1969c
Vernomenin		
Vernodalin		

For the sake of completeness it may be mentioned that the sesquiterpene lactone amaralin, from *Helenium amarium* (Raf.) H. Roch, when applied subcutaneously causes a strong analgesic action which is not antagonized by nalorphin (Lucas *et al.*, 1964).

Another cytotoxic principle has been found in the flavone and flavonol di- and trimethyl ethers of *Eupatorium perfoliatum*, *E. semiserratum* and *Baccharis sarothroides* (Kupchan *et al.*, 1965b; Kupchan and Bauerschmidt, 1970). The active flavonols are derived from quercetagetin, 6-hydroxyluteolin or 6-hydroxyapigenin and they all have a methoxy group in position 6. Eupatorin, 5,3'-dihydroxy-6,7,4'-trimethoxyflavone (**27**), is active against nasopharynx carcinoma. Activity is also shown by 3,4'-dimethoxy-5,7,3'-trihydroxyflavone and centaureidin, 3,6,4'-trimethoxy-5,7,3'-trihydroxyflavone.

(27) Eupatorin

In Table II are listed the drugs additionally used in folk medicine. Table III enumerates the crude drugs of the family which have been listed in certain individual pharmacopeias.

TABLE II Drugs used in folk medicine

Anthelmintic activity	Santonin (sesquiterpene lactone) essential oil	*Artemisia cina* (Flores Cinae) *Chrysanthemum vulgare* (Flores Tanaceti)
Expectorant activity	Polysaccharides	*Tussilago farfara* L. (Fol. Farfarae)
Sedative activity hypoglycaemic (Blood) activity	Lactucin Lactupicrin (sesquiterpene lactone)	*Lactuca virosa* L.
Diuretic activity	Flavonol glycosides Saponins	*Solidago virgaurea* L. (Herba Virgaureae)
Spasmolytic activity	Saponins	*Grindelia* rob. (Herba Grindeliae)

TABLE III. Drugs in European pharmacopeias

Crude Drug	DAB 7 BRD	DAB 7 DDR	Helv. VI	ÖAB 9	CF 65	Hisp. IX
Herba Absinthii	+	+	+	+		
Herba Cardui bened.			+	+		+
Herba Millefolii		+	+	+		
Radix Arnicae				+		+
Radix Bardanae c. herba					+	+
Radix Cynoglossae						+
Folia Cynarae				+		
Folia Farfarae	+		+			
Flores Arnicae	+	+	+	+	+	+
Flores Chamomillae	+	+	+	+	+	+
Flores Cinae				+	+	
Flores Calendulae		+				
Flores Cyani			+			
Flores Farfarae			+		+	
Flores Pyrethri				+	+	
Flores Stoechados			+			

FOOD INDUSTRY

Sweet substances

Whereas bitter substances are fairly widely distributed in the Compositae, we know only one plant in this family that produces a sweet material. The plant, *Stevia rebaudiana* Bertoni. earlier known as *Eupatorium rebaudianum*, grows in Paraguay (Fletcher, 1955). It has been used for a long time by the natives for sweetening bitter drinks. Interest was renewed in this plant as a result of sugar shortage during the second world war. The sweetening action is due to a diterpene glycoside, which contains a glucose bound to the carboxyl group, and two glucose moieties, in the form of 1-2 linked sophorose, bound to the alcoholic group (Wood et al., 1955). Known as stevioside (**28**), this compound is 300 times sweeter than sucrose and only one fifth less sweet than saccharine. Nothing is known so far about side effects. Nowadays, many synthetic replacements for sugar have been discredited as potential carcinogenic agents. Since stevioside also has certain advantages over saccharine with respect to taste, the study of this plant is again very topical. 1 kg of dried leaves will yield up to 65 g of stevioside, but the large scale cultivation of this plant in tropical countries would be necessary for an adequate and economic supply. Otherwise, we must wait until a cheap and convenient laboratory synthesis of stevioside has been devised.

In this connection, the 4-β, 10-α-dimethyl-1,2,3,4,5,10-hexahydrofluorene-4α, 6-dicarboxylic acid (**29**), synthesized from common pine resin, is of theoretical interest, because its sodium salt is about 1600–2000 times sweeter than sucrose (Tahara *et al.*, 1971).

(**28**) Stevioside

(**29**) 4-β,10α-diMe-1,2,3,4,5,10-hexahydrofluorene-4α,6-dicarboxylic acid

Bitter substances

Although bitter substances occur fairly widely in the Compositae, only *Artemisia absinthium* and *Cnicus benedictus* are important in pharmacy and in the food industry. The bitter taste of *Artemisia absinthium* is due to the guaianolide lactones, absinthin (**30**) (bitter value = 70 000) and anabsinthin (**31**).

(30) Absinthin   (31) Anabsinthin

Cnicin, the bitter principle of *Cnicus benedictus*, is an ester of dihydroxymethyl-acrylic acid with a sesquiterpene dihydroxylactone of the germacrane type. Whereas the absinthe drug with a bitter value of 10–25 000, together with the gentian drug (bitter value 10–30 000), is one of the most bitter drugs, the *Cnicus* drug, with an average bitter value of 800–1500, is qualitatively about the same as Cortex Chinae. The pharmaceutical usefulness of these bitter drugs is due primarily to their stimulation of stomach secretion, which is the result of reflex nervous activity. Furthermore, the release of gastrin causes an increase in stomach acidity. The motor action of the intestine and the bile and pancreatic secretion are therefore also increased (Schmid, 1966). Recent physiological studies by Glatzel (1968) have shown that there can be considerable differences between the individual bitter drugs with respect to their action on the salivary flow, amylase activity and the hexosamine concentration of the saliva. On this basis, absinthe is classed with gentian as a bitter drug that causes a marked increase in salivary flow, with a simultaneous small decrease in the amylase activity and hexosamine concentration (Blumberger and Glatzel, 1966). The great popularity of absinthe for the preparation of aperitifs (e.g. vermouth) is largely due to the aromatic substances that it also contains.

Vermouth wines are prepared predominantly from *Artemisia pontica* L. Since absinthe liqueurs are prepared from Oleum Absinthii, they contain considerable quantities of thujone. In large doses this is very toxic, and can lead to chronic poisoning. For this reason their preparation is prohibited in Germany and Switzerland.

Spices and flavouring agents

The Compositae contains only a small proportion of spice plants. Apart from absinthe, only *Artemisia dracunculus* (estragon) and *Artemisia vulgaris* (mugwort) are of any importance, owing to their aromatic smell and taste. They are used to season salads, mayonnaises, gravies, fish dishes, and pickled gherkins, and in the preparation of estragon vinegar.

As a coffee supplement and as a substitute for coffee, use has been found for the ground roasted roots of chicory, *Cichorium intybus*, especially during the second world war.

Topinambur, the rhizome of the ground artichoke, *Helianthus tuberosus*, when stewed or boiled is used similarly to potato as a food, or as animal fodder; and roasted it serves as a coffee substitute. It is also used for the technical preparation of fructose.

SUBSTANCES OF GENERAL ECONOMIC IMPORTANCE

Insecticides

Now that DDT has been sacrificed for the sake of environmental protection, the old insecticides have become important again. From the scientific standpoint, the pyrethroids of the long known *Chrysanthemum cinerariaefolium* Vis. are interesting. Its dried and powdered flowers have been used for a long time as fly powder. The insecticidal activity is due to the presence of the following six constituents: pyrethrin I (**32**) and II, cinerin I (**33**) and II, and jasmolin I (**34**) and II. Pyrethrins are esters of pyrethrolone, which is a 2-cyclopentane-4-ol-1-one, with (+)*trans*-chrysanthemummonocarboxylic acid, or (+)*trans*-pyrethrum acid. The cinerins and jasmolins are analogous esters of cinerolone and jasmolone (Crombie and Elliot, 1961).

14. PHARMACEUTICAL AND ECONOMIC USES

$H_3C-C\equiv CH$ — with CH_3, CH_3 groups on cyclopropane ring, connected via $-C(=O)-O-$ to cyclopentenone bearing H_3C and R_2 substituents.

(32) Pyrethrin I $R_1 = CH_3$ $R_2 = -CH_2-CH=CH-CH=CH_2$

(33) Cinerin I $R_1 = CH_3$ $R_2 = -CH_2-CH=CH-CH_3$

(34) Jasmolin I $R_1 = CH_3$ $R_2 = -CH_2-CH=CH-CH_2-CH_3$

(35) Allethrin $R_1 = CH_3$ $R_2 = -CH_2-CH=CH_2$

Of these compounds, pyrethrin I and II show the strongest activity against *Musca domestica* L. and *Phaedon cochleariae* Fab., whereas both cinerins and both jasmolins all have about the same activity, which is markedly less than that of pyrethrin (Table IV).

TABLE IV. Relative average toxicity of the pyrethroids against *Musca domestica* and *Phaedon cochleariae* Fab.

	Pyrethrin I	Pyrethrin II	Cinerin I	Cinerin II
Musca	100	120	40	45
Phaedon	100	38	50	20

It is essential for activity that the acid moiety contains a cyclopropane ring with gem-dimethyl-groups. Hydrogenation of the double bond decreases the toxicity. On their own pyrethrolone and chrysanthemum-monocarboxylic acid are inactive. A review of their structure–activity relationships has been published by Matsui and Yamamoto (1971). Pyrethroids have the advantage that they are practically non-toxic to warm-blooded animals, and the development of resistance against them is far less than for the synthetic insecticides.

These studies have stimulated many chemical syntheses, in which the chrysanthemum acid is esterified with a non-naturally occurring alcohol, and in which the natural alcohol components are esterified with a modified acid. The most active synthetic compounds belong to the first group. For example, allethrin (**35**), developed by Schlechter *et al.* in 1949, is used commercially as its isomeric mixture. Allethrin (**35**) has about the same activity as pyrethrin I (**32**), and differs from the latter only by the side chain on the 5-ring.

Pyrethroids rapidly lose activity by oxidative degradation. Antioxidants, such as hydroquinone or tannins, so-called activators such as ethylene glycol ether and pinene, or synergistic agents are therefore added. Suitable synergistic agents are piperonyl butoxide, piperonyl cyclones or sesamin. The extracts are used as aerosols or dusting preparations.

Other compounds with insecticidal activity have been obtained from species of *Echinacea, Chrysanthemum, Heliopsis* and *Anacyclus*. They are isobutylamides of long chain simple unsaturated fatty acids or acetylene fatty acids (Jacobson, 1967). They possibly act as natural protective agents or resistance factors for the plant against insect attack. The bitter substance absinthin (**30**) from *Artemisia absinthium*, which is a dimeric sesquiterpene lactone, also has protective properties against insect attack (Wada and Munakata, 1971).

Oil drugs
Of the oil drugs, only the fat oils of the fruits of *Helianthus annuus* L. and *Carthamus tinctorius* have achieved any industrial importance. The annual world production of sunflower oil is at present about 3 million tons. The safflower plant used to be cultivated as a dye plant. Safflower oil is a drying oil; its highly unsaturated character is due chiefly to its content of linoleic acid (up to 75% of the total fatty acids). The oil is used as a salad oil, and more rarely it is used pharmaceutically in vitamin preparations.

Polysaccharides
Roots and rhizomes, in particular those of *Inula* species, the tubers of *Dahlia* and *Helianthus* species, roots of *Taraxacum officinale, Arctium lappa, Pyrethrum* and *Cichorium* species are characterized by a great abundance of inulin. Inulin consists of approximately 20–30 fructose units and has a molecular weight of 3000–5000. The fructofuranose units are joined by 1,2-linkages. Since inulin is cleaved by the body into D-fructose (fructofuranose), it is tolerated better than other carbohydrates by diabetics. Plants containing inulin, or inulin itself are therefore used for the preparation of diabetic bread. Furthermore, inulin is the starting material for the technical preparation of fructose.

Caoutchouc
Two plants of the family Compositae, *Taraxacum bicorne* (= Kok Saghyz) Rodin and *Parthenium argentatum* L. yield caoutchouc (rubber) and they are economically important (Ulmann, 1951). The first plant grows wild in Russian central Asia and yields the so-called Koksagis caoutchouc. The useful part of the plant is the tap-root, from which up to 50% of latex can be obtained. The second plant is a dwarf tree of the mexican uplands. Unlike the other known caoutchouc-yielding plants, in which the caoutchouc is contained in latex vessels, the caoutchouc of this tree is found in the cells of the medulla. This is the source of guayul caoutchouc. Qualitatively it is the same as tropical caoutchouc.

With the development of synthetic rubber both these types of caoutchouc have ceased to be so important. A certain area of application still remains, however, because the elasticity of synthetic rubber can be increased by the addition of Koksagis and guayul caoutchouc (Ulmann, 1951).

Toxic substances

Systematic, chemical and pharmacological research within a family, tribe or genus can occasionally disclose the fact that plants used in herbal medicine for many years are in fact dangerously toxic. This is the case with the genus *Senecio*. The vegetative parts of the flowers of *Senecio* were approved in antiquity as a folk remedy for inflamed wounds, stomach trouble, worms, and illnesses of the liver and bile system. Nothing concerning side effects was recorded in the earlier literature. The turning point came with the isolation of toxic pyrrolizidine alkaloids from *Senecio* plants towards the end of the nineteenth century. This discovery led to the explanation of the aetiology of an illness that was responsible for the annual loss of many cattle and horses, which ate *Senecio* plants when grazing. The illness was known in Canada as Pictou disease, in New Zealand as Winton disease, in South Africa as Malteno disease and in southern Bohemia as Zdar disease. Later it was found that acute liver poisoning and liver cirrhosis in humans was associated with the use of so-called "bush teas", which are based in part on leaves of *Senecio* species.

These toxic effects can all be attributed to the pyrrolizidine alkaloids, which have a ubiquitous occurrence in *Senecio* species (McLean, 1970). Monoesters, acyclic diesters and macrocyclic diester alkaloids with 11, 12 and 13 membered rings are now also known, in addition to the simple pyrrolizidine alkaloids (Warren, 1955; Bull *et al.*, 1968; Klasek and Weinbergova, 1975). Representative formulae (**36**)–(**39**) are shown below.

The many animal experiments performed with pure *Senecio* alkaloids have shown that the primary effect is liver poisoning, which is manifested as megalocytosis, fatty liver, necroses, haemorrhages and cirrhosis. In addition, damage to the kidneys, lungs, coronary vessels and other organs is also observed. Furthermore, the pyrrolizidine alkaloids also possess hepatocarcinogenic activity (Schoental, 1954). An atropine-like activity has been observed for platyphyllin from species of *Senecio*, *Petasites* and *Cynoglossum* (Goldenhershel, 1943). It would seem that the antitumour activity of some alkaloids observed by Culvenor in 1968 can never be exploited for treating cancer, because of their hepatotoxic properties.

In addition to *Senecio* species, other composites are reported as being toxic especially to animals; in most cases the active principles responsible are unknown. The list given in relevant books (Watt and Breyer-Brandwijk 1962; Muenscher, 1939) includes *Geigeria*, *Eupatorium*, *Baccharis*, *Baileya*, *Helenium*, *Iva* and *Xanthium* species.

(36) Retronecine: $8\alpha, \Delta^{1,2}$, R = OH
(*Crotalaria retusa*)

(37) Strigosine (*Heliotropium strigosum*)

(38) Echiumine (*Echium lycopsis*)

(39) Jacobine (*Senecio jacobaea*)

REFERENCES

ABRAHAM, D. J., TAKAGI, S., ROSENSTEIN, R. D., RYONOSUKE, S., WAGNER, H., HÖRHAMMER, L., SELIGMANN, O. and FARNSWORTH, N. R. (1970). The structure of Silydianin, an isomer of Silymarin (Silybin) by X-ray analysis. *Tetrahedron Lett.* **31**, 2675.

AEBI, A., WAALER, T. and BÜCHI, J. (1958). Petasin and S-Petasin, the spasmolytic active principle in *Petasites officinalis* (*P. hybridus*). *Pharm. Weekblad* **93**, 397.

AEBI, A. and WAALER, T. (1959). "Über die Inhaltsstoffe von *Petasites hybridus*", (**2**) Fl. Welt-Verlag Haelbing und Lichtenhan, Basel.

BARTON, H., HEINE, U., GRAFFI, A. and JUNG, F. (1954). Versuche zur Beeinflussung der Cancerogenese durch 1-Isopropyl-5-methyl-azulen. *Naunyn-Schmiedebergs Arch. exp. Pathol. Pharmacol* **223**, 443.

BLUMBERGER, W. and GLATZEL, H. (1966). In "Pflanzliche Bitterstoffe", p. 52. Planta Medica Suppl., Hippokrates-Verlag, Stuttgart.

BOHLMANN, F., BORNOWSKI, H. and v. KÖHN, S. (1964). Polyacetylenverbindungen, Polyine der Gattung *Cosmos*. *Chem. Ber.* **97**, 2583.

BOHLMANN, F., BURKHARDT, T. and ZDERO, C. (1973). "Naturally Occurring Acetylenes". Academic Press, London and New York.

BREINLICH, J. (1970). En-In-Wirksubstanzen in einigen arzneilich gebrauchten Compositen-Drogen, unter besonderer Berücksichtigung von *Achillea millefolium*. *Pharm. Ztg* **45**, 1699.

BREINLICH, J. and SCHARNAGEL, K. (1968). Pharmakologische Eigenschaften des EN-IN-Dicycloäthers aus *Matricaria chamomilla*. *Arzneim.-Forsch.* (*Drug Res.*) **18**, 429.
BULL, L. B., CULVENOR, C. D. J. and DICK, A. T. (1968). "The Pyrrolizidine Alkaloids". North-Holland, Amsterdam.
CROMBIE, L. and ELLIOTT, M. (1961). Chemistry of the natural pyrethrins. In "Fortschritte der Chemie Organischer Naturstoffe", (L. Zechmeister ed.), Bd. XIX, p. 120. Springer-Verlag, Wien.
DOSKOTCH, R. W. and HUFFORD, C. D. (1969). Damsin, the cytotoxic principle of *Ambrosia ambrosioides* (Cav.) Payne. *J. pharm. Sci.* **58**, 186.
DOSKOTCH, R. W. and HUFFORD, C. D. (1970). The structure of damsinic acid, a new sesquiterpene from *Ambrosia ambrosioides* (Cav.) Payne. *J. Org. Chem.* **35**, 486.
EICHLER, O. and HAHN, M. (1949). Versuche zum Schutz gegen leberschädigende Gifte. *Naunyn-Schmiedebergs Arch. exp. Pathol. Pharmacol.* **206**, 674.
FLETCHER, H. G. JR. (1955). The sweet herb of Paraguay. *Chemurgic Digest* **14**.
FRIMMER, M. and KROKER, R. (1975). Phalloidin-Antagonisten 1. Mitteilung. Wirkung von Silybin-Derivaten an der isoliert perfundierten Rattenleber. *Arzneim. Forsch.* (*Drug Res.*) **25**, 394.
GLATZEL, H. (1968). "Die Gewürze". Nicolaische Verlagsbuchhandlung Manualia Nicolai, Herford.
GNECCO, S., POYSER, J. PH., SILVA, M., SAMMER, P. G. and TYLER, TH. W. (1973). Sesquiterpene lactones from *Podanthus ovatifolius*. *Phytochemistry* **12**, 2469.
GOLDBERG, A. S., MUELLER, E. C., EIGEN, F. and DESALRA, S. J. (1969). Isolierung einer entzündungshemmenden Substanz aus *Achillea millefolii*. *J. pharm. Sci.* **58**, 938.
GOLDENHERSHEL, T. I. (1943). A new Atropin-like substance called *Platyphyllin*. *Klinicheskaya meditsina* **31**(3), 56. *Am. Rev. Soviet Med.* **1**, 155.
HAHN, G., LEHMANN, H. D., KÜRTEN, M., UEBEL, K. and VOGEL, G. (1968). Zur Pharmakologie und Toxikologie von Silymarin, des antihepatotoxischen Wirkprinzipes aus *Silybum marianum* (L.) Gaertn. *Arzneim. Forsch.* (*Drug Res.*) **18**, 698.
HEROUT, V. and SORM, F. (1959). Isolation and structure of Costunolide from *Artemisia balchanorum*. *Chem. Ind.*, 1067.
HERZ, W. and WAHLBERG, I. (1973). Provincialin, a cytotoxic Germacradienolide from *Liatris provincialis* Godfrey with an unusual ester side chain. *J. org. Chem.* **38**, 2485.
HÖRHAMMER, L., WAGNER, H. and SALFNER, B. (1963). Neue Flavonglykoside aus der Kamille (*Matricaria chamomilla* L.). *Arzneim. Forsch.* (*Drug Res.*) **13**, 33.
ISAAC, O. (1974). Fortschritte in der Kamillenforschung. *Dt. apoth. Ztg* **114**, 255.
JACOBSON, M. (1967). Structure of echinacein, the insecticidal component of American cornflower roots. *J. org. Chem.* **32**, 1646.
JAKOVLEV, V. and SCHLICHTEGROLL, A. VON (1969). Zur entzündungshemmenden Wirkung von ($-$)-α-Bisabolol, einem wesentlichen Bestandteil des Kamillenöls. *Arzneim. Forsch.* (*Drug Res.*) **19**, 615.

Klasek, A. and Weinbergova, O. (1975). "Recent Developments in the Chemistry of Natural Carbon Compounds" p. 37. Ed. Akadémiai Kiadó, Hung. Academy of Sciences, Budapest.

Kraul, M. A. and Schmidt, F. (1957). Über die wachstumshemmende Wirkung bestimmter Extrakte aus Flores Chamomillae und eines synthetischen Azulenpräparates auf experimentelle Mäusetumoren. *Arch. Pharmaz.* **290**, 66.

Kupchan, S. M. (1970). Recent advances in the chemistry of terpenoid tumor inhibitors. *Pure appl. Chem.* **21**, 227.

Kupchan, S. M. and Bauerschmidt, E. (1970). Cytotoxic flavonols from *Baccharis sarothroides*. *Phytochemistry* **10**, 664.

Kupchan, S. M., Cassady, J. M., Bailey, J. and Knox, J. R. (1965a). Tumor inhibitors XII. Gaillardin, a new cytotoxic sesquiterpene lactone from *Gaillardia pulchella*. *J. pharm. Sci.* **54**, 1703.

Kupchan, S. M., Knox, J. R. and Udayamurthy, M. S. (1965b). Tumor inhibitors VIII. Eupatorin, new cytotoxic flavone from *Eupatorium semiserratum*. *Pharmac. Sci.* **54**, 929.

Kupchan, S. M., Aynehchi, Y., Cassady, J. M., McPhail, A. I., Sim, G. A., Schnoes, H. K. and Burlingame, A. L. (1966). The isolation and structural elucidation of two novel sesquiterpenoid tumor inhibitors from *Elephantopus elatus*. *J. Am. chem. Soc.* **88**, 3674.

Kupchan, S. M., Hemingway, J. C., Cassady, J. M., Knox, J. R., McPhail, A. T. and Sim, G. A. (1967). The isolation and structural elucidation of Euparotin acetate, a novel Guaianolide tumor inhibitor from *Eupatorium rotundifolium*. *J. Am. chem. Soc.* **89**, 465.

Kupchan, S. M., Aynehchi, Y., Cassady, J. M., Schnoes, H. K. and Burlingame, A. L. (1969a). The isolation and structural elucidation of Elephantin and Elephantopin. Two novel sesquiterpenoid tumor inhibitors from *Elephantopus elatus*. *J. org. Chem.* **34**, 3867.

Kupchan, S. M., Hemingway, J. C., Karim, A. and Werner, D. (1969b). Tumor inhibitors XLVII. Vernodalin and Vernomygdin, two new cytotoxic sesquiterpene lactones from *Vernonia amygdalina* Del. *J. org. Chem.* **34**, 3908.

Kupchan, S. M., Hemingway, R. J., Werner, D. and Karim, A. (1969c). Vernolepin, a novel sesquiterpene dilactone tumor inhibitor from *Vernonia hymenolepsis* A. Rich. *J. org. Chem.* **34**, 3903.

Kupchan, S. M., Davies, V. H., Fujita, T., Cox, M. R. and Bryan, R. F. (1971a). Liatrin, a novel antileukemic sesquiterpene lactone from *Liatris chapmanii*. *J. Am. chem. Soc.* **93**, 4916.

Kupchan, S. M., Maruyama, M., Hemingway, R. J., Hemingway, J. C., Shibuya, S., Fujita, T., Cradwick, P. D., Hardy, A. D. U. and Sim, G. A. (1971b). Eupacunin a novel antileukemic sesquiterpene lactone from *Eupatorium cuneifolium*. *J. Am. chem. Soc.* **93**, 4914.

Kupchan, S. M., Fujita, T., Maruyama, M. and Britton, R. W. (1973a). The isolation and structural elucidation of Eupaserrin and deacetyleupaserrin, new antileukemic sesquiterpene lactones from *Eupatorium semiserratum*. *J. org. Chem.* **38**, 1260.

KUPCHAN, S. M., MARUYAMA, M., HEMINGWAY, R. J., HEMINGWAY, J. C., SHIBUYA, S. and FUJITA, T. (1973b). Structural elucidation of novel tumor inhibitory sesquiterpene lactones from *Eupatorium cuneifolium*. *J. org. Chem.* **38**, 2189.

LEE, K. H., FURUKAWA, H., KOZUKA, M., HUANG, H. C., LUHAN, P. A. and MCPHAIL, A. T. (1973). Molephatin, a novel cytotoxic germacranolide from *Elephantopus mollis* X-ray crystal structure. *J. chem. Soc.* D 476.

LEE, K. H., IBUKA, T., HUANG, H. CH. and HARRIS, D. L. (1975). Antitumor. Agents XIV: Molephantinin, a new potent antitumor sesquiterpene lactone from *Elephantopus mollis*. *Pharmac. Sci.* **64**, 1077.

LUCAS, R. A., ROVINSKI, S., KIESEL, R. J., DORFMAN, L. and MACPHILLAMY, H. B. (1964). A new sesquiterpene lactone with analgesic activity from *Helenium amarum* (Raf.) H. Rock. *J. org. Chem.* **29**, 1549.

MAROS, T., RACZ, G., KATONAI, B. and KOVACS, V. V. (1966). Wirkungen der *Cynara scolymus*-Extrakte auf die Regeneration der Rattenleber. 1. Mitteilung. *Arzneim. Forsch. (Drug Res.)* **16**, 127.

MAROS, T., SERES-STURM, L., RACZ, G., RETTEGI, C., KOVACS, V. V. and HINTS, M. (1968). Wirkungen der Cynara Scolymus-Extrakte auf die Regeneration der Rattenleber. 2. Mitteilung. *Arzneim. Forsch. (Drug Res.)* **18**, 884.

MATSUI, M. and YAMAMOTO, I. 1971. In "Naturally Occurring Insecticides" (M. Jacobson and G. Crosby, eds). Marcel Decker, New York.

MAYER, F. and MENGE, F. (1949). *Carduus marianus*. *Arzt und Patient* **62**, 256.

MCLEAN, E. K. (1970). Toxic actions on pyrrolizidine (*Senecio*) alkaloids. *Pharmac. Rev.* **22**, 429.

MCPHAIL, A. T., ONAN, K. D., LEE, K. H., IBUKA, T., KOZUKA, M., SHINGU, T. and HUANG, H. C. (1974a). Structure and stereochemistry of the Epoxide of Phantomolin, a novel cytotoxic sesquiterpene lactone from *Elephantopus mollis*. *Tetrahedron Lett.* 2739.

MCPHAIL, A. T., ONAN, K. D., LEE, K. H., IBUKA, T. and HUANG, H. C. (1974b). Structure and stereochemistry of Eupaformonin, a novel cytotoxic sesquiterpene lactone from *Eupatorium formosanum* Hay. *Tetrahedron Lett.* **3203**.

MUENSCHER, W. C. (1939). "Poisonous Plants of the United States". Macmillan, New York.

ORINDA, D., DIEDERICH, J. and WACKER, A. (1973). Antivirale Aktivität von Inhaltsstoffen der Composite *Echinacea purpurea*. *Arzneim. Forsch. (Drug Res.)* **23**, 1119.

PANIZZI, L. and SCARPATI, M. L. (1954a). Isolation and constitution of the active principle of the artichoke. *Gazz. chim. ital.* **84**, 792.

PANIZZI, L. and SCARPATI, M. L. (1954b). Synthesis of cinarine, the active principle of the artichoke. *Gazz. chim. ital.* **84**, 812.

PANIZZI, L. and SCARPATI, M. L. (1965). 1,4- and 1,5-Dicaffeoylquinic acids. *Gazz. chim. ital.* **95**, 71.

PELTER, A. and HÄNSEL, R. (1968). The structure of silybin (Silybum substance E_6), the first flavonolignane. *Tetrahedron Lett.* **25**, 2911.

PETTIT, G. R., BUDZINSKI, J. C., CRAGG, G. M., BROWN, P. and JOHNSTON LA REA, D. (1974). Antineoplastic agents. 34. *Helenium autumnale* L. *J. med. Chem.* **17**, 1013.

PETZINGER, E., HOMANN, J. and FRIMMER, M. (1975). Phalloidin-Antagonisten. 2. Mitteilung. Protektive Wirkung von Disilybin bei der Vergiftung isolierter Hepatozyten mit Phalloidin. *Arzneim. Forsch. (Drug Res.)* **25**, 571.

POPLAWSKI, J., HOLUB, M., SAMEK, Z. and HEROUT, V. (1971). Arnicolides-sesquiterpenic lactones from the leaves of *Arnica montana*. *Collect. Czech. Chem. Commun.* **36**(8), 2189.

PROKOPENKO, O. P., SPIRIDONOV, V. N., LITVINENKO, V. I., CHORNOBAI, V. T., OBOLENTSEVA, G. V., KHADZHAI, YA.I. and TATARKO, Z. I. (1972). Phenol. compounds of *Helichrysum* and their biological activity. *Farm. Zh. (Kiev)* **27**, 3.

RAO, A. S., KELKAR, G. R. and BHATTACHARYYA, S. C. (1960). The structure of Costunolide, a new sesquiterpene lactone from *Costus* root oil. *Tetrahedron* **8**, 275.

REISCH, J., SPITZNER, W. and SCHULTE, K. E. (1967). Zur Frage der mikrobiologischen Wirksamkeit einfacher Acetylen-Verbindungen. *Arzneim. Forsch. (Drug Res.)* **17**, 816.

RÜCKER, G. (1963). Polyacetylene als Inhaltsstoffe von Arzneipflanzen. *Pharm. Ztg* **108**, 1169.

SCHLECHTER, M. S., GREEN, N. and LA FORGE, F. B. (1949). Constituents of Pyrethrum flowers. XXIII. Cinerolone and the synthesis of related cyclopentenolones. *J. Am. chem. Soc.* **71**, 3165.

SCHMID, W. (1966). *In* "Pflanzliche Bitterstoffe" p. 34 Planta Medica Suppl., Hippokrates-Verlag, Stuttgart.

SCHMIDT-THOMÉ, J. (1950). Über die antibakterielle Wirkung der *Silberdistelwurzel. Z. Naturf.* **5b**, 409.

SCHOENTAL, R. (1954). *Senecio* alkaloids and cancer. *Br. med. J.* **1**, 335.

SCHREIBER, J., ERB, W., WILDGRUBE, J. and BÖHLE, E. Z. (1970). Die fäkale Ausscheidung von Gallensäuren und Lipiden des Menschen bei normaler und medikamentös gesteigerter Cholerese. *Gastroenterologie* **8**, 230.

SCHULTE, K. E. and RÜCKER, G. (1966). Polyacetylene und einige andere Inhaltsstoffe der *Arnica*-Blüten. *Arch. Pharm.* **299**, 468.

SCHULTE, K. E. and RÜCKER, G. (1970). Polyacetylene als Inhaltsstoffe von Arzneipflanzen. *In* "Fortschritte der Arzneimittelforschung", Bd. 14, p. 517. Birkhäuser-Verlag, Basel and Stuttgart.

SCHULTE, K. E., REISCH, J. and RÜCKER, G. (1963). Einige neue Inhaltsstoffe der Wurzel von *Arnica montana*. *Arch. Pharmaz.* **296**, 273.

SCHULTE, K. E., RÜCKER, G. and PERLICK, J. (1967). Das Vorkommen von Polyacetylen-Verbindungen in *Echinacea purpurea Moench* und *Echinacea angustifolia* DC. *Arzneim. Forsch. (Drug Res.)* **17**, 825.

SMOLENSKI, S. J., BELL, C. L. and BAUER, L. (1967). The isolation of Achillin from *Achillea millefolium*. *Lloydia* **30**, 144.

SONNENBICHLER, J. (1975). Max-Planck-Institut für Biochemie, Martinsried, München. Private communication.

STOLL, A., RENZ, J. and BRACK, A. (1950). Isolierung und Konstitution des Echinacosids, eines Glykosids aus den Wurzeln von *Echinacea angustifolia* DC. *Helv. chim. Acta* **33**, 1877.

STOLL, A., MORF, R., RHEINER, A. and RENZ, J. (1956). Über Inhaltsstoffe aus

Petasites officinalis Moench. 1. Petasin und die Petasolester Bund. C. *Experientia* **12**, 360.

STRUPPLER, A. and RÖSSLER, H. (1957). Über die choleretische Wirkung des Artischockenextraktes. *Med. Monatsschr.* **11**, 221.

SZADOWSKA, A. (1962). Pharmacology of galenic preparations and flavonoids from *Helichrysum arenarium*. *Acta polon. pharm.* **19**, 465.

TAHARA, A., NAKATA, T. and OHTSUKA, Y. (1971). New type of compound with strong sweetness. *Nature Lond.* **233**, 619.

THIELE, K., JAKOVLEV, V., ISAAC, O. and SCHULLER, W. A. (1969). Äther und Ester von (−)-α-*Bisabolol* und analogen Mono- und Sesquiterpenoiden mit antiphlogistischer Wirkung. *Arzneim. Forsch. (Drug Res.)* **19**, 1878.

ULMANN, M. (1951). "Wertvolle Kautschukpflanzen des gemäßigten Klimas". Akademie-Verlag, Berlin.

VANHAELEN-FASTRÉ, R. (1973). Antibiotische und cytotoxische Wirkung des Cnicins aus *Cnicus benedictus*. *Zentbl. Pharm.* **112**, 624.

VOGEL, G., TROST, W., BRAATZ, R., ODENTHAL, K. P., BRÜSEWITZ, G., ANTWEILER, H. and SEEGER, R. (1975). Untersuchungen zur Pharmakokinetik, Angriffspunkt und Wirkungsmechanismus von Silymarin, dem antihepatotoxischen Prinzip aus *Silybum marianum* (L.) Gaertn. *Arzneim. Forsch. (Drug Res.)* **25**, 82.

WADA, K. and MUNAKATA, K. (1971). Insect-feeding inhibitors in plants. Feeding inhibitory activity of terpenoids in plants. *Agr. biol. Chem.* **35**, 115.

WAGNER, H., HÖRHAMMER, L. and MÜNSTER, R. (1968). Zur Chemie des Silymarins (Silybin), des Wirkprinzips der Früchte von *Silybum marianum* (L.) Gaertn. (*Carduus marianus* L.). *Arzneim. Forsch. (Drug Res.)* **18**, 688.

WAGNER, H., SELIGMANN, O., HÖRHAMMER, L., SEITZ, M. and SONNENBICHLER, J. (1971). Zur Struktur von Silychristin, einem zweiten Silymarin-Isomeren aus *Silybum marianum*. *Tetrahedron Lett.* **22**, 1895.

WARREN, F. H. (1955). The pyrrolizidine alkaloids. *In* "Fortschritte der Chemie organischer Naturstoffe" (L. Zechmeister, ed.), p. 198. Springer-Verlag, Vienna.

WATT, J. M. and BREYER-BRANDWIJK, M. G. (1962). "The Medicinal and Poisonous Plants of Southern and Eastern Africa". E. and S. Livingstone, Edinburgh and London.

WOOD, H. B. Jr., ALLERTON, R., DIEHL, H. W. and FLETCHER, H. G. Jr. (1955). Stevioside. The structure of the glucose moieties. *J. org. Chem.* **20**, 875.

Section II

Chapter 15
Eupatorieae—systematic review

H. ROBINSON and R. M. KING

National Museum of Natural History
Smithsonian Institution
Washington D.C., U.S.A.

Abstract. The Eupatorieae have been well delimited by workers at the tribal level. The only recent changes accepted here are the inclusion of *Isocarpha* and *Microspermum* and the exclusion of *Adenostyles* and *Dyscritothamnus*. In contrast, previous concepts at the subtribal and generic level are regarded as highly artificial. The characters of the tribe are reviewed in detail with some indication of distribution in the tribe. The present treatment presents results of a revision that is still in progress. The tribe contains approximately 2000 species. At present 160 genera are recognized, allowing placements of all but a few mostly South American species. The previous series of four subtribes is replaced with a series of 19 groups many of which are potential subtribes: Adenostemmatinae, Eupatoriinae, *Disynaphia* group, *Gyptis* group, *Acritopappus* group, *Piqueria* group, *Trichocoronis* group, *Ayapana* group, Alomiinae, *Liatris* group, *Fleischmannia* group, *Critonia* group, *Praxelis* group, *Hebeclinium* group, *Neomirandea* group, *Mikania* group, *Ageratina* group, *Hofmeisteria* group and *Oaxacania* group. Correlations with cytology suggest a base number of $n=10$ for the tribe with important derived groups having $n=9$, $n=11-12$ and $n=16-18$.

CONTENTS

Introduction	438
Features of the Eupatorieae	441
Cytology	452
Distribution of characters	453
Subdivisions of the tribe	454
Adenostemmatinae	454
Eupatoriinae	454
Disynaphia group	455
Gyptis group	455
Acritopappus group	456
Piqueriinae	456
Trichocoronis group	456
Ayapana group	456

Alomiinae 457
Liatris group 457
Fleischmannia group 458
Critonia group 458
Praxelis group 459
Hebeclinium group 459
Neomirandea group 459
Mikania group 460
Ageratina group 460
Hofmeisteria group 461
Oaxacania group 461
Acknowledgements 468
References 469

INTRODUCTION

The history of the largely New World Eupatorieae began with the description and establishment of the genus *Eupatorium* by Tournefort (1700) on the basis of the single European member of the tribe. The tribe has continued to suffer since its centers of diversity were remote from the early centers of study. The nomenclatural history of the tribe began with Linnaeus (1753) who recognized the genera *Eupatorium* and *Ageratum*. Linnaeus also started the taxonomic and nomenclatural confusion in the tribe by naming both a *Eupatorium altissimum* and an *Ageratum altissimum* from eastern North America, but the latter species actually belonged in the Linnaean *Eupatorium* with the closely related *E. aromaticum* L. *Ageratum altissimum* has been treated in *Eupatorium* under the various names, *E. ageratoides* L.f., *E. urticaefolium* Reich. and *E. rugosum* Houtt. Both *Ageratum altissima* and *Eupatorium aromaticum* are now placed in *Ageratina* and others of the original *Eupatorium* species of Linnaeus are now in the genera *Mikania*, *Conoclinium* and *Vernonia*.

By 1818 many presently recognized genera of the Eupatorieae had been described, including *Critonia* P. Browne (1756), *Kuhnia* Linnaeus (1763, =*Brickellia* nom. cons.), *Adenostemma* Forst. & Forst. (1776), *Liatris* Gaertn. ex Schreber (1791), *Piqueria* Cav. (1794), *Stevia* Cav. (1797), *Mikania* Willd. (1803), *Carphephorus* Cass. and *Sclerolepis* Cass. (1816b), *Isocarpha* R. Brown (1817) and *Alomia* H.B.K. (1818).

The Eupatorieae were recognized as a natural group by various authors including Cassini (1813) and Humboldt, Bonpland and Kunth (1818). It was formally established, along with most others in the family, by Cassini in 1819. The tribe and many of the genera were treated in detail in Cassini's contributions to the "Dictionnaire des Sciences naturelles" (1817–30) and in the Cassini "Opuscules" (1826–34). These writings might have been the foundation for further advances in the field but even before Cassini's death the family was subjected by Lessing (1832) to a reorganization outrageously lacking in perception. Cassini had placed the Adenostylieae and Tussilagineae between the Senecioneae and Eupatorieae, which was wrong, but Lessing did far worse in placing them in the Eupatorieae. Other genera included in the tribe on the basis of very superficial similarities were *Paleolaria* Cass. (=*Palafoxia*) which is Heliantheae and *Shawia* Forst. (=*Olearia*) which is Astereae. The members of the Eupatorieae were placed in four subtribes: Alomieae for genera lacking a pappus (*Alomia* and *Piqueria*), Agerateae for genera with a squamose, awned or clavate pappus (*Ageratum, Adenostemma*, etc.), Eupatorieae for genera with a capillary pappus, and Tussilagineae for genera with heterogamous often subdioicous heads. De Candolle (1836) was the last to summarize the whole tribe completely to species level, but most of the errors of Lessing were maintained.

In 1873 Bentham produced a classification of the family that survived with minor alterations for the next 100 years. The treatment of the tribe Eupatorieae was improved over that of Lessing by removal of such groups as the Tussilagineae and *Palafoxia*, but *Adenostyles* was retained. The system introduced a few new characters and distinguished three subtribes: Piquerieae, having anthers truncate apically and exappendiculate, and achenes 5-costate (including *Adenostemma* and *Gymnocoronis* with *Piqueria*, etc.); Agerateae, having anthers appendaged and achenes 5-costate; and Adenostyleae, having anthers appendaged and achenes 8-10-costate. The Bentham system was decidedly aphyletic and showed no insight into any peculiarities of the tribe, but it was perpetuated in its entirety by Hoffmann (1894). The system was modified in minor parts by B. L. Robinson (1913a) with some doubts regarding certain relationships. Four subtribes were recognized by Robinson and they were nearly as artificial as the three subtribes of Bentham and Hooker from which they were derived. The smallest subtribe, Adenostemmatinae, was new with Robinson and has proven to be with the exclusion of *Hartwrightia*, the most natural. The Kuhniinae was what remained of the Adenostyleae after the transfer of *Adenostyles* to the Senecioneae and, as noted by Gaiser (1954), four separate basic elements were involved. The Piqueriinae continued to be based on the reduced anther appendage but this character recurs in many different groups of the Eupatorieae. The residual subtribe, Ageratinae, of Robinson was unnatural to the extent that it was very diverse and that genera had been removed from it and placed in other artificial subtribes. With the exception of the Adenostemmatinae all the subtribes must be regarded as completely unnatural with only nomenclatural significance.

B. L. Robinson made no attempt to alter most of the artificial generic limits but various comments indicate his insight into some of the real relationships. In his discussion of the new species, *Eupatorium rivulorum*, Robinson (1926a) clearly indicated that the generic distinction between *Eupatorium* and *Fleischmannia* violated obvious relationships. In 1913 (b), for technical reasons, he placed in *Alomia* two species which he related actually to *Trichogonia*. With the publication of his key Robinson (1913a) stated, "Further study, especially of the larger genera, may well reveal profitable generic segregations not as yet clear. This is especially likely to be the case among the numerous and as yet very imperfectly known South American members of the tribe. It is also by no means improbable that when these are more satisfactorily represented in herbaria some new and more convincingly natural re-adjustment of generic lines will become possible." It seems to have been this feeling more than anything else that caused B. L. Robinson to maintain such distinctive-looking genera as *Symphyopappus* although he could furnish no workable technical difference.

At the species level the regional treatments of B. L. Robinson have continued to be the most useful for Mexico and western South America. Other useful works include Baker (1876), Barroso (1950, 1951, 1957, 1958), and Cabrera and Vittet (1963) for Brazil, Standley (1938) for Costa Rica, and Aristeguieta (1964) for Venezuela. A preliminary study has been published for Nuevo Galicia, Mexico by McVaugh (1972) and a generic key for the Eupatorieae of the area has been written by King and Robinson (1972jj). A complete treatment of the Eupatorieae of Panama using revised genus and species concepts has been prepared by King and Robinson (1976, 1975z).

Since the work of B. L. Robinson the number of genera in the Eupatorieae has continued to grow as a result of the works of Mattfeld (*Stylotrichium* and *Arrojadocharis* (as *Arrojadoa*) 1923; *Sciadocephala*, 1938), Barroso (*Praxeliopsis*, 1949), Blake (*Ferreyrella* and *Iltisia*, 1957), Cuatrecasas (*Ellenbergia*, 1964; *Ascidiogyne*, 1965), King (*Piqueriopsis*, 1965; *Carterothamnus*, 1967a, *Cronquistia*, 1968), Barroso and King (*Monogerion*, 1971) and King and Robinson (1967–76).

A few recent studies have made minor adjustments in the tribal limits. *Dyscritothamnus* B. L. Robinson has been transferred to the Senecioneae by Paray (1958) or to the Heliantheae by Turner and Powell (Chapter 25). *Adenostyles* has been firmly established in the Senecioneae (Toman *et al.*, 1968). *Microspermum* has been transferred from the Helenieae to the Eupatorieae, a position first suggested by Rzedowski (1970), and accepted by Turner and Powell (Chapter 25). *Isocarpha* and *Lepidesmia* which have been placed in the Heliantheae have been restored to the Eupatorieae (King and Robinson, 1970r), although not without disagreement (cf. Stuessy, Chapter 23).

Two major phyletic interpretations of the family have been presented in this century. Both maintain the basic Bentham and Hooker classification and both relegate the Eupatorieae to a peripheral position. Small (1917–19) placed the Senecioneae at the base of his system and suggested a relatively late derivation of the Eupatorieae from the Astereae in the Middle Miocene. Cronquist (1955), like Bentham, derived all tribes more or less directly from the Heliantheae and suggested the Eupatorieae were closest to the Vernonieae.

The present study reflects a more basic position of the Eupatorieae in the family. From all indications, the tribe is very isolated without true intermediates with other tribes. Such features as the expanded peripheral corollas of *Microspermum* are viewed as developments within a specialized group of the tribe having no direct relationship to members of the family with ray florets. The Heliantheae, however, do have some significant characters in common with the Eupatorieae which indicate that it is the most closely related of the other tribes.

Regarding the Compositae in general, one must regret that the work of Cassini was not fully appreciated in its time. The Bentham system, while an improvement on Lessing, should never have persisted as long as it has. Progress toward a phyletic system in the Asteraceae requires abandonment of the Bentham system, as has occurred recently, in part, in treatments of the Helenieae (see Chapter 25) Heliantheae (Chapter 23), Liabeae and Senecioneae (see Chapter 29). In the case of the Eupatorieae the resulting classification is complex, but this is the inevitable result of the previously unrecognized complexity and diversity of the tribe.

FEATURES OF THE EUPATORIEAE

Members of the tribe have a variety of habits but are mostly herbs or shrubs. A few species such as *Fleischmannia microstemon* (Cass.) K. & R., *F. sinclairii* (Benth. ex Oerst.) K. & R. and *Brickellia diffusa* (Vahl) Gray are annuals. Some species of *Critonia* are small trees while *Neomirandea panamensis* K. & R. is an arborescent perennial herb up to 13 m tall. Four genera (*Mikania, Tuberostyles, Neomirandea* and *Gongrostylus* have developed an epiphytic habit. In *Neomirandea* the more highly evolved members of the genus have reverted to deep humus substrates but they have extensive prop-root systems.

The anatomy of the stems has been reviewed by Carlquist (1965, 1966) who found no unique features in the vascular tissue of the tribe. Vessel elements were short and broad with large intervascular pits in the liana, *Mikania cordifolia* (L.f.) Willd. This species also had strands of axial parenchyma cells as long as wide. Narrow vessels and prominent growth rings are found in such xeric types as *Brickellia*. Most of the tribe had strands of 2–6 elongate cells. Solereder (1908) had suggested possible anomalous secondary growth in *Mikania*. Carlquist (1965) indicated that width of rays, presence or absence of procumbent cells in addition to erect ones in rays, and relative height of rays, notably short and narrow or wide and long vessel elements, might be useful in distinguishing species. The Eupatorieae can be distinguished from other tribes by resin ducts which occur between the vascular bundles rather than directly outside them (Col, 1904). The pith may be solid but is often fistulose. The condition is often stable within genera (*Matudina* versus *Eupatoriastrum*). Chambering of the pith is often prominent and the septae can persist in considerably enlarged stems (*Neomirandea*).

The tribe is one of the few in the Asteraceae that is basically opposite-leaved. Various isolated species have strictly alternate leaves (most of *Decachaeta*, one species of *Guevaria*) and most of the genera centering about *Liatris*. Among opposite-leaved genera the upper leaves and bracts of the inflorescence often become alternate. Almost all the upper leaves of species such as *Neohintonia monantha* (Sch.-Bip.) K. & R. are alternate with opposite leaves at the base. The almost strictly opposite condition in *Chromolaena* contrasts with the alternate inflorescence branches of such genera as *Austroeupatorium*. The inflorescence branching is variable between species in some genera such as *Heterocondylus*.

Whorled leaves are present in isolated members of the tribe. Direct assumption of the whorled condition is seen in such species as *Neomirandea costaricensis* K. & R. The whorls of *Eupatoriadelphus* and one species of *Eupatorium* are related to the nearly sessile tripartite leaves of some other species of *Eupatorium*.

Prominent in the *Liatris* relationship is the basal rosette of leaves. Rosettes are found to a lesser extent in other isolated genera such as *Eupatorina, Ciceronia, Antillia* and *Gyptis*.

In basic structure the petioles are very narrowly winged in all members of the tribe and grooved on the upper surface. This basic structure is not evident, however, and most members of the tribe would be described as unwinged. Long petioles are particularly notable in members of the *Hebeclinium* complex, especially in *Bartlettina*. In marked contrast, a sessile leaf or broadly winged petiole is rather characteristic of the *Ayapana–Isocarpha* group. Most groups of the tribe are more variable in their petiolar development.

Leaf blades vary from entire to strongly dissected with a predominance of simple ovate or elliptical types. Dissection may involve broad lobes (*Carterothamnus*, some *Hofmeisteria*, one *Lomatozoma*), linear lobes (*Acanthostyles* and some *Eupatorium*) or bipinnatifid blades (*Gyptis pinnatifida* Cass.). The most extensively dissected lobes in the tribe occur in the calcicolous *Eupatorina* of Hispaniola. The blade varies from finely serrate to bipinnatifid in *Grazielia gaudichaudiana* (DC.) K. & R. (Malme, 1933, p. 42, Fig. 7). Venation of the blades may be palmate or pinnate but there is a strong tendency for a trinervate condition from near the base. The venation varies within many genera. Laticifers are found only along the veins in most members of the tribe, but they form distinct vesicles in the areoles of the leaf in the genus *Critonia*. In some members of the genus these become prominent translucent or even lens-like spots.

The inflorescence of the Eupatorieae is usually a flat-topped or dome-shaped corymbose panicle. Central heads may mature slightly after the peripheral ones but more often they mature first. The cymose sequence is often reflected in the structure of the branching and is either restricted to the branches as in *Condylidium* or may extend to the whole inflorescence as in *Ageratum*. Heads in extreme cases may be sessile in rounded clusters (*Neohintonia*, *Mexianthus*, *Sphaereupatorium*) or may be single on long peduncles (*Hofmeisteria*, and *Brickellia monocephala* B. L. Robins.). The heads are borne in a spicate manner in most species of *Liatris*. Major groups of *Mikania* have been classically distinguished by their corymbose, thyrsoid and racemose, or spicate inflorescence branches.

The involucre of the Eupatorieae is traditionally characterized as imbricate, subimbricate or eximbricate. The imbricate form is characteristic of what has been called the "Cylindrocephalae" with the cylindrical involucre of strongly overlapping bracts in many series of gradually increasing lengths. The character is most evident in dry plants because of the failure of the bracts to spread in that condition. In living condition many other members of the tribe show the same cylindrical appearance of the involucre. Within the imbricate group is a truly natural element in which all the involucral bracts are completely deciduous. The relationship includes a few genera such as *Praxelis* which are not in the traditional Cylindrocephalae. The subimbricate involucre has bracts progressively longer in overlapping series but at least the outer bracts spread at maturity and do not fall. The eximbricate form has bracts of subequal length which spread at maturity and at most only a few inner ones are deciduous. The involucral types may vary considerably within genera, although general trends may be evident through whole groups at the subtribal level. The involucre of the Adenostemmatinae is distinct from all others by the indistinct nonarticulate bases of the bracts and by partial basal fusion in many of the members.

The receptacle varies in shape and cellular differentiation, the most distinctive form occurring in the Adenostemmatinae where the tissue between the individual achene scars is not sclerified and where the scars are able to shift in position as the receptacle matures. In most other Eupatorieae the surface between the achene scars is completely sclerified and often forms distinct ridges or short spines. In the extreme opposite condition, most species of *Hebeclinium* have the receptacle sclerified throughout and have no central pith. Receptacles of many Eupatorieae have paleae arising from junctures between the flowers. The paleaceous groups are rarely closely related (*Eupatoriastrum*, one species of *Gyptidium*, several species of *Chromolaena*, *Sphaereupatorium*, some *Ageratum*, *Jaliscoa*, *Blakeanthus*, *Ferreyrella*, *Isocarpha*). Such forms seem to represent an erratic capacity in the Eupatorieae for revival of this suppressed structure. The paleae of two genera, *Oaxacania* and *Carterothamnus*, seem to represent a very different form where each palea is paired with a flower and when pulled from the head carries a flower with it. This seems most likely to be the primitive condition in the tribe. Receptacles of many Eupatorieae have hairs or small chaff. These are more common, but not entirely consistent, throughout the *Hebeclinium–Bartlettina* complex. Some other comparatively unrelated genera such as *Urolepis* and *Polyanthina* also have prominent hairs. In *Neomirandea* and *Neocabreria* the presence or absence of receptacle hairs is correlated with hairs on the inner surface of the corolla. The receptacle shape is usually flat or slightly convex. A low-conical shape occurs in many relatives of *Gyptis* and *Ageratum* and in one species of *Aristeguietia*. A high-conical to columnar shape occurs in *Praxelis*, *Eupatoriopsis* and *Isocarpha*.

The heads of the Eupatorieae may contain from one to many hundred flowers. The genera having clusters of single-flowered heads (*Mexianthus* and *Neohintonia*) are closely related to each other but still retain many differences. Other genera with small numbers of flowers may have the number fixed, and in *Stevia* with five, *Mikania* with four and *Piqueria* with three to five, the number is identical to the number of involucral bracts. The highest number with such floret-stability seems to be 20 which occurs in six of the seven species of *Lourteigia*. The extremely high flower numbers occur in various distantly related groups such as *Polyanthina*, *Hofmeisteria* and *Eupatoriastrum*.

Corollas are whitish, reddish or bluish but, so far as known, never truly yellowish in the Eupatorieae. All flowers are perfect and fertile. True rays are lacking but peripheral flowers may be differentiated. The peripheral corollas are slightly zygomorphic in *Bartlettina tuerckheimii* (Klatt) K. & R. and *Ferreyrella peruviana* Blake and the outer lobes are greatly expanded in *Praxeliopsis* and *Microspermum*. The peripheral flowers are differentiated by lack of a pappus in some species of *Trichogonia* and *Fleischmannia*. B. L. Robinson (1930d) has shown that there is a definite sequence in the five involucral bracts and five flowers in the heads of *Stevia* and that this is reflected in heterocarpy. He noted the usual pattern of heterocarpy where the idiocarp with a partly or completely reduced pappus occurs in the flower subtended by the outermost bract. The remaining adelphocarps are usually alike, but sometimes differences occur in the second or third achene following a 2/5 sequence around the circle of flowers.

Corollas in the tribe are funnelform to tubular with or without a distinct basal tube. There are five lobes except in the minute species of the monotypic genera *Iltisia* and *Piqueriopsis* which have four lobes. The lobes are characteristically short with exceptions of deeply cut narrow lobes in some species of *Mikania*, *Neomirandea* and *Steyermarkina*; they are usually without stomata but these occur in some species of *Stomatanthes* of Africa and South America and in the four species of *Eupatoriadelphus* of North America. The cells of most corollas are narrow with obvious sinuous walls, but some species of *Mikania* and most species of *Neomirandea* have large quadrate cells with mostly straight walls. The species with such cells seem to be restricted to moist and often epiphytic habitats. Corolla cells never contain obvious druses or other complex raphides but rarely very simple crystals are seen. The inside surface of the corolla is usually glabrous but *Stevia, Steyermarkina, Neocabreria,* some *Neomirandea,* some *Hebeclinium* and various others have hairs. Hairs on the inner surface in *Eitenia* are along the veins. In *Cronquistianthus korthalsianus* (B. L. Robins.) K. & R. the inner surface of the corolla near the base of the anther has a pair of crests formed by invaginations of the inner surface.

The filaments are usually inserted above a slightly to strongly narrowed tube. In *Urbananthus* the insertion is very near the base. In *Praxeliopsis* the filaments are reduced to short collars inserted just below the bases of the spreading lobes. In *Polyanthina* the corollas are so narrow that the insertions of the filaments are staggered at various levels.

The lower part of the filament is usually elongate and smooth. In various genera such as *Phania, Praxeliopsis* or some *Cronquistianthus* the collar is mounted almost directly on the corolla. In a few genera such as *Piqueria* and *Ellenbergia* the lower filament is papillose or pubescent.

The anther collar shows more variation in the Eupatorieae than in any other tribe. Cells in the lower part are usually short and those of the upper elongate; the walls are usually thin or show slight nodular thickenings. In *Ageratina, Hebeclinium* and *Neomirandea* and various relatives the collars are elongate with great numbers of quadrate cells and there are usually few thickenings in the walls (Fig. 1). *Fleischmannia* shows the opposite extreme with few or no quadrate cells and with very prominent annular thickenings (Fig. 4). In *Chromolaena* and its relatives, the lower part of the collar is usually enlarged with very short cells where the direction of the thickenings is often oblique or vertical (Figs 2, 3). The pattern of thickenings of the collar cells in its extreme forms is taxonomically useful. members of the *Gyptis, Fleischmannia* and *Piqueria* groups having the strongly annulate thickenings and relatives of *Ageratina* and *Hebeclinium* having almost none. In the *Critonia* group there is variation in some genera. Prominent annular thickenings in the collar are known in a few genera outside of the Eupatorieae such as *Adenocaulon* of the Mutiseae and some members of the Liabeae.

The stamens remain included in the corolla in most members of the tribe but are characteristically exserted in *Mikania* and are partially exposed in the deeply lobed species of *Neomirandea* and *Steyermarkina.*

Exothecial cells of the tribe are almost all subquadrate with thickenings evenly distributed on horizontal and vertical walls (Fig. 5). The cells tend to be more elongate in a few genera such as *Ageratina. Carphochaete* and *Hofmeisteria* (Fig. 6) where some of the horizontal walls have fewer thickenings. Only the Vernonieae and a few species of Liabeae, among other

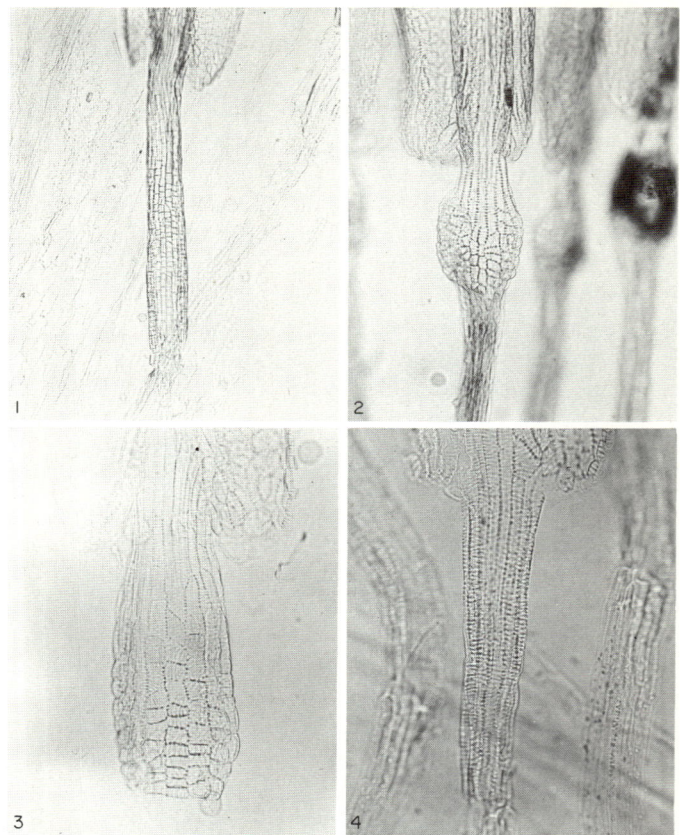

FIGS 1–4. Anther collars. Eupatorieae. 1. *Bartlettina breedlovei* K & R., × 140 2. *Praxelis asperulacea* (Baker) K. &. R., × 220. 3. *Chromolaena chasei* (B. L. Robinson) K. &. R., × 250. 4. *Fleischmannia gentryi* K. & R., × 270. Photos by V. Krantz of U. S. National Museum of Natural History and A. Guevara.

tribes, have a similar distribution of thickenings on the exothecial cells and in these there are subtle differences by which they can be distinguished from the Eupatorieae.

The anther appendage is flat with two layers of cells (Figs 9–11). It is often hollow in the mature state as the internal structure disappears. A pair of flanges on the inner surface may occur as slight continuations of the inner valves of the thecae and between these there is sometimes a prominent groove that may end in a distinct apical emargination. In some genera such as *Diacranthera* the appendage may be divided into two completely separate parts. The appendage is usually oblong-ovate but seems to show strong predisposition to reduction in the tribe. The appendage is shortened or lacking in such genera as *Adenostemma* of the Adenostemmatinae, *Ciceronia*, *Ophryosporus* and *Koanophyllon* of the *Critonia* group, *Praxeliopsis* of the *Praxelis* group, *Decachaeta* of the *Hebeclinium* group, *Ageratella* of the *Brickellia* group, *Gongrostylus* of the *Ayapana* group, *Diacranthera* of the *Gyptis* group, many genera such as *Phania*, *Ellenbergia*, *Piqueria*, *Phalacraea* and *Ascidiogyne* (Fig. 12) of the *Piqueria* group, and two species of the isolated genus *Hofmeisteria*. The appendage is unusually indurated in one species of *Microspermum* and unusually lobed and narrowed in some *Praxelis*. The appendage is characteristically rather obovate with a crenulate distal margin in *Stevia* and *Metastevia*.

The nectary surrounds the style base and in most genera of the Eupatorieae the style is fully immersed. There is some elevation of the style in

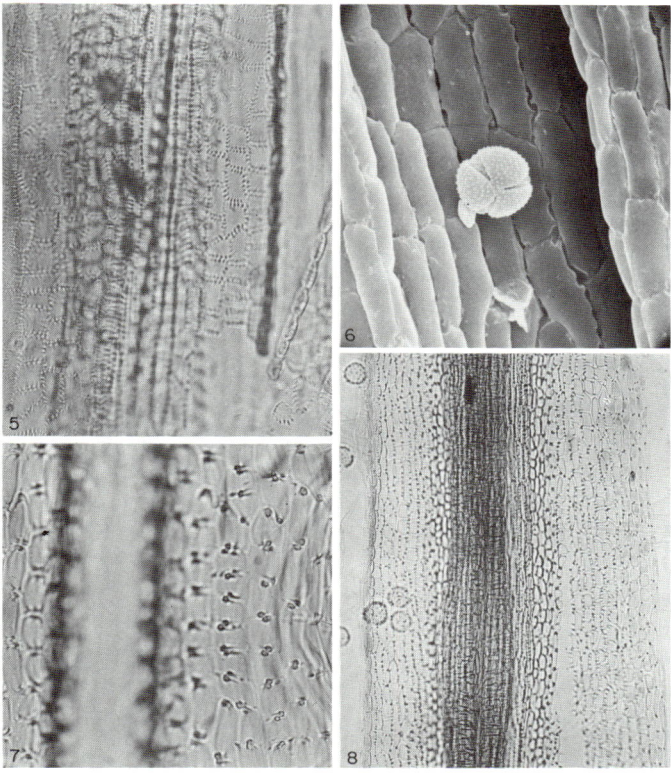

FIGS 5–8. Exothecial cells. 5–6. Eupatorieae. 5. *Praxelis capillaris* Sch. Bip., × 400. 6. *Hofmeisteria schaffneri* (A. Gray) K. & R., × 900. 7. Heliantheae, *Palafoxia arida* Turner, × 500. 8. Senecioneae, *Pittocaulon velatum* (Greenm.) R. & B., × 150. SEM photo (6) by Walter Brown of U. S. National Museum of Natural History.

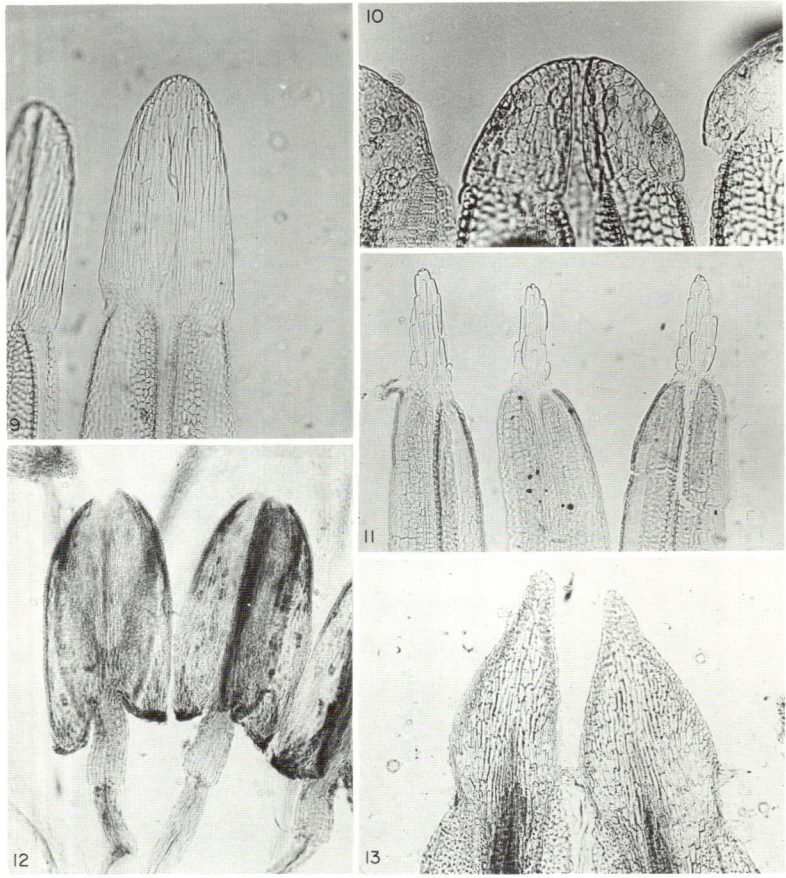

FIGS 9–13. Anther appendages. 9–12. Eupatorieae. 9. *Brickellia peninsularis* Rose & Standley, × 160. 10. *Eupatoriastrum nelsonii* Greenm., × 250. 11. *Praxelis capillaris* Sch. Bip., × 140. 12. *Ascidiogyne wurdackii* Cuatr., × 100. 13. Heliantheae, *Palafoxia integrifolia* (Nutt.) Torr. & A. Gray, × 100.

a few genera such as *Praxeliopsis*, *Ageratina* and *Carphochaete*. This contrasts with such tribes as the Heliantheae where the style is never completely immersed, or the Astereae or Senecioneae where the style is seated on top of the nectary. The nectary has stomates in all cases, but in *Sciadocephala amazonica* K. & R. there are hairs at the tip of the nectary, a feature not reported for any other species in the family.

The style furnishes the most important distinguishing characters of the tribe. Except for *Arnica cordifolia* Hook., we have found no pubescence on the style base in any other tribe. Also, in other tribes enlargements in the style base are usually not taxonomically reliable. In the Eupatorieae the basal nodes are often sharply defined and usually constant within genera or groups of genera. Hairs on the style base are characteristic of some groups such as *Eupatorium* and its relatives and they occur in many of the *Brickellia* group where the character is constant for genera. It varies within some genera such as *Ayapanopsis, Heterocondylus* and *Isocarpha* in the *Ayapana* group. The occurrence of hairs on the style is widely distributed in the tribe and the potential seems basic to the tribe. The hairs occur on the shaft of the style in some *Adenostemma* species and in *Stylotrichium* and both hairs and glands occur on the style shaft in *Sartorina*, which is a relative of *Fleischmannia*.

The style branches have stigmatic lines in two rows usually restricted to the lower half (Cassini, 1813; Chamberlain, 1891). The stigmatic lines are close together on the inner surface and reach near the tip in one genus, *Carphochaete*. In most Eupatorieae the lines are widely separated and in many genera from various groups in the tribe there are glands on the inner surface between the lines, or above. Glands are not known on the inner surface of the style branches in any other tribe. The style appendage is quite short in *Brickellia diffusa* (Vahl) Gray and in most species of *Ayapana* and *Isocarpha*. The reduced appendages of the latter two has led to some confusion with those in the Heliantheae. In all other Eupatorieae the appendage is at least as long as the stigmatic region and in some it is strongly pigmented lending much of the visible color of the inflorescence. Styles in the Eupatorieae are often at least as prominent as rays of some Astereae and they apparently function in a manner equivalent to colored rays or bracts.

The achenes of the Eupatorieae are usually prismatic with five ribs. In *Brickellia, Liatris* and the *Kanimia* part of *Mikania* there are up to 10 ribs but these genera do not form a related group as treated by Bentham (1873). The genera with flattened, mostly two-ribbed achenes such as *Macvaughiella, Oaxacania* and *Eupatoriopsis* are also of diverse relationships in the tribe. The genus *Lourteigia* is unique in the tribe by the extreme constriction of the achene below the pappus. The surface of the achene may have biseriate hairs, capitate glands, various combinations of these structures, or may be glabrous. The characteristically biseriate hairs may be parted from the base or even uniseriate in members of the *Liatris* group.

Internally the walls of the achenes lack crystals, which are prominent in many other tribes of the family. The walls have carbonized resin-like deposits in the mature state in most genera of the Eupatorieae and the deposits are formed around minute projections which appear as small pores (Figs 14-16). The punctations that result are usually randomly distributed but are often in transverse rows in members of the *Gyptis* and *Disynaphia* groups (Fig. 15). The punctations are unusually sparse in *Piqueria* (Fig. 16) and are totally lacking in some genera such as *Sartorina*.

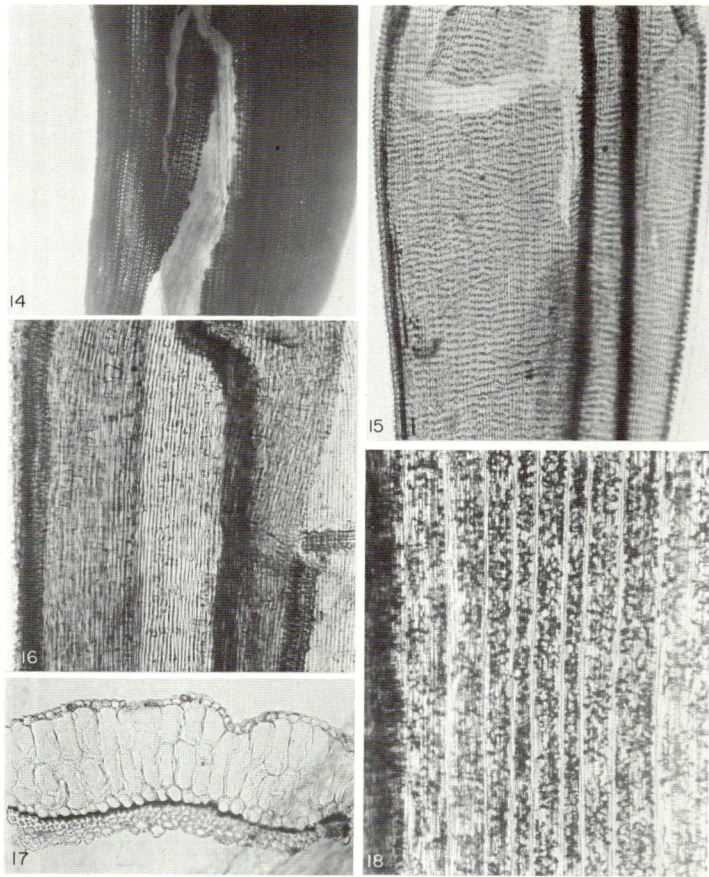

FIGS 14–18. Achene walls. 14–17. Eupatorieae. 14. *Ageratina aschenborniana* (Schauer) K. & R., ×175. 15. *Disynaphia radula* (Chod.) K. & R., ×115. 16. *Piqueria laxiflora* B. L. Robinson & Seaton, ×150. 17. *Tuberostyles axillaris* Blake, cross-section, ×220. 18. Heliantheae, *Palafoxia arida* Turner, ×160.

The only other tribe with such carbonized deposits associated with pitting is the Heliantheae (including most Helenieae and some Senecioneae of other systems). Most of these latter can be distinguished by the presence of numerous clear longitudinal striations in the wall (Fig. 18). The outer layers of the achene walls expand at maturity to form a rind in the genus *Tuberostyles* (Fig. 17) and in some species of *Mikania*. In *Ascidiogyne* the outer layer is expanded to form a large fluid-filled sac around the achene.

The carpopodium represents, in part, the abscission zone at the base of the achene and it has a great variety of structure in the Eupatorieae (Figs 19–26). It is a particularly useful taxonomic character as shown in the study of *Hofmeisteria* and associated genera by King and Robinson (1966). Many genera have distorted bases, such as *Piqueria* (Fig. 24), *Ageratum, Alomia, Flyriella, Condylidium* and *Cronquistianthus*, but in the first two genera the character varies between species. *Guayania, Brickellia* and *Praxelis* are examples with the opening turned to one side. *Eupatorium Conoclinium* and *Sartorina* have the carpopodium poorly differentiated or undifferentiated. Cell shapes produce elongate surface patterns in *Pleurocoronis* (Fig. 19) or many series of short cells as in *Hofmeisteria* (Fig. 26). The cell walls may be thin as in *Ageratina* (Fig. 22) or thick as in *Fleischmannia* (Fig. 25). In *Ayapana, Polyanthina* and *Ayapanopsis* (Fig. 21) the lowermost row of cells is much enlarged and differentiated. The cells

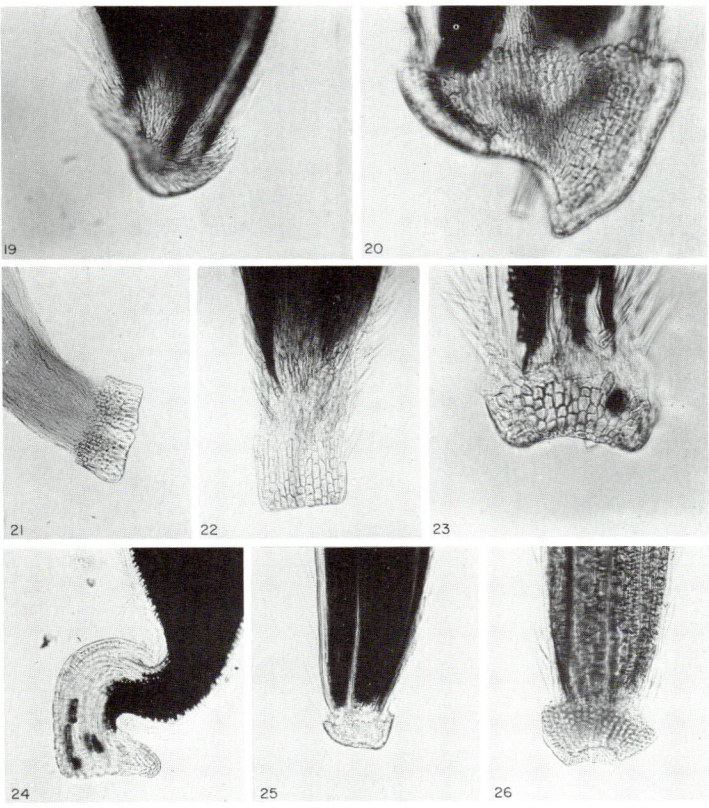

FIGS 19–26. Carpopodia, Eupatorieae. 19. *Pleurocoronis pluriseta* (A. Gray) K. & R., ×140. 20. *Phanerostylis coahuilensis* (A. Gray) K. & R., ×250. 21. *Ayapanopsis vargasii* K. & R., ×100. 22. *Ageratina aschenborniana* (Schauer) K. & R., ×160. 23. *Trichogonia villosa* (Spreng.) Sch. Bip., ×250. 24. *Piqueria trinervia* Cav., ×130. 25. *Fleischmannia guatemalensis* K. & R., ×130. 26. *Hofmeisteria schaffneri* (A. Gray) K. & R., ×165.

above the carpopodium may collapse at maturity leaving a marked upper edge on the carpopodium as in *Fleischmannia* and *Campuloclinium* or the carpopodium may be structurally continuous with the ribs of the achene as in *Hebeclinium* and *Bartlettina*.

The ovule in the Eupatorieae has a single lobe at the lower end. The surface cells of the lobe are usually sclerified, but the ovule is unsclerified in some genera such as *Ayapana* and *Isocarpha*.

The pappus of the Eupatorieae is usually formed of many capillary bristles in one series. The pappus elements may be reduced to 10 or five setae, as in *Hofmeisteria* or *Fleischmannia*. The pappus is reduced to five scales, a crown or may be completely absent in *Ageratum*. Individual species of some genera such as *Koanophyllon* and *Spaniopappus* may have the pappus reduced or lacking. Pappus elements of *Pleurocoronis* and *Malperia* are differentiated with scales above the flat surfaces of the achene and awns above the ribs. There is a tendency for pappus setae to be congested and to form a partial second row in some genera such as *Mikania* and *Aristeguietia*. In the tribe only one species of *Ascanthus* has the pappus characteristically multiseriate with over a hundred setae.

The outer surfaces of the pappus setae are sometimes flattened as in *Brickellia* and some closely related genera. *Carminatia*, *Liatris*, some species of *Brickellia* and a few species of *Helogyne* have the pappus plumose. In *Helogyne* the setae of some species have vascular traces reaching about half the length of the setae. The tips of the pappus setae are usually narrow with sharp apical cells. In various members of the *Eupatorium*, *Disynaphia* and *Gyptis* groups the apical cells are rounded to bulbous. The tips of the pappus setae are extremely enlarged in the genus *Amboroa*.

Pollen in the Eupatorieae is echinate, becoming almost smooth in one genus, *Hofmeisteria* (Fig. 6). Some members of the tribe have notably short spines and seem to be primarily wind-pollinated (Grashoff and Beaman, 1970; Sullivan, 1975). Pollen size on the average is perhaps smaller than that of any other tribe. In the Eupatorieae the pollen of most genera is 18–22 μm in diameter. A number of unrelated genera such as *Phalacraea* and *Aristeguietia* have pollen nearer 25 μm. Size may vary between related species in the Adenostemmatinae and is as much as 25 μm in diameter in some *Sciadocephala*. In one of the distinctive genus-pairs in the tribe, pollen of 13–17 μm diameter has been recorded in *Carterothamnus* and of 25 μm diameter in the related *Oaxacania* (King and Robinson, 1970e). The largest pollen in the tribe, reaching *c.* 35 μm, is found in *Carphochaete*.

CYTOLOGY

Numerous chromosome counts have accumulated during the last 50 years and the most notable efforts at synthesis have been the papers by Gaiser

(1949–54) and Grant (1953). A detailed review and discussion are being published separately (King, et al., 1977), but a general accounting of these is included in Table I of genera (p. 460) and the over-all pattern may be summarized as follows.

The majority of the groups in the Eupatorieae, including the Adenostemmatinae, *Eupatorium*, *Disynaphia*, *Gyptis*, *Ayapana*, *Fleischmannia*, *Liatris*, *Critonia* and large parts of the *Brickellia* and *Piqueria* groups have a chromosome base number of $x=10$. Variations from this base include $n=5$ in a single species of *Adenostemma*, $n=4$ in two species of *Fleischmannia*, $n=9$ in most of the advanced members of the *Brickellia* group, $n=11$ and 12 in many specialized members of the *Piqueria* group and indications of $n=17$ or 18 in members of the genus *Ayapana*. There are other groups in which the base $x=10$ is rare or unknown and these show little relation structurally to any of the preceding except perhaps the *Critonia* group. The base $x=16$ occurs in many genera of the *Hebeclinium* group and $n=16$ and 17 are characteristic of the *Ageratina* group. *Mikania* has mostly $n=18$ and 19, while *Neomirandea* has two subgenera, one with $n=17$ and the other with $n=20$ and $c.\ 25$. All the latter groups share other characters besides their higher chromosome numbers and they may represent a related though very diverse series.

DISTRIBUTION OF CHARACTERS

Intermittent recurrence of many characters is evident among various members of the Eupatorieae. There is evidence of at least two of the mechanisms that are involved in such characters.

In the genera *Eupatorium* and *Austroeupatorium* the corollas and achenes characteristically have glands without the hairs or setae found in most other genera of the tribe. In a few specimens an isolated hair or seta is present, indicating that the genetic potential is not absent but only suppressed. The distribution of many other characters in the tribe such as the pubescent style bases also suggests that their absence is the result of suppression rather than loss of genetic potential.

The two genera *Koanophyllon* and *Chromolaena* are markedly different in a series of corolla, anther, and style characters. Many of the characters of *Koanophyllon* are shared by related genera concentrated in Mexico and Central America while many characters of *Chromolaena* are shared by its relatives which are concentrated in Brazil. However, in some West Indian members of the tribe the characters of these two complexes are intermixed in ways that could only be explained by old or ancient hybridization since there is no evidence of recent occurrences.

Another suspected intergeneric hybrid situation is represented by the type and only known collection of *Eupatorium atrescens* B. L. Robins. of Ecuador. In this taxon glands occur on the tips of otherwise normal setae of the achene and a number of other combinations of *Aristeguietia* and *Badilloa* characters occur. These two examples of putative hybridization in the tribe are between genera that are in or close to the *Critonia* group, but other more distant genera may occasionally hybridize.

The present treatment recognizes 160 genera in the tribe.

SUBDIVISIONS OF THE TRIBE

This study of the Eupatorieae is incomplete, especially in reorganization at the subtribal level. While presenting what is known about the tribe at the present time it has been necessary to avoid forcing premature concepts. No new formal subtribes are established and some groups may not ultimately be given subtribal status. Some groups of genera recognized here have been dimly perceived since early in our joint study, but the present refined concepts are mostly derived from our recent effort to correlate the anatomical and morphological studies with the cytology of the tribe. There are about 100 described species not yet assigned to genera and there seem to be an unusually large number of undescribed species, but these are not expected to alter significantly the over-all concepts of the tribe as we preview them.

Adenostemmatinae

The group is one of the most isolated in the tribe and can be distinguished by receptacles with soft tissue separating the areolae, and by the non-articulated bases of the involucral bracts. The pappus of glanduliferous knobs distinguishes two of the genera from all other Eupatorieae but the third genus, *Gymnocoronis*, has no pappus.

The genus *Adenostemma* had not received any real taxonomic treatment since the superficial summary by De Candolle in 1836. Grierson (1972) in his study of Ceylon material gave an initial indication of the need for narrower interpretation of some species, in particular *A. lavenia* (L.) O. Kuntze. King and Robinson (1974l) clarify species concepts, especially in the South American members of the genus. The study reveals three Andean species having elongate parallel non-overlapping distributions that might be the result of distribution by animals. The sticky knobs of the pappus seem particularly suited for animal dispersal and these may have been a factor in the extensive representation of the genus in the Eastern Hemisphere.

Eupatoriinae

This typical element of the tribe is characterized by subimbricate involucral bracts and by pubescent style bases. Some of the species in various genera have apical cells of the pappus rounded or bulbous at the tip. The corolla lobes are smooth and sometimes have stomata. The three southern genera of the group have distinct annular thickenings in the cells of the anther collar, the two northern genera lack these thickenings.

The group is mostly limited to eastern North America and eastern South America and has probably never occurred extensively in the western parts of these continents. The two natural migrations into the Eastern Hemisphere are notable, one by *Eupatorium* apparently through Alaska reaching temperate Asia and Europe (King and Robinson, 1970u), and the other by *Stomatanthes* across the South Atlantic giving rise to three species in Africa (King and Robinson, 1975y). In recent times *Austroeupatorium inulaefolium* (H.B.K.) K. & R. has become widely adventive in Indomalaya and Ceylon.

Disynaphia group

The group is characterized by subimbricate involucral bracts and by a glabrous unenlarged style base. The number of flowers per head is 5 in all but one species where it is 10. The anther collars are strongly annulated and some species have enlarged round-tipped apical cells on the pappus setae. The corolla lobes are smooth.

The group is restricted geographically to eastern South America in the region from the hump of Brazil south to Uruguay. The generic limits in the group have been summarized by King and Robinson (1971s).

Gyptis group

This diverse group is most notable for the eximbricate to weakly subimbricate involucral bracts, the heads with more than five flowers and the pappus of many capillary or sometimes plumose bristles. The pappus is rarely short as in *Stylotrichium* or lacking as in a few species of *Trichogonia*. The anther collar has prominent annular thickenings, the apical cells of the pappus sometimes have rounded or bulbous tips, the corolla lobes are often papillose and the style bases are, with few marked exceptions, glabrous and unenlarged. Many of the genera have a characteristic low-conical receptacle of a type seen elsewhere in the tribe only in the *Ageratum–Phalacraea* group and in one species of *Aristeguietia* of the *Critonia* group.

Of the 20 genera listed for the *Gyptis* group, 17 are concentrated in the eastern parts of South America. Two additional genera are concentrated in eastern North America and eastern Mexico. One genus is endemic to the eastern edge of the Andes in Peru and Bolivia. The *Gyptis* group shares the general geographic distribution and the tendency for blunt or bulbous-tipped pappus setae with the two preceding groups. Of the three, the *Gyptis* group is the largest and most diverse.

Acritopappus group

The genera *Acritopappus* and *Radlkoferotoma* have most of the characters of the *Gyptis* group, but they have a reduced pappus and a more subimbricate involucre. The two genera are from adjacent parts of eastern South America and are probably closely related.

Piqueriinae

A number of genera with reduced pappus can be placed together in a perhaps artificial group having eximbricate involucral bracts, the corolla lobes and style branches papillose, strongly annulate cells of the anther collars and sometimes a low-conical receptacle. Most of the characters indicate relation to the *Gyptis* group and the reduced pappus alone would not be particularly significant, but the distinction is reinforced by the predominantly western South American and Central American distribution of almost all the genera. There are tendencies for reduced anther appendages in the group and the style bases are glabrous except in a few species of *Stevia*.

The two species, *Ageratum conyzoides* L. and *A. houstonianum* Mill. are widely distributed in tropical and subtropical areas of the World and are commonly adventive. *Ageratum houstonianum* is notable for its larger corollas and broader style branches which usually have a prominent blue color.

Trichocoronis group

Trichocoronis, *Shinnersia* and *Sclerolepis* are aquatic or semiaquatic and show sessile leaves (whorled in *Sclerolepis*), eximbricate or weakly subimbricate involucral bracts, reduced pappus, papillose corolla lobes, unenlarged glabrous style bases and linear papillose style branches. The three genera have most of these features in common with the *Ageratum* group but the anther collars have few or no annulate thickenings in the cell walls. The anther collars, the more aquatic nature, and the North American distribution would suggest that the three genera are related in spite of many differences in detailed structure.

Ayapana group

The relatives of *Ayapana* are notable for the subimbricate persistent nature of their involucre, the usually slender style branches, the smooth corolla lobes, the usually annulated anther collars, and the enlarged sometimes hirsute base of the style. A number of the most highly developed members have the basal row of cells in the carpopodium prominently enlarged. The combination of a reduced pappus, presence of paleae, and a reduced style appendage in *Isocarpha* and in one species of *Ayapana* has led to their placement in the Heliantheae by some authors (e.g. Steussy, Chapter 23).

The group is primarily South American with a few widely distributed species such as *Ayapana amygdalina* (Lam.) K. & R., *A. triplinervis* (Vahl) K. & R., *Condylidium iresinoides* (H.B.K.) K. & R. and *Isocarpha oppositifolia* (L.) R.Br.

Alomiinae

The relatives of *Brickellia* share many of the characters of the allied *Ayapana* group, the subimbricate persistent involucre, the usually smooth corolla lobes, the usually annulated cells of the anther collar, and the often enlarged hirsute base of the style. Almost all members of the group have a characteristic tubular corolla constricted above to scarcely allow passage of the broadly clavate style branches. One exception to the broad style branches is the most widely distributed species in the group. *Brickellia diffusa* (Vahl) Gray. A conventional capillary pappus occurs in most of the genera, but the setae are flattened externally in *Brickella*, *Brickelliastrum* and *Barroetia*, they are plumose in *Carminatia* and parts of *Brickellia* and *Helogyne*, the pappus is squamate in *Ageratella* and *Pleurocoronis*, the setae are of mixed lengths in *Dissothrix* and deciduous in *Leptoclinium*, and pappus is lacking in *Alomia* and *Planaltoa*. The ten ribbed achenes of *Brickellia* have been the basis of its distinction from *Eupatorium* but most members of the group as delimited here have five ribs.

The *Brickellia* group is widely distributed through the range of the tribe in the Western Hemisphere, but seems less developed in western South America where it is represented by *Helogyne* and is almost completely lacking in the West Indies.

Liatris group

The *Liatris* group has usually subimbricate persistent involucral bracts, linear style branches, smooth or scarcely papillose corolla lobes, and an unenlarged glabrous style base. The hairs of the achene are usually of a distinctive form with the normally biseriate structure separated from near the base and sometimes completely uniseriate. Gaiser (1954) noted the alternate leaves of this group, which characteristically form a basal rosette at least in the initial stages of their development. The pappus is usually capillary or plumose and the achene is mostly 10-ribbed, a combination of characters that has led to an erroneous association with *Brickellia*.

The anomalous genus *Hartwrightia* of Florida has only glands on the achene, has a reduced anther appendage and has no pappus, but the habit, alternate leaves with a basal rosette, and the geographical distribution suggest relation to the *Liatris* group.

The group is distributed in eastern North America with a concentration in the southeastern United States.

Fleischmannia group

Fleischmannia and *Sartorina* seem isolated in the tribe. They have a combination of subtle characters that result in a distinctive appearance. The involucre is usually strongly subimbricate with persistent bracts, the corolla has a short basal tube, the corolla lobes have both surfaces papillose by the projecting upper ends of the cells, the anther collars are narrow, the collar cells are elongate with strongly annulated walls and the style base is usually not enlarged. *Fleischmannia* has distinctive carpopodia with projecting upper margins and thickened walls. *Sartorina* has distinctive terete achenes and has scattered glands and hairs on the shaft of the style. The pappus in the group is capillary but sometimes with a reduced number of setae, five in *Fleischmannia arguta* (H.B.K.) B. L. Robins. and 10 in *F. capillipes* (Benth. ex Oerst.) K. & R.

The group is concentrated in Mexico and Central America and in the western parts of South America. One species, *Fleischmannia microstemon* (Cass.) K. & R., is weedy and has become widely adventive.

Critonia group

Critonia and the numerous associated genera seem unspecialized in most of their basic features. The pappus is usually capillary, rarely squamate or lacking, the corolla lobes are smooth, and the base of the style lacks hairs or enlargements. The anther collars vary, sometimes showing differences of cell shape and annulation in a single genus. The group is most notable for the imbricate or strongly subimbricate usually partially deciduous involucral bracts. Similar types of involucre are found in the related *Praxelis*, *Hebeclinium* and *Neomirandea* groups. Most members of the *Critonia* group have style tips that are flat and somewhat broadened. Two genera of the group, *Mexianthus* and *Neohintonia*, have single-flowered heads borne in globose clusters.

The genus *Tuberostyles* has a specialized epiphytic habit in the mangrove areas of eastern Panama and western Colombia and further resembles some species of *Mikania* by the cortication of the mature achenes. In other characters, *Tuberostyles* proves thoroughly distinct from other epiphytic members of the tribe. The anther collars of *Tuberostyles* are broad and have mostly elongate cells with slight but distinct annular thickenings in the walls.

The *Critonia* group is concentrated in the Mexican, Central American and West Indian region south into the Andes. There are comparatively few members of the group in Brazil.

Praxelis group

Praxelis and its relatives form a distinctive group having imbricate or subimbricate bracts that are all deciduous at maturity, corolla lobes that are usually papillose on the whole inner surface and anther collars that are usually broader below with prominent annulations on the cell walls. Some of the genera have fewer or shorter pappus setae, have flattened achenes, or have high-conical to columnar receptacles.

In spite of the useful character differences, the *Praxelis* group seems closely related to the *Critonia* group. Some Mexican and West Indian species seem to be intermediate between the two groups but more complete evidence indicates these latter are probably the result of intergeneric hybridization.

The *Praxelis* group is concentrated in Brazil with some species of *Praxelis* and *Chromolaena* more widely distributed. The three most widely distributed species are *Chromolaena ivaefolia* (L.) K. & R., *C. laevigata* (Lam.) K. & R., and the weedy often adventive *C. odorata* (L.) K. & R.

Hebeclinium group

The *Hebeclinium* group has mostly Critonioid features including the subimbricate partially deciduous involucral bracts, the smooth corolla lobes and the unenlarged glabrous style base. A number of trends help separate the groups. The receptacles, with many isolated exceptions, are hirsute in the *Hebeclinium* group, and the anther collars are of an elongate type with mostly subquadrate cells. Also the carpopodia usually are continuous with the lower parts of the achene ribs, a feature seen best in *Bartlettina*. The group is notable for the occurrence of chromosome numbers of both $n=10$ and $n=16$ or $c.$ 17, sometimes within the same species.

The group is found mostly from central Mexico southward through Central America to the central Andes. One widely distributed species is *Hebeclinium macrophyllum* (L.) DC.

Neomirandea group

Neomirandea has the subimbricate, partially deciduous, involucre and the smooth corolla lobes of the *Critonia* group. The anther collars are elongate with mostly subquadrate cells as in *Hebeclinium*. The genus has some species with unenlarged style bases and others with distinct enlargements.

All the members of the group are either epiphytic or grow in deep humus supported by prop-root systems.

The chromosome base numbers in *Neomirandea* follow subgeneric lines (King and Robinson, 1970g). The subgenus *Neomirandea* with higher numbers, with enlarged style bases, often with hairs inside the corolla and with some terrestrial species, seems to be the more advanced subgenus.

The *Neomirandea* group is restricted to the moist tropical regions between central Mexico and Ecuador with the greatest concentration in Costa Rica and western Panama.

Mikania group

Mikania is one of the most distinctive genera in the Compositae but due to the idiocies that have prevailed in the Eupatorieae even this genus has not escaped the suggestion that it be combined with *Eupatorium* (Correll and Johnston, 1970, p. 1551). *Mikania* has four subequal principle involucral bracts and four flowers in each head. The stamens are exserted partially beyond the corolla lobes at anthesis and the anther collars are broad with mostly subquadrate non-annulated cells. The style shaft is thick but the base is not nodular. The style base is usually glabrous but is papillose in a few species. Most of the species are vines and some are epiphytic but a number of terrestrial species occur in Brazil where the genus is concentrated. *Kanimia* has been artificially separated from *Mikania* and associated with *Brickellia* on the basis of having 10-ribbed achenes. The possible value of the genus name for a more natural segregate needs further study. The anther collar cells, the epiphytic habit, the higher chromosome number and the quadrate straight-walled corolla cells suggest possible relation to the *Neomirandea* group.

Mikania is distributed throughout tropical America with one endemic species, *M. scandens* (L.) Willd., in eastern North America. Members of the *M. scandens* relationship are pantropical, being found in Africa, Ceylon, southeastern Asia and the Indonesian area as well as the Americas.

Ageratina group

The relatives of *Ageratina* and *Oxylobus* are distinct among the groups with higher chromosome numbers by the eximbricate to weakly subimbricate involucral bracts and by the papillose inner surfaces of the corolla lobes. The corolla lobes are smooth only in parts of the specialized genus *Piptothrix*. Style bases are always glabrous; and they are usually enlarged, except in *Pachythamnus*, *Standleyanthus*, *Spaniopappus* and a few species of *Ageratina*. The pappus is usually capillary and often easily deciduous, but it is reduced in a few taxa such as *Oxylobus* and one species of *Spaniopappus*. The tube of the corolla is usually long and often very slender. The anther collars are elongate with many subquadrate non-annulated cells. This latter character and the existence of such a transitional genus as *Standleyanthus* suggests relationship to the *Neomirandea* group.

Ageratina and its relatives are almost totally restricted to the western parts of the Americas. Three species of *Ageratina* extend into eastern North America (Clewell and Wooten, 1971), the genus *Spaniopappus* and a few *Ageratina* species are in the West Indies, and a single *Ageratina* has recently been found in Brazil. Two species, *A. adenophora* (Spreng.) K. & R. and *A. riparia* (Regel) K. & R. are weedy and widely adventive in many tropical and subtropical parts of the World.

Hofmeisteria group

Hofmeisteria seems to be an isolated genus in the tribe, with its long monocephalic peduncles arising from condensed parts of the leafy stems and with its scarcely roughened pollen. The genus has subimbricate involucres with narrowly pointed bracts, narrow glabrous corollas, smooth lobes, anther collars with mostly elongate cells and glabrous usually slightly enlarged style bases. The carpopodium is characteristically short-tapering with the cells multiseriate and subquadrate at the surface. The pappus and anther appendages are notably variable in the genus (King and Robinson, 1966).

Hofmeisteria is mostly restricted to the regions surrounding the Gulf of Lower California with one species extending to southern Mexico.

Oaxacania group

Oaxacania and *Carterothamnus* are evidently related by shrubby habit, partly dissected leaves, long-pedunculate heads, prominently subimbricate involucral bracts, prominent paleae, narrow corollas, narrow anther collars with many subquadrate non-annulated cells, enlarged glabrous style bases, tapering carpopodia with many rows of subquadrate cells and a reduced pappus with mostly short or vestigial setae. Still, there are marked differences in style tips, achene and pappus form, corolla lobe inner surface, and possibly also in pollen size (King and Robinson, 1970e). The most important character of the group is the paleaceous receptacle where the paleae can be extracted each with a flower in the axil. This form of palea is most suggestive of the primitive type seen in many Heliantheae.

The two genera of the group are both Mexican but from widely separated localities. *Oaxacania* is from a local area in northern Oaxaca while *Carterothamnus* is from the Sierra de la Giganta in southern Baja California.

TABLE I. Groups and genera of the Eupatorieae

Genera	Number of species	Habit	Distribution	Chromosome number
I ADENOSTEMMATINAE				
Sciadocephala Mattf.	4	P	N. S. America	
Adenostemma J. R. & G. Forster	20	P	pantropical	5, 10
Gymnocoronis DC.	5	P	Cent. & S. Amer.	
II EUPATORIINAE				
Eupatorium L.	38	P	E. U.S., Asia, Europe	10, 15, 20
Eupatoriadelphus K. & R.	4	P	E. N. Amer.	10, 20
Austroeupatorium K. & R.	11	P, S?	S. Amer., adventive	10
Stomatanthes K. & R.	15	P, S	Brazil, Africa	10
Hatchbachiella K. & R.	2	P	E. Brazil to N. Argentina	
III DISYNAPHIA GROUP				
Acanthostyles K. & R.	1	P, S	Bolivia, Brazil, N. Argentina	
Raulinoreitzia K. & R.	2	P, S	Brazil to N. Argentina	10
Disynaphia Hook. & Arn.	12	P, S	Brazil, Uruguay	10
Campovassouria K. & R.	1	P, S	Brazil, Paraguay, Uruguay, N. Argentina	10
Grazielia K. & R.	9	P, S	Brazil	10
Symphyopappus Turcz.	13	P, S	Brazil	10
IV GYPTIS GROUP				
Gyptis (Cass.) Cass.	7	P	S. Brazil to Argentina	30
Gyptidium K. & R.	2	P	S. Brazil, Argentina	
Urolepis (DC.) K. & R.	1	P	Bolivia, Brazil, Argentina	
Barrosoa K. & R.	11	P	S. America	10
Dasycondylus K. & R.	7	P	Peru to Brazil	
Conocliniopsis K. & R.	1	P	S. America	10, 30
Platypodanthera K. & R.	1	P	Brazil	
Trichogonia (DC.) Gardn.	22	P	S. America	
Trichogoniopsis K. & R.	2	P	Brazil	10
Neocuatrecasia K. & R.	7	P	Peru, Bolivia	
Diacranthera K. & R.	2	P	Brazil	

TABLE I. Groups and genera of the Eupatorieae—*continued*

Genera	Number of species	Habit	Distribution	Chromosome number
IV GYPTIS GROUP—*continued*				
Vittetia K. & R.	1	P	Brazil	10
Bahianthus K. & R.	1	P, S	Brazil	
Agrianthus Mart. ex DC.	6	S	Brazil	
Lasiolaena K. & R.	2	P, S	Brazil	
Stylotrichium Mattf.	2	P, S	Brazil	
Campuloclinium DC.	20	P	Cent. America to Brazil	10
Macropodina K. & R.	3	P	S. Brazil, Paraguay	
Conoclinium DC.	3	P	E. U.S., Mexico	10
Tamaulipa K. & R.	1	P	Texas, N. Mexico	10
Lourteigia K. & R.	7	P	Colombia, Venezuela	10
V ACRITOPAPPUS GROUP				
Acritopappus K. & R.	3	S	Brazil	9
Radlkoferotoma O. Kuntze	3	S	S. Brazil, Uruguay	
VI PIQUERIA GROUP				
Ageratum L.	43	P	Trop. America, adventive	9, 10, 11, 15, 20
Phania DC.	5	P	W. Indies	
Phalacraea DC.	5	P	Colombia–Peru	20
Blakeanthus K. & R.	1	P	Guatemala	
Ascidiogyne Cuatr.	2	P	Peru	10
Ellenbergia Cuatr.	1	P	Peru	
Guevaria K. & R.	4	P	Colombia–Peru	10
Ferreyrella Blake	2	P	Peru	
Piqueriella K. & R.	1	P	Brazil	
Piqueriopsis King	1	A	Mexico	
Arrojadocharis Mattf.	1	P	Brazil	
Piqueria Cav.	7	P	Cent. America, W. Indies	12, 25
Stevia Cav.	150–200	A, P, S	Trop. America	11, 12, *c.* 30, $2n = 33$I, 34I, 36I
Metastevia Grashoff	1	P	Mexico	
Cronquistia King	1	P	Mexico	12
Revealia K. & R.	1	S	Mexico	

TABLE I. Groups and genera of the Eupatorieae—*continued*

Genera	Number of species	Habit	Distribution	Chromosome number
VI PIQUERIA GROUP—*continued*				
Carphochaete A. Gray	5	S	SW. U.S., Mexico	11
Macvaughiella K. & R.	1	S	Mexico, Guatemala	
Iltisia Blake	1	P	Costa Rica	
Microspermum Lag.	7	P	Mexico	12
VII TRICHOCORONIS GROUP				
Trichocoronis A. Gray	2	P	Texas, Mexico	15
Shinnersia K. & R.	1	P?	Texas, Mexico	*c.* 30
Sclerolepis Cass.	1	P	E. U.S.	15, *c.* 30
VIII AYAPANA GROUP				
Ayapana Spach.	15	P	Trop. America, adventive	10, 17, 18
Ayapanopsis K. & R.	16	P	W.S. America	
Isocarpha R. Brown	12	P	Trop. America	10
Polyanthina K. & R.	1	P	Costa Rica to Peru	10, 12 ± 1
Heterocondylus K. & R.	12	P	Cent. & S. America	*c.* 20
Condylidium K. & R.	2	P	Trop. America	10
Gongrostylus K. & R.	1	P, S	Costa Rica to Ecuador	
Gymnocondylus K. & R.	1	P	Brazil	
Alomiella K. & R.	1	P	Brazil	
Monogerion Barroso & King	1	P	Brazil	
IX ALOMIINAE				
Brickellia Elliott	100	A, P, S	U.S., Mexico 1 sp. S. America	9
Barroetea A. Gray	6	P	Mexico	9
Phanerostylis (A. Gray) K. & R.	3	sS	N. Mexico	9
Brickelliastrum K. & R.	1	P	S.W. U.S.	10
Flyriella K. & R.	4	P	Texas, N. Mexico	
Ageratella A. Gray ex Watson	1	S	Mexico	
Asanthus K. & R.	3	S	S.W. U.S., Mexico	

TABLE I. Groups and genera of the Eupatorieae—*continued*

Genera	Number of species	Habit	Distribution	Chromosome number
IX ALOMIINAE—*continued*				
Malperia Watson	1	P	S.W. U.S., Mexico	10
Pleurocoronis K. & R.	3	sS	S.W. U.S., Mexico	9
Alomia H.B.K.	4	P	Mexico	
Dyscritogyne K. & R.	2	P	Mexico	
Kyrsteniopsis K. & R.	4	P	Mexico	
Pseudokyrsteniopsis K. & R.	1	P	Guatemala	
Steviopsis K. & R.	3	P	Mexico	
Carminatia Moc. ex DC.	2	P	Mexico	10
Dissothrix A. Gray	1	P	Brazil	
Austrobrickellia K. & R.	3	P	Brazil, Bolivia, Argentina	10
Pseudobrickellia K. & R.	3	P, S	Brazil	
Crossothamnus K. & R.	1	S	Peru	10
Helogyne Nutt.	c. 12	S	Peru, Bolivia, Chile	
Planaltoa Taubert	2	P	Brazil	
Leptoclinium Bentham	1	P, S?	Brazil	
Condylopodium K. & R.	4	S	Colombia	
X LIATRIS GROUP				
Liatris Gaertn. ex Schreb.	34	P	E.N. America	10, 20, 30
Litrisa J. K. Small	1	P	S.E. U.S.	10
Trilisa (Cass.) Cass.	2	P	S.E. U.S.	10
Carphephorus Cass.	5	P	S.E. U.S.	10
Garberia A. Gray	1	S	S.E. U.S.	10
Hartwrightia A. Gray ex Watson	1	P	S.E. U.S.	
XI FLEISCHMANNIA GROUP				
Fleischmannia Sch. Bip.	75	A, P	N. America, W.S. America advent.	4, 10, 20, 29, 30
Sartorina K. & R.	1	P	Mexico	
XII CRITONIA GROUP				
Mexianthus B. L. Robins	1	P	Mexico	
Neohintonia K. & R.	1	P	Mexico	
Critonia P. Browne	33	P,S,T	Cent. America, W. Indies, N. S. America	10

TABLE I. Groups and genera of the Eupatorieae—*continued*

Genera	Number of species	Habit	Distribution	Chromosome number
XII CRITONIA GROUP—*continued*				
Critoniadelphus K. & R.	2	S	Cent. America	
Urbananthus K. & R.	2	S	W. Indies	
Adenocritonia K. & R.	1	S	Jamaica	
Antillia K. & R.	1	P	W. Indies	
Ciceronia Urban	1	P	W. Indies	
Eupatorina K. & R.	1	P	W. Indies	
Fleischmanniopsis K. & R.	3	P	Cent. America	
Verieckia K. & R.	1	S	N. Mexico	
Koanophyllon Arruda da Camara	109	P, S	Trop. America	10, 20
Eupatoriastrum Greenm.	3	P	Cent. America	
Sphaereupatorium (Hoffm.) O. Kuntze ex B. L. Robins	1	P	Brazil, Bolivia	
Critoniella K. & R.	5	P	Colombia Venezuela	10
Aristeguietia K. & R.	20	S	Andes	10
Asplundianthus K. & R.	9	S	N.S. America	10
Austocritonia K. & R.	3	S	Brazil	10
Badilloa K. & R.	9	S	N. Andes	
Grosvenoria K. & R.	3	S	Ecuador, Peru	
Grisebachianthus K. & R.	8	P	Cuba	
Lorentzianthus K. & R.	1	P	Bolivia, Argentina	
Chacoa K. & R.	2	P	Paraguay, Argentina	
Idiothamnus K. & R.	4	S	S. America	
Imeria K. & R.	1	S	Venezuela	
Ophryosporus Meyen	38	P, S	S. America	10
Cronquistianthus K. & R.	16	S	Andes	
Steyermarkina K. & R.	4	S	Brazil, Venezuela	
Neocabreria K. & R.	3	P	S. Brazil	
Uleophytum Hieron.	1	S	Peru	
Amboroa Cabrera	2	sS	Peru, Bolivia	
Tuberostyles Steetz	2	S	Panama to Ecuador	
XIII PRAXELIS GROUP				
Praxelis Cass.	13	P	S. America	20
Chromolaena DC.	over 130	P	Trop. America, advent.	10, 29, *c*. 80[1]
Eupatoriopsis Hieron.	1	P	Brazil	

TABLE I. Groups and genera of the Eupatorieae—*continued*

Genera	Number of species	Habit	Distribution	Chromosome number
XIII PRAXELIS GROUP—*continued*				
Lomatozoma Baker	2	P	Brazil	
Praxeliopsis Barroso	1	P	Brazil	
Eitenia K. & R.	1	P	Brazil	
Osmiopsis K. & R.	1	P	Haiti	
XIV HEBECLINIUM GROUP				
Hebeclinium DC.	18	P, S	Trop. America	10
Bartlettina K. & R.	20	P, S	Trop. N. & S. America	10, 16
Decachaeta DC.	7	S	Mexico, Cent. America	16
Amolinia K. & R.	1	S	Mexico, Guatemala	
Erythradenia (B. L. Robins.) K. & R.	1	S?	Mexico	
Guayania K. & R.	4	P	Venezuela	
Matudina K. & R.	1	P	Mexico	16
Peteravenia K. & R.	4	P	Mexico, Cent. America	10, *c.* 17
XV NEOMIRANDEA GROUP				
Neomirandea K. & R.	24	S	Mexico to Ecuador	17, 20, *c.* 24, *c.* 25
XVI MIKANIA GROUP				
Mikania Willd.	300	P, Li, S	Trop. America, Africa, Trop. Asia	*c.* 16, 17, 18, 19, 20, $2n = 34-38$!
XVII AGERATINA GROUP				
Ageratina Spach.	230	P, S	N. & Cent. America, W. Indies, W.S. America, advent.	17, 18, 20, *c.* 40, 48, 50, 51
Oxylobus (Moc. ex DC.) A. Gray	5	S		16
Piptothrix A. Gray	6	S	Mexico	
Jaliscoa Watson	3	S	Mexico	
Spaniopappus B. L. Robins.	5	P, S	Cuba	
Pachythamnus (K. & R.) K. & R.	1	S	Mexico, Cent. America	
Standleyanthus K. & R.	1	P, S	Costa Rica	

TABLE I. Groups and genera of the Eupatorieae—*continued*

Genera	Number of species	Habit	Distribution	Chromosome number
XVIII HOFMEISTERIA GROUP				
Hofmeisteria Walp.	8	P, S	Mexico	18, 19
XIX OAXACANIA GROUP				
Oaxacania B. L. Robins. & Greenm.	1	S	Mexico	
Carterothamnus King	1	S	W. Mexico	

GENERIC SYNONYMY

Addisonia Rusby = *Helogyne*
Ageratiopsis Sch. Bip. nom. nud. = *Ageratina*
Ammopursus J. K. Small = *Liatris*
Arrojadoa Mattf. = *Arrojadocharis*
Batschia Moench. = *Ageratina*
Biolettia Greene = *Trichocoronis*
Brachyandra Philippi = *Helogyne*
Bulbostylus DC. = *Brickellia*
Calostelma D.Don = *Liatris*
Caradesia Raf. = *Eupatorium*
Carelia Juss. ex Cav. = *Mikania*
Carelia Less. = *Radlkoferotoma*
Carelia Pont. ex Fabr. = *Ageratum*
Chone Dulac = *Eupatorium*
Clavigera DC. = *Brickellia*
Coelestina Cass. = *Ageratum*
Coleosanthus Less. = *Brickellia*
Corynanthelium Kunze = *Mikania*
Cunigunda Bub. = *Eupatorium*
Dichaeta Sch. Bip. nom. nud. = *Macvaughiella*
Dimorpholepis K. & R. = *Grazielia*
Eriopappus Loud. = *Brickellia*
Eutrochium Raf. = *Eupatoriadelphus*?
Haberlea Pohl nom. nud. = *Praxelis*
Heterolaena Sch. Bip. nom. nud. = *Chromolaena*
Ismaria Raf. = *Brickellia*
Isocarpha Less. = *Ageratum*
Kallophyllon Pohl nom. nud. = *Symphyopappus*

Kanimia Gardn. = *Mikania*
Kuhnia L. = *Brickellia*
Kyrstenia Neck. ex Greene = *Ageratina*
Lacinaria Hill = *Liatris*
Lavenia Swartz = *Adenostemma*
Lepidesmia Klatt = *Ayapana*
Leptoclinium (Nutt.) A. Gray = *Garberia*
Leto Philippi = *Helogyne*
Mallinoa Coult. = *Ageratina*
Margacola Buckl. = *Trichocoronis*
Microconia Dus. nom. nud. = *Stomatanthes*
Mustelia Spreng. = *Stevia*
Neobartlettia K. & R. = *Bartlettina*
Nothites Cass. = *Stevia*
Ooclinium DC. = *Praxelis*
Osmia Sch. Bip. = *Chromolaena*
Rosalesia Llave & Lex. = *Brickellia*
Schaetzellia Sch. Bip. = *Macvaughiella*
Suprago Gaertn. = *Liatris*
Tamayoa Badillo = *Ayapana*
Traganthes Wallr. = *Eupatorium*
Trichinolepis B. L. Robinson = *Ophryosporus*
Uncasia Greene = *Eupatorium*
Wikstroemia Spreng. = *Critonia*
Willoughbya Neck. ex O. Kuntze = *Mikania*
Xetoligus Raf. = *Stevia*

ACKNOWLEDGEMENTS

The research that gave rise to this chapter was supported in part by the following grants to the junior author: National Science Foundation—BMS 70-00537; Penrose Fund of the American Philosophical Society; and the National Geographic Society.

REFERENCES

Adams, C. D. (1971). Miscellaneous additions and revisions to the flowering plants of Jamaica III. *Phytologia* **21**, 405–410.
Alain, H. (1960). Novedades en la flora Cubana. XII. *Contr. ocas. Mus. Hist. nat. La Salle* **18**, 1–16.
Aristeguieta, L. (1964). Compositae. *In Flora de Venezuela* **10**, 91–243 (Eupatorieae), 523, 525–531 (*Isocarpha* and *Lepidesmia*).
Badillo, V. M. (1943). Dos nuevas compuestas de Venezuela. *Boln Soc. venez. Cienc. nat.* **8**, 237–239.
Badillo, V. M. (1944a). Compuestas venezolanas notables o nuevas. *Boln Soc. venez. Cienc. nat.* **9**, 131–137.
Badillo, V. M. (1944b). *Tamayoa*, género nuevo de las compuestas (Asteraceae). *Boln Soc. venez. Cienc. nat.* **9**, 139–140.
Badillo, V. M. (1946). Contribución al conocimiento de la sistemática y distribución geografica de las compuestas en Venezuela. *Boln Soc. venez. Cienc. nat.* **10**, 279–320.
Badillo, V. M. (1952). *Mikania araguensis*, una nueva compuesta del parque nacional de Rancho Grande. Novedades Científicas. *Contr. ocas. Mus. Hist. nat. La Salle, Bot.* **2**, 1–4.
Baker, H. G. (1967). The evolution of weedy taxa in the *Eupatorium microstemon* species aggregate. *Taxon* **16**, 293–300.
Baker, J. G. (1876). Compositae II. Eupatoriaceae *in* Martius, *Flora Brasiliensis*. **6**, 181–374.
Barroso, G. M. (1949). *Praxeliopsis*—Um novo gênero de Compositae. *Arq. Jard. bot.* **9**, 175–178.
Barroso, G. M. (1950). Considerações sôbre o gênero *Eupatorium*. *Arq. Jard. bot.* **10**, 13–116.
Barroso, G. M. (1951). Estudos das espécies brasileiras de "*Trichogonia*" Gardn. *Arq. Jard. bot.* **11**, 7–18.
Barroso, G. M. (1957). Compositae—O gênero *Stylotrichium* Mattfeld. *Arq. Jard. bot.* **15**, 23–25.
Barroso, G. M. (1958). Mikaniae do Brasil. *Arq. Jard. Botânico* **16**, 239–333.
Barroso, G. M. (1965). De compositarum novitatibus. *Sellowia* **17**, 79–83.
Barroso, G. M. (1969). Novitates compositarum, II. *Loefgrenia* **36**, 1–3.
Barroso, G. M. and King, R. M. (1971). New Taxa of Compositae (Eupatorieae) from Brazil. *Brittonia* **23**, 118–121.
Bentham, G. (1873). Notes on the classification, history, and geographical distribution of Compositae. *J. Linn. Soc. Lond. (Bot.)* **13**, 335–577.
Bentham, G. and Hooker, J. D. (1873). "Genera Plantarum", Vol. 2. London.
Blake, S. F. (1924). New American Asteraceae. *Contr. U.S. Natn. Herb.* **22**, 587–661.
Blake, S. F. (1941). Note on the name *Eupatorium rugosum*. *Rhodora* **43**, 557–558.
Blake, S. F. (1957). Two new genera of Compositae from Peru and Costa Rica. *J. Wash. Acad. Sci.* **47**, 407–410.
Borhidi, A. and Muñiz, O. (1971). New plants in Cuba I. *Acta bot. Acad. Sci. Hungar.* **17**, 1–36.

Borhidi, A. and Muñiz, O. (1973). New plants in Cuba II. *Acta bot. Acad. Sci. Hungar.* **18**, 29–48.

Brown, R. (1817–1818). Some observations on the natural family of plants called Compositae. *Trans. Linn. Soc. Lond.* **12**, 76–142.

Browne, P. (1756). "The Civil and Natural History of Jamaica in Three Parts". London.

Cabrera, A. L. (1941). Compuestas Bonaerenses. II. *Adenostemma* Forst.—VI. *Mikania* Willd. 21–49.

Cabrera, A. L. (1945). Cuatro compuestas nuevas del Perú. *Rev. Univ. Cuzco* **33**, 117–122.

Cabrera, A. L. (1956). Un nuevo genero de Eupatorieas (Compositae) de Bolivia. *Boln Soc. argent. Bot.* **6**, 91–93.

Cabrera, A. L. (1957). El genero *Carelia* (Compositae). *Boln Soc. argent. Bot.* **6**, 239–242.

Cabrera, A. L. (1959a). Compositae Catarinenses novae. *Boln Soc. argent. Bot.* **7**, 187–200.

Cabrera, A. L. (1959b). Notas sobre tipos de compuestas subamericanas en herbarios europeos. I. *Boln Soc. argent. Bot.* **7**, 233–246.

Cabrera, A. L. (1959c). Notas sobre tipos de compuestas sudamericanas en herbarios europeos. II. *Boln Soc. argent. Bot.* **8**, 26–35.

Cabrera, A. L. (1959d). Ocho compuestas sudamericanas nuevas. *Univ. Nac. La Plata, Not. Mus.* **19**, 191–210.

Cabrera, A. L. (1962). Compuestas andinas nuevas. *Boln Soc. argent. Bit.* **10**, 21–45.

Cabrera, A. L. (1973). Notas sobre tipos de compuestas sudamericanas en herbarios europeos. IV. *Boln Soc. argent. Bot.* **15**, 113–125.

Cabrera, A. L. (1974). Especies nuevas o criticas de la flora Jujeña. VI. *Boln Soc. argent. Bot.* **15**, 319–339.

Cabrera, A. L. and Vittet, N. (1954). Catalogo de las Eupatorieas Argentinas. *Revta Mus. Univ. Eva Peron n.s.* **8**, 179–263.

Cabrera, A. L. and Vittet, N. (1963). Compositae Catharinenses II. Eupatorieae. *Sellowia* **15**, 149–258.

Candolle, A. P. De (1836). Eupatoriaceae *in* "Prodromus" **5**, 103–211.

Carlquist, S. (1965). Wood anatomy of Eupatorieae (Compositae). *Aliso* **6**, 89–103.

Carlquist, S. (1966). Wood anatomy of Compositae: a summary with comments on factors controlling wood evolution. *Aliso* **6**, 25–44.

Cassini, H. (1812). Extrait d'un premier Mémoire de M. Henri Cassini, sur les Synanthérées. *Bull. Soc. Philom.* **3**, 189–191.

Cassini, H. (1813). Observations sur le style, et le stigmate des Synanthérées. *J. Phys.* **76**, 97–128, 181–201, 249–275.

Cassini, H. (1814a). Second Mémoire de M. Henri Cassini, sur les Synanthérées. *Bull. Soc. philomath. Paris* 1814: 9–11.

Cassini, H. (1814b). Précis d'un second Mémoire sur les Synanthérées. *J. Phys.* **78**, 272–291.

Cassini, H. (1815). Extrait d'un troisième Mémoire de M. Henri Cassini, sur les Synanthérées. *Bull. Soc. philomath. Paris* **1815**, 171–175.

Cassini, H. (1816a). Troisième Mémoire sur les Synanthérées, analyse de la corolle. *J. Phys.* **82**, 116–146.
Cassini, H. (1816b). Aperçu des genres nouveaux formés par M. Henri Cassini, dans la famille des Synanthérées. *Bull. Soc. philomath. Paris* **1816**, 198–200.
Cassini, H. (1817a). Extrait d'un quatrieme Mémoire de M. Henri Cassini, sur les Synanthérées. *Bull. Soc. philomath. Paris* **1817**, 115–118.
Cassini, H. (1817b). Quatrième mémoire sur la famille des Synanthérées, contenant l'analyse de l'ovaire et de ses accessoires. *J. Phys.* **85**, 5–21.
Cassini, H. (1817–1830). In "Dictionaire des Sciences Naturelles" (G. Cuvier, ed.). (1817). *Caelestina.* **6** (suppl.), **8**; Carphephore. **7**, 148–150; (1818). *Coleosanthus.* **10**, 36–37; Composées. **10**, 131–159; (1819). *Critonia.* **12**, 1–2; (1820). Eupatoire. **16**, 2–4; Eupatorées. **16**, 5–8; Eupatoriées. **16**, 9–10; (1821). Gyptide. **20**, 177–179; Hélianthées (with discussion of tribes). **20**, 354–385; (1822). Isocarphe. **24**, 18–20; Kuhnie. **24**, 515–520; Lavenie. **25**, 360–365; (1823). Comments on Catalogue des plantes du Jardin médical de Paris. **26**, 223–235; *Liatris.* **26**, 235–239; (1825). Nothite. **35**, 163–167; (1826). Piquérie. **41**, 115–119; Praxelide. **43**, 261–262; (1827). Sclerolèpe. **48**, 155–156; Suprage. **51**, 384–388; Synanthérologie. **51**, 443–455; (1828). *Traganthes.* **55**, 129–131; Trilise. **55**, 310–311; (1829). Wikstromia. **59**, 60; (1830). Zyégée (with tableau synoptique des Synanthérées). **60**, 560–619.
Cassini, H. (1818). Cinquième Mémoire, sur la famille des Synanthérées, contenant les fondemens de la Synanthérographie. *J. Phys.* **86**, 120–129, 173–189.
Cassini, H. (1819). Sixième Mémoire sur la famille des Synanthérées, contenant les caractères des tribus. *J. Phys.* **88**, 150–169, 189–204.
Cassini, H. (1826–1834). "Opuscules Phytologiques", 3 vols. Paris.
Cavanilles, A. J. (1794–1797). "Icones et Descriptiones Plantarum, Vols 3, 4. Madrid.
Chamberlain, J. S. (1891). A comparative study of the styles of Compositae. *Bull. Torrey bot. Club* **18**, 175–209.
Clewell, A. F. and Wooten, J. W. (1971). A revision of *Ageratina* (Compositae: Eupatorieae) from eastern North America. *Brittonia* **23**, 123–143.
Col, M. A. (1904). Recherches sur l'appareile secreteur interne des Composées. *J. Bot. (Paris)* **18**, 110–133, 153–175.
Correa, M. D. and Wilbur, R. L. (1969). A revision of the genus *Carphephorus* (Compositae-Eupatorieae). *J. Elisha Mitchell scient. Soc.* **85**, 79–91.
Correll, D. S. and Johnston, M. C. (1970). "Manual of the Vascular Plants of Texas". Texas Research Foundation, Renner.
Cronquist, A. (1955). Phylogeny and taxonomy of the Compositae. *Am. Midl. Nat.* **53**, 478–511.
Cuatrecasas, J. (1935a). Plantae Isernianae, 1. *An. Univ. Madrid* **4**, 206–265.
Cuatrecasas, J. (1935b). Plantae novae colombianae: serie altera. *Trab. Mus. nac. Cienc. nat. Jard. bot.* **29**, 1–46.
Cuatrecasas, J. (1964). Studies on Andean Compositae: VI *Proc. biol. Soc. Wash.* **77**, 127–156.
Cuatrecasas, J. (1965). Some new Compositae from Peru. *Ann. Mo. bot. Gdn* **52**, 304–313.

FERREIRA, A. G. (1968). Contribuição ao estudo da nervação foliar das Compositae dos cerrados—IV—tribo Eupatoriae. *Arq. bot. Estado São Paulo* **4**, 153–170.
FLYR, D. (1968). New names and records in *Brickellia* (Compositae). *Sida* **3**, 252–256.
FORSTER, J. R. and FORSTER, G. (1776). "Characteres Generum Plantarum". London.
GAISER, L. O. (1946). The genus *Liatris*. *Rhodora* **48**, 165–183, 216–263, 273–326, 331–382, 393–412.
GAISER, L. O. (1949). Chromosome studies in *Liatris*. I. Spicatae and Pycnostachyae. *Am. J. Bot.* **36**, 122–135.
GAISER, L. O. (1950a). Chromosome studies in *Liatris*. II. Graminifolia and Pauciflorae. *Am. J. Bot.* **37**, 414–423.
GAISER, L. O. (1950b). Chromosome studies in *Liatris*. III. Punctatae. *Am. J. Bot.* **37**, 763–777.
GAISER, L. O. (1953). Chromosome studies in Kuhniinae (Eupatorieae). I. *Brickellia*. *Rhodora* **55**, 253–267, 269–288, 297–321, 328–345.
GAISER, L. O. (1954). Studies in the Kuhniinae (Eupatorieae). II. *J. Arnold Arboretum* **35**, 87–133.
GODFREY, R. K. (1950). Studies in the Compositae of North Carolina. III. *J. Elisha Mitchell scient. Soc.* **66**, 186–194.
GODFREY, R. K. (1961). *Liatris provincialis*, sp. nov., (Compositae), endemic in western Florida. *Am. Midl. Nat.* **66**, 466–470.
GRANT, W. F. (1953). A cytotaxonomic study in the genus *Eupatorium*. *Am. J. Bot.* **40**, 729–742.
GRASHOFF, J. L. (1972). "A systematic Study of the North and Central American Species of *Stevia*". Ph.D. dissertation, University of Texas.
GRASHOFF, J. L. (1974). Novelties in *Stevia* (Compositae: Eupatorieae). *Brittonia* **26**, 347–384.
GRASHOFF, J. L. (1975). *Metastevia* (Compositae: Eupatorieae): a new genus from Mexico. *Brittonia* **27**, 69–73.
GRASHOFF, J. L. and BEAMAN, J. H. (1969a). Studies in *Eupatorium* (Compositae), I. Revision of *Eupatorium bellidifolium* and allied species. *Rhodora* **71**, 566–576.
GRASHOFF, J. L. and BEAMAN, J. H. (1969b). Studies in *Eupatorium* (Compositae), II. A new species of *Eupatorium* (section *Hebeclinium*). *Rhodora* **71**, 577–579.
GRASHOFF, J. L. and BEAMAN, J. H. (1970). Studies in *Eupatorium* (Compositae), III. apparent wind pollination. *Brittonia* **22**, 77–84.
GRAY, A. (1852). Compositae *in* Plantae Wrightianae Texano—Neo-mexicanae. I. *Smithson. Contr. Knowl.* **3**, 82–129.
GRAY, A. (1884). Compositae. *In Synoptical flora of North America*. **1**, 48–449.
GREENE, E. L. (1889–1892). *Biolettia*, a new genus of Compositae. *Pittonia* **2**, 215–216.
GREENE, E. L. (1893). Observations on the Compositae—Tribe II. Eupatoriaceae. *Erythea* **1**, 41–45.
GREENE, E. L. (1903). Neglected Eupatoriaceous genera. *Leafl. Bot. Obs. Crit.* **1**, 7–13.
GRIERSON, A. J. C. (1972). Critical notes on the Compositae of Ceylon. *Ceylon J. Sci.* **10**, 42–60.

HARCOMBE, P. A. and BEAMAN, J. H. (1967). Transfer of two Mexican species from *Eupatorium* to *Brickellia* (Compositae). *Southwestern Nat.* **12**(2), 127–133.

HERBERT, H. J.-C. (1968). Generic considerations concerning *Carphephorus* and *Trilisa* (Compositae). *Rhodora* **70**, 474–485.

HOFFMANN, O. (1894). Compositae. *In* "Die naturlichen Pflanzenfamilien" A. Engler and K. Prantl, eds. **4**(5), 87–387.

HUMBOLDT, F. H. A. von, BONPLAND, A. J. and KUNTH, C. S. (1818). Sectio III. Eupatoreae. *In* "Nova Genera et Species Plantarum", **4**, 82–120.

JAMES, C. W. (1958). Generic considerations concerning *Carphephorus*, *Trilisa* and *Litrisa* (Compositae). *Rhodora* **60**, 117–122.

JOHNSON, M. F. (1971a). A monograph of the genus *Ageratum* L. (Compositae-Eupatorieae). *Ann. Mo. bot. Gdn* **58**, 6–88.

JOHNSON, M. F. (1971b). The genera *Carphephorus*, *Mikania* and *Kuhnia* (Eupatorieae-Asteraceae) in Virginia. *Virginia J. Sci.* **22**, 38–41.

JOHNSON, M. F. (1971c). The genus *Liatris* in Virginia. *Castanea* **36**, 137–147.

JOHNSON, M. F. (1972). Eupatorieae (Asteraceae) in Virginia: *Eupatoriadelphus*, *Ageratina*, *Fleischmannia* and *Conoclinium*. *Virginia J. Sci.* **23**, 48–55.

JOHNSON, M. F. (1974). Eupatorieae (Asteraceae) in Virginia: *Eupatorium* L. *Castanea* **39**, 205–228.

KING, R. M. (1965). *Piqueriopsis*, a new genus of Compositae from southwestern Mexico. *Brittonia* **17**, 352–353.

KING, R. M. (1967a). Studies in the Eupatorieae (Compositae), I–III. *Rhodora* **69**, 35–47, 240.

KING, R. M. (1967b). Studies in the Compositae-Eupatorieae V, notes on the genus *Piqueria*. *Sida* **3**(2), 107–109.

KING, R. M. (1967c). Studies in the Compositae-Eupatorieae IV. *Rhodora* **67**, 352–371.

KING, R. M. (1967d). Studies in the Compositae-Eupatorieae VII. *Sida* **3**(3), 163–164.

KING, R. M. (1968). Studies in the Compositae-Eupatorieae VI. *Brittonia* **20**, 11–12.

KING, R. M. and ROBINSON, H. (1966). Generic limitations in the *Hofmeisteria* complex (Compositae-Eupatorieae). *Phytologia* **12** 465–476.

KING, R. M. and ROBINSON, H. (1967). Multiple pollen forms in two species of the genus *Stevia* (Compositae). *Sida* **3**(3), 165–169.

KING, R. M. and ROBINSON, H. (1968a). Studies in the Compositae-Eupatorieae VIII. Observations on the microstructure of *Stevia*. *Sida* **3**(4), 257–269.

KING, R. M. and ROBINSON, H. (1968b). *Macvaughiella* King & Robinson, nomen novum for *Schaetzellia* Sch.-Bip., not Klotzsch (Compositae). *Sida* **3**(4), 282.

KING, R. M. and ROBINSON, H. (1969a). Studies in the Compositae-Eupatorieae, IX. A review of the genus *Eupatorium* section *Hebeclinium* in Colombia. *Sida* **3**(5), 321–326.

KING, R. M. and ROBINSON, H. (1969b). Studies in the Compositae-Eupatorieae, X. A new species of *Helogyne* Nuttall. *Sida* **3**(5), 327–328.

KING, R. M. and ROBINSON, H. (1969c). Studies in the Compositae-Eupatorieae, XI. Typification of genera. *Sida* **3**(5), 329–342.

KING, R. M. and ROBINSON, H. (1969d). Studies in the Eupatorieae (Compositae). XVI. A monograph of the genus *Decachaeta* DC. *Brittonia* **21**, 275–284, 397.

KING, R. M. and ROBINSON, H. (1969e). Studies in the Eupatorieae (Compositae). XVII. The genus *Erythradenia* (B. L. Robinson) R. M. King and H. Robinson. *Brittonia* **21**, 285.

KING, R. M. and ROBINSON, H. (1970a). Studies in the Eupatorieae (Compositae). XVIII. New combinations in *Fleischmannia*. *Phytologia* **19**, 201–207.

KING, R. M. and ROBINSON, H. (1970b). Studies in the Eupatorieae (Compositae). XIX. New combinations in *Ageratina*. *Phytologia* **12**, 208–229.

KING, R. M. and ROBINSON, H. (1970c). Studies in the Eupatorieae (Compositae). XII. A new genus, *Shinnersia*. *Phytologia* **19**, 297–298.

KING, R. M. and ROBINSON, H. (1970d). Studies in the Eupatorieae (Compositae). XIII. The genus *Conoclinium*. *Phytologia* **19**, 299–300.

KING, R. M. and ROBINSON, H. (1970e). Studies in the Eupatorieae (Compositae). XIV. Another example of dimorphic pollen? *Phytologia* **19**, 301–302.

KING, R. M. and ROBINSON, H. (1970f). Studies in the Eupatorieae (Compositae). XX. New combinations in *Spaniopappus*. *Phytologia* **19**, 303–304.

KING, R. M. and ROBINSON, H. (1970g). Studies in the Eupatorieae (Compositae). XXI. A new genus, *Neomirandea*. *Phytologia* **19**, 305–310.

KING, R. M. and ROBINSON, H. (1970h). Studies in the Compositae-Eupatorieae, XV. *Jaliscoa*, *Macvaughiella*, *Oaxacania*, and *Planaltoa*. *Rhodora* **72**, 100–105.

KING, R. M. and ROBINSON, H. (1970i). The new Synantherology. *Taxon* **19**, (1), 6–11.

KING, R. M. and ROBINSON, H. (1970j). Studies in the Eupatorieae (Compositae). XXII. The genus *Piptothrix*. *Phytologia* **19**, 425–426.

KING, R. M. and ROBINSON, H. (1970k). Studies in the Eupatorieae (Compositae). XXIII. New combinations in *Jaliscoa*. *Phytologia* **19**, 427–428.

KING, R. M. and ROBINSON, H. (1970l). Studies in the Eupatorieae (Compositae). XXIV. A new genus *Stomatanthes*. *Phytologia* **19**, 429–430.

KING, R. M. and ROBINSON, H. (1970m). Studies in the Eupatorieae (Compositae). XXV. A new genus *Eupatoriadelphus*. *Phytologia* **19**, 431–432.

KING, R. M. and ROBINSON, H. (1970n). Studies in the Eupatorieae (Compositae). XXVI. A new genus *Austroeupatorium*. *Phytologia* **19**, 433–435.

KING, R. M. and ROBINSON, H. (1970o). Studies in the Eupatorieae (Compositae). XXVII. A monograph of the genus, *Trichocoronis*. *Phytologia* **19**, 497–500.

KING, R. M. and ROBINSON, H. (1970p). Studies in the Eupatorieae (Compositae). XXVIII. The genus *Praxelis*. *Phytologia* **20**, 193–195.

KING, R. M. and ROBINSON, H. (1970q). Studies in the Eupatorieae (Compositae). XXIX. The genus *Chromolaena*. *Phytologia* **20**, 196–209.

KING, R. M. and ROBINSON, H. (1970r). Studies in the Eupatorieae (Compositae). XXX. The genus *Ayapana*. *Phytologia* **20**, 210–212.

KING, R. M. and ROBINSON, H. (1970s). Studies in the Eupatorieae (Compositae). XXXI. A new genus, *Polyanthina*. *Phytologia* **20**, 213–214.

KING, R. M. and ROBINSON, H. (1970t). Studies in the Eupatorieae (Compositae). XXXII. A new genus, *Neocuatrecasia*. *Phytologia* **20**, 332–333.

KING, R. M. and ROBINSON, H. (1970u). *Eupatorium*, a composite genus of arcto-tertiary distribution. *Taxon* **19**(5), 769-774.

KING, R. M. and ROBINSON, H. (1971a). Studies in the Eupatorieae (Compositae). XXXIII. The genus *Gyptis*. *Phytologia* **21**, 22-25.

KING, R. M. and ROBINSON, H. (1971b). Studies in the Eupatorieae (Compositae). XXXIV. A new genus, *Barrosoa*. *Phytologia* **21**, 26-27.

KING, R. M. and ROBINSON, H. (1971c). Studies in the Eupatorieae (Compositae). XXXV. A new genus, *Lourteigia*. *Phytologia* **21**, 28-30.

KING, R. M. and ROBINSON, H. (1971d). Studies in the Eupatorieae (Compositae). XXXVI. A new genus, *Neobartlettia*. *Phytologia* **21**, 294-297.

KING, R. M. and ROBINSON, H. (1971e). Studies in the Eupatorieae (Compositae). XXXVII. The genus *Hebeclinium*. *Phytologia* **21**, 298-301.

KING, R. M. and ROBINSON, H. (1971f). Studies in the Eupatorieae (Compositae). XXXIX. A new genus, *Guayania*. *Phytologia* **21**, 302-303.

KING, R. M. and ROBINSON, H. (1971g). Studies in the Eupatorieae (Compositae). XL. The genus, *Urolepis*. *Phytologia* **21**, 304-305.

KING, R. M. and ROBINSON, H. (1971h). Studies in the Eupatorieae (Compositae). XLI. The genus, *Eupatoriastrum*. *Phytologia* **21**, 306-307.

KING, R. M. and ROBINSON, H. (1971i). Studies in the Eupatorieae (Asteraceae). XXXVIII. A new genus, *Peteravenia*. *Phytologia* **21**, 394-395.

KING, R. M. and ROBINSON, H. (1971j). Studies in the Eupatorieae (Asteraceae). XLII. A new genus, *Eupatorina*. *Phytologia* **21**, 396-397.

KING, R. M. and ROBINSON, H. (1971k). Studies in the Eupatorieae (Asteraceae). XLIII. A new genus, *Antillia*. *Phytologia* **21**, 398-399.

KING, R. M. and ROBINSON, H. (1971l). Studies in the Eupatorieae (Asteraceae). XLIV. The genus, *Radlkoferotoma*. *Phytologia* **21**, 400-401.

KING, R. M. and ROBINSON, H. (1971m). Studies in the Eupatorieae (Asteraceae). XLV. A new genus, *Fleischmanniopsis*. *Phytologia* **21**, 402-404.

KING, R. M. and ROBINSON, H. (1971n). Studies in the Eupatorieae (Asteraceae). XLVI. A new genus, *Standleyanthus*. *Phytologia* **22**, 41-42.

KING, R. M. and ROBINSON, H. (1971o). Studies in the Eupatorieae (Asteraceae). XLVII. A new genus, *Steyermarkina*. *Phytologia* **22**, 43-45.

KING, R. M. and ROBINSON, H. (1971p). Studies in the Eupatorieae (Asteraceae). XLVIII. The genus, *Critonia*. *Phytologia* **22**, 46-51.

KING, R. M. and ROBINSON, H. (1971q). Studies in the Eupatorieae (Asteraceae). XLIX. A new genus, *Critoniadelphus*. *Phytologia* **22**, 52-53.

KING, R. M. and ROBINSON, H. (1971r). Studies in the Eupatorieae (Asteraceae). L. A new genus, *Urbananthus*. *Phytologia* **22** 54-55.

KING, R. M. and ROBINSON, H. (1971s). Studies in the Eupatorieae (Asteraceae). LI. The Disynaphioid complex. *Phytologia* **22**, 109-110.

KING, R. M. and ROBINSON, H. (1971t). Studies in the Eupatorieae (Asteraceae). LII. A new genus, *Acanthostyles*. *Phytologia* **22**, 111-112.

KING, R. M. and ROBINSON, H. (1971u). Studies in the Eupatorieae (Asteraceae). LIII. A new genus, *Raulinoreitzia*. *Phytologia* **22**, 113-114.

KING, R. M. and ROBINSON, H. (1971v). Studies in the Eupatorieae (Asteraceae). LIV. The genus, *Symphyopappus*. *Phytologia* **22**, 115-117.

KING, R. M. and ROBINSON, H. (1971w). Studies in the Eupatorieae (Asteraceae). LV. The genus, *Dimorpholepis*. *Phytologia* **22**, 118-120.

KING, R. M. and ROBINSON, H. (1971x). Studies in the Eupatorieae (Asteraceae). LVI. A new genus, *Campovassouria*. *Phytologia* **22**, 121–122.
KING, R. M. and ROBINSON, H. (1971y). Studies in the Eupatorieae (Asteraceae). LVII. The genus, *Disynaphia*. *Phytologia* **22**, 123–125.
KING, R. M. and ROBINSON, H. (1971z). Studies in the Eupatorieae (Asteraceae). LXII. A new genus, *Neohintonia*. *Phytologia* **22**, 143–144.
KING, R. M. and ROBINSON, H. (1971aa). Studies in the Eupatorieae (Asteraceae). LXIII. A new genus, *Kyrsteniopsis*. *Phytologia* **22**, 145–146.
KING, R. M. and ROBINSON, H. (1971bb). Studies in the Eupatorieae (Asteraceae). LXIV. The genus, *Koanophyllon*. *Phytologia* **22**, 147–152.
KING, R. M. and ROBINSON, H. (1971cc). Studies in the Eupatorieae (Asteraceae). LVIII. A new genus, *Tamaulipa*. *Phytologia* **22**, 153–155.
KING, R. M. and ROBINSON, H. (1971dd). Studies in the Eupatorieae (Asteraceae). LIX. A new genus, *Steviopsis*. *Phytologia* **22**, 156–157.
KING, R. M. and ROBINSON, H. (1971ee). Studies in the Eupatorieae (Asteraceae). LX. A new genus, *Dyscritogyne*. *Phytologia* **22**, 158–159.
KING, R. M. and ROBINSON, H. (1971ff). Studies in the Eupatorieae (Asteraceae). LXI. Additions to the *Hebeclinium* complex with *Bartlettina*, a new generic name. *Phytologia* **22**, 160–162.
KING, R. M. and ROBINSON, H. (1972a). Studies in the Eupatorieae (Asteraceae). LXV. A new genus, *Neocabreria*. *Phytologia* **23**, 151–152.
KING, R. M. and ROBINSON, H. (1972b). Studies in the Eupatorieae (Asteraceae). LXVI. The genus, *Pachythamnus*. *Phytologia* **23**, 153–154.
KING, R. M. and ROBINSON, H. (1972c). Studies in the Eupatorieae (Asteraceae). LXVII. *Grazielia* nom. nov. for *Dimorpholepis*. *Phytologia* **23**, 305–306.
KING, R. M. and ROBINSON, H. (1972d). Studies in the Eupatorieae (Asteraceae). LXVIII. A new genus, *Conocliniopsis*. *Phytologia* **23**, 307–309.
KING, R. M. and ROBINSON, H. (1972e). Studies in the Eupatorieae (Asteraceae). LXIX. A new genus, *Gyptidium*. *Phytologia* **23**, 310–311.
KING, R. M. and ROBINSON, H. (1972f). Studies in the Eupatorieae (Asteraceae). LXX. A new genus, *Bahianthus*. *Phytologia* **23**, 312–313.
KING, R. M. and ROBINSON, H. (1972g). Studies in the Eupatorieae (Asteraceae). LXXI. A new genus, *Hatschbachiella*. *Phytologia* **23**, 393–394.
KING, R. M. and ROBINSON, H. (1972h). Studies in the Eupatorieae (Asteraceae). LXXII. Notes on the genus *Koanophyllon*. *Phytologia* **23**, 395–396.
KING, R. M. and ROBINSON, H. (1972i). Studies in the Eupatorieae (Asteraceae). LXXIII. The genus, *Ophryosporus*. *Phytologia* **23**, 397–400.
KING, R. M. and ROBINSON, H. (1972j). Studies in the Eupatorieae (Asteraceae). LXXIV. New species of *Critonia*, *Fleischmannia* and *Hebeclinium*. *Phytologia* **23**, 405–408.
KING, R. M. and ROBINSON, H. (1972k). Studies in the Eupatorieae (Asteraceae). LXXV. A new genus, *Cronquistianthus*. *Phytologia* **23**, 409–412.
KING, R. M. and ROBINSON, H. (1972l). Studies in the Eupatorieae (Asteraceae). LXXVI. Additions to the genus *Kyrsteniopsis*. *Phytologia* **24**, 57–59.
KING, R. M. and ROBINSON, H. (1972m). Studies in the Eupatorieae (Asteraceae). LXXVII. Additions to the genus *Steviopsis*. *Phytologia* **24**, 60–62.
KING, R. M. and ROBINSON, H. (1972n). Studies in the Eupatorieae (Asteraceae). LXXVIII. A new genus, *Brickelliastrum*. *Phytologia* **24**, 63–64.

KING, R. M. and ROBINSON, H. (1972o). Studies in the Eupatorieae (Asteraceae). LXXIX. A new genus, *Asanthus*. *Phytologia* **24**, 65–66.
KING, R. M. and ROBINSON, H. (1972p). Studies in the Eupatorieae (Asteraceae). LXXX. A new genus, *Flyriella*. *Phytologia* **24**, 67–69.
KING, R. M. and ROBINSON, H. (1972q). Studies in the Eupatorieae (Asteraceae). LXXXI. The genus, *Phanerostylis*. *Phytologia* **24**, 70–71.
KING, R. M. and ROBINSON, H. (1972r). Studies in the Eupatorieae (Asteraceae). LXXXII. A new genus, *Austrobrickellia*. *Phytologia* **24**, 72–73.
KING, R. M. and ROBINSON, H. (1972s). Studies in the Eupatorieae (Asteraceae). LXXXIII. A new genus, *Pseudobrickellia*. *Phytologia* **24**, 74–76.
KING, R. M. and ROBINSON, H. (1972t). Studies in the Eupatorieae (Asteraceae). LXXXIV. A new genus, *Crossothamnus*. *Phytologia* **24**, 77–78.
KING, R. M. and ROBINSON, H. (1972u). Studies in the Eupatorieae (Asteraceae). LXXXV. Additions to the genus *Ageratina* with a key to the Costa Rican species. *Phytologia* **24**, 79–104.
KING, R. M. and ROBINSON, H. (1972v). Studies in the Eupatorieae (Asteraceae). LXXXVI. Additions to the genus, *Neocuatrecasia*. *Phytologia* **24**, 105–107.
KING, R. M. and ROBINSON, H. (1972w). Studies in the Eupatorieae (Asteraceae). LXXXVII. The genus, *Alomia*. *Phytologia* **24**, 108–111.
KING, R. M. and ROBINSON, H. (1972x). Studies in the Eupatorieae (Asteraceae). LXXXVIII. Additions to the genus, *Ageratum*. *Phytologia* **24**, 112–117.
KING, R. M. and ROBINSON, H. (1972y). Studies in the Eupatorieae (Asteraceae). LXXXIX. A new genus, *Blakeanthus*. *Phytologia* **24**, 118–119.
KING, R. M. and ROBINSON, H. (1972z). Studies in the Eupatorieae (Asteraceae). XC. The genus, *Campuloclinium*. *Phytologia* **24**, 170–172.
KING, R. M. and ROBINSON, H. (1972aa). Studies in the Eupatorieae (Asteraceae). XCI. A new genus, *Macropodina*. *Phytologia* **24**, 173–175.
KING, R. M. and ROBINSON, H. (1972bb). Studies in the Eupatorieae (Asteraceae). XCII. The genus, *Trichogonia*. *Phytologia* **24**, 176–179.
KING, R. M. and ROBINSON, H. (1972cc). Studies in the Eupatorieae (Asteraceae). XCIII. A new genus, *Trichogoniopsis*. *Phytologia* **24**, 180–181.
KING, R. M. and ROBINSON, H. (1972dd). Studies in the Eupatorieae (Asteraceae). XCIV. A new genus, *Platypodanthera*. *Phytologia* **24**, 182–183.
KING, R. M. and ROBINSON, H. (1972ee). Studies in the Eupatorieae (Asteraceae). XCV. Additions to the genus *Barrosoa*. *Phytologia* **24**, 184.
KING, R. M. and ROBINSON, H. (1972ff). Studies in the Eupatorieae (Asteraceae). XCVI. A new genus, *Lasiolaena*. *Phytologia* **24**, 185–186.
KING, R. M. and ROBINSON, H. (1972gg). Studies in the Eupatorieae (Asteraceae). XCVII. A new genus, *Dasycondylus*. *Phytologia* **24**, 187–191.
KING, R. M. and ROBINSON, H. (1972hh). Studies in the Eupatorieae (Asteraceae). XCVIII. A new genus, *Diacranthera*. *Phytologia* **24**, 192–194.
KING, R. M. and ROBINSON, H. (1972ii). Studies in the Eupatorieae (Asteraceae). XCIX. A new genus, *Amolinia*, and a new combination in *Bartlettina*. *Phytologia* **24**, 265–266.
KING, R. M. and ROBINSON, H. (1972jj). Studies in the Eupatorieae (Asteraceae). C. A key to the genera of Nueva Galicia, Mexico. *Phytologia* **24**, 267–280.
KING, R. M. and ROBINSON, H. (1972kk). Studies in the Eupatorieae (Asteraceae). CI. New species of *Fleischmannia* and *Neomirandea*. *Phytologia* **24**, 281–284.

KING, R. M. and ROBINSON, H. (1972ll). Studies in the Eupatorieae (Asteraceae). CII. A new genus, *Condylidium*. *Phytologia* **24**, 380–381.

KING, R. M. and ROBINSON, H. (1972mm). Studies in the Eupatorieae (Asteraceae). CIII. A new genus, *Ayapanopsis*. *Phytologia* **24**, 382–386.

KING, R. M. and ROBINSON, H. (1972nn). Studies in the Eupatorieae (Asteraceae). CIV. A new genus, *Gongrostylus*. *Phytologia* **24**, 387–388.

KING, R. M. and ROBINSON, H. (1972oo). Studies in the Eupatorieae (Asteraceae). CV. A new genus, *Heterocondylus*. *Phytologia* **24**, 389–392.

KING, R. M. and ROBINSON, H. (1972pp). Studies in the Eupatorieae (Asteraceae). CVI. A new genus, *Gymnocondylus*. *Phytologia* **24**, 393–394.

KING, R. M. and ROBINSON, H. (1972qq). Studies in the Eupatorieae (Asteraceae). CVII. A new genus, *Alomiella*. *Phytologia* **24**, 395–396.

KING, R. M. and ROBINSON, H. (1972rr). Studies in the Eupatorieae (Asteraceae). CVIII. A new genus, *Condylopodium*. *Phytologia* **24**, 397–400.

KING, R. M. and ROBINSON, H. (1972ss). Studies in the Eupatorieae (Asteraceae). CIX. A new genus, *Acritopappus*. *Phytologia* **24**, 401–403.

KING, R. M. and ROBINSON, H. (1972tt). Studies in the Eupatorieae (Asteraceae). CX. Additions to the genus, *Campuloclinium*. *Phytologia* **24**, 404–406.

KING, R. M. and ROBINSON, H. (1972uu). Studies in the Eupatorieae (Asteraceae). CXI. Additions to the genus, *Ophryosporus*. *Phytologia* **25**, 65–67.

KING, R. M. and ROBINSON, H. (1972vv). *Neomirandea allenii*, a new epiphytic composite of the American rain forest. *Rhodora* **74**, 272–275.

KING, R. M. and ROBINSON, H. (1973a). Studies in the Eupatorieae (Asteraceae). CXII. A new species of *Ferreyrella*. *Phytologia* **26**, 167–169.

KING, R. M. and ROBINSON, H. (1973b). Studies in the Eupatorieae (Asteraceae). CXIII. A new genus, *Matudina*. *Phytologia* **26**, 170–173.

KING, R. M. and ROBINSON, H. (1973c). Studies in the Eupatorieae (Asteraceae). CXIV. The genera of Barro Colorado Island, Panama. *Phytologia* **27**, 233–240.

KING, R. M. and ROBINSON, H. (1973d). Studies in the Eupatorieae (Asteraceae). CXV. A new genus and species, *Pseudokyrsteniopsis perpetiolata*. *Phytologia* **27**, 241–244.

KING, R. M. and ROBINSON, H. (1973e). Studies in the Eupatorieae (Asteraceae). CXVI. New species of *Neomirandea*. *Phytologia* **27**, 245–251.

KING, R. M. and ROBINSON, H. (1974a). Studies in the Eupatorieae (Asteraceae). CXVII. A new species of *Oxylobus* from Oaxaca, Mexico. *Phytologia* **27**, 385–386.

KING, R. M. and ROBINSON, H. (1974b). Studies in the Eupatorieae (Asteraceae). CXVIII. New species of *Ageratum*, *Fleischmannia* and *Hebeclinium* from northern South America. *Phytologia* **27**, 387–394.

KING, R. M. and ROBINSON, H. (1974c). Studies in the Eupatorieae (Asteraceae). CXIX. Additions to the genera *Cronquistianthus*, *Helogyne* and *Neocuatrecasia* from Peru. *Phytologia* **27**, 395–401.

KING, R. M. and ROBINSON, H. (1974d). Studies in the Eupatorieae (Asteraceae). CXX. Additions to the genus *Koanophyllon* in Panama. *Phytologia* **28**, 67–72.

KING, R. M. and ROBINSON, H. (1974e). Studies in the Eupatorieae (Asteraceae). CXXI. Additions to the genus *Fleischmannia*. *Phytologia* **28**, 73–96.

KING, R. M. and ROBINSON, H. (1974f). Studies in the Eupatorieae (Asteraceae). CXXII. A new genus, *Sartorina*. *Phytologia* **28**, 97–100.

KING, R. M. and ROBINSON, H. (1974g). Studies in the Eupatorieae (Asteraceae). CXXIII. Additions to the genus *Mikania*. *Phytologia* **28**, 272–281.

KING, R. M. and ROBINSON, H. (1974h). Studies in the Eupatorieae (Asteraceae). CXXIV. A new genus, *Eitenia*. *Phytologia* **28**, 282–285.

KING, R. M. and ROBINSON, H. (1974i). Studies in the Eupatorieae (Asteraceae). CXXV. Additions to the genus, *Bartlettina*. *Phytologia* **28**, 286–293.

KING, R. M. and ROBINSON, H. (1974j). Studies in the Eupatorieae (Asteraceae). CXXVI. A new species of *Ageratum*. *Phytologia* **28**, 491–493.

KING, R. M. and ROBINSON, H. (1974k). Studies in the Eupatorieae (Asteraceae). CXXVIII. Four additions to the genus *Ageratina* from Mexico and Central America. *Phytologia* **28**, 494–502.

KING, R. M. and ROBINSON, H. (1974l). Studies in the Eupatorieae (Asteraceae). CXXVII. Additions to the American and Pacific Adenostemmatinae. *Adenostemma, Gymnocoronis* and *Sciadocephala*. *Phytologia* **29**, 1–20.

KING, R. M. and ROBINSON, H. (1974m). Studies in the Eupatorieae (Asteraceae). CXXIX. A new genus, *Vittetia*. *Phytologia* **29**, 121–122.

KING, R. M. and ROBINSON, H. (1974n). Studies in the Eupatorieae (Asteraceae). CXXX. Notes on *Campuloclinium, Koanophyllon, Mikania* and *Symphyopappus*. *Phytologia* **29**, 123–129.

KING, R. M. and ROBINSON, H. (1974o). Studies in the Eupatorieae (Asteraceae). CXXXII. The genus, *Phalacraea*. *Phytologia* **29**, 251–256.

KING, R. M. and ROBINSON, H. (1974p). Studies in the Eupatorieae (Asteraceae). CXXXI. A new genus, *Guevaria*. *Phytologia* **29**, 257–263.

KING, R. M. and ROBINSON, H. (1974q). Studies in the Eupatorieae (Asteraceae). CXXXIII. A new genus, *Piqueriella*. *Phytologia* **29**, 264–266.

KING, R. M. and ROBINSON, H. (1975a). Studies in the Eupatorieae (Asteraceae). CXXXIV. A new species of *Sciadocephala* from Panama. *Phytologia* **29**, 343–346.

KING, R. M. and ROBINSON, H. (1975b). Studies in the Eupatorieae (Asteraceae). CXXXV. A new species of *Ageratina* from Panama. *Phytologia* **29**, 347–350.

KING, R. M. and ROBINSON, H. (1975c). Studies in the Eupatorieae (Asteraceae). CXXXVI. Four new species of *Neomirandea*. *Phytologia* **29**, 351–361.

KING, R. M. and ROBINSON, H. (1975d). Studies in the Eupatorieae (Asteraceae). CXXXVII. Two new species of *Neomirandea*. *Phytologia* **30**, 9–14.

KING, R. M. and ROBINSON, H. (1975e). Studies in the Eupatorieae (Asteraceae). CXXXIX. A new genus, *Aristeguietia*. *Phytologia* **30**, 217–220.

KING, R. M. and ROBINSON, H. (1975f). Studies in the Eupatorieae (Asteraceae). CXL. A new genus, *Grosvenoria*. *Phytologia* **30**, 221–222.

KING, R. M. and ROBINSON, H. (1975g). Studies in the Eupatorieae (Asteraceae). CXLI. A new genus, *Asplundianthus*. *Phytologia* **30**, 223–228.

KING, R. M. and ROBINSON, H. (1975h). Studies in the Eupatorieae (Asteraceae). CXLII. A new genus, *Badilloa*. *Phytologia* **30**, 229–234.

KING, R. M. and ROBINSON, H. (1975i). Studies in the Eupatorieae (Asteraceae). CXXXVIII. A new genus, *Critoniella*. *Phytologia* **30**, 284–285.

KING, R. M. and ROBINSON, H. (1975j). Studies in the Eupatorieae (Asteraceae). CXLV. A new species of *Bartlettina*. *Phytologia* **31**, 62–65.

King, R. M. and Robinson, H. (1975k). Studies in the Eupatorieae (Asteraceae). CXLIII. A new genus, *Austrocritonia*. *Phytologia* **31**, 115–117.

King, R. M. and Robinson, H. (1975l). Studies in the Eupatorieae (Asteraceae). CXLIV. A new genus, *Viereckia*. *Phytologia* **31**, 118–121.

King, R. M. and Robinson, H. (1975m). Studies in the Eupatorieae (Asteraceae). CXLVI. Two new species of *Fleischmannia* from Central America. *Phytologia* **31**, 305–310.

King, R. M. and Robinson, H. (1975o). Studies in the Eupatorieae (Asteraceae). CXLVII. Additions to the genera *Amboroa*, *Ayapanopsis*, and *Hebeclinium* in South America. *Phytologia* **31**, 311–316.

King, R. M. and Robinson, H. (1975p). Studies in the Eupatorieae (Asteraceae). CXLVIII. A new species of *Lomatozoma*. *Phytologia* **32**, 246–249.

King, R. M. and Robinson, H. (1975q). Studies in the Eupatorieae (Asteraceae). CXLIX. A new genus, *Osmiopsis*. *Phytologia* **32**, 250–251.

King, R. M. and Robinson, H. (1975r). Studies in the Eupatorieae (Asteraceae). CL. Limits of the genus *Koanophyllon*. *Phytologia* **32**, 252–267.

King, R. M. and Robinson, H. (1975s). Studies in the Eupatorieae (Asteraceae). CLI. A new genus, *Grisebachianthus*. *Phytologia* **32**, 268–270.

King, R. M. and Robinson, H. (1975t). Studies in the Eupatorieae (Asteraceae). CLII. A new genus, *Imeria*. *Phytologia* **32**, 271–272.

King, R. M. and Robinson, H. (1975u). Studies in the Eupatorieae (Asteraceae). CLIII. A new genus, *Lorentzianthus*. *Phytologia* **32**, 273–274.

King, R. M. and Robinson, H. (1975v). Studies in the Eupatorieae (Asteraceae). CLIV. A new genus, *Chacoa*. *Phytologia* **32**, 275–276.

King, R. M. and Robinson, H. (1975w). Studies in the Eupatorieae (Asteraceae). CLV. A new genus, *Idiothamnus*. *Phytologia* **32**, 277–282.

King, R. M. and Robinson, H. (1975x). Studies in the Eupatorieae (Asteraceae). CLVI. Various new combinations. *Phytologia* **32**, 283–285.

King, R. M. and Robinson, H. (1975y). New species of *Stomatanthes* from Africa (Eupatorieae, Compositae). *Kew Bull.* **30**, 463–465 (1976).

King, R. M. and Robinson, H. (1975z). Eupatorieae. *In* "Flora of Panama" (R. E. Woodson and R. W. Schery, eds). *Ann. Mo. bot. Gdn.* **62**, 888–1004.

King, R. M. and Robinson, H. (1976a). Studies in the Eupatorieae (Asteraceae). CLVII. A new genus, *Revealia* from Mexico. *Phytologia* **33**, 277–280.

King, R. M. and Robinson, H. (1976b). Studies in the Eupatorieae (Asteraceae). CLVIII. A new genus, *Adenocritonia* from Jamaica. *Phytologia* **33**, 281–284.

King, R. M. and Robinson, H. (1976c). Studies in the Eupatorieae (Asteraceae). CLIX. Additions to the genus, *Ayapana*. *Phytologia* **34**, 57–66.

King, R. M., Kyhos, D. W., Powell, M., Raven, R. H. and Robinson, H. (1976). Chromosome numbers in Compositae. XIII Eupatorieae.

Lessing, C. F. (1832). "Synopsis Generum Compositarum". Berlin.

Linnaeus, C. (1753). "Species Plantarum", 2 vols. Stockholm.

Linnaeus, C. (1763). "Species Plantarum" (2nd edition), 2 vols. Stockholm.

Malme, G. O. (1933). Eupatorieae *In* Compositae Paranenses Dusenianae. *K. Svenska Vetensk.-Akad. Handlingar* **12**(2), 28–62.

Mattfeld, J. (1923). Compositae. *In* "Plantae Lützelburgianae brasiliensis" (R. Pilger, ed.), *Notizbl. bot. Gart. Mus. Berl.* **8**, 428–451.

MATTFELD, J. (1938). Compositae. *In* Neue Arten aus Ecuador. *Notizbl. bot. Gart. Mus. Berl.* **14**, 41–44.

McVAUGH, R. (1972). Tribe III. Eupatorieae *in* Compositarum Mexicanarum Pugillus. *Contr. Univ. Mich. Herb.* **9**(4), 378–408.

MONTGOMERY, J. D. and FAIRBROTHERS, D. E. (1970). A biosystematic study of the *Eupatorium rotundifolium* complex (Compositae). *Brittonia* **22**, 134–150.

PARAY, L. (1953). Las Compuestas del Valle Central de Mexico. *Boln Soc. bot. México* **15**, 1–12.

PARAY, L. (1954). Nuevas Fanerógamas de México. *Boln Soc. bot. México* **16**, 20–25.

PARAY, L. (1956). El genero *Eupatorium* en la Valle Central de Mexico. *Boln Soc. bot. México* **19**, 1–15.

PARAY, L. (1958). Nuevas Compuestas de Mexico. *Boln Soc. bot. México* **22**, 1–12.

ROBINSON, B. L. (1892). Descriptions of new plants collected in Mexico by C. G. Pringle in 1890 and 1891, with notes on a few other species. *Proc. Am. Acad. Arts Sci.* **27**, 165–185.

ROBINSON, B. L. (1900). New phanerogams, chiefly Gamopetalae, from Mexico and Central America. *Proc. Am. Acad. Arts Sci.* **35**, 323–342.

ROBINSON, B. L. (1901). New species and newly noted synonymy among the Spermatophytes of Mexico and Central America. *Proc. Am. Acad. Arts Sci.* **36**, 471–488.

ROBINSON, B. L. (1903). *Eupatorium hypomalacum in* Smith, J. D., Undescribed plants from Guatemala and other Central American Republics. *Bot. Gaz.* **35**, 4.

ROBINSON, B. L. (1904a). Diagnoses and synonymy of some Mexican and Central American Eupatoriums. *Proc. Boston Soc. nat. Hist.* **31**, 247–254.

ROBINSON, B. L. (1904b). Synopsis of the Mikanias of Costa Rica. *Proc. Boston Soc. nat. Hist.* **31**, 254–257.

ROBINSON, B. L. (1905). Diagnoses and notes relating to American Eupatorieae. *Proc. Am. Acad. Arts Sci.* **41**, 271–278.

ROBINSON, B. L. (1906). Studies in the Eupatorieae. I. Revision of the genus *Piqueria*. II. Revision of the genus *Ophryosporus*. III. The genus *Helogyne* and its synonyms. IV. Diagnoses and synonymy of Eupatorieae and of certain other Compositae which have been classed with them. *Proc. Am. Acad. Arts Sci.* **42**, 3–48.

ROBINSON, B. L. (1907). New or otherwise noteworthy Spermatophytes chiefly from Mexico. *Proc. Am. Acad. Arts Sci.* **43**, 21–48.

ROBINSON, B. L. (1909). Diagnoses and transfers of tropical American phanerogams. *Proc. Am. Acad. Arts Sci.* **44**, 613–626.

ROBINSON, B. L. (1911a). On the classification of certain Eupatorieae. *Proc. Am. Acad. Arts Sci.* **47**, 191–202.

ROBINSON, B. L. (1911b). Revision of the genus *Barroetea*. *Proc. Am. Acad. Arts Sci.* **47**, 202–206.

ROBINSON, B. L. (1913a). A key to the genera of the Compositae-Eupatorieae. *Proc. Am. Acad. Arts. Sci.* **49**, 429–437.

ROBINSON, B. L. (1913b). Revisions of *Alomia*, *Ageratum* and *Oxylobus*. *Proc. Am. Acad. Arts Sci.* **49**, 438–491.

ROBINSON, B. L. (1916). New, reclassified or otherwise noteworthy spermatophytes. *Proc. Am. Acad. Arts Sci.* **51**, 527–540.

ROBINSON, B. L. (1917). A monograph of the genus *Brickellia*. *Mem. Gray Herb.* **1**, 3–151.

ROBINSON, B. L. (1918a). Diagnoses and notes relating to tropical American Eupatorieae. *Proc. Am. Acad. Arts Sci.* **54**, 235–263.

ROBINSON, B. L. (1918b). A descriptive revision of the Colombian Eupatoriums. *Proc. Am. Acad. Arts Sci.* **54**, 264–330.

ROBINSON, B. L. (1918c). Keyed recensions of the Eupatoriums of Venezuela and Ecuador. *Proc. Am. Acad. Arts Sci.* **54**, 331–367.

ROBINSON, B. L. (1919a). On tropical American Compositae, chiefly Eupatorieae. *Proc. Am. Acad. Arts Sci.* **55**, 3–41.

ROBINSON, B. L. (1919b). A recension of the Eupatoriums of Peru. *Proc. Am. Acad. Arts Sci.* **55**, 42–88.

ROBINSON, B. L. (1920a). Further diagnoses and notes on tropical American Eupatorieae. *Contr. Gray Herb. n.s.* **61**, 3–30.

ROBINSON, B. L. (1920b). The Eupatoriums of Bolivia. *Contr. Gray Herb. n.s.* **61**, 30–80.

ROBINSON, B. L. (1922a). Records preliminary to a general treatment of the Eupatorieae, I. *Contr. Gray Herb. n.s.* **64**, 3–21.

ROBINSON, B. L. (1922b). The Mikanias of Northern and Western South American. *Contr. Gray Herb. n.s.* **64**, 21–116.

ROBINSON, B. L. (1922c). *Dyscritothamnus*, a new genus of Compositae. *Contr. Gray Herb. n.s.* **65**, 24–28, pl.

ROBINSON, B. L. (1922d). Records preliminary to a general treatment of the Eupatorieae, II. *Contr. Gray Herb. n.s.* **65**, 46–54.

ROBINSON, B. L. (1923). Records preliminary to a general treatment of the Eupatorieae, III. *Contr. Gray Herb. n.s.* **68**, 3–43.

ROBINSON, B. L. (1924). Records preliminary to a general treatment of the Eupatorieae, IV. *Contr. Gray Herb. n.s.* **73**, 3–31.

ROBINSON, B. L. (1925). Records preliminary to a general treatment of the Eupatorieae, V. *Contr. Gray Herb. n.s.* **75**, 3–15.

ROBINSON, B. L. (1926a). Records preliminary to a general treatment of the Eupatorieae, VI. *Contr. Gray Herb. n.s.* **77**, 3–62.

ROBINSON, B. L. (1926b). The woody species of *Eupatorium* and *Ophryosporus* occurring in Mexico. *In* "Trees and Shrubs of Mexico" (P. C. Standley, ed.). *Contr. U.S. nat. Herb.* **23**, 1432–1470.

ROBINSON, B. L. (1928). Records preliminary to a general treatment of the Eupatorieae, VII. *Contr. Gray Herb. n.s.* **80**, 3–42.

ROBINSON, B. L. (1930a). *Ageratum* (Coelestina) *Standleyi in* Standley, P. C., Woody Plants of Siguatepeque, Honduras. *J. Arnold Arbor.* **11**, 44.

ROBINSON, B. L. (1930b). *Eupatorium* (Subimbricata) *hondurense*. *In* "Woody plants of Siguatepeque, Honduras" (P. C. Standley, ed.). *Jour. Arnold Arbor.* **11**, 44–45.

ROBINSON, B. L. (1930c). Records preliminary to a general treatment of the Eupatorieae, VIII. *Contr. Gray Herb. n.s.* **90**, 3–36.

ROBINSON, B. L. (1930d). Observations on the genus *Stevia*. *Contr. Gray Herb. n.s.* **90**, 36–58, pl.

ROBINSON, B. L. (1930e). The Stevias of the Argentine Republic. *Contr. Gray Herb.* n.s. **90**, 58–79.
ROBINSON, B. L. (1930f). The Stevias of Paraguay. *Contr. Gray Herb.* n.s. **90**, 79–90.
ROBINSON, B. L. (1930g). The Stevias of North America. *Contr. Gray Herb.* n.s, **90**, 90–160.
ROBINSON, B. L. (1931a). Compositae-Eupatorieae (of the Tyler-Duida Expedition). *Bull. Torrey bot. Club* **58**, 482–485.
ROBINSON, B. L. (1931b). Records preliminary to a general treatment of the Eupatorieae, IX. *Contr. Gray Herb.* n.s. **96**, 3–27.
ROBINSON, B. L. (1931c). The Stevias of Colombia. *Contr. Gray Herb.* n.s. **96**, 28–36.
ROBINSON, B. L. (1931d). The Stevias of Venezuela. *Contr. Gray Herb.* n.s. **96**, 37–43.
ROBINSON, B. L. (1931e). The Stevias of Ecuador. *Contr. Gray Herb.* n.s. **96**, 43–49.
ROBINSON, B. L. (1932a). Records preliminary to a general treatment of the Eupatorieae, X. *Contr. Gray Herb.* n.s. **100**, 3–19.
ROBINSON, B. L. (1932b). The Stevias of Peru. *Contr. Gray Herb.* n.s. **100**, 20–36.
ROBINSON, B. L. (1932c). The Stevias of Bolivia. *Contr. Gray Herb.* n.s. **100**, 36–39.
ROBINSON, B. L. (1933). Taxonomic notes on several South American Eupatoriums. *Ostenia* 349–358.
ROBINSON, B. L. (1934a). Records preliminary to a general treatment of the Eupatorieae, XI. *Contr. Gray Herb.* n.s. **104**, 3–49.
ROBINSON, B. L. (1934b). The variability of two wide-ranging species of *Mikania*. *Contr. Gray Herb.* n.s. **104**, 49–55.
ROBINSON, B. L. (1934c). *Mikania scandens* and its near relatives. *Contr. Gray Herb.* n.s. **104**, 55–71.
ROBINSON, B. L. (1934d). New Compositae-Eupatorieae from Brazil. *Candollea* **5**, 170–174.
ROBINSON, B. L. and GREENMAN, J. M. (1896). Synopsis of the Mexican and Central American species of the genus *Mikania*. *Proc. Am. Acad. Arts Sci.* **32**, 10–13.
ROBINSON, B. L. and GREENMAN, J. M. (1899). Supplementary notes upon *Calea*, *Tridax*, and *Mikania*. *Proc. Boston Soc. nat. Hist.* **29**, 105–108.
ROBINSON, B. L. and SEATON, H. E. (1893). Additions to the phanerogamic flora of Mexico discovered by C. G. Pringle in 1891–1892. *Proc. Am. Acad. Arts Sci.* **28**, 103–115.
ROBINSON, H. (1970). South American species of *Stomatanthes* (Eupatorieae, Compositae). *Phytologia* **20**, 334–338.
RZEDOWSKI, J. (1970). Estudio sistematico del genero *Microspermum* (Compositae). *Boln. Soc. bot. México* **31**, 49–107.
RZEDOWSKI, J. (1972). Dos especies nuevas del genero *Microspermum* (Compositae) del Estado de Jalisco (Mexico). *Boln Soc. bot. México* **32**, 77–86.
SCHREBER, J. C. D. VON (1791). "Genera Plantarum" (8th edition), Vol. 2. Frankfurt.

SCHULTZ-BIPONTINUS, C. H. (1850). *Fleischmannia* novum plantarum genus. *Flora, Jena* **27**, 417–418.

SCHULTZ-BIPONTINUS, C. H. (1856). Tribus II. Eupatoriaceae. *In* "The Botany of the Voyage of H.M.S. Herald" (B. C. Seemann, ed.), pp. 298–301. Flora of Northwestern Mexico.

SHINNERS, L. H. (1943). A revision of the *Liatris scariosa* complex. *Am. Midl. Nat.* **29**, 27–41.

SHINNERS, L. H. (1946). Revision of the genus *Kuhnia* L. *Wrightia* **1**, 122–144.

SHINNERS, L. H. (1951). Notes on Texas Compositae—VII. *Fld. Lab.* **19**, 74–82.

SHINNERS, L. H. (1955). Notes on Compositae—X. *Fld Lab.* **23**, 34–36.

SHINNERS, L. H. (1971). *Kuhnia* L. transferred to *Brickellia* Ell. (Compositae). *Sida* **4**, 274.

SMALL, J. (1917–1919). The origin and development of the Compositae. *New Phytol.* **16**, 157–177, 198–221, 253–276; **17**, 13–40, 69–94, 114–142, 200–230; **18**, 1–35, 65–89, 129–176, 201–234.

SOLEREDER, H. (1908). "Systematic Anatomy of the Dicotyledons" (translation). Clalendon Press, Oxford.

SPACH, E. (1841). "Histoire naturelle des végétaux. Phanérogames", Vol. 10. Paris.

STANDLEY, P. C. 1938. Compositae. *In* Flora of Costa Rica. *Field Mus. nat. Hist. (Bot.)* **18**(4), 1418–1538.

STEETZ, J. (1854). Tribus II. Eupatoriaceae. *In* "The Botany of the Voyage of H.M.S. Herald". (B. C. Seemann, ed.), pp. 142–151. Flora of the isthmus of Panama.

STEYERMARK, J. A. (1953). Compositae *in* Botanical exploration in Venezuela—III. *Fieldiana, Bot.* **28**(3), 620–678.

SULLIVAN, V. I. (1975). Pollen and pollination in the genus *Eupatorium* (Compositae). *Can. J. Bot.* **53**(6), 582–589.

TAUBERT, P. (1896). Beiträge zur Kenntnis der Flora des centralbrasilianischen Staates Goyaz mit einer pflanzengeographischen Skizze von E. Ule. Engler. *Bot. Jahrb.* **21**, 402–457.

TOMAN, J., HARMATHA, J. and NOVOTNY, L. (1968). Verwandschaftsbeziehungen der Gattung *Adenostyles* und die Berechtigung ihrer Einreihung in die Tribus Senecioneae. *Preslia* **40**, 122–132.

TOURNEFORT, J. P. DE (1700). "Institutiones Rei Herbariae", 3 vols. Paris.

TURCZANINOW, N. (1847–1848). Decades 3, 4, et 5 generum adhuc non descriptorum, adjectis descriptionibus nonnullarum specierum Myrtacearum xerocarpicarum atque Umbelliferarum imperfectarum. *Moscou Soc. nat. Bull.* **20**, 148–174; **21**, 570–591.

URBAN, I. (1925). *Ciceronia in* Sertum antillarum. XXIII. *Fedde Repert.* 224–225.

URBAN, I. (1931). Compositae *in* Plantae Haitienses et Domingenses novae vel rariores IX. a cl. E. L. Ekman 1924–1930 lectae. *Ark. Bot.* **23A**(11), 1–103.

WATSON, T. J. (1973). Chromosome numbers in Compositae from the Southwestern United States. *S. West. Nat.* **18**, 117–124.

WIEGAND, K. M. (1920). *Eupatorium purpureum* and its allies. *Rhodora* **22**, 57–70.
WIEGAND, K. M. and WEATHERBY, C. A. (1937). The nomenclature of the verticillate Eupatoria. *Rhodora* **39**, 297–306.
WILLDENOW, C. L. (1803). Caroli a Linné Species plantarum. . . . ed. 4. Classis XIX. *Syngenesia.* 3(3), 1475–2409.
WOOTEN, J. W. and CLEWELL, A. F. (1971). *Fleischmannia* and *Conoclinium* (Compositae, Eupatorieae) in eastern North America. *Rhodora* **73**, 566–574.

References added in proof:
CABRERA, A. L. (1976). Una segunda especie del genero *Ascidiogyne* (Compositae). *Hickenia* **1**, 1–3.
KEIL, D. J. and PINKAVA, D. J. (1976). Chromosome counts and taxonomic notes for Compositae from the United States and Mexico. *Am. J. Bot.* **63**, 1393–1403.
KEIL, D. J. and PINKAVA, D. J. (1977). Reinstatement of *Carminatia* DC. (Compositae: Eupatorieae). *Phytologia* **35**, 323.

Chapter 16
Eupatorieae—chemical review

XORGE A. DOMÍNGUEZ
Department of Chemistry, Institute of Technology, Monterrey, Mexico

Abstract. About 10% of the species of this tribe have been examined chemically; for more than half of the 50 genera there are no reports on the chemistry. *Ageratum, Brickellia, Carphephorus, Eupatorium, Liatris, Mikania, Piqueria, Stevia* are the genera that have yielded the most interesting chemical substances. Some species of *Ageratum, Carphephorus* and *Eupatorium* contain essential oils, from which benzofuran and coumarin derivatives have been isolated. Euparin, a benzofuran derivative, has been found in the roots of *Eupatorium* and *Liatris* species. Ursene, oleanene and friedalene triterpenoids have been isolated from *Eupatorium* and other genera. A high percentage of *Eupatorium, Liatris* and *Mikania* species are the sources of germacranolide and guaianolide-like sesquiterpene lactones. These three genera, together with *Stevia* and *Brickellia*, contain flavonoids polyoxygenated in the A-ring. Polyacetylenes have been isolated from 14 species belonging to eight genera. From five *Eupatorium* species, pyrrolizidine alkaloids have been isolated.

CONTENTS

Chemical pattern of the tribe	487
Low molecular weight constituents	488
Terpenoids	488
Flavonoids and other phenolics	494
Alkaloids	498
Other compounds	498
Economic and pharmaceutical uses	499
References	500

CHEMICAL PATTERN OF THE TRIBE

For almost 80% of the members of this tribe there is no record of any chemical study. Some members produce essential oils and biogenetically related resins. The essential oils from some *Ageratum, Liatris* and *Carphephorus* species contain mainly aromatic terpene derivatives like ageratochromene or coumarin derivatives. From some *Eupatorium*, thymolhydroquinone dimethyl ether, *p*-cymol, limonene and neryl acetate have been isolated or detected.

Several benzofuran derivatives of euparin have been found in the roots and aerial parts of *Eupatorium* and *Liatris*. From one species of *Adenostyles* three keto-furan eremophilane derivatives were isolated. Among the non-volatile sesquiterpene components are germacranolide and guaianolide-type lactones isolated from *Eupatorium*, *Mikania* and *Liatris* species. Only one pseudoguaianolide-type lactone has been reported, in a *Stevia* sp. Diterpenes are rare; the only ones reported are a kauren-16-en-19-oic acid glycoside from *Mikania*, two kaurene derivatives in *Stevia* and geranylnerol derivatives from *Liatris*.

Triterpenes occur frequently, but rarely in high concentration; most are found as esters, a few as glycosides or free. Most of the triterpenes are oleanene- and ursene-type and are constituents of *Eupatorium* and some *Mikania* species. In the latter genus, friedelene and dammarene types have also been found. There are no reports on tetraterpenoids.

Flavonoids are the most widely studied phenolic constituents of the Eupatorieae. Fifteen *Eupatorium*, 10 *Liatris*, three *Brickellia* and one *Mikania* species contain polyhydroxylated flavonols, flavones and their methyl ethers; one dihydroflavone has been reported. The flavone *C*-glycosides vicenin-1 and -2 have been isolated from several *Eupatorium* and *Liatris* species. The presence of alkaloids has been reported in *Ageratum*, *Alomia*, *Adenostemma*, *Liatris*, and *Eupatorium* species. From three *Eupatorium* species five related pyrrolizidine ester alkaloids have been isolated. Of the 33 Eupatorieae species so far studied 14 contain polyacetylenes, some of which are present as thiophenes. The acetylene-containing plants are widely distributed at the generic level. There are only few reports on the alkanes present in the tribe and the only polyalcohol found so far is inositol.

LOW MOLECULAR WEIGHT CONSTITUENTS

Terpenoids

Essential oils. Some *Ageratum*, *Eupatorium*, *Liatris* and *Carphephorus* species, on steam distillation or extraction, give volatile oils containing aromatic terpene derivatives. Ageratochromene (**1**) has been isolated from *Ageratum mexicanum* and *A. conyzoides* (Alertsen, 1955) and a benzofuran derivative (**2**) has been obtained from *Liatris graminifolia* and *L. spicata* (Karlsson *et al.*, 1973). The volatile compounds from *Carphephorus odoratissimus* (deertongue) are predominantly coumarin, dihydrocoumarin, 2,3-benzofuran and the common mono- and sesquiterpenes (Karlsson *et al.*, 1972). The essential oils of *C. corymbosus* and *C. paniculatus* contain traces of coumarin (Karlsson *et al.*, 1972). The extracts from *Alomia fastigiata*, contain 6,7-methylenedioxycoumarin (Pozzitti, 1966). *Eupatorium stoechadosmum* has yielded coumarins and the only reported quinones (Shimoda and Sawada, 1957).

The essential oil from *Eupatorium capillifolium* is rich in limonene (Domínguez *et al.*, 1969). The terpenes from *E. perfoliatum* were examined (Schindler, 1953). In *E. triplinerve*, thymol hydroquinone dimethyl ether was found as the main component, in *E. fortunei* p-cymol, neryl acetate and the methyl ether of thymol were identified (Hegnauer, 1964). Thymol methyl ether and thymolhydroquinone have been found in three *Carphephorus* species and in *Liatris spicata*, *L. elegans* and *L. gracilis*, which also contained thymolhydroquinone dimethyl ether and benzofuran derivatives (Karlsson *et al.*, 1973). The benzofuran euparin (**2**) has been

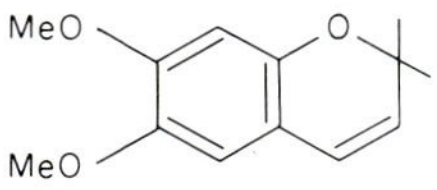

(1) Ageratochromene

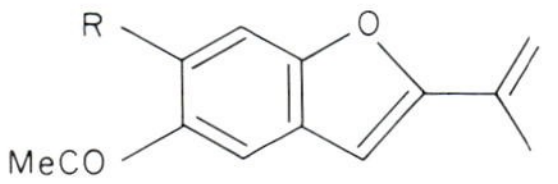

(2) Euparin R = OH
(3) Dehydrotremetone R = H

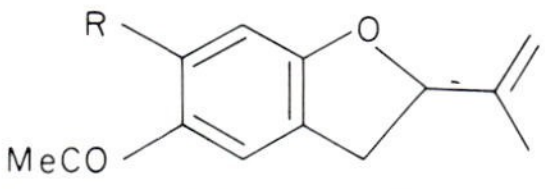

(4) Tremetone R = H
(5) Hydroxytremetone R = OH

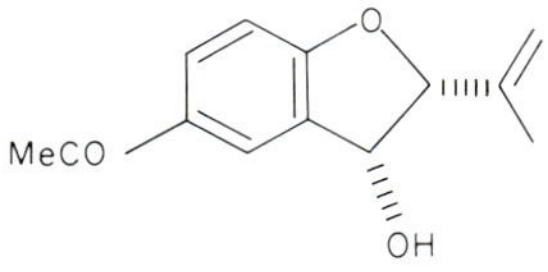

(6) Toxol

isolated from roots of five different *Eupatorium* species, *E. fortunei* and from aerial parts of *Liatris graminifolia* and *L. spicata* (Herz *et al.*, 1975), and it may be a chemical marker of the tribe. Besides euparin, the related toxic compounds, dehydrotremetone (**3**), tremetone (**4**), hydroxytremetone (**5**) and toxol (**6**) have been isolated from *Eupatorium rugosum* (Bonner *et al.*, 1961, 1963); *E. urticaefolium* also afforded tremetone (Bonner and DeGraw, 1962).

In *Piqueria trinervia*, Bohlmann and Zdero (1968) isolated (−)-santalal (**7**) and in *Carelia cistifolia* (Bohlmann and Zdero, 1971) they isolated a benzofuran closely related to euparin. In *Brickellia guatemalensis*, Bohlmann and Zdero (1969) found derivatives of dehydronerolidol. From the roots of *Adenostyles alliaria*, Harmatha *et al.*, (1969) isolated derivatives of eremophilene, including adenostylone (**8**).

Sesquiterpene lactones. Many *Eupatorium* species, several *Mikania*, *Liatris* and even a *Stevia* species contain sesquiterpene lactones (Table I). Other *Eupatorium* species, such as *E. azureum*, *E. havanense*, *E. capillifolium*, *E. perfoliatum*, lack these lactones (Domínguez *et al.*, 1971, 1972, 1974).

TABLE I. Sesquiterpene lactones isolated from tribe Eupatorieae

Genus	Species total/with	Germacrano- lides	Guaianolides	Pseudoguaiano- lides	Total
Eupatorium	400/6	8	9		17
Liatris	15/6	3	5		8
Mikania	120–150/5	5	1		
Stevia	100/1			1	1

Germacranolides have been isolated from several members of the tribe. *E. cannabinum* afforded eupatoriopicrin (**9**) (Dolejs and Herout, 1962) and eucannabinolide (**10**) (Drozdz, 1972). *E. formosanum* contains eupaformonin (**11**) (McPhail *et al.*, 1974).

Geissman and Atala (1971) reported the presence of eupatoriopicrin in substantial amounts in four members of the tribe Helenieae. *E. cuneifolium*

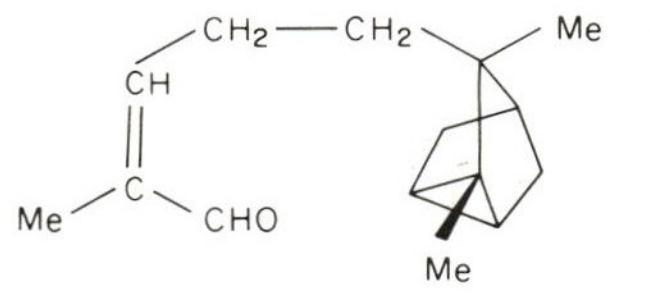

(7) (−)-Santalal

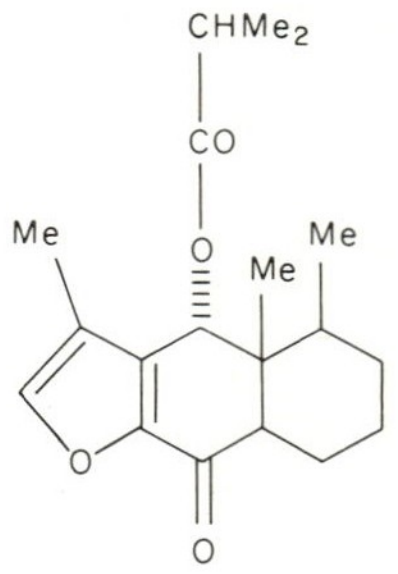

(8) Adenostylone

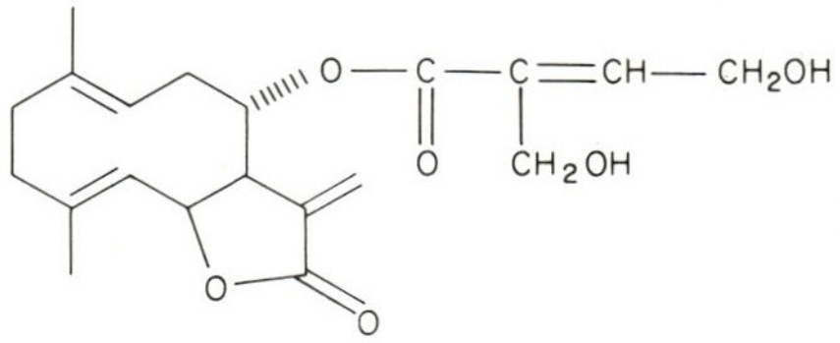

(9) Eupatoriopicrin

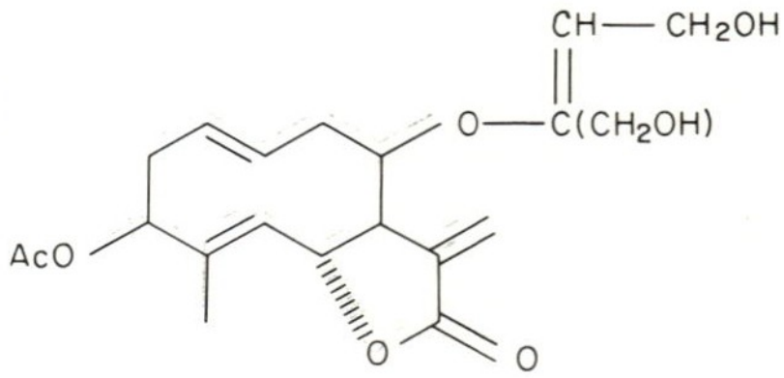

(10) Eucannabinolide

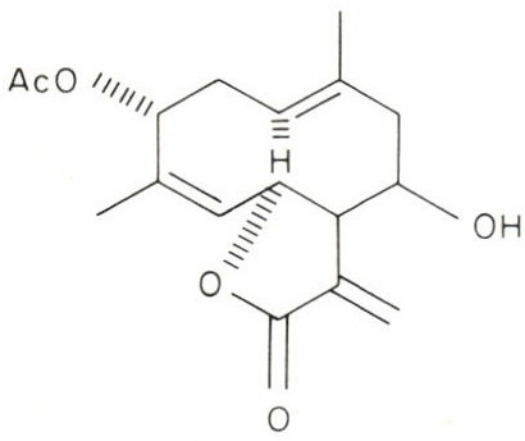

(11) Eupaformonin

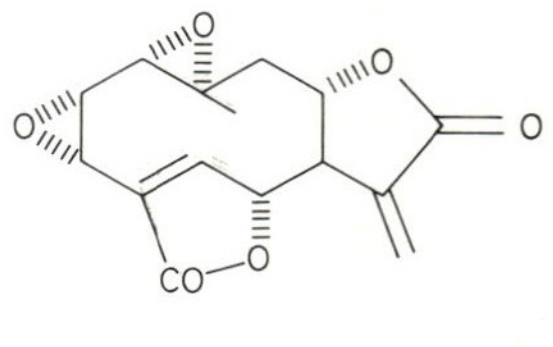

(12) Mikanolide

afforded an antileukemic germacranolide and two other cytotoxic germacranolides: eupacunin, eupacunoxin and eupatocunoxin (Kupchan et al., 1973). From *E. semiserratum* two antileukemic germacranolides were isolated (Kupchan et al., 1973), eupaserrin and deacetyleupaserrin. In *Mikania scandens* (Herz et al., 1967, 1970) five germacranolides were isolated: mikanolide (**12**), desoxymikanolide (**13**) dihydromikanolide (**14**), miscandenin (**15**), scandenolide (**16**). *M. batatifolia* only contains (**12**) and (**14**).

Mikanolide (**12**) and dihydromikanolide are present in *M. micrantha* (Herz et al., 1975). Dihydromikanolide (**14**) has also been isolated from *M. mongasensis* (Mathur and Fermin, 1973). From *Liatris* several cytotoxic

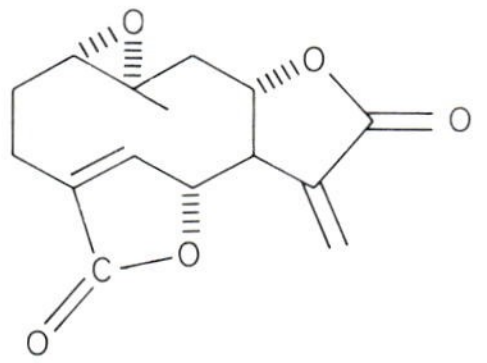

(13) Desoxymikanolide

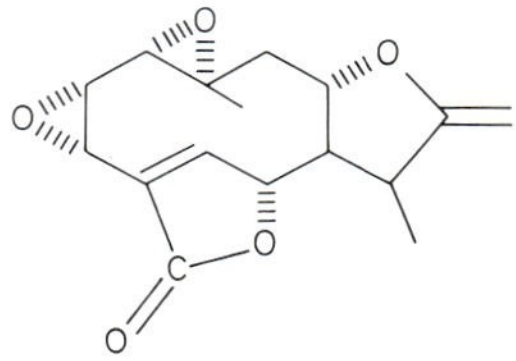

(14) Dihydromikanolide

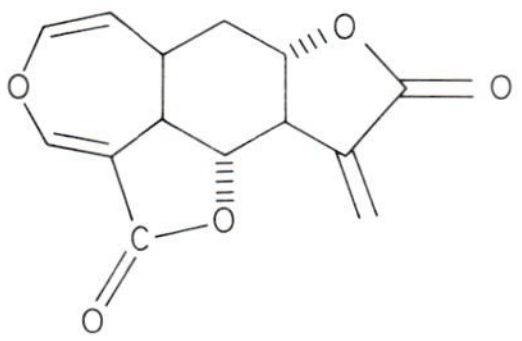

(15) Miscandenin

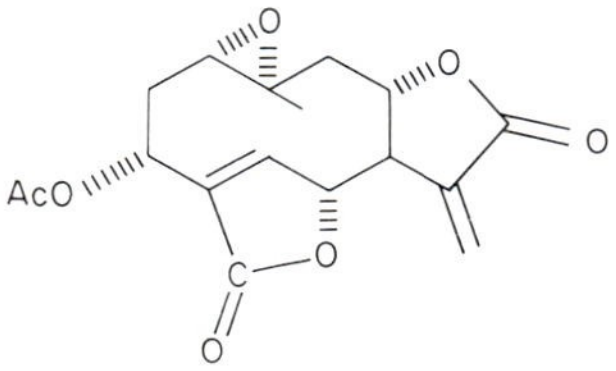

(16) Scandenolide

and antileukemic germacranolides have been found; among them is liatrin (**17**) isolated from *L. chapmanii*. *L. punctata* contains punctatin (**18**) (Herz and Wahlber, 1972) and *L. provincialis*, the cytotoxic provincialin (**19**) (Herz and Wahlber, 1973).

Several guaianolides have been obtained from *Eupatorium* and *Liatris* species; some are cytotoxic. *E. rotundifolium* contains euparotin (**20**), its acetate (**21**), eupachlorin (**22**) and its acetate (**23**), eupatoroxin (**24**), *epi*-eupatoroxin (**25**) eupachloroxin (**26**) and eupatundin (**27**) (Kupchan et al., 1969). In *E. ligustrinum*, ligustrin (**28**) was found by Romo et al., (1968). The new guaianolide mikanokryptin has been isolated from a new species of *Mikania* (Herz et al., 1975). *Liatris graminifolia* affords graminiliatrin (**29**), deoxygraminiliatrin (**30**) and graminichlorin (**31**). *L. spicata* yields spicatin (**32**) and from *L. pycnostachya* spicatin (**32**) and epoxyspicatin were isolated (Herz et al., 1975).

Pseudoguaianolides (ambrosanolides) are considered to be unique to the Heliantheae, so the isolation of stevin (**33**) from *Stevia rhombifolia*

(Rios et al., 1967) in the Eupatorieae is anomalous from the taxonomic point of view.

Diterpenoids. Stevia rebaudiana contains stevioside (**34**), a glycoside that on hydrolysis affords the kaurene diterpene steviol, glucose and sophorose (Vis and Fletcher, 1957). From *Mikania mongasensis*, Mathur and Fermin (1973) isolated kaur-16-en-19-oic acid. From *Liatris elegans*, two new geranylnerol derivatives have been isolated namely ligantrol (**35**) and ligantrol acetate (**36**) (Herz and Sharma, 1975).

Triterpenoids and steroids. The flowers of *Eupatorium cannabinum* and *E. cannabinum* var. *syriacum* contain taraxasterol (Grzybowska et al., 1954). From the whole plant dammaradienyl acetate (**37**), taraxasterol and stigmasterol were isolated (Talapatra et al., 1974). From *E. azureum*, epifriedelinol and taraxasterol acetate were isolated (Domínguez et al., 1973).

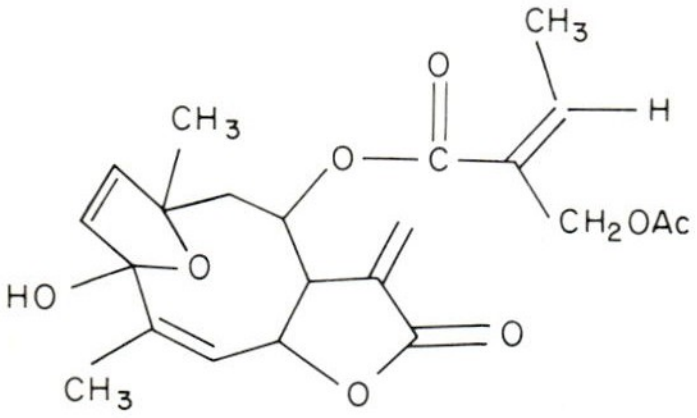

(17) Liatrin

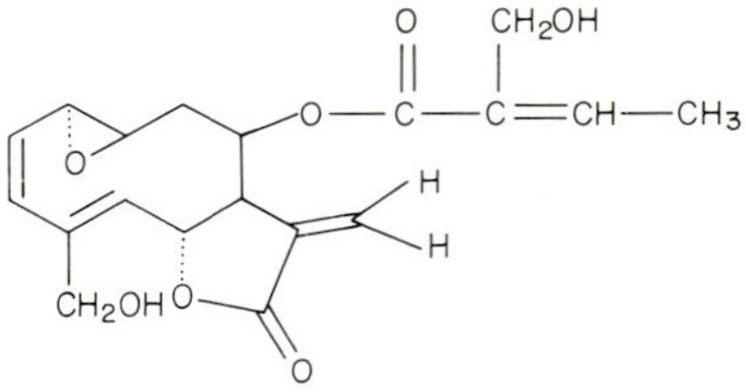

(18) Punctatin (16)

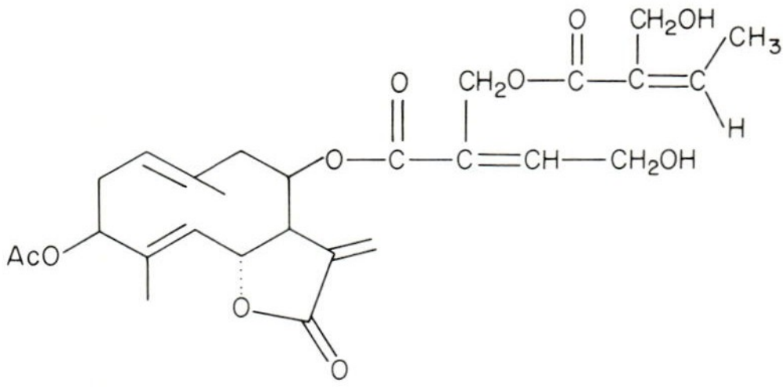

(19) Provincialin (17)

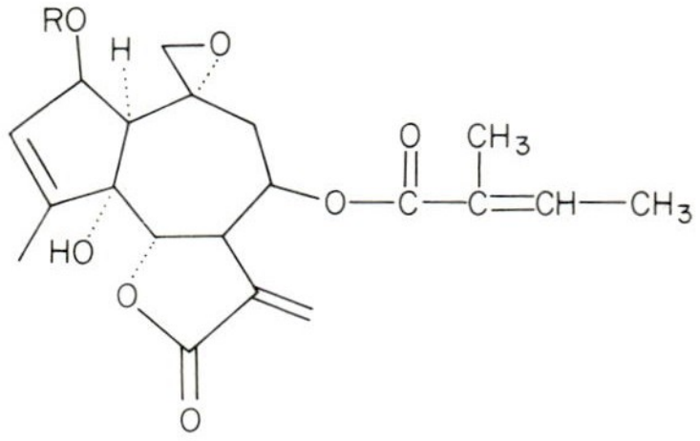

(20) Euparotin R=H
(21) Acetate R=Ac

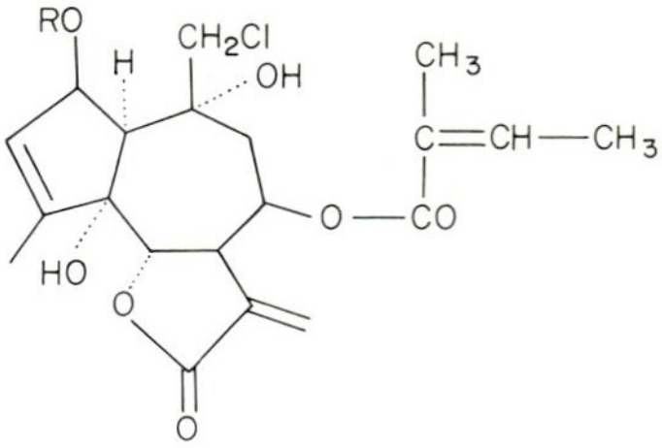

(22) Eupachlorin R=H
(23) Acetate R=Ac

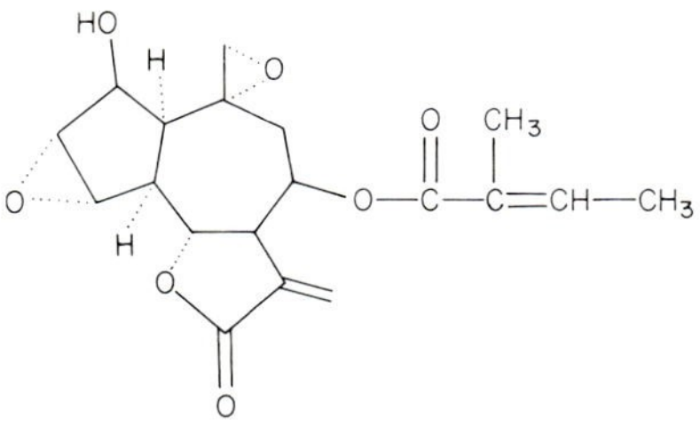

(24) Eupatoroxin

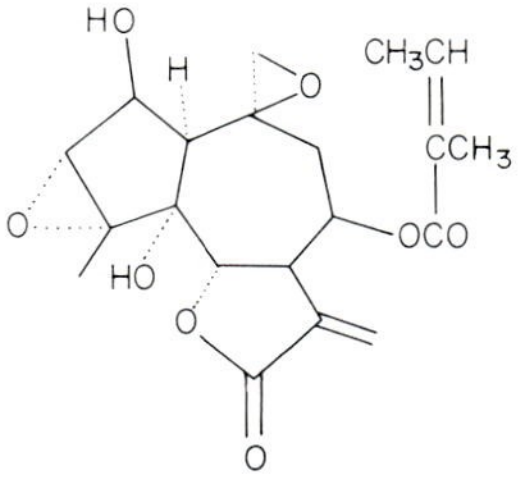

(25) *epi*-eupatoroxin

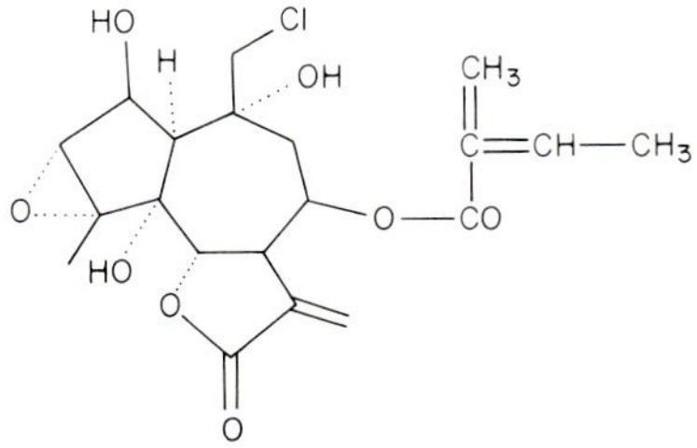

(26) Eupachloroxin

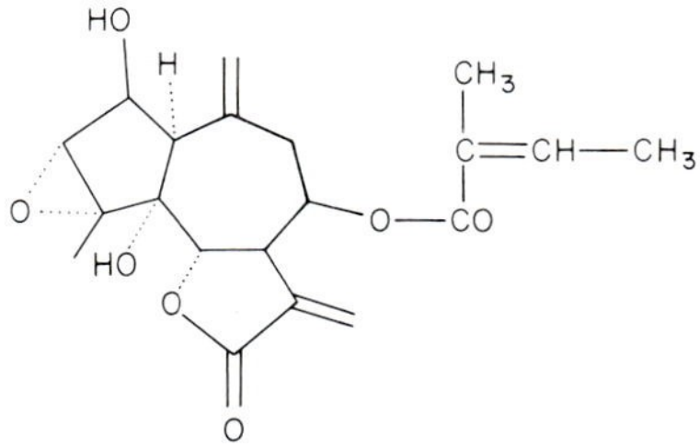

(27) Eupatundin

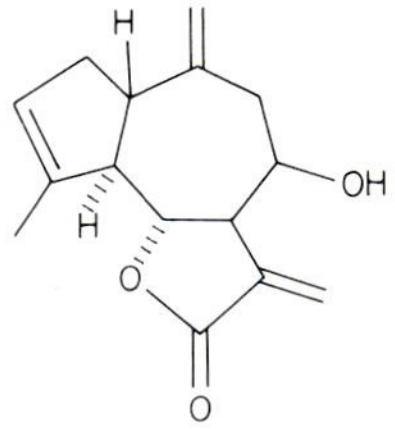

(28) Ligustrin

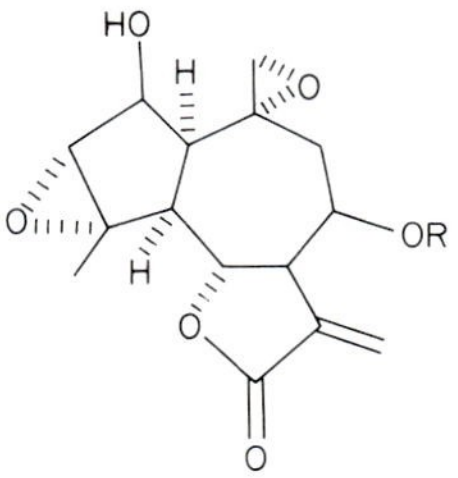

(29) Graminiliatrin

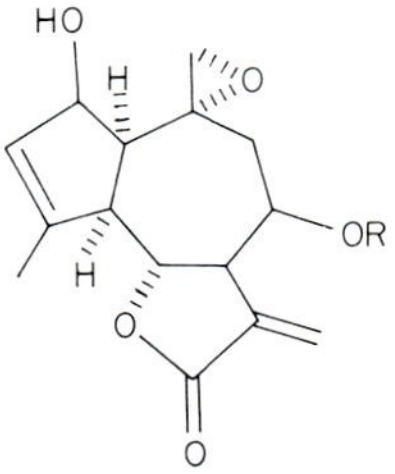

(30) Deoxygraminiliatrin

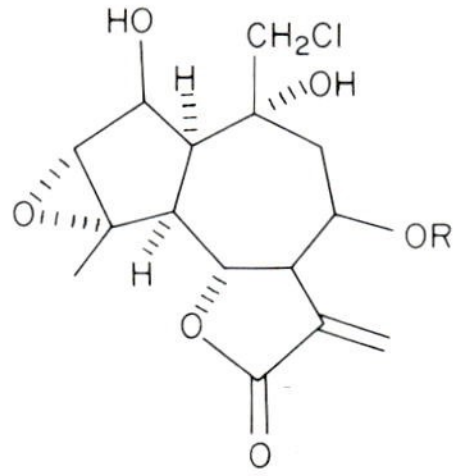

(31) Graminichlorin

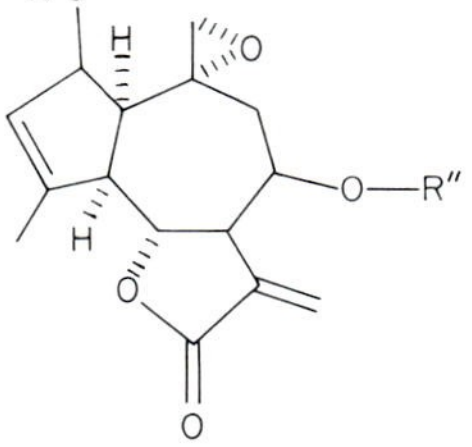

(32) Spicatin

R = COCMe=CHCH$_2$OAc
R' = COC(CH$_2$OH)=CHMe
R" = COC(CH$_2$OAc)=CHMe

E. havanense contains pulcherryl acetate (Domínguez and Roehl, 1973); the aerial part of *E. perfoliatum* contains two ursene derivatives (Domínguez et al., 1974). In *Mikania hirsutissima* a triterpene saponin was isolated (Gomez da Cruz and Liberalli, 1938).

In *Mikania batatifolia* taraxasterol acetate has been found (Herz et al., 1967). *M. cordata* affords epifriedelanol (Kiang, 1965). It has been reported (Wahlberg et al., 1972; Appleton, 1971), that *Carphephorus odoratissimus* leaves and aerial parts contain lupeol, α- and β-amyrin, together

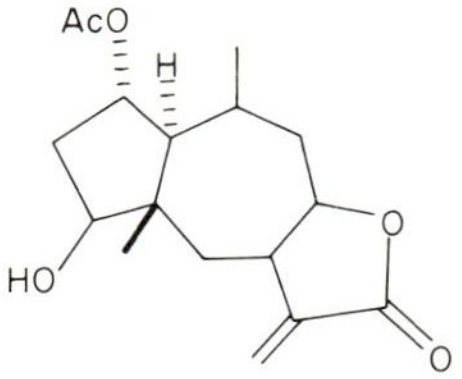

(33) Stevin

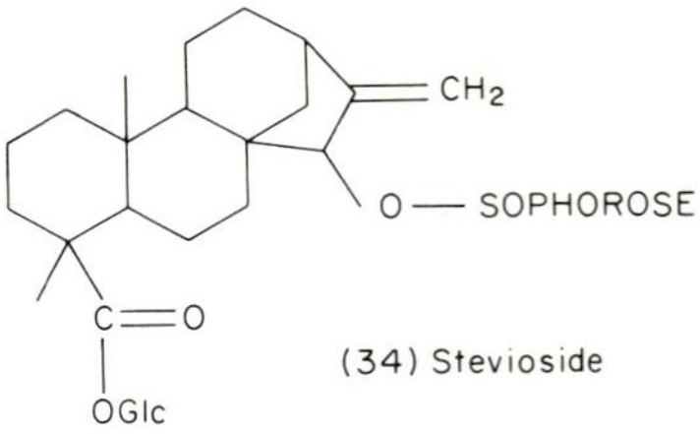

(34) Stevioside

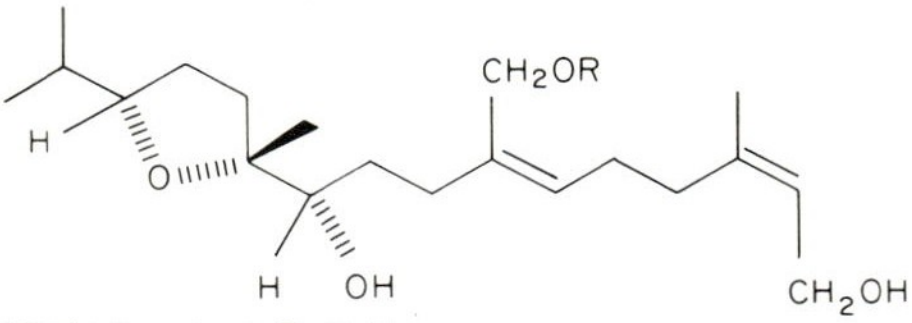

(35) Ligantrol, R = H
(36) Acetate, R = Ac

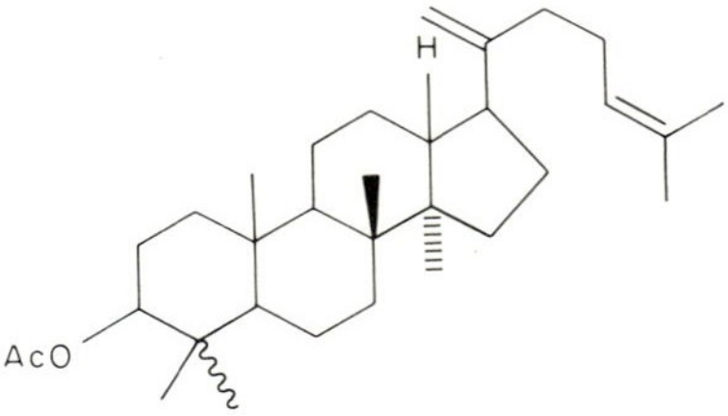

(37) Dammaradienyl acetate

with their corresponding palmitates, cycloartenyl palmitate, β-amyrin acetate, α- and β-amyrenone and 11-oxo-β-amyrin and 11-oxo-α-amyrin, sitosterol and stigmasterol. In *Ageratum houstonianum* there is friedelan-3-3β-ol and friedelin and in *A. conyzoides* only friedelin (Hui and Lee, 1971). *Eupatorium fortunei* contains taraxasteryl palmitate and acetate (Masao, 1974), which are also present in *E. chinense* var. *simplicifolium*. In *E. odoratum* Talapatra et al. (1974) reported the presence of lupeol and β-amyrin.

Flavonoids and other phenolics

The Eupatorieae are rich in polymethoxylated flavones and flavonols and their glycosides, particularly apigenin and quercetin derivatives (see Table II). Several flavonoids show cytotoxic activity. Kupchan et al. (1969) found seven cytotoxic flavones in two *Eupatorium* species. *E. cuneifolium* afforded two cytotoxic flavones, the known hispidulin (**38**), also present in

Table II. Distribution of flavonoids among genera of the Eupatorieae

Genera	Species total/with	Flavones	3-OCH₃ flavones	Flavonols	O-glycosyl flavonoids	C-glycosyl flavones	Flavanone	Total flavones
Brickellia	60/3		1		2			3
Eupatorium	400/15	5		2	9	1	2	19
Kuhnia	3/1	1						1
Liatris	15/10				4	2		6
Mikania	120–150/1		1					1
Stevia	100/1	1						1

E. rotundifolium, and eupafolin (6-methoxy-5,7,3'4'-tetrahydroxyflavone). *E. semiserratum* was the source of the other five flavones, the known pectolinarigenin (**39**) and the new compounds, eupatorin (**40**), eupatilin (**41**), eupatoretin (3,3'-dihydroxy-5,6,7,4'-tetramethoxyflavone) and eupatin (3,5,3'-trihydroxy-6,7,4'-trimethoxyflavone). Eupatin has been also isolated from *E. stoechadosmum* (Nakaobi and Moreta, 1958). Wagner *et al.* (1973) found eupatilin (**41**) in *Liatris punctata*.

E. ligustrinum yields two rhamnosides, eupalin (3-O-rhamnoside of 3,5,4'-trihydroxy-6,7-dimethoxyflavone) and eupatolin (3-O-rhamnoside of 3,5,3',4'-tetrahydroxy-6,7-dimethoxyflavone) (Quijano *et al.*, 1970).

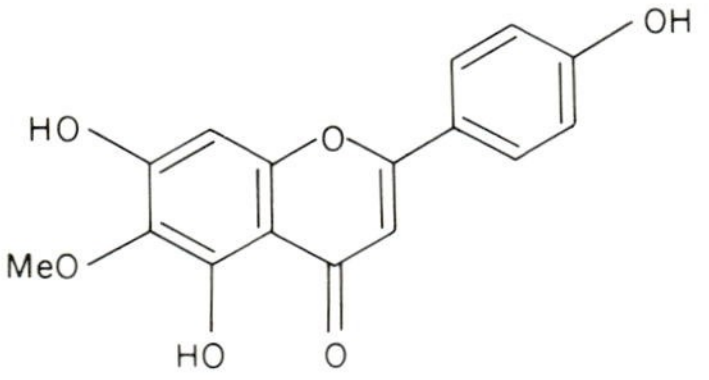

(38) Hispidulin

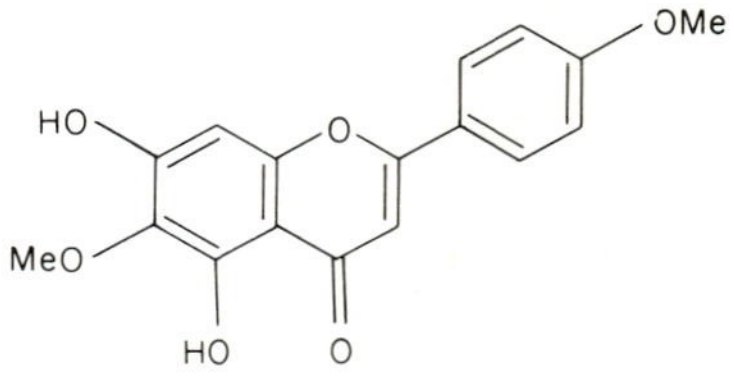

(39) Pectolinaringenin

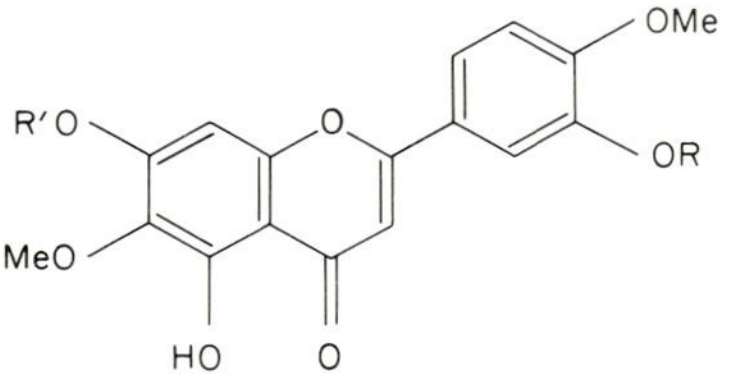

(40) Eupatorin R = H, R' = Me
(41) Eupatilin R = Me, R' = H

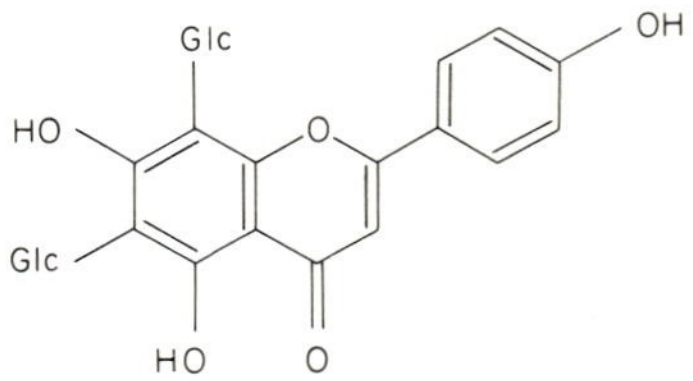

(42) Vicenin-2

Several *Eupatorium* spp. contain more common flavonol glycosides (Wagner *et al.*, 1972). *E. capillifolium* and *E. cuneifolium* contain hyperoside and astragalin, the last species also containing quercetin and kaempferol, while *E. hyssopifolium*, *E. alba*, *E. recurvans*, *E. rugosum*, and *E. cannabinum* var. *syriacum* contain rutin and kaempferol-3-rutinoside. In *E. subhastatum*, besides eupafolin, Ferraro and Cussio (1973) found eriodictyol (5,7,3'4'-tetrahydroxyflavanone) and three quercetin glycosides (hyperoside, quercitrin and rutin).

In *E. serotinum*, vicenin-2 (**42**) a flavone C-diglycoside has been isolated besides the known hyperoside and astragalin (Wagner *et al.*, 1972b). In *E. odoratum* Talapatra *et al.* (1974) reported the isolation of salvigenin (**43**). *E. havanense* contains sakuranetin (Domínguez and Roehl, 1973). In

Brickellia pendula, the flavone glycoside, pendulin (**44**) has been found (Flores and Herran, 1958). From *B. guatemaliensis* its aglycone penduletin has been isolated (Bohlmann and Zdero, 1969). *B. squarrosa* also contains pendulin and a 3-methoxyflavone, atanasin (**45**).

From *Mikania cordata* a flavonol, mikanin (**46**) has been obtained and in *M. batatifolia* the flavone, batatifolin (**47**) (Herz *et al.*, 1970). From *Kuhnia eupatorioides* Herz *et al.* (1961) isolated artemetin. Extracts from *Stevia berlandieri* (Domínguez *et al.*, 1974) contain 5,6-dihydroxy-7,8,4'-trimethoxyflavone. *Liatris spicata* contains two glycosides of quercetin (Kagan, 1968).

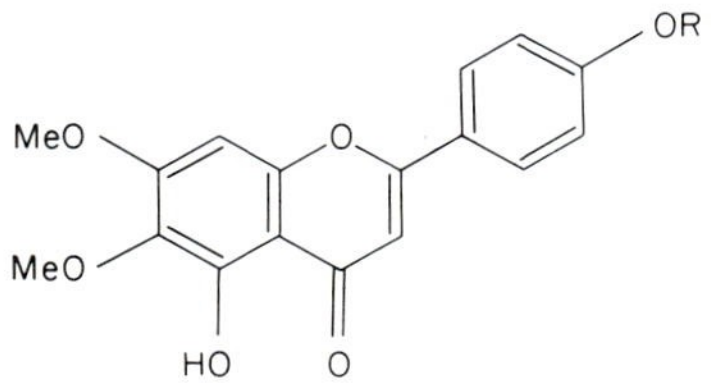

(43) Salvigenin R = Me
(44) Pendulin R = Glc

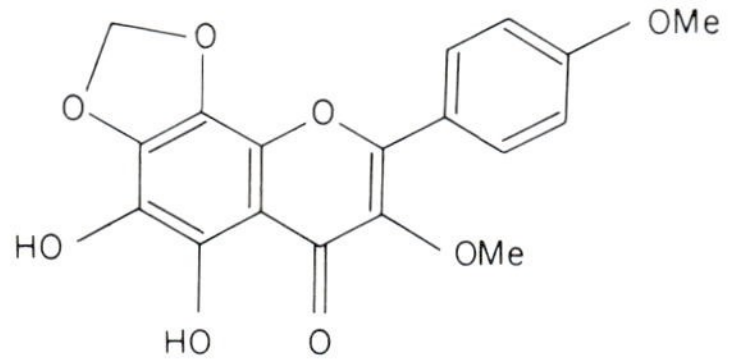

(45) Atanasin

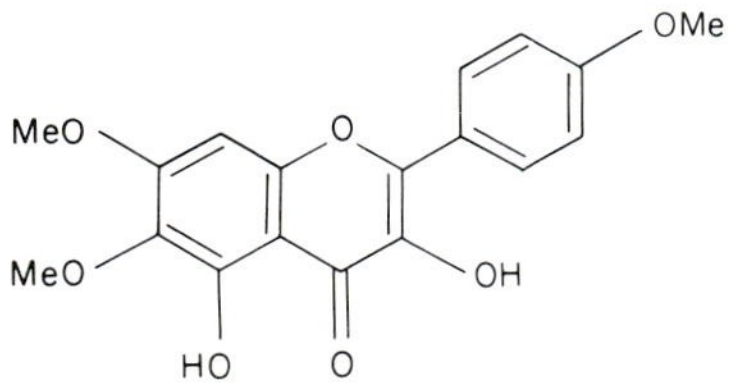

(46) Mikanin

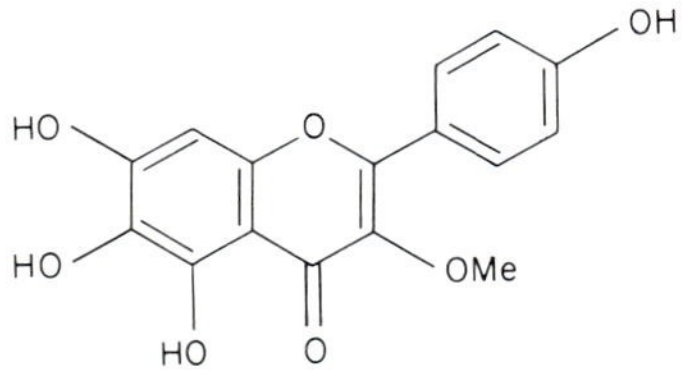

(47) Batatifolin

Wagner *et al.* (1973) reported on the flavonoids in 10 species of *Liatris* collected in the U.S.A. *L. provincialis, L. punctata, L. chapmanii, L. secunda*, contained kaempferol and kaempferitrin (kaempferol-3,7-dirhamnoside) only once reported elsewhere in the Compositae. Kaempferitrin was also present in *L. graminifolia, L. tenuifolia*, and *L. elegans*. The last two *Liatris* species contained vicenin-1 (apigenin-6-xyloside-8-glucoside) and vicenin-2 (apigenin-6, 8-diglucoside). The last diglycoside also occurred in *L. chapmanii, L. secunda* and *L. gracilis*. Traces of rutin were found only in *L. gracilis* and *L. spicata*. Racemates of the lignans eudesmin and epiedesmin are present in the *Carphephorus odoratissimus* (Wahlberg *et al.*, 1972). Caffeic acid and chlorogenic acid were found in the flowers of *Eupatorium cannabinum* var. *syriacum*. The presence of *p*-anisic acid has been reported in *E. odoratum* (Ahmad, 1969). 7-Methoxycoumarin has been isolated from the leaves of *Eupatorium trinerve* syn. *E. ayapana* (Karrer, 1959). Methylriparichromene A and four other new 2,2-dimethylchromenes have been found in *E. riparium* (Anthonsen, 1969; Taylor and Wright, 1971).

Alkaloids

An early report (Righini, 1811) that the leaves and flowers of *Eupatorium cannabinum* contain the alkaloid, eupatorine, has not been further investigated. More recently some species of *Eupatorium* have been suspected of causing liver damage in cattle; as a result Tsuda and Marion (1963) searched for alkaloids in the roots of *E. maculatum* and found at least four alkaloids, two being the pyrrolizidine alkaloids, trachelanthamidine (**48**) and echinatine (**49**). The other two, eupatorium base-C and eupatorium

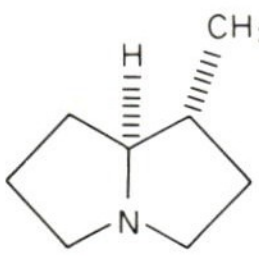

(48) Trachelanthamidine

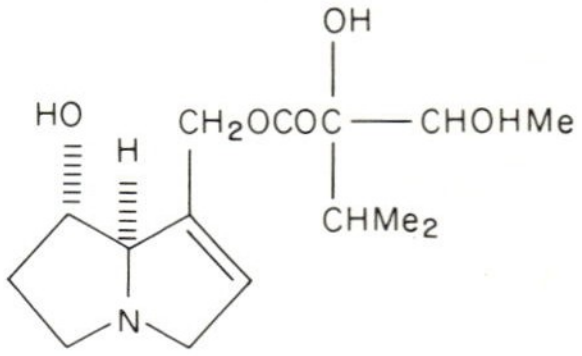

(49) Echinatine

base D are probably also of the pyrrolizidine type, but the elucidation of their structure remains to be completed.

Furaya and Hikichi (1973) isolated from *E. stoechadosmum*, two pyrrolizidine alkaloids, supinine (**50**) and lindefoline. In a survey of nine *Eupatorium* species native to Ohio, U.S.A. Locock et al. (1966) found that *E. serotinum* contained two pyrrolizidine ester alkaloids, supinine (**50**) and rinderine (**51**). *E. capillifolium* also contains alkaloids (Domínguez et al., 1969).

Extracts from *Alomia fastigiata* gave a strong reaction for alkaloids with Dragendorff's reagent (Pozzitti, 1966). Positive alkaloid tests have been reported for *Adenostemma lavenia*, *Ageratum conyzoides*, (Arthur, 1954) and *Liatris laevigata* (Wall et al., 1959).

Other compounds

The seed oils of some *Liatris* species have been examined (Earle et al., 1960). Fumaric acid was isolated from *Mikania cordata* (Kiang, 1965). Inositol was extracted from *Eupatorium cannabinum* (Plouvier, 1962).

Bohlmann et al. (1973) mention that 33 members of the Eupatorieae have been searched for acetylene derivatives, and that only 14 contain them. The most widespread compound in the roots is the pentaynene $CH_3-(C{\equiv}C)_5-CH{=}CH_2$, which appears in small amounts ($10^{-3}-10^{-5}\%$) of the fresh root weight. Because of the small amount of material available, it is not clear whether it is present in all the examined species or not. The plants that contain the acetylene compounds are: *Piqueria trinervia, P. lavenia, Adenostyles alliariae, Carelia cistifolia, Ageratum mexicanum, Stevia ovata, Eupatorium altissimum, E. cannabinum, Mikania officinalis, Symphyopappus halschboachii*. Two acetylenic thiophene derivatives (52), (53) are also present in *Mikania scandens*. In *Liatris pycnostachya* and *L. spicata*, Atkinson and Curtis (1968) found besides the pentaynene the thiophene ketone (54); in *L. scariosa* the thiophene ketone only was detected.

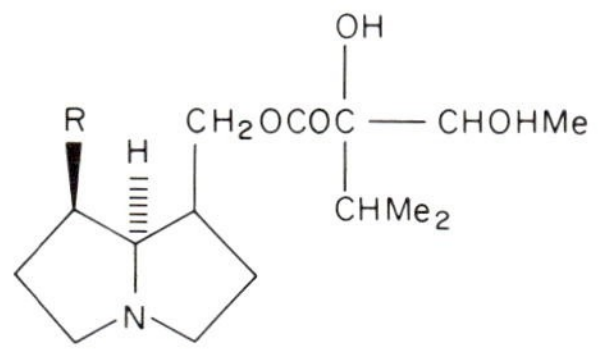

(50) Supinine R = H
(51) Rinderine R = OH

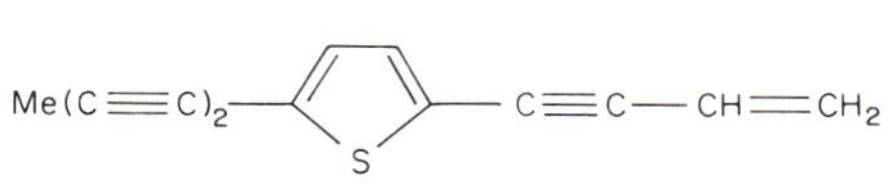

(52)

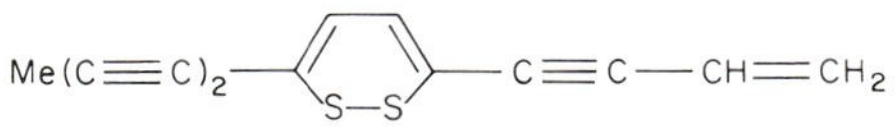

(53)

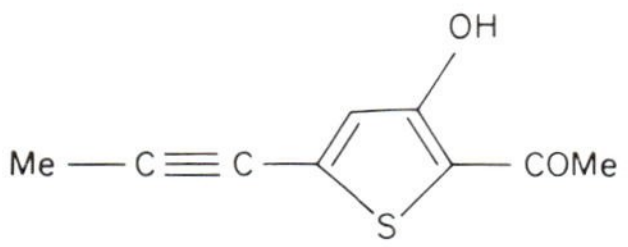

(54)

ECONOMIC AND PHARMACEUTICAL USES

The oleoresin obtained from *Carphephorus odoratissimus* (deertongue) is used as a fixative in perfumery and the dried leaves as a flavour additive to tobacco. *Eupatorium odoratum* is used in India as fish poison; in Mexico it is used for medicinal purposes, as are *E. collinum* and several *Brickellia* species. *Stevia rebaudiana* is the source of stevioside, a glycoside that is 300 times sweeter than sucrose.

Many members of the tribe find application in folk medicine. *Eupatorium cannabinum* has been used in India to cure jaundice and scurvy and fomenting sores and ulcers. In Italy, *E. cannabinum* var. *syriacum* has been frequently used in homeopathic medicine. *E. azureum* is used as an astringent for poultices. *Mikania cordata* and *M. capenses* are used to cure snake bites in South America and South Africa. *Ageratum conyzoides* is used for the relief of abdominal pain.

Recently, cytotoxic and antileukemic properties have been demonstrated in several sesquiterpene lactones from *Eupatorium rotundifolium* (euparotin), *E. cuneifolium* (eupacunin, eupacunoxin and eupatocunin), *E. semiserratum* (eupaserrin and desacetyleupaserrin) and *E. formosanum* (eupaformonin), *Liatris provincialis*, (provincialin); the polyhydroxylated flavonoids present in *E. semiserratum* and *E. cuneifolium*, have also been found to be cytotoxic. These results have opened up a new field of interesting physiological research.

REFERENCES

AHMAD, M. (1969). *Sci. Res. (Dacca, Pak.)* **6**, 37.
ARTHUR, H. R. (1954). *J. Pharm. Pharmacol.* **6**, 66.
ALERTSEN, A. R. (1955). *Acta chem. scand.* **9**, 1725.
APPLETON R. A. and ENZEL, C. R. (1971). *Phytochemistry* **10**, 447.
ANTHONSEN, T. (1969). *Acta chem. scand.* **23**, 3605
ATKINSON, R. E. and CURTIS, R. F. (1971). *Phytochemistry* **10**, 454.
BOHLMANN, F. and ZDERO, C. (1968). *Tetrahedron Lett.* 1533.
BOHLMANN, F. and ZDERO, C. (1969). *Tetrahedron Lett.* 5109.
BOHLMANN, F. and ZDERO, C. (1971). *Chem. Ber.* **104**, 964.
BOHLMANN, F., BURKHARDT, T. and ZDERO, C., 1973, "Naturally Occurring Acetylenes". Academic Press, New York and London.
BONNER, W. A. (1963). *Tetrahedron Lett.* 1295.
BONNER, W. A. and DE GRAW, J. J. (1962). *Tetrahedron* **18**, 1295.
BONNER, W. A., DEGRAW, J. J., BOWEN, D. M. and SHAH, V. R. (1961). *Tetrahedron Lett.* 417.
DOLEJS, L. and HEROUT, V. (1962). *Collect Czech chem. Commun.*, **27**, 2654.
DOMÍNGUEZ, X. A. and ROEHL, E. (1973). *Phytochemistry* **12**, 2060.
DOMÍNGUEZ, X. A., GÓMEZ, M. E., GÓMEZ, P. A., VILLARREAL, A. N. and ROMBOLD, C. (1970). *Planta Med.* **19**, 52.
DOMÍNGUEZ, X. A., ROJAS, P., DUEÑAS, MA. DEL C. and ESCARRIA, S. (1973). *Phytochemistry* **12**, 224.
DOMÍNGUEZ, X. A., GONZÁLEZ Q., J. A. and ROJAS, P. (1974a). *Phytochemistry* **13**, 673.
DOMÍNGUEZ, X. A., GONZÁLEZ, A., ZAMUDIO, M. A. and GARZA, A. (1974b). *Phytochemistry* **13**, 2001.
DROZDZ, B., SAMEK, Z., HOLUB, M., HEROUT, M. and SORM, F. (1972). *Collect Czech. chem. commun.* **37**, 1546.
EARLE, F. R., WOLFF, I. A. and JONES, Q. (1960). *J. Am. Oil Chemists' Soc.* **37**, 254.
FERRARO, G. E. and CUSSIO, J. D. (1973). *Phytochemistry* **12**, 1825.
FLORES, S. E. and HERRAN, J. (1958). *Tetrahedron* **2**, 308.
FLORES, S. E. and HERRAN, J. (1960). *Chemy Ind.* 291
FURUYA, T. and HIKICHI, M. (1973). *Phytochemistry* **12**, 225.
GEISSMAN, T. A. and ATALA, S. (1971). *Phytochemistry* **10**, 1075.
GÓMEZ DA CRUZ, J. P. and LIBERALLI, C. H. (1938). *Bol. Chim. Farm. Brasil* **17**, 693.

CRYZBOWSKA, J., JERZMANOWSKA, Z. and WITOKOWSKI, H. (1954). *Roczniki Chemi* **28**, 197.
HARMATHA, J., SMAEK, Z. and NOVOTNY, L. (1969). *Collect., Czech. chem. Commun* **34**, 1739.
HEGNAUER, R. (1964). "Chemotaxonomie der Pflanzen" Vol. III, p. 448. Birkhauser Verlag, Basel.
HERBERT, H. J. (1968). *Rhodora* **70**, 474.
HERZ, W. (1961). *J. org. Chem.* **26**, 3014.
HERZ, W. and SHARMA, R. P. (1975). *J. org. Chem.* **40**, 192.
HERZ, W. and WAHLBERG, I. (1973a). *Phytochemistry* **12**, 1421.
HERZ, W. and WAHLBERG, I. (1973b). *J. org. Chem.* **38**, 2485.
HERZ, W., SANTHANAM, P. S., SUBRAMANIAN, P. and SCHMID, J. J. (1967). *Tetrahedron Lett.* 3111.
HERZ, W., SANTHANAM, P. S., WAGNER, H., HOER, R., HORHAMMER, L. and FARKAS, L. (1970). *Chem. Ber.* **103**, 1822.
HERZ, W., POPLAWSKI, J. and SHARMA, R. P. (1975a). *J. org. Chem.* **40**, 199.
HERZ, W., POPLAWSKI, J. and SHARMA, R. P. (1975b). *J. org. Chem.* **44**, 199.
HERZ, W., SRINIVASAN, A. and KALYANARAMAN, P. S. (1975c). *Phytochemistry* **14**, 233.
HUI, W. H. and LEE, W. K. (1971). *Phytochemistry* **10**, 899.
KAGEN, J. (1968). *Phytochemistry* **7**, 1205.
KARLSSON, K., WAHLBERG, I. and ENZELL, C. R. (1972a). *Acta. chem. scand.* **26**, 3839.
KARLSSON, K., WAHLBERG, I. and ENZELL, C. R. (1972b). *Acta chem. scand.* **26**, 2837.
KARLSSON, K., WAHLBERG, I. and ENZELL, C. R. (1973a). *Acta chem. scand.* **27**, 1630.
KARLSSON, K., WAHLBERG, I. and ENZELL, C. R., (1973b). *Acta chem. scand.* **27**, 1613.
KARRER, W., "Konstitution und Vorkommen der organischen Pflanzenstoffe". Birkhauser, Basel.
KIANG, A. K. (1965). *J. chem. Soc.* 6371.
KUPCHAN, S. M., KELSEY, J. E., MURAYAMA, M., CASSADY, J. M., HEMINGWAY, J. C. and KNOW, J. R. (1969a). *J. org. Chem.* **34**, 3876.
KUPCHAN, S. M., SIGEL, C. W., HEMINGWAY, R. J., KNOX, J. R. and UDAGAMURTHY, M. S. (1969b). *Tetrahedron* **25**, 1603.
KUPCHAN, S. M., FUJITA, T., MARUYAMA, M. and BRITTON, R. W. (1973a). *J. org. Chem.* **38**, 1260.
KUPCHAN, S. M., MURAYAMA, M., HEMINGWAY, R. J., HEMINGWAY, J. C., SHIBUYA, S. and FUJITA, T. (1973b). *J. org. Chem.* **38**, 1853.
KUPCHAN, S. M., MURAYAMA, M., HEMINGWAY, R. J., HEMINGWAY, J. C., SHIBUYA, S. and FIJUTA, T. (1973c). *J. org. Chem.* **38**, 2189.
LOCOCK, R. A., BEAL, J. L. and DOSKOTOCH, R. W. (1966). *Lloydia* **29**, 201.
MASAO, Y., LAN-SO and ZE-IAN, I. (1974). *Yakugaku Zasshi* **94**, 338.
MATHUR, S. B. and FERMIN, C. M. (1973). *Phytochemistry* **12**, 226.
MCPHAIL, A. T., ONAN, K. D., LEE, K. H., ISUKA, T. and HUONG, H. C. (1974). *Tetrahedron Lett.* 3203.

NAKAOKI, T. and MORITA, N. (1958). *Yakugaku Zasshi* **78**, 557.
PLOUVIER, V. (1962). *C. r. hebd. Séanc. Acad. Sci., Paris.* **256**, 1397.
POZZITTI, G. L. (1966). *Rev. Fac. Farm. Bioquim. Univ. Sao Paulo* **4**, 137.
QUIJANO, L., MALANCO, F. and RÍOS, R. (1970). *Tetrahedron* **26**, 2851.
RIGHINI, A. (1811). *J. pharm. chim.* **14**, 623.
RÍOS, T., ROMO DE VIVAR, A. and ROMO, J. (1967). *Tetrahedron* **23**, 4265.
ROMO, J., RÍOS, T. and QUIJANO, L. *Tetrahedron* **24**, 6087.
SCHINDLER, H. (1953). *Arzneimittel-Forsch.* **3**, 541.
SHIMADA, G. and SAWADA, T. (1967). *Yakugaku Zasshi* **77**, 1246.
TALAPATRA, S. K., BHAR, D. S. and TALAPATRA, B. (1974a). *Aust. J. Chem.* **27**, 1137.
TALAPATRA, S. K., BHAR, D. S. and TALAPATRA, B. (1974b). *Phytochemistry* **13**, 284.
TAYLOR, D. R. and WRIGHT, J. A. (1971). *Phytochemistry* **10**, 1665.
TSUDA, Y. and MARION, L. (1963). *Can. J. Chem.* **41**, 1919.
WAGNER, H., IYENGAR, M. A., HORHAMMER, L. and HERZ, W. (1972a). *Phytochemistry* **11**, 1504.
WAGNER, H., IYENGAR, M. A., DULL, P. and HERZ, W. (1972b). *Phytochemistry* **11**, 1506.
WAGNER, H., IYENGAR, M. A. and HERZ, W. (1973). *Phytochemistry* **12**, 2063.
WAHLBERG, I., KARLSSON, K. and ENZELL, C. R. (1972c). *Acta chem. scand.* **26**, 1383.
WALL, M. E., GARVIN, J. W., WILLAMAN, J. J., JONES, Q., SCHUBERT, B. G. and GENTRY, H. S. (1959). *Am. J. pharm. Assoc. Sci. Ed.* **48**, 695.
VIS, E. and FLETCHER, H. G. (1956). *J. Am. chem. Soc.* **78**, 4709.

Chapter 17

Vernonieae—systematic review

SAMUEL B. JONES

Botany Department, University of Georgia, Athens, Georgia, U.S.A.

Abstract. The tribe is reviewed and reported to have *c.* 1456 species, *c.* 70 genera, of which 37 are monotypic. *Vernonia*, the largest genus in the tribe has *c.* 1000 species. It is concluded that our knowledge of the Vernonieae is still largely in the alpha state; there is no comprehensive treatment of the tribe at the subtribal or generic level nor have several of the larger genera been carefully examined. The need is stressed for a systematic revision of the tribe and its genera on a worldwide basis rather than piece-meal treatments of taxa from limited geographic areas. Comparative pollen morphology, phytochemistry, and cytogenetics should yield information useful in developing an improved classification of the tribe.

CONTENTS

Introduction	503
Description of the Tribe	504
Division into Subtribes	505
Genera of the Tribe	506
Geographical Distributional Patterns	506
Biology of the Vernonieae	509
Comparative Anatomy and Morphology	509
Cytotaxonomy	510
Cytogenetics	512
Herbivore interactions	513
Ecotypes	514
Coevolution with rusts	514
Evolutionary Considerations	514
Conclusion	515
Acknowledgements	516
References	516

INTRODUCTION

The tribe was established by Cassini (1817, 1819) and delimited by Lessing (1829, 1831a, b). Lessing's organization formed the basis of De Candolle's (1836) classification of the tribe. The next major revision of the tribe was that of Bentham (1873b). Hoffmann (1894) made a few rearrangements and additions; however, his classification is essentially identical with that of Bentham.

The tribe has attracted the attention of numerous workers who described species, revised genera, or considered portions of the group in floristic treatments. Among the major contributions are: Martius (1822), *Lychnophora*; Lessing (1829, 1831a, b); Schultz Bip (1861, 1863); Baker (1873), Brazil; Oliver and Hiern (1877), Africa; Hieronymus (1897), southern South America; Moore (1902, 1918), Africa; Gleason (1906, 1922, 1923a), North America and Bolivia; Ekman (1914), West Indies: Blake (1926), woody Vernonias of Mexico; Malme (1931, 1932, 1933), Brazil; Koster (1935), Malaysia; Markötter (1939), *Corymbium*; Toledo (1941), *Heterocoma*; Cabrera (1944), Argentina; Cuatrecasas (1956), section *Critoniopsis* of *Vernonia*; Humbert (1960), Madagascar; Aristeguieta (1963), *Pollalestra*; Smith (1971), African Stengelioid Vernonias; Clonts (1972), *Elephantopus*; Barroso (1969, 1970), Brazil; my colleagues and I published a series of papers on the Vernonias of North America (Jones, 1964, 1966, 1967, 1968, 1970a, 1972a, b, 1973; Jones et al., 1970; Faust, 1972, Urbatsch, 1972; Faust and Jones, 1973; Chapman, 1973; Mabry et al., 1974; Jones and Faust, in press); and Jeffrey and Wild (pers. comm.) have prepared treatments for certain African taxa.

DESCRIPTION OF THE TRIBE

Vernonieae Cass. (Cassini, 1817). Perennial herbs, rarely annuals, shrubs, vines or trees. Leaves alternate, rarely opposite or whorled, sessile or petiolate, occasionally lobed. Heads homogamous, 1–many flowered, sometimes reduced and syncephalous. Florets normally bisexual and fertile, rarely unisexual. Involucre usually ovoid or globular, phyllaries many, closely or loosely imbricated in several series, or rarely few in one series. Receptacle flat or subconvex, either smooth or pitted, rarely alveolate, sometimes with pales. Pappus usually elongate and setose, sometimes flattened in two series, outer series often reduced; rarely absent. Corollas tubular, usually regular (sub-ligulate in *Stokesia*) tube elongate, with five narrow lobes to the limb, rarely 3–4 lobed, or somewhat bilabiate, deep purplish-red to white or blue (rarely cream, yellow or orange in some African and Madagascan taxa), often glandular. Anthers with terminal appendage, sagittate at the base, the auricles obtuse, acute or rarely tailed; filaments inserted high above the base. Style branches semi-cylindrical, long slender, tips acute, or obtuse, usually short-hirsute throughout, rarely glabrate, stigmatic papillae on the inner surface near the base. Achenes variable, terete to slightly flattened, often 10-ribbed, or 4- or 5-angled, occasionally smooth, rarely dimorphic. Pollen grains lophate, the ridges often spiny. Type genus: *Vernonia* Schreb. Gen. **2**, 541 (1791).

DIVISION INTO SUBTRIBES

Both Bentham (1873b) and Hoffmann (1894) recognized two subtribes: the Vernoninae, with separate heads and distinct involucre; and the Lychnophoreae, with few-flowered heads aggregated into secondary heads or glomerules, sometimes with the florets clustered forming a dense globular or oblong compound head with or without a common involucre.

Bentham noted that subtribe Vernoninae consists principally of one large genus, *Vernonia*, with a number of smaller genera closely connected and clustered around the central core formed by *Vernonia*. He noted extensive intergradation of the diagnostic features which distinguished many of these genera; Ekman (1914) concurred with this observation and noted further that delimitation of all taxonomic units in *Vernonia* was extremely difficult, both at the specific and generic levels. Another problem is the rather large number of genera in the Vernonieae which are small or monotypic. A number of these smaller genera, including *Stokesia*, *Corymbium*, *Struchium*, *Pacourina*, and *Heterocoma*, have the essential features of the tribe but are not especially close to *Vernonia*, making classification difficult. For example *Stokesia* is anomalous in having the corollas of at least the outer flowers ligulate and rather deeply 5-lobed, while the innermost flowers, likewise deeply 5-lobed are essentially regular. Thus *Stokesia* approaches the tribe Mutisieae in features of the corolla, yet it has the anther, style and pollen of the Vernonieae. Among the Lychnophorinae, a large group of Old and New World genera, the heads are few-flowered, and either closely packed and sessile or syncephalous, giving the compound inflorescence the general appearance of a single head. However, since the successive development of the heads in the Compositae is usually centrifugal, whereas the opening of the florets is constantly centripetal, (Cronquist, Chapter 1), these do not pose recognition problems.

Philipson (1938) expressed dissatisfaction with the subtribal arrangement of Bentham, noting that the subtribe Elephantopeae, first proposed by Cassini (1817) on the basis of their glomerules of secondary heads and asymmetric corollas and maintained by Lessing (1829) and De Candolle (1836), seems preferable to placing *Elephantopus* in the subtribe Lychnophorinae. He further believed that the older arrangement of Cassini (1817) in which the Rolandreae was also separated from the subtribe Lychnophorinae, provided for better classification.

Our understanding of the Vernonieae is far from satisfactory. Bentham's two subtribes, which were recognized over a century ago, were based upon presence or absence of secondary aggregation of heads, but several problems arise with this classification. Examination of the Vernonieae in some detail, however, has convinced me that taxonomic distinction of those taxa with separate heads from those with small clustered capitula arranged in secondary heads is artificial. Some genera which typically have separate heads have species that are almost syncephalous.

GENERA OF THE TRIBE

The genera of the tribe are listed in Table I. Except where noted, I have examined personally one or more species of each genus. The tribe has c. 70 genera, including c. 37 monotypic genera, and c. 1456 species. About 1000 species are found in the large genus *Vernonia* which seems to form the central core of the tribe. The numerous small monotypic genera need to be carefully evaluated and probably many could be included in other genera. Some genera, such as *Centratherum*, appear to be heterogeneous and are in need of revision. Certain of the African genera such as *Erlangea, Gutenbergia* and *Triplotaxis* are not well delimited. *Hoehnelia* may not be distinct from *Ethulia* (C. Jeffrey, personal communication). The cream, yellow, and orange-flowered species of *Vernonia* from Africa and Madagascar were placed in *Gongrothamnus* by Merxmüller (1954), but at least for the present time I would retain them in *Vernonia*. Several genera, especially certain of the monotypic South American ones, have rarely been collected and little is known of them.

Since there has been no comprehensive review of the Vernonieae since that of Bentham (1873a, b), subsequent revisionary work and changes have been restricted to the generic level or below. The current delimitation of most of the genera in the Vernonieae is nebulous; additional collections over the past 100 years have erased certain of the boundaries between formerly distinct taxa. The traditional sections of *Vernonia* are clearly artificial, being based upon relatively few, poorly understood, characters (Smith, 1971; C. Jeffrey, pers. comm.; S. B. Jones, unpublished observations).

GEOGRAPHICAL DISTRIBUTION PATTERNS

There is little doubt that the Vernonieae originated in the tropics, since that is its center of diversity, the area where its primitive species occur, and the region in which the majority of its genera are located. The tribe has two centers of distribution, one in southern Brazil (Baker, 1873) and the second in tropical Africa (Oliver and Hiern, 1877; Hutchinson and Dalziel, 1963; Smith, 1971). Vernonieae are commonly found in certain regions of Southeast Asia and associated archipelagos (Hooker, 1881; Gagnepain, 1924; Koster, 1935; Backer and van den Brink, 1965; Merrill, 1967; Ridley and Hutchinson, 1967), and in the West Indies, Central America and North America (Gleason, 1923b). Twenty-five endemic genera occur in South America; Brazil alone has about 16 endemic genera. In southern Brazil, Angely (pers. comm.) lists over 200 species of *Vernonia*, and 13

17. VERNONIEAE—SYSTEMATIC REVIEW 507

TABLE I. List of accepted genera in the tribe Vernonieae with approximate number of species, general distribution, habit, and chromosome number

Genera[a]	Number of Species[b]	Distribution[b]	Habit[c]	Chromosome Number (n)[d]
Acanthodesmos C. D. Adams	1	Jamaica	Shrub	
Adendöon Dalz.	1	Indo-Malaysia	Perennial	
Aedesia O. Hoffm.	3	Tropical West Africa	Perennial	10
Ageratinastrum Mattf.	5	Tropical Africa	Perennial	
Albertinia Spreng.*	1	Brazil	Shrub*	
Alcantara Glaziou*	2	Brazil	Shrub*	
Bipontia Blake	1	Brazil	Perennial	
Blanchetia DC.	1	Brazil	Shrub	
Bolanosa Gray	1	Mexico	Perennial	
Bothriocline Oliv. ex Benth.	14	Tropical Africa	Annual Perennial	20
Camchaya Gagnep.	4	Indochina	Annual	
Centauropsis Boj. ex DC.	10	Madagascar	Shrub	
Centratherum Cass.	20	Tropical	Annual Perennial	9, 16, 32
Chronopappus DC.	1	Brazil	Shrub	
Corymbium L.	17	South Africa	Perennial	
Decastylocarpus Humbert*	1	Madagascar	Perennial*	
Dewildemania O. Hoffm.	3	Tropical Africa	Perennial	
Diaphractanthus Humbert*	1	Madagascar	Annual*	
Dipterocypsela Blake*	1	Columbia		
Ekmania Gleason	1	Cuba	Shrub	
Elephantopus L.[e]	32	Tropical	Perennial	11, 13, 22
Eremanthus Less.	25	Brazil	Perennial Shrub	15
Erlangea Sch. Bip.	60	Tropical Africa	Annual Perennial Shrub	10, 20
Ethulia L.f.	1	Africa Southeast Asia	Annual	10, 20
Glaziovianthus Barroso*	1	Brazil		
Gorceixia Baker	1	Brazil	Shrub	
Gossweilera S. Moore	2	Angola	Perennial	
Gutenbergia Sch. Bip.	20	Tropical Africa	Annual Perennial	
Haarera Hutch. & E. A. Bruce*	1	Tropical East Africa	Annual*	
Haplostephium Mart. ex DC.	3	Brazil	Shrub	
Harleya Blake	1	Mexico	Perennial	

[a] One or more species in each genus was examined personally for Vernonieae features except for those marked with an asterisk.

[b] The number of species and distribution is from Shaw (1973) except where a more recent treatment has been used.

[c] The habit was determined from an examination of material at Kew and the British Museum (Natural History). Relatively few labels included habit information. In a few genera, the habit was determined from the literature and is indicated by an asterisk.

[d] From Bolkhovskikh *et al.* (1969); Moore (1973, 1974); Jones (1974); Olorode (1974); Powell *et al.* (1974); Jones (unpubl.), Turner (unpubl.).

[e] Includes *Orthopappus* Gleason and *Pseudelephantopus* Rohr. (Clonts, 1972).

Genera[a]	Number of Species[b]	Distribution[b]	Habit[c]	Chromosome Number (n)[d]
Herderia Cass.	6	Tropical Africa	Annual Perennial	
Heterocoma DC.*	1	Brazil	Shrub*	
Hoehnelia Schweinf.	2	Tropical East Africa	Perennial	
Hoplophyllum DC.	25	South Tropical Africa	Shrub	
Hystrichophora Mattf.	1	Tropical East Africa	(fragment at BM)	
Iodocephalus Thorel ex Gagnep.	3	Southeast Asia	Annual	
Lachnorhiza A. Rich	2	Cuba	Perennial	
Lamprachaenium Benth.	1	India	Annual	
Lepidonia Blake*	1	Central America	Perennial*	
Lychnophora Mart.	23	Brazil	Shrub	
Lychnophoriopsis Sch. Bip.*	1	Brazil	Shrub*	
Msuata O. Hoffm.	1	Tropical Africa	Shrub	
Muschleria S. Moore	2	Tropical Africa	Perennial	
Neurolakis Mattf.	1	West Tropical Africa	Perennial	
Oiospermum Less.*	1	Brazil	Annual*	
Oliganthes Cass.	9	Madagascar	Shrub	
Omphalopappus O. Hoffm.	1	Angola	Perennial	
Pacourina Aubl.	1	South America	Perennial	
Paurolepis S. Moore	1	South Tropical Africa	Perennial	
Piptocarpha R. Br.	50	Central America South America West Indies	Perennial Shrub Tree Liana	
Piptocoma Cass.	2	West Indies	Shrub	
Piptolepis Sch. Bip.	8	Brazil	Shrub	
Pithecoseris Mart. ex DC.	1	Brazil	Annual	
Pleurocarpaea Benth.	1	Australia	Perennial	
Pollalestra Kunth	24	Central America South America West Indies	Shrub Tree	
Proteopsis Mart. & Zucc. ex DC.	5	Brazil	Perennial	
Rastropoyllum Wild* (in ms.)	1	Africa		
Rolandra Rottb.	1	South America Malaysia	Perennial	
Spiracantha Kunth	1	Central America Columbia	Annual	
Stephanolepis S. Moore*	1	Tropical Africa	Perennial*	
Stilpnopappus Mart. ex DC.	20	South America	Annual Perennial Shrub	
Stokesia L'Hér	1	Southeast United States	Perennial	
Struchium P. Br.	1	Tropical America Tropical Africa	Perennial	
Telmatophila Mart. ex Baker*	1	Brazil	Perennial*	

Genera[a]	Number of Species[b]	Distribution[b]	Habit[c]	Chromosome Number (n)[d]
Trichospira (Kunth) Robin. & Brettell*	1	Tropical America	Annual* Perennial*	
Triplotaxis Hutch.	3	Tropical Africa	Annual Perennial	8, 10, 20
Vanillosmopsis Sch. Bip.	7	Brazil	Tree	
Vernonia Schreb.[f]	1000	America, Africa Southeast Asia	Annual Perennial Shrub Tree Liana	8, 9, 10, 14, 16, 17, 18, 20, 26–27, 28–30, 33–34, 36–39, 51, 68
Volkensia O. Hoffm.	8	Tropical East Africa	Perennial Shrub	

[f] Includes *Gongrothamnus* Steetz.

other genera with 60 species. Nineteen endemic genera are found in Africa, and four in Madagascar. Several small endemic genera occur in such diverse regions as Australia, Indo-Malaysia, Mexico, the West Indies and the south eastern United States (Shaw, 1973). The general distribution for each genus is shown in Table I.

The paucity of recent monographs on the tribe precludes a meaningful analysis of species diversity but a few trends are apparent in *Vernonia*. This large genus has *c.* 100 species in Southeast Asia, *c.* 200 in Africa, a surprisingly large number (100) in Madagascar (Humbert, 1960), at least 250 in South America, *c.* 24 in Central America, *c.* 40 in Mexico, *c.* 50 in the West Indies, and 17 species in the eastern United States. *Vernonia* is limited to the tropical and semitropical areas of Southeast Asia; in Africa it grows in savannas south of the Sahara into South Africa. In the New World, *Vernonia* can be found from central Argentina north into southern Canada but is absent from the central and southern Andes, the lower Amazon Basin, and the western United States.

BIOLOGY OF THE VERNONIEAE

Comparative Anatomy and Morphology

Wodehouse (1928) observed a wide range of variation in the types of pollen surfaces in species of *Vernonia* from North America; this was confirmed by Jones (1970b) using scanning electron microscopy (SEM). Light microscopy of pollen grains of 64 species of *Vernonia* from Africa indicated to Smith (1969) that pollen morphology and diameter confirmed traditional groupings in the genus. Ueno (1972) published SEM micrographs of *Stokesia*; the grains were similar to others in the tribe. Cabrera

(1944) described the surface features of pollen from several genera. Kingham (in press) examined the external pollen morphology of tropical African and certain other Vernonieae. Her work suggests that some taxonomic re-arrangement may be necessary and that several of the genera appear heterogeneous. She demonstrated that the pollen types exhibited by the segregate genera are also found in *Vernonia*. Comparative pollen morphology using a combination of SEM and TEM (transmission electron microscopy) promises to yield valuable information regarding arrangements in the tribe.

Relatively little work has been published on the comparative anatomy and morphology of the Vernonieae. Metcalfe and Chalk (1950) mentioned selected wood anatomy features of *Lychnophora*, *Piptocarpha* and *Vernonia*. Carlquist (1964, 1966) described wood anatomy for 14 Old and New World *Vernonia* species, samples from *Lychnophora*, *Oliganthes*, *Piptocarpha*, *Proteopsis* and *Vanillosmopsis*. The Vernonieae sampled have relatively long vessel elements and mostly uniseriate rays.

A number of investigators have examined the anatomy or morphology of selected Vernonieae. Cabrera (1944), in an excellent paper on the Vernonieae of Argentina, compared the leaf anatomy, glands, trichomes, pollen grains, corollas, stamens, styles, achenes, pappus, and inflorescences of selected species from *Centratherum*, *Elephantopus*, *Pacourina*, *Piptocarpha* and *Vernonia*. He demonstrated the systematic value of inflorescence types, confirming Gleason's (1923b) observations. Jones (1974) examined the trichomes, venation and epidermal features of cleared leaves of *Vernonia* Section *Eremosis* and found taxonomically useful features. Trichomes can be diagnostic in *Vernonia* particularly when they are interpreted with reference to other lines of evidence (Faust and Jones, 1973; Hunter and Austin, 1967). Wild has found that trichomes are useful characters with the African Vernonias (c. Jeffrey, pers. comm.). Gleason (1919) concluded that environmental and genetic factors interact in *V. missurica* Raf. to control the number of flowers per head. He observed the greatest number of flowers in the terminal head of each cyme; the number is relatively constant in the other heads on each plant.

More restricted studies have been made: McCarty and Scifres (1969), anatomy and life history of *Vernonia baldwinii* Torr.; Cozzo (1946). anatomy of *V. nudiflora* Less.; Petriella (1966), stem anatomy of *V. fulta* Griseb; Alencastro (1973), foliar anatomy of *V. oppositifolia* Less.; Wagner (1915), inflorescence of *V. rubicaulis* H. & B.; Schaffner (1918), dichotomous branching in *V. baldwinii*; Manilal (1966), floral anatomy of *V. cinerea* (L.) Less. and *V. patula* Mart. ex DC.; Tiagi and Taimni (1960, 1963), micro- and megasporogensis in *V. cinerascens* Sch. Bip. and *V. cinerea*; Misra (1972), floral morphology of *V. anthelmintica* (L). Wild; and Tiagi and Kshetrapal (1967), morphology and vascular anatomy of *Elephantopus scaber* L. In general, however, these papers provide little information of taxonomic value because the genera of the tribe have not yet received the necessary synoptic treatments with which to evaluate these characters.

External features of the pappus and achenes when examined by both light microscopy and SEM provide good markers for certain African genera (C. Jeffrey, pers. comm.). A synoptical study of these features is badly needed in *Vernonia* and should yield useful information.

Cytotaxonomy

Relatively few chromosome numbers have been reported for the tribe; nine of the 50 genera are known chromosomally and *c.* 80 of the 800–1000 species of *Vernonia* have been counted (Table II). From this limited

TABLE II. A summary of generic cytotaxonomy in the tribe Vernonieae.[a]

Genus	Chromosome number
Aedesia	$n=10$ (1)[b]
Bothriocline	$n=20(1)$
Centratherum	$n=9(2)$, 16(2)
Elephantopus	$n=11(7)$, 22(1), 13(1)
Eremanthus	$n=15(1)$
Erlangea	$n=10(4)$, 20(3)
Ethulia	$n=10(1)$, 20(1)
Stokesia	$n=7(1)$
Triplotaxis	$n=8(1)$

Vernonia[c]

$n=\overset{**}{8(1)}\ \overset{*}{9(19)}\ \overset{*}{10(20)}\ 14(1)\ \overset{**}{16(4)}\ \overset{***}{17(48)}\ \overset{***}{18(6)}\ \overset{**}{20(5)}\ 26\text{--}30(2)\ \overset{**}{33\text{--}39(6)}\ \overset{**}{51(1)}$
$\overset{**}{68(1)}$

* Old World species; ** New World species; *** Both Old and New World species

[a] From Bolkhovskikh *et al.* (1969); Moore (1973, 1974); Jones (1974); Olorode (1974); Powell, Khyos, and Raven (1974); Jones (unpublished); Turner (unpublished). A summary of the chromosome numbers for each species in the tribe is available in xeroxed form from the author.

[b] Number of species counted in parentheses.

[c] The majority of the counts for *Vernonia* are $n=9$, 10, 17, 18, 20, and 34. The other numbers are not frequent and some may be questionable. The Old World species have $n=9$, 10, 18, 20; whereas the New World Vernonias have $n=c$ 8, 17, 34, 51 and 68. The latter two numbers of $n=51$ and 68 are unpublished and were obtained after this manuscript was prepared, as were several of the additional numbers listed here.

sample, some generalizations are apparent: first, the chromosomal variation in the tribe is of taxonomic and evolutionary value; second, several basic numbers appear in the tribe, namely $x=7$, 8, 9, 10, 11 and 13 (Table II).

Significantly, chromosome counts of *Vernonia* in the Old World have been $n=9$ or 10 and a few polyploids of these, whereas in the New World $n=17$ and multiples thereof are found. The few exceptions to this may be questionable counts. If the genus *Vernonia* is a natural unit, the ancestral base number is likely to have been $x=9$ or 10 since those numbers are widely distributed in Africa and Southeast Asia. The New World species can be accounted for as older aneuploids. Bentham (1873a) first suggested that the Compositae might have first appeared in northern South America (Turner, Chapter 2), hence it seems probable the New World Vernonias represent an older evolutionary group than do those of the Old World.

Cytogenetics

Many species of *Vernonia* grow in disturbed habitats. This has led to hybridization and the production of intermediates which in some instances has tended to obscure species boundaries or has resulted in the naming of hybrids (Jones, 1967, 1972b; Jones *et al.*, 1970; Smith, 1971). That the potential for hybridization is high is shown by the nearly complete fertility of artificial hybrids between quite different species (Faust, 1972; Jones, 1972a; Urbatsch, 1972). I have recently confirmed Smith's (1971) hypothesis that hybridization can occur among certain African Vernonias. The artificial first-generation hybrids were vigorous and fertile, but some hybrid breakdown was noted in the second-generation progeny. Since natural hybridization often occurs in the tribe, and since much sympatry occurs, one or more phenomena must account for the numerous taxa which preserve their specific integrity. The isolating mechanisms are undoubtedly ecological, seasonal, and geographical, coupled with hybrid inviability or sterility.

Intra-subsectional crosses (Gleason, 1906) in subsections *Paniculatae verae* of eastern North America and *Paniculatae umbelliformes* of the Mexican highlands yielded fertile first-generation hybrids, but inter-subsectional crosses produced weak and sterile hybrids (S. B. Jones, unpublished). These Vernonias seem to fit what Grant (1971, p. 100) called the "*Geum* Pattern" of reproductive isolation. The plants are perennial herbs with an outcrossing breeding system and similar floral mechanisms; closely related species are interfertile and chromosomally homologous, but with strong incompatibility barriers developed between higher groupings.

Both cross-fertilization and autogamy are known in *Vernonia*. The pantropical weed, *V. cinerea*, is self-fertile; *V. anthelmintica* is largely selfing but does outcross; about 45 New World and about eight African species are self-incompatible (S. B. Jones, unpublished). Wide experimental hybridization in *Vernonia* sometimes causes a normally self-incompatible species to self (Jones, 1970 and unpublished).

Herbivore Interactions

Duviard (1969, 1970a, b) reported an extensive study on the geographic distribution, life history, floral morphology, habit, and insect fauna of *Vernonia guineensis* in the Ivory Coast. *V. guineensis* occurs in fire-maintained savannas, dying back to the ground during the dry season, then resprouting very quickly following fire. Several ant species, notably *Camponotus acvapimensis*, feed extensively on the *Vernonia* plants, obtain nectar from the heads, and exudates from the leaves and stems. They also "milk" aphids present on the plants, building chambers or "houses" for the aphids out of leaves or shoots cemented together or in hollowed-out heads where they cultivate the aphids. Interestingly, Duviard (1970a) observed that the "houses" are of varying design and appear to represent taxonomic differences among the ant species.

The larvae of several genera and species of tephritid flies infest the heads of *Vernonia* in eastern North America (M. D. Huttel, pers. comm.) greatly reducing the number of mature achenes. Duviard (pers. comm.) noted heavy insect predation on the achenes of *V. guineensis*. Schwitzgebel and Wilbur (1942a, b, 1943) published lists of Coleoptera, Lepidoptera, Hemiptera, Homoptera and Diptera associated with *V. baldwinii* in Kansas, U.S.A.

Larval feeding tests were undertaken to determine the preference of six Lepidoptera species for two *Vernonia* species which contain the sesquiterpene lactone glaucolide-A against one species which lacked it (Burnett, 1974; Burnett *et al.*, 1974). Significant differences were found between the feeding preferences of the insect species. Glaucolide-A incorporated into a *Vernonia* powder-agar medium reduced larval feeding and adversely affected the insect's life cycle (Burnett, 1974). Furthermore, this compound may influence oviposition by gravid female Lepidoptera (Burnett, 1974). Glaucolide-A was shown to deter the feeding of rabbits and deer on *Vernonia* (Burnett, 1974). In short, the abundance of sesquiterpene lactones in this tribe may be related to the ability of these compounds to serve as herbivore feeding deterrents rather than insect deterrents (Burnett *et al.*, 1974; Mabry *et al.*, 1974). Habitat correlation also suggests this for species are commonly found in open, grassy habitats with high-density mammal populations.

Insect pollinators are attracted to *Vernonia* in large number and variety. The insects gather both nectar and pollen from the *Vernonia* flowers. Some species of *Vernonia* are highly fragrant, complementing their showy heads and flower clusters. Certain African species have petaloid phyllaries which are probably adaptations to attract pollinators; to the human eye, they make the head seem much larger and more visible. The corolla tubes of the African Stengelioid Vernonias are much longer than the corolla tubes of other *Vernonia* taxa. This adaptation suggests some relationship to pollinators.

Ecotypes

Using reciprocal cloned transplants, Urbatsch (1973) demonstrated differences in flowering responses between southern and northern populations of *V. gigantea* (Walt.) Trel. Jain (1972) reported genetically distinct populations of erect and decumbent ecotypes of *V. patula* in India. Sesquiterpene lactone races have been found in several species of *Vernonia* from Mexico (Mabry et al., 1974); it is likely that these represent adaptations to different environments.

Coevolution with rusts

Urban (1973) investigated 17 autoecious species of the rust *Puccinia* parasitizing Vernonieae in North America. These rusts show remarkably close distributions to the systematics of their hosts. For example, *P. longipes* Lagerh. occurs on all *Vernonia* species in subsection *Paniculatae verae* (Gleason, 1906); whereas the Vernonias of subsection *Paniculatae umbelliformes* (Gleason, 1906) are parasitized by *P. semiinsculpta* Arth.

EVOLUTIONARY CONSIDERATIONS

The tribe presents an impressive array of form, habit, and ecology demonstrating the radiation and diversification that occurred during its evolution. Their habit includes shrubs, trees, perennial herbs and annuals. It has one of the largest woody Compositae, *Vernonia arborea* Buch.-Ham., said to reach a height of 33 m in the tropical forests of the Malay archipelago and southern part of the Asiatic continent (Good, 1974). Several African grassland species, such as *V. acrocephala* Klatt and *V. chthonocephala* O. Hoffm., are acaulescent perennials of fire habitats. The tribe also includes annuals such as *V. cinerea* and at least one aquatic species with edible leaves, *Pacourina edulis* Aubl. from South America. Good (1974) lists *Vernonia* among the 14 largest genera of flowering plants. One cannot but be impressed with the success of *Vernonia* in exploiting both habitats and geographical regions.

The Vernonieae as a group are characterized by discoid heads, anthers sagittate at the base, and a long slender style with slender almost acute branches that are nearly hirsute throughout and have the stigmatic surface on their inner flattened side. The Vernonioid style is found among all genera of the tribe, but since this style is also found elsewhere it must be considered in combination with other features (Bentham, 1873). Cronquist (1955) suggested that the style may be the only character in which the Vernonieae are more primitive than the Heliantheae. Augier and DuMerac (1951) suggested that the Vernonieae is primitive within the family but their hypothesis was rejected by Cronquist. Carlquist (1966) noted that three tribes (Vernonieae, Mutisieae, and Heliantheae) have relatively unspecialized wood with primitive features, but he cautioned that a search for the most primitive tribe is pointless, since characters, not tribes, are primitive.

The tribe is remarkable for the small size of its genera: c. 37 genera are monotypic and c. 51 genera have five or fewer species (Table I); this is a product of the evolutionary history of the tribe, or an artifact of classification, or both.

The sesquiterpene lactone data summarized by Harborne and Williams (Chapter 18) provide considerable insight into phylogenetic relations in *Vernonia*. They suggest that the New World *Vernonia* belong to the same phyletic line which also includes the genera *Stokesia* and *Piptocarpa*. The sesquiterpene lactones elaborated by the New World species of *Vernonia* are sufficiently different from those isolated from Old World species to support the earlier hypothesis (Gleason, 1923b) that, on a world-wide basis, *Vernonia* has two centers, one in Africa and another in South America (Mabry *et al.*, 1974). I believe the flavonoid evidence (Harborne, and Williams, Chapter 18) also supports the hypothesis of a common ancestral line for both the North and South American species of *Vernonia* and provides evidence linking the Old and New World species.

Supporting evidence is also available from chromosomal studies. *Vernonia* in the New World has a chromosome base number of $x=17$, whereas the Old World Vernonias are dibasic with $x=9$, or 10 (Table II) supporting Gleason's "two centers" hypothesis.

CONCLUSION

The tribe is abundantly represented in the New World, particularly in Brazil where the variety of genera and species is remarkable. In the Old World numerous species are found in Africa, a secondary center of distribution, and Madagascar, southern Asia, and the Malay Archipelago. Several species are troublesome weeds including *V. baldwinii* of North America and the pan-tropical weeds *V. cinerae*, *Elephantopus scaber*, and *E. spicatus* Aubl. *Vernonia anthelmintica*, *Stokesia laevis* (Hill) Greene and *Erlangea* are sources of seed oil; several species are used locally as drug plants; cancer inhibitors have been found in several genera: *Stokesia laevis*, several species of *Vernonia*, and *Pitocarpha sellowii* Baker are valued as ornamentals.

Studies of comparative pollen morphology using both SEM and TEM are needed and should help resolve critical taxonomic problems at higher levels in the tribe. The basic chromosome numbers for the genera are $x = 7, 8, 9, 10, 11, 13$ and 17. Natural hybrids among species are common in *Vernonia*; this often makes classification difficult at the species level.

The general classification of the Vernonieae was established from 1817 to 1873 but no modern synopsis of the tribe is available and few revisions or generic monographs have been published in the last 100 years. Several new genera and many species, however, have been described in recent years. Our knowledge of the Vernonieae is still in the alpha state. Chromosome numbers are known for only c. 100 of 1456 species and nine of the c. 70 genera in the tribe. The present attempt to survey the tribe reveals that the Vernonieae is composed of about 1456 species, *Vernonia* alone having about 1000 of them. Thirty-four of the genera are monotypic. From regional manuals, early nineteenth century diagnoses and herbarium studies, it is difficult to obtain an in-depth overview of the tribe such as attempted here. Only a few generalizations can be drawn; this is especially true of *Vernonia* because of its size and wide distribution. Finally, the need is stressed for a systematic revision of the tribe, subtribes, and genera on a world-wide basis rather than piece-meal treatments in limited geographic areas.

ACKNOWLEDGEMENTS

I wish to thank Mrs Nancy Coile for her help in preparing this chapter and Dr B. L. Turner for encouraging me to obtain an overview of *Vernonia* and the Vernonieae. The preparation of this paper was supported by grants from the National Science Foundation and by the University of Georgia.

REFERENCES

ALENCASTRO, F. M. M. R. DE. (1973). Contribuicão ão estudo da anatomia folia das Vernonias do Brasil I—*Vernonia oppositifolia* Less. *Arquivos.* **19**, 109–123.

ARISTEGUIETA, L. (1963). El genero *Oliganthes* de Madagascar y su equivalente Americano *Pollalestra*. Bolm *Soc. venez. Cienc. nat., Caracas* **23**, 255–288.

AUGIER, J. and DUMERAC, M. (1951). La phylogénie des Composées. *Revue scient.* **3311**, 167–182.

BACKER, C. A. and VAN DEN BRINK, R. C. (1965). *Flora of Java* **2**, 369–375.

BAKER, J. G. (1873). Compositae. I. Vernoniaceae. *Flora of Brazil* **6**, 1–180.

BARROSO, G. M. (1969). Novitates Compositarum. II. *Loefgrenia* **36**, 1–3.

BARROSO, G. M. (1970). Sôbre o colorido vermelho-purpúreo de *Vernonia erythrophila* DC. *Bolm. Mus. paraense Emilio Goeldi* **33**, 1–7.

BENTHAM, G. (1873a). Notes on the classification, history and geographical distribution of Compositae. *J. Linn. Soc. (Bot)* **13**, 335–577.
BENTHAM, G. (1873b). *In* "Genera Plantarum" (G. Bentham and J. D. Hooker, eds), Vol. 2, pp. 165–238.
BLAKE, S. F. (1926). Compositae. *In* "Trees and Shrubs of Mexico" (P. C. Standley, ed.). *Contr. U.S. natn. Herb.* **23**, 1412–1417.
BOLKHOVSKIKH, Z., GRIF, V., MATVEGEVA, T. and ZAKHARYEVA, O. (1969). "Chromosome Numbers of Flowering Plants". V. L. Komarov Botanical Institute, Acad. Sci. U.S.S.R.
BURNETT, W. C. (1974). "Sesquiterpene Lactones—Herbivore Feeding Deterrents in *Vernonia* (Compositae)". Ph.D. dissertation, University of Georgia, Athens, Georgia, U.S.A.
BURNETT, W. C., JONES, S. B., MABRY, T. J. and PADOLINA, W. G. (1974). Sesquiterpene lactones—Insect feeding deterrents in *Vernonia*. *Biochem. Syst. Ecol.* **2**, 25–29.
CABRERA, A. L. (1944). Vernonieas Argentinas (Compositae). *Darwiniana* **6**, 19–379.
CANDOLLE, A. P. DE. (1836). "Prodromus Systematis Naturalis Regni Vegetabilis", Vol. 5, pp. 15–66.
CARLQUIST, S. (1964). Wood anatomy of Vernonieae (Compositae). *Aliso* **5**, 451–467.
CARLQUIST, S. (1966). Wood anatomy of Compositae: A summary, with comments on factors controlling wood evolution. *Aliso* **6**, 25–44.
CASSINI, H. (1817). Apercu des genres nouveaux formés par M. Henri Cassini dans la famille des Synanthérées. *Bull. scient. Soc. phil.* **4**, 66.
CASSINI, H. (1819). Sur la famille des Synanthérées contenant les caractéres des tribus. *J. Phys. Chim. Hist. nat. Arts.* **88**, 190–204.
CHAPMAN, G. C. 1973. "Biosystematic Study of the Texanae Species-group of *Vernonia* (Compositae)". Ph.D. dissertation, University of Georgia, Athens, Georgia, U.S.A.
CLONTS, J. A. (1972). "A Revision of the genus *Elephantopus* including *Orthopappus* and *Pseudoelephantopus* (Compositae)." (Ph.D. dissertation, Mississippi State University, State College, U.S.A.
COZZO, D. (1946). Los géneros de famerogamas arbentinas con radios leñosos altos en sui lenõ secundario. *Revta argent. Agron.* **13**, 207–330.
CRONQUIST, A. (1955). Phylogeny and taxonomy of the Compositae. *Am. Midl. Nat.* **53**, 478–511.
CUATRECASAS, J. (1956). Neue *Vernonia*-arten und synopsis der Andinen-arten der sektion *Critoniopsis*. *Bot. Jb.* **77**, 52–84.
DUVIARD, D. (1969). Importance de *Vernonia guineensis* Benth. dans l'alimentation de quelques fourmis de savane. *Insectes Soc.* **16**, 115–134.
DUVIARD, D. (1970a). Place de *Vernonia guineensis* Benth. (Composées) dans la biocénose d'une savane préforestiére de Côte d'Ivoire. *Annls Univ. Abidjan*, Serie E. (*Ecologie*) **3**, 7–174.
DUVIARD, D. (1970b). Recherches ecologiques dans la savane de lamto (Côte D'Ivoire): L'Entomocoenose de *Vernonia guineensis* Benth. (Composées). *Terre et Vie* **1**, 62–79.
EKMAN, E. L. (1914). West Indian Vernonieae. *Ark. Bot.* **13**, 1–106.

FAUST, W. Z. (1972). A biosystematic study of the *Interiores* species group of the genus *Vernonia* (Compositae). *Brittonia* **24**, 363–378.

FAUST, W. Z. and JONES, S. B. (1973). The systematic value of trichome complements in a North American group of *Vernonia* (Compositae). *Rhodora* **75**, 517–528.

GAGNEPAIN, F. (1924). *In* Lecomte, M. H. "Flore generale de L'Indo-Chine" (M. H. Lecomte, ed.), Vol. 4, pp. 456–487.

GLEASON, H. A. (1906). A revision of the North American Vernonieae. *Bull. N.Y. bot. Gdn* **4**, 144–243.

GLEASON, H. A. (1919). Variability in flower-number in *Vernonia missurica* Raf. *Am. Nat.* **53**, 526–534.

GLEASON, H. A. (1922). Vernonieae. *N. Am. Flora* **33**, 52–95.

GLEASON, H. A. (1923a). The Bolivian species of *Vernonia*. *Am. J. Bot.* **10**, 297–309.

GLEASON, H. A. (1923b). Evolution and geographical distribution of the genus *Vernonia* in North America. *Am. J. Bot.* **10**, 187–202.

GOOD, R. (1974). "The Geography of the Flowering Plants". Longman, London.

GRANT, V. (1971). "Plant Speciation". Columbia University Press, New York.

HIERONYMUS, G. (1897). Erster beitrag zur Kenntnis der Siphonogamenflora der Argentina und der angrenzenden Länder, besonders von Uruguay, Paraguay, Brasilien und Bolivian. *Bot. Jb.* **22**, 672–701.

HOFFMANN, O. (1894). Vernonieae. *In* "Die Naturlichen Pflanzenfamilien" (A. Engler and K. Prantl, eds), Vol. 4, pp. 120–129.

HOOKER, J. D. (1881). *Flora of British India*. **3**, 226–242.

HUMBERT, H. (1960). *Flore de Madagascar. Composées*. pp. 5–198. Paris.

HUNTER, G. E. and AUSTIN, D. F. (1967). Evidence from trichome morphology of interspecific hybridization in *Vernonia*: Compositae. *Brittonia* **19**, 38–41.

HUTCHINSON, J. and DALZIEL, J. M. (1963). (Second edition edited by J. N. Hepper). *Flora of West Tropical Africa* **2**, 268–285.

JAIN, N. K. (1972). The effect of clipping on the erect type of *Vernonia patula* (Dryandl) Merrill. *J. Indian bot. Soc.* **51**, 374–378.

JONES, S. B. (1964). Taxonomy of the narrow-leaved *Vernonia* of the southeastern United States. *Rhodora* **66**, 382–401.

JONES, S. B. (1966). Experimental hybridizations in *Vernonia* (Compositae). *Brittonia* **18**, 39–44.

JONES, S. B. (1967). *Vernonia georgiana*—species or hybrid? *Brittonia* **29**, 161–164.

JONES, S. B. (1968). An example of a *Vernonia* hybrid in a disturbed habitat. *Rhodora* **70**, 486–491.

JONES, S. B. (1970a). The taxonomy of *Vernonia acaulis*, *V. glauca*, and *V. noveboracensis*. *Rhodora* **72**, 145–163.

JONES, S. B. (1970b). Scanning electron microscopy of pollen as an aid to the systematics of *Vernonia* (Compositae). *Bull. Torrey bot. Club* **97**, 325–335.

JONES, S. B. (1972a). A systematic study of the Fasciculatae group of *Vernonia* (Compositae). *Brittonia* **24**, 28–45.

JONES, S. B. (1972b). Hybridization of *Vernonia acaulis* and *V. noveboracensis* (Compositae) in the Piedmont of North Carolina. *Castanea* **37**, 244–253.

[K]es, S. B. (1973). Revision of *Vernonia* Section *Eremosis* (Compositae) in North America. *Brittonia* **25**, 86–115.

[K]es, S. B. (1974). Vernonieae (Compositae) chromosome numbers. *Bull. Torrey bot. Club* **101**, 31–34.

[K]es, S. B. and Faust, W. Z. (1976). Vernonieae (North of Mexico). *N. Am. Flora* in press.

[K]es, S. B., Faust, W. Z. and Urbatsch, L. E. (1970). Natural hybridization between *Vernonia crinita* and *V. baldwinii* (Compositae). *Castenea* **35**, 61–67.

[King]ham, D. L. (1976). A study of the pollen morphology of tropical African and certain other Vernonieae (Compositae). *Kew Bull*, **31**, 9–26.

K[oster], J. T. (1935). The Compositae of the Malay Archipelago. I. Vernonieae and Eupatorieae. *Blumea* **1**, 351–536.

L[ess]ing, C. F. (1829). De synanthereis herbarii regii berolinensis dissertationes, Vernonieae. *Linnaea* **4**, 240–356.

L[ess]ing, C. F. (1831a). De synthereis dissertatio quarta. *Linnaea* **4**, 240–288; 295–339.

L[ess]ing, C. F. (1831b). De synanthereis herbarii regii berolinensis dissertationes, V, Vernoniearum mantissa. *Linnaea* **6**, 624–721.

M[ab]ry, T. J., Abdel-Baset, Z., Padolina, W. G. and Jones, S. B. (1974). Systematic implications of flavonoids and sesquiterpene lactones in species of *Vernonia*. *Biochem. Syst. Ecol.* **2**, 185–192.

Mc[C]arty, M. K. and Scifres, C. J. (1969). Western ironweed: research on anatomy, physiology, life history and control. *Univ. Neb. agric. Exp. Stn. Res. Bull.* **231**.

M[al]me, G. O. A. N. (1931). Die Compositen der zweiten regnellschen reise. I. Rio Grande do Sul. *Ark. Bot.* **24**, 15–23.

M[al]me, G. O. A. N. (1932). Die Compositen der zweiten regnellschen reise. I. Mato Grosso. *Ark. Bot.* **24**, 5–20.

M[a]lme, G. O. A. N. (1933). Compositae Paranenses Dusenianae. *K. svensk. Vetensk. Akad. Handl.* **12**, 8–26.

M[a]nilal, K. S. (1966). Studies in the floral anatomy of Compositae—I the tribes—Vernonieae and Eupatorieae. *Proc. natn. Acad. Sci. (India)* **36**, 513–526.

M[a]rkötter, E. L. (1939). Eine revision der gattung *Corymbium* L. *Bot. Jb.* **70**, 54–372.

M[a]rtius, C. F. P. (1822). Novum planatarum genus. *Lychnophora*. *Denkschr. K. bayer, bot. Ges. Regensb.* **2**, 148–159.

M[e]rrill, E. D. (1967: reprint). An enumeration of Philippine flowering plants. **3**, 591–596.

M[e]rxmüller, H. (1954). Compositen—Studien IV: Die Compositen-Gattungen Südwestafrikas. *Mitt. bot. StSamml., Munch.* **1**, 357–443.

M[e]tcalfe, C. R. and Chalk, L. (1950). "Anatomy of the Dicotyledons", Vol. 2. Clarendon Press, Oxford.

M[i]sra, S. (1972). Floral morphology of the family Compositae. IV. Tribe Vernonieae—*Vernonia anthelmintica*. *Bot. Mag., Tokyo* **85**, 187–199.

M[o]ore, R. J., ed. (1973). "Index to Plant Chromosome Numbers, 1967–1971". International Bureau for Plant Taxonomy and Nomenclature, Utrecht. Netherlands.

MOORE, R. J., ed. (1974). "Index to Plant Chromosome Numbers for 1972", International Bureau for Plant Taxonomy and Nomenclature, Utrecht, Netherlands.

MOORE, S. L. M. (1902). A contribution to the Compositae flora of Africa. *J. Linn. Soc. (Bot.)* **35**, 305–367.

MOORE, S. L. M. (1918). Compositae (Vernonieae) africanae novea vel rariores. His Alabastra divera XXIX, 2. *J. Bot., Lond.* **56**, 204–212.

OLIVER, D. and HIERN, W. P. (1877). *Flora of Tropical Africa* **3**, 266–297.

OLORODE, O. (1974). Chromosome numbers in Nigerian Compositae. *J. Linn. Soc. (Bot.)* **68**, 329–335.

PETRIELLA, B. (1966). Estudio anatómico del tallo de *Vernonia fulta* (Compositae). *Boll. Soc. argent. Bot.* **11**, 19–25.

PHILIPSON, W. R. (1938). An enumeration of the African species of *Elephantopus* L. *J. Bot., Lond.* **76**, 299–305.

POWELL, A. M., KYHOS, D. W. and RAVEN, P. H. (1974). Chromosome numbers in Compositae. X. *Am. J. Bot.* **61**, 909–913.

RIDLEY, H. N. and HUTCHINSON, J. (1967, reprint). *The Flora of the Malay Peninsula* **2**, 186–189.

SCHAFFNER, J. H. (1918). Unusual dichotomous branching in *Vernonia*. *Ohio J. Sci.*, **18**, 487–490.

SCHULTZ BIP., C. H. (1861). Cassiniaceae uniflorae, oder verzeichniss der cassiniaceen mit l-blüthigen kopfchen. *Jber. Pollichia* **18–19**, 157–190.

SCHULTZ BIP., C. H. (1863). *Lychnophora* Martius und einige benachbarte gattungen. *Jber. Pollichia* **20–21**, 321–439.

SCHWITZGEBEL, R. B. and WILBUR, D. A. (1942a). Coleoptera associated with ironweed, *Vernonia interior* Small, in Kansas. *Kans. ent. Soc. J.* **15**, 37–44.

SCHWITZGEBEL, R. B. and WILBUR, D. A. (1942b). Lepidoptera, Hemiptera, and Homoptera associated with ironweed, *Vernonia interior* Small, in Kansas. *Trans. Kans. Acad. Sci.* **45**, 195–202.

SCHWITZGEBEL, R. B. and WILBUR, D. A. (1943). Diptera associated with ironweed, *Vernonia interior*, in Kansas. *Kans. ent. Soc. J.*, **16**, 4–13.

SHAW, H. K. A., ed. (1973). Eighth edition of Willis, J. C. "A Dictionary of the Flowering Plants and Ferns". Cambridge University Press.

SMITH, C. E. (1969). Pollen characteristics of African species of *Vernonia*. *J. Arnold Arbor.* **50**, 469–477.

SMITH, C. E. (1971). Observations on stengelioid species of *Vernonia*. *U.S.D.A. A.R.S. Agr. Handbook.* **396**.

TIAGI, B. and TAIMNI, S. (1960). Embryo-sac development in *Vernonia cinerascens* Schult. and seed development in *V. cinerea* Less. *Curr. Sci.* **29**, 406.

TIAGI, B. and TAIMNI, S. (1963). Floral morphology and embryology of *Vernonia cinerascens* Schult. and *V. cinerea* Less. *Agra* **12**, 123–138.

TIAGI, Y. D. and KSHETRAPAL, S. (1967). Studies on the vascular anatomy of the family Compositae. *Natn. Acad. Sci. (India)* **37**, 111.

TOLEDO, J. F. (1941). Notas sôbre o gênero monotípico *Heterocoma* DC. Compositae-Vernonieae. *Archos Bot. Est. S. Paulo* **1**, 71–73.

UENO, J. (1972). The fine structure of pollen surface. III. *Gazania* and *Stokesia*. *Shizuoka Univ.* **7**, 103–116.

URBAN, Z. (1973). The autoecious species of *Puccinia* on Vernonieae in North America. *Acta Univ. carol. biol.* **1971**, 1–84.
URBATSCH, L. E. (1972). Systematic study of the Altissimae and Giganteae species groups of the genus *Vernonia* (Compositae). *Brittonia* **24**, 229–238.
URBATSCH, L. E. (1973). A study of ecotypes in *Vernonia gigantea* ssp. *gigantea* (Compositae). *Trans. Neb. Acad. Sci.* **2**, 182–189.
WAGNER, R. (1915). Verzweigungsanomalien bei *Vernonia rubricaulis* H.B.K. Ser. Akad. Wiss. Wien. **124**, 547–565.
WODEHOUSE, R. P. (1928). The phylogenetic value of pollen-grain characters. *Ann. Bot.* **42**, 891–934.

Chapter 18
Vernonieae—chemical review

JEFFREY B. HARBORNE AND CHRISTINE A. WILLIAMS
Botany Department, University of Reading, England.

Abstract. The Vernonieae have not been as widely surveyed for chemical constituents as have other composite tribes. Although most emphasis has been placed on the very large genus *Vernonia*, some 100 species of the tribe representing 31 genera have been examined to some extent.

Amongst the taxonomically interesting constituents are the sesquiterpene lactones. In *Vernonia* spp. these are mostly of the primitive germacranolide type but three elemanolides have also been found and in *Elephantopus* species germacranolide dilactones have been characterized Sesquiterpene lactones of the guaianolide type are present in *Eremanthus elaeagnus*, *Vanillosmopsis erythropappa* and *Vernonia sublutea* and a heliangolide has been reported in *Eremanthus goyazensis*. New and Old World species of *Vernonia* can be distinguished on the basis of their sesquiterpene lactones, which are of a simpler type in New World species.

The flavonoids of New World *Vernonia* species have been examined in some detail and it appears that the complex flavonoid profiles found in these plants, when compared with the profiles of Old World species, might represent a "primitive" character within the genus. A wider survey for flavonoids in *Vernonia* and 30 other genera has indicated that African and Indian species generally produce only flavones, whilst New World species produce flavones, or flavonols or both types. This evidence, with the sesquiterpene lactone data, supports the hypothesis that the genus *Vernonia* has two centres of origin: in Africa and in South America. Other chemical constituents reported in the tribe include triterpenes, sterols, alkaloids, lipids, cyclitols and acetylenes but too little is known of their distribution for them to be useful taxonomic markers. The fatty acid, vernolic acid, a common constituent of the Vernonieae has also been reported from the tribes Cichorieae and Cynareae.

CONTENTS

Introduction	524
Low Molecular Weight Constituents	524
Terpenoids	524
Flavonoids	529
Alkaloids	532
Other compounds	532
Economic and Pharmaceutical Uses	533

Conclusion	.	.	.	.	.	.	.	.	.	.	533
Acknowledgements	.	.	.	.	.	.	.	.	.	534
References.	.	.	.	.	.	.	.	.	.	534

INTRODUCTION

The Vernonieae, a tribe of some 1400 species and 50 genera, is dominated by a single large genus *Vernonia* which alone has 800–1000 species. It is mainly tropical, with many species growing in Brazil and tropical Africa. For this reason, the chemistry has not been as widely explored as in most other composite tribes. Indeed, until relatively recently, little was known of the chemical constituents. Thus Hegnauer (1964) gives only a few references to the tribe and most of these concern *Vernonia*. The situation has improved in the last few years, partly due to recent screening programmes of tropical plants for anti-tumour agents. One important chemotaxonomic study has just been published on North American *Vernonia* species (Mabry *et al.*, 1975).

The present account of the chemistry is therefore based on Hegnauer (1964) but has taken into consideration as far as possible more recent references.

LOW MOLECULAR WEIGHT CONSTITUENTS

Terpenoids

Essential oils. The essential oil of one species of the tribe Vernonieae has been studied: *Vanillosmopsis erythropappa* Sch. Bip. In the wood of this plant, Gottlieb and Magalhaes (1958) identified ($-$)-α-bisabolol (**1**) as the major oil constituent, with traces of isovaleric acid and β-bisabolene.

Sesquiterpene lactones. Some 19 sesquiterpene lactones have been identified in four genera of the Vernonieae: *Vernonia*, *Elephantopus*, *Eremanthus* and *Vanillosmopsis*.

In *Vernonia* species, the lactones are mostly of the simple germacranolide type: five compounds have been identified in Old World species and three of nine sesquiterpene lactones isolated from New World species have been characterized. Three elemanolides, vernodalin (**2**), vernolepin (**4**) and vernomenin (**5**), have also been found in Old World species (Table I) and one guaianolide (**10**) in *Vernonia sublutea*. On the basis of sesquiterpene lactone evidence, Mabry *et al.* (1975) suggest that all the New World *Vernonia* species and the genus *Stokesia* belong to the same phyletic line. They find that the more primitive species contain glaucolide-A(**12**) or -B(**13**), whilst the most recently evolved species contain structurally simpler compounds or else lack sesquiterpene lactones. They also report

TABLE I. The distribution of sesquiterpene lactones in *Vernonia*

Sesquiterpene lactone	Source	Reference
From Old World Species		
Vernodalin (**2**)	*V. amygdalina* Delile	Kupchan *et al.*, 1969a
Vernomygdin (**3**)		
Vernolepin (**4**)	*V. hymenolepis* A. Richard	Kupchan *et al.*, 1969b
Vernomenin (**5**)		
Vernolide (**6**)	*V. colorata* Drake	Toubiana and Gaudemer, 1967; Pascard, 1970.
Hydroxyvernolide (**7**)	*V. colorata* Drake	Toubiana, 1969; Ho and Toubiana, 1970
Pectorolide (**8**)	*V. pectoralis* Baker	Mompon *et al.*, 1973
Confertolide (**9**)	*V. conforta* Bentham	Toubiana *et al.*, 1970
Subluteolide (**10**)	*V. sublutea* Scott Eliott	Mompon *et al.*, 1974
From New World Species		
Marginatin (**11**)	*V. marginata* Oliver & Hiern.	Padolina, 1973; Padolina *et al.*, 1974
	V. fasiculata Michx.	
	V. arkansana DC.	
Glaucolide-A (**12**)	*V. acaulis* (Walt.) Gl.	Padolina, 1973;
	V. angustifolia Michx.	Padolina *et al.*, 1974
	V. gigantea (Walt.) W. Trelease ex Bronner & Colville	
	V. glauca (L.) Willd.	
	V. lettermanii Engelm. ex Gray	
	V. missurica Rafin	
	V. noveboracensis (L.) Willd.	
	V. texana (Gray) Small	
	V. greggii Gray	
	V. alamanii DC.	
	V. capreaefolia (Sch. Bip.) Gl.	
	V. liatroides DC.	
	V. incana Less.	
	V. nidiflora Less.	
	V. unifolia Sch. Bip.	
Glaucolide-B (**13**)	*V. baldwinii* Torrey	Padolina, 1973;
	V. divaricata Sw.	Padolina *et al.*, 1974
	V. fruticosa (L.) Sw.	
	V. canescens H.B.K.	
	V. brevifolia Less.	
	V. duncanii S. B. Jones	

that the sesquiterpene lactones of the New World species of *Vernonia* are sufficiently different from those isolated from Old World species to support the hypothesis (Gleason, 1923) that *Vernonia* has two centres of distribution, one in Africa and the other in South America.

In *Elephantopus* species, five germacranolide dilactones have been characterized: elephantin (**14**), elephantopin (**15**) (Kupchan *et al.*, 1966, 1969c) and elephantol (**16**) (McPhail and Sim, 1972) from *E. elatus* Bertol; deoxyelephantopin (**17**) (Kurkawa *et al.*, 1970) and isodeoxyelephantopin (Govindachari *et al.*, 1972) from *E. scaber* L. and molephantin (**18**) from *E. mollis* H.B.K. (Lee *et al.*, 1973). In heartwood oils of *Eremanthus elaeagnus* Sch. Bip. and *Vanillosmopsis erythropappa* Sch. Bip., a sesquiterpene lactone of the guaianolide type is present, namely eremanthine (**19**) (Vichnewski and Gilbert, 1972). The substance from *Vanillosmopsis* was first called vanillosmin (Corbella *et al.*, 1974) until it was realized that it was identical with the substance from *Eremanthus* (Vichnewski *et al.*, 1976). Goyazensolide (**20**), a sesquiterpene lactone of yet another type, a schistosomicidal heliangolide, has recently been characterized in *Eremanthus goyazensis* Sch. Bip. These reports of a guaianolide and a heliangolide in the Vernonieae are of phylogenetic interest in that previously only the "primitive" germacranolides and a few elemanolides had been found in the tribe, evidence that had been used to support the suggestion that the Vernonieae is amongst the most "primitive" tribes in the Compositae. To clarify the phyletic position of the Vernonieae, a survey of other genera of the Vernonieae for sesquiterpene lactones is obviously needed.

Triterpenes and Sterols. Only three *Vernonia* and *Elephantopus* species have been examined for triterpenes and/or sterols. The triterpenes: lupeol, lupeol acetate and β-amyrin together with the common phytosterols: sitosterol, stigmasterol and α-spinasterol were identified in *V. cinerea* Less. (Rao, 1962). In the seed of *V. anthelmintica* (L.) Willd. six sterols were characterized. $\triangle^7$-Avenasterol (**21**), first isolated from *Avena sativa* L. (Knights and Laurie, 1967) represents 70% of the sterol mixture in *V. anthelmintica* (Frost and Ward, 1968). The same workers have since confirmed this structure and reported the presence of stigmasterol, stigmastanol and spinasterol together with a new phytosterol, 5α-stigmasta-8,14Z-24(28)-trien-3β-ol (**22**) in the seeds of this species (Frost and Ward, 1970). In the same year Fioretti *et al.* (1970) characterized yet another new phytosterol, vernosterol, 5α-stigmasta-8(14),15,Z,-24,28-trien-3β-ol (**23**) in seeds of the same plant; this is a simple isomer of compound (**21**). A third new sterol has more recently been reported in the stem of *Vernonia amygdalina* Delile: 7,24(28)-stigmastadien-3β-ol (Arene, 1972).

In *Elephantopus scaber* L. epifriedelinol, lupeol, stigmasterol, triacontan-1-ol and diotriacontan-1-ol were isolated by Sim and Lee (1969).

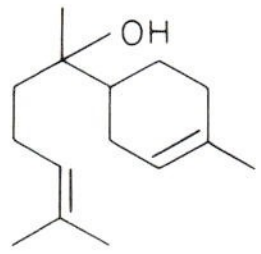

(1) Bisabolol

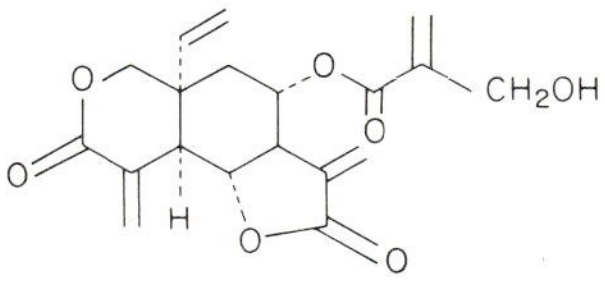

(2) Vernodalin

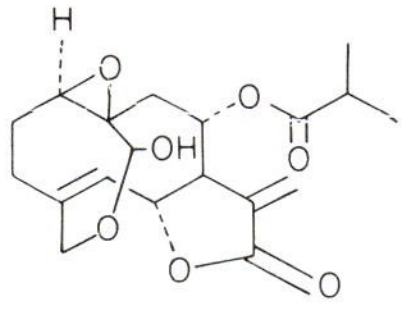

(3) Vernomygdin

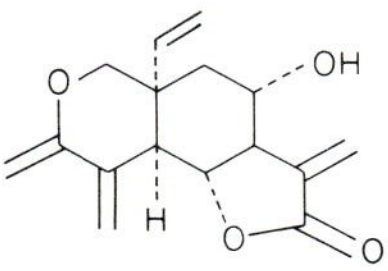

(4) Vernolepin

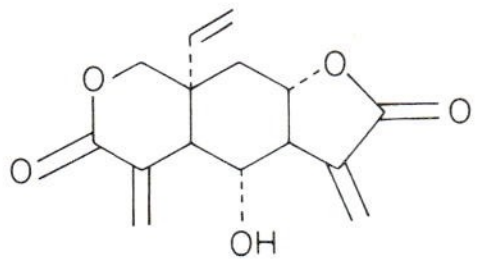

(5) Vernomenin

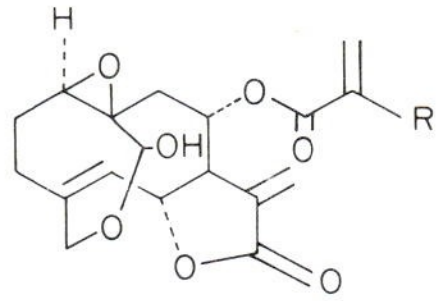

(6) Vernolide, R = CH₃
(7) Hydroxyvernolide R = CH₂OH

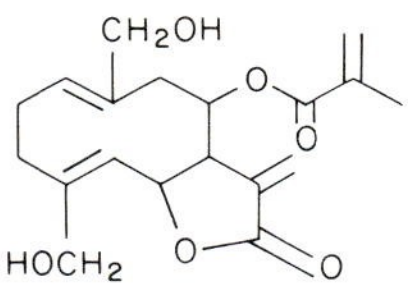

(8) Pectorolide

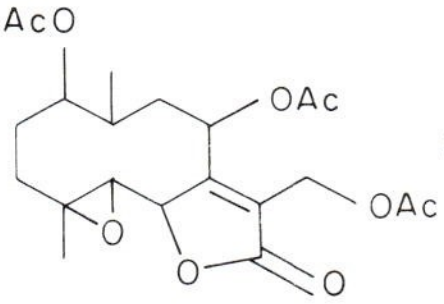

(9) Confertolide

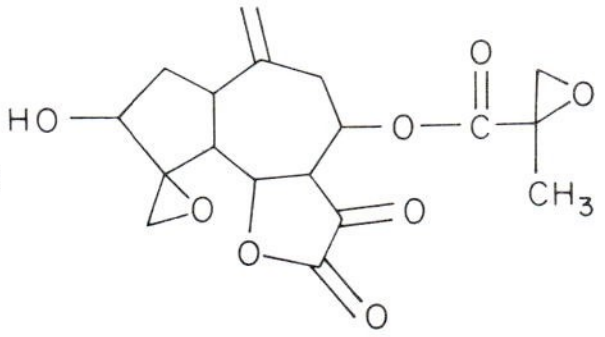

(10) Subluteolide

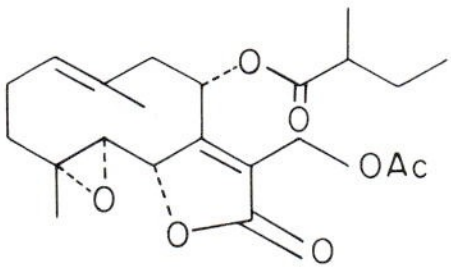

(11) Marginatin

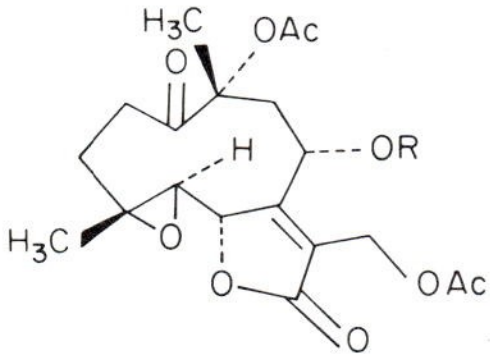

(12) R = $\underset{O}{\underset{\parallel}{-C}}-\underset{\text{CH}_2}{\overset{\text{CH}_3}{\underset{\parallel}{C}}}$, Glaucolide-A

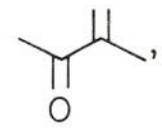

(13) R = Ac , Glaucolide-B

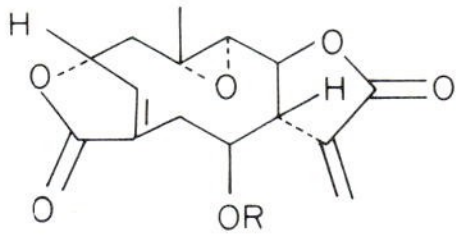

(14) Elephantin R = COCH=C(CH₃)₂

(15) Elephantopin R = COC(=CH₂)CH₃

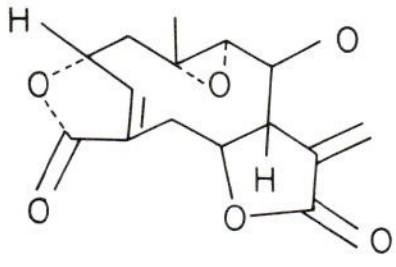

(16) Elephantol

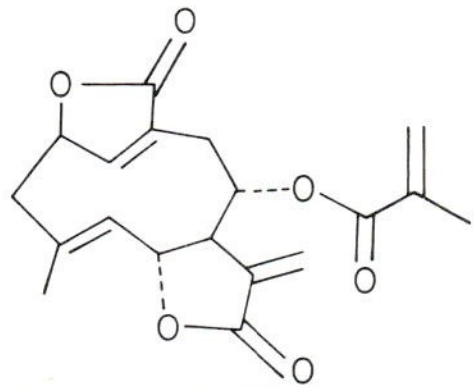

(17) Deoxyelephantopin

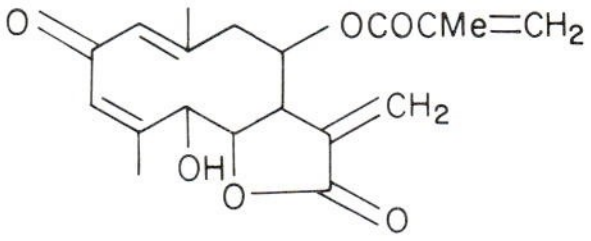

(18) Molephantin

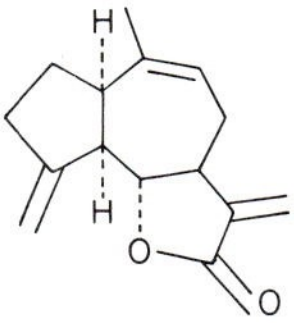

(19) Eremanthine

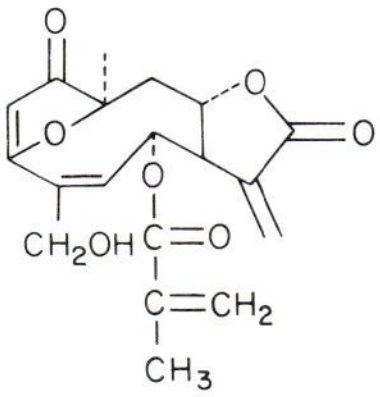

(20) Goyazensolide

(21) Δ^7-Avenasterol; 5α-stigmasta-7,Z-24(28)-dien-3β-ol

(22) 5α-Stigmasta-8,14,Z-24(28)-trien-3β-ol

(23) Vernosterol; 5α-stigmasta-8(14),15,Z-24(28)-trien-3β-ol

(24) Apigenin 7-methyl ether

(25) R = H, Luteolin 7-methyl ether
(26) R = Me, Luteolin 7,3'-dimethyl ether

(27) Kaempferol 3-methyl ether

$CH_3 \cdot [CH_2]_4 \cdot CH \overset{O}{\frown} CH \cdot CH_2 - CH:CH \cdot [CH_2]_7 \cdot CO_2H$
(28) Vernolic acid

$H_3C-(C{\equiv}C)_5-CH{=}CH_2$
(31) Pentaynene

(29) L-Inositol   (30) Scyllitol

$H_3C-CH{=}CH-(C{\equiv}C)_4-CH{=}CH_2$
(32) Tetrayne

$H_3C-(C{\equiv}C)_2$—[thiophene]—$C{\equiv}C-CH{=}CH_2$ (33) Thiophene acetylene

Flavonoids

Flavonoid reports in the Vernonieae have been largely restricted to the large genus *Vernonia*, although in 1963 Ghanim *et al.* did record the presence of luteolin 7-glucoside in flowers of *Elephantopus scaber* L. This glycoside was later identified in *Vernonia cinerea* Less. by Wagner *et al.* (1972), who also recorded the presence of quercetin, luteolin, kaempferol and quercetin 3-*O*-methyl ether in *V. patens* H.B.K.

In a detailed study of New World and some Old World species of *Vernonia* Padolina (1973) and Mabry *et al.* (1975) have identified 26 flavonoids from 41 species (see Table II). Several of the methylated flavonoids detected in this survey, for example, apigenin 7-methyl ether (**24**) luteolin 7-methyl ether (**25**), luteolin 7,3'-dimethyl ether (**26**), and kaempferol 3-methyl ether (**27**) are relatively rare plant constituents and do not appear to have been reported in other tribes of the Compositae.

Within *Vernonia* the South American species show the most complex and the Old World species the simplest flavonoid patterns. Mabry *et al.* (1975) suggest, because some *Vernonia* species, which are recognized as "advanced" on other than chemical grounds contain only a small number (one to four) flavonoids, that the presence of a relatively large number (10-17) of flavonoid constituents might be considered as a "primitive" character within the genus. They give as an example two closely related North American species *V. greggii* A. Gray and *V. texana* (Gray) Small, considered to be primitive members of the subsection *Paniculata verae* and which both contain 17 flavonoids, i.e. one of the most complex patterns observed in North American *Vernonia* species. These workers also suggest the grouping of *V. acaulis* (Walt.) Gl. with *V. glauca* Willd., *V. angustifolia* Michx. with *V. blodgettii* Small, *V. lindheimeri* Gray & Engelm. with *V. larsenii* King, *V. noveboracensis* (L.) Willd. with *V. gigantea* (Walt) W. Trelease ex Branner & Colville ssp. *gigantea* and *V. marginata* Oliver & Hierr. with *V. fasciculata* Michx. on both flavonoid and sesquiterpene lactone evidence. The data also serve to distinguish between two subspecies of *V. gigantea* ssp. *gigantea* and *ovalifolia*. The latter occupies a geologically younger area in Florida than does the former and contains fewer methylated aglycones. Two subspecies of *V. baldwinii* Torrey, *baldwinii* and *interior*, may also be separated on chemical grounds and Jones (in Mabry *et al.*, 1975) suggests that these two subspecies should be treated as two separate species. Mabry *et al.* (1975) found that the Old World species of *Vernonia*, section *Stengelia*, had fewer but similar flavonoid constituents to the New World species.

In a later survey, of 52 species (representing 31 genera) of the Vernonieae, Williams and Harborne (unpublished results) also found that Old World species of *Vernonia* and indeed most other genera are depauperate in flavonoid constituents. It is however, interesting that in all six African

TABLE II. Distribution of flavonoids in *Vernonia*

Taxa and geographical distributions	Flavonoids
Section: Lepidaploa (New World):	
Subsection: Paniculatae verae (U.S. & Sierra Madre Oriental, Mexico)	
V. acaulis (Walt.) Gl.	1, 2, 3, 4, 13, 16, 17, 19, 20, 21, 22
V. angustifolia Michx.	5, 19, 20, 21, 22
V. arkansana DC.	1, 2, 3, 4, 5, 6, 13, 17, 19, 20, 21, 22, 23
V. baldwinii Torr. ssp. *baldwinii*	1, 2, 3, 4, 5, 6, 13, 17, 18, 20, 21, 22, 23
V. baldwinii ssp. *interior* (Small) Faust	1, 2, 3, 4, 5, 6, 13, 14, 15, 16, 17, 18, 21, 23
V. blodgetti Small	1, 3, 4, 19, 20, 21, 22
V. fasciculata Michx.	1, 2, 3, 4, 5, 6, 13, 14, 15, 16, 17
V. flaccidifolia Small	19, 20, 21, 22
V. gigantea (Walt) Trel, ex Branner & Colville ssp. *gigantea*	1, 2, 3, 4, 5, 6, 13, 17, 19, 20, 21, 22
V. gigantea ssp. *ovalifolia* (T. & G.) Urbatsch	3, 4, 5, 6, 13, 17, 19, 20, 21, 22
V. glauca (L.) Willd.	1, 2, 3, 4, 13, 16, 17, 19, 20, 21, 22
V. greggii Gray	1, 2, 3, 4, 5, 6, 13, 14, 15, 16, 17, 18, 19, 20, 21, 22, 23
V. larsenii King	1, 2, 3, 4, 5, 6, 13, 14, 16, 17, 19, 20, 21, 22
V. lettermannii Engelm. ex Gray	1, 3, 4, 6, 19, 20, 21, 22
V. lindheimeri Engelm. & Gray	1, 2, 3, 4, 5, 6, 13, 14, 16, 17, 19, 20, 21, 22
V. marginata Oliver & Hiern	1, 2, 3, 4, 5, 6, 13, 14, 15, 16, 17
V. missurica Rafin	1, 2, 3, 4, 6, 13, 17, 19, 20, 21, 22, 23
V. noveboracensis (L.) Willd.	1, 2, 3, 4, 5, 6, 13, 14, 17, 19, 20, 21, 22
V. pulchella Small	5
V. texana (Gray) Small	1, 2, 3, 4, 5, 6, 13, 14, 15, 16, 17, 18, 19, 20, 21, 22, 23
Subsection: Paniculatea umbelliformes: (Mexico)	
V. alamanii DC.	1^a, 2, 3, 4, 5, 6, 13, 14, 16^a, 17, 21, 22
V. capreaefolia (Sch. Bip.) Gl.	5, 13, 14, 16, 17, 19, 20, 21, 22, 26
V. liatroides DC.	1, 2, 3, 4, 5, 6, 13, 14, 16, 17, 19, 20, 21, 22, 26
Subsection: Scorpioideae foliatae	
V. diavaricata Sw. (Jamaica)[b]	1^a, 2^a, 5, 13^a, 14^a, 17, 19^a, 20^a, 21^a, 22^a
V. arbuscula Less. (Bahamas)	5, 12, 22, 23
Subsection: Scorpioideae aphyllae (Central America)	
V. canescens H.B.K.	5, 13, 14, 17, 21, 22, 23
V. patens H.B.K.	No detectable amounts of flavonoids
Subsection: Not known	
V. rubricaulis H. & B. (Argentina)	3, 4, 6, 21, 22
V. brasiliana Druce (Brasil)	1, 2, 3, 4, 5, 10, 13, 14, 17, 19, 20, 21, 22
V. brevifolia Less. (Argentina)	1, 2, 5, 6, 7, 8, 9, 11, 12, 13, 19, 20, 21, 22, 23, 24, 25, 26, 27
V. nudiflora Less. (Argentina)	1, 3, 4, 5, 6, 10, 19^a, 20^a, 21, 22^a, 23, 26
V. incana Less. (Argentina)	1, 3, 4, 5, 6, 10, 19, 20, 21, 22, 23, 26
Section: Eremosis (Mexico)	
V. salicifolia (DC.) Sch. Bip.	5, 13, 14, 17
V. duncanii S. B. Jones	2, 5, 6, 10, 13, 14, 17
V. paniculata D.C.	3, 4, 5, 6, 13, 14
Section: Stengelia (Old World)	
V. anthelmintica (L.) Willd. (Pakistan)	13, 14, 16, 17
V. abyssinica Sch. Bip. ex Walp. (Africa)	5, 10, 13, 14, 17, 21, 22, 23
V. glabra (Steetz ex Peters) Vatke (Africa) (=melleri)	13, 14, 17
V. afromontana R. E. Fries (Africa)	6, 10, 19, 20, 21, 22
V. adoensis Sch. Bip. ex Walp. (Africa)	3, 4, 6, 13, 14, 16, 19, 20, 21, 22
Section: Tephrodes	
V. conerea (L.) Less. (Pan-tropical)	14, 16, 17, 21, 22, 23

[a] Geographical variation observed for the occurrence of this compound in this taxon.
[b] One sample used here may have been *V. fruticosa*; this matter requires re-investigation.

Key:
1. Luteolin 7,3'-dimethyl ether
2. Apigenin 7-methyl ether
3. Luteolin 3'-methyl ether
4. Luteolin 7-methyl ether
5. Apigenin
6. Luteolin
7. Quercetin 7,3,4'-trimethyl ether
8. 3-O-Acyl quercetin 7,3'-dimethyl ether*
9. Kaempferol 3-methyl ether
10. Unknown
11. Quercetin
12. Kaempferol

13. Apigenin 7-O-glucoside
14. Luteolin 7-O-glucoside
15. Luteolin 7-O-arabinoglucoside
16. Luteolin 7-O-galactoglucoside
17. Luteolin 7-O-glucoglucoside
18. Luteolin 4'-O-glucoside 7-O-galactoside
19. Kaempferol 3-O-glucoside
20. Kaempferol 3-O-rhamnoglucoside
21. Quercetin 3-O-glucoside
22. Quercetin 3-O-rhamnoglucoside
23. Quercetin 7-O-glucoside 3-O-rhamnoside
24. Quercetin 7-O-diglycoside 3-O-glycoside
25. Quercetin 7-O-diglucoside 3-O-diglycoside
26. 3,3'-O-Diacyl quercetin 7-O-glucoside*
27. Hesperidin

* Tentative structure.

From Mabry et al. (1975).

Vernonia species examined only flavone glycosides (luteolin and apigenin) were found, whereas in South, Central and North American species flavones, flavonols or both flavones and flavonols were detected. These results reinforce the sesquiterpene lactone evidence that *Vernonia* has two centres of distribution, one in Africa and another in South America. Also in all but one of the Old World genera examined only flavone glycosides were found. The following species of the Vernonieae were surveyed:

1. *From India*: *Adenoön indicum* Dalz. flavonol, *Centratherum phyllolaenum* Bentham, flavone.

2. *From Africa*: *Aedesia glabra* O. Hoffm., *Centauropsis lanuginosa* Boj ex DC., *Corymbium glabrum* L., *Ethulia conyzoides* L.fil., *Erlangia cordifolia* S. Moore, *Gutenbergia pembensis* S. Moore, *Herderia truncata* Cass., *Vernonia glabra* (Steetz ex Peters) Vatke, *V. jugulas* Oliver and Hiern var. *lanuginosa*, *V. leptalepis* Baker, *V. perrottetii* Sch. Bip., *V. senegalensis* Lam. and *V. theophrastifolia* Schweinf., which all contain flavones.

3. *From South, Central and North America*: Flavone-containing species: *Alcantara petroana* Glaziou, *Bolanosa coulteri* A. Gray, *Piptocarpha eleagnoides* Baker, *P. sellowiana* Baker var. *balnusiana* Hieron, *Stokesia laevis* (Hill) Greene, *Vernonia balansae* Hieron, *V.* aff. *nitidula* Less. and *V. squamulosa* Hooker & Arn.

Flavonol-containing species: *Blanchetia heterotricha* DC., *Ekmania leipidota* Griseb. Gleason, *Oliganthes condensata* Sch. Bip., *Piptocarpha rotundifolia* Baker, *Piptocoma ekmanii* Alain, *P. antillaria* Urb., *Piptolepis gardneri* Baker, *Vanillosmopsis erythropappa* Sch. Bip., *Vernonia noveboracensis* (L.) Willd., *V. patens* H.B.K., *V. praealta* DC., *V. rubicaulis* Humb. & Bonpl., *V. saltensis* Hieron, *V. vernicosa* Klatt, *Haplostephium passerinia* Mart. B. var. *sublatum Lychnophora staavioides* Michx. ex Baker, *L. brunioides* Mart.

Species with both flavones and flavonols: *Oliganthes discolor* Sch. Bip., *Pacourina spinissima* Britton, *Stilpnopappus scaposus* Dec., *Hysterionica jasionoides* Willd.

No flavonoids detected: *Elephantopus carolinianus* Willd., *E. mollis* H.B.K., *Spiracantha cornifolia* H.B.K.

Alkaloids

The alkaloids of the Vernonieae have not been well studied. Pernet (1959) in a survey of 12 *Vernonia* species from Madagascar found that these plants were mostly rich in (probably indole-based) alkaloids. On the other hand, Patel *et al.* (1964) found no alkaloids in the four species they examined: *Vernonia amygdalina* Delile, *V. colorata* Drake, *V. guineensis* Bentham and *V. nigritiana* Oliver & Hiern but did find toxic cardenolides in all four taxa.

Other compounds

Lipids. A number of *Vernonia* seed oils have been shown to contain oxygenated acids (8–90%); mainly (+)-12,13-epoxyoleic acid, vernolic (**28**). This epoxy acid was first identified in *Vernonia anthelmintica* Willd. (Gunstone, 1954), where it represents 26·9% of the seed oil. Vernolic acid has also been found in *V. colorata* Drake (Chisholm and Hopkins, 1957), *V. amygdalina* Delile, *V. cinerea* Less., *V. biafrae* Oliver & Hiern, *V. nigritiana* Oliver & Hiern and *V. camporum* A. Cheval. (Badami and Gunstone, 1963) but with the exception of *V. camporum* this epoxy acid is present in small amount when compared with the unsaponifiable fraction of the seed oil (Badami and Gunstone, 1963). Other species of *Vernonia*, for example, *V. baldwinii* Torr. and *V. missurica* Rafin. produce no vernolic acid in their seed oils (Hegnauer, 1964). Scott *et al.* (1962) also report the presence of 12, 13-dihydroxyoleic acid in the seed oil of *V. anthelmintica* Willd. No other genera of the Vernonieae appear to have been surveyed for lipids. Vernolic acid has been found in two other tribes of the Compositae—the Cichorieae and the Cynareae.

Cyclitols. The only reports of these cyclohexane derivatives in the Vernonieae are of L-inositol (**29**) and scyllitol (**30**) in *Vernonia altissima* Nutt. (Rowe *et al.*, 1955).

Acetylenes. Acetylenes have been found in small amount in the roots of 18 species of the Vernonieae (Bohlmann, 1973). Most widespread is the pentaynene (**31**), which is present in 14 taxa: *Sparganophorus sparganophora* Boehm, *Ethulia conyzoides* L., *Centratherum camporum* Malme, *C. punctatum* Cass., *Vanillosmopsis discolor* Bentham & Hooker, *Vernonia anthel-*

mintica Willd., *V. brevifolia* Less., *V. cinerea* Less., *V. echioides* Less., *V. megapotamica* Sprengel., *V. noveboracensis* (L.) Willd., *V. senegalensis* Less., *Corymbium villosum* L. fil., and *Elephantopus corilianus* Willd. The tetrayne (**32**) appears in only five species: *Ethulia conyzoides* L., *Corymbium villosum* L. fil., *Hoplophyllum spinosum* DC., *Elephantopus corilianus* Willd. and *E. mollis* H.B.K. and a thiophene acetylene (**33**) is present in only one species: *Ethulia conyzoides* L.

ECONOMIC AND PHARMACEUTICAL USES

Only two species of the tribe appear to have been examined as possible crop plants. One is *Vanillosmopsis erythropappa* which has been suggested as a possible commercial forest tree in Brazil (Arauja, 1944). The other is *Vernonia anthelmintica* or Indian ironweed, a plant which has been used as a source of an anthelmintic against thread worms and for the treatment of asthma and kidney troubles. Apart from its medicinal value, the plant has potential as an industrial oil-seed crop, with yields of up to 1 t/ha (Higgins and White, 1968). The achenes contain vernolic acid, which is mainly present in the oil as the triglyceride, trivernolin. Epoxidized *Vernonia* oil and trivernolin have potential as plasticizers with improved heat and light stability for polyvinyl chloride (Krewson *et al.*, 1966). Further plant breeding work is needed before the plant is likely to reach crop status (Krewson, 1968).

Many *Vernonia* species have had wide use medicinally in native cultures for treating a variety of diseases. That active substances are present in these plants has been demonstrated by Kupchan *et al.* (1966, 1968) with the isolation of vernolepin from *Vernonia hymenolepis* and of elephantopin and elephantin from *Elephantopus*. These sesquiterpene lactones (for structures, see p. 525) show significant activity *in vitro* against cells derived from human carcinoma of the nasopharynx and *in vivo* against rat carcinomas. Vernolepin is also active as an inhibitor of plant growth (e.g. in wheat coleoptile tests), an inhibition which is reversed by indoleacetic acid (Sequiera *et al.*, 1968).

Perhaps the most important pharmaceutical activity of any Vernonieae constituent discovered to date is that of eremanthine, the sesquiterpene lactone of *Eremanthus elaeagnus*. This inhibits the penetration of cercarieae of the trematode *Schistosoma mansonii* (Baker *et al.*, 1972) and it promises to be of value in treatment of this human parasite, which is a serious pest in Brazil and other South American countries.

CONCLUSION

So little is known of the chemistry of the Vernonieae that it is not possible to draw any firm conclusions about the utility of secondary compounds as systematic markers in the tribe. Indeed, the taxonomy of the Vernonieae is still poorly understood (see Jones, Chapter 17) and no comprehensive modern treatment of the whole group is available. Most of the chemical data pertain to species of *Vernonia*, as might be expected. Other genera which have received some detailed attention include *Centratherum, Corymbium, Elephantopus, Eremanthus, Ethulia, Haplophyllum, Stokesia* and *Vanillosmopsis*. This is a total of only nine in a tribe containing 50 genera.

Undoubtedly the most fascinating chemical constituents are the sesquiterpene lactones. A number of structurally distinctive compounds have been found in the tribe and several have significant biological activity. Clearly further exploration of these lactones is the first priority in future chemosystematic studies in the tribe. The flavonoids also show some promise as being of value in relation to the systematics of *Vernonia*. However, there is also evidence that the tribe as a whole is depauperate and lacks the structurally complex flavonoids present in many other of the composite tribes. Again, from what is known of the polyacetylenes, there is less variation in structural types than elsewhere in the Compositae.

The most characteristic single compound of the Vernonieae is the lipid constituent vernolic acid, but although this compound is common in *Vernonia*, it is not present in every species. Its value as a marker for the tribe is also limited by the fact that it has also been found in the Cichorieae and the Cynareae. Finally, three new unsaturated sterols have recently been identified in *Vernonia*, so that this is yet another chemical type worthy of further exploration in the tribe.

ACKNOWLEDGEMENTS

The authors thank Dr C. Humphries and the Head of the Botany Division, British Museum for provision of herbarium material for a flavonoid survey of the Vernonieae. They are grateful to Dr S. B. Jones for supplying references to the economic and pharmaceutical literature on the tribe.

REFERENCES

ARAUJA, L. C. (1944). *Vanillosmopsis erythropappa* Sua exploracão florestal. *Bolm Soc. bras. Agron.* **7**, 101–151.

ARENE, E. O. (1972). 7, 24(28)-Stigmastadien-3β-ol from *Vernonia amygdalina*. *Phytochemistry* **11**, 2886–2887.

BADAMI, R. C. and GUNSTONE, F. D. (1963). Vegetable oils. XII. *Vernonia* seed oils. *J. Sci. Fd Agric.* **14**, 481–484.

BAKER, P. M., FORTES, C. C., FORTES, E. G., GAZZINELLI, G., GILBERT, B., LOPES, J. N. C., PELLEGRINA, J., TOMASSINI, T. C. B. and VICHNEWSKI, W. (1972). Chemoprophylactic agents in schistosomiasis: eremanthine, castunolide, α-cyclocostunolide and bisabolol. *J. Pharm. Pharmacol.* **24**, 853–857.

BOHLMANN, F. (1973). In "Naturally Occurring Acetylenes" (BOHLMANN, F., BURKHARDT, T. and ZDERO, C., eds), p. 340–342. Academic Press, London and New York.

CHISHOLM, M. J. and HOPKINS, C. Y. (1957). An oxygenated fatty acid from the seed oil of *Hibiscus esculentus*. *Can. J. Chem.* **35**, 358–364.

CORBELLA, A., GARIBOLDI, P. and JOMMI, C. (1974). Structure and absolute stereochemistry of vanillosmin, a guaianolide from *Vanillosmopsis erythropappa*. *Phytochemistry* **13**, 459–465.

FIORETTI, J. H., KOLOR, M. G. and MCNAUGHT, R. P. (1970). A new sterol from *Vernonia anthelmintica* seed oil. *Tetrahedron Lett.* No. 34, 2971–2974.

FROST, D. J. and WARD, J. P. (1968). Stereochemistry of 7, 24(28)-stigmastion-3β-ol and the fucosterols. *Tetrahedron Lett.* No. 34, 3779–3782.

FROST, D. J. and WARD, J. P. (1970). Sterols in *Vernonia anthelmintica* seed. 8,14,(Z)-24(28) Stigmastatrienol, a new phytosterol. *Recl Trav. Chem. Pays-Bas*, **89**, 1054–1056.

GARCIA, M., DA SILVA, A. J. R., BAKER, P. M., GILBERT, B. and RABI, J. A. (1976). Absolute stereochemistry of eremanthine—a schistosomicidal sesquiterpene lactone from *Eremanthus elaeagnus*. *Phytochemistry* **15**, 331.

GHANIM, A., PRAKASH, L. and KIDWAI, A. R. (1963). Chemical examination of *Elephantopus scaber*. *Indian J. Chem.* **1**, 320–321.

GLEASON, H. A. (1923). Evolution and geographical distribution of the genus *Vernonia* in North America. *Am. J. Bot.* **10**, 187–200.

GOTTLIEB, O. R. and MAGALHAES, M. T. (1958). Essential oil of the wood of *Vanillosmopsis erythropappa* Schultz-Bip. *Perfumery Essent. Oil Record* **49**, 711–714.

GOVINDACHARI, T. R., VISWANATHAN, N., FUHRER, H. and CIBA-GEIGY, A. G. (1972). Isodeoxyelephantopin, a new germacranediolide from *Elephantopus scaber*. *Indian J. Chem.* **10**, 272–273.

GUNSTONE, F. D. (1954). Fatty acids. Part II. The nature of the oxygenated acid present in *Vernonia anthelmintica* (Willd.) seed oil. *J. chem. Soc.*, 1611–1616.

HEGNAUER, R. (1964). "Chemotaxonomie der Pflanzen", Vol. III. Birkhauser Verlag, Basle and Stuttgart.

HIGGINS, J. J. and WHITE, G. A. (1968). *Vernonia anthelmintica*—a potential seed oil source of epoxy acid. *Agron. J.* **60**, 59–60.

HO, C. M. and TOUBIANA, R. (1970). Lactones sesquiterpeniques. Determination de la position du cycle lactonique dans le vernolide et l'hydroxyvernolide. *Tetrahedron* **26**, 941–949.

KNIGHTS, B. A. and LAURIE, W. (1967). Application of combined gas–liquid chromatography-mass spectrometry to the identification of sterols in oat seed. *Phytochemistry* **6**, 407–416.

KREWSON, C. F. (1968). Naturally occurring epoxy oils. *J. Am. Oil Chem. Soc.* **45**, 250–260.

KREWSON, C. F., RISER, G. R. and SCOTT, W. E. (1966). *Euphorbia* and *Vernonia* seed oil products as plasticiser stabilizers for polyvinyl chloride. *J. Am. Oil Chem. Soc.* **43**, 377–379.

KUPCHAN, S. M., AYNEHCHI, Y., CASSADY, J. M., MCPHAIL, A. T., SIM, G. A., SHNOES, H. K. and BURLINGAME, A. L. (1966). The isolation and structural elucidation of two novel sesquiterpenoid tumor inhibitors from *Elephantopus elatus*. *J. Am. chem. Soc.* **88**, 3674-3676.

KUPCHAN, M. S., HEMINGWAY, R. J., WERNER, D., KARIM, A., MCPHAIL, A. T. and SIM, G. A. (1968). Vernolepin, a novel elemanolide dilactone tumor inhibitor from *Vernonia hymenolepis*. *J. Am. chem. Soc.* **90**, 3586-3597.

KUPCHAN, S. M., HEMINGWAY, R. J., KARIM, A. and WERNER, D. (1969a). Tumor inhibitors. XLVII. Vernodalin and vernomygdin, two new cytotoxic sesquiterpene lactones from *Vernonia amygdalina*. *J. org. Chem.* **34**, 3908-3911.

KUPCHAN, S. M., HEMINGWAY, R. J., WERNER, D. and KARIM, A. (1969b). Tumor inhibitors. XLVI. Vernolepin, a novel sesquiterpene dilactone tumor inhibitor from *Vernonia hymenolepis*. *J. org. Chem.* **34**, 3903-3908.

KUPCHAN, S. M., AYNEHCHI, Y., CASSADY, J. M., SHNOES, H. K. and BURLINGAME, A. L. (1969c). Tumor inhibitors. XL. Isolation and structures elucidation of elephantin and elephantopin, two novel sesquiterpenoid tumor inhibitors from *Elephantopus elatus*. *J. org. Chem.* **34**, 3867-3875.

KURKAWA, T., NAKANISHI, K., WU, W., HAU, H. Y., MARUYAMA, M. and KUPCHAN, S. M. (1970). Deoxyelephantopin and its interrelation with elephantopin. *Tetrahedron Lett.* 2863-2866.

LEE, K., FURUKAWA, H., KOZUKA, M., HUANG, H., LUHAN, P. A. and MCPHAIL, A. T. (1973). Molephantin, a novel cytotoxic germacranolide from *Elephantopus mollis*. X-ray crystal structure. *J. chem. Soc. chem. Commun.* 476-477.

MABRY, T. J., ABDEL-BASET, Z. and PADOLINA, W. G. (1975). Systematic implications of flavonoids and sesquiterpene lactones in species of *Vernonia*. *Biochem. Syst. Ecol.* **2**, 185-192.

MCPHAIL, A. T. and SIM, G. A. (1972). Sesquiterpenoids, XIII. Constitution and absolute stereochemistry of elephantol. X-ray analysis of elephantol *p*-bromobenzoate. *J. chem. Soc., Perkin Trans.* **2**, 1313-1316.

MOMPON, B., HO, C. M. and TOUBIANA, R. (1973). Sesquiterpene lactones. 6. Structure of pectorolide, a new sesquiterpenoid lactone from *Vernonia pectoralis*. *C. r. hebd. Séanc. Acad. Sci., Paris. Ser. A.* **276**, 1799-1801.

MOMPON, B., MASSIOT, G. and TOUBIANA, R. (1974). Structure due subluteolide nouveau guaianolide isole du *Vernonia sublutea* Scott Elliott (Composées). *C. r. hebd. Séanc. Acad. Sci., Paris, Ser. C. T.* **279**, No. 22, 907-909.

PADOLINA, W. G. (1973). Ph.D. thesis, University of Texas at Austin.

PADOLINA, W. G., NAKATANI, N., YOSHIOKA, M., MABRY, T. J. and MONTI, S. A. (1974). Marginatin, a new germacranolide from *Vernonia* species. *Phytochemistry* **13**, 2225-2229.

PASCARD, G. (1970). Determination par les rayons x de la structure d'un ester sesquiterpenique: le vernolide. *Tetrahedron Lett.* 4131-4134.

PATEL, M. B. and ROWSON, J. M. (1964). Investigations of certain Nigerian medicinal plants. *Planta Med.* **12**, 33-41.

PERNET, R. (1959). Les plantes medicinales Malgaches (suites de notes analytiques). *Mém. Inst. Sci. Madagascar, Sér. B.* **9**, 217-248.

RAO, K. V. (1962). Chemical constituents of *Vernonia cinerea* Less. *J. Indian chem. Soc.* **39**, 749-752.

ROWE, J., HARWOOD, A. A. and MYERS, D. B. (1955). The isolation of three inositols from *Vernonia altissima*. *J. Am. pharm. Assoc.* **44**, 308-310.

SCOTT, W. E., KREWSON, C. F. and RIEMENSCHNEIDER, R. W. (1962). *Vernonia anthelmintica*: (+)- and (−)-three-12, 13-dihydroxyoleic acid. *Chemy Ind.*, 2038–2039.

SEQUEIRA, L., HEMINGWAY, R. L. and KUPCHAN, S. M. (1968). Vernolepin, a new reversible plant growth inhibitor. *Science, N.Y.* **161**, 789–790.

SIM, K. Y. and LEE, H. T. (1969). Constituents of *Elephantopus scaber*. *Phytochemistry* **8**, 933–934.

TOUBIANA, R. (1969). Structure of hydroxyvernolide, a new sesquiterpene ester from *Vernonia colorata*. *C. r. hebd. Séanc. Acad. Sci., Paris, Sér. C.* **268**, 82–85.

TOUBIANA, R. and GAUDEMER, A. (1967). Structure du vernolide, nouvel ester sesquiterpenique isole de *Vernonia colorata*. *Tetrahedron Lett.* 1333–1336.

TOUBIANA, R., TOUBIANA, M. J. and BAS, B. C. (1970). Sesquiterpene lactones, IV. Structure of confertolide, a new sesquiterpene lactone isolated from *Vernonia conforta*. *C. r. hebd. Séanc. Acad. Sci., Paris, Sér. C.* **270**, 1033–1035.

VICHNEWSKI, W. and GILBERT, B. (1972). Schistosomicidal sesquiterpene lactone from *Eremanthus elaeagnus*. *Phytochemistry* **11**, 2563–2566.

VICHNEWSKI, W., SARTI, S. J., GILBERT, B. and HERZ. W. (1976). Goyazensolide a schistosomicidal heliangolide from *Eremanthus goyazensis*. *Phytochemistry* **15**, 191–194.

WAGNER, H., IYENGAR, M. A., SELIGMANN, O., HORHAMMER, L. and HERZ, W. (1972). Flavonoides in *Vernonia* arten. *Phytochemistry* **11**, 3086–3087.

CHAPTER 19
Astereae—systematic review

J. GRAU

Institut für Systematische Botanik, München, Germany

Abstract. The total number of Astereae today is 135 genera with about 2500 species. The Astereae can be characterized by the styles of the bisexual disc florets which have distinct, linear to triangular, sterile appendages. The main centres of distribution of this tribe are both Americas and Southern Africa. Current subtribal delimitations are highly artificial and need much additional study. Additional characters will have to be found so as to construct a new and better system.

CONTENTS

Description of the tribe	539
Geographical distribution	542
Division into subtribes	543
Delimitation of the genera	544
Cytotaxonomy	556
Appendix 1	558
Appendix 2	559
References	560

DESCRIPTION OF THE TRIBE

The Astereae has a predominantly extratropical distribution. Genera based on fewer differences are more frequent in the northern hemisphere. They often form large groups and frequently occur on more than one continent. Genera in the southern hemisphere seem to be more readily distinguished and form geographically smaller units. Moving in the direction South America–Africa–Australia with New Zealand, there is an increasing generic differentiation with a greater deviation from the basic type of Astereae as it is known in the northern hemisphere.

The circumscription of the tribe has been unchanged since the days of Bentham (1873) and Hoffmann (1890). Four characters mainly typify the tribe:

540   J. GRAU

(1) *The style.* The typical style of the Astereae is to be found in the bisexual disc florets. Style branches are in cross section semi-orbicular and tipped with subulate to triangular appendages covered with collecting hairs. The basal part of these branches is margined with stigmatic lines.
(2) *The anthers.* The anthers are obtuse at base and bear a terminal, more or less, triangular appendage.
(3) *The pollen.* The spheroid spinous pollen grains are normally 3-colporate. They have a one-layered sexine fusing with the nexine at the margins of the colpi only. They correspond (Skvarla, Chapter 8) to the helianthoid pattern.
(4) *The anatomy of the fruit.* The achenes show in cross-section a very typical testa. Its epidermis seems to be always constituted by a single layer of cells which are thickened on three sides (u-cells).

Other characters of importance may be mentioned as follows. In habit they include perennial or sometimes annual herbs, quite often shrubs and rarely, on insular outskirts, small- to medium-sized trees (*Apodocephala, Psiadiella* and *Vernoniopsis* on Madagascar, *Commidendron* and *Melanodendron* on St Helena and *Olearia* in Australia and South-East Asia). The leaves are seldom opposite (e.g. some groups of *Felicia*), entire or toothed or occasionally divided (this occurs in different groups and especially in annual herbs). In genera of drier areas the leaves are often resiniferous.

Heads are mostly heterogamous, more rarely homogamous (*Chrysocoma, Nardophyllum* and others) and in a few cases the heads are dioecious (*Archibaccharis, Baccharis* and *Heterothalamus* — all American ones). The involucral bracts are normally imbricated and in several rows, seldom in only two rows (some species of *Felicia*). Genera of quite different areas may have receptacular bracts (e.g. *Erodiophyllum* in Australia, *Amellus* in Africa, and *Chiliotrichium* in South America), but normally the receptacle is naked. The outer florets are usually ligulate and female, the central ones are tubular and bisexual. Sometimes the corolla is reduced in the outer florets (e.g. in *Nolletia* with a short ligule; in *Cyathocline* with only the tubular part left and in *Thespis* where the corolla of florets is completely lacking). Sometimes the pistillate or ray florets are more tubular with the corolla teeth of different lengths as in *Hinterhubera*, or the ligules may have a small inner appendix as in *Remya*. The colour of the ligules may be yellow (especially in the subtribe Solidagininae or Homochrominae of Bentham) or blue to violet (in the subtribe Asterinae or "Heterochromeae" of Bentham), rarely more or less white, which is certainly derived.

The disc florets are almost exclusively actinomorphic. In the genus *Colobanthera* (Madagascar) the disc florets are sometimes curved in one direction and therefore slightly zygomorphic. Normally the disc florets are pentamerous, but in some, mainly Pacific and Madagascar genera, there is a reduction to a tetramerous state (*Keysseria, Myriactis*, sometimes also in *Calotis, Dicrocephala, Egletes, Grangeopsis, Psiadiella* and *Remya*). The colour is normally yellow, but sometimes whitish and seldom reddish (but this sometimes occurs in the same species which also has yellow corollas). The tetramerous florets also possess four anthers. In some Pacific and Madagascar genera the appendages of the anthers are absent (e.g. in *Piora, Colobanthera* and in one section of *Brachycome*); reduced appendages are present in *Keysseria* and *Myriactis*. In *Cyathocline* the appendages bear some few-celled glands. Finally, the anthers may be auriculate to sagittate at the base as in *Apodocephala, Celmisia, Cyathocline, Myriactis, Olearia, Phacellothrix, Vernoniopsis* and less distinctly so in *Pleurophyllum, Olivaea* and in occasional species of *Grindelia*.

The style of fertile disc florets is, so far as is known, basically transformed into a nectarium. There is, essentially, no deviation among styles in the tribe. The only significant variation is in the length of the apical appendage, which may vary from shortly triangular to filiform. The extreme appendage types may occur within one genus (e.g. *Felicia*); usually however, the appendages are relatively uniform within a genus. In some genera, mainly from Madagascar (*Colobanthera, Grangea* and *Grangeopsis*), the branches of the style are thickened, but they are otherwise typical.

The style of the ray florets is very uniform, having two slender branches without hairs. When the disc florets have a sterile gynoeceum, two types of styles seem to be present. In the typical case (e.g. *Polyarrhena* and *Sommerfeltia*) the branches are ovoid, flat at the inner side and rounded and hairy on the outer. In some Australian genera (*Calotis, Erodiophyllum* and *Minuria*) the sterile style has short branches, but has hairs far below the point of bifurcation. This type resembles somewhat the fertile styles of the Anthemideae. In *Cyathocline* the style of the disc florets is entire.

The achenes of the Astereae are very polymorphic as regards their external morphology. They may be flat and two-nerved (e.g. *Felicia*) to round and many-nerved (e.g. *Heteromma*) and they may have lateral protuberances (e.g. *Brachycome*) or apical ones (e.g. *Calotis*). These protuberances are formed of sclerenchymatic tissue. Sometimes the achene has an apical sterile (*Lagenophora, Ixiochlamys, Podocoma*) rarely a basal sterile (*Vittadinia*) part. The achenes are often pubescent with basically two types of hairs: (1) the typical "twin-hairs" and (2) hairs with few- to many-celled glands. The latter type can be found among diverse genera (e.g. European and African Asters, *Hinterhubera* and many others).

The usual pappus consists of several rows of shortly-toothed bristles. It may exist as plumose bristles, short scales or reduced to a cartilaginous crown or, especially in North American genera, be absent. Quite often the disc florets may be male in function without any visible differentiation, but in other cases the stigmas are lacking (see above) and the ovary reduced. Occasionally there is differentiation among the ray florets. In the Australian genus *Erodiophyllum* the outer ray florets are sterile, the inner ones having a corolla reduced to a short ring, these being the only fertile ones (compare also *Erigeron* sect. *Trimorphaea*). The disc florets, at least, are only male in function. Apomixis is known from the genera *Brachycome, Colotis, Erigeron* and *Townsendia*. The wood of the Astereae shows advanced characters according to Carlquist (1960).

The Astereae are very uniform without regard to the genera *Geissolepis* and *Isoetopsis*. These genera, though having much in common with the other genera, differ clearly in the anatomy of the fruit. Further investigations have yet to show where they really belong. *Rhamphogyne* seems to be related to *Apodocephala* from Madagascar. There seems to be no real relationship to *Abrotanella*, which certainly does not belong in the Astereae (see page 554). The nearest relationship of the Astereae is with several groups of the Inuleae.

There are about 136 genera in the tribe, that is, if these are not too finely circumscribed as might be the case for North American genera. The number of known species is around 2520. That means that since the time of Bentham (who described 90 genera and 1400 species) the number of species has increased almost twice as much as the number of genera, from an average of about 15 species per genus to about 18 species today.

GEOGRAPHICAL DISTRIBUTION

Genera of Astereae which are more or less tropical include *Conyza, Egletes, Gundlachia* and a relatively few species of *Baccharis, Aster* and *Erigeron* in the New World; in the Old World, *Dichrocephala, Grangea* and *Microglossa*, may be singled out. The genera that occur in more than one continent are few and they are relatively fewer yet in the southern hemisphere. The number of such genera will probably be reduced as more detailed monographic study is pursued.

More or less primary cosmopolitan genera are *Aster, Conyza* and *Erigeron*. These genera are largely confined to the northern hemisphere, presumably originating there. The connections between South and North America are quite clear. Genera common to both are *Baccharis, Egletes, Grindelia, Gutierrezia, Haplopappus* (*sensu* Hall, 1928) and *Psilactis*. There is also an obvious connection between Australia and New Zealand by the genera *Brachycome, Celmisia, Lagenophora, Olearia* and *Vittadinia*, which are also partly connected by some conspicuous characters (see pp. 538 and 539). Moreover, *Lagenophora* shows a more or less circumpacific distribution. Some of the Australian genera like *Brachycome* and *Olearia* also reach New Guinea and even Japan (e.g. *Solenogyne*). The genus *Vittadinia*, thought to be represented by one species in South America, now seems to be confined to Australia since the South American species is quite different in fruit structure and is perhaps only remotely related to that genus. It is separated as *Microgynella*. The same is true of the Australian *Podocoma*: its South American species are quite different and placed in

Ixiochlamys. The genera *Keysseria* and *Tetramolopium* connect New Guinea and Hawaii. In addition to *Aster*, *Conyza* and *Erigeron*, there also exist connections between Asia and Africa shown by *Cyathocline, Grangea, Microglossa, Nidorella* and *Psiadia*. Species common to both Europe and America or Europe and Africa are very few. The reason for this is that the Astereae are poorly represented in Europe. The endemic Madagascan genera are, in some respects, more closely related to the pacific than to other African genera.

Most genera of the Astereae have a more or less restricted distribution and are typical for certain regions. Remarkable disjunctions do not exist. The statement of Hoffmann (1890) that the subtribes, with the exception of the Baccharinae, do not show typical distribution patterns may not be generally valid. His opinion seems to be based mainly on an inadequate construction of subtribal groupings. Summing up succinctly then, most of the genera in Astereae are distributed in and about land areas peripheral to the Pacific and Indian oceans.

DIVISION INTO SUBTRIBES

One of the main problems in the Astereae is still that associated with the delimitation of subtribes. Bentham (1873a) says that "the Astereae are not divisible into distinct subtribes" and he thus mentioned only seven different "types" named after characteristic genera. But he (1873b) nevertheless divided the tribe into six subtribes, a system which is still in use today. Hoffmann accepted this subdivision, but he noted that this was perhaps the weakest part of Bentham's treatment. According to Hoffmann the transitions between several genera (e.g. *Erigeron* and *Aster* on the one side, and *Erigeron* and *Conyza* on the other) make the recognition of tribes difficult. Thus *Erigeron* and *Aster* belong to one subtribe (Asterinae) while *Erigeron* and *Conyza* are included in another (Conyzinae). Hoffmann obviously accepted the homochromous-heterochromous breakdown, and he only mentioned the genus *Pentachaeta* as an exception to the characters which distinguish between them. Today, however, we know more genera which possess both homochromous and heterochromous heads, as for example, *Felicia, Mairia* and *Machaeranthera*.

Ray colour, therefore, while expressive of a trend within the tribe, does not permit absolute distinction of these groups. Moreover, for discoid genera other criteria are needed.

The pappus has also been used for tribal recognition. Presence or absence of this character serves to distinguish the subtribes Asterinae and Bellidinae. In some genera (e.g. *Felicia*) it is difficult to distinguish between a reduced pappus and its absence. In other cases (e.g. *Calotis*) it is uncertain whether the part of the fruit referred to as a pappus is homologous with a true pappus. Thus, subtribal division on the basis of this particular character is not very satisfying.

No doubt, new characters must be sought, either singly or in combination, by which to circumscribe subtribes. Some previously recognized characters, however, do appear to suffice. The derived subtribe Baccharinae is one such natural grouping which should be kept. Dioecious heads are without doubt a good tribal character in this instance but this is the rare exception. A second natural group may be the Grangeinae with a slightly changed circumscription, comprising the genera *Grangea*, *Dichrocephala*, *Cyathocline*, *Ceruana*, *Egletes*, *Plagiocheilus* and perhaps *Floscaldasia* and *Gyrodoma*. Quite related to this group are some of the members of the *Bellidinae* which are situated around *Lagenophora* as *Myriactis*, *Solenogyne*, *Laestadia*, *Keysseria* and probably *Rhynchospermum*.

Cuatrecasas (1969) has made an effort to clarify subtribal groupings in his recognition of the subtribe Hinterhuberinae with only one genus. Monotypic subtribal erections without consideration of the tribe as a whole are not recommended, but there is no doubt that the characters used in the erection of this new subtribe are of sufficient value to warrant "testing" elsewhere in the tribe.

A character heretofore not widely used is that of the fruiting bodies. Comparative studies among genera might show correlations with other floral features such as tetramerous flowers, appendaged anthers, etc. The second type of sterile styles (*Calotis*, *Erodiophyllum*, *Minuria*) may also be characters which might be used at the subtribal level. It is to be hoped that this will permit a better phyletic arrangement of the genera, especially those of the northern hemisphere such as *Erigeron* and *Conyza* which surely belong to the same subtribe if equal treatment of subtribes on a worldwide basis is attempted.

Because of the above, and the need to examine in detail the large number of genera that are listed in Table I, I have not tried to arrange these in any sort of subtribal breakdown. The historical subtribes are as follows: (1) Solidagininae Hoffm. (Homochrominae B. & H.); (2) Grangeinae B. & H.; (3) Bellidinae B. & H.; (4) Asterinae Dum. (Heterochromeae B. & H.); (5) Conyzinae B. & H.; (6) Baccharinae Less. (Heterothalaminae Endl.). From the Asterinae, Cuatrecasas (1969) has split off Hinterhuberinae, placing this between the Grangeinae and Bellidinae.

DELIMITATION OF THE GENERA

The characters that mark the subtribes have been discussed in the above. The use of flower colours does not serve to distinguish natural taxa, even at the generic level (*Homochroma* with yellow ligules from *Mairia* with violet ligules). Even recently such differences have been used to distinguish *Luteidiscus*, as a new genus, from *Tetramolopium* (St. John, 1974). Such a

TABLE I. List of genera of the Astereae arranged geographically

Genus	Number of species	Habit	Chromosome numbers	Distribution
COSMOPOLITAN GENERA				
Aster L.	c. 250	annual, perennial herbs, shrubs	5, 7, 8, 9, 10, 13, 16, 18, 19, 20, 23, 24, 25, 27, 32, 33, 36, 72	mainly in the northern hemisphere
included here:				
Asterothamnus Novopokr.	*Kalimeris* Cass.			
Biotia DC.	*Kemulariella* Tamamsch.			
Brachyactis Ledeb.	*Krylovia* Schisch.			
Doellingeria Nees	*Linosyris* Cass.			
Galatella Cass.	*Oreastrum* Greene			
Gymnaster Kit.	*Pseudaster* Tamamsch.			
Heleastrum DC.	*Pseudolinosyris* Novopokr.			
Ionactis Greene	*Sericocarpus* Nees			
Bellidiastrum Micheli	*Tripolium* Nees			
Conyza Less.	c. 50	annuals, perennial herbs	9, 27	mainly in the tropical regions
Erigeron L.	c. 200	annuals, perennial herbs, shrubs	9, 16, 18, 20, 27, 32, 36	mainly in northern hemisphere, only a few in Africa, Australia and New Zealand
NORTH AMERICA INCLUDING MEXICO				
Acamptopappus Gray	2	shrubs	9	S. North America
Achaetogeron Gray	10	perennial herbs	9, 27	Mexico to North America
Amphiachyris (DC.) Nutt.	2	annuals	5	North America

Table I.—*continued*

Genus	Number of species	Habit	Chromosome numbers	Distribution
NORTH AMERICA INCLUDING MEXICO—*continued*				
Amphipappus Torrey & Gray	1	shrub	9	S. North America
Aphanostephus DC.	c. 6	annuals, perennial herbs	3, 4, 5	Mexico to North America
Archibaccharis Heering	22	shrubs	9	Mexico to Costa Rica
Astranthium Nutt.	12	annuals, perennial herbs	3, 4, 5, 6, 8, 9, 10, 12, 18	Mexico and southeastern North America
Benitoa Keck.	1	annual	5	California
Bigelovia DC.	2	perennial herbs	9	North America
Boltonia L.'Hérit.	8	perennial herbs	9, 18, 27	North America, 1 East Asia
Bradburia Torrey & Gray	1	annual	3	S. North America
Chaetopappa DC. (including *Pentachaeta*)	9	annuals, perennial herbs	8, 9, 16	Mexico, S.W. North America
Chrysoma Nutt.	1	shrub	9	North America
Chrysopsis Nutt.	20	annuals, perennial herbs, shrubs	4, 5, 9, 12, 18	North America
Chrysothamnus Nutt.	14	shrubs	9, 18, 36	W. North America
Corethrogyne DC.	3	perennial herbs	5	North America (California)
Dichaetophora Gray	1	annual	3	Southern North America
Eastwoodia Brandegee	1	shrub	9	North America (California)
Eremiastrum Gray	2	annuals	?	North America
Euthamia Nutt.	19	perennial herbs	9, 18, 27	North America
Geissolepis B. L. Robinson	1	annual	?	Mexico

19. ASTEREAE—SYSTEMATIC REVIEW

Greenella Gray	2	annuals, perennial herbs	4	S. North America
Gymnosperma Less.	1	shrub	7, 8	Texas to C. America
Heterotheca Cass. (including *Pityopsis*)	c. 15	perennial herbs	9, 18	North America to Mexico
Lessingia Cham.	7	annuals	5, 6	North America (California)
Leucelene Greene	1	perennial herbs	8, 16	North America
Machaeranthera Nees	26	annuals, perennial herbs	4, 5, 6, 8, 9	North America
Monoptilon Torrey & Gray	2	annuals	8, 27	S. North America
Olivaea Sch. Bip. ex Benth.	2	annuals	6	Mexico
Petradoria Greene	2	shrubs	9, 18	North America
Rigiopappus Gray	1	annual	?	North America
Solidago L.	c. 100	shrubs, perennial herbs	9, 18, 27	few species also in Eurasia and South America
Stephanodoria Greene	1	perennial herb	?	Mexico
Thurovia Rosc.	1	annual	ca. 5	S. North America
Townsendia Hooker	21	annual, perennial herbs	9, 18	W. North America to Mexico
Tracyina Blake	1	annual	9	North America (California)
Vanclevea Greene	1	perennial herbs	9	S. North America
Xanthisma DC.	1	annual	4, 8	S. North America
Xanthocephalum Willd.	8	annuals, perennial herbs, shrubs	4, 6	Mexico to S. North America

NORTH AND SOUTH AMERICA

Baccharis L.	c. 400	perennial herbs, shrubs	9, 18, 25	Centre in South America

Table I.—continued

Genus	Number of species	Habit	Chromosome numbers	Distribution
NORTH AND SOUTH AMERICA—continued				
Egletes Cass.	12	annuals, perennial herbs	9	Mexico, Central America to Texas and to Ecuador, Columbia, Brazil
Grindelia Willd. (including *Chrysophtalmum* Phil.)	c. 60	perennial herbs, shrubs	6, 12	Centre in North America, in South America south of the tropics
Guterrezia Lag.	20	annuals, perennial herbs, shrubs	4, 8, 12, 16, 20, 28	North America and South America south of the tropics
Haplopappus Cass. included here: *Chroilema* Bernh. *Croptilon* Raf. *Ericameria* Nutt. *Hazardia* Greene *Hesperodoria* Greene *Isopappus* Torrey & Gray *Prionopsis* Nutt. *Stenotopsis* Rydb.	c. 160	annuals, perennial herbs, shrubs	2, 3, 4, 5, 6, 7, 8, 9, 10, 12, 18, 45	North and South America
Psilactis Gray	7	annuals, perennial herbs	4, 5, 9	Mexico, one species Peru and Columbia
SOUTH AMERICA (including West Indies)				
Aylacophora Cabrera	1	shrub	?	Argentina

19. ASTEREAE—SYSTEMATIC REVIEW

Baccharidastrum Cabrera	2	perennial herbs shrubs	?	Argentina, Uruguay, Paraguay, S. Brazil, Columbia
Blakiella Cuatrec.	1	perennial herb	?	Venezuela
Chiliophyllum Philippi	3	shrubs	?	South America south of the tropics
Chiliotrichium Cass.	2	shrubs	?	South America south of the tropics
Chiliotrichiopsis Cabrera	3	shrubs	?	Argentina
Darwiniothamnus Harling	2	shrubs	?	Galapagos
Diplostephium H.B.K.	c. 90	shrubs	?	Venezuela to Northern Chile
Floscaldasia Cuatrec.	1	perennial herb	?	Columbia
Gundlachia Gray	9	shrubs	?	Bahamas to Curacao, Haiti
Heterothalamus Less.	8	shrubs	?	Brazil and Argentina to Chile and Guyana
Hinterhubera Sch. Bip.	8	shrubs	?	Venezuela and Columbia
Hysterionica Willd.	10	perennial herbs, shrubs	?	Brazil, Uruguay and Argentina
Laestadia Kunth	6	perennial herbs, shrubs	?	Columbia, Venezuela to Bolivia
Lepidophyllum Cass.	1	shrubs	?	Patagonia
Llerasia Triana	11	shrubs, trees	?	Columbia, Bolivia
Microgynella Grau	1	shrub	?	Brazil, Argentina
Nardophyllum Hooker & Arn.	7	shrubs	?	South America south of the tropics
Noticastrum DC.	12	perennial herbs	9	South America south of the tropics
Oritrophium (H.B.K.) Cuatrec.	c. 15	perennial herbs	?	Venezuela, Peru, Bolivia

Table I.—continued

Genus	Number of species	Habit	Chromosome numbers	Distribution
SOUTH AMERICA (including West Indies)—continued				
Paleaepappus Cabrera	1	shrub	?	Patagonia
Parastrephia Nutt. (including most of *Lepidophyllum*)	5	shrubs	?	South America south of the tropics
Plagiocheilus Arn. ex DC.	c. 8	annuals	?	Ecuador to Peru
Podocoma Cass. (including *Inulopsis* O. Hoffm.)	c. 12	perennial herbs	?	Brazil, Argentina
Sommerfeltia Less.	2	shrubs	?	Brazil, Uruguay, Argentina
SOUTH AMERICA, AUSTRALIA AND NEW ZEALAND				
Lagenophora Cass.	15	perennial herbs	9	9 Australia and New Zealand, 6 South or Central America
AUSTRALIA AND NEW ZEALAND				
Achnophora F. v. Müller	1	perennial herb	?	America to Venezuela Australia
Brachycome Cass.	66	annuals, perennial herbs	2, 3, 4, 5, 6, 7, 8, 9, 10, 11, 12, 13, 14, 15, 16, 18, 27, 36, 45	62 in Australia, 3 in New Zealand, 1 in New Guinea
Calotis R. Br.	22	annuals, perennial herbs	4, 7, 8, 9, 14	New Zealand

19. ASTEREAE—SYSTEMATIC REVIEW

Celmisia Cass. (with *Damnamenia* Given)	61	perennial herbs	9, 27.	58 New Zealand, 2 Australia, 1 Tasmania
Erodiophyllum F. v. Müll.	2	perennial herbs	8	Australia
Isoetopsis Turcz.	3	annual	17	Australia
Ixiochlamys F. v. Müll. & Sond.	2	annual, perennial herbs	?	Australia
Minuria DC. (with *Minuriella* Tate)	6	annuals, perennial herbs, shrubs	9, *c*. 14, 18	Australia
Olearia Mnch.	*c*. 130	perennial herbs, shrubs, trees	9, 27, 54	mainly Australia, *c*. 30 New Zealand, few New Guinea and Lord Howe Island
Pachystegia Cheesem.	1	shrub	?	New Zealand
Phacellothrix F. v. Müll.	1	annual	?	North Australia, Malesia
Pleurophyllum Hook.	3	perennial herbs	?	New Zealand
Solenogyne Cass.	3	perennial herbs	?	Australia, 1 up to Japan
Vittadinia A. Rich.	15	perennial herbs, shrubs	9	Australia and New Zealand, 2 in New Guinea

NEW GUINEA AND HAWAII

Keysseria Lauterbach	15	perennial herbs, shrubs	?	New Guinea, Malesia, Fiji and Hawaii
Myriactis Less.	*c*. 12	perennial herbs	13, 18	South Asia to New Guinea
Piora Koster	1	shrub	?	New Guinea
Remya Hillebr.	2	shrubs	?	Hawaii
Tetramolopium Nees.	32	perennial herbs, shrubs	7, 9	New Guinea, Hawaii

Table I.—continued

Genus	Number of species	Habit	Chromosome numbers	Distribution
EUROPE AND ASIA				
Asteromoea Blume	2	perennial herbs	9	East Asia
Bellis L.	7	annuals, perennial herbs	9	Europe to Asia
Bellium L.	3	annuals, perennial herbs	9	Mediterranean Europe
Callistephus Cass.	1	annual	?	East Asia
Heteropappus Less.	5	annuals, perennial herbs	18	East Asia
Lachnophyllum Bunge	2	annuals	?	South-West to South-Central Asia
Nannoglottis Maxim.	c. 8	perennial herbs	?	China
Psychrogeton, Boiss.	20	perennial herbs	?	South-Central Asia
Rhynchospermum Reinw.	1	perennial herb	9	South-East Asia
Thespis DC.	1	annual	?	South-East Asia
Tolbonia O. Kuntze	1	perennial herb	?	South-East Asia
AFRICA AND ASIA (AUSTRALIA pp.)				
Cyathocline Cass.	4	annuals	11	Ethiopia, Southern Asia
Dichrocephala DC.	13	annuals	9	Southern and tropical Africa, Madagascar, S.E. Asia and Australia
Grangea Adans.	6	annuals, perennial herbs	9	tropical Africa and Asia

19. ASTEREAE—SYSTEMATIC REVIEW 553

Microglossa DC.	*c.* 10	shrubs	9, *c.* 18	Southern to tropical Africa, Eastern Asia
Nidorella Cass.	11	perennial herbs	9	Southern Africa to the Middle East
Psiadia Jacq.	*c.* 60	shrubs	9	Southern to tropical Africa, Madagascar, tropical Asia
AFRICA				
Amellus L.	12	annuals, perennial herbs	6, 8, 9	Southern Africa
Apodocephala (Bak.) H. Humb.	8	shrubs, trees	?	Madagascar
Chrysocoma L.	18	perennial herbs, shrubs	9	Southern Africa
Ceruana Forsk.	1	annual	?	Egypt, tropical Africa
Colobanthera H. Humb.	1	annual, perennial herb	?	Madagascar
Commidendron DC.	4	shrubs, trees	?	St Helena
Dacryotrichia Wild	1	annual	?	Zambia
Engleria O. Hoffm.	2	annuals, perennial herbs	?	Southern Africa
Felicia Cass.	83	annuals, perennial herbs, shrubs	5, 6, 8, 9, 18	Southern Africa up to Ethiopia
Grangeopsis H. Humb.	1	annual	?	Madagascar
Gymnostephium Less.	6	perennial herbs	?	Southern Africa
Gyrodoma Wild	1	annual	?	Mozambique
Heteromma Benth.	3	perennial herbs, shrubs	?	Eastern South Africa
Jeffreya Wild	1	annual	?	Tanzania, Zambia
Mairia Nees	*c.* 10	perennial herbs, shrubs	9	Southern Africa

Table I.—*continued*

Genus	Number of species	Habit	Chromosome numbers	Distribution
AFRICA—*continued*				
Melanodendron DC.	1	tree	?	St Helena
Microtrichia DC.	1	annual	?	tropical Africa
Nolletia Cass.	4	shrubs	?	Southern and Northern Africa
Polyarrhena Cass.	4	perennial herbs, shrubs	?	Southern Africa
Psednotrichia Hiern	1	annual	?	Southern Africa
Psiadiella H. Humb.	1	shrub	?	Madagascar
Pteronia L.	79	shrubs	?	Southern Africa
Rochonia DC.	4	shrubs	?	Madagascar
Rhamphogyne	1	perennial herb		Rodriguez Island
Vernoniopsis H. Humb.	1	small tree	?	Madagascar

Excluded genera
  Adelostigma Steetz = Inuleae
  Bellidia A. J. Ewert = Inuleae
  Cratystylis Moore = Inuleae
  Haastia Hook. fil. = Inuleae
  Pseudoconyza Cuatrec. = Inuleae

Table I.—*continued: Comments on selected genera*

Aster L. It is still questionable whether this cosmopolitan genus should be treated as one unit. However, the more regional attempts made so far to divide *Aster* into more practical units are not yet sufficient. A treatment on a world-wide scale based on fruit characters might give better results.

Benitoa Keck. This genus, which is sometimes considered as a member of *Haplopappus*, is clearly distinct.

Bigelowia DC. This has to be restricted according to current circumscriptions to only two species (Anderson, 1970).

Celmisia Cass. The Australian genus *Celmisia* has been regarded sometimes as having some members in South America. Cuatrecasas (1969) treats these species as the separate genus *Oritrophium*.

Chrysocoma L. *Chrysocoma* is restricted to South Africa. All New World species belong to other genera of the *Astereae* (*Baccharis*, *Bigelowia*) or even to other tribes.

Haplopappus Cass. This genus is much more uniform in South America than in North America. In South America it includes *Chroilema* Bernh.

Geissolepis B. L. Robinson This genus, transferred from the Heliantheae (Robinson and Brettel, 1972) is problematical in the same way as *Isoetopsis*. The fruit anatomy is very distinct from all other genera of Astereae investigated so far.

Isoetopsis Turcz. This genus does not fit very well in the Astereae as Robinson and Brettell (1973) thought. The squamose pappus and the type of dry involucral bracts are not very typical of the Astereae but the main objection is the very different anatomy of the fruits.

Ixiochlamys F. Muell. This genus should be separated from *Podocoma* (Grau, 1975).

Mairia Nees. This South African genus is still artificial and seems to comprise two different units held together only by a superficial pappus character.

Microgynella Grau. This South American genus is different from the Australian genus *Vittadinia* with which it has been united to date (Grau, 1975).

Nannoglottis Maxim. This genus from eastern Asia, so far not mentioned as belonging to the Astereae, is a good member of this tribe and belongs into the neighbourhood of *Erigeron*.

Plagiocheilus Arn. ex DC. This former member of the Anthemideae is a very typical member of the Astereae, and should be placed in the Grangeinae, more or less between *Egletes* and *Grangea*.

Podocoma Cass. *Podocoma* is restricted to South America. It includes both *Asteropsis* and *Inulopsis*. The typical character of *Podocoma*, the beaked achene, is very unequally developed.

Psednotrichia Hiern. This is a very doubtful genus and could well be reduced to synonymy on further investigation.

Pteronia L. This genus is reported to have a species in Australia too but this seems very unlikely and needs a careful investigation.

Rhamphogyne S. Moore The illustrations in the original publication show the curved beak of the young achene. This unique character is very reminiscent of some species of *Apodocephala*. This genus is endemic to Madagascar and it is quite probable that the ecologically highly specialized genus *Rhamphogyne* is related to it. This would fit in quite well with the geographical pattern. The supposed connections with *Abrotanella* seem to be based only on their similar habit.

Thurovia Rose. The inadequate original description caused some doubts as to whether this genus really belongs to the Astereae, but these doubts have now been resolved.

---

procedure seems out of place today because it leads to unnatural groupings. Also the presence or absence of a pappus should be used with care as a diagnostic feature, and only in connection with other characters.

Another problem in classification has been the variation in evaluation of the same or equal characters by workers in different countries. This is only partly an expression of the different phylogenetic situations on different continents. It seems that this is largely a question of character assessment, reflecting some degree of "overtreatment" of the tribe on some continents (notably North America and Eurasia). Thus, Cronquist and Keck (1957) noted that the genus *Machaeranthera* is separated from *Aster* principally by its "taprooted habit", but the South African Asters, also characterized by different root types, are still included in *Aster*. Other genera have similar variation in other characters such as *Brachycome* spp. which may or may not possess appendages on their anthers, yet are united into one genus. Regionally different treatments of the Astereae are reflected, apparently, in the number of species per endemic genus, and the number of monotypic genera in different continents (Table II).

CYTOTAXONOMY

As well as several papers dealing with special groups there are four papers which treat the Astereae as a whole (Raven *et al.*, 1960, Solbrig *et al.*, 1964, Solbrig *et al.*, 1969, Anderson *et al.*, 1974).

TABLE II. Average number of species per endemic genus according to region.

Region	Number of endemic genera	Number of monotypic genera	Average number of species per endemic genus
Eurasia	11	4	4·6
North America	39	15	8·4
South America	25	7	8·8
Africa	25	12	10·0
New Guinea and Hawaii	5	1	12·4
Australia and New Zealand	14	4	22·4

These papers make it quite clear that $x=9$ is the basic number of the whole tribe. One finds polyploidy particularly in *Brachycome*, *Haplopappus* (decaploid species) and *Olearia* (dodecaploid species). There are hexaploid species in *Aster*, *Chrysothamnus* and *Erigeron*. On the other hand there are several descending aneuploid series, frequently with a gap at $x=7$, this number being found only in five genera (*Aster*, *Brachycome*, *Calotis*, *Haplopappus* and *Tetramolopium*). The lowest chromosome number, $x=2$, is found in the two genera *Brachycome* and *Haplopappus*. *Brachycome* and *Haplopappus* are the most cytologically variable genera in the tribe. Both show polyploidy and descending aneuploid series from $x=9$ down to $x=2$ as well.

Descending aneuploidy seems to be widely correlated with an annual habit or occurs at least in regions where derived annuals are frequent. North America is the most important region from this point of view, containing nearly 20 genera with chromosome numbers of $x=5$ down to $x=2$. Australia, as far as is known, has three genera with species with chromosome numbers less than $x=9$ (*Brachycome* from $x=9$ down to $x=2$, *Calotis* $x=4$, $x=7$ and 8, *Erodiophyllum* $x=8$). In Africa *Felicia* ($x=9$, 8, 6 and 5) and *Amellus* ($x=9$, 8 and 6) are representatives of this development. In the two latter continents further investigations may show more examples. Eurasia and South America (with the exception of *Cyathocline* in Eurasia and of *Haplopappus* in the latter, which even here has some species with low chromosome numbers) do not have any genera with a basic number other than $x=9$. The reason for this may be that Eurasia is rather poor in Astereae and there are no areas with high concentrations of derived annual species. South America, on the other hand, has quite a large number of Astereae but these are mostly all shrubby, perhaps more primitive, genera; this needs further investigations. An exception to the correlation of herbaceous habit and chromosome numbers less than $x=9$ is the subtribe Grangeinae. As far as is known (counts in three genera of total seven are known) all the annuals belonging to this subtribe are based on $x=9$ or $x=11$ (*Cyathocline*). For this and other reasons they may represent an older, or at least a different development, which retained partly the original basic number, and do not show descending aneuploidy.

APPENDIX I: SELECTED SYNONYMY

Agathaea Cass. = *Felicia*
Aphanochaeta Gray = *Chaetopappa*
Asterigeron Rydb. = *Aster*
Asteropsis Less. = *Podocoma*
Asterothamnus Novopokr. = *Aster*
Bellidiastrum Micheli = *Aster*
Biotia DC. = *Aster*
Bourdonia Greene = *Chaetopappa*
Brachyactis Ledeb. = *Aster*
Chaetophora Nutt. = *Chaetopappa*
Charieis Cass. = *Felicia*
Chroilema Bernh. = *Haplopappus*
Chrysophthalmum Phil. = *Grindelia*
Croptilon Raf. = *Haplopappus*
Doellingeria Nees = *Aster*
Detridium Nees = *Felicia*
Detris Adans. = *Felicia*
Diplostelma Gray = *Chaetopappa*
Distasis DC. = *Chaetopappa*
Elcismia B. L. Robinson = *Celmisia*
Ericameria Nutt. = *Haplopappus*
Eyselia Reichenb. = *Egletes*
Fresenia DC. = *Felicia*
Galatella Cass. = *Aster*
Golionema Watson = *Olivaea*
Grindeliopsis Sch. Bip. = *Xanthocephalum*
Guenthera Regel = *Xanthocephalum*
Gymnaster Kitamura = *Aster*
Haenelia Walpers = *Amellus*
Hazardia Greene = *Haplopappus*
Hecatactis F. v. Muell. = *Keysseria*
Heleastrum DC. = *Aster*
Hemibaccharis Blake = *Archibaccharis*
Herrickia Wott. & Small = *Aster*
Hesperodoria Greene = *Haplopappus*
Inulopsis O. Hoffm. = *Podocoma*
Ionactis Greene = *Aster*

Isopappus T. & G. = *Haplopappus*
Kalimeris Cass. = *Aster*
Kaulfussia Nees = *Felicia*
Keerlia Gray = *Chaetopappa*
Kemulariella Tamamsch. = *Aster*
Kraussia Sch. Bip. = *Amellus*
Krylovia Schischk. = *Aster*
Lagenifera Cass. = *Lagenophora*
Lasallea Greene = *Aster*
Leptilon Raf. = *Conyza*
Leucopsis (DC.) Baker = *Noticastrum*
Leucopsidium DC. = *Aphanostephus*
Linochilus Benth. = *Diplostephium*
Luteidiscus St. John = *Tetramolopium*
Lynosyris Cass. = *Aster*
Martinia Vaniot = *Asteromoea*
Microcalia A. Rich. = *Lagenophora*
Microgyne Less. = *Microgynella*
Minuriella Tate = *Minuria*
Molina Ruiz & Pavon = *Baccharis*
Munychia Cass. = *Felicia*
Neja D. Don = *Hysterionica*
Neosyris Greene = *Llerasia*
Ocyroe Phil. = Nardophyllum
Oreastrum Greene = *Aster*
Oreochrysum Rydb. = *Solidago*
Oreostemma Greene = *Aster*
Pentachaeta Nutt. = *Chaetopappa*
Pentheriella O. Hoffm. = *Heteromma*
Phyllochilium Cabr. = *Chiliophyllum*
Phyllostelidium Beauv. = *Baccharis*
Piofontia Cuatrec. = *Diplostephium*
Pityopsis Nutt. = *Heterotheca*
Platystephium Gardener = *Egletes*
Prionopsis Nutt. = *Haplopappus*
Pseudaster Tamamsch. = *Aster*
Pseudobaccharis Cabr. = *Baccharis*
Pseudolynosyris Novopokr. = *Aster*

19. ASTEREAE—SYSTEMATIC REVIEW

Psila Phil. = *Baccharis*
Schaetzelia Sch. Bip. = *Hinterhubera*
Sericocarpus Nees = *Aster*
Stenotopsis Rydb. = *Haplopappus*
Stereosanthus Franch. = *Nannoglottis*
Susanna Phillips = *Amellus*
Tonestus Nelson = *Haplopappus*
Tripolium Nees = *Aster*
Tumionella Greene = *Haplopappus*
Tursenia Cass. = *Baccharis*
Vierhapperia Hand. Mazz. = *Nannoglottis*
Wardaster J. Small = *Aster*
Wyomingia Nelson = *Erigeron*
Xanthocoma H.B.K. = *Xanthocephalum*
Xerobius Cass. = *Egletes*

APPENDIX 2

The most important publications dealing with the different genera of Astereae. Treatments in Floras are only mentioned when of predominantly monographic character.

Aphanostephus (Shinners, 1946a)
Apodocephala (Humbert, 1960)
Archibaccharis (Heering, 1904; Blake, 1924; Jackson, 1975)
Aster (Cronquist, 1943a, 1948 North America; Grierson, 1964 South Asia; Lippert, 1973 Africa; Onno, 1932 Eurasia)
Astranthium (Larsen, 1933; De Jong, 1965)
Aylacophora (Cabrera, 1953)
Baccharis (Cuatrecasas, 1967, 1969 Columbia; Heering, 1914; Luis, 1958)
Baccharidastrum (Cabrera, 1951)
Bigelowia (Anderson, 1970, 1972)
Blakiella (Cuatrecasas, 1969)
Brachycome (Davis, 1948, 1949)
Calotis (Davis, 1952)
Chaetopappa (Shinners, 1946b)
Colobanthera (Humbert, 1960)
Chrysothamnus (Hall and Clements, 1923; Anderson, 1964, 1966)
Conyza (Wild, 1969 Africa; Cronquist, 1943b)
Corethrogyne (Canby, 1927)
Dacryotrichia (Wild, 1973)
Darwiniothamnus (Harling, 1962)
Dichaetophora (Shinners, 1946c)
Diplostephium (Cuatrecasas, 1969 Columbia)
Egletes (Shinners, 1949)
Erigeron (Cronquist, 1947 North America; Solbrig, 1962 South America; Vierhapper, 1906 Asia, Europe)
Felicia (Grau, 1973)
Floscaldasia (Cuatrecasas, 1969)
Geissolepis (Robinson & Brettell, 1972)
Grangeopsis (Humbert, 1960)
Grindelia (Cabrera, 1931 South America; Steyermark, 1934 North America)
Gutierrezia (Solbrig, 1965 North America; Solbrig, 1966 South America)
Gyrodoma (Wild, 1974)
Haplopappus (Hall, 1928; Cabrera, 1934 Argentina; Smith, 1965 North America)
Heteromma (Hilliard and Burtt, 1973)
Hysterionica (Cabrera, 1946)
Ixiochlamys (Grau, 1975)
Jeffreya (Wild, 1974)
Keysseria (Koster, 1966; Mattfeld, 1938)
Laestadia (Cuatrecasas, 1969 Columbia)
Lagenophora (Cabrera, 1966; Davis, 1950)
Lessingia (Howell, 1928)
Machaeranthera (Cronquist and Keck, 1957; Turner and Horne, 1964)
Microgynella (Grau, 1975)
Nardophyllum (Cabrera, 1954)

Nidorella (Wild, 1969)
Olearia (Koster, 1966 New Guinea)
Olivaea (De Jong and Beaman, 1963)
Paleaepappus (Cabrera, 1969)
Parastrephia (Cabrera, 1954)
Pentachaeta (Van Hoorn, 1973)
Petradoria (Anderson, 1964)
Piora (Koster, 1966)
Polyarrhena (Grau, 1970)
Psiadia (Humbert, 1960 Madagascar)
Psiadiella (Humbert, 1960)
Psychrogeton (Grierson, 1967)
Pteronia (Hutchinson and Phillips, 1917)
Rigiopappus (Robinson and Brettell, 1973)
Rochonia (Humbert, 1960)
Solenogyne (Davis, 1950b)
Tetramolopium (Koster, 1966 New Guinea; Sherff, 1935)
Townsendia (Beaman, 1957; Larsen, 1927)
Vernoniopsis (Humbert, 1960)
Xanthisma (Semple, 1975)
Xanthocephalum (Solbrig, 1961)
Xylorhiza (Watson, 1973)

REFERENCES

ANDERSON, L. C. (1964). Studies on *Petradoria* (Compositae): anatomy, cytology, taxonomy. *Trans. Kansas Acad. Sci.* **66**, 632–684.

ANDERSON, L. C. (1964). Taxonomic notes on the *Chrysothamnus viscidiflorus* complex (Astereae, Compositae). *Madroño* **17**, 222–227.

ANDERSON, L. C. (1966). Cytotaxonomic studies in *Chrysothamnus* (Astereae, Compositae). *Am. J. Bot.* **53**, 204–212.

ANDERSON, L. C. (1970). Studies on *Bigelowia* (Astereae, Compositae). 1. Morphology and taxonomy, *Sida.* **3(7)**, 451–465.

ANDERSON, L. C. (1971). Embryology of *Chrysothamnus* (Astereae, Compositae). *Madroño* **20**, 337–342.

ANDERSON, L. C. (1972). Studies on *Bigelowia* (*Asteraceae*), II. Xylary comparisons, woodiness, and paedomorphosis. *J. Arnold Arbor.* **53**, 499–514.

ANDERSON, L. C., KYHOS, D. W., MOSQUIN, T., POWELL, A. M. and RAVEN, P. H. (1974). Chromosome numbers in *Compositae*. IX. *Haplopappus* and other Astereae. *Am. J. Bot.* **61**, 665–671.

BEAMAN, J. H. (1954). Chromosome numbers, apomixis and interspecific hybridization in the genus *Townsendia*. *Madroño* **12**, 169–180.

BEAMAN, J. H. (1957). The systematics and evolution of *Townsendia* (Compositae) *Contr. Gray Herb.* **183**, 1–138.

BLAKE, S. F. (1924). *Hemibaccharis*, a new genus of Baccharidinae. *Contr. U.S. natn. Herb.* **20**, 543–554.

CABRERA, A. L. (1931). Revisión de las especies sudamericanos del género *Grindelia*. *Rev. Mus. La Plata* **33**, 207–249.

CABRERA, A. L. (1934). Las especies argentinas del género *Haplopappus*. *Notas prelim. Mus. La Plata* **2**, 233–257.

CABRERA, A. L. (1941). *Compuestas* Bonaerenses. *Rev. Mus. La Plata* (N. Ser.) **4**, 1–450.

CABRERA, A. L. (1946). El género *Hysterionica* en el Uruguay y en la Republica Argentina. *Not. Mus. La Plata* **11**, 349–358.

CABRERA, A. L. (1953). Un nuevo género de *Astereas* de la Republica Argentina. *Bol. Soc. arg. not.* **4**, 261–271.

CABRERA, A. L. (1954). Las especies del género *Nardophyllum*. *Not. Mus. Buenos Aires* **18**, 55–66.

CABRERA, A. L. (1966). The genus *Lagenophora* (Compositae). *Blumea* **14**, 285–307.

CABRERA, A. L. (1969). *Compuestas* nuevas de Patagonia. *Bol. soc. arg. bot.* **11**, 271–291.

CANBY, M. L. (1927). The genus *Corethrogyne* in Southern California. *Bull. Soc. Calif. Acad.* **26**, 8–16.

CARLQUIST, S. (1960). Wood anatomy of Astereae (Compositae). *Trop. Woods* **113**, 54–84.

CARTER, C. R., SMITH-WHITE, S. and KYHOS, D. W. (1974). The cytology of *Brachycome lineariloba*. 4. The 10-chromosome quasi-diploid. *Chromosoma* **44**, 439–456.

CRONQUIST, A. W. N. (1943a). American species of *Aster* centering about *Aster foliaceus*. *Am. Midl. Nat.* **29**, 429–468.

CRONQUIST, A. (1943b). The separation of *Erigeron* from *Conyza*. *Bull. Torrey bot. Club*, **70**, 629–632.

CRONQUIST, A. (1947). Revision of the North American species of *Erigeron*, North of Mexico. *Brittonia* **6**, 121–302.

CRONQUIST, A. (1948). Revision of the Oreastrum group of *Aster*. *Leafl. West. Bot.* **5**, 73–82.

CRONQUIST, A. (1955). Phylogeny and taxonomy of the Compositae. *Am. Midl. Nat.* **53**, 478–511.

CRONQUIST, A. and KECK, D. D. (1957). A reconstruction of the genus *Machaeranthera*. *Brittonia* **9**, 231–240.

CUATRECASAS, J. (1967). Revisión de las especies columbianas del género *Baccharis*. *Rev. Acad. Colomb. Cienc. ex., fis. y nat.* **8**, 5–12.

CUATRECASAS, J. (1969). Prima Flora Colombiana. 3. Compositae-Astereae. *Webbia* **24**, 1–335.

DAVIS, G. L. (1948). A revision of the genus *Brachycome*. *Proc. Linn. Soc. N.S.W.* **73**, 132–241.

DAVIS, G. L. (1949). Revision of the genus *Brachycome* Cass. II The New Zealand species. *Proc. Linn. Soc. N.S.W.* **74**, 97–106.

DAVIS, G. L. (1950a). A revision of the Australian species of the genus *Lagenophora* Cass. *Proc. Linn. Soc. N.S.W.* **75**, 122–132.

DAVIS, G. L. (1950b). Revision of the genus *Solenogyne* Cass. *Proc. Linn. Soc. N.S.W.* **75**, 188–194.

DAVIS, G. L. (1952). Revision of the genus *Calotis*. *Proc. Linn. Soc. N.S.W.* **77**, 146–188.

DAVIS, G. L. (1963). Generative apospory and diploid parthenogenesis in *Brachycome ciliaris* var. *lanuginosa*. *Aust. J. Sci.* **26**, 90.

DAVIS, G. L. (1964). The embryology of *Minuria integerrima*: a somatic apomict. *Phytomorphology* **14**, 231–239.

DAVIS, G. L. (1964). Development of the female gametophyte of *Minuria cunninghamii* (DC.) Benth. (Compositae). *Aust. J. Bot.* **12**, 152–156.

DAVIS, G. L. (1964). Embryological studies in the *Compositae* IV. Sporogenesis, gametogenesis, and embryogeny in *Brachycome ciliaris* (Labill.) Less. *Aust. J. Bot.* **12**, 142–151.

DAVIS, G. L. (1968). Apomixis and abnormal anther development in *Calotis lappulacea* Benth. (Compositae). *Aust. J. Bot.* **16**, 1–17.

DE JONG, D. C. D. (1965). A systematic study of the genus *Astranthium*. *Publs. Mich. State Univ. Biol., Ser. 2*, 429–538.

DE JONG, D. C. D. and BEAMAN, J. H. (1963). The genus *Olivaea* (Compositae–Astereae). *Brittonia* **15**, 86–92.

DRURY, D. G. (1968). A clarification of the generic limits of *Olearia* and *Pleurophyllum* (Astereae–Compositae). *N.Z. Jl Bot.* **6**, 459–466.

GIVEN, D. R. (1971). *Damnamenia* gen. nov. A new subantarctic genus allied to *Celmisia* Cass. (Astereae–Compositae). *N.Z. Jl Bot.* **11**, 785–786.

GRAU, J. (1970). Die Gattung *Polyarrhena* Cass (Asteraceae–Asterinae). *Mitt. Bot. München* **7**, 347–368.

GRAU, J. (1973). Revision der Gattung *Felicia* (Asteraceae). *Mitt. Bot. München* **9**, 195–705.

GRAU, J. (1975). *Podocoma* und *Vittadinia*—zwei vermeintlich bikontinentale Gattungen. *Mitt. Bot. München* **12**, 181–194.

GRIERSON, A. J. C. (1964). A revision of the *Asters* of the Himalayan area. *Not. R. bot. Gdn Edinb.* **26**, 67–163.

GRIERSON, A. J. C. (1966). A note on *Aster* ageratoides and observations on *Aster* in the Philippine islands. *Not. R. bot. Gdn Edinb.* **28**, 227–231.

GRIERSON, A. J. C. (1967). The genus *Psychrogeton* (Compositae). *Not. R. bot. Gdn Edinb.* **27**, 101–148.

HALL, H. M. (1928). The genus *Haplopappus*. A phylogenetic study in the Compositae. *Carnegie Instn Publs* **389**, 1–391.

HALL, H. M. and CLEMENTS, F. E. (1923). The phylogenetic method in Taxonomy. The north american species of *Artemisia*, *Chrysothamnus* and *Atriplex*. *Carnegie Instn Publs* **326**, 157–234.

HARLING, G. (1951). Embryological Studies in the Compositae III, Astereae. *Acta hort. berg.* **16**, 73–120.

HARLING, G. (1951). The embryo-sac development of *Vittadinia triloba* (Gaud.) DC. *Svensk. bot. Tidskr.* **48**, 490–496.

HARLING, G. (1962). On some *Compositae* endemic to the Galapagos Islands. *Acta hort. berg.* **20**, 63–120.

HEERING, W. (1904). Die *Baccharis*-Arten des Hamburger Herbars. *Jb. Hamb. wiss. Anst.* **21**, (Beih. 3), 39–42

HEERING, W. (1914). Systematische und pflanzengeographische Studien über die *Baccharis*-Arten des außertropischen Südamerikas. *Jb. Hamb. wiss. Anst.* **31**, 65–173.

HILLIARD, O M. and BURTT, B. L. (1973). Notes on some plants of southern Africa chiefly from Natal: III. *Not. R. bot. Gdn Edinb.* **32**, 303–388.

HOFFMANN, O. (1890). Astereae. *In* "Die natürlichen Pflanzenfamilien" (Engler and Prantl, eds.) **4** (5), 142–172.

HOWELL, J. T. A. (1928). A systematic study of the genus *Lessingia* Cham. *Univ. Calif. Publs Bot.* **16**, 1–44.

HUMBERT H. (1960). "Flore de Madagascar et des Comores", Composées I, Asterées, 204–338.

HUTCHINSON, J. and PHILLIPS, E. P. (1917). A revision of the genus *Pteronia*. (Compositae). *Ann. S. Afr. Mus.* **9**, 277–329.

HUZIWARA, Y. (1958). Karyotype analysis in some genera of Compositae V. The chromosomes of American *Aster* species. *Jap. J. Genet.* **33**, 129–137.

HUZIWARA, Y. (1958). Karyotype analysis in some genera of Compositae IV. The karyotypes within the genera *Gymnaster*, *Kalimeris* and *Heteropappus*. *Cytologia* **23**, 33–45.

HUZIWARA, Y. (1959). Chromosomal evolution in the subtribe Asterinae. *Evolution* **13**, 188–193.

HUZIWARA, Y. (1965). Chromosome analysis in the tribe Astereae. *Jap. J. Genet.* **40**, 63–71.

JACKSON, J. D. (1975). A revision of the genus *Archibaccharis* Heering (Compositae–Astereae). *Phytologia* **32**, 81–194.

JACKSON, R. C. and CROVELLO, T. J. (1971). A comparison of numerical and biosystematic studies in *Haplopappus*. *Brittonia*. **23**, 54–70.

KITAMURA, S. (1937). Compositae Japonicae. *Mem. Coll. Sci. Ser. 8* **13**, 299–399.

KOSTER, J. TH. (1966). The Compositae of New Guinea I. *Nova Guinea Bot.* **24**, 497–614.

LARSEN, E. L. (1927). A revision of the genus *Townsendia*. *Ann. Mo. bot. Gdn* **14**, 1–47.

LARSEN, E. L. (1933). *Astranthium* and related genera. *Ann. Mo. bot. Gdn* **20**, 23–44.

LIPPERT, W. (1973). Revision der Gattung *Aster* in Afrika. *Mitt. bot. München.* **11**, 153–258.

LUIS, J. T. (1958). Novum index *baccharidinarum*. *Contr. Inst. Geobiol. Canoes* **9**, 1–35.

MATTFELD, J. (1938). Einige neue oder bemerkenswerte *Compositen* aus Neuguinea. *Bot. Jb.* **68**, 248–268.

ONNO, M. (1932). Geographisch-morphologische Studien über *Aster alpinus* L. und verwandte Arten. *Biblthca bot.* **106**, 1–83.

RAVEN, P. H., SOLBRIG, T., KYHOS, D. W. and SNOW, R. (1960). Chromosome numbers in Compositae. I. Astereae. *Am. J. Bot.* **47**, 124–132.

RECHINGER, K. M. (1950). Die ausdauernden iranischen Arten von *Erigeron* sectio *Conyzastrum* Boiss. *Phyton* **2**, 124–133.

ROBINSON, H. and BRETTELL, B. D. (1972). Tribal revisions in the Asteraceae. I. The relationship of *Geissolepis*. *Phytologia* **24**, 299–301.

ROBINSON, M. and BRETTELL, R. D. (1973). Tribal revisions in the Asteraceae. V. The relationships of *Rigiopappus*. *Phytologia* **26**, 69–70.

SEMPLE, J. C. (1976). The cytogenetics of *Xanthisma texonium* DC. (Asteraceae) and its B-chromosomes. *Am. J. Bot.* **63**, 388–398.

ST JOHN, H. (1974). *Luteidiscus*, new genus (Compositae). *Bot. Jb.* **94**, 549–555.

SHERFF, E. E. (1935). Revision of *Tetramolopium*, *Lipochaeta*, *Dubautia* and *Raillardia*. *Bull. Bernice P. Bishop Mus.* **135**, 1–136.

SHINNERS, H. (1946a). Revision of the genus *Aphanostephus* DC. *Wrightia* **1**, 95–121.

SHINNERS, L. H. (1946b). Revision of the genus *Chaetopappa* DC. *Wrightia* **1**, 63–81.

SHINNERS, L. H. (1946c). The genus *Dichaetophora* A. Gray and its relationships. *Wrightia* **1**, 90–94.

SHINNERS, L. H. (1946). Revision of the genus *Leucelene* Greene. *Wrightia* **1**, 82–89.

SHINNERS, L. H. (1949). Revision of the genus *Egletes* Cassini North of South America, *Lloydia* **12**, 239–250.

SMITH, E. B. (1965). Taxonomy of *Haplopappus*, section *Isopappus* (Compositae). *Rhodora* **67**, 217–238.

SMITH, E. B. (1966). Cytogenetics and phylogeny of *Haplopappus* section *Isopappus* (Compositae). *Can. J. Genet. Cytol.* **8**, 14–36.

SMITH-WHITE, S. and CARTER, C. R. (1970). The cytology of *Brachycome lineariloba*. *Chromosoma* **30**, 129–153.

SMITH-WHITE, S , CARTER, C. R. and STACE, H. M. (1970). The cytology of *Brachycome*. I. The subgenus *Eubrachycome*: a general survey. *Aust. J. Bot.* **18**, 99–125.

SOLBRIG, O. T. (1965). The California species of *Gutierrezia* (Compositae–Astereae). *Madroño* **18**, 75–84.

SOLBRIG, O. T. (1960). The status of the genera *Amphipappus, Amphiachyris, Greenella, Gutierrezia, Gymnosperma* and *Xanthocephalum* (Compositae). *Rhodora* **62**, 43–54.

SOLBRIG, O. T. (1960). The South American sections of *Erigeron* and their relation to *Celmisia*. *Contr. Gray Herb.* **188**, 65–86.

SOLBRIG, O. T. (1961). Synopsis of the genus *Xanthocephalum* (Compositae). *Rhodora* **63**, 151–164.

SOLBRIG, O. T. (1962). The South American species of *Erigeron*. *Contr. Gray Herb.* **191**, 3–82.

SOLBRIG, O. T. (1965). The typification of *Xanthocephalum* (Compositae). *Rhodora* **67**, 182–184.

SOLBRIG, O. T. (1966). The South American species of *Gutierrezia*. *Contr. Gray Herb.* **197**, 3–42.

SOLBRIG, O. T., ANDERSON, L. C., KYHOS, D. W., RAVEN, P. H. and RÜDENBERG, L. (1964). Chromosome numbers in Compositae V. Astereae II. *Am. J. Bot.* **51**, 513–519.

SOLBRIG, O. T., ANDERSON, L. C., KYHOS, D. W. and RAVEN, P. H. (1969). Chromosome numbers in Compositae VII: Astereae III. *Am. J. Bot.* **56**, 348–353.

STEYERMARK, J. A. (1934). The North American species of the genus *Grindelia*. *Ann. Mo. bot. Gdn* **21**, 433–608.

TURNER, B. L. (1970). Chromosome numbers in the *Compositae*. XII. Australian species. *Am. J. Bot.* **57**, 383–389.

TURNER, B. L. and HORNE, D. (1964). Taxonomy of *Machaeranthera* sect. *Psilactis* (Compositae–Astereae). *Brittonia* **16**, 316–331.

VAN HORN, G. S. (1973). The taxonomic status of *Pentachaeta* and *Chaetopappa* with a revision of *Pentachaeta*. *Univ. Calif. Publs. Bot* **65**, 1–41.

VIERHAPPER, F. (1906). Monographie der alpinen *Erigeron*-Arten Europas und Vorderasiens. *Beih. bot. Zbl.* **19**, 385–560.

VIERHAPPER, F. (1916). Analytische Übersicht über einige patagonische und feuerländischer *Erigeron*-Formen. *Bot. Notiser* **1916**, 241–250.

Watson, T. J. (1973). Systematics of *Xylorhiza* (Compositae–Astereae). *Diss. Abstr. Int., Bot.* **33**, 1, 4155.

Withaker, T. W. and Steyermark, J. A. (1935). Cytological aspects of *Grindelia* species. *Bull. Torrey bot. Club* **62**, 69–75.

Wild, H. (1969). The genus *Nidorella* Cass. *Bolm Soc. broteriana* **43**, 209–240.

Wild, H. (1969). The species of *Conyza* L. with ligulate or lobed ray florets in Africa, Madagascar and the Cape Verde Islands. *Bolm. Soc. broteriana* **43**, 247–277.

Wild, H. (1973). A new genus of *Compositae* (*Astereae*) from the Flora Zambesiaca area. *Garcia de orto.* **1**, 67–68.

Wild, H. (1974). New and interesting Compositae from south central Africa, 2. *Kirkia* **9**, 293–300.

Wodehouse, R. P. (1930). Pollen grains in the identification and classification of plants. V. *Haplopappus* and other Astereae. *Bull. Torrey bot. Club.* **57**, 21–46.

ADDENDA

Barroso (1975) separated out a new genus from *Aster* or *Baccaris*. It has to be included in the South American species: *Baccharidiopsis* Barroso 1 species, perennial, from Brazil.

Also in 1975, Stucky and Jackson, in their investigations of the DNA content of related American Astereae with different chromosome numbers (*Machaeranthera* with $n=4$, 5 and 9, *Aster* with $n=5$ and 9), showed that the species with a low chromosome number resulted from an aneuploid reduction from 9 to 4 (to 5 in *Aster*). In these cases the basic number $x=9$ is quite probable.

References

Barroso, G. M. (1975). Baccharidiopsis—un género novo da Subtribo Baccharidinae Hoffmann (Tribo Astereae). *Sellowia* **26**, 95–101.

Stucky, J. and Jackson, R. C. (1975). DNA content of seven species of Astereae and its significance to theories of chromosome evolution in the Tribe. *Am. J. Bot.* **62**, 509–518.

Chapter 20

Astereae—chemical review

WERNER HERZ

Department of Chemistry,
Florida State University,
Tallahassee, Florida U.S.A.

Abstract. At this date the very limited information available on constituents of herbaceous parts of Astereae precludes any attempts at relating chemistry to systematics. Knowledge of root constituents is more advanced, particularly in subtribe Asterinae where variations in polyacetylene and coumarin composition reflect to some extent divisions based on classical techniques. Diterpenoids appear to be more characteristic of Solidaginae.

CONTENTS

Introduction	567
Low-molecular-weight constituents	568
Terpenoids	568
Flavonoids and other phenolics	571
Alkaloids	572
Acetylenes	572
Discussion	573
References	574

INTRODUCTION

At the time of writing, somewhat more than 10% of the approximately 2000 species in this tribe have been subjected to chemical scrutiny. However, work has been confined largely to root constituents. Among these, polyacetylenes, polyenes and related substances and, in certain groups, coumarins are very characteristic. Less widely distributed, perhaps because they are less intensively searched for, are diterpenoids, so far largely of the labdane and clerodane type.

    Epigeal parts have been studied only sporadically. As a result, the information on the distribution of flavonoids and other low-molecular-weight secondary metabolites in these parts is so spotty that it cannot be used to adumbrate a chemical pattern of the tribe and practically nothing is known of macromolecular constituents. However, the rarity or absence of sesquiterpene lactones and alkaloids appears to distinguish Astereae from several other tribes of Compositae.

In the following, the literature since 1963 will be reviewed with occasional reference to earlier work cited in Vol. 3 of "Chemotaxonomie der Pflanzen" (Hegnauer, 1964) and in "Naturally Occurring Acetylenes" (Bohlmann et al., 1973).

LOW-MOLECULAR-WEIGHT CONSTITUENTS

Terpenoids

Monoterpenoids and Sesquiterpenoids. As the essential oils of Astereae are of no commercial significance, little is known of their composition. In subtribe Solidaginae *Haplopappus rigidifolius* E. B. Smith (≡ *Croptilon divaricatum* (Nutt) Raf. var. *hirtellum* Shinners) contains (Dominguez and Jimenez, 1973) a stereoisomer of cumambrin B (**1**). This is the only sesquiterpene lactone so far found in Astereae. The composition of the oil of *Cyathocline lyrata* Cass. (subtribe Bellidinae) is discussed by Hegnauer (1964, p. 455) as is that of *Olearia paniculata* Druce (subtribe Asterinae) and *Solidago canadensis* L. (Solidaginae). The sesquiterpene fraction of the latter contains the relatively rare aromadendrane derivative cyclocolorenone. The oil from *Baccharis genistilloides* Pers. (Baccharinae) contains, *inter alia*, two sesquiterpene alcohols of the aromadendrane type and the abnormal monoterpene carquejol (Hegnauer, 1964, pp. 454-455), whereas roots of *Baccharis timera* (sic! = *B. trimera* DC.? ≡ *B. genistelloides* Pers.) yielded two esters **2a** and **2b** of the carquejol type (Bohlmann and Zdero, 1969) and no acetylenes. The terpene fraction of *Psiadia salviifolia* Bak. (Conyzinae) contains *inter alia* the unusual monoterpene **3** (Dennis, 1973).

Diterpenoids. The occurrence of diterpenes in members of the subtribe Solidaginae is well documented. Epigeal parts of *Haplopappus rigidifolia* contain the pimarane derivative justintetrol (**4**) (Dominguez and Jimenez, 1973), while *cis*-clerodanes such as **5** have been isolated from *H. angustifolius* Reiche and *H. foliosus* DC. (Silva and Sammes, 1973). An unusual chlorine-containing *ent-cis*-clerodane gutierolide (**6**) has been found in *Gutierrezia dracunculoides* (DC.) Blake (Cruse *et al.*, 1971), while labdanes have been isolated from *Grindelia robusta* Nutt. and *G. squarrosa* (Pursh.) Dunal (Hegnauer, 1964, p. 481).

Roots of *Solidago* species have proved to be a rich source of clerodanes; other types of diterpenoids are occasionally found as will be evident from the following listings:

  S. altissima L.: *trans*-Clerodanes (Kusumoto *et al.*, 1969; Ohsuka *et al.*, 1973; Okazaki *et al.*, 1973)

S. arguta Ait.: *cis*- and *trans*-Clerodanes (Anderson *et al.*, 1974)
S. canadensis L.: Labdanes (Anthonsen 1966; Anthonsen *et al.*, 1967; 1969; 1970); modified labdanes (**7**) (McCabe *et al.*, 1969)
S. elongata Nutt.: *trans*-Clerodanes (Anthonsen and McCrindle, 1969)
S. flexicaulis L.: No diterpenoids (Anthonsen and Bergland, 1971)
S. juncea Ait.: *trans*-Clerodanes, abietanes, *ent*-kauranes (Henderson *et al.*, 1973)
S. missouriensis Nutt.: *ent-trans*-Clerodanes, abietanes (Anthonsen and Bergland, 1970; 1973)

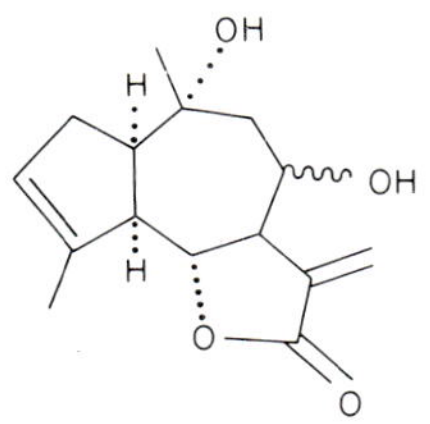

(1)

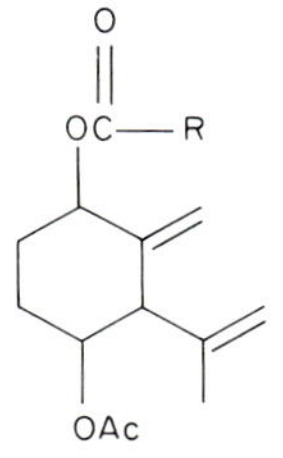

(2a) R = $CHMe_2$
(b) R = CHMeEt

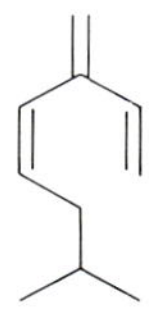

(3)

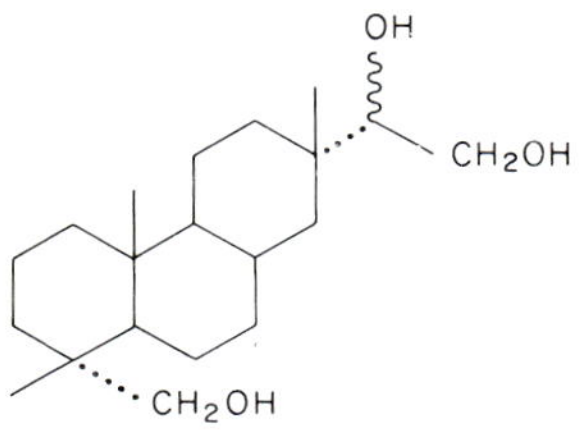

(4) Justintetrol

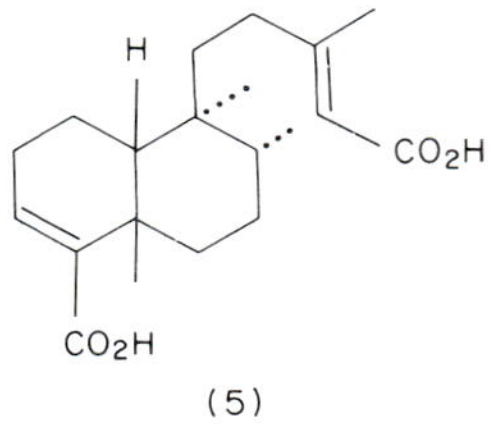

(5)

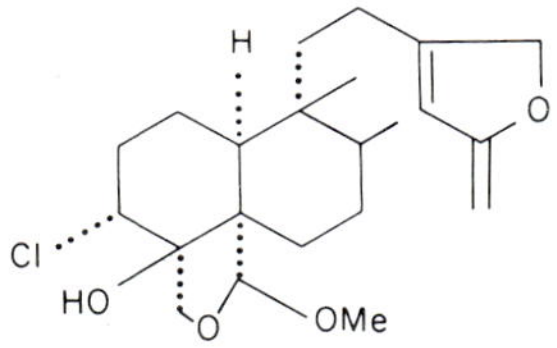

(6) Gutierolide

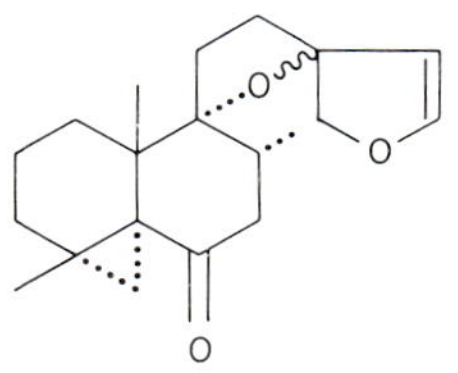

(7)

S. rigida L.: *ent*-Kauranes (Anthonsen and Bergland, 1971)
S. serotina Ait. (≡ *S. gigantea* Air. var. *serotina*): *ent-cis*-Clerodanes (Anthonsen *et al.*, 1973; Henderson *et al.*, 1973); *trans*-clerodanes (McCrindle and Nakamura, 1974)
S. shortii T. & G.: *trans*-Clerodane (Anthonsen and Bergland, 1971)
S. virgaurea L.: No diterpenoids (Gerlach, 1965; Anthonsen and Bergland, 1971)

Hinterhubera imbricata Cuatr. (Bohlmann *et al.*, 1973) and *Olearia heterocarpa* S. T. Blake (Pinhey *et al.*, 1971) are the only members of tribe Asterinae from whose epigeal parts diterpenes (*trans*-clerodane type) have so far been reported. *Conyza ivaefolia* Less. (Bohlmann and Grenz, 1972) and *Psiadia altissima* B. & H. (Canonica *et al.*, 1969), both in subtribe Conyzinae, have yielded *trans-* and *ent-trans*-clerodanes, respectively. Lastly, a labdane derivative has been isolated from *Baccharis glutinosa* Pers. (referred to as *Gymnosperma glutinosa*, Miyakado *et al.*, 1974) in subtribe Baccharinae.

Triterpenoids and sterols. The following is a listing of triterpenes and sterols, mostly well distributed through the entire plant kingdom. Exceptions are baccharis oxide (**8**) which seems to be a characteristic constituent of some *Baccharis* species, shionone (**9**), found in some asters, and two sterols of *Haplopappus heterophyllus*.

Solidaginae:
  Haplopappus angustifolius Reiche: Friedelanone, epi-friedelanol (Silva and Sammes, 1973)
  H. foliosus DC.: Friedelanone, epi-friedelinol, stigmasterol (Silva and Sammes, 1973)
  H. heterophyllus (Gray) Blake: 5α-Androstan-3β-16α,17α-triol (Zalkow *et al.*, 1964), stigma-8(14),22-dien-3β-ol (Zalkow *et al.*, 1968)
  Solidago elongata Nutt. and *S. serotina* Ait.: Stigmasterol (Anthonsen and McCrindle, 1969)
  S. virgaurea L.: Steroidal diols (Kasprzyk and Kozierowska, 1966)
Asterinae:
  Aster baccharoides Steetz; Friedelin, friedelanol, β-amyrin and acetate, lupeol, α-spinasterol, shionone (Hui *et al.*, 1971)
  A. scaber Thunb.: Friedelin, friedelanol α-spinasterol, squalene (Tada *et al.*, 1974)
  A. tataricus L.: Shionone (Hegnauer, 1964), shionone, friedelin, epi-friedelanol (Kamisake and Takahashi, 1964)
  Hinterhubera imbricata Cuatr.: Oleanolic acid (Bohlmann *et al.*, 1973b)
Conyzinae:
  Conyza filaginoides DC.: β-amyrin, α-spinasterol (Dominguez *et al.*, 1972a)

Baccharinae:
> *Baccharis glutinosa* Pers. and *B. halimifolia* L. (roots): Baccharis oxide (Anthonsen *et al.*, 1970, Mo *et al.*, 1972)
> *B. rhomboidalis* Remj.: Brein, oleanolic acid (Silva *et al.*, 1971)
> *B. salicifolia* Pers.: Friedoolean-3β-ol, stigmasterol, baccharis oxide (Dominguez *et al.*, 1972b)
> *B. salicina* Shinner (roots): Friedelanol (Anthonsen, Bruun *et al.*, 1970)
> *B. viminea* DC. (pollen): sitosterol (Standifer *et al.*, 1968)

Flavonoids and other Phenolics

Flavonoids. There is no thorough, published study of flavonoid distribution in any genus of the Astereae, although there is much unpublished data on *Haplopappus* and its relatives in process of publication (B. L. Turner, personal communication). The scattered results tabulated below indicate the relatively common occurrence of apigenin, kaempferol and quercetin derivatives.

Solidaginae:
> *Grindelia squarrosa* (Purs.) Dunal: Chrysoeriol 7-glucuronide (Wagner *et al.*, 1972)
> *Haplopappus bailahuen* Remy: Quercetin, kaempferol, quercetin 3-*O*-methyl ether (Hörhammer *et al.*, 1973)
> *Solidago canadensis* L.: Quercetin, kaempferol, rutin, isoquercitrin, isorhamnetin, isorhamnetin-3-glucosylrhamnoside (Batyuk and Koltsova 1968; 1969)

Bellidinae:
> *Bellis perennis* L.: Apigenin 7-glucoside (Harborne, 1967)

Asterinae:
> *Aster altaicus* Willd.: 5-Hydroxy-6,7,8,3′,4′-pentamethoxyflavone (Troshchenko and Limasova, 1966)
> *A. ageratoides* var. *ovatus* pollen: Quercetin, carotenoids, galactose (Hisamichi, 1961)
> *A. pilosus* Willd.: Hyperoside (Farnsworth *et al.*, 1968)
> *A. tataricus* L. and *A. yomena* (Kitamura) Honda pollen: Quercetin, carotenoids, galactose (Hisamichi, 1961)
> *Erigeron annuus* (L.) Pers.: Apigenin 7-glucuronide (Harborne, 1967)
> *Olearia heterocarpa* S. T. Blake: Xanthomicrol (Pinhey *et al.*, 1971)

Conyzinae:
> *Conyza ivaefolia* Less.: Quercetin (Bohlmann and Grenz, 1972)
> *Psiadia altissima* Benth. + Hook.: Apigenin, aromadendrin, 7-*O*-methylaromadendrin (Canonica *et al.*, 1969)

Baccharinae:
  Baccharis angustifolia Mich.: Isoquercitrin, astragalin (Wagner *et al.*, 1972)
  B. glutinosa Pers. (≡ *Gymnosperma glutinosa*): 5,7-Dihydroxy-6,8,3′,4′,5′-pentamethoxyflavone, 5,7-dihydroxy-3,6,8,3′-tetramethoxyflavone (Dominguez and Torre, 1974)
  B. rhomboidalis Remj.: 5-Hydroxy-7,4′-dimethoxyflavone (Silva *et al.*, 1971)
  B. sarothroides A. Gr.: Centaureidein, 3,4′-Dimethoxy-3′,5,7-trihydroxyflavone (Kupchan and Bauerschmidt, 1971)

Other Phenolics. Geranyl ethers of umbelliferone of type **10** and variations thereof have been found in the roots of 23 *Aster* species all belonging to section *Aster* (Bohlmann *et al.*, 1968, 1973a). Coumarin itself has been isolated from the roots of *A. depauperatus* Fernald (Bohlmann *et al.*, 1973a). Roots of *Solidago rigida* L. and *S. virgaurea* L. gave an ester **11a** (Anthonsen and Bergland, 1971) which appears to be identical with one of two esters **11a** and **11b** found in the roots of *Aster ptarmicoides* T. & G. (Anthonsen and Bergland, 1971; Bohlmann *et al.*, 1969).

The toxic dihydrobenzofurans toxol (**12**,5-acetyl-2,3-dihydrobenzofuran) and tremetone (5-acetyl-2,3-dihydrobenzofuran, **13**) also found in *Eupatorium rugosum* Houtt. (Hegnauer 1964, p. 453), as well as 2-isopropenyl-5-acetylbenzofuran (dehydrotremetone) and 2,5-diacetylbenzofuran were isolated from the epigeal parts of *Haplopappus heterophyllus* (Gray) Blake (Zalkow *et al.*, 1962, 1964, 1968).

Alkaloids

Chemical tests have suggested the presence of alkaloids in several species (Hegnauer, 1964, p. 509; Raffauf, 1970), but no well-characterized substances have been obtained. Recent reports claim the isolation of several alkaloids of unknown structure from *Baccharis coridifolia* DC. (Moreira, 1966) (for a study of the wax constituents of this species, see Barbará *et al.*, 1974) and *B. linearis* Pers. (Montes *et al.*, 1971).

Acetylenes

The main acetylenic constituents and related compounds in the root extracts of more than 160 species of Astereae in 36 genera have been listed by Bohlmann *et al.* (1973a, pp. 343–350).* Additional reports not included in the tables deal with *Solidago flexicaulis* L. (*cis-trans* and *trans-cis*-matricaria and dehydromatricaria esters; Anthonsen and Bergland, 1971),

---

* These tables contain a few taxonomic and chemical ambiguities.

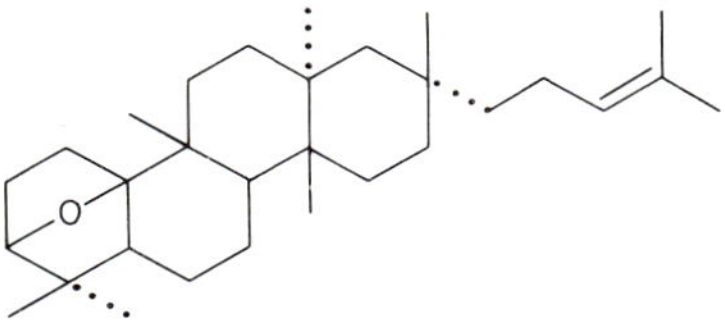

(8) *Baccharis* oxide

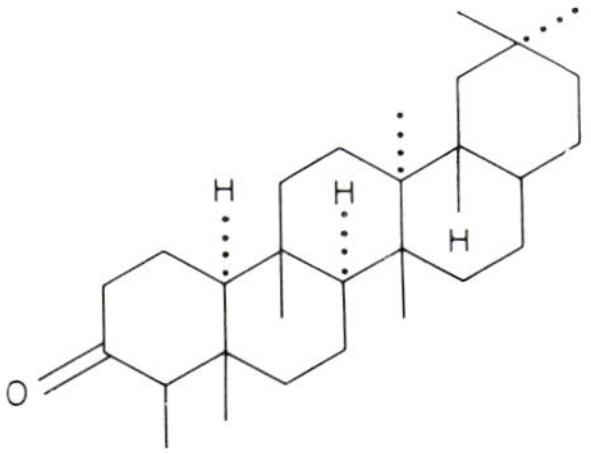

(9) Shionone

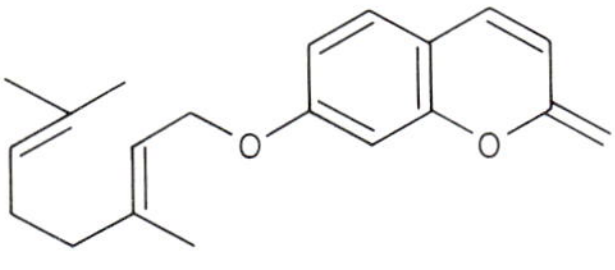

(10)

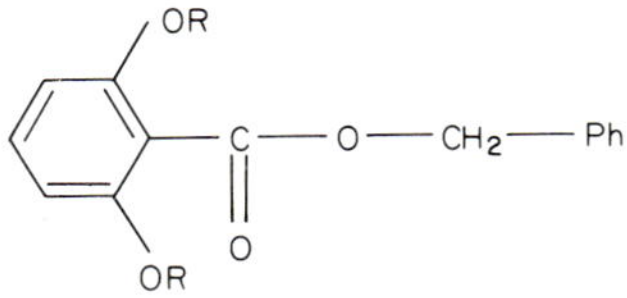

(11a) R = Me, Me
(11b) R = H, Me

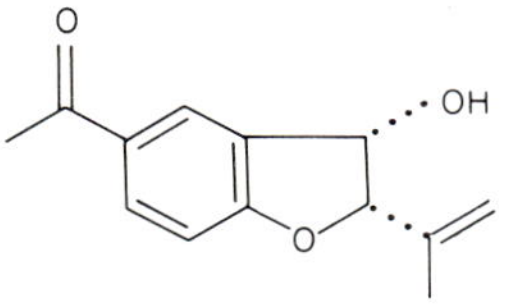

(12)

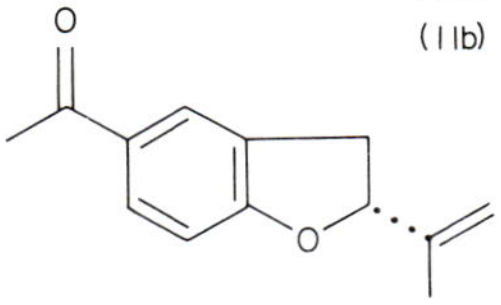

(13) Tremetone

(14) $CH_3CH=CH-(C\equiv C)_2-(CH=CH)_2-(CH_2)_4-CO_2Me$

(15) $CH_3CH=CH-(C\equiv C)_2-CH=CH-CO_2Me$
Lachnophyllum ester

(16) $CH_3CH_2CH_2-(C\equiv C)_2-CH=CH-CO_2Me$
Matricaria ester

Chrysocoma peduncularis DC. and *C. tenuifolia* Berg. (Conyzinae) which contain unusual C_{16} esters such as **14** (Bohlmann and Zdero, 1972), and details on the constituents of *Diplopappus filifolius* DC., *Felicia tenella* Nees. (Bohlmann and Rao, 1972) and *Chrysocoma coma-aurea* L. (Bohlmann and Zdero, 1972). Approximately 10% of the species contained no acetylenes, prominent among which were six out of the nine *Felicia* (Asterineae) and both of the *Pteronia* (Solidaginae) species examined.

DISCUSSION

As has been pointed out in the introduction, information on the chemistry of the epigeal parts of Asterinae is so limited that its relevance to the systematics within the tribe cannot be assayed, although the apparent absence of sesquiterpene lactones would appear to differentiate Asterinae sharply from Vernonieae, Eupatorieae, Inuleae (in part), Heliantheae, Helenieae, Anthemideae, Cynareae and Cichorieae.

Root chemistry offers greater opportunities. Subtribe Solidaginae contains almost exclusively C_{10}-acetylenes (25 species) except for *Pteronia* which contains no acetylenes (two species) and *Solidago* (17 species) where acetylenes are frequently accompanied or replaced by diterpenes. Because of very limited information no generalizations are possible in subtribes Grangeinae and Baccharinae; two species of the former contained no acetylenes of interest, whereas two species of the latter yielded C_{10}- and C_{17}-acetylenes. Two species of Bellidinae afforded C_{17}- and five C_{10}-acetylenic alcohols, with angeloyl derivatives of 3-hydroxymatricaria ester or matricarianol apparently typical of *Bellis*. In Conyzinae, the presence of unusual C_{16}-esters in two *Chrysocoma* species (Bohlmann and Zdero, 1972) has already been commented upon; five other Conyzinae afforded C_{10}- and/or C_{17}-compounds.

The most thoroughly studied subtribe has been the Asterinae (approximately 135 species examined). Generalizations appear possible for *Felicia*, from which acetylenes are largely absent, and *Erigeron* whose root constituents typically are lachnophyllum ester **15**, matricaria ester **16** and three closely related C_{10}-lactones as well as cosmene. Within the large genus *Aster*, the variations are said to be partially in agreement with systematics (Bohlmann *et al.*, 1973a, p. 350). The presence of C_{17}-acetylenes is typical for sections *Alpigena* and *Biota*. For section *Aster*, the presence of **15**, its 8-angeloyloxy derivative and, in many instances, coumarins of type **10** seems to be very characteristic, while sections *Lynoseris* and *Machaeranthera* appear to elaborate primarily matricarianol derivatives typical also of *Grindelia* (Solidaginae). Finally, in view of the poorly defined subtribes of this tribe (Grau, Chapter 19), it should be noted that chemical characters might prove most helpful in future infrageneric classificatory schemes.

REFERENCES

ANDERSON, A. B., MCCRINDLE, R. and NAKAMURA, E. (1974). *J. chem. Soc. chem. Comm.*, 453.
ANTHONSEN, T. (1966). *Acta chem. scand.* **20**, 904.
ANTHONSEN, T. and BERGLAND, G. (1970). *Acta chem. scand.* **24**, 1860.
ANTHONSEN, T. and BERGLAND, G. (1971). *Acta chem. scand.* **25**, 1924.
ANTHONSEN, T. and BERGLAND, G. (1973). *Acta chem. scand.* **27**, 1073.
ANTHONSEN, T. and MCCRINDLE, R. (1969). *Acta chem. scand.* **23**, 1068.
ANTHONSEN, T., MCCABE, P. H., MCCRINDLE, R. and MURRAY, R. D. H. (1967). *Acta chem. scand.* **21**, 2289.
ANTHONSEN, T., MCCABE, P. H., MCCRINDLE, R. and MURRAY, R. D. H. (1969). *Tetrahedron* **25**, 2233.

ANTHONSEN, T., MCCABE, P. H., MCCRINDLE, R., MURRAY, R. D. H. and YOUNG, G. A. R. (1970a). *Tetrahedron* **26**, 3091.
ANTHONSEN, T., BRUUN, T., HEMMER, E., HOLME, D., LAMVIK, A., SUNDE, E. and SORENSEN, N. A. (1970b). *Acta chem. scand.* **24**, 2479.
ANTHONSEN, T., HENDERSON, M. S., MARTIN, A., MURRAY, R. D. H., MCCRINDLE, R. and MCMASTER, D. (1973). *Can. J. Chem.* **51**, 1332.
BARBARÁ, N. H., CADENAS, R. A. and GARCIA, P. T. (1974). *Phytochemistry* **13**, 671.
BATYUK, V. S. and KOLTSOVA, L. F. (1968). *Khim. Prir. Soedin.* **4**, 381.
BATYUK, V. S. and KOLTSOVA, L. F. (1969). *Khim. Prir. Soedin.* **5**, 121.
BOHLMANN, F. and GRENZ, M. (1972). *Chem. Ber.* **105**, 3123.
BOHLMANN, F. and RAO, N. (1972). *Chem. Ber.* **105**, 2421.
BOHLMANN, F. and ZDERO, C. (1969). *Tetrahedron Lett.*, 2419.
BOHLMANN, F. and ZDERO, C. (1972). *Chem. Ber.* **105**, 3587.
BOHLMANN, F., ZDERO, C. and KAPTEYN, H. (1968). *Liebigs Ann.* **717**, 186.
BOHLMANN, F., ZDERO, C. and KAPTEYN, H. (1969). *Chem. Ber.* **102**, 1689.
BOHLMANN, F., BURKHARDT, T. and ZDERO, C., eds (1973a). "Naturally Occurring Acetylenes". Academic Press, London and New York.
BOHLMANN, F., GRENZ, M. and SCHWARZ, H. (1973b). *Chem. Ber.* **106**, 2479.
CANONICA, L., RINDONE, B., SCOLASTICO, C., FERRARI, G. and CASAGRANDE, C. (1969). *Gazz. chim. ital.* **99**, 260, 276.
CRUSE, W. T. B., JAMES, M. N. G., AL-SHAMMA, A. A., BEAL, J. K. and DOSKOTCH, R. W. (1971). *Chem. Commun.* 1278.
DENNIS, R. (1973). *Phytochemistry* **12**, 2705.
DOMINGUEZ, X. A. and JIMENEZ, S. (1973). *Rev. latinoamer. Quim.* **3**, 179.
DOMINGUEZ, X. A. and TORRE, B. (1974). *Phytochemistry* **13**, 1624.
DOMINGUEZ, X. A., QUINTERO, G. and BUTRUILLE, D. (1972a). *Phytochemistry* **11**, 1855.
DOMINGUEZ, X. A., SANCHEZ, H., MERIJANIAN, B. A. and ROJAS, M. P. (1972b). *Phytochemistry* **11**, 2628.
FARNSWORTH, N. R., WAGNER, H., HÖRHAMMER, L. and HÖRHAMMER, H.-P. (1968). *J. pharm. Sci.* **57**, 1059.
GERLACH, H. (1965). *Pharmazie* **20**, 523.
HARBORNE, J. B., ed. (1967). "Comparative Biochemistry of the Flavonoids". Academic Press, London and New York.
HEGNAUER, R. (1964). "Chemotaxonomie der Pflanzen", Vol. 3. Birkhäuser Verlag, Basel and Stuttgart.
HENDERSON, M. S., MCCRINDLE, R. and MCMASTER, D. (1973a). *Can. J. Chem.* **51**, 1346.
HENDERSON, M. S., MURRAY, R. D. H., MCCRINDLE, R. and MCMASTER, D. (1973b). *Can. J. Chem.* **51**, 1322.
HISAMICHI, S. (1961). *Yakugaku Zasshi* **81**, 446.
HÖRHAMMER, L., WAGNER, H. WILKOMIRSKY, M. T. and IYENGAR, M. A. (1973). *Phytochemistry* **12**, 2068.
HUI, W. H., LAM, W. K. and TYE, S. M. (1971). *Phytochemistry* **10**, 903.
KAMISAKE, W. and TAKAHASHI, M. (1964). *Yakugaku Zasshi* **84**, 318.
KASPRZYK, Z. and KOZIEROWSKA, T. (1966). *Bull. Acad. Pol. Sci., Ser. Sci. Biol.* **14**, 645.

Kupchan, S. M. and Bauerschmidt, E. (1971). *Phytochemistry* **10**, 664.
Kusumoto, S., Okazaki, T., Ohsuka, A. and Kotake, M. (1969). *Bull. chem. Soc. Japan* **42**, 812.
McCabe, P. H., McCrindle, R. and Murray, R. D. H. (1969). *Tetrahedron* **25**, 2133.
McCrindle, R. and Nakamura, E. (1974). *Can. J. Chem.* **52**, 2029.
Miyakado, M., Ohno, N., Yoshioka, H., Mabry, T.-J. and Whiffin, T. (1974). *Phytochemistry* **13**, 189.
Mo, R., Anthonsen, T. and Bruun, T. (1972). *Acta chem. scand.* **26**, 1287.
Montes, G. M., Wilkomirski, F. T., Valenzuela, R. R. and Neira, M. R. (1971). *Rev. Real Acad. Cienc. Exactas, Fis. Natur. Madrid* **65**, 499. *Chem. Abstr.* **75**, 11588k.
Moreira, A. A. (1966). *Tribuna farm.* **34**, 27.
Ohsuka, A., Kusumoto, S. and Kotake, M. (1973). *Nippon Kagaku Kaishi* **590**, 631.
Okasaki, T., Ohsuka, A. and Kotake, M. (1973). *Nippon Kagaku Kaishi*, 584.
Pinhey, J. T., Simpson, R. F. and Baley, I. L. (1971). *Aust. J. Chem.* **24**, 2621.
Raffauf, R. F. (1970). "A Handbook of Alkaloids and Alkaloid-Containing Plants". Wiley-Interscience, New York, London, Sydney and Toronto.
Silva, M. and Sammes, P. G. (1973). *Phytochemistry* **12**, 1755.
Silva, M., Mundaco, J. M. and Sammes, P. G. (1971). *Phytochemistry* **10**, 1942.
Standifer, L. N., Devys, M. and Barbier, M. (1968). *Phytochemistry* **7**, 1361.
Tada, M., Takahashi, T. and Koyama, H. (1974). *Phytochemistry* **13**, 670.
Troshchenko, A. T. and Limasova, T. I. (1966). *Khim. Prir. Soedin.* **2**, 437.
Wagner, H., Iyengar, M. A. and Herz, W. (1972). *Phytochemistry* **11**, 444.
Wagner, H., Iyengar, M. A., Seligmann, O. and Herz, W. (1972). *Phytochemistry* **11**, 2350.
Zaikow, L. H., Burke, N., Kabat, G. and Grula, E. A. (1962). *J. Med. Chem.* **5**, 1342.
Zalkow, L. H., Burke, N. and Keen, G. (1964). *Tetrahedron Lett.*, 217.
Zalkow, L. H., Cabat, G. A., Chetty, G. L., Ghosal, M. and Keen, G. (1968). *Tetrahedron Lett.*, 5727.

ADDENDUM

Reference to a derodane lactone isolated from *Baccharis conferta* H.B.K. by C. Guerrero and A. Romo de Vivar. *Rev. latinoqmer. Quim.* **4**, 178 (1973) was omitted from the section on diterpenoids.

Chapter 21
Inuleae—systematic review

H. MERXMÜLLER*, P. LEINS† and H. ROESSLER*

Abstract. The tribe Inuleae (*c.* 180 genera with some 2100 spp.) has been reinvestigated. Pollen and style characters, together with karyological data, enable a clear-cut distinction to be made between the two enlarged subtribes, Inulinae *sensu amplo* and Gnaphaliinae *sensu amplo*, whereas the third one, Athrixiinae *sensu amplo*, takes a somewhat intermediate position. The problems of establishing a better generic concept are still unsolved but a putatively natural arrangement of generic groups is discussed. A revised list of all genera described in or transferred to the Inuleae is added; 23 genera are excluded.

CONTENTS

Introduction	577
List of genera	578
Important characters	587
Discussion	589
Inulinae	590
Gnaphaliinae	592
Athrixiinae	597
References	598

INTRODUCTION

The Inuleae is a rather well defined group whose delimitation has not been altered essentially since Bentham (1873). Their main character is considered to be the combination of usually caudiculate, or at least sagittate, anthers with styles belonging to the "*Vernonia*", *Senecio* or *Inula* type, respectively; in addition, the predominance of heterogamy, yellow flowers and setose or plumose pappus elements facilitates recognition when the main characters show some ambiguity. Defined in this way the tribe comprises *c.* 180–200 genera with some 2100 species. The centres of differentiation and distribution are S. Africa and Australia and to some extent

* State Herbarium, Munich, Germany.
† Botanical Institute, University of Bonn, Germany.

also S. America and the Mediterranean, whereas N. America, Eurasia and Indomalaysia play only a minor role.

The delimitation of subtribes has also been considered to be relatively simple until today. Again following Bentham, one distinguishes four subtribes with filiform female florets, i.e. Tarchonanthinae (dioecious trees and shrubs), Plucheinae ("*Vernonia*"-type styles), Filagininae (*Senecio*-type styles and paleae) and Gnaphaliinae (*Senecio*-type styles and epaleaceous); then there are the homogamous Angianthinae (pseudocephalia) and lastly four subtribes with mostly ligulate female florets, i.e. Relhaniinae (*Senecio*-type styles, ericoid), Athrixiinae (*Senecio*-type styles, not ericoid), Inulinae (*Inula*-type styles, epaleaceous) and Buphthalminae (*Inula*-type styles and paleae). Most difficulties have always been found in generic delimitation, which in many groups has remained unsolved to the present. Therefore, in the following "List of Genera" we have found it necessary to add a category of "semi-accepted" genera (bold type, indented) to the "accepted" ones (bold type at the beginning of the line); it would be dishonest to pretend here to have a better degree of knowledge.

Since the catalogue of Dalla Torre and Harms (1905) gives the last complete enumeration of known and accepted genera, the following list is based on it; genera described meanwhile are inserted in the places indicated by their authors. We have tried to include also genera hitherto placed in other tribes but which in our opinion belong to the Inuleae, as well as to exclude presumed "Inuleae" which we believe to have been placed wrongly here. Such excluded genera should by no means vanish into space but rather they merit, in many cases, further investigation. The numbers added to each genus at the end refer to the grouping in our "Discussion".

LIST OF GENERA

8755   *Thysanurus* O.Hoffm. → 9090 *Geigeria* (see Merxmüller, 1953).
8867/1   *Poilania* Gagnepain (1924) → 8953 *Epaltes divaricata* (L.) Cass., syn. nov.
8922 Haastia Hooker fil. (quoad *H. pulvinaris* Hooker fil., **Lectotypus** generis).—1 sp. (the two others seem to represent quite another genus), New Zealand.—?Near 8985 *Pterygopappus*.   **II (9/16?)**
8924 Adelostigma Steetz.—2 spp., trop. Africa. Type spec. not seen.—Near 8940 *Laggera* and 8954 *Porphyrostemma*.   **I (4)**
8926/1   *Pseudoconyza* Cuatr. (1961), syn.: *Ernstia* Badillo (1947), nomen.— 1 sp., Mexico to Northern S. America.—Included in 8940 *Laggera* by Leins (1971a), in 8939 *Blumea* by Badillo (1974), but isolated geographically.   **I (4)**
8934/1 Bellida Ewart (1907).—2 spp., Australia.—*n* = 9.—Near 8999 *Schoenia* (see Moore 1917).   **II (16)**

8935   *Synchodendron* Bojer ex DC. → 8936 *Brachylaena*.—Not Inuleae.

8936 *Brachylaena* R.Br.—Not Inuleae (see Leins, 1971a).
8937 *Tarchoanthus* L.—Not Inuleae (see Leins, 1971a).
8938 Stenachaenium Bentham.—4 spp., S. America. **I (5a)**
8939 Blumea DC. (incl. *Leveillea* Vaniot, *Bi-Leveillea* Vaniot). *c.* 75 spp., Africa, Asia and Australia.—*n* = 9, 10, 11, 18, 22, 27.—Rev.: Randeria (1960).
I (4)
8939/1 **Merrittia** Merrill (1910).—1 sp., Philippines.—Rather near to 8939 *Blumea*. **I (4)**
8939/2 **Blumeopsis** Gagnepain (1920).—1 sp., S.E. Asia. **I (4)**
8940 Laggera Schultz-Bip. ex Hochst.—*c.* 15 spp., Africa, Asia and Australia.—*n* = 10.—United with 8939 *Blumea* by Phillips (1951) and Wild (1968/69), but maintained by Randeria (1960) and Merxmüller (1967) **I (4)**
8940/1 *Petrollinia* Chiov. (1911) → 9061 *Inula* (see Cufodontis, 1967).
8941 Pluchea Cass. (incl. 8941/4 *Eremohylema* A. Nelson).—*c.* 40 spp., warmer parts of both hemispheres.—*n* = 10, 15. **I (4)**
8941/1 **Karelinia** Less.—1 sp., Western C. Asia and adj. European part of U.S.S.R.—*n* = 10.—Often included in *Pluchea*, but maintained by Russian authors. **I (4)**
8941/2 *Cavea* W.W.Smith & Small (1917).—Not Inuleae (pollen grains with infrategillar bacules: Leins, unpubl.).
8941/3 *Stera* Ewart (1912) → 9100a *Cratystylis*.
8941/4 *Eremohylema* A. Nelson (1924) → 8941 *Pluchea*.
8942 Tessaria Ruiz & Pavon.—1 sp., S. America.—See Robinson and Cuatrecasas (1973). **I (4)**
8943 Pechuel-Loeschea O.Hoffm.—1 sp., S.W. Africa to Rhodesia.—Included by Brown (1909), Phillips (1951) and Wild (1968/69) in 8941 *Pluchea*, but maintained by Merxmüller (1954a, 1967) and Leins (1971a). **I (4)**
8944 Sachsia Griseb.—3–4 spp., Cuba and the Bahamas. **I (5)**
8945 Rhodogeron Griseb.—1 sp., Cuba.—Not seen. **I (5?)**
8946 Pterigeron (DC.) Bentham.—9 spp., Australia. *n* = 10. **I (4)**
8947 Thespidium F. Mueller.—1 sp., Australia. **I (4)**
8948 Coleocoma F. Mueller.—1 sp., Australia. **I (4)**
8949 Denekia Thunb.—1 sp., Southern Africa.—Near 8972 *Amphidoxa* (see Leins 1971a). **II (9)**
8950 Delamerea S. Moore.—1 sp., E. Africa. **I (4)**
8951 Nicolasia S. Moore.—6 spp., Africa.—Rev.: Merxmüller 1954b. **I (4)**
8952 Nanothamnus Thomson.—1 sp., India. **I (4)**
8953 Epaltes Cass. (incl. 8867/1 *Poilania* Gagnepain).—*c.* 15 spp., warmer parts of both hemispheres.—*n* = 10.—Heterogeneous (see Leins, 1971a). **I (4)**
8954 Porphyrostemma Bentham ex Oliver.—4 spp., tropical Africa. **I (4)**
8954/1 *Aostea* Buscalioni & Muschler (1913) → *Vernonia* (see Volkens in *Bot. Jahrb.* (1915) **53**, 374).
8955 Sphaeranthus L. (incl. 9140/1 *Tisserantia* Humbert).—40 spp., trop. Africa, S. Asia to Australia.—*n* = 10.—Rev.: Ross-Craig (1955). **I (8)**
8956 Triplocephalum O.Hoffm.—1 sp., E. Africa. **I (8)**
8957 Pterocaulon Elliot.—*c.* 25–30 spp., trop. America, Madagascar and Mauritius, trop. Asia and Australia.—*n* = 10. **I (8)**
8958 Monarrhenus Cass.—3 spp., Mascarenes. **I (8)**

8959 Cylindrocline Cass.—1 sp., Mauritius. **I (8)**
8960 Blepharispermum Wight ex DC.—*c.* 10 spp., trop. Africa, Madagascar, Arabia and India. **I (8)**
8961 Athroisma DC. (incl. 9034/1 *Aetheocephalus* Gagnepain, 9035 *Polycline* Oliver).—8 spp., Africa and S.E. Asia.—See Mattfeld (1936). **I (8)**
8962 Symphyllocarpus Maxim.—Not Inuleae (pollen grains with infrategillar bacules: Leins, unpubl.).
8963 Micropus L.—1 sp., Mediterranean to S.W. Asia. **II (13)**
8963/1 Bombycilaena (DC.) Smoljan. (1955) (=*Micropus* sect. *Bombycilaena*).—3 spp., Mediterranean to C. Asia, pacif. N. America. **II (13)**
8963/2 Cymbolaena Smoljan. (1955) (=*Micropus* sect. *Cymbolaena* (Smoljan.) Wagenitz).—1 sp., W. Asia to C. Asia.—Near 8964 *Stylocline* (see Wagenitz, 1971). **II (13)**
8964 Stylocline Nutt.—6 spp., Western U.S.A. and Mexico—Lacking in the Old World (see Wagenitz, 1971). **II (13)**
8965 Psilocarphus Nutt.—5 spp., Western N. America, Chile.—$n=14$. **II (13)**
8966 Evax Gaertner.—*c.* 15–20 spp., Mediterranean, N. America.—$n=13$. —Included in 8969 *Filago* by Wagenitz (1969). **II (13)**
8967 Ifloga Cass. (incl. 8996 *Petalactella* N.E.Br.).—11 spp., most in S. Africa, one from the Canary Islands to India.—$n=7$. **II (15)**
8968 Micropsis DC.—5 spp., S. America (see Beauverd, 1913b). **II (13)**
8969 Filago L. (sensu lato)—*c.* 50 spp., Eurasia, N. Africa and southwestern N. America.—$n=13$, 14.—The following scheme gives the taxonomic treatment by Wagenitz (1969) and includes the taxa sometimes given generic rank: *Filago* subgen. *Filago* ($n=14$) with sect. *Filago*, sect. *Gifolaria* Cosson & Kralik (=*Gifolaria* (Cosson & Kralik) Pomel), sect. *Evacopsis* (Pomel) Batt. (=*Evacopsis* Pomel, *Evax* sect. *Evacella* Smoljan.), sect *Filaginoides* (Smoljan.) Wagenitz (=*Evax* sect. *Filaginoides* Smoljan.);—subgen. *Evax* (Gaertner) Wagenitz ($n=13$), see 8966;—subgen. *Oglifa* (Cass.) Gren. ($n=14$), see 8969/1.
 II (13)
8969/1 Logfia Cass. (=*Oglifa* Cass.)—*c.* 10 spp., Eurasia, N. Africa, N. America.—$n=14$. **II (13)**
— *Pseudevax* Pomel (1888) (=*Filagopsis* (Batt.) Rouy) and *Giflifa* Chrtek & Holub (1963) are "generic" hybrids *Filago* × *Evax* and *Filago* × *Logfia*, respectively.
8969/2 Evacidium Pomel.—1 sp., Sicily and N. Africa. **II (13)**
8969/3 Lifago Schweinf. & Muschler (1911) (=9085/2 *Niclouxia* Batt. 1915). —1 sp., N. Africa.—Certainly not a member of the Filago group, rather near 9085 *Anvillea*.
8970 Gymnarrhena Desf.—Not Inuleae (see Leins, 1973).
8971 Stuartina Sonder.—2 spp., Australia.—On the relationship of "*Epaltes* *tatei* F.Mueller see Leins (1971a). **II (18)**
8972 Amphidoxa DC.—*c.* 6 spp., Southern Africa and Madagascar.— Near, if not belonging, to the "*Gnaphalium declinatum*—group" ("*Helichrysum* sect. *Declinata*"), see Hilliard and Burtt (1973) **II (9)**
8973 Artemisiopsis S.Moore.—1 sp., trop. and S.W. Africa. **II (9)**
8974 Chiliocephalum Bentham → 9006 *Helichrysum* (see Hilliard and Burtt, 1973).

8975 *Chevreulia* Cass.—*c.* 6 spp., S. America. **II (11)**
8976 *Loricaria* Wedd. (= *Tafalla* D.Don).—17 spp., Andean S. America.—
$n = 14$.—Rev.: Cuatrecasas (1954a). **II (12)**
8977 *Luciliopsis* Wedd.—4 spp., S. America. **II (11)**
8978 *Antennaria* Gaertner.—*c.* 40–50 spp. (with extreme splitting of the apomicts more than 100), Holarctic and a few spp. in the Andean S. America.—
$n = 7$, 12, 13, *14*, *21*, 24, *28*, 30, *35*, 40, *42*, 50, 52, *63*. **II (10)**
8978/1 *Parantennaria* Beauverd (1911).—1 sp., Australia.—Not seen.
II (9?)
8979 *Mniodes* (A.Gray) Bentham & Hooker.—5 spp., Andean Peru.—see Cuatrecasas (1954b) **II (11)**
8980 *Chionolaena* DC.—*c.* 8 spp., Mexico, Colombia, Brazil.—Very near 9004 *Leucopholis* (see Cufodontis, 1933).—Under revision (Stockholm). **II (11)**
8981 *Oligandra* Less.—3 spp., S. America. **II (11)**
8982 *Leontopodium* R.Br. ex Cass.—*c.* 30–40 spp., mountains of Europe, C. and E. Asia, Japan; 2 spp. (not seen) in Andean S. America?—$n = 12$, 13, 26, 52.—Rev.: Handel-Mazzetti (1927). **II (10)**
8983 *Anaphalis* DC. (incl. 8984 *Nacrea* A.Nelson).—*c.* 60—100 spp., most in Asia (in the S. until New Guinea), a few in N., (C. and S.?) America.—
$n = 14$, ?28.—Problematic, see Beauverd (1913a) and Grierson (1972). Cf. 8992/1 *Anaphalioides*, too. **II (10)**
8984 *Nacrea* A. Nelson → 8983 *Anaphalis* (see Rydberg, 1954).
8985 *Pterygopappus* Hooker fil.—1 sp., Tasmania. **II (9?)**
8986 *Facelis* Cass.—4 spp., S. America.—see Beauverd (1913b). **II (11)**
8987 *Lasiopogon* Cass.—2 spp., S. and S.W. Africa, one of them in the Saharo-Sindian area, too.—see Nordenstam (1964). **II (9)**
8987/1 *Comptonanthus* B.Nord. (1964).—3 spp., S. and S.W. Africa. **II (9)**
8988 *Phagnalon* Cass.—*c.* 30 spp., Macaronesia to C. Asia.—$n = 9$.
?III (25)
8988/1 *Pseudocrupina* Velen. (1923) → 9052 *Leysera leyseroides* (Desf.) Dandy, **syn. nov.**
8989 *Psila* Phil.—Astereae (near *Baccharis*: see Cabrera, 1955).
8990 *Achyrocline* Less.—*c.* 20–30 spp., C. and S. America, ?Africa, Madagascar.—$n = 14$.—Probably heterogeneous, cf. 9003 *Stenocline* **II (9, 14)**
8991 *Lucilia* Cass. (incl. 9010 *Pachyrhynchus* DC.).—*c.* 20 spp., S. America.
II (11)
8991/1 *Belloa* Remy.—11 spp., S. America.—see Cabrera (1958). **II (11)**
8991/2 *Berroa* Beauverd (1913).—1 sp., S. America. **II (11)**
8992 *Gnaphalium* L. (sensu lato)—*c.* 150 spp., world-wide.—$n = 7$, 14, 21, 28.—A badly understood complex, see Cabrera (1961a), Drury (1970, 1971, 1972), Kirpichnikov and Kuprijanova (1950), Wagenitz (1965). The following scheme enumerates the sections accepted by Wagenitz (1965; Drury's "gnaphalioid cudweeds"), given generic rank by some authors: *Gnaphalium* sect. *Gnaphalium* (= *Filaginella* Opiz), sect. *Omalotheca* (Cass.) Endl. (= **Omalotheca** Cass.), sect. *Synchaeta* (Kirp.) Wagenitz (= **Synchaeta** Kirp.), sect. *Gamochaeta* (Wedd.) Bentham (= **Gamochaeta** Wedd.), sect. *Euchiton* (Cass.) DC. (= **Euchiton** Cass.).—A reinvestigation of the African and many American spp. is

urgently needed.—The sect. *Calolepis* Kirp., based on *G. luteo-album* L., should be excluded ("achyroclinoid cudweeds", Drury 1970). **II (9)**

8992/1 ***Anaphalioides*** (Bentham) Kirp. (1950).—Several spp., Australia and New Zealand.—"Anaphalioid cudweeds" Drury's (1970), near 8983 *Anaphalis*. **II (10)**

8992/2 ***Gnaphaliothamnus*** Kirp. (1950).—1 sp., Mexico.—Like the following monotypic splits Kirpichnikov's to be reconsidered in an overall investigation of *Gnaphalium*. **II (11 ?)**

8992/3 ***Helichrysopsis*** Kirp. (1950).—1 sp., S.E. Africa. **II (9)**

8992/4 ***Homognaphalium*** Kirp. (1950).—1 sp., N. Africa. **II (9)**

8992/5 ***Hypelichrysum*** Kirp. (1950).—1 sp., S. America.—Not seen. **II (?)**

8992/6 ***Pseudognaphalium*** Kirp. (1950).—1 sp., C. and S. America. **II (9)**

8992/7 ***Stuckertiella*** Beauverd (1913).—2 spp., S. America. **II (9)**

8993 ***Raoulia*** Hooker fil. (incl. 8993/1 *Psychrophyton* Beauverd).—20 spp., New Zealand.—See Beauverd (1912), Koster (1972). **II (9)**

8993/1 *Psychrophyton* Beauverd (1910) → 8993 *Raoulia* (subgen. *Psychrophyton*), see Beauverd (1912), Allan (1961).

8993/2 ***Leucogenes*** Beauverd (1910).—2 spp., New Zealand.—$n=14$. **II (9)**

8993/3 ***Ewartia*** Beauverd (1910).—5 spp., Australia, Tasmania and New Zealand. **II (9)**

8993/4 ***Raouliopsis*** S.F.Blake (1938).—2 spp., Andean Colombia.—Not seen. **II (11 ?)**

8994 ***Cassinia*** R.Br.—*c.* 20 spp., Australia and New Zealand. **II (14)**

8994/1 ***Rhynea*** DC.—1 sp., S. Africa.—see Hilliard and Burtt (1973). **II (14)**

8994/2 ***Basedowia*** E.Pritzel (1918).—1 sp., Australia.—Included in "Angianthinae" by Black (1943/57) and Besold (1971). **II (18)**

8995 ***Petalacte*** D.Don.—1 sp., S. Africa.—Rev.: Lundgren (1974). **?III (24)**

8996 *Petalactella* N.E.Br. → 8967 *Ifloga* (see Hilliard and Burtt, 1971).

8997 ***Phaenocoma*** D.Don.—1 sp., S. Africa.—$n=8$. **?III (24)**

8998 ***Anaxeton*** Gaertner.—9 spp., S. Africa.—Rev.: Lundgren (1972). **?III (24)**

8999 ***Schoenia*** Steetz.—1 sp., Australia.—$n=12$. **II (16)**

9000 ***Helipterum*** DC. (incl. 9000/1 *Griffithia* Black).—*c.* 60 spp., S. Africa and Australia.—$n=7$ (S. Africa); $n=5, 7, 8, 10, 11, 12, 14, 24$ (Australia).—Perhaps heterogeneous, partly very near 9006 *Helichrysum*, partly near Schoenia group; reinvestigation needed. **II (14/16)**

9000/1 *Griffithia* Black (1913) → 9000 *Helipterum* (see Black, 1943/57).

9001 ***Gratwickia*** F.Mueller.—1 sp., Australia.—Included in 9006 *Helichrysum* by Eichler (1963). **II (14)**

9002 ***Eriosphaera*** Less.—1 sp., S. Africa. **II (14)**

9003 ***Stenocline*** DC.—6 spp., Brazil, Madagascar, Mauritius ("S. Africa" goes back to a printing error in *Ind. Kew*.—The species named there belongs to *Stenoglottis*, Orchid.).—Probably heterogeneous, cf. the closely related 8990 *Achyrocline*. **II (9, 14)**

9004 ***Leucopholis*** Gardner.—4 spp., Brazil.—Very near 8980 *Chionolaena* (see Cufodontis, 1933).—Under revision (Stockholm). **II (11)**

9005 *Ixiolaena* Bentham.—6 spp., Australia.—*n*=21.   **II (14/16)**
9006 *Helichrysum* Miller corr. Pers. (sensu lato; incl. 8974 *Chiliocephalum* Bentham).—*c*. 500 spp., Eurasia, Africa, Madagascar, Australia and New Zealand.—*n*=7, 14, 28 (old world); *n*=8, 10, 11, 12, 14, 15, 38 (Austral.).—Together with *Calomeria, Cassinia, Helipterum, Leontonyx* and *Rhynea* (as well as with some members of *Gnaphalium* sensu lato and of the *Schoenia* group) a poorly understood complex which urgently needs reinvestigation. The same is true for the 3 following splits, based on ±identical pappus aberrations.—see Hedberg (1957), Hilliard and Burtt (1973), Moeser (1909).   **II (14/16)**
9006/1   *Cladochaeta* DC.—2 spp., Russia.—Mostly included in 9006, but maintained in Fl.U.S.S.R.   **II (14)**
9006/2   *Acanthocladium* F.Mueller.—1 sp., Australia.—Not seen.—Mostly included in 9006, but see Burbidge (1958)   **II (14?)**
9006/3   *Argyroglottis* Turcz.—1 sp., Australia.—Not seen.—Mostly included in 9006, but see Burbidge (1958).   **II (14?)**
9007 *Phacellothrix* F. Mueller.—cf. Astereae.
9008   *Leontonyx* Cass.—*c*. 5–8 spp., S. Africa.—Scarcely separable from 9006 *Helichrysum* (see Moeser, 1909; Merxmüller, 1954a).   **II (14)**
9009 *Podotheca* Cass., **nom. conserv.** (=*Podosperma* Labill.)—6 spp., Australia.—*n*=13, 26.   **II (16)**
9010   *Pachyrhynchus* DC.→ 8991 *Lucilia acutifolia* (Poiret) Cass.—see Nordenstam, *Bot. Notiser* (1977) **129**, 428.
9011 *Leptorhynchos* Less.—9 spp., Australia.—*n*=12, 19, 20.   **II (14/16)**
9011/1 *Gilruthia* Ewart (1909).—1 sp., Australia.—Allegedly connecting "Angianthinae" and "Gnaphaliinae".—Not seen.   **II (14?)**
9012 *Waitzia* Wendl.—6 spp., Australia.—*n*=10, 12.   **II (16)**
9013 *Millotia* Cass.—4 spp., Australia.—*n*=8, 10, 11, 13.—Rev.: Schodde (1963)   **II (16)**
9014 *Quinetia* Cass.—1 sp., Australia.—*n*=12.   **II (16)**
9015 *Rutidosis* DC.—7 spp., Australia.—*n*=19.   **II (14/16)**
9016 *Scyphocoronis* A.Gray.—2 spp., Australia.   **II (16)**
9017 *Ammobium* R.Br.—2–3 spp., Australia.—*n*=12 or 13.   **II (14/16)**
9018 *Toxanthes* Turcz.—2 spp., Australia.   **II (16)**
9019 *Calomeria* Vent. (=*Humea* Sm., see Heine in *Adansonia* (1967) (ser. 2) **7**, 137).—10 spp., E. and S. Africa, Madagascar, Australia.—Near 9006 *Helichrysum* (especially the African ones), see Hilliard and Burtt (1973).
    II (14/16)
9020   *Astephanocarpa* Baker → 9040 *Syncephalum* (see Humbert, 1962).
9021 *Eriochlamys* Sonder & F.Mueller.—2 spp., Australia.—Included in "Angianthinae" by Black (1943/57) and Besold (1971).   **II (18)**
9022 *Acomis* F.Mueller.—3 spp., Australia.   **II (16)**
9022a *Neotysonia* Dalla Torre & Harms (1905) (=*Tysonia* F.Mueller non Bolus; *Swinburnia* Ewart (1907)).—1 sp., Australia.—Probably near 9017 *Ammobium*.   **II (14/16)**
9023 *Pithocarpa* Lindley.—4 spp., Australia.—Rev.: Lewis and Summerhayes (1951).   **II (14/16)**
9023a *Thiseltonia* Hemsley.—1 sp., Australia.—Not seen.   **II (16)**
9024 *Ixodia* R.Br.—2 spp., Australia.   **II (14/16)**

9024/1 *Tugarinovia* Iljin (1928).—Not Inuleae (pollen grains with infrategillar bacules: Leins, unpubl.).
9025 Caesulia Roxb.—1 sp., India.—$n=7$.—see Besold (1971), Leins (1971a)
    I (8)

9026 *Dimeresia* A.Gray.—Heliantheae-Helenieae (see Johnston, 1923; Skvarla and Turner, 1966; Besold, 1971).
9027 Myriocephalus Bentham.—10 spp., Australia.—$n=8$, 10.   **II (18)**
9028 Angianthus Wendl.—c. 30 spp., Australia.—$n=6$.   **II (18)**
9029 Gnephosis Cass.—c. 12 spp., Australia.—$n=4$, 12.   **II (18)**
9030 Calocephalus R.Br.—c. 15 spp., Australia.—$n=7$, 14.   **II (18)**
9031 Actinobole Fenzl ex Endl. (=*Gnaphalodes* A.Gray, non Miller; see Eichler, 1963).—3 spp., Australia.—$n=10$.   **II (18)**
9032 Cephalipterum A.Gray.—1 sp., Australia.—$n=12$, 14.   **II (18)**
9033 Craspedia Forster fil.—6 spp., Australia, Tasmania and New Zealand.—$n=11$, 70.—"Near Heliantheae" (Skvarla and Turner, 1966), but see Besold (1971).   **II (18)**
9034 Chthonocephalus Steetz.—4 spp., Australia.   **II (18)**
9034/1   *Aetheocephalus* Gagnepain (1920) → 8961 *Athroisma laciniatum* DC., **syn. nov.**
9035   *Polycline* Oliver → 8961 *Athroisma* (see Mattfeld, 1936).
9036 Decazesia F.Mueller.—1 sp., Australia   **II (18)**
9037 Stoebe L. (incl. 9038 *Perotriche* Cass.).—34 spp., C. and S. Africa, Madagascar, Reunion.—Rev.: Levyns (1937).   **III (21)**
9038   *Perotriche* Cass. → 9037 *Stoebe* (see Levyns, 1937).
9039 Disparago Gaertner.—7 spp., S. Africa. $n=9$.—Rev.: Levyns (1936).
    III (21)

9040 Syncephalum DC. (incl. 9020 *Astephanocarpa* Baker).—5 spp., Madagascar.   **II (14)**
9040/1 Catatia H.Humb. (1923).—2 spp., Madagascar.—Not seen.   **II (14)**
9041 Elytropappus Cass.—8 spp., S. Africa.—Rev.: Levyns (1935).   **III (21)**
9042 Pterothrix DC.—4 spp., S. Africa.   **III (21)**
9043 Metalasia R.Br.—33 spp., S. Africa.—Rev.: Pillans (1954)   **?II (14)**
9044   *Cullumiopsis* Drake del Castillo → *Dicoma* (Mutisieae; see Humbert, 1962).
9045 Lachnospermum Willd.—1 or 2 spp., S. Africa.   **?II (14)**
9046 Amphiglossa DC.—4 spp., S. Africa.   **III (21)**
9047 Bryomorphe Harvey.—1 sp., S. Africa.   **III (21)**
9048   *Nestlera* Sprengel → 9050 *Relhania* (see Bremer, 1976a).
9049   *Anaglypha* DC. (quoad speciem typicam) → *Gibbaria scabra* (Thunb.) T. Norl.; the remaining species are near 9055 *Athrixia* and 9096 *Anisopappus*—see Hilliard and Burtt (1976).
9050 Relhania L'Hér. (incl. 9048 *Nestlera* Sprengel pro parte, 9099 *Osmites* L. quoad typum).—29 spp., S. Africa.—$n=5$, 7.—Rev.: Bremer (1976a).   **III (20)**
9051 Rosenia Thunb. (incl. 9048 *Nestlera* Sprengel pro parte).—4 spp., S. Africa.—$n=7$, 14, 28.—Rev.: Bremer (1976b).   **III (20)**
9051/1   *Zoutpansbergia* Hutch. (1946) → 9094 *Callilepis* (see Leins, 1971b).
9052 Leysera L. (=*Astropterus* Adanson, see Rothmaler in *Feddes Repert.* (1944), **53**, 4; both names July/Aug. 1763, see Stafleu (1967), "Taxonomic

Literature", pp. 2 & 287.—Incl. 8988/1 *Pseudocrupina* Velen.).—4–5 spp., one saharo-sindian, the others in S. Africa.—*n*=4. **III (20)**

9053 Macowania Oliver (incl. 9062 *Homochaete* Bentham).—11 spp., S. Africa, Ethiopia and Arabia (see Burtt and Grau, 1972).—Rev.: Hilliard and Burtt (1976). **III (20)**

9054 Podolepis Labill.—18 spp., Australia.—*n*=3, 7, 8, 9, 10, 11, 12, 30.—Rev.: Davis (1957). **II (17)**

9055 Athrixia Ker-Gawler.—*c.* 20 spp., E. and S. Africa, Madagascar; Australia (=*Asteridea* Lindley).—*n*=9 (Austral.).—Under revision (Munich). **III (20)** The Australian species only seem to be near 9054 *Podolepis* (unpubl.). **II (17)**

9056 Antithrixia DC.—1 sp., S. Africa, see Burtt and Grau (1972). **III (20)**

9057 Heterolepis Cass.—Not Inuleae (?Mutisieae: see Merxmüller, 1950; Arctotideae: see Besold, 1971).

9058 Arrowsmithia DC.—1 sp., S. Africa. **III(20)**

9058/1 Alatoseta Compton (1931).—1 sp., S. Africa. **III (22)**

9059 Printzia Cass. (incl. *Bojeria* pro parte, see Leins, 1971a). —*c.* 5 spp., S. Africa. **III (19)**

9060 *Codonocephalum* Fenzl → 9061 *Inula* (see Grierson, 1974).

9061 Inula L. (sensu lato; incl. 8940/1 *Petrollinia* Chiov., 9060 *Codonocephalum* Fenzl, *Monactinocephalus* Klatt).—*c.* 80–100 spp., Eurasia (Canary Islands to Japan), Africa and Madagascar.—*n*=8, 9, 10, 16, 20.—A badly delimited complex needing a thorough revision. The following splits have been given generic rank by several authors and seem to merit recognition. **I (1)**

9061/1 **Limbarda** Adanson (*Inula* sect. *Enula* pro parte).—1 sp., Mediterranean.—*n*=9.—see Briquet and Cavillier (1917). **I (1)**

9061/2 **Schizogyne** Cass. (*Inula* sect. *Cappa* pro parte).—2 spp., Canary Islands.—*n*=9. **I (1)**

9061/3 **Pentanema** Cass. (*Inula* sect. *Vicoa* pro parte).—*c.* 10 spp., from W. Africa to India.—*n*=9. **I (1)**

9061/4 Pentatrichia Klatt (*Inula* sect. *Vicoa* pro parte).—4 spp., Southern Africa.—See Merxmüller (1954a), Leins (1971a). **III (19)**

9061/5 Dittrichia Greuter (1973) (=*Cupularia* Godron & Gren. non Link; *Inula* sect. *Cupularia*).—2 spp., Mediterranean.—*n*=9, 10. **I (1)**

9062 *Homochaete* Bentham → 9053 *Macowania* (see Phillips (1950), *J. S. Afr. Bot.* **16**, 21).

9063 Varthemia DC.—*c.* 6 spp., Saharo-Sindian area.—*n*=8.—Some species may belong to 9075 *Jasonia*, see Botschantzev (1964). **I (1)**

9064 *Minurothamnus* DC. → 9057 *Heterolepis*.—Not Inuleae, see Merxmüller (1950).

9065 Iphiona Cass.—*c.* 10 spp., from N. E. Africa to C. Asia; ? Madagascar. **I (1)**

9065/1 Antiphiona Merxm. (1954).—2 spp., S. W. Africa. **I (1)**

9066 Hirschia Baker.—1 sp., S. Arabia. **I (1)**

9067 *Cypselodontia* DC. → *Dicoma picta* (Thunb.) Druce (Mutisieae), **syn. nov.**

9068 Anisochaeta DC.—1 sp., S. Africa.—Wrongly placed in Mutisieae by Bentham (1873) and Jeffrey (1967), see Leins (1971a), Marticorena and Parra (1974) **III (19)**

9069 *Calostephane* Bentham.—4 spp., Southern Africa. **I (1)**
9069/1 **Sclerostephane** Chiov. (1929).—2 spp., Somalia.—Near 9078 *Pulicaria*, not to *Calostephane*. **I (1)**
9070 *Grantia* Boiss.—6 spp., Arabia to Iran. **I (1)**
9071 *Perralderia* Cosson.—1 sp., N. Africa.—$n=9$. **I (1)**
9071/1 *Fontquera* Maire (1931).—1 sp., N. Africa. **I (1)**
9072 **Anisothrix** O.Hoffm.—1 sp., S. Africa.—Very near 9073 *Pegolettia*; *P. dentata* Bolus (1906) is a later synonym. **I (1)**
9073 *Pegolettia* Cass.—6 spp., Africa, one of them also Madagascar, subtrop. Asia to Java.—$n=10$. **I (1)**
9074 *Vieraea* Schultz-Bip.—1 sp., Canary Islands. **I (1)**
9075 *Jasonia* Cass.—4–5 spp., Mediterranean.—$n=9$. **I (1)**
9076 *Pelucha* S.Watson.—1 sp., California. **I (1)**
9077 *Allagopappus* Cass.—2 spp., Canary Islands.—$n=10$. **I (1)**
9078 *Purlicaria* Gaertner.—c. 40 spp., Eurasia and Africa.—$n=7$ (sect. *Poloa*), 9, 10. **I (1)**
9078/1 **Francoeuria** Cass. (*Pulicaria* sect. *Francoeuria*).—Several spp., Cape Verde Islands to Tibet.—$n=9, 10$. **I (1)**
9078/2 *Platychaete* Boiss. (*Pulicaria* sect. *Platychaete*).—c. 8 spp., Arabia to India. **I (1)**
9079 *Mollera* O.Hoffm.—4 spp., trop. Africa and Madagascar. **I (1)**
9080 *Amblyocarpum* Fischer & Meyer.—1 sp., Caspian Region **I (1)**
9081 *Carpesium* L.—20 spp., from Europe to Japan, Java and New Guinea.—$n=10, 18, 20$. **I (3)**
9082 *Adenocaulon* Hooker.—Not Inuleae (see Leins, 1971a).
9083 *Philyrophyllum* O.Hoffm.—1 sp., Southern Africa. **III (19)**
9084 *Rhanterium* Desf. (probably incl. 9093/2 *Musilia* Velen.).—2 spp., N. Africa to C. Asia.—$n=12$. **I (1)**
9085 *Anvillea* DC.—2 spp., N. Africa to Iran. **I (1)**
9085/1 **Anvilleina** Maire (1940).—1 sp., N. Africa.—Not seen. **I (1 ?)**
9085/2 **Niclouxia** Batt. (1915) → 8969/3 *Lifago* (see Jahandiez and Maire, 1934).
9086 **Astephania** Oliver → 9096 *Anisopappus* (see Taylor (1933) in *J. Bot., Lond.* **71**, 165; Wild, 1964).
9087 **Sphacophyllum** Bentham → 9096 *Anisopappus* (see Taylor (1933) in *J. Bot., Lond.* **71**, 165; Wild, 1964).
9088 **Eenia** Hiern & S.Moore → 9096 *Anisopappus* (see Merxmüller, 1954a).
9089 *Ondetia* Bentham.—1 sp., S. W. Africa. **I (2)**
9090 *Geigeria* Griesselich (incl. 8755 *Thysanurus* O.Hoffm.).—28 spp., mainly Southern Tropical and S. Africa, a few extending to N. Africa and S. Arabia.—$n=10$.—Rev.: Merxmüller (1953). **I (2)**
9091 *Pallenis* Cass.—1 sp., Mediterranean.—$n=5$. **I (1)**
9092 *Buphthalmum* L.—2 spp., Europe.—$n=10$. **I (1)**
9092/1 *Telekia* Baumg.—2 spp., S. Europe to Caucasus.—$n=10$. **I (1)**
9093 *Asteriscus* Miller (=*Odontospermum* Necker pro parte).—3 spp., Mediterranean.—$n=6$. **I (1)**
9093/1 **Bubonium** Hill (=*Odontospermum* Necker pro parte).—c. 8 spp., Mediterranean.—$n=7, 8, 9$.—Mostly included in *Asteriscus*, but maintained by

Briquet and Cavillier (1917) and some other authors. **I (1)**
9093/2 *Musilia* Velen. (1923).—1 sp., Arabia.—cf. 9084 *Rhanterium epapposum* Oliver (Burtt, in litt.).
9094 Callilepis DC. (incl. 9051/1 *Zoutpansbergia* Hutch.).—5 spp., S. Africa.
III (23)
9095 Chrysophthalmum Schultz-Bip.—3 spp., W. Asia. **I (1)**
9096 Anisopappus Hooker & Arn. (incl. 9086 *Astephania* Oliver, 9087 *Sphacophyllum* Bentham, 9088 *Eenia* Hiern & S.Moore, 9176 *Temnolepis* Baker, 9194 *Epallage* DC.).—c. 30 spp., Africa and Madagascar, one of them in Asia, too.—n = 14.—Somewhat heterogenous.—Rev.: Wild (1964). **I (1)**
9097 Postia Boiss. & Blanche.—4 spp., W. Asia. **I (1)**
9098 *Osmitopsis* Cass. emend. Bremer—Anthemideae (see Bremer, 1972).
9099   *Osmites* auct. non L.→ 9098 *Osmitopsis* (see Bremer, 1972). See also 9050 *Relhania*.
9100 Nablonium Cass.—1 sp., Tasmania.—See Leins (1971a). **II (18)**
9100/1   *Symphipappus* Klatt (1896) → *Cadiscus* (Senecioneae).
9100a Cratystylis S.Moore (=8941/3 *Stera* Ewart).—3 spp., Australia.—See Black (1943/57). **I (7)**
9140/1   *Tisserantia* Humbert (1927) → 8955 *Sphaeranthus*, **syn. nov.** (see also Cronquist, 1955, p. 480).
9176   *Temnolepis* Baker → 9096 *Anisopappus* (see Taylor (1933) in *J. Bot., Lond.* **71**, 165; Wild, 1964).
9194   *Epallage* DC. → 9096 *Anisopappus* (see Taylor (1933) in *J. Bot., Lond.* **71,** 165, Wild, 1964).
9354   Sphaeromorphaea DC.—1 sp., E. and S. E. Asia to Australia.—Identical with 8953 *Epaltes australis* Less. (see Gagnepain, 1924; Kitamura, 1937); aff. 8947 *Thespidium* (see Leins, 1971a). **I (4)**
9373 *Eremothamnus* O.Hoffm.—"Inulinae" (see Moore, 1929), "near *Ondetia*" (see Merxmüller, 1954a)—cf. Arctotideae (see Leins, 1970), "Eremothamneae" (see Robinson and Brettell, 1973b).
9378 *Nannoglottis* Maxim.—Wrongly described as an anomalous Inulea by Maximowicz, later treated as senecionid.—cf. Astereae.
9430 *Eriachaenium* Schultz-Bip.—"Inuleae-Adenocauline" (see Cabrera, 1961b), "Mutisieae" (see Robinson and Brettell, 1973a).
9485/1 Feddea Urban (1925).—1 sp., Cuba.—Wrongly placed in Mutisieae by Urban, attributed to Inuleae by Marticorena and Parra (1974). **I (6)**

Excluded taxa:

Adenocauline, Symphyllocarpinae, Tarchonanthinae; *Adenocaulon, Anaglypha, Aostea, Brachylaena, Cavea, Cullumiopsis, Cypselodontia, Dimeresia, Eremothamnus, Eriachaenium, Gymnarrhena, Heterolepis, Minurothamnus, Nannoglottis, Osmites, Osmitopsis, Phacellothrix, Psila, Symphipappus, Symphyllocarpus, Synchodedron, Tarchonanthus, Tugarinovia.*

Important characters

The well known contributions of Wodehouse, Erdtman, Stix and Skvarla have revealed the taxonomically relevant diversity of pollen grain structure in Compositae. The investigations on Inuleae pollen, made by Leins and

Besold in recent years, have covered nearly all described genera and some 800 spp. It was somewhat surprising to find that throughout the whole tribe the structure is fairly uniform, exhibiting two variants only of the same basic type. All "true" Inuleae show spinous (very rarely spinulose) pollen grains and spines with a baculate base and a more solid top; the sexine of the (acetolysed) grain is joined with the nexine at the margins of the colpi only. In three of the conventional subtribes the sexine is one-layered, baculate and mostly with dome-like arrangement of thickenings in the spine-bases; in all the others it is two-layered with an outer baculate layer and an inner granular or irregular one. As we did not find any transitions between these types we attached great importance to them; on the other hand, the basic uniformity of the tribe seems to require the elimination of taxa with a quite aberrant exine structure (see the "excluded taxa" named above).

There is a sharp contrast between this basic uniformity of the pollen grains and the diversity of stylar structures (in hermaphrodite florets) which always carry weight in the Compositae but seem to be poorly scrutinized in many groups—one needs series of cross-sections, at least in critical cases. In the Inuleae, the only common character is two marginal stigmatic rows. They fuse at the top of the stylar arm in the Inulinae/Buphthalminae (*Inula* type) as well as in the Plucheinae ("*Vernonia*"-type); the difference in the latter is confined to a more gradual tapering of the stylar arm (occurring in some Inulinae, too) and the presumed similarity with the *Vernonia*-type is a superficial one. In the aberrant *Callilepis* and *Cratystylis*, the stigmatic rows fuse near the base. In most of the other subtribes the stigmatic arms are parallel and remain quite separated at the apex, but in such cases, the apices are not always truncate with a terminal hair brush (*Senecio*-type): in several groups we find, above the stigmatic part, some sterile tissue, always bearing sweeping hairs. Here, the stylar arm therefore has a rounded or elongated, obtuse or even acute appendix though the existence of such appendices has mostly been neglected in Inuleae. In our opinion, these structures are taxonomically relevant, and the same holds good for the arrangement of the sweeping hairs which are not always terminal.

The results of chromosome counts look rather confusing, at first sight, though we suggested even 20 years ago (Merxmüller, 1954c) that some parts of Inuleae cluster at about $x=10$ and others about $x=7$. The counts available today corroborate this suggestion to a surprising extent. The three subtribes with inuloid styles and one-layered sexine are based on $x=10$ and 9 except for very few odd cases; most of the others are based almost monotonously, on $x=7$. The remarkable long dysploid series, shown by some Australian groups, could be interpreted, with some speculation, as descendent lines from $n=7$, 14 and 21, as it is commonly admitted in the apomictic Antennarias.

A very remarkable feature is the formation of pseudocephalia which are quite frequently represented in the Inuleae. Their distribution does not look irregular, as most of the taxa concerned can be arranged in well delimitated taxonomic groups. This has been accepted for a long time in the *Angianthus* group which is nearly related to some Gnaphaliinae, and has been convincingly shown to be in the plucheoid *Sphaeranthus* group by Leins (1971a). Moreover, we agree with Burtt's earlier view on the pseudocephaloid nature of the strange capitula of the Filagininae as well as of the isolated genus *Ifloga*. In some cases these progressions go as far as to develop heads looking like quite simple ones, e.g. in *Nablonium* (*Angianthus* group) and *Evacidium* (*Filago* group), where one can speculate only on their presumed complexity.

Other biological trends have some importance in minor groups only, like apomixis (well known in *Antennaria*, combined with high ploidy levels), autogamy (*Filago* group and probably other pseudocephaloid groups, Wagenitz in litt.) synaptospermy (*Geigeria* group, less distinctly also in the more or less related groups around *Asteriscus* and *Anvillea*), anemogamy (*Stoebe*, *Elytropappus*; *Calomeria*, *Cassinia*; *Loricaria*—in some cases deduced only from the spinulose pollen grains), and eventually imperfect or perfect dioecism (*Anaphalis* and *Lucilia* groups). On the other hand, pseudohermaphrodite, i.e. female-sterile disc florets are scattered throughout the tribe though not quite irregularly.

Finally one has to add three character complexes whose investigation is urgently needed in Inuleae. The importance of studies on the fruit anatomy has been shown by Briquet's excellent results in the *Inula* and *Asteriscus* groups; they certainly ought to be resumed on a larger scale. More detailed studies of the pappus elements (e.g. with the scanning electron microscopy) could be promising, too, as the different structures of pappus scales (true scales always being rather unusual in the Inuleae), the relations between some scales and hair tufts, or the diversity of plumose setae may reveal taxonomic implications not yet understood. The most surprising fact is the nearly absolute lack of biosystematic studies in the whole tribe.

DISCUSSION

If one combines the results of the above-mentioned investigations with the facts already known and the classification used hitherto the first impression is that of a surprisingly clear-cut bipartition of the tribe.

On the one hand we have the Plucheinae, Inulinae and Buphthalminae, all three exhibiting the basic pattern of few-spined pollen grains with one-layered sexine, stigmatic rows fusing at the apex, and a predominance of the basic numbers $x = 10$ and 9. The characters allegedly separating the named subtribes (form of the female florets, shape of the stylar arms, presence of paleae) show a rather reticulate distribution, though more fixed in certain groups. Therefore, we find it useful to subsume all three under a single subtribe which then must be called Inulinae sensu amplo. This enlarged subtribe comprises c. 65 genera with some 570 species; it has a mainly old-world distribution and reaches America and Australia with a few genera only.

Inulinae sensu amplo
In this enlarged subtribe one can first distinguish three major groups. Two are more or less equivalent to the old Inulinae/Buphthalminae and the Plucheinae, respectively, delimited by a typical combination of minor characters; both are so coherent as to make generic distinction the major problem. The third main group, a little more heterogeneous, is characterized by its pseudocephalia. Beside these there are less than 10 more isolated genera either combining some characters of the main groups or deviating markedly in special characters.

A first series of genera (1. *Inula* group) is characterized by mostly ligulate (rarely miniligulate-filiform or lacking) female florets, styles with rounded or obtuse arms and sweeping hairs in the upper part. The spine tips of the pollen grains are more or less distinctly set off from their bases and the bacular thickenings mostly have a dome-like arrangement. This group comprises firstly the genera *Allagopappus*, *Amblyocarpum*, *Anvillea*, *Buphthalmum*, *Chrysophthalmum*, *Dittrichia*, *Fontquera*, *Grantia*, *Hirschia*, *Inula* sensu lato (incl. *Codonocephalum*, *Limbarda*, *Pentanema* and *Schizogyne*), *Iphiona*, *Jasonia*, *Pegolettia* (incl. *Anisothrix*), *Perralderia*, *Platychaete*, *Postia*, *Pulicaria* sensu lato (incl. *Francoeuria*), *Sclerostephane*, *Telekia*, *Varthemia* and *Vieraea*. It stretches from the Cape Verde Islands to Japan and from Europe to S. Africa, with a distinct centre in the Mediterranean and the adjacent regions; the single New World genus *Pelucha* (from S.W. North America) seems to be near this Mediterranean stock. All these genera are based, as far as is known, on $x = 10$ and 9; very rare are 8 (and 7?). The few genera with even lower basic numbers *Astericus*, *Bubonium*, *Rhanterium* and *Pallenis* ($x = 9, 8, 7, 6, 5$) stick close together and may be related to a predominantly African group including *Anisopappus* ($x = 7$), *Antiphiona*, *Calostephane* and *Mollera*, whose chromosome numbers are still unknown.

There are three genera which clearly combine the structure of this main group with some characters of the following one. The (2.) *Geigeria* group with the African *Geigeria* and *Ondetia*, possibly related to the genera just mentioned, has distinctly narrowed, obtuse to nearly acute stylar arms and sweeping hairs reaching down below the partition. On the other hand, the (3.) *Carpesium* group (from Europe to New Guinea) possesses typical *Inula*-type styles, but pluriseriate, narrowly tubular to filiform female florets and robust conical pollen spines.

Such pluriseriate, mostly filiform female florets characterize the (4.)

Pluchea group. The stylar arms are always distinctly narrowed, obtuse to nearly acute; but often the styles of the (then pseudohermaphrodite) disc florets are bidentate or undivided. The more or less robust pollen spines have bases gradually tapering towards the top and the bacular thickenings in the spine-bases are often nearer to the bacule base or (rarely) even lacking. This group comprises *Adelostigma, Blumea, Blumeopsis, Coleocoma,*

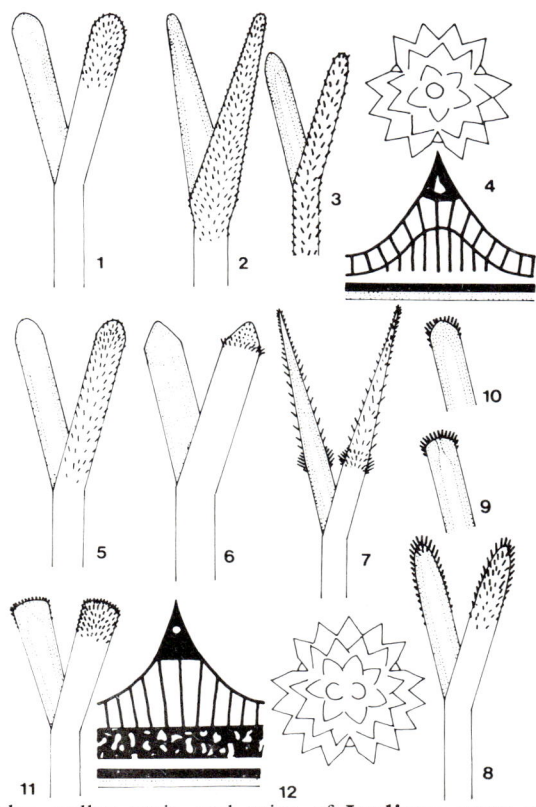

FIGS 1-4. Styles, pollen grain and exine of **Inulinae** sensu amplo: 1. *Inula*, 2. *Geigeria*, 3. *Blumea*, 4. pollen schemes.
FIGS 5-12. Styles, pollen grain and exine of **Athrixiinae** sensu amplo: 5. *Printzia* 6. *Callilepis*, 7. *Alatoseta*, 8. *Relhania quinquenervis*, 9. *Athrixia elata*, 10. *Athrixia capensis*, 11. *Phagnalon sordidum*, 12. pollen schemes.

Delamerea, Epaltes, Laggera, Merrittia, Nanothamnus, Nicolasia, Pechuel-Loeschea, Pluchea (incl. *Karelinia*), *Porphyrostemma, Pseudoconyza, Pterigeron* (mini-ligulate), *Sphaeromorphaea, Tessaria* and *Thespidium*; it is distributed over the warmer parts of both hemispheres, reaching in the north as far as the European part of USSR. The basic number is $x = 10$, a few weedy species of *Blumea* being dysploid with 9 and 11. The generic limits are especially weak due to several phenomena of reduction which may have arisen independently in various taxa, i.e. the above-mentioned sterilization of the disc florets, the loss of the anther-tails and the gradual reduction of the number of pappus elements to zero.

Here we also find several more deviating genera whose inclusion in group 4 does not seem justified. The Cuban (5.) *Sachsia* group including *Sachsia* and, presumably, the mini-ligulate *Rhodogeron,* has sexine bacules branching towards the base, the even more aberrant South American (5a.) *Stenachaenium* has bacules anastomosing towards the base and submarginal stigmatic rows. The homogamous, Cuban (6.) *Feddea* combines the style of group 1 with a sexine whose bacules show elongate basal thickenings; it looks extremely isolated as does the strange, subdioecious Australian (7.) *Cratystylis* whose stigmatic rows fuse near the base and cover nearly the entire surface.

In the last main group, the (8.) *Sphaeranthus* group, we put together all taxa whose single heads are contracted to very dense glomerules or into true pseudocephalia. The heads are homogamous or heterogamous with filiform female florets, the pollen grains identical with those of group 4, but the shape of the style arms and the arrangement of the sweeping hairs varies greatly. This may be the reason why the members of this group have hitherto been included in Plucheinae, Filagininae and Angianthinae, respectively. But most of the genera like *Athroisma* (incl. *Polycline*), *Blepharispermum, Cylindrocline, Monarrhenus, Sphaeranthus* and *Triplocephalum* form a sufficiently homogeneous assemblage, distributed over Africa and the adjacent isles, Arabia, S. and S.E. Asia to New Guinea and Australia. Only *Pterocaulon* (with mostly less contracted heads) reaches with a few species, both Americas. The Indian *Caesulia* is rather aberrant, not only in having $x=7$ but also in its inflorescence structure (Leins, 1971a).

On the other hand we find the so-called Filagininae, Gnaphaliinae and Angianthinae which constantly exhibit the "opposite" basic pattern, i.e. frequently plurispinous pollen grains with a two-layered sexine, parallel stigmatic rows completely separated at the apex and an exclusive basic number of $x=7$ and its eu- or dysploid polyploids. The female florets, if present, are nearly always filiform. Here also the recognition of several subtribes does not seem to reflect adequately the natural relationships; we have here subsumed all the taxa concerned under one subtribe, to be called Gnaphaliinae sensu amplo. This enlarged subtribe comprises c. 95 genera with some 1350 species and is distributed all over the world, with distinct centres in Australia and South America and less prominent ones in South Africa and the Mediterranean.

(II) *Gnaphaliinae sensu amplo*

As far as we can see today, a classification of this subtribe can be based on little more than stylar characters, the smooth style-arm with truncate apex, parallel stigmatic rows and exclusively terminal hairs ("*Senecio*-style") having always been considered as "normal". But, as already mentioned, there are so many other forms with rounded, obtuse or appendiculate arms and with other arrangements of sweeping hairs as to suggest a derived position of the allegedly basic *Senecio*-type. However this may be, a classification based on style-types is possible, but is much complicated by the unusually strong trends to pseudohermaphroditism, subdioecism and even dioecism. These features, with their associated aberrations in the style-shapes, are superimposed on the basic patterns to such an extent that an unambiguous grouping has not yet been entirely successful. A major problem is still the collective genus *Gnaphalium* with its "achyroclinoid" relatives whose styles all belong to the *Senecio*-type. Although we do not consider them a primitive complex we shall refer back to this "Gnaphalium group" several times.

In this (9.) *Gnaphalium* group we include *Gnaphalium* sensu Drury (with the sections *Euchiton, Gamochaeta, Gnaphalium, Omalotheca* and *Synchaeta*), the still obscure sect. *Calolepis* (Drury's "achyroclinoid cudweeds") and all the doubtful African cudweeds including some *Helichrysum* species, as well as *Helichrysopsis* and *Homognaphalium*. Within this complex one can certainly include the major part of the few-flowered groups named "*Achyrocline*" and "*Stenocline*", respectively. Moreover, we add to the *Gnaphalium* group in America *Pseudognaphalium*, as well as the central-sterile *Stuckertiella* and "*Anaphalis*" *chilensis*, in Africa *Amphidoxa, Artemisiopsis, Comptonanthus, Lasiopogon* and, possibly, the somewhat deviating *Denekia*. In the Australian region, *Euchiton* seems to be connected closely with *Leucogenes, Raoulia* and the subdioecious *Ewartia* whereas such relations can be supposed only in the perfectly dioecious genera *Parantennaria* and *Pterygopappus*. Perhaps one ought to add here *Haastia* (the type-species only) which deviates greatly, certainly, by the typical stylar appendices of the (16.) *Schoenia* group but has the same kind of sagittate anthers as found in *Pterygopappus*.

A second line (10. *Anaphalis* group) may begin in this last region with *Anaphalioides* whose stylar arms are covered with sweeping hairs far down. This aberrant arrangement may be connected with the tendency to pseudohermaphroditism and dioecism rapidly developing in this group and leading through *Anaphalis* (with styles becoming less and less typical) and *Leontopodium* (with $x = 13$, 12) to *Antennaria* (with polyploidy and apomixis). With this development, the group reaches over the whole Northern Hemisphere going as far south in America as the Andes.

The other Gnaphaliinae from Central and South America, better known from the excellent work of Cabrera and Cuatrecasas, never show really truncate stylar arms but rounded or obtuse to elongate-subacute ones; the sweeping hairs reach down further even in the wholly fertile taxa. In this (11.) *Lucilia* group we include certainly, *Belloa, Berroa, Chevreulia*,

Facelis, *Leucopholis*, *Lucilia* and possibly *Gnaphaliothamnus* and *Raouliopsis*. Once more we can only suggest a relationship with the central-sterile, subdioecious or dioecious genera *Chionolaena*, *Luciliopsis*, *Mniodes* and *Oligandra*, as in these cases connections with the northern *Anaphalis* or the southern *Raoulia* group cannot yet be excluded. The equally dioecious (12.) *Loricaria* has spinulose pollen grains which may point to a tendency towards anemogamy.

There is a strange similarity of these *Lucilia* styles with the admittedly always smaller styles of the (13.) *Filago* group, whose inflorescence structure still offers as many problems as does their geographical distribution, reaching over South and North America on the one side, through the Mediterranean and adjacent Eurasia on the other. The conical receptacle as well as the often curiously arranged "paleae" (lacking in *Evacidium*!) could indicate reduced pseudocephalia, but this certainly requires further investigation. This group comprises *Bombycilaena*, *Cymbolaena*, *Evacidium*, *Evax* ($x=13$), *Filago* sensu lato (incl. *Logfia*), *Micropsis*, *Micropus* sensu lato, *Psilocarphus* and *Stylocline*. The Asian *Symphyllocarpus*, treated as a "subtribe" of its own by Russian authors, seems not to belong to the Inuleae at all.

Thus, we come back to the Old World. The still hopeless situation around *Gnaphalium* and *Helichrysum* in Africa (which perhaps can really be improved only by some splitting) makes a separate (14.) *Helichrysum* group somewhat doubtful if not ridiculous. This name shall be used here only provisionally; as far as we know there is no clear-cut difference between this and the *Gnaphalium* group. Even if we exclude, temporarily, the difficult problems of the Australian taxa, there can be no doubt whatever that the huge *Helichrysum*/*Helipterum* complex (incl. *Calomeria*, *Eriosphaera*, *Leontonyx* and *Rhynea*) ought to be reinvestigated thoroughly. The generic separation of some arbitrarily chosen isolated units does not help any more here than in *Gnaphalium*. But this complex apart, there remain in the African region only a few genera which might be included in 14. We may mention the Madagascan few-flowered group including *Catatia*, *Syncephalum* and the taxa called there, once more, *Achyrocline* and *Stenocline*; in S. Africa perhaps the relhanioid genera *Metalasia* and *Lachnospermum* as well as the rather odd, central-sterile group of *Anaxeton*, *Petalacte* and *Phaenocoma* (the last one with $x=8$!). But we have to return once more to both South African groups in our final discussion. If we accept *Metalasia* as helichrysoid, then we could affiliate here the isolated genus (15.) *Ifloga* (incl. *Petalactella*) whose heads we interpret with Hilliard & Burtt as reduced pseudocephalia; but we cannot agree with these authors who regard it as near the *Filago* group from which it is clearly separated by the typical *Senecio*-style.

In all (amphimictic) Gnaphaliinae treated until now the basic number of $x=7$ has been found almost exclusively; the numbers 13 in *Evax*, 13 and

12 in *Leontopodium* are undoubtedly to be interpreted as hypotetraploid. Quite another picture offers itself in Australia where the prevalent karyological pattern is formed by large dysploid series with $n = 21, 20, 19; 14, 13, 12, 11, 10, 9, 8, 7, 6, 5, 4$ and 3. It is mere speculation if we interpret these as descendent series too, down from 21, 14 and 7, but if so this might be correlated with the numerous annual growth forms in this region.

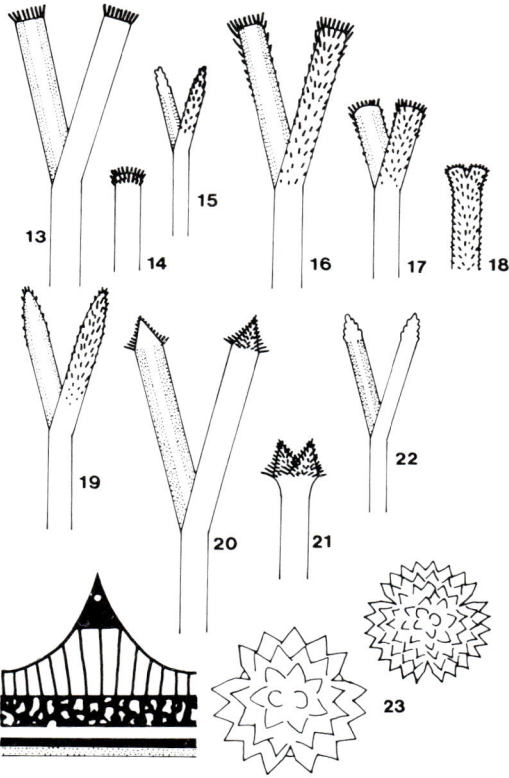

FIGS 13–23. Styles, pollen grains and exine of **Gnaphaliinae** sensu amplo: 13. *Gnaphalium*, 14. *Petalacte canescens*, 15. *Filago gallica*, 16. *Anaphalis* pro parte, 17. *Lucilia araucana*, 18. *Antennaria*, 19. *Lucilia longifolia*, 20–21. *Schoenia*, 22. *Toxanthes*, 23. pollen schemes.

There are several other new trends in Australia. Filiform female florets are decidedly less frequent than in other regions; sometimes they are replaced either by tubular florets which strangely overhang the involucre, or even by irregularly tubulate-ligulate florets. The pappus elements are strikingly variable. But the most remarkable feature is the frequent appearance of a capitate or triangular hairy appendix on the otherwise smooth stylar arm which then shows similarity to the so-called Aster type.

Some genera could, at least technically, still be included in our (14.)

Helichrysum group, especially the genera with truncate stylar arms like *Helichrysum* pro parte and *Helipterum* pro parte, *Cassinia* and *Calomeria* (both partly with spinulose pollen grains) as well as *Ammobium, Ixodia, Neotysonia* and *Pithocarpa*. But here already the exclusively dysploid numbers are somewhat disturbing and this disturbance increases with *Ixiolaena, Leptorhynchos* and *Rutidosis* which approach the next group with respect to other characters, too.

This (16.) *Schoenia* group is characterized by their peculiar stylar appendices mentioned above. To this group belongs the other part of the Australian *Helichrysum* and *Helipterum*, furthermore *Acomis, Bellida, Millotia, Podotheca, Schoenia* and *Waitzia*; especially slender are the appendices in *Quinetia, Scyphocoronis* and *Toxanthes*, all being nearly related to *Millotia*. We hope we are correct in adding here also the two alleged Athrixiinae of Australia, (17.) *Podolepis* and *Athrixia* (pro parte, Australian) which fit extremely well in this group from which they differ solely by the (partial) transformation of the marginal tubular florets into irregularly ligulate ones.

The last Australian complex is the (18.) *Angianthus* group which once again shows an aggregation of their heads into very dense glomerules or into pseudocephalia. *Actinobole, Angianthus, Calocephalus, Cephalipterum, Chthonocephalus, Craspedia, Decazesia, Eriochlamys, Gnephosis* and *Myriocephalus* are homogamous, dysploid as far as known and have truncate stylar arms (except a few appendiculate *Myriocephalus* spp.). Not very far are *Basedowia* with naked female florets and most probably also the heterogamous *Stuartina* with filiform female florets. The Tasmanian genus *Nablonium* whose capitula are interpreted as extremely simplified pseudocephalia by Leins (1971a) may represent the last member of this series.

Although our distinction of two clear-cut subtribes seems to work very well, there remain—as always in such cases—about 20 genera with some 200 species whose inclusion in either subtribe would severely disrupt their delimitation. The groups concerned are a few genera of the Inulinae/Buphthalminae ("*Printzia* group": Leins, 1971a), nearly all Athrixiinae and a part at least of the Relhaniinae. Whereas we initially attributed these inconsistencies to convergent developments, a more detailed investigation revealed surprising similarities. The pollen grains are often sparsely-spinous, the sexines constantly two-layered, the female florets if present ligulate; the only pertinent chromosome counts are based on $x=9$ (*Disparago*), $x=7, 5$ (*Relhania*) and $x=4$ (*Leysera*). A disappointing feature is the presence of diverse styles, some genera exhibiting the *Inula*-type or the *Senecio*-type, respectively, whereas the larger part shows transitional stylar forms with stigmatic rows converging towards the apex. The taxa concerned are restricted to South and East Africa (extending to Arabia).

In such situations one always wonders whether this may be a primitive group, not yet fixed in stylar characters, or a group of transitional taxa, or even whether one is creating a "dustbin group" in the well known sense of Davis and Heywood. We do not feel able to decide on the first possibility but we can now exclude the third possibility. We have, therefore, decided to group all these taxa under a third subtribe to be called *Athrixiinae* sensu amplo.

(III) *Athrixiinae sensu amplo.*

Firstly, we put together provisionally the inuloid genera always with distinctly fusing stigmatic rows, i.e. *Anisochaeta, Philyrophyllum, Pentatrichia* and *Printzia*. In this (19.) *Printzia* group the sweeping hairs are either confined to the upper part or reach down to the partition, perhaps reflecting some heterogeneity.

At one extreme, members of the (20.) *Athrixia* group are practically indistinguishable from group 19, but here we also find every transition to the other extreme, namely having the styles with convergent or parallel stigmatic rows. According to our concept, this group comprises *Antithrixia, Arrowsmithia, Athrixia, Leysera, Macowania, Relhania* and *Rosenia*, genera hitherto allotted to Athrixiinae sensu stricto or Relhaniinae somewhat vaguely. Current investigations on *Macowania* (Edinburgh), *Relhania* and *Rosenia* (Stockholm) as well as on *Athrixia* and *Printzia* (Munich) give us hope for some clarification of this group.

The (21.) *Amphiglossa* group can be distinguished from the *Gnaphalium* or *Helichrysum* groups by the ligulate flowers only, if one disregards the vague character of an "ericoid habit" and the possible difference in chromosome number. *Amphiglossa, Bryomorphe* and *Disparago* obviously belong to this group; but if this grouping is a natural one, then the homogamous genera *Pterothrix* as well as *Elytropappus* and *Stoebe* (both of these partly with spinulose pollen grains and then probably anemogamous) cannot be excluded. *Lachnospermum* and *Metalasia* (together with *Ifloga*) seem to be closer to the *Helichrysum* group, but these uncertainties show clearly what problems have yet to be solved.

There remain two more aberrant genera which differ greatly from each other, whose stigmatic rows fuse near the base and cover the larger part of the surface (as in the inuloid *Cratystylis*); the stylar arm, moreover, bears at the base of its conical apex a more or less distinct crown of hairs ("Fegehaarkranz"). This crown of hairs is very conspicuous in (22.) *Alatoseta*, a genus attributed to Athrixiinae but unique in its extremely elongated, nearly acuminate stylar arms, but less conspicuous in (23.) *Callilepis* (incl. *Zoutpansbergia*) included hitherto in Buphthalminae because of its remarkable paleae. In both cases the styles are very similar to those of some African Mutrisieae (e.g. *Passacardoa* or *Erythrocephalum*) which remain clearly separated by their pollen grains only, showing the characteristic infrategillar bacula. This strange similarity reminds us of *Anisochaeta* (*Printzia* group) which has been treated as mutisioid by several authors.

It would be dishonest to conceal that this third subtribe disturbs us still to a considerable degree. We have very little information on chromosome numbers; we have not yet succeeded in finding better characters to justify the inclusion, for example, of the homogamous Relhaniinae. But most of all we still hesitate as to whether we could "allow" our Athrixiinae sensu amplo to include also taxa with filiform female florets. A positive answer would allow a reconsideration of the strange (24.) *Phaenocoma* group with female-sterile disc florets and undivided styles, including *Phaenocoma* ($x=8$) as well as *Anaxeton* and *Petalacte*. We then could find a place, eventually, for the as yet unmentioned genus (25.) *Phagnalon* with its "inuloid" distribution in the northern hemisphere (from the Cape Verde Islands to Central Asia). It can certainly not be included in our Gnaphaliinae sensu amplo as in most points it agrees perfectly (apart from the filiform florets and the tailless anthers) with our Athrixiinae. The stigmatic rows in this genus are either convergent or even fusing; the chromosome number, known for many species, is uniformly $n=9$. We are not yet prepared to make a positive decision and we therefore treat these last groups as an "informal appendix" to subtribe III.

REFERENCES

ALLAN, H. H. (1961). "Flora of New Zealand", Vol. 1, Government Printer, Wellington.

BADILLO, V. M. (1974). *Blumea viscosa* y *Piptocarpa cuatrecasasiana*, dos nuevas combinaciones en Compositae. *Revta Fac. Agron.* (*Maracay*) **7**, 9–16.

BANERJEE, A. K. and A. SHARMA, A. (1974). Chromosome studies on some Indian Members of Compositae. I. Tribe Inuloideae. *Bróteria* **43**, 15–32.

BEAUVERD, G. (1910). Contribution à l'étude des Composées. IV: Recherches sur la tribu des Gnaphaliées. *Bull Soc. bot. Genève* ser. 2 **2**, 207–252.

BEAUVERD, G. (1912). Contribution à l'étude des Composées. VI: Nouveaux *Leontopodium* et *Raoulia*. *Bull. Soc. bot. Genève*, ser. 2 **4**, 12–55.

BEAUVERD, G. (1913a). La constante générique des *Anaphalis* DC. *Bull. Soc. bot. Genève*, ser. 2 **5**, 146–147.

BEAUVERD, G. (1913b). Contribution à l'étude des Composées (Suite VIII). *Bull. Soc. bot. Genève*, ser. 2 **5**, 205–228.

BENTHAM, G. (1873). Notes on the Classification, History and Geographical Distribution of Compositae. *J. Linn. Soc.* (*Bot.*) **13**, 335–577.

BENTHAM, G. and HOOKER, J. D. (1873). "Genera Plantarum", Vol. **2(1)** London.

BERGMAN, B. (1951). On the formation of reduced and unreduced gametophytes in the females of *Antennaria carpatica*. *Hereditas* **37**, 501–518.
BESOLD, B. (1971). Pollenmorphologische Untersuchungen an Inuleen (Angianthinae, Relhaniinae, Athrixiinae). *Dissertationes Botanicae* **14**, J. Cramer, Lehre.
BLACK, J. M. (1943/57). "Flora of South Australia" (2nd edition). Government Printer, Adelaide.
BORISSOVA, A. (1960). De genere *Antennaria* Gaertn. notulae systematicae (russ.). *Notul. Syst. Herb. Inst. Bot. Acad. Sci. URSS*, **20**, 289–295.
BOTSCHANTZEV, V. (1964). Additamenta ad floram Aegypti, *Novit. Syst. Pl. Vasc.*, *Acad. Sci. URSS* 1964, 349–378.
BREMER, K. (1972). The Genus *Osmitopsis* (Compositae). *Bot. Notiser* **125**, 9–48.
BREMER, K. (1976a). The genus *Relhania* (Compositae). *Opera bot.* **40**, 1–86.
BREMER, K. (1976b). The genus *Rosenia* (Compositae). *Bot. Notiser* **129**, 97–111.
BRIQUET, J. and CAVILLIER, F. (1917). *In* "Flore des Alpes Maritimes" (E. Burnat, ed.) Vol. VI, pp. 221–314.
BROWN, N. E. (1909). List of plants collected in Ngamiland and the northern part of the Kalahari Desert, chiefly in the neighbourhood of Kwebe and along the Botletle and Lake Rivers. *Kew Bull.* **1909**, 89–146.
BURBIDGE, N. T. (1958). A monographic study of *Helichrysum* subgenus *Ozothamnus* (Compositae) and of two related genera formerly included therein. *Aust. J. Bot.* **6**, 229–284.
BURTT, B. L. and GRAU, J. (1972). An extension of the genus *Macowania* (Compositae). *Not. R. bot. Gdn. Edinb.* **31**, 373–376.
CABRERA, A. L. (1955). La identidad del genero *Psila* Philippi. *Bol. Soc. argent. bot.* **5**, 209–211.
CABRERA, A. L. (1958). El genero "*Belloa*" Remy. *Bol. Soc. argent. bot.* **7**, 79–85.
CABRERA, A. L. (1961a). Observaciones sobre las Inuleae-Gnaphalineae (Compositae) de América del sur. *Bol. Soc. argent. bot.* **9**, 359–386.
CABRERA, A. L. (1961b). Compuestas Argentinas. *Revista Mus. Argent. Ci. Nat. Bernardino Rivadavia, Ci.Bot.* **2**, 291–362.
CLARKE, C. B. (1876). "Compositae Indicae". Calcutta.
CRONQUIST, A. (1955). Phylogeny and Taxonomy of the Compositae. *Am. Midl. Nat.* **53**, 478–511.
CUATRECASAS, J. (1954a). Synopsis der Gattung *Loricaria* Wedd. *Feddes Repert.* **56**, 149–172.
CUATRECASAS, J. (1954b). El género *Mniodes*. *Folia Biol. Andina* **1**, 1–7.
CUFODONTIS, G. (1933). *Leucopholis capitata* (Baker) comb. nov. *Feddes Repert.* **31**, 329–330.
CUFODONTIS, G. (1966/67). Enumeratio Plantarum Aethiopiae. Spermatophyta. *Bull. Jard. bot. nat. belg.* **36**, Suppl., 1091–1114 (1966); and **37**(3), Suppl., 1115–1124 (1967).
DALLA TORRE, C. G. DE and HARMS, H. (1905). "Genera Siphonogamarum". Engelmann, Leipzig.
DAVIS, G. L. (1957). Revision of the genus *Podolepis* Labill. *Proc. Linn. Soc. N.S. Wales* **81**, 245–286.
DIELS, L. and PRITZEL, E. (1904). Fragmenta Phytographiae Australiae occidentalis. *Bot. Jahrb.* **35**, 55–662
DRURY, D. G. (1970). A fresh approach to the classification of the genus *Gnapha-*

lium with particular reference to the species present in New Zealand (Inuleae-Compositae). *N.Z. Jl. Bot.* **8**, 222–248.

DRURY, D. G. (1971). The American Spicate Cudweeds Adventive to New Zealand: (*Gnaphalium* Section *Gamochaeta*-Compositae). *N.Z. Jl Bot.* **9**, 157–185.

DRURY, D. G. (1972). The Cluster and Solitary-headed Cudweeds Native to New Zealand: (*Gnaphalium* Section *Euchiton*-Compositae). *N.Z. J. Bot.* **10**, 112–179.

EICHLER, H. (1963). Some new names and new combinations relevant to the Australian Flora. *Taxon* **12**, 295–297.

EICHLER, H. (1965). "Supplement to J. M. Black's Flora of South Australia" (Second Edition, 1943–1947). Government Printer, Adelaide.

GAGNEPAIN, F. (1924). Composées. In (H. Lecomte, H. Humbert and F. Gagnepain, eds), "Flore générale del'Indo-Chine" Vol. 3. Paris.

GRIERSON, A. J. C. (1972). A new species of *Anaphalis* (Compositae) from Mexico *Not. R. bot. Gdn Edinb.* **31**, 389–392.

GRIERSON, A. J. C. (1974). *Inula.* In Davis, P. H., Materials for a Flora of Turkey. XXX: Compositae, I. *Not. R. bot. Gdn Edinb.* **33**, 248–251.

HANDEL-MAZZETTI, H. (1927). Systematische Monographie der Gattung *Leontopodium. Beih. bot. Centralbl.* **44** (II) ,1–178.

HEDBERG, O. (1957). Afroalpine vascular plants. *Symb. bot. upsal.* **15**, 1.

HILLIARD, O. M. and BURTT, B. L. (1971). Notes on some plants of Southern Africa chiefly from Natal: II. *Not. R. bot. Gdn. Edinb.* **31**, 1–33.

HILLIARD, O. M. and BURTT, B. L. (1973). Notes on some plants of Southern Africa chiefly from Natal: III. *Notes R. bot. Gdn. Edinb.* **32**, 303–387.

HILLIARD, O. M. and BURTT, B. L. (1976). Notes on some plants of Southern Africa chiefly from Natal: V. *Not. R. bot. Gdn Edinb.* **34**, 260–276.

HOFFMANN, O. (1894). Compositae. *In* "Die Natürlichen Pflanzenfamilien" (Engler and Prantl, eds) **4**(5). Engelmann, Leipzig.—Nachträge I 320–330 (1897); II: 75–78 (1900); III: 337–349 (1908); IV: 315–327 (1915).

HUMBERT, H. (1962). "Flore de Madagascar et des Comores", 189. Composées. IV. Inulées. Paris.

JAHANDIEZ, E. and MAIRE, R. (1934). "Catalogue des Plantes du Maroc", Vol. 3, Alger.

JEFFREY, C. (1967). Notes on Compositae. II. The Mutisieae in East Tropical Africa. *Kew Bull.* **21**, 41–114.

JOHNSTON, J. M. (1923). Diagnoses and Notes relating to the Spermatophytes chiefly of North America. *Contr. Gray Herb.*, **68**, 80–104.

JUEL, O. (1900). Vergleichende Untersuchungen über typische und parthenogenetische Fortpflanzung bei der Gattung *Antennaria. Kongl. Svenska Vetensk. Akad. Handl.* **33**, (5), 1–59.

KIRPICHNIKOV, M. E. and KUPRIJANOVA, L. A. (1950). Morphological, geographical and palynological contributions to the understanding of the genera of the subtribe Gnaphaliinae. (russ.). *Acta Inst. Bot. Acad. Sci. URSS*, Ser. 1, Fasc. **9**, 7–37.

KITAMURA, S. (1937). Compositae Japonicae. Pars Prima. *Mem. College Sci., Kyoto Imp. Univ.*, ser B. **13**.

KOSTER, J. T. (1972). The Compositae of New Guinea III. *Blumea* **20**, 193–226.

Leins, P. (1970). Die Pollenkörner und Verwandtschaftsbeziehungen der Gattung *Eremothamnus* (Asteraceae). *Mitt. bot. Münch.* **7**, 369–376.

Leins, P. (1971a). Pollensystematische Studien an Inuleen. I. Tarchonanthinae, Plucheinae, Inulinae, Buphthalminae. *Bot. Jahrb.* **91**, 91–146.

Leins, P. (1971b). Neukombinationen einiger Inuleen. *Mitt. bot. Münch.* **9**, 107–108.

Leins, P. (1973). Pollensystematische Studien an Inuleen. II. Filagininae. *Bot. Jahrb.* **93**, 603–611.

Levyns, M. R. (1935). A revision of *Elytropappus* Cass. *J. S. Afr. Bot.*, **1**, 89–103.

Levyns, M. R. (1936). A revision of *Disparago* Gaertn. *J. S. Afr. Bot.* **2**, 95–103.

Levyns, M. R. (1937). A revision of *Stoebe* L. *J. S. Afr. Bot.* **3**, 1–35.

Lewis, P. and Summerhayes, V. S. (1951). *Pithocarpa* Lindl. *Kew Bull.* **1950**, 435–440.

Lundgren, J. (1972). Revision of the genus *Anaxeton* Gaertn. (Compositae). *Opera Bot.* **31**, 1–59.

Lundgren, J. (1974). The Genus *Petalacte* D. Don (Compositae). *Bot. Notiser* **127**, 119–124.

Marticorena, C. and Parra, O. (1974). Morfologia de los granos de polen y posición sistematica de *Anisochaeta* DC., *Chionopappus* Benth., *Feddea* Urb. y *Gochnatia glomeriflora* Gray (Compositae). *Bol. Soc. biol. Concepcion* **47**, 187–197.

Mattfeld, J. (1936). Compositae II. *Notizbl. bot. Gart. Mus. Berlin*, **13**, 287–303.

Merxmüller, H. (1950). Compositen-Studien I. *Mitt. bot. Münch.* **1**, 33–46.

Merxmüller, H. (1953). Compositen-Studien III. Revision der Gattung *Geigeria* Griesselich. *Mitt. bot. Münch.* **1**, 239–316.

Merxmüller, H. (1954a). Compositen-Studien IV. Die Compositen-Gattungen Südwestafrikas. *Mitt. bot. München.* **1**, 357–443.

Merxmüller, H. (1954b). Compositen-Studien V. Revision der Gattung *Nicolasia* S. Moore. *Mitt. bot. Münch.* **2**, 1–10.

Merxmüller, H. (1954c). Beiträge zur Taxonomie der Compositen. *Ber. dt. bot. Ges.* **67** (61. Gen.-Vers.-Heft): (23)–(24).

Merxmüller, H. (1967). Asteraceae. In "Prodromus einer Flora von Südwestafrika **139**. (H. Merxmüller, ed.) J. Cramer, Lehre.

Moeser, W. (1909). Über die systematische Gliederung und geographische Verbreitung der afrikanischen Arten von *Helichrysum* Adans. *Bot. Jahrb.* **43**, 420–460.

Moore, S. (1917). *Bellida* Ewart. *J. Bot. Lond.* **55**, 100–101.

Moore, S. (1929). Notes on African Compositae. *J. Bot. Lond.* **67**, 273–274.

Nordenstam, R. B. (1964). *Comptonanthus*, a new genus of the Compositae with notes on *Lasiopogon* in South Africa. *J. S. Afr. Bot.* **30**, 53–65.

Phillips, E. P. (1951). "The Genera of South African Flowering Plants". (2nd edition) Bot. Surv. Mem. No. 25. Government Printer, Pretoria.

Pillans, N. S. (1954). A revision of *Metalasia*. *J. S. Afr. Bot.*, **20**, 47–87.

Poljakov, P. P. (1967). "Sistematika i proischoždenie složnocvetnych". Alma-Ata.

Powell, A. M., Kyhos, D. W. and Raven, P. H. (1974). Chromosome numbers in Compositae. X. *Am. J. Bot.* **61**, 909–913.

Pritzel, E. (1918). *Basedowia*, eine neue Gattung der Compositen aus Zentral-Australien. *Ber. dt. bot. Ges.* **36**, 332.

Randeria, A. J. (1960). The Composite Genus *Blumea*, a taxonomic Revision. *Blumea* **10**, 176–317.

Robinson, H. and Brettell, R. D. (1973a). Tribal revisions in the Asteraceae. VI. The relationship of *Eriachaenium*. *Phytologia* **26**, 71–72.

Robinson, H. and Brettell, R. D. (1973b). Tribal revisions in the Asteraceae. XI. A new tribe, Eremothamneae. *Phytologia* **26**, 163–166.

Robinson, H. and Cuatrecasas. J. (1973). The generic limits of *Pluchea* and *Tessaria* (Inuleae, Asteraceae). *Phytologia* **27**, 277–285.

Ross-Craig, S. (1955). A revision of the genus *Sphaeranthus*. *Hooker's Icon. Pl.* **36**, t.3501–3525.

Rydberg, P. A. (1954). "Flora of the Rocky Mountains and Adjacent Plains". (2nd edition), New York.

Schischkin, B. K. (1959). "Flora SSSR", Vol. 25. Moscow and Leningrad.

Schodde, R. (1963). A taxonomic revision of the genus *Millotia* Cassini (Compositae). *Trans. R. Soc. S. Aust.* **87**, 209–241.

Skvarla, J. J. and Turner, B. L. (1966). Systematic implications from electron microscopic studies of Compositae pollen. *Ann. Mo. bot. Gdn.* **53**, 220–256.

Stix, E. (1960). Pollenmorphologische Untersuchungen an Compositen. *Grana Palynol.* **2**, 41–114.

Urbanska-Worytkiewicz, K. (1961/62). Embryological investigations in *Antennaria* Gaertn. I–III. *Acta Biol. Cracov. Ser. Bot.*, **4**, 49–64; **5**, 97–102, 103–115.

Wagenitz, G. (1964). Compositae. In "Englers Syllabus der Pflanzenfamilien" (H. Melchior, ed.), vol. II. Ed. 12. Borntraeger, Berlin-Nikolassee.

Wagenitz, G. (1965). Compositae (Korbblütler). In "Illustrierte Flora von Mitteleuropa" (G. Hegi, ed., 2nd edition). Vol. VI, 3, Lief. 2. C. Hanser, München.

Wagenitz, G. (1969). Abgrenzung und Gliederung der Gattung *Filago* L. s.l. (Compositae-Inuleae). *Willdenowia* **5**, 395–444.

Wagenitz, G. (1971). Zur taxonomischen Stellung und Nomenklatur von *Micropus longifolius* (Compositae-Inuleae). *Öst. bot. Z.* **119**, 399–403.

Wild, H. (1964). A revision of the genus *Anisopappus* Hook. & Arn. (Compositae) *Kirkia* **4**, 45–73.

Wild, H. (1968/69). The Compositae of the Flora Zambesiaca Area, 2. *Kirkia* **7**, 121–135.

Chapter 22
Inuleae—chemical review

JEFFREY B. HARBORNE
Plant Science Laboratories, University of Reading, England

Abstract. While something is known of the chemistry of between 5 and 10% of the 2000 species which constitute the Inuleae, detailed chemical investigations have been limited to a handful of species belonging mainly to the genera *Blumea*, *Gnaphalium*, *Helichrysum* and *Inula*. At least 16 sesquiterpene lactones have been identified variously in these plants; the pattern is similar to those of the Heliantheae and Anthemideae in that all four major types of lactone are represented. Although no unusual monoterpenes have been found in the Inuleae, a unique series of diterpene lactones have been uncovered in the roots of *Inula royleana*. Flavonoid pigments have been isolated, especially from species of *Gnaphalium* and *Helichrysum* and the similarity in pattern in these two genera correlates with their close morphological affinities. The most characteristic feature of Inuleae flavonoids is the presence of flavonols lacking B-ring hydroxylation. 6- and/or 8-hydroxyflavonols and their methyl ethers occur regularly in the tribe. By contrast, anthochlor pigments occur only in *Gnaphalium* and *Helichrysum*. Flavonoid sulphates have been detected in eight of 80 species surveyed. A number of structurally diverse phenolic compounds have been found in the tribe, and especially in *Blumea* species. Alkaloids appear to be uncommon and only one has been fully characterized, an aconitine-type structure from *Inula royleana*. Simple polyacetylenes are widespread. Mono- and di-thiophenic acetylenes occur in three subtribes and acetylenes with pyran and furan attachments are found only in the Gnaphalinae. The results of work to date indicate the presence of a distinctive chemistry in the Inuleae and suggest that detailed chemotaxonomic surveys for flavonoids, sesquiterpene lactones and polyacetylenes would be especially rewarding.

CONTENTS

Introduction	604
Terpenoids	604
Simple terpenes	604
Sesquiterpene lactones	605
Diterpenes	607
Triterpenoids	608
Flavonoids and other phenolics	608
Flavonoids	608
Simple phenolics	612

Other secondary constituents 614
    Alkaloids 614
    Polyols and cyclitols 614
    Lipids 615
    Polyacetylenes 616
Conclusion 617
References 617

INTRODUCTION

The Inuleae are a cosmopolitan tribe of the Compositae, well represented in the European flora (27 genera, 116 species) but especially abundant in South Africa and Australia. There are some 200 genera and 2000 species. While Bentham divides these plants into some nine subtribes, Merxmüller *et al.* (Chapter 21) arrange them into three major groupings, two of which, the Inulinae and Gnaphaliinae, are natural while the third, the Arthrixiinae, is a more artificial assemblage.

The Inuleae include a selection of plants which have been prized by man for their useful properties. Three of the best known which were used medicinally in the past are elecampane, *Inula helenium* (for treating chest diseases), fleabane *Pulicaria dysenterica* (in herbal remedies) and cat's foot *Antennaria dioica* (for throat infections). Elecampane roots have been used as sweetmeats and they have given their name to inulin, the characteristic composite storage polysaccharide based on fructose. Leaves of *Inula* and *Pulicaria* species have also been employed as insecticidal sources.

In spite of the practical interest in plants of the Inuleae, the chemistry of the tribe has not been extensively explored. The only representative survey seems to be of the polyacetylenes in roots and leaves (Bohlmann *et al.*, 1973). The sesquiterpene lactones, which have been so thoroughly explored in other composite tribes, have only been identified in relatively few members. The flavonoid pigments have been studied in great detail in *Helichrysum* by Hänsel and his co-workers, but otherwise relatively little work has been done on them. Fragmentary information on a range of other constituents is also available (Hegnauer, 1964). The present account of the chemistry of the Inuleae is perforce of limited taxonomic interest. Clearly, there is an interesting chemistry in these plants but more surveys are urgently needed.

TERPENOIDS

Simple terpenes

In the few cases where essential oil fractions have been examined, they have usually been found to be more interesting from the point of view of methylated phenolic constituents (see section on phenolics) than for the normal terpenes. Commonly occurring mono- and sesquiterpenes have been identified in several plants. Pinene, limonene, p-cymol and caryophyllene have been found in *Achyrocline satureioides*; borneol and camphor in *Blumea balsamifera*; and methylchavicole, p-hydroxycinnamaldehyde, α-ionone and cadinene in root oil of *Sphaeranthus indicus* (for references, see Hegnauer, 1964).

Sesquiterpene lactones

At least 16 sesquiterpene lactones have been identified in 12 species drawn from five genera of the Inuleae (Table I). A rich variety of structures have

TABLE I. Distribution of Sesquiterpene lactones in the Inuleae

Compound	Structural Type[a]	Source(s)	References
Alantolactone (**1**)	EU	*Codonocephalum grande* O. et B. Fedtsch root, *Inula helenium* L., *I. racemosa* Hook f., *I. magnifica* Lipsky root, *Telekia speciosa* (Schreb.) Baumg.	Hegnauer, 1964; Nikonova, 1973
Isoalantolactone (**2**)	EU		
Dihydroalantolactone (**3**)	EU		
Carpesia lactone (**4**)	GU	*Carpesium abrotanoides* L. seeds	
Carabrone (**5**)			
Geigerin (**6**)	GU	*Geigeria africana* Gries and *G. aspera* Harv. plant	Anderson et al., 1967
Geigerinin (**7**)	PG		
Gafranin (**8**)	GM		
Vermeerin (**9**)	R		
Graveolide (**10**)	EU	*Inula graveolens* Desf.	Alcontres et al., 1973
Granilin (**11**)	EU	*I. grandis* Shrenk aerial parts	Nikonova and Nikonov, 1972
Ivalin (**12**)	EU		
Carabrone	R		
Inulicin (**13**)	R	*I. japonica* Thunb.	Kiseleva et al., 1971
Britannin (**14**)	PG	*I. britannica* Bieb.	Chugunov, et al., 1971
Telekin (**15**)	EU	*Telekia speciosa* aerial parts	Benesova and Herout 1961
Isotelekin (**16**)	EU		

[a] Key: EU = eudesmanolide; GM = germacranolide; GU = guaianolide; PG = pseudoguaianolide; R = rearranged carbon skeleton type derived from any of the others.

been uncovered (**1–16**) and compounds representing all but one of the five main "biogenetic" types of lactone have been found. Most characteristic of the tribe is alantolactone (**1**), a eudesmanolide present in three genera; it is accompanied in some sources by the isomer (**2**) or the dihydro derivative (**3**). Only one germacranolide, gafranin (**8**), has apparently so far been reported and this occurs in *Geigeria*. This genus is a rich source of structures; geigerin (**6**) is a guaianolide from the same plant. Finally, the genus *Inula* has also been found to be rich in these compounds; nine structures have been obtained from seven species. Typical of recent Russian research in this field is the identification of the rearranged guaianolide inulicin (**13**), from *I. japonica* (Kiseleva *et al.*, 1971).

The pattern in the Inuleae is similar to those found in the Heliantheae, Helenieae and Anthemideae in that all four major types of lactone are

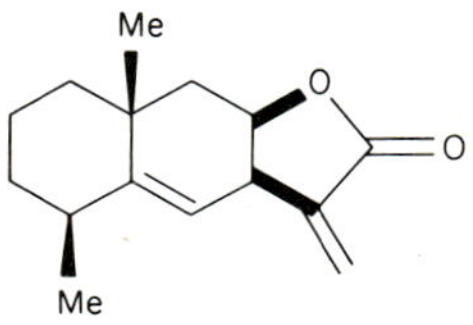

(1) Alantolactone

(2) Isoalantolactone

(3) Dihydroalantolactone
(single bond at C5 and C6)

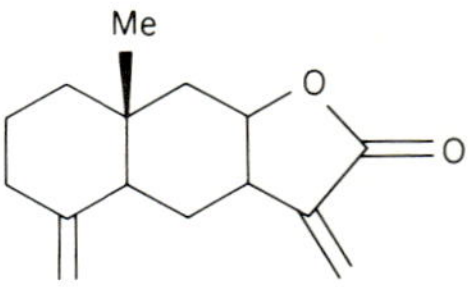

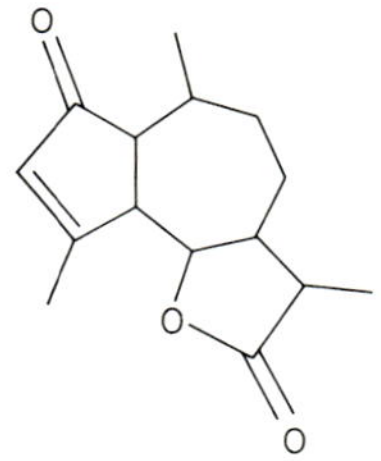

(4) Carpesialactone

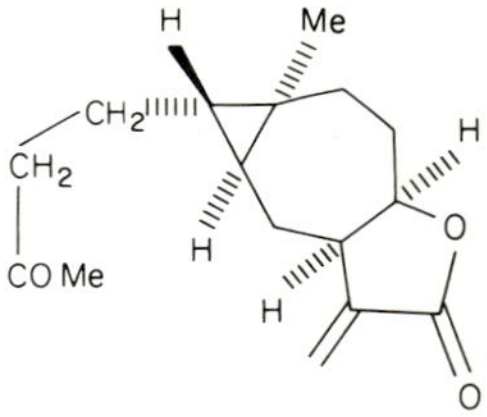

(5) Carabrone

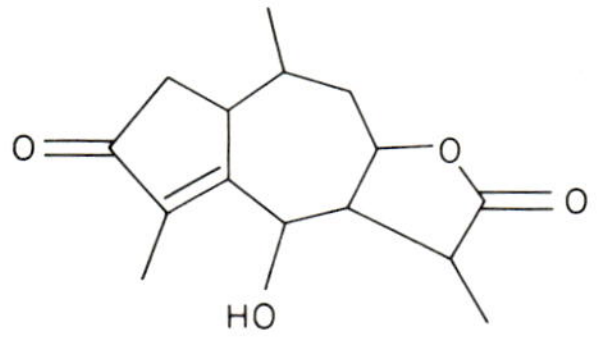

(6) Geigerin

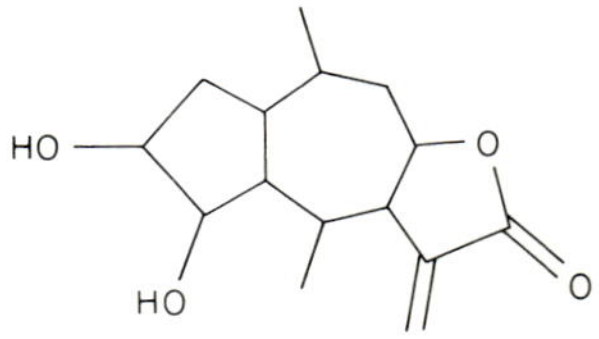

(7) Geigerinin

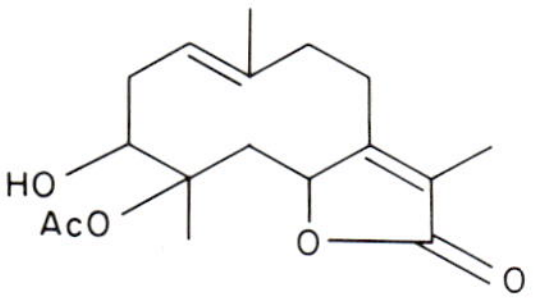

(8) Gafranin

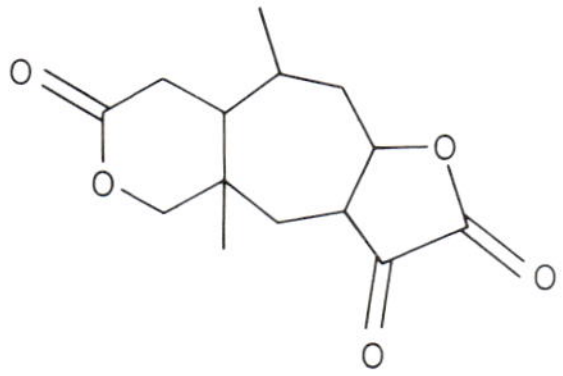

(9) Vermeerin

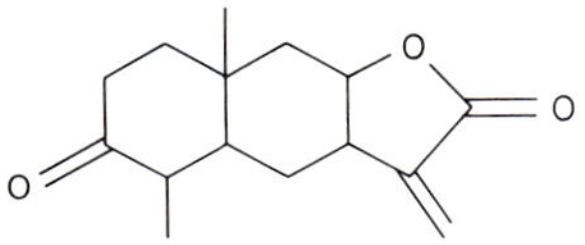

(10) Graveolide

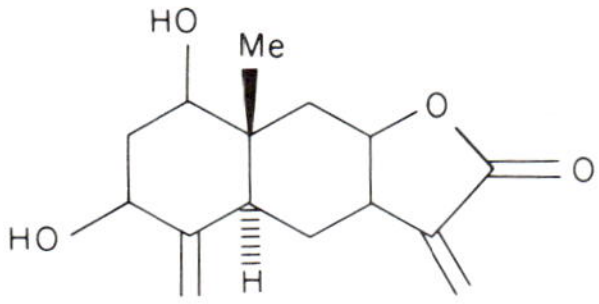

(11) Granilin

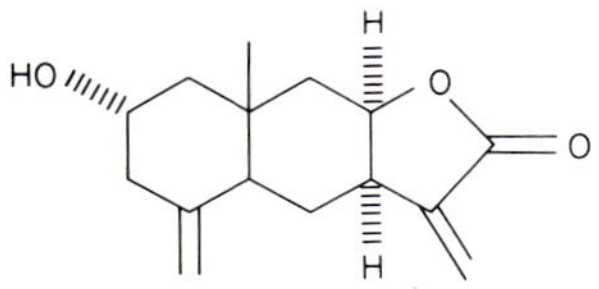

(12) Ivalin

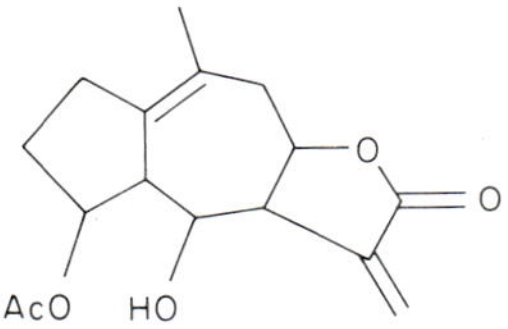

(13) Inulicin

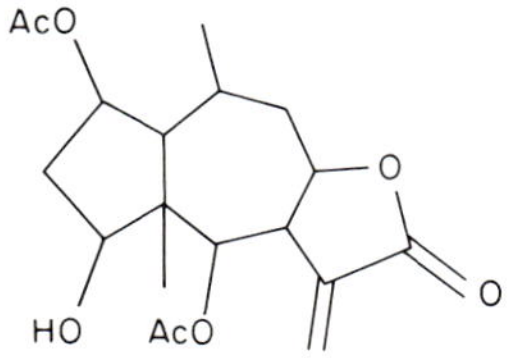

(14) Britannin

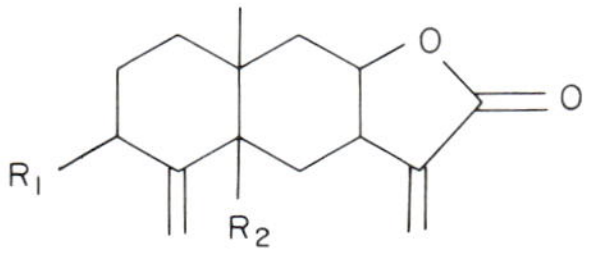

(15) $R_1=H, R_2=OH$ Telekin
(16) $R_1=OH, R_2=H$ Isotelekin

present. The only missing type, the eremophilanolide skeleton, is restricted in any case to the Senecioneae (Herout and Sorm, 1969). Unfortunately there are not enough lactone data to draw, as yet, any chemical conclusions regarding classification within the tribe.

Diterpenes

A remarkable range of yellow diterpene quinones (**17–19**) have recently been isolated from *Inula royleana* roots. They are all derived from the

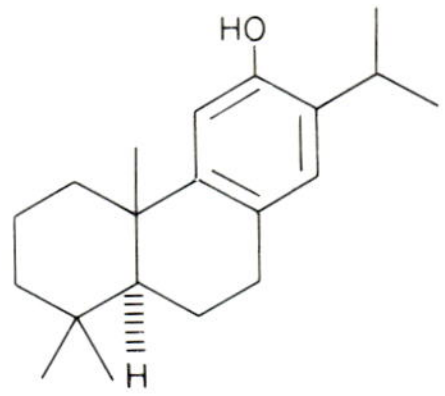

(17) Ferruginol

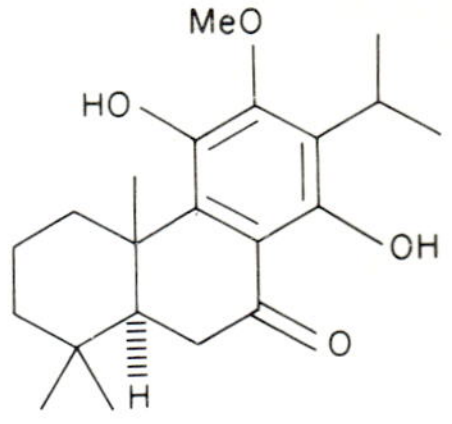

(18) Inuroyleanol

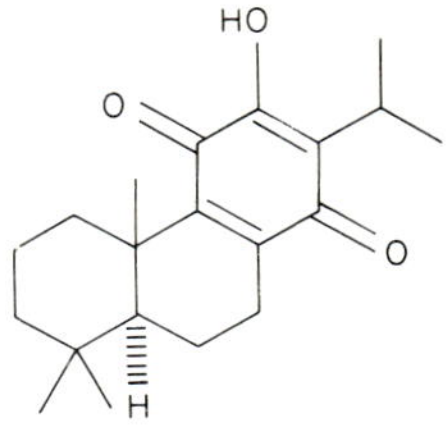

(19) Royleanone

phenolic diterpene ferruginol (**17**) which is also present. An intermediate between (**17**) and the basic quinone (**19**) has recently been isolated, namely inuroyleanol (**18**). Royleanone (**19**) itself is accompanied by its 6,7-dehydro, 7-acetoxy and 7-keto derivatives (Bhat *et al.*, 1975).

Triterpenoids

Triterpene patterns may eventually be of systematic interest in the Inuleae but the present data are quite sparse. *Blumea lacerata* has campesterol (Pal *et al.*, 1972); *Helichrysum italicum* has sitosterol and ursolic acid, *H. steochas* ursolic and oleanolic acids, uvaol, erythrodiol, sitosterol and stigmasterol (Quesada *et al.*, 1972); *Inula chinensis* root has taraxasterol, *I. helenium* root friedelin, dammaradienol and stigmasterol; and *Pluchea odorata* has β-amyrin acetate and campesterol (Dominguez and Zamudio, 1972).

FLAVONOIDS AND OTHER PHENOLICS

Flavonoids

There seems to be no information on anthocyanins in the Inuleae, but this is perhaps not surprising since yellow is the predominant flower colour in the tribe. A range of other flavonoids have, however, been isolated from these plants (**20–28**), particularly from flowerheads but also from leaves and stems. Most work has been done on a handful of species from two genera, *Gnaphalium* and *Helichrysum*. The results of these investigations

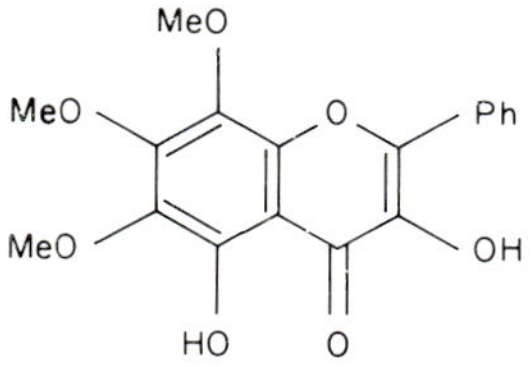

22. INULEAE—CHEMICAL REVIEW

(20) 3,5-Dihydroxy-6,7,8-trimethoxyflavone

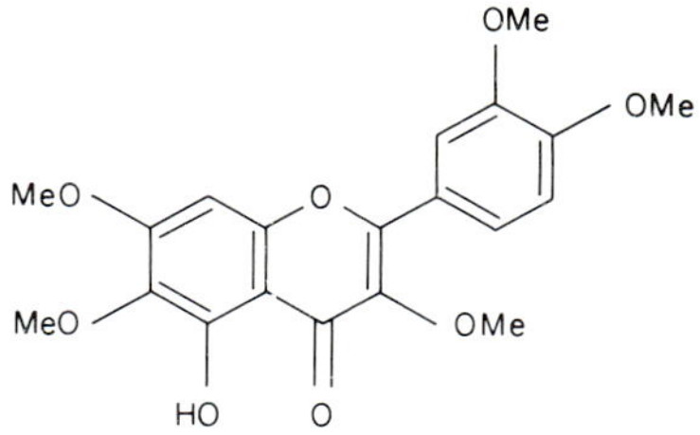

(21) Artemetin

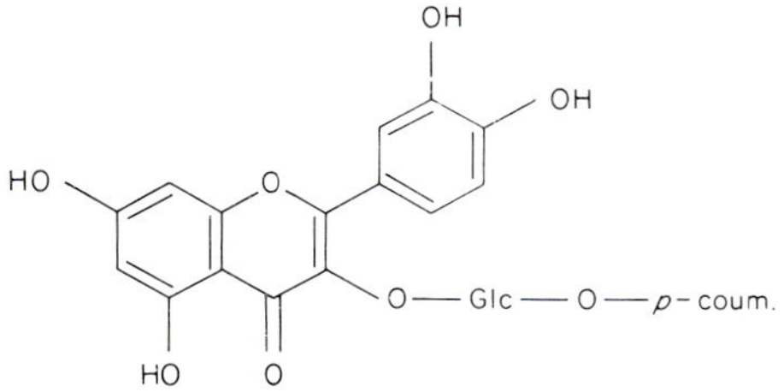

(22) Quercetin 3-(p-coumarylglucoside)

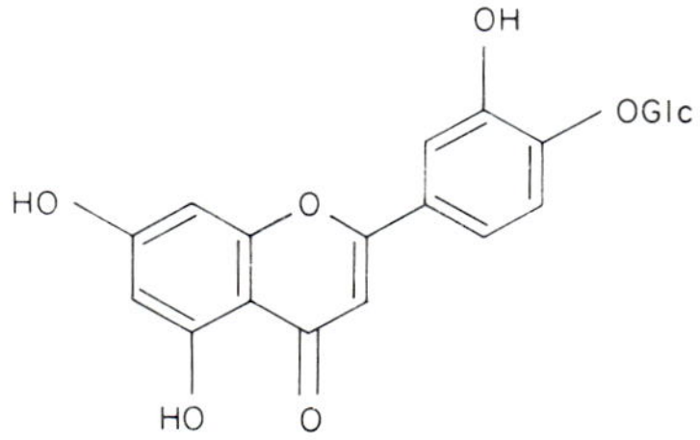

(23) Luteolin 4'-glucoside

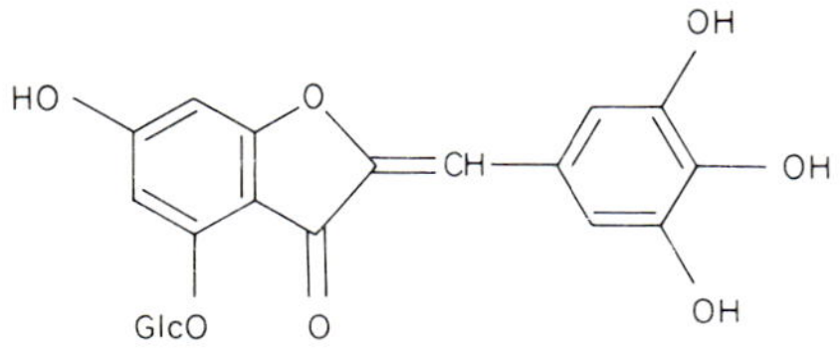

(24) Bractein

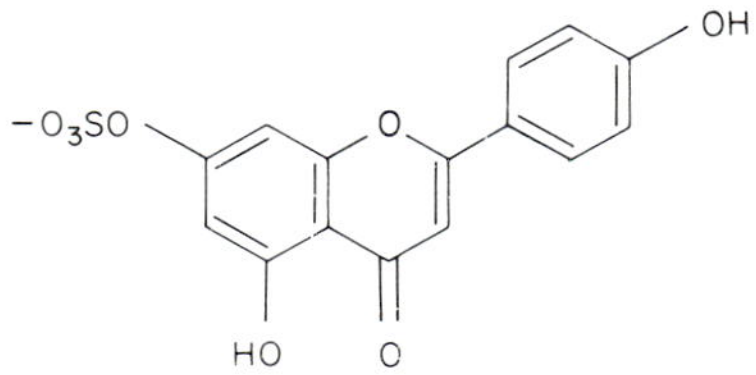

(26) Apigenin 7-sulphate

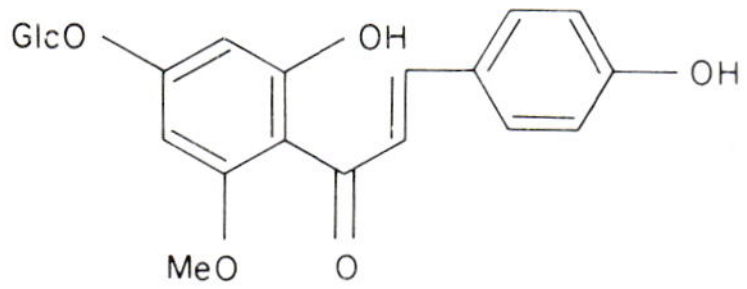

(25) 4,2'-Dihydroxy-4'-glucosyloxy-6'-methoxychalcone

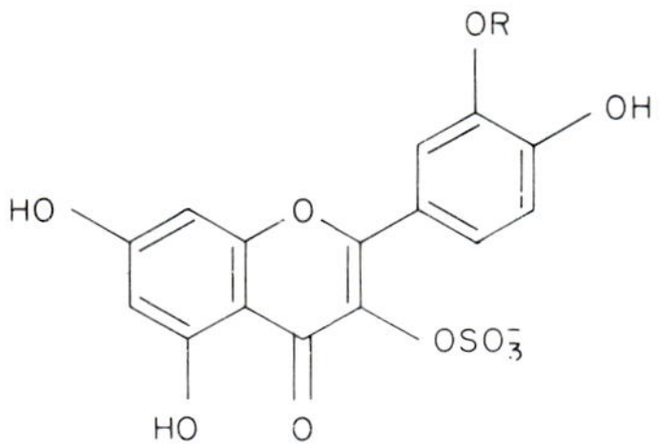

(27) R = H Quercetin 3-sulphate

(28) R = Me Isorhamnetin 3-sulphate

are summarized in Table II. A brief survey of leaf flavonoids has been carried out recently in our laboratories and the results of this are shown in Table III.

With regard to flavonoid type, perhaps the most characteristic feature distinguishing members of the Inuleae from those of other composite tribes is the presence of flavonols lacking B-ring hydroxylation. One such compound is 3,5-dihydroxy-6,7,8-trimethoxyflavone (**20**) isolated from flowerheads of *H. bracteatum*, *H. graveolens* and *H. kraussii*. Similar compounds have been found in *Achyrocline* and *Gnaphalium* species (see Table II). Other 6- or 8-hydroxylated flavonols also occur in the tribe. The presence of quercetagetin 3,6,7,3′,4′-pentamethyl ether (**21**) in *Blumea eriantha* provides a link with *Artemisia* of the Anthemideae, the plant from which it was first isolated. Our own survey (Table III) suggests that methylated flavonols such as (**21**) may occur with some frequency in leaves of this tribe. Our data also indicate that quercetagetin and/or 6-hydroxykaempferol glycosides may occur regularly in the leaves. The free 6-hydroxyflavonols, however, seem to be more frequent in Merxmüller's subtribes Gnaphaliinae and Arthrixiinae than in the subtribe Inulinae. Thus they are probably abundant in *Antennaria*, *Gnaphalium*, *Helichrysum*, *Leontopodium* and *Phagnalon*.

The two widely occurring flavonols of the angiosperms, kaempferol and quercetin, do not appear to be very common in the Inuleae. Kaempferol was found in only two of the 14 species that have been investigated in detail (Table II). In our survey of leaves of 80 species, quercetin was only identified in 19. An interesting quercetin derivative has recently been isolated from *Helichrysum kraussii*—the 6-(*p*-coumarylglucoside) (**22**)—and its structure proved by X-ray crystallography (Candy *et al.*, 1975).

The two common flavones, i.e. apigenin and luteolin, were found in leaves of only 10 of the 80 Inuleae species surveyed (Table III). They are probably more common in flowers than in leaves of the Inuleae (see Table II), as in composites generally (cf. Harborne, 1967). Of taxonomic interest is the occurrence of the relatively rare 4′-glucoside of luteolin (**23**) in flowers of four related species: *Antennaria dioica*, *Gnaphalium affine*, *G. multiceps* and *Leontopodium alpinum* (all in the subtribe Gnaphaliinae). By contrast, *C*-glycosylflavones appear to be rare in the tribe; there is only one report of them, in the leaves of *Helichrysum bracteatum*.

Anthochlor pigments, which are so characteristic of the Coreopsidinae (Heliantheae) occur in two plants of the Inuleae. Flowerheads of *Helichrysum bracteatum* contain several chalcones and the aurone bractein (**24**) while flowers of *Gnaphalium multiceps* have the anthochlor 4,2′-dihydroxy-4′-glucosyloxy-6′-methoxychalcone (**25**). Chalcones are often accompanied in plants by the structurally and biogenetically related flavanones and this is true of *Helichrysum*, from which naringenin 5-glucoside has been isolated. Geissman *et al.* (1967), in examining an Australian form of *H.*

TABLE II. Flavonoids identified in the Inuleae

Plant Species	Compounds identified	Reference
Achyrocline satureioides DC. flowering stems	3,7-Dimethoxy-5,8-dihydroxyflavone	Hänsel and Ohlendorf, 1971
Antennaria dioica Gaertn. flowers	Luteolin 7- and 4'-glucosides	Tira *et al.*, 1969
Blumea eriantha DC.	Quercetagetin 3,6,7,3,',4'-pentamethyl ether (artemetin)	Bose *et al.*, 1968
Gnaphalium affine Urv.	Luteolin 4'-glucoside	Aritomi *et al.*, 1964
G. multiceps Wall.	Luteolin 4'-glucoside and 4,2'-dihydroxy-4'-glucosyloxy-6'-methoxy chalcone	Maruyama *et al.*, 1974
G. obtusifolium L. aerial parts	3,5,7,triOH-6,8-diOMe flavone 3,5-diOH-7,8-diOMe flavone 5-OH,-3,7,8-triOMe flavone 5,7-diOH 8-subst-flavanone (obtusifolin)	Hänsel and Ohlendorf, 1969; Ohlendorf *et al.*, 1971 Narayanam *et al.*, 1970
Helichrysum affine D. Don flowers	2',6'-Dihydroxy-4-glucosyloxy-4'-methoxychalcone	Aritomi and Kawasaki, 1974
Helichrysum arenium L. roots	Kaempferol 3-glucoside, galangin and naringenin	Vrkoc *et al.*, 1975b
Helichrysum bracteatum Andr. leaves flower heads (European forms)	orientin, iso-orientin bractein, naringenin 5-glucoside, Kaempferol 3-glucoside, 3,5-diOH-6,7,8-triOMe flavone, 3,4,2',4',6'-pentaOH chalcone 2'-glucoside, 3,4,5,2',4',6'-hexaOHchalcone 3'-glucoside.	Hänsel *et al.*, 1962, 1967; Rimpler and Hänsel, 1965
aerial parts (Australian form)	naringenin, eriodictyol, homoeriodictyol, 5,7,4'-triOH-6,3'-diOMe flavone, 5,4'-diOH-6,7-diOMe flavone.	Geissman *et al.*, 1967
H. graveolens Sweet flower heads	3,5-diOH-6,7,8-triOMe flavone	Hänsel and Cubuken, 1972
H. kraussii Sch. Bip. flower heads	Quercetin 3-(6-*p*-coumaryl-glucoside), quercetin 3-methyl ether, 3,5-diOH-6,7,8-triOMe flavone	Candy *et al.*, 1975

TABLE II.—*continued*

Plant Species	Compounds identified	Reference
H. italicum G. Don	5,7-diOH-3,8-dioMe flavone	Opitz *et al.*, 1971
H. polyphyllum Ledeb.	Naringenin 5-glucoside	Zapesochnaya *et al.*, 1972
Inula grandis Shrenk	5,6,4'-triOH-3,7 diOMe flavone	Nikonova and Nikonov 1975
Leontopodium alpinum Cass.	Luteolin 7 and 4'-glucosides	Tira *et al.*, 1970
Pulicaria dysenterica Gaertn. flowers	Kaempferol 3-glucoside, 5,6,3'-triOH-3,7,4'-triOMe flavone	Schulte *et al.*, 1968
Telekia speciosa (Schreb.) Baumg.	Luteolin 7-glucoside	Bandyukova *et al.*, 1970

bracteatum, found two different flāvanones, eriodictyol and homoeriodictyol, as well as two methylated flavones (see Table II). Chemical variation is thus obviously pronounced within this particular species.

Representatives of an unusual class of flavonoid conjugate, those with sulphate, have been reported in two composite genera, *Flaveria* and *Lasthenia* (both Helenieae) (see Harborne, 1975). It was of interest to see if these compounds also occur elsewhere in the family and indeed our survey of 80 species revealed them in 8 species (Table IV). Sulphates of both flavones (e.g. **26**) and flavonols (**27** and **28**) were detected. However, sulphates occur in unrelated taxa in the Inuleae and their occurrence seems to be more closely related to an aquatic or saline habitat than to taxonomy. Indeed, the correlation between sulphate distribution and the halophyte condition seems to be particularly pronounced (see Table IV).

As in most other tribes of the Compositae, too little is known as yet regarding the distribution of flavonoids to make much use of them as taxonomic markers. Two points, however, may be noted. Firstly, *Gnaphalium* and *Helichrysum*, which are regarded as being particularly closely related taxonomically (see Chapter 21), have a very similar and complex pattern of flavonoids in their tissues. Secondly, the distribution of flavonoid types is generally in accord with the revised subtribal arrangement proposed by Merxmüller *et al.* (Chapter 21). On the other hand, no particular flavonoid structure has yet been found as a specific subtribal character.

Simple phenolics

A rich variety of phenolic compounds based on a C_6, C_6-C_2 or C_6-C_3 skeleton have been isolated from individual members of the Inuleae but it is not known whether any of these compounds are at all widespread in the tribe. Phloracetophenone trimethyl ether (**29**), for example, has been isolated from *Blumea balsamifera* DC. and *Pulicaria dysenterica* (L.) Bernth, and 4-hydroxy-3-(isopent-2-yl)acetophenone (**30**) from *Helichrysum stoechas*; the second compound also occurs in *Helianthella uniflora* (Helantheae). Several structurally interesting chromans (e.g. **31**) occur also in *H. stoechas* (Quesada *et al.*, 1972). Some unusual antibacterial substances based on 5,7-dihydroxyphthalide (**32**) have been

TABLE III. Frequency of different flavonoid types in leaves of Inuleae

Classification after Merxmüller		Genus	No. of species examined	No. of species containing:[a]			
				Methylated flavones	Ordinary flavones	Ordinary flavonols	6-Hydroxy flavonols
I	1	*Allagopappus*	2	2	2	0	0
		Buphthalmum	1	1	1	1	1
		Bubonium	1	1	0	0	0
		Fontquera	1	1	0	1	0
		Grantia	1	1	0	0	0
		Inula	12	9	5	6	2
		Jasione	2	2	2	2	0
		Pulicaria	9	8	1	7	0
		Odontospermum	8	4	5	2	0
		Perralderia	1	1	0	1	0
		Schizogyne	2	1	0	1	0
		Vieraea	1	0	1	1	0
I	2	*Geigeria*	1	0	1	1	0
I	3	*Carpesium*	1	1	0	0	0
I	4	*Blumea*	1	1	0	0	0
Inulineae totals			44	33	18	23	3
II	9	*Gnaphalium*	6	0	1	0	3
		Lasiopogon	1	1	1	0	0
II	10	*Leontopodium*	2	2	0	0	1
		Antennaria	3	3	2	0	2
II	14	*Helichrysum*	15	9	3	6	9
II	15	*Ifloga*	1	0	0	1	0
Gnaphaliinae totals			28	15	7	7	15
III	23	*Callolepis*	1	0	0	1	0
	25	*Phagnalon*	7	7	7	0	4
Arthrixiinae totals			8	7	7	1	4
INULEAE totals			80	55(69%)	32(40%)	34(42%)	22(28%)

[a] Flavones assumed to be methylated from very high R_fs in all solvents, includes methylated flavonols; ordinary flavones are mainly luteolin, occasionally apigenin; ordinary flavonols are mainly quercetin; 6-hydroxyflavonols refer to spots with R_fs corresponding to 6-hydroxykaempferol and/or quercetagetin.

TABLE IV. Species of Inuleae containing flavonoid sulphates

Plant species	Samples examined	Collecting site and country	Sulphate present
Callilepis laureola DC.	1	sandy waste ground, Swaziland	6-hydroxy-flavone[a]
Epaltes gariepina (DC.) Steez	1	paddy banks, Swaziland	Quercetin and isorhamnetin
Helichrysum angustifolium DC.	4	coastal sand, Spain	unidentified
Inula crithmoides L.	3	salt marsh or sea Spain	Patuletin
Phagnalon graecum Boiss. & Heldr.	1	unrecorded site Germany	unidentified
Pulicaria burchardii Hutch.	1	coastal rocks, Canary Islands	6-hydroxy-flavone[a]
Pulicaria scabra (Thunb.) Druce	1	saline rice paddy, Swaziland	apigenin
Sphaeranthus incisus Robyns	1	boggy ground, Swaziland	luteolin

[a] Not fully identified; could be alternatively a sulphate of a 8-hydroxyflavone.

found in *Helichrysum arenarium*. One of the latest to have been characterized is a derivative with a 4-isopentanyl sidechain with xylose attached (Vrkoc et al., 1975a). *Bis*-hydroxypyran-2-ones also occur in the roots of this plant, e.g. compound (33) (Vrkoc et al., 1975b). A final example of bizarre structural variation in this tribe is the substituted coniferyl alcohol derivative (34) isolated by Bohlmann and Zdero (1969) from *Blumea lacerata*.

OTHER SECONDARY CONSTITUENTS

Alkaloids

Alkaloids have been detected in at least eight species (Hegnauer, 1964) but in most cases little is known of their chemistry. The only alkaloid to have been fully investigated seems to be that of *Inula royleana*, a plant from which a complex diterpene alkaloid of the *Aconitum* type has been characterized.

Polyols and cyclitols

L-Inositol, unlike the widely distributed D-isomer, has only been so far reported in two families: the Euphorbiaceae and the Compositae (Plouvier,

1963). In the Compositae, it occurs in nine of 27 species examined. Among those positive were three Inuleae: *Helichrysum arenarium*, *Inula helenium* and *Pulicaria dysenterica*.

Lipids

Where they have been investigated, the lipids of the Inuleae seem to be of the usual type. For example, the seed oil of *Inula grandis* gave, on saponification, linoleic (68%), oleic (14%), palmitic (12%) and stearic (3·8%), together with traces of myristic, palmitoleic and lauric acids (Nikonova, 1972). More unusual fatty acid derivatives have occasionally been reported. Thus, β-diketones derived from branched chain organic acids have been found in *Helichrysum italicum* G. Don stems. Two of the five structures identified are EtCO-CH-Me-CO-$(CH_2)_3$-Me and Et-CO-CH-Me-CO-CH-Me-Et (Manitto *et al.*, 1972).

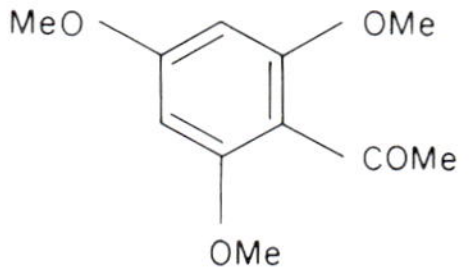

(29)

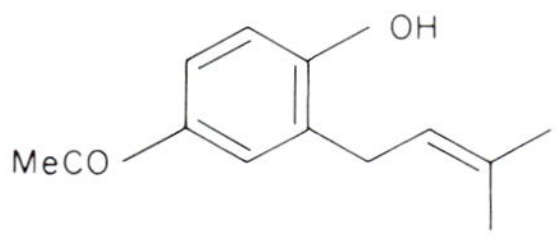

(30)

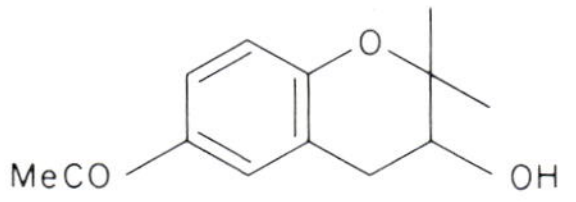

(31) 2,2-Dimethyl-3-hydroxy-6-acetylchroman

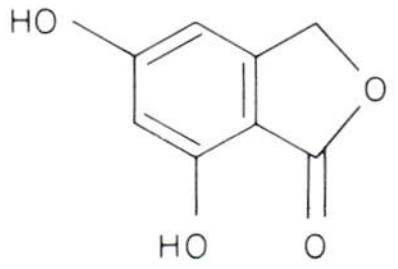

(32) 5,7-Dihydroxyphthalide

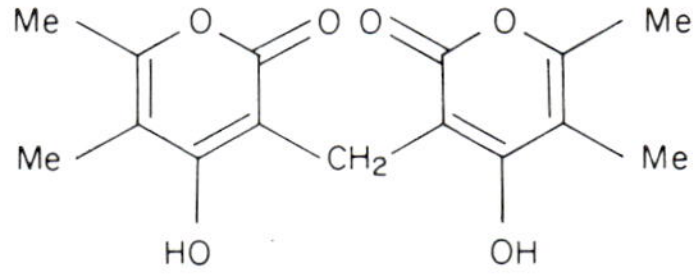

(33)

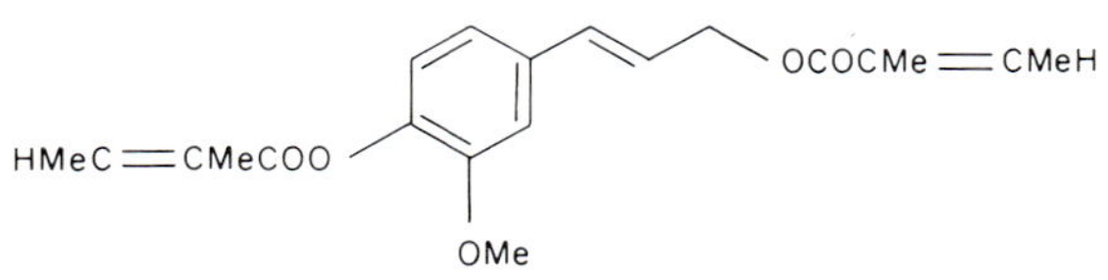

(34)

Polyacetylenes

Bohlmann *et al.* (1973) have reported the results of a survey of 108 species, representing all the Bentham subtribes, for polyacetylenes (**35–42**) and these are summarized in Table V. The simple pentayn-ne (**35**) is universally distributed and two related structures (**36, 37**) are also widespread. Monothiophene acetylenics (e.g. **38**) occur in three groups, while dithiophene acetylenics (**39**) are found only in the Buphthalminae. Acetylenics with pyran (**40**) and furan (**41**) attachments, some with epoxy and/or chlorine (**42**) substitution, occur characteristically in the Gnaphalinae. Members of this group contain the most structurally complex polyacetylenes of the Inuleae.

The distribution of polyacetylene types fits in as well with the Bentham system as with the more recent Merxmüller revision (see Table V). Taxonomically, the polyacetylenes do not as yet seem to add a great deal to present classification, but this is largely because only a relatively small percentage of species and genera have been examined. From the structural variation already revealed, there is undoubtedly much potential in this chemical character in future work on the systematics of the tribe.

$Me(C{\equiv}C)_5CH{=}CH_2$
(35)

$MeCH{=}CH(C{\equiv}C)_4CH{=}CH_2$
(36)

$Me(C{\equiv}C)_3CH{=}CH(CH_2)_3OH$
(37)

$Me(C{\equiv}C)$—[thiophene]—$(C{\equiv}C)_2CH{=}CH_2$
(38)

Me—[thiophene]—[thiophene]—$C{\equiv}C{-}CH{=}CH_2$
(39)

$Me(C{\equiv}C)_3CH{=}CH$—[pyran]—CH_2Cl
(40)

$Me(C{\equiv}C)_3CH{=}CH$—[pyran]—OH
(41)

$Me(C{\equiv}C)_3CH{=}CH$—[epoxy-pyran]—Cl
(42)

TABLE V. Distribution of polyacetylenes in the Inuleae

Merxmüller grouping	Bentham Subtribe	Pentaynne (35)	Dihydro compound (36)	Distribution of: Alcohol (37)	Monothiophene (e.g. 38)	Dithiophene (e.g. 39)	Pyran and furans
—	Tarchonanthinae (1)a	−	−	−	+	−	−
A	Plucheinae (6)	+	−	−	+	−	−
A	Inulinae (21)	+	+	+	−	−	−
A	Buphthalminae (6)	+	+	−	−	+	−
B	Filagininae (5)	+	+	+	−	−	−
B	Gnaphalinae (60)	+	−	+	−	−	+
B	Agianthinae (4)	+	−	+	+	−	−
C	Relhaninae (3)	+	−	−	−	−	−
C	Athrixinae (2)	+	−	−	−	−	−

a Removed from the Inuleae by Merxmüller.

CONCLUSION

Chemical investigations have indicated that the Inuleae contain a considerable range of secondary constituents, some of which are undoubtedly unique to the tribe. Because of lack of information on distribution, it is not known how useful these compounds are as systematic markers. It is clear, however, that the most promising constituents to examine for in any chemosystematic programme would be sesquiterpene lactones, polyacetylenes and flavonoids. Apart from the presence of inulin, practically nothing is known of the macromolecular components of the Inuleae and this remains an important area for future investigation.

REFERENCES

ALCONTRES, G. S., GATTUCO, M., AVERSA, M. C. and CARISTI, C. (1973). Graveolide, a new sesquiterpene lactone of *Inula graveolens*. *Gazz. Chim. ital.* **103**, 239–246.

ANDERSON, L. A. P., KOCK, W. T., PACHLER, K. G. R. and BRINK, C. M. (1967). Vermeerin, a sesquiterpenoid lactone from *Geigeria africana*. *Tetrahedron* **23**, 4153–4160.

ARITOMI, M. and KAWASAKI, T. (1974). Dehydro-*para*-asebotin, a new chalcone glucoside in the flowers of *Helichrysum affine* D. Don. *Chem. pharm. Bull. (Tokyo)* **22**, 1800–1805.

ARITOMI, M., SHIMOJO, M. and MAZAKI, T. (1964). Constituents in flowers of *Gnaphalium affin*. *J. pharm. Soc. Japan* **84**, 895–896.

BANDYUKOVA, V. A., SERGEERA, N. V. and DZHUMISRKO, S. T. (1970). Luteolin 7-glucoside from *Telekia speciosa*. *Khim Prir. Soedin.* 470.

BENESOVA, V., HEROUT, V. and SORM, F. (1961). Structure of telekin and isotelekin, sesquiterpene lactones from *Telekia speciosa*. *Coll. Czech. Comm.* **26**, 2916–2923.

BHAT, S. V., KALYANARAME, P. S., KOHL, H., DE SOUSA, N. J. and FEHLHABER, H. W. (1975). Inuroyleanol and 7-ketoroyleanine, diterpenoids of *Inula royleana* roots. *Tetrahedron* **31**, 1001.

BOHLMANN, F. and ZDERO, C. (1969). A new coniferyl alcohol derivative from *Blumea lacerata. Tetrahedron Lett.* 69–70.

BOHLMANN, F., BURKHARDT, T. and ZDERO, C. (1973). "Naturally Occurring Acetylenes", pp. 350–356. Academic Press, London and New York.

BOSE, P. K., BARNA, A. K. and CHAKRABORTI, P. (1968). A revised structure for erianthin, a flavonol of *Blumea eriantha. J. Indian chem. Soc.* **45**, 851.

CANDY, H. A., LAING, M., WEEKS, C. M. and KRUGER, G. J. (1975). Helichrysoside, a new acylated quercetin glycoside from *Helichrysum kraussii* flowerheads. *Tetrahedron Lett.* 1211–1212.

CHUGUNOV, P. V., SHEICHENKO, V. I. and BANKOVSKII, A. I. (1971). Britannin, a sesquiterpene lactone from *Inula britannica. Khim. Prir. Soedin.* **7**, 276–280.

DOMINGUEZ, X. A. and ZAMUDIO, J. A. (1972). β-Amyrin acetate and campesterol from *Pluchea odorata. Phytochemistry* **11**, 1179.

GEISSMAN, T. A., MUKHERJEE, R., and SIM, K. Y. (1967). Constituents of *Helichrysum viscosum* var. *bracteatum. Phytochemistry* **6**, 1575–1582.

HÄNSEL, R. and CUBUKEN, R. (1972). 3,5-Dihydroxy-6,7,8-trimethoxyflavone from *Helichrysum graveolens. Phytochemistry* **11**, 2632.

HÄNSEL, R. and OHLENDORF, D. (1971). A new flavone from *Achyrocline satureioides. Arch. Pharm.* **304**, 893–896.

HÄNSEL, R. and OHLENDORF, D. (1969). B-ring unsubstituted flavones from *Gnaphalium obtusifolium. Tetrahedron Lett.* 431.

HÄNSEL, R., LANGHAMMER, L. and ALBERT, A. G. (1962). A new aurone glycoside from *Helichrysum bracteatum. Tetrahedron Lett.* 599–601.

HÄNSEL, R., RIMPLER, H. and SCHWARTZ, R. (1967). The structure of "Helichrysum-auronol', a substance from *Helichrysum arenarium. Tetrahedron Lett.* 735–739.

HARBORNE, J. B. (1967). "The Comparative Biochemistry of the Flavonoids". Academic Press, London and New York.

HARBORNE, J. B. (1975). Flavonoid sulphates—a new class of sulphur compounds in plants. *Phytochemistry* **14**, 1147–1155.

HEGNAUER, R. (1964). "Chemotaxonomie der Pflanzen", Vol. 4, pp. 447–544. Birkhauser-Verlag, Basle.

HEROUT, V. and SORM, F. (1969). Chemotaxononomy of the sesquiterpenoids of the Compositae. In "Perspectives in Phytochemistry" (J. B. Harborne and T. Swain, eds), pp. 139–166. Academic Press, London and New York.

KISELEVA, E. Y., SHEICHENKO, V. I., RYBALKO, K. S. and BANKOVSKII, A. I. (1971). Inulicin, a new sesquiterpene lactone from *Inula japonica. Khim. Prir. Soed.* **7**, 263–270.

MANITTO, P., MONTI, D. and COLOMBO, E. (1972). Two new β-diketones from *Helichrysum italicum. Phytochemistry* **11**, 2112.

MARUYAMA, M., HAYASAKA, K., SASAKI, S., HOSOKAWA, S. and UCHIYAMA, H. (1974). A new chalcone glucoside from *Gnaphalium multiceps* flowers. *Phytochemistry* **13**, 286–288.

NARAYANAM, P., ZECHMEISTER, K., ROHRL, M. and HOPPE, W. (1970). The crystal structure analysis of obtusifolin, a flavanone. *Tetrahedron Lett.* 2643–3644.

NIKONOVA, L. P. (1972). Oil from *Inula grandis* seed. *Khim Prir. Soed.* **8**, 664.

NIKONOVA, L. P. (1973). Alantalactone and isoalantalactone from *Inula magnifica* root. *Khim. Prir. Soed.* **9**, 558.

NIKONOVA, L. P. and NIKONOV, G. K. (1972). Granilin, a new lactone from *Inula grandis. Khim. Prir. Soed.* **8**, 289.

NIKONOVA, L. P. and NIKONOV, G. K. (1975). 5,6,4'-Trihydroxy-3,7-dimethoxyflavone from *Inula grandis* leaves. *Khim. Prir. Soed.* **11**, 96.

OHLENDORF, D., SCHWARZ, R. and HANSEL, R. (1971). 3,5,7-Trihydroxy-6,8-dimethoxyflavone from *Gnaphalium obtusifolium. Arch. Pharm.* **304**, 213–215.

OPITZ, L., OHLENDORF, D. and HANSEL, R. (1971). 5,7-Dihydroxy-3,8-dimethoxyflavone from *Helichrysum italicum. Phytochemistry* **10**, 1948.

PAL, R., MOITRA, S. K., CHAKRAVORTI, N. M., and ADHYA, R. M. (1972). Campesterol from *Blumea lacera. Phytochemistry* **11**, 1855.

PLOUVIER, V. (1963). The distribution of aliphatic polyols and cyclitols. In "Chemical Plant Taxonomy" (T. Swain, ed.), pp. 313–336. Academic Press, London and New York.

QUESADA, T. G., RODRIGUEZ, B. and VALVERDE, S. (1972). The constituents of *Helichrysum stoechas. Phytochemistry* **11**, 446.

RIMPLER, H. and HANSEL, R. (1965). Two new chalcones from *Helichrysum bracteatum. Arch Pharm.* **298**, 838.

SCHULTE, K. E., RUCKER, G. and MULLER, F. (1968). Flavonoids of *Pulicaria dysenterica* blossoms. *Arch. Pharm.* **301**, 115.

TIRA, S., GALEFFI, C. and MONACHE, E. M. (1969). NMR of flavonoids, flavones from *Antenarria dioica. Ann. chim. Roma* **59**, 284–294.

TIRA, S., GALEFFI, C. and MODICA, G. di (1970). Flavonoids of *Leontopodium alpinum. Experientia* **26**, 1192.

VRKOC, J., BUDESINSKY, M., DOLEJS, L. and VASICKOVA, S. (1975a). Arenophthalide A—a new phthalide glycoside from *Helichrysum arenarium* roots. *Phytochemistry* **14**, 1845–1848.

VRKOC, J., DOLEJS, L. and BUDESINSKY, M. (1975b). Methylene-bis-2H-pyran-2-ones and phenolic constituents from *Helichrysum arenarium* roots. *Phytochemistry* **14**, 1383–1384.

ZAPESOCHNAYA, G. G., BANKOVSKII, A. I. and NAKAIDZE, A. K. (1972). Flavonoids of *Helichrysum polyphyllum. Khim. Prir. Soed.* **8**, 804–806.

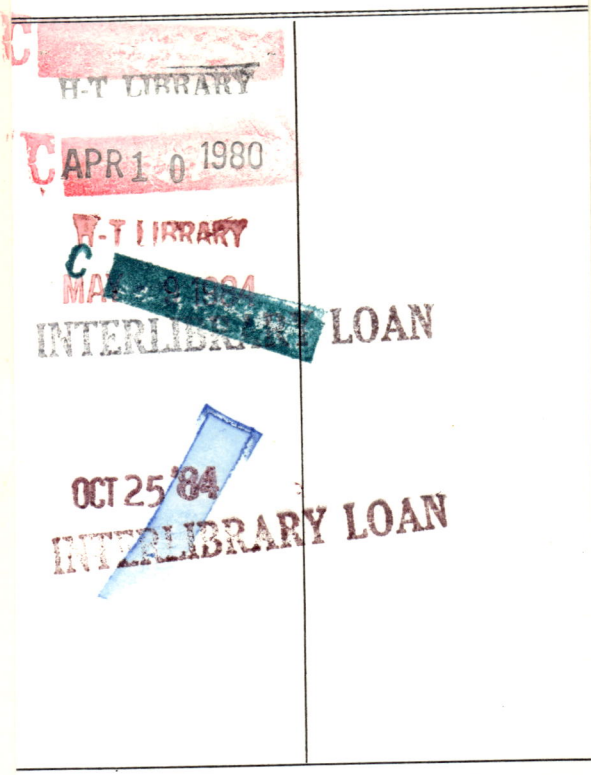